S E C O N D E

Gas Turbine Heat Transfer and Cooling Technology

Je-Chin Han ▪ Sandip Dutta ▪ Srinath Ekkad

CRC Press
Taylor & Francis Group
Boca Raton   London   New York

CRC Press is an imprint of the
Taylor & Francis Group, an **informa** business

First Indian Reprint, 2015

CRC Press
Taylor & Francis Group
6000 Broken Sound Parkway NW, Suite 300
Boca Raton, FL 33487-2742

© 2012 by Taylor & Francis Group, LLC
CRC Press is an imprint of Taylor & Francis Group, an Informa business

Printed in India by Manipal Technologies Ltd, Manipal

International Standard Book Number-13: 978-1-4398-5568-3

Library of Congress Cataloging-in-Publication Data

Han, Je-Chin, 1946-
  Gas turbine heat transfer and cooling technology / Je-Chin Han, Sandip Dutta, Srinath Ekkad. -- 2nd ed.
    p. cm.
  Includes bibliographical references and index.
  ISBN 978-1-4398-5568-3 (hardback)
  1. Gas-turbines. 2. Heat--Transmission. 3. Gas-turbines--Cooling. I. Datta, Sandipa. II. Ekkad, Srinath, 1958- III. Title.

TJ778.H24 2012
621.43'3--dc23   2012019250

Visit the Taylor & Francis Web site at
http://www.taylorandfrancis.com

and the CRC Press Web site at
http://www.crcpress.com

For sale in India, Pakistan, Nepal, Bhutan, Bangladesh and Sri Lanka only.

Contents

Preface to the Second Edition

This book, *Gas Turbine Heat Transfer and Cooling Technology*, was first published in 2000. There have been many new technical papers available in the open literature over the last 10 years. These new published data provide gas turbine researchers, designers, and engineers with advanced heat transfer analysis and cooling technology development references. There is a need to revise the first edition by including the latest information in order to keep this book relevant for users. The second edition provides information on state-of-the-art cooling technologies such as advanced turbine blade film cooling and internal cooling schemes. It updates modern experimental methods for gas turbine heat-transfer and cooling research as well as advanced computational models for gas turbine heat-transfer and cooling performance predictions.

ASME Turbo Expo (IGTI International Gas Turbine Institute) has made conference CDs available to every year's attendees since 2000 (GT2000–GT2010). These conference CDs contain all gas turbine heat-transfer papers presented in each year's IGTI conference. The number of heat transfer–related conference papers has increased from about 100 in the year 2000 to about 200 in the year 2010. These reviewed technical papers are widely used in gas turbine heat transfer and added new knowledge to this field after the publication of our first edition.

This text is a revision of the first edition, not a new book. The major contents and framework have been based on the first edition. To keep the same book format, the revised second edition adds new information at the end of each chapter, mainly based on selected papers from the open literature published in 2000–2010. The open literature has many excellent articles available on this subject; however, we cannot use all of them in this book. To reduce the book size, we have mainly used our own published results in the second edition. We hope this book will be useful for the gas turbine community. We would be happy to receive constructive comments and suggestions on the material in the book.

Je-Chin Han
Sandip Dutta
Srinath Ekkad

Preface to the First Edition

Gas turbines are used for aircraft propulsion and in land-based power generation or industrial applications. Modern development in turbine-cooling technology plays a critical role in increasing the thermal efficiency and power output of advanced gas turbines. Research activities in turbine heat transfer and cooling began in the early 1970s; since then, many research papers, state-of-the-art review articles, and book chapters have been published. However, there is no book focusing entirely on the range of gas turbine heat-transfer issues and the associated cooling technologies.

This book is intended as a reference book for researchers and engineers interested in working with gas turbine heat-transfer and cooling technology. Specifically, it is for researchers and engineers who are new to the field of turbine heat-transfer analysis and cooling design; it can also be used as a textbook or reference book for graduate-level heat-transfer and turbomachinery classes.

In the beginning, we thought of covering all aspects of gas turbine–related heat-transfer and cooling problems. After careful survey, however, we decided to focus on the heat-transfer and cooling issues related to turbine airfoils only, because a vast amount of information on this subject alone is available in the published literature. Assembling all the scattered information in a single compilation requires a great deal of effort. The book does not include combustor liner cooling and turbine disk-cooling problems although they are important to gas turbine hot gas path component designs. The book is divided into eight chapters:

Chapter 1 *Fundamentals*. Discusses the need for turbine cooling, gas turbine heat-transfer problems, and cooling methodology

Chapter 2 *Turbine Heat Transfer*. Discusses turbine rotor and stator heat-transfer issues, including endwall and blade tip region under engine conditions as well as under simulated engine conditions

Chapter 3 *Turbine Film Cooling*. Includes turbine rotor and stator blade film cooling and a discussion of the unsteady high free-stream turbulence effect on simulated cascade airfoils

Chapter 4 *Turbine Internal Cooling*. Includes impingement cooling, rib-turbulated cooling, pin-fin cooling, and compound and new cooling techniques

Chapter 5 *Turbine Internal Cooling with Rotation*. Discusses the effect of rotation on rotor coolant passage heat transfer

Chapter 6 *Experimental Methods*. Includes heat-transfer and mass-transfer techniques, liquid crystal thermography, optical techniques, flow and thermal field measurement techniques

Chapter 7 *Numerical Modeling*. Discusses governing equations and turbulence models and their applications for predicting turbine blade heat transfer and film cooling and turbine blade internal cooling

Chapter 8 *Final Remarks*. Provides suggestions for future research in this area

The open literature has many excellent articles available on this subject; however, we cannot use all of them in this book. We do not claim any new ideas in this book, but we do attempt to present the topic in a systematic and logical manner. We hope this book is a unique compilation and is useful for the gas turbine community. We would be happy to receive constructive comments and suggestions on the material in the book.

Je-Chin Han
Sandip Dutta
Srinath Ekkad

Authors

Je-Chin Han received his BS from National Taiwan University, MS from Lehigh University, and his ScD from MIT in 1976, all in mechanical engineering. He joined Texas A&M University in 1980 and has been doing research on gas turbine heat transfer and cooling technology for aircraft propulsion and in land-based power generation or industrial applications for more than 35 years. Currently he is a Texas A&M University Distinguished Professor and holder of the Marcus C. Easterling Endowed Chair.

Sandip Dutta received his BS from IIT, Kharagpur, MS from Louisiana State University, and PhD from Texas A&M University in 1995, all in mechanical engineering. He has worked in both academia and industry. Currently he is with General Electric in the power generation gas turbines division located in Greenville, South Carolina.

Srinath Ekkad received his BS from Jawaharlal Nehru Technological University, India, MS from Arizona State University, and PhD from Texas A&M University in 1995, all in mechanical engineering. Since then, he has worked at Rolls-Royce and Louisiana State University. He is currently a professor of mechanical engineering at Virginia Tech, leading gas turbine–related efforts as director of Commonwealth Center for Aerospace Propulsion Systems.

1

Fundamentals

1.1 Need for Turbine Blade Cooling

1.1.1 Recent Development in Aircraft Engines

Gas turbines are used for aircraft propulsion and in land-based power generation or industrial applications. Thermal efficiency and power output of gas turbines increase with increasing turbine rotor inlet temperatures (RIT). This is illustrated in Figure 1.1, which plots specific core power production (which can be related to specific thrust) as a function of turbine RIT. Aircraft engines tend to track fairly close to the ideal performance line, which represents a cycle power output with 100% efficient turbines with no leakage or cooling flows. Clearly, increasing RIT is one of the key technologies in raising gas turbine engine performance. Figure 1.2 shows that the RIT in advanced gas turbines are far higher than the softening point of the blade material: therefore, turbine blades need to be cooled. To double the engine power in aircraft gas turbines, the RIT should increase from today's 2500°F to 3500°F using the same amount of cooling air (3%–5% of compressor bleed air). Meanwhile, the compressor pressure ratio should increase from today's 20 times the compression ratio to 40 times the compression ratio, or even higher, as shown in Figure 1.3. This means that future aircraft gas turbines would have a higher turbine inlet temperature with the same amount of hotter cooling air from high-pressure compressor bleed. Therefore, high-temperature material development such as thermal barrier coating (TBC) or highly sophisticated cooling schemes is an important issue that needs to be addressed to ensure high-performance, high-power gas turbines for the next century. To reach this goal, the U.S. Department of Defense (DOD), NASA, and U.S. aircraft gas turbine manufacturers established the long-range R&D program known as integrated high-performance turbine engine technology (IHPTET). Begun in 1993, it was targeted with doubling the engine power by the year 2003 (Daly, 1993). Research and Development funds are provided by the U.S. Air Force, Navy, Army, and NASA, and by the U.S. gas turbine manufacturers such as GE Aircraft Engines, Pratt & Whitney, Allison, and Allied Signal. Research is performed at U.S. government laboratories, industrial

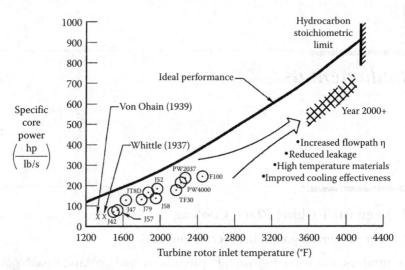

FIGURE 1.1
Increased turbine inlet temperature dramatically improves cycle power output. (Courtesy of Pratt & Whitney, East Hartford, CT; Sautner, M. et al., Determination of surface heat transfer and film cooling effectiveness in unsteady wake flow conditions, *AGARD Conference Proceedings*, Vol. 527, pp. 6-1–6-12, 1992. With permission.)

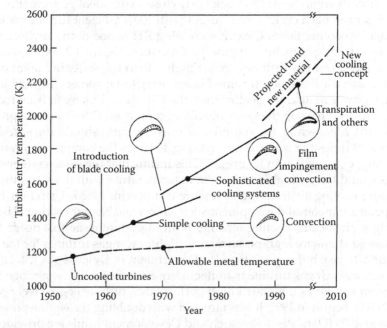

FIGURE 1.2
Variation of turbine entry temperature over recent years. (From Clifford, R.J., Rotating heat transfer investigations on a multipass cooling geometry, AGARD CP 390, 1985; collected in Lakshminarayana, B.: *Fluid Dynamics and Heat Transfer of Turbomachinery.* Chapter 7, pp. 597–721. 1996. Copyright Wiley-VCH Verlag GmbH & Co. KGaA. With permission.)

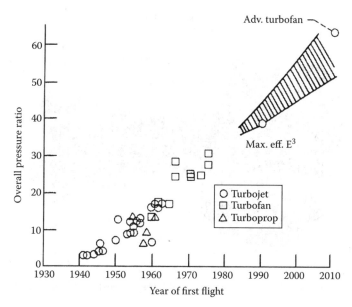

FIGURE 1.3

Progress in compressor pressure ratio. (From Rohlik, H.E., Current and future technology trends in radial and axial gas turbines, NASA TM 83414, 1983; collected in Lakshminarayana, B.: *Fluid Dynamics and Heat Transfer of Turbomachinery*. Chapter 7, pp. 597–721. 1996. Copyright Wiley-VCH Verlag GmbH & Co. KGaA. With permission.)

laboratories, and university laboratories. Figure 1.4 is a diagram of the F117 turbofan engine developed by Pratt & Whitney. All R&D activities are aimed at doubling the capability of turbine engines through (1) improved cooling effectiveness, (2) high-temperature materials with TBC, and (3) increased flow path efficiency with reducing leakage.

1.1.2　Recent Development in Land-Based Gas Turbines

For land-based gas turbines, including power generation (300 MW combined cycles), marine propulsion, and industrial applications such as pumping and cogeneration (less than 30 MW), the RIT should maintain today's level of 2500°F–2600°F due to the constraint of the NO_x pollution problem. Therefore, the main issue for land-based gas turbines is how to further improve the thermal efficiency at the current RIT level—for example, how to move the efficiency for the stand-alone gas turbine from today's 35% up to 40%, and for the large gas turbine combined cycle from today's 55% up to 60% by the year 2003. To reach this goal, the U.S. Department of Energy (DOE) and the U.S. land-based gas turbine manufacturers established a long-range R&D program known as advanced turbine systems (ATS). Begun in 1992, the ATS program was targeted to increase the combined cycle efficiency to 60% by the year 2002 (Davis and Randolph, 1993). Research and development funds are

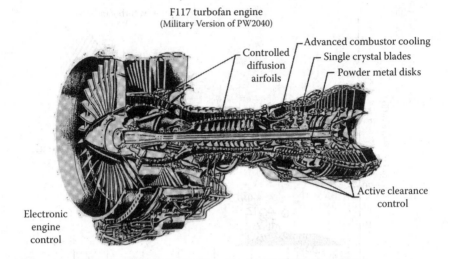

FIGURE 1.4
F117 turbofan engine developed by Pratt & Whitney. (Courtesy of Pratt & Whitney, East Hartford, CT.)

provided by the U.S. DOE and by the gas turbine manufacturers such as GE Power Systems, Westinghouse Electric, Allison, Solar Turbines, and Allied Signal. Research is performed at government laboratories, industrial laboratories, and university laboratories. For example, Figure 1.5 is a schematic of the H-Engine Hot Gas Parts developed by GE Power Systems under the U.S. DOE ATS program. They used a closed-loop steam-cooled nozzle with

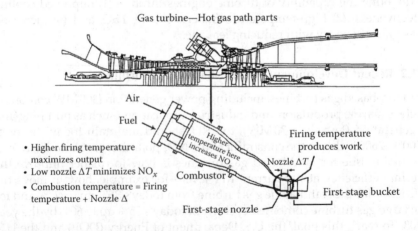

FIGURE 1.5
Relationship-combustion temperature to fire temperature for H-Engine Hot Gas Parts developed by GE Power Systems under U.S. DOE ATS Program. (From Corman, J.C. and Paul, T.C., Power systems for the 21st century "H" gas turbine combined cycles, GE Power Systems, Schenectady, NY, GER-3935, pp. 1–12, 1995. With permission.)

TBC in order to reduce the hot-gas temperature drop through the first-stage nozzle. Therefore, the RIT (or firing temperature) can be higher to produce more power at the same combustion temperature. The GE H-Engine uses the most advanced combustor design incorporating fuel/air premixing and lean combustion to reduce the environmental pollution problem (dry low NO_x). All research activities are aimed at (1) combustion and combustion instability issues because of the use of lean-premixed dry low NO_x combustors, (2) material development such as a single crystal blade with TBC and ceramic blades, and (3) advanced turbine blade cooling such as a closed-loop steam-cooled blade with TBC.

1.2 Turbine-Cooling Technology

1.2.1 Concept of Turbine Blade Cooling

Advanced gas turbine engines operate at high temperatures (2500°F–2600°F) to improve thermal efficiency and power output. As the turbine inlet temperature increases, the heat transferred to the turbine blades also increases. The level and variation in the temperature within the blade material (which causes thermal stresses) must be limited to achieve reasonable durability goals.

The operating temperatures are far above the permissible metal temperatures. Therefore, there is a need to cool the blades for safe operation. The blades are cooled by extracted air from the compressor of the engine. Since this extraction incurs a penalty to the thermal efficiency, it is necessary to understand and optimize the cooling technique, operating conditions, and turbine blade geometry. Gas turbine blades are cooled internally and externally. Internal cooling is achieved by passing the coolant through several enhanced serpentine passages inside the blades and extracting the heat from the outside of the blades. Both jet impingement and pin-fin cooling are also used as a method of internal cooling. External cooling is also called film cooling. Internal coolant air is ejected out through discrete holes or slots to provide a coolant film to protect the outside surface of the blade from hot combustion gases.

The engine cooling system must be designed to ensure that the maximum blade surface temperatures and temperature gradients during operation are compatible with the maximum blade thermal stress for the life of the design. Too little coolant flow results in hotter blade temperatures and reduced component life. Similarly, too much coolant flow results in reduced engine performance. The engine cooling system must be designed to minimize the use of compressor bleed air for cooling purposes to achieve maximum benefits of the high inlet gas temperature.

Highly sophisticated cooling techniques in advanced gas turbine engines include film cooling, impingement cooling, and augmented convective

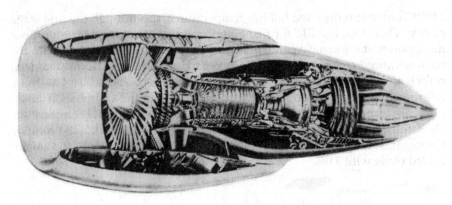

FIGURE 1.6
Cutaway view of GE CF6 turbofan engine. (From Treager, I.E., General electric CF6, in *Aircraft Gas Turbine Engine Technology*, 2nd edn., McGraw-Hill, New York, Chapter 25, pp. 469–525, 1979. With permission.)

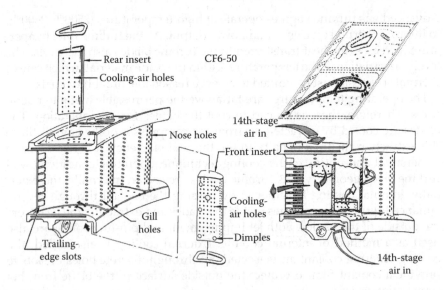

FIGURE 1.7
Stage-1 high-pressure turbine nozzle vane for the GE CF6 engine. (From Treager, I.E., General electric CF6, in *Aircraft Gas Turbine Engine Technology*, 2nd edn., McGraw-Hill, New York, Chapter 25, pp. 469–525, 1979. With permission.)

cooling. Figures 1.6 and 1.7 show the cutaway view of the General Electric CF6 turbofan engine and the stage-1 high-pressure nozzle guide vane (NGV) for the GE CF6 turbofan engine, respectively. The cooling air comes from the 14th-stage compressor bleed and impinges on the inner walls of the NGV. After impingement cooling, the spent air provides film cooling through the leading-edge holes, gill holes, midchord holes, and trailing-edge

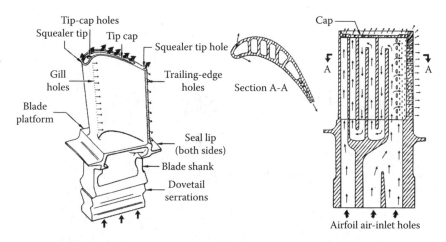

FIGURE 1.8
Stage-1 high-pressure turbine rotor blade for the GE CF6 engine. (From Treager, I.E., General electric CF6, in *Aircraft Gas Turbine Engine Technology*, 2nd edn., McGraw-Hill, New York, Chapter 25, pp. 469–525, 1979. With permission.)

slots. Figure 1.8 shows the stage-1 internally cooled high-pressure turbine rotor blade for a GE CF6 turbofan engine. The cooling system is based on the use of convective cooling in the leading-edge region and film cooling through the gill holes, augmented convective cooling with rib turbulators in the midchord region, and squealer-tip-cap cooling and augmented convective cooling with pin fins in combination with film cooling at the trailing edge. The optimum combination of these cooling techniques to meet the highly complex design requirements is the key to designing air-cooled turbine blades and vanes.

1.2.2 Typical Turbine-Cooling System

Gas turbine-cooling technology is complex and varies from engine manufacturer to engine manufacturer. Even the same engine manufacturer uses different cooling systems for various engines. Most turbine-cooling systems are proprietary in nature and also are not available in open literature. However, most cooling system designs are quite similar regardless of engine manufacturer and models. The following paragraph will discuss a typical turbine-cooling system using NASA's energy efficient engine (E^3), developed by GE Aircraft Engines (Halila et al., 1982). This is used as an example since it is available in the public domain. Note that the cooling systems for today's advanced gas turbine engines have improved beyond the E^3 engine.

Figure 1.9 is an overall view of the rotor, stator, and casing cooling supply system. The stage-1 nozzle is cooled by air extracted from the inner and outer combustion liner cavities, and the stage-1 rotor is cooled by air extracted at

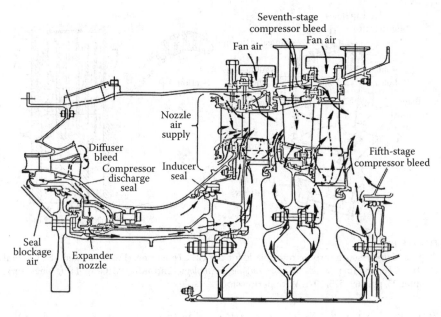

FIGURE 1.9
Overall view of the rotor, stator, and casing cooling supply system for E³ engine. (From Halila, E.E. et al., Energy efficient engine, General Electric Company, Fairfield, CT (prepared for NASA CR-167955), 1982. With permission.)

the compressor diffuser midspan. The stage-2 nozzle coolant comes from the stage-7 compressor bleed, and the stage-2 rotor coolant comes from the stage-1 rotor inducer system. Figures 1.10 through 1.12 show the cooling air supply for the nozzle pitch line and the inner-band and outer-band design. This design includes two separate impingement inserts and trailing-edge pressure side bleed slots. The design also uses both impingement cooling and film cooling at the nozzle leading-edge and midchord region with two rows of compound angle holes on the pressure side and two rows of diffusion-shaped holes on the suction side. The vane inner band (endwall) is cooled by impingement, then film-cooled through shaped holes. The vane outer band is film-cooled through diffusion-shaped holes. The goal is to minimize the coolant consumption, maximize the cooling effect, and produce an acceptable temperature level and distribution on the vane surface, as shown in Figure 1.13.

Figure 1.14 shows the detailed cooling air supply for the stage-1 rotor blade cooling system. The stage-1 rotor blade cooling system uses a two-circuit augmented convection and film-cooling design. In the forward circuit, leading-edge impingement holes are supplied by a three-pass serpentine passage with rib turbulators. The leading edge is cooled by a combination of impingement and film cooling through three rows of radial angle holes. A single row of round, axial angle holes provides the pressure side

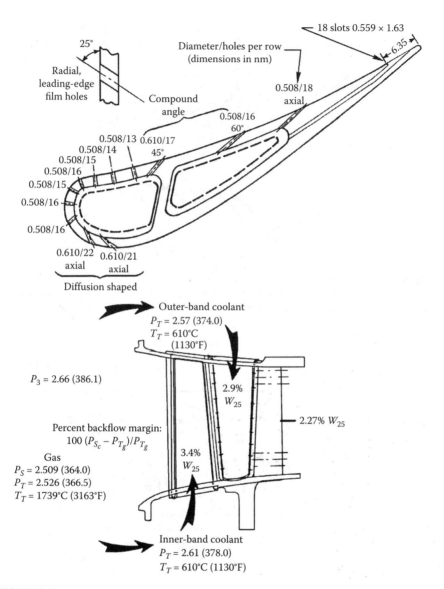

FIGURE 1.10

E³ stage-1 HPT vane-cooling geometry and cooling air supply. (From Halila, E.E. et al., *Energy efficient engine*, General Electric Company, Fairfield, CT (prepared for NASA CR-167955), 1982. With permission.)

film-cooling air, while a single row of diffusion-shaped, axial angle holes provides the suction side film-cooling air. The second circuit has a three-pass forward-flowing serpentine passage with rib turbulators. This second loop also provides air for the trailing-edge impinged-pin film-cooling design. Trailing-edge spent cooling air exits through the pressure side bleed slots

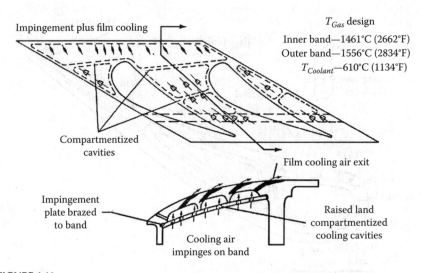

FIGURE 1.11
E[3] stage-1 HPT vane, inner-band-cooling design. (From Halila, E.E. et al., Energy efficient engine, General Electric Company, Fairfield, CT (prepared for NASA CR-167955), 1982. With permission.)

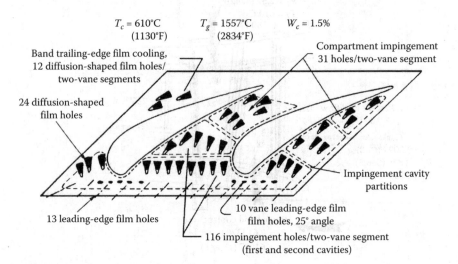

FIGURE 1.12
E[3] stage-1 HPT vane, outer-band-cooling design. (From Halila, E.E. et al., Energy efficient engine, General Electric Company, Fairfield, CT (prepared for NASA CR-167955), 1982. With permission.)

and produces external film cooling for the remainder of the trailing edge. The third pass of the second loop exits through a single row of pressure side, midchord, compound angle holes for reinforcing the pressure side film cooling from the upstream gill holes. Figure 1.15 shows the stage-1 rotor blade-tip-cap cooling design. The tip-cap and squealer-tip cooling are achieved

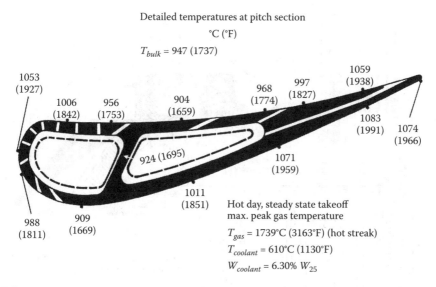

Detailed temperatures at pitch section
°C (°F)

T_{bulk} = 947 (1737)

1053 (1927)
1006 (1842)
956 (1753)
904 (1659)
968 (1774)
997 (1827)
1059 (1938)
1083 (1991)
1074 (1966)
924 (1695)
1071 (1959)
1011 (1851)
988 (1811)
909 (1669)

Hot day, steady state takeoff
max. peak gas temperature

T_{gas} = 1739°C (3163°F) (hot streak)
$T_{coolant}$ = 610°C (1130°F)
$W_{coolant}$ = 6.30% W_{25}

FIGURE 1.13

E^3 stage-1 HPT vane pitch section detailed temperature distribution. (From Halila, E.E. et al., Energy efficient engine, General Electric Company, Fairfield, CT (prepared for NASA CR-167955), 1982. With permission.)

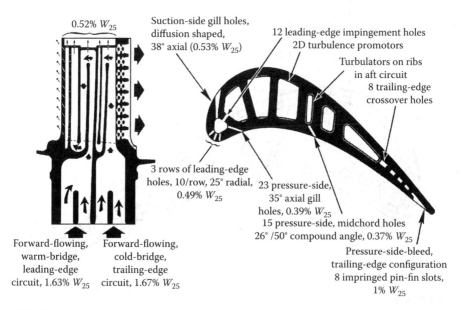

0.52% W_{25}

Suction-side gill holes, diffusion shaped, 38° axial (0.53% W_{25})

12 leading-edge impingement holes 2D turbulence promotors

Turbulators on ribs in aft circuit 8 trailing-edge crossover holes

3 rows of leading-edge holes, 10/row, 25° radial, 0.49% W_{25}

23 pressure-side, 35° axial gill holes, 0.39% W_{25}

15 pressure-side, midchord holes 26° /50° compound angle, 0.37% W_{25}

Forward-flowing, warm-bridge, leading-edge circuit, 1.63% W_{25}

Forward-flowing, cold-bridge, trailing-edge circuit, 1.67% W_{25}

Pressure-side-bleed, trailing-edge configuration 8 impringed pin-fin slots, 1% W_{25}

FIGURE 1.14

E^3 stage-1 HPT rotor blade cooling system. (From Halila, E.E. et al., Energy efficient engine, General Electric Company, Fairfield, CT (prepared for NASA CR-167955), 1982. With permission.)

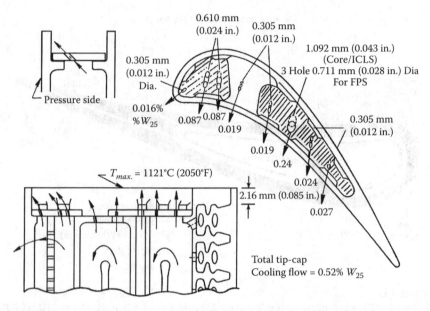

FIGURE 1.15
E³ stage-1 HPT rotor blade-tip-cap cooling design. (From Halila, E.E. et al., Energy efficient engine, General Electric Company, Fairfield, CT (prepared for NASA CR-167955), 1982. With permission.)

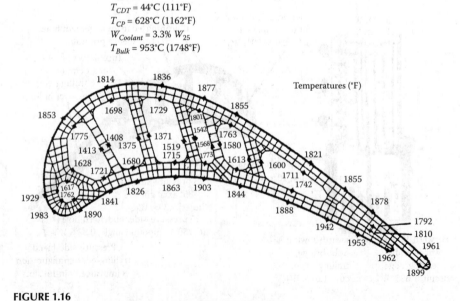

FIGURE 1.16
E³ stage-1 HPT rotor blade pitch-line temperature distribution at steady-state takeoff. (From Halila, E.E. et al., Energy efficient engine, General Electric Company, Fairfield, CT (prepared for NASA CR-167955), 1982. With permission.)

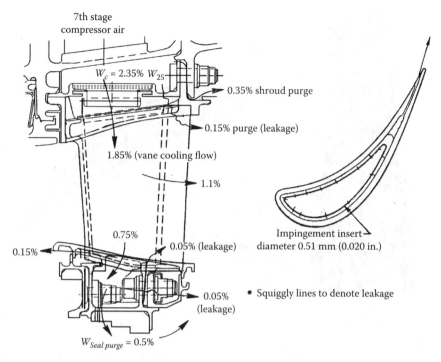

FIGURE 1.17

E^3 stage-2 nozzle-cooling flows. (From Halila, E.E. et al., Energy efficient engine, General Electric Company, Fairfield, CT (prepared for NASA CR-167955), 1982. With permission.)

by bleeding a small portion of the cooling air through holes in the tip-cap region. The goal is to minimize the coolant consumption, maximize the cooling effect, and produce an acceptable temperature level and distribution on the rotor blade surface, as shown in Figure 1.16.

Figure 1.17 shows the cooling system for the stage-2 vane. The design uses a single impingement insert in the vane. It uses no film cooling but uses pressure side bleed slots to cool the trailing-edge region and inject the spent coolant (after impingement cooling) back into the hot-gas flowpath with low aerodynamic mixing loss. After impingement, portions of spent air are vented through the inner diameter of the vane to provide purging for the interstage seal. Figure 1.18 shows the stage-2 rotor blade cooling system with a two-circuit, augmented internal convection design. In the forward circuit, cooling air is forced from the leading edge through a three-pass serpentine passage with rib turbulators. Similarly, in the second circuit, cooling air is brought from the blade dovetail through a serpentine passage with rib turbulators located near the trailing edge. The spent air from the forward and second circuit exits through a single slot on the pressure side near the tip. No film-cooling or trailing-edge holes are used.

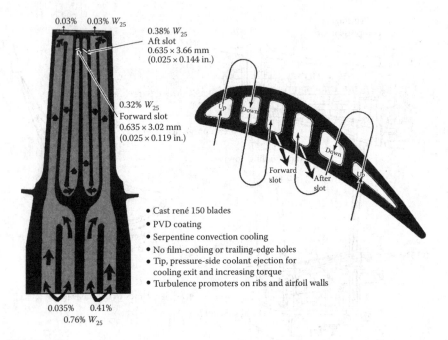

FIGURE 1.18

E³ stage-2 rotor blade design feature. (From Halila, E.E. et al., Energy efficient engine, General Electric Company, Fairfield, CT (prepared for NASA CR-167955), 1982. With permission.)

1.3 Turbine Heat Transfer and Cooling Issues

1.3.1 Turbine Blade Heat Transfer

Modern gas turbines are designed to run at high turbine inlet temperatures well in excess of current metal temperature limits. In addition to improved temperature capability materials and TBCs, highly sophisticated cooling techniques such as augmented internal cooling and external film cooling must be used to maintain acceptable life and operational requirements under such extreme heat load conditions. To design a system that most efficiently cools the turbine hot-gas flowpath components, it is necessary to better comprehend the detailed hot-gas flow physics within the turbine itself. There is a great need to increase the understanding of heat transfer within this unsteady high-turbulence and highly 3-D complex flow field. Note that the blade life may be reduced by half if the blade metal temperature prediction is off by only 50°F. Therefore, it is critical to predict accurately the local-heat-transfer coefficient as well as the local blade temperature in order to prevent local hot spots and increase turbine blade life. Current turbine designs are characterized by an inability to predict accurately heat-transfer coefficient distributions under turbomachinery flow conditions. This results in a nonoptimized design using inordinate amounts of cooling air, which ultimately causes significant penalties to the cycle in terms of thrust and specific fuel consumption.

Hot-gas path components include turbine stator vanes and turbine rotor blades. Turbine first-stage vanes are exposed to high-temperature, high-turbulence hot gases from the exit of the combustor. It is important to determine the heat load distributions on the first-stage vanes and their endwalls under engine flow Reynolds and Mach numbers for a typical gas turbine engine. An accurate estimate of the heat-transfer distributions can help in designing an efficient cooling system and prevent local hot-spot overheating. Figure 1.19 shows both typical annular and can-annular type combustion chambers and their cross sections. Figure 1.20 shows both typical radial and circumferential temperature distributions, as well as the average radial

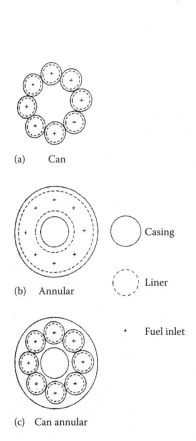

(a) Can

(b) Annular

(c) Can annular

Casing

Liner

Fuel inlet

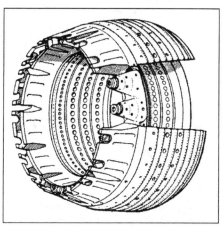

Typical annular-type combustion chamber

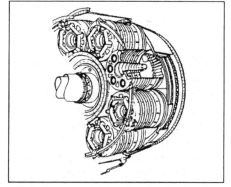

Typical can-annular-type combustion chamber

FIGURE 1.19
Typical annular type and can-annular type combustion chambers and their cross-sections. (From Mattingly, J.D., *Elements of Gas Turbine Propulsion*, McGraw-Hill, New York, p. 229, 1996. With permission.)

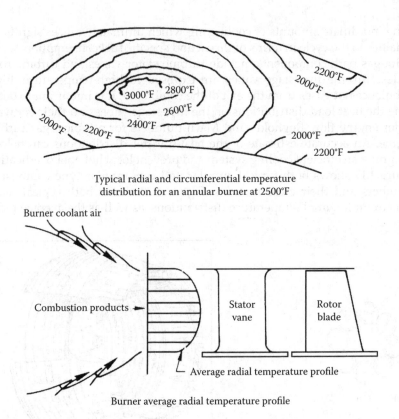

Typical radial and circumferential temperature
distribution for an annular burner at 2500°F

Burner coolant air

Combustion products ▸

Stator
vane

Rotor
blade

└ Average radial temperature profile

Burner average radial temperature profile

FIGURE 1.20
Typical annular burner outlet temperature profile. (From Suo, M., Turbine cooling,
in *Aerothermodynamics of Aircraft Gas Turbine Engines*, Oates, G., Ed., Air Force Aero
Propulsion Laboratory, Wright-Patterson Air Force Base, OH, AFAPL TR 78-5, Chapter 19,
pp. 19-1–19-23, 1978. With permission.)

temperature profile for an annular burner at 2500°F. These temperature profiles, as well as the associated turbulence levels (both intensity and scale), are
important in determining the stage-1 nozzle heat load distributions. Besides
the inlet temperature profile and high turbulence effect (turbulence intensity
up to 20%), the secondary flows produced near the endwalls make the turbine stator vane heat-transfer prediction even harder.

After accelerating from the first-stage vanes, hot gases move into the first-
stage rotor blades to produce turbine power. At the inlet of the first-stage
rotor blade, both the temperature and turbulence levels are lower compared
to the inlet of the first-stage vane. However, the inlet velocity could be two
to three times higher. Besides, the blade receives unsteady wake flows from
the upstream vane trailing edge. More importantly, blade rotation causes hot
gases to leak from the pressure side through a tip gap to the suction side. This
often causes damage on the blade tip near the pressure side trailing-edge

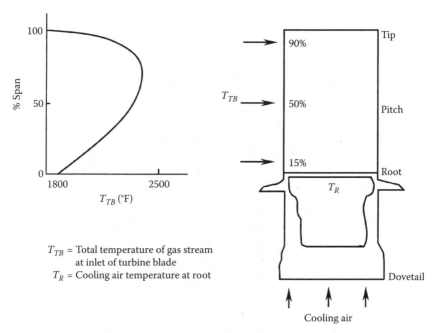

FIGURE 1.21
Typical average radial temperature profile at inlet of stage-1 turbine rotor blade.

region. It is important to understand the complex 3-D flow physics and associated heat-transfer distributions on the rotor blade, particularly near the tip region, under typical engine flow conditions. This can aid in designing a better cooling system and prevent blade failure due to insufficient cooling under high rotating thermal stress conditions.

Figure 1.21 shows typical average radial temperature profiles at the inlet of the stage-1 turbine rotor blade. It is important to note that rotation causes the peak temperature to shift from the blade pitch line toward the tip region. It is also important to predict correctly the RIT profile as well as the associated unsteady velocity profile and turbulence levels. Figure 1.22 depicts a complex flow phenomenon in a turbine rotor hot gas passage including secondary flows, tip flows, wakes, and rotation flows. Figure 1.23 shows a typical strategic cooling scheme for a turbine rotor blade, including film cooling, impingement cooling, and augmented convection cooling with ribs and pins. Figure 1.24 shows the variation of the heat-transfer rate around a turbine blade. The heat-transfer distributions for blades not film-cooled are higher than those for film-cooled blades at the same engine flow conditions. These heat-transfer distributions could differ for varied engine flow conditions; therefore, it is critical for a designer to be able to accurately predict these distributions for film-cooled or no film-cooled blades in order to design an efficient cooling scheme.

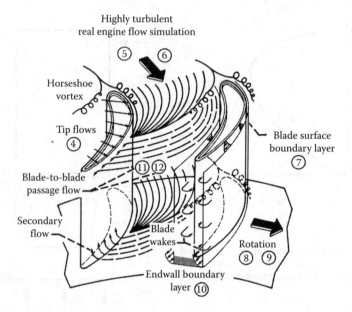

FIGURE 1.22

Complex flow phenomena in a turbine rotor hot-gas passage, including secondary flows, tip flows, wakes, and rotation flows. (From Gladden, H.J. and Simoneau, R.J., Review and assessment of the database and numerical modeling for turbine heat transfer, in *Toward Improved Durability in Advanced Aircraft Engine Hot Sections*, Sokolowski, D.E., Ed., IGTI, Norcross, GA, Vol. 2, pp. 39–55, 1988; collected in Sokolowski, D.E., Toward improved durability in advanced aircraft engine hot sections, *The 1988 ASME Turbo Expo*, Amsterdam, the Netherlands, June 5–9, 1988. With permission.)

1.3.2 Turbine Blade Internal Cooling

Jet impinging on the inner surfaces of the airfoil through tiny holes in the impingement insert is a common, highly efficient cooling technique for first-stage vanes. Part of the spent air moves through rows of distributed film holes on the necessary locations for film-cooling surface protection. The other part of the spent air moves toward the trailing-edge region and exits through tiny holes for trailing-edge region cooling. Impingement cooling is very effective because the cooling air can be delivered to impinge on the hot region. However, a spent-air cross flow effect can reduce the impingement cooling effect. It is important to determine the impingement heat-transfer coefficient distributions inside the turbine vane under typical jet hole configurations and distributions and engine cooling flow conditions.

Jet impingement cooling can be used only in the leading-edge region of the rotor blade, due to structure constraints on the rotor blade under high-speed rotation and high centrifugal loads. Figure 1.25 depicts the cooling concepts of a modern, multipass turbine rotor blade. Currently, a blade mid-chord region uses serpentine coolant passages with rib turbulators on the

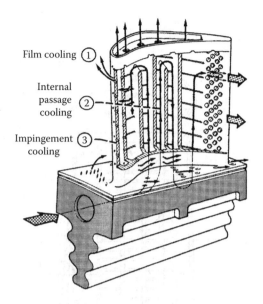

FIGURE 1.23

Typical cooled aircraft gas turbine blade. (From Gladden, H.J. and Simoneau, R.J., Review and assessment of the database and numerical modeling for turbine heat transfer, in *Toward Improved Durability in Advanced Aircraft Engine Hot Sections*, Sokolowski, D.E., Ed., IGTI, Norcross, GA, Vol. 2, pp. 39–55, 1988; collected in Sokolowski, D.E., Toward improved durability in advanced aircraft engine hot sections, *The 1988 ASME Turbo Expo*, Amsterdam, the Netherlands, June 5–9, 1988. With permission.)

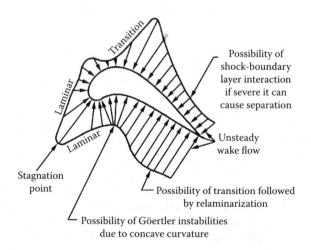

FIGURE 1.24

Variation of heat-transfer rate around a turbine blade. (From Daniels, L.C. and Schultz, D.L., Heat transfer rate to blade profiles—Theory and measurement in transient facilities, VKI Lecture Series, Vols. 1 and 2, 1982; Lakshminarayana, B.: *Fluid Dynamics and Heat Transfer of Turbomachinery*. Chapter 7, pp. 597–721. 1996. Copyright Wiley-VCH Verlag GmbH & Co. KGaA. With permission.)

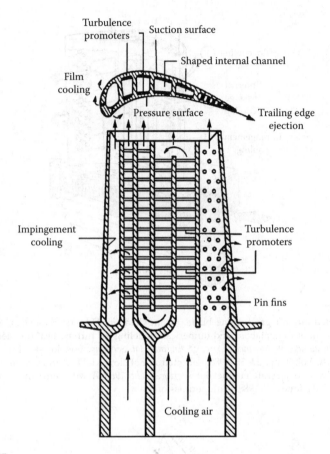

FIGURE 1.25
Cooling concepts of a modern multipass turbine blade. (From Han, J.C. et al., *Heat Transfer and Pressure Drop in Blade Cooling Channels with Turbulence Promoters*, Texas A&M University, College Station, TX (prepared for NASA CR-3837), 1984. With permission.)

inner walls of the rotor blades, while the blade trailing-edge region uses short pins due to space limitation and structure integration. Heat transfer in rotating coolant passages is very different from that in stationary coolant passages. Both Coriolis and rotating buoyancy forces can alter the flow and temperature profiles in the rotor coolant passages and affect their surface-heat-transfer coefficient distributions. It is very important to determine the local-heat-transfer distributions in the rotor coolant passages with impingement cooling, rib-turbulated cooling, or pinned cooling under typical engine cooling flow, coolant-to-blade temperature difference (buoyancy effect), and rotating conditions. It is also important to determine information on the associated coolant passage pressure losses for a given internal cooling design. This can help in designing an efficient cooling system and prevent local hot-spot overheating of the rotor blade.

1.3.3 Turbine Blade Film Cooling

Film cooling depends primarily on the coolant-to-hot-mainstream pressure ratio (p_c/p_t), temperature ratio (T_c/T_g), and the film-cooling hole location, configuration, and distribution on a film-cooled airfoil. The coolant-to-mainstream pressure ratio can be related to the coolant-to-mainstream mass flux ratio (blowing ratio), while the coolant-to-mainstream temperature ratio can be related to the coolant-to-mainstream density ratio. In a typical gas turbine airfoil, the p_c/p_t ratios vary from 1.02 to 1.10, while the corresponding blowing ratios vary approximately from 0.5 to 2.0. Whereas the T_c/T_g values vary from 0.5 to 0.85, the corresponding density ratios vary approximately from 2.0 to 1.5. Both the pressure (p_c/p_t) and temperature (T_c/T_g) ratios are probably the most useful measure in quantifying film-cooling effectiveness, since these ratios essentially give the ratio of the coolant-to-hot-mainstream thermal capacitance. In general, the higher the pressure ratio, the better the film-cooling protection (i.e., reduced heat transfer to the airfoil) at a given temperature ratio, while the lower the temperature ratio, the better the film-cooling protection at a given pressure ratio. However, a too high pressure ratio (i.e., blowing too much) may reduce the film-cooling protection because of jet penetration into the mainstream (jet liftoff from the surface). Therefore, it is important to optimize the amount of coolant for airfoil film cooling under engine operating conditions (Reynolds number ~10^6; Mach number $\approx$ 0.9 at exit conditions). It is also important to determine the effects of free-stream turbulence and unsteady wakes on the airfoil film-cooling performance (i.e., film-cooling effectiveness and heat-transfer coefficient distributions). As mentioned earlier, turbine-cooling system designers need to know where the heat is transferred from the hot mainstream to the airfoil in order to design better film-cooling patterns for airfoils. These film-hole patterns (i.e., film-hole location, distribution, angle, and shape) affect film-cooling performance.

1.3.4 Thermal Barrier Coating and Heat Transfer

For safer operation, the turbine blades in current engines use nickel-based superalloys at metal temperatures well below 2000°F for safe operations. For higher RIT, the advanced casting techniques, such as directionally solidified and single crystal blades with TBC coating, have been proposed for advanced gas turbines. TBC coating serves as insulation for the turbine airfoils and allows a 200°F–300°F higher RIT, thereby enhancing turbine efficiency. There are two types of coating techniques: (1) air plasma spray with plate structure/porosity/low thermal conductivity and (2) electron beam physical vapor deposition with column structure/dense/high thermal conductivity (Nelson et al., 1995). The performance of TBC coatings, the zirconia-based ceramics, depends on the aforementioned coating techniques and the coating thickness (5–50 miles). The U.S. government laboratories and gas turbine manufacturers, as well as industry, are conducting research to identify better coating materials, better coating techniques, controllable coating thicknesses, good bonding coats, and hot corrosion tests for TBC life prediction. It is important to determine the effects of TBC roughness and the potential TBC spallations on turbine aerodynamic and heat-transfer performance.

1.4 Structure of the Book

Chapter 1: Fundamentals. This chapter discusses the need for turbine cooling, turbine-cooling technology, turbine heat-transfer and cooling issues, over-all structure of the book, and review articles and book chapters on turbine blade cooling and heat transfer. This book focuses on heat-transfer and cooling technology in turbine hot-gas path components, including turbine stator vanes and rotor blades. However, the book does not include the combustor liner cooling and turbine disk cooling, although they are important to hot-gas path component designs.

Chapter 2: Turbine Heat Transfer. This chapter includes information on turbine stage heat transfer; cascade vane and blade heat transfer; endwall, blade-tip, and leading-edge heat transfer; and flat-surface heat transfer. It also discusses turbine rotor and stator heat transfer, including endwall and blade-tip region under engine conditions, and cascade vane and blade heat transfer under simulated engine conditions. This follows the simulated endwall, blade-tip, and leading-edge region heat transfer. Finally, the simplest flat-plate heat transfer is included.

Chapter 3: Turbine Film Cooling. This chapter discusses turbine heat transfer, which includes film cooling on rotating blades, on cascade vane and blade simulations, film cooling on endwall, blade-tip, leading-edge regions, and flat surfaces. Turbine rotor and stator blade film cooling under engine conditions are reviewed first, followed by a discussion on the high free-stream turbulence effect on simulated cascade vanes and the unsteady wakes effect on simulated cascade blades. Then the simulated endwall, blade-tip, and leading-edge region film cooling under low-speed flow conditions are presented, followed by flat-plate film cooling.

Chapter 4: Turbine Internal Cooling. This chapter includes impingement cooling, rib-turbulated cooling, pin-fin cooling, and compound and new cooling techniques. It also deals with heat-transfer and pressure drop correlations for ranges of cooling flow and geometric conditions applicable for turbine stator blade designs.

Chapter 5: Turbine Internal Cooling with Rotation. This chapter discusses the effects of rotation. It includes rotating effect on coolant passage heat transfer with smooth and rib-turbulated walls, rotating effect on impingement cooling with smooth and rib-turbulated walls under simulated engine cooling flow, and geometric conditions.

Chapter 6: Experimental Methods. This chapter addresses heat-transfer measurements, mass-transfer analogy techniques, liquid crystal thermography, optical techniques, flow, and thermal-field measurement techniques. Sample cases are given and discussed for each measurement technique. Experimental methods are important and useful for determining turbine blade heat-transfer and film-cooling performance. These methods can provide detailed heat-transfer distributions and help understand how heat is transferred from the turbine blade hot-gas path and how to remove it from the coolant passages.

Chapter 7: Numerical Modeling. Experimental results are useful but expensive, especially for engine condition tests. CFD simulations on turbine blade heat transfer and cooling are useful and are becoming more feasible in the future. This chapter discusses governing equations and turbulence models, hot-gas path heat-transfer predictions, film-cooling predictions, and coolant passage heat-transfer predictions. CFD for turbine heat-transfer and cooling predictions is relatively new. Although progress has been made in the past 10 years, there is plenty of room for improvement in this area. It is important to have the predicted results validated with the experimental data under engine flow and geometric conditions. It is hoped that the CFD can be a feasible design tool for further turbine-cooling applications.

Chapter 8: Final Remarks. This chapter provides some suggestions regarding the next step in research in order to gain significant heat-transfer and cooling information for the next century's gas turbine blades.

1.5 Review Articles and Book Chapters on Turbine Cooling and Heat Transfer

Several publications are available that address state-of-the-art reviews of turbine blade cooling and heat transfer. These include "Film cooling" by Goldstein (1971), "Turbine cooling" by Suo (1978), *Heat Transfer and Fluid Flow in Rotating Coolant Channels* by Morris (1981), "Cooling techniques for gas turbine airfoils" by Metzger (1985), and "Turbine blade heat transfer" by Moffat (1987). Others are "Some considerations in the thermal design of turbine airfoil cooling systems" by Elovic and Koffel (1983), "Heat transfer problems in aero-engines" by Hennecke (1984), "Aero-thermal aspects of gas turbine flows: Turbine blading internal cooling and external effects" by Harasgama (1995), "Internal convection heat transfer and cooling: An experimental approach" by Han and Dutta (1995), and "Turbine cooling and heat transfer" by Lakshminarayana (1996). Several review articles related to gas turbine heat-transfer problems (Graham, 1979, 1990; Taylor, 1980; Gladden and Simoneau, 1988; Jones, 1988; Simoneau and Simon, 1993) are also available.

1.6 New Information from 2000 to 2010

This *Gas Turbine Heat Transfer and Cooling Technology* book has been published since 2000. There have been many new technical papers available in open literature over the past 10 years. These new published data provide gas turbine researchers, designers, and engineers with advanced heat-transfer analysis and cooling technology references. There is a need to revise the first edition book by including these amounts of available new information in order to keep this book updated for users. The revised book provides information on state-of-the-art cooling technologies such as advanced turbine blade film cooling and internal cooling schemes as sketched in Figures 1.26 (Han and Wright, 2007) and 1.27 (Han and Rallabandi, 2010). The revised book updates modern experimental methods for gas turbine heat transfer and cooling research, as well as advanced computational models for gas turbine heat transfer and cooling performance predictions.

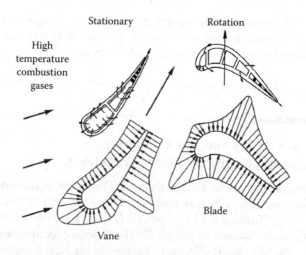

FIGURE 1.26
Gas turbine blade thermal loading schematic. (From Han, J.C. and Wright, L.M., Enhanced internal cooling of turbine blades and vanes. In *The Gas Turbine Handbook*, U.S. DOE, National Energy Technology Laboratory, Morgantown, WV, pp. 321–352, 2007. With permission.)

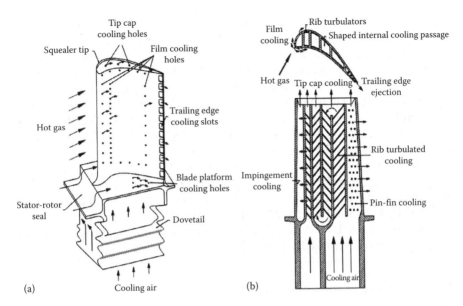

FIGURE 1.27
Gas turbine blade cooling schematic. (a) External cooling; (b) internal cooling. (From Han, J.C. and Rallabandi, A.P., *Front. Heat Mass Transfer*, 1(1), 1, 2010. With permission.)

1.6.1 ASME Turbo Expo Conference CDs

ASME Turbo Expo (IGTI International Gas Turbine Institute) has made conference CDs available to every year's attendees since 2000 (GT2000–GT2010). These conference CDs contain all gas turbine heat-transfer papers presented in each year's IGTI conference. The numbers of heat-transfer-related conference papers have increased from about 100 in the year 2000 to about 200 in the year 2010. Approximately 25%–30% of each year's conference heat-transfer papers have subsequently published in the *ASME Journal of Turbomachinery*. These tremendous amounts of conference and journal papers are the main research sources of gas turbine heat transfer and cooling technology for the interested readers.

1.6.2 Book Chapters and Review Articles

In addition, several books have been published since 2000: *Heat Transfer in Gas Turbines* edited by Sunden and Faghri (2001), *Heat Transfer in Gas Turbine Systems* edited by Goldstein (2001), *The Gas Turbine Handbook* edited by DOE-National Energy Technology Laboratory (2006), "Special section: Turbine science and technology" (included 10 review papers) edited by Shih (2006), *Heat Transfer in Gas Turbine Systems* (included 10 keynote papers) edited by Simon and Goldstein (2009), "Advanced Turbine Program Annual Report" edited by DOE-National Energy Technology Laboratory (2010). Meanwhile, many review papers related to gas turbine heat transfer and cooling problems are available: "Convection heat transfer and aerodynamics in axial flow turbines" by Dunn (2001), "HSystem technology update" by Pritchard (2003), "A review of W501G engine design" by Diakunchak et al. (2003), "Application of H gas turbine design technology" by Kallianpur et al. (2003), "A review of turbine shaped film cooling technology" by Bunker (2005), "Gas turbine heat transfer: 10 remaining hot gas path challenges" by Bunker (2006), "Gas turbine film cooling" by Bogard and Thole (2006), "Enhanced internal cooling of turbine blades and vanes" by Han and Wright (2007), "The challenges facing the utility gas turbine" by Kiesow and McQuiggan (2007), "The effects of manufacturing tolerances on gas turbine cooling" by Bunker (2008), "Turbine cooling systems design—Past, present and future" by Downs and Landis (2009), "Aerothermal challenges in syngas hydrogen-fired and oxy-fuel turbines" by Chyu et al. (2009), "Recent studies in turbine blade internal cooling" by Han and Huh (2009), "Film cooling: Breaking the limits of diffusion shaped holes" by Bunker (2009), "Turbine blade film cooling using PSP technique" by Han and Rallabandi (2010).

1.6.3 Structure of the Revised Book

This is a revision for the second edition, not a new book. Major book contents and framework have already been done in the 1st edition book. To keep the same book format, the revised second edition adds new information, mainly based on selected 2000–2010 published papers in open literature, at the end of each chapter. Chapter 1—Fundamentals; Chapter 2—Turbine Heat Transfer: turbine vane and blade interactions, deposition and roughness effects, combustor effects on turbine components, transition fundamentals, turbine tip and casing heat transfer, end-wall heat transfer; Chapter 3—Turbine Film Cooling: airfoil film cooling, endwall film cooling, shaped hole film cooling, tip and trailing edge film cooling; Chapter 4—Turbine Internal Cooling: pin fin and dimple cooling, impingement and compound cooling, rib turbulated cooling, combustor liner cooling; Chapter 5—Turbine Internal Cooling with Rotation: rotating heat transfer; Chapter 6—Experimental Methods: heat flux gauges, liquid crystals, IR camera, PSP, TSP, LDV and PIV; Chapter 7—Numerical Modeling: RANS and Unsteady RANS, LES, and conjugate heat transfer; Chapter 8—Final Remarks.

References

ASME Turbo Expo 2000. *IGTI Conference CDs*, May 8–11, 2000, Munich, Germany.
ASME Turbo Expo 2001. *IGTI Conference CDs*, June 4–8, 2001, New Orleans, LA.
ASME Turbo Expo 2002. *IGTI Conference CDs*, June 3–6, 2002, Amsterdam, the Netherlands.
ASME Turbo Expo 2003. *IGTI Conference CDs*, June 16–19, 2003, Atlanta, GA.

ASME Turbo Expo 2004. *IGTI Conference CDs*, June 14–17, 2004, Vienna, Austria.

ASME Turbo Expo 2005. *IGTI Conference CDs*, June 6–9, 2005, Reno, NV.

ASME Turbo Expo 2006. *IGTI Conference CDs*, May 8–11, 2006, Barcelona, Spain.

ASME Turbo Expo 2007. *IGTI Conference CDs*, May 14–17, 2007, Montreal, Quebec, Canada.

ASME Turbo Expo 2008. *IGTI Conference CDs*, June 9–13, 2008, Berlin, Germany.

ASME Turbo Expo 2009. *IGTI Conference CDs*, June 8–12, 2009, Orlando, FL.

ASME Turbo Expo 2010. *IGTI Conference CDs*, June 14–18, 2010, Glasgow, U.K.

Bogard, D.G. and Thole, K.A., 2006. Gas turbine film cooling. *Journal of Propulsion and Power*, 22(2), 249–270.

Bunker, R.S., 2005. A review of turbine shaped film cooling technology. *Journal of Heat Transfer*, 127, 441–453.

Bunker, R.S., 2006. Gas turbine heat transfer: 10 remaining hot gas path challenges. ASME Paper GT2006-90002.

Bunker, R.S., 2008. The effects of manufacturing tolerances on gas turbine cooling. ASME Paper GT2008-50124.

Bunker, R.S., 2009. Film cooling: Breaking the limits of diffusion shaped holes, *Turbine09 Symposium on Heat Transfer in Gas Turbine Systems*, August 9–14, 2009, Antalya, Turkey.

Chyu, M.K., Mazzotta, D.W., Siw, S.C., Karaivanov, V.G., Slaughter, W.S., and Alvin, M.A., 2009. Aerothermal challenges in syngas hydrogen-fired and oxyfuel turbines. *ASME Journal of Thermal Science and Engineering Applications*, 1, 011002, 011003.

Clifford, R.J., 1985. Rotating heat transfer investigations on a multipass cooling geometry. AGARD CP 390.

Corman, J.C. and Paul, T.C., 1995. Power systems for the 21st century "H" gas turbine combined cycles. GE Power Systems, Schenectady, New York, GER-3935, pp. 1–12.

Daly, K., 1993. How the products of the integrated high performance turbine engine technology initiative are entering into the real world. *Flight International*, 14, 28–30.

Daniels, L.C. and Schultz, D.L., 1982. Heat transfer rate to blade profiles—Theory and measurement in transient facilities. VKI Lecture Series, Vols. 1 and 2.

Davis, J.M. and Randolph, J.G., 1993. Comprehensive program plan for advanced turbine systems. U.S. Department of Energy, Office of Fossil Energy, Office of Energy Efficiency and Renewable Energy, DOE/FE-0279, pp. 1–44.

Diakunchak, I., McQuiggan, G., and Bancalari, E., 2003. A review of W501G engine design: Development and field operating experience. ASME Paper GT2003-38843.

Downs, J.P. and Landis, K.K., 2009. Turbine cooling systems design—Past, present and future. ASME Paper GT2009-59991.

Dunn, M.G., 2001. Convection heat transfer and aerodynamics in axial flow turbines. *Journal of Turbomachinery*, 123(4), 637–686.

Elovic, E. and Koffel, W.K., 1983. Some considerations in the thermal design of turbine airfoil cooling systems. *International Journal of Turbo and Jet-Engines*, 1(1), 45–65.

Gladden, H.J. and Simoneau, R.J., 1988. Review and assessment of the database and numerical modeling for turbine heat transfer. In *Toward Improved Durability in Advanced Aircraft Engine Hot Sections*, Sokolowski, D.E., Ed. IGTI, Norcross, GA, Vol. 2, pp. 39–55.

Goldstein, R.J., 1971. Film cooling. In *Advancement in Heat Transfer*, Irvine, Jr. and Hartnett, Eds. Academic Press, New York, Vol. 7, pp. 321–379.

Glodstein, R.J., Ed., 2001. *Heat Transfer in Gas Turbine Systems*. Annals of the New York Academy of Sciences, New York, Vol. 934.

Graham, R.W., 1979. Fundamental mechanisms that influence the estimate of heat transfer to gas turbine blades. ASME Paper 79-HT-43.

Graham, R.W., 1990. Recent progress in research pertaining to estimates of gas side heat transfer in an aircraft gas turbine. ASME Paper 90-GT-100.

Halila, E.E., Lenahan, D.T., and Thomas, T.T., 1982. Energy efficient engine. General Electric Company, Fairfield, CT (prepared for NASA CR-167955).

Han, J.C. and Dutta, S., 1995. Internal convection heat transfer and cooling: An experimental approach, Arts, T., Ed. VKI Lecture Series, 1995-05.

Han, J.C. and Huh, M., 2009. Recent studies in turbine blade internal cooling. *Turbine-09*, August 9–14, Antalya, Turkey.

Han, J.C., Park, J.S., and Lie, C.K., 1984. *Heat Transfer and Pressure Drop in Blade Cooling Channels with Turbulence Promoters*. Texas A&M University, College Station, TX (prepared for NASA CR-3837).

Han, J.C. and Rallabandi, A.P., 2010. Turbine blade film cooling using PSP technique. *Frontiers in Heat and Mass Transfer*, 1(1), 1–21.

Han, J.C. and Wright, L.M., 2007. Enhanced internal cooling of turbine blades and vanes. In *The Gas Turbine Handbook*, U.S. National Energy Technology Laboratory, Morgantown, WV, Section 4-2-2-2, pp. 321–352.

Harasgama, S.P., 1995. Aero-thermal aspects of gas turbine flows: Turbine blading internal cooling and external effects, Arts, T., Ed. VKI Lecture Series 1995-05.

Hennecke, D.K., 1984. Heat transfer problems in aero-engines. In *Heat and Mass Transfer in Rotating Machinery*, Metzger, D.E. and Afgan, N.H., Eds. Hemisphere, Washington, DC, pp. 353–379.

Jones, T.V., 1988. Gas turbine studies at Oxford 1969–1987. ASME Paper 88-GT-112.

Kallianpur, V., Iwasaki, Y., and Fukuizumi, Y., 2003. Application of H gas turbine design technology to increase thermal efficiency and output capability of the Mitsubishi M701G2 gas turbine. ASME Paper GT2003-38956.

Kiesow, H.J. and McQuiggan, G., 2007. The challenges facing the utility gas turbine. ASME Paper GT2007-27180.

Lakshminarayana, B., 1996. Turbine cooling and heat transfer. In *Fluid Dynamics and Heat Transfer of Turbomachinery*. John Wiley, New York, Chapter 7, pp. 597–721.

Mattingly, J.D., 1996. *Elements of Gas Turbine Propulsion*. McGraw-Hill, New York, p. 229.

Metzger, D.E., 1985. Cooling techniques for gas turbine airfoils. AGARD CP 390, pp. 1–12.

Moffat, R.J., 1987. Turbine blade heat transfer. In *Heat Transfer and Fluid Flow in Rotating Machinery*, Yang, W.J., Ed. Hemisphere, Washington, DC, pp. 1–24.

Morris, W.D., 1981. *Heat Transfer and Fluid Flow in Rotating Coolant Channels*. Research Studies Press, Hertfordshire, England, pp. 1–228.

Nelson, W.A., Orenstein, R.M., Dimascio, P.S., and Johnson, C.A., 1995. Development of advanced thermal barrier coatings for severe environments. ASME Paper 95-GT-270.

Pritchard, J., 2003. HSystem technology update. ASME Paper GT2003-38711.

Rohlik, H.E., 1983. Current and future technology trends in radial and axial gas turbines. NASA TM 83414.

Sautner, M., Clouser, S., and Han, J.C., 1992. Determination of surface heat transfer and film cooling effectiveness in unsteady wake flow conditions. *AGARD Conference Proceedings*, Vol. 527, pp. 6-1–6-12.

Shih, T.I.-P., 2006. Special section: Turbine science and technology. *Journal of Propulsion and Power*, 22(2), 225–396.

Simon, T.W. and Goldstein, R.J., 2009. Heat transfer in gas turbine systems. *Turbine-09* (edited), August 9–14, Antalya, Turkey.

Simoneau, R.J. and Simon, F.F., 1993. Progress towards understanding and predicting convection heat transfer in the turbine gas path. *International Journal of Heat Fluid Flow*, 14(2), 106–127.

Sunden, B. and Faghri, M., 2001. *Heat Transfer in Gas Turbines* (edited). WIT Press, Boston, MA.

Suo, M., 1978. Turbine cooling. In *Aerothermodynamics of Aircraft Gas Turbine Engines*, Oates, G., Ed. Air Force Aero Propulsion Laboratory, Wright-Patterson Air Force Base, OH, AFAPL TR 78-5, Chapter 19, pp. 19-1–19-23.

Taylor, J.R., 1980. Heat transfer phenomena in gas turbines. ASME Paper 80-GT-172.

Treager, I.E., 1979. General Electric CF6. In *Aircraft Gas Turbine Engine Technology*, 2nd edn. McGraw-Hill, New York, Chapter 25, pp. 469–525.

U.S. DOE, 2006. National Energy Technology Laboratory (edited). *The Gas Turbine Handbook*. DOE/NETL-2006/1230, Pittsburgh, PA.

U.S. DOE, 2010. Turbines for coal based power systems that capture carbon. Advanced Turbine Program Annual Report (edited). DOE/NETL-2010/1437, Pittsburgh, PA.

2

Turbine Heat Transfer

2.1 Introduction

In this chapter, the focus will be on the effect of various parameters on turbine component external heat-transfer characteristics. Advances in combustor design have resulted in higher turbine inlet temperatures, which in turn affect the heat load to the hot-gas-path components. It is important for the gas turbine designer to know the effects of increased heat load to gas-path components to design efficient cooling schemes to protect the components. The combustor exit gases are highly turbulent, with turbulence levels up to 20%–25% at the first-stage stator vane leading edges. Primary hot-gas-path components are the stationary nozzle guide vanes (NGV) and the rotating blades. Turbine shrouds, blade tips, and platforms and stator endwalls also represent critical areas in the hot-gas path. Fundamental and applied studies relating to all the aforementioned components have helped in providing better understanding and predicting heat load more accurately. Most heat-transfer studies relating to hot-gas-path components are large-scale models run at simulated conditions to provide a fundamental understanding of the phenomena. The components have been simulated using flat and curved surfaces, leading-edge models, and scaled airfoil cascades. In this chapter, the focus will be on the experimental heat-transfer results obtained by various researchers on gas-path components. Heat transfer to first-stage vanes is primarily influenced by such parameters as combustor outlet temperature profile, high free-stream turbulence, and hot streaks. Heat transfer to first-stage rotor blades is affected by medium to low free-stream turbulence, unsteady wakes shed by upstream vane trailing edges, hot streaks, and, of course, rotation.

2.1.1 Combustor Outlet Velocity and Temperature Profiles

Turbulence levels in a combustor are very important due to the significant effect on the convective heat transfer to hot-gas-path components in the turbine. Turbulence affecting the heat transfer in turbines is generated in the combustor due to the mixing of fuel with the compressor gases. Knowledge of combustor-generated turbulence intensity is important for designers in estimating heat-transfer levels in the turbine. Reduced combustor turbulence can lead to reduction in thermal load to turbine components, and that will lead to longer life and also reduce cooling requirements. There are, however, few studies that have focused on measuring combustor outlet velocity and turbulence profiles.

Goldstein et al. (1983) presented exit velocity and turbulence profiles for model combustors. Moss and Oldfield (1991) presented turbulence spectra at combustor exits. Both the studies, however, were at atmospheric pressure and low temperature. Although it is difficult to obtain measurements under real conditions, it is essential for a gas turbine designer to understand the gas profiles exiting the combustor. This information will help in the improvement of combustor geometry, which will have significant effects on turbine-cooling requirements.

Recently, Goebel et al. (1993) measured combustor velocity and turbulence profiles downstream of a small-scale combustor using a laser Doppler velocimeter (LDV) system. They presented normalized velocity, turbulence, and exit temperature profiles for all combustion tests. They used a single can-type combustor for extensive measurements. Figure 2.1 shows a typical can-type combustor configuration used in modern gas turbine engines (Ames, 1997). The flow from the compressor enters into the combustor through holes and mixes with the burning fuel at different locations downstream. Combustor design requires a minimal pressure drop through the combustor to the turbine inlet. The combustor process is controlled by gradual mixing of compressed air with the fuel in the can combustor. Modern combustor designers also focus on swirl problems and fuel-air mixing process. Clean combustion is also a focus for designers due to stringent environmental standards required by the U.S. federal government and the EPA. However, combustor design is not a focus in this book.

Figure 2.2 illustrates the effect of combustion on axial velocity, axial turbulence intensity, swirl velocity, and swirl turbulence intensity. All velocities

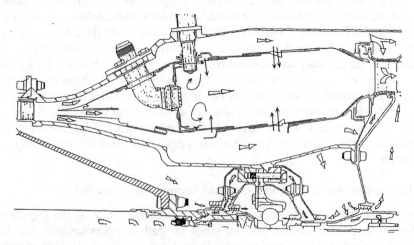

FIGURE 2.1
Typical can combustor configuration. (From Ames, F.E., Film cooling: A short course, Conducted by DOE-AGTSR, Clemson, SC, 1997a.)

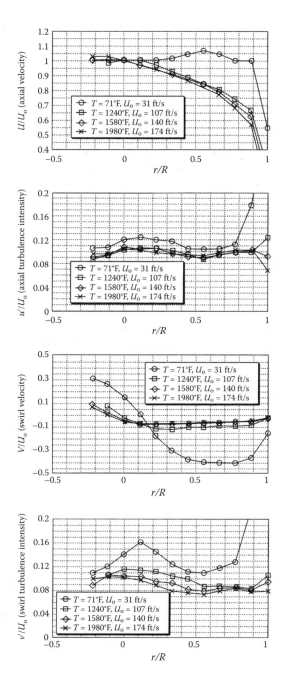

FIGURE 2.2
Effect of combustion on axial velocity, axial turbulence intensity, swirl velocity, and swirl turbulence intensity. (From Goebel, S.G. et al., Measurements of combustor velocity and turbulence profiles, ASME Paper 93-GT-228, 1993.)

are normalized by the measured centerline velocity and plotted against normalized radius. The pressure and mass flow of air were kept the same for different strengths of combustion. Increasing the fuel flow increased combustion strength. The adiabatic flame temperature was varied. It is important to note here that the compressed air in a gas turbine engine is pre-heated due to the compression process. However, in this study, the air is not preheated. The mass flow and the pressure were 0.45 kg/s (1.0 lbm/s) and 6.8 atm, respectively. Firing temperatures ranged from 71°F (unfired-295 K) to 1980°F (1356 K). The effect of combustion is strongly evident when comparing the unfired case with the rest of the fired cases. The axial velocity and the swirl velocity are strongly influenced by combustion. The levels of swirl are greatly reduced by combustion. The reduction in swirl can also be seen in the swirl turbulence intensity. The peak levels in swirl turbulence intensity were reduced from 16% to 10% from the unfired to the fully fired case.

Temperature profiles were also measured for the combustion cases. Figure 2.3 compares the effect of dilution holes for similar firing temperatures (1200°F). The temperature profile appears fairly flat and uniform without dilution holes, with peak levels at the centerline. However, the addition of dilution holes decreases the temperature levels between centerline and the edges. Knowing the combustor outlet temperature profile is a must for gas-path heat-transfer calculations. Temperature exit profile measurements are a routine for gas turbine manufacturers. The inlet gas temperature profiles are needed for gas-path component heat-transfer calculations for estimating component temperatures.

It is difficult to compare combustor temperature profiles due to uniqueness of design. The combustor configuration dictates the exit temperature profile. However, the aforementioned studies provide some insights into the velocity, turbulence intensity, and temperature profiles, and the effects of

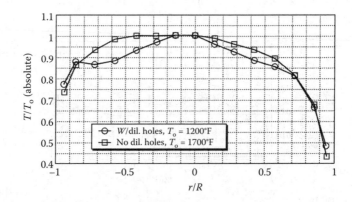

FIGURE 2.3
Effect of dilution holes on spanwise temperature distributions for similar firing temperatures. (From Goebel, S.G. et al., Measurements of combustor velocity and turbulence profiles, ASME Paper 93-GT-228, 1993.)

combustion on them. Turbulence length scale is also an important factor for gas-path heat transfer. However, the aforementioned studies do not provide any length scale information.

2.2 Turbine-Stage Heat Transfer

2.2.1 Introduction

A turbine stage consists of a row of NGV or stators and a row of rotating blades called rotors. The working fluid enters the vane passage and is deflected in the direction of the rotor blades' leading edge. A part of the fluid energy is converted into mechanical energy due to the rotational motion of the rotor blades. The rotor blades are attached to the turbine shaft. The rotating motion transferred to the shaft is used to run the compressor. Figure 2.4 shows a typical turbine stage comprised of an NGV passage and a rotor blade passage. The velocity diagram for the stage is also indicated.

2.2.2 Real Engine Turbine Stage

It is important to understand the heat-transfer aspects to all turbine components under actual conditions. Typically, measurements on a single-stage turbine under engine conditions can be used to provide all the heat-transfer

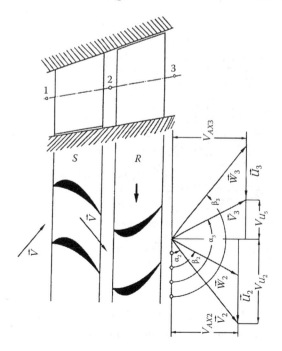

FIGURE 2.4
Typical turbine stage comprised of NGV and rotor passages.

information on gas-path components. Instrumentation and experimentation on actual turbine stages under engine conditions are very rare. The lack of accurate high-temperature measurement tools and difficulty in instrumentation of the turbine stage with temperature and pressure measuring devices are some of the reasons for the few attempts to study turbine heat transfer on an actual stage under actual engine conditions.

Most of the primary results on real rotor/stator heat transfer have been provided by Dunn et al. at Calspan Advanced Technology Center. Dunn et al. (Dunn and Hause, 1979; Dunn, 1984, 1986; Dunn et al., 1984, 1986a,b) provided a considerable amount of information on heat flux measurements for NGV (endwall and airfoil), the rotor blade, rotor tip, platform, and the shroud. Dunn et al. used a full-stage rotating turbine of the Garrett TFE 731-2 engine. They reported heat flux measurements on the NGV, rotor, and shroud of the turbine. A shock-tunnel facility was intended to provide well-defined flow conditions and duplicate sufficient number of parameters to validate and improve confidence in design data and predictive techniques under development. Static pressure measurements were obtained using pressure transducers over the entire turbine section. Thin-film heat flux gauges were installed in the turbine stage on the NGV tip endwall, suction and pressure surfaces, leading edge, rotor platform, shroud, and tip and rotor suction and pressure surfaces. Figure 2.5 shows the shock-tunnel facility used by Dunn et al.

Dunn et al. (1986b) also provided heat flux and pressure measurements for a "low-aspect-ratio turbine stage." In the aforementioned studies, they investigated a high-pressure turbine stage with an aspect ratio of approximately 1.5.

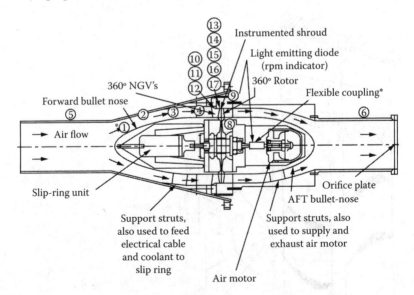

FIGURE 2.5

Shock-tunnel facility used by Dunn et al. (From Dunn, M.G. et al., *ASME J. Eng. Gas Turbines Power*, 106, 229, 1984.) Numbers in circles denote location of pressure transducers.

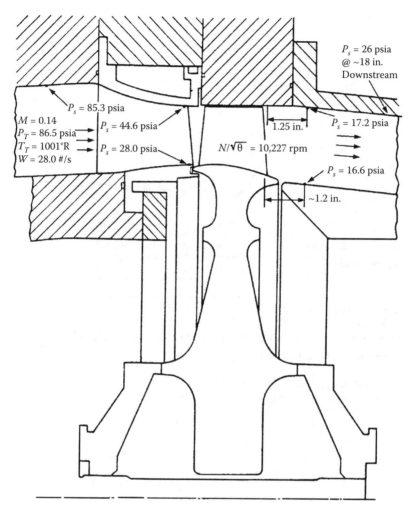

FIGURE 2.6
Sketch of the LART stage used by Dunn et al. (From Dunn, M.G. et al., *ASME J. Turbomach.*, 108, 108, 1986b.)

An Air Force/Garrett low-aspect-ratio turbine (LART) with an aspect ratio of approximately 1.0 was used for this study. A similar shock tube wind tunnel facility was used as in the earlier studies. Figure 2.6 shows a sketch of the LART stage. The inlet Mach number, total pressure, total temperature, and weight flow are all shown on the figure. Measurements were obtained on the NGV, NGV hub, and tip endwalls, and the rotor blade for the stage.

Figure 2.7 shows the measured pressure distributions on the NGV and the rotor meanlines. The pressure distributions clearly show the highest and the lowest velocity locations on the NGV and blade surfaces along the centerline. Figure 2.8 presents the Stanton number distribution for the

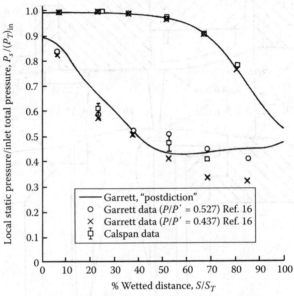

Pressure distribution for LART IIB NGV meanline

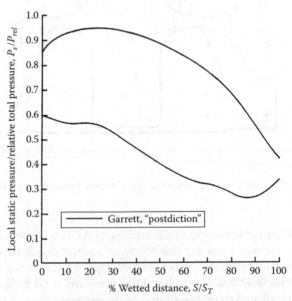

Pressure distribution for LART IIA rotor meanline

FIGURE 2.7

Measured pressure distributions on the LART NGV and rotor meanlines. (From Dunn, M.G. et al., *ASME J. Turbomach.*, 108, 108, 1986b.)

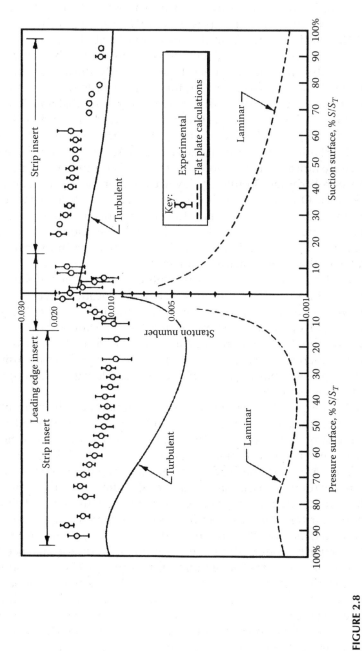

FIGURE 2.8
Stanton number distribution on the LART NGV meanline. (From Dunn, M.G. et al., *ASME J. Turbomach.*, 108, 108, 1986b.)

NGV meanline. The solid and dashed lines on the plot indicate the calculated Stanton number distributions based on turbulent and laminar flat-plate correlations, respectively. The highest Stanton number occurs at about 2% surface distance on the pressure surface. Then the Stanton number drops rapidly on the pressure surface to about half the value at 11% surface distance, and then increases gradually over the entire pressure surface to the trailing edge. From the pressure distributions, it is evident that the flow on the pressure surface is very slow over the initial 50% of the surface distance and then accelerates rapidly toward the trailing edge. On the suction surface, the Stanton number falls rapidly from the leading edge to about 6% surface distance. The pressure distribution indicates a highly accelerated flow in this region. The pressure distribution indicates that the peak velocity on the suction surface is at 50% surface distance. The Stanton number increases from 6% surface distance to about 25% surface distance. This behavior in the Stanton number is attributed to the laminar–turbulent boundary layer transition occurring about 6% surface distance. Once transition is completed at 25% surface distance, the Stanton number decreases gradually toward the trailing edge. From the correlations, it appears that both the pressure and suction surfaces have strongly turbulent boundary layers, and the predicted Stanton numbers are much lower than the measured values.

Dunn et al. (1986b) also made measurements on the NGV hub and tip endwalls. Figure 2.9 presents Stanton number data for locations near pressure surface endwall, mid-endwall, and the near suction surface endwall regions. Both the hub and tip endwalls show similar trends. Stanton numbers are unaffected for about 60% surface distance from leading edge to trailing edge and then increase toward trailing edge. The higher Stanton numbers near the trailing edges may be due to the accelerated flow going through the throat of the vane passage.

Figure 2.10 presents the Stanton number distribution on the rotor blade. The resolution of the data is not as good as for the NGV due to the additional problem of obtaining data on a rotating component. The rotation of the airfoil appears to nullify some of the effects seen on the NGV. Similar Stanton number distributions on the blade suction and pressure surfaces can be attributed to the rotation of the blade. Dunn et al. indicate that they observed that rotation effect reduces the variations of the Stanton number distribution over the entire airfoil. The peak Stanton number occurred at approximately 3.5% surface distance on the pressure side. The suction surface Stanton number falls rapidly from leading edge to about 30% surface distance. The pressure distribution for the blade indicates that the flow becomes sonic at about 37% surface distance on the suction side. At this point, the Stanton number increases over the surface and reaches another maximum around 70% surface distance. Beyond 70% surface distance, the Stanton numbers are expected to decrease toward the trailing edge. However, Dunn et al. do not have any measurements near the trailing-edge region except for a single

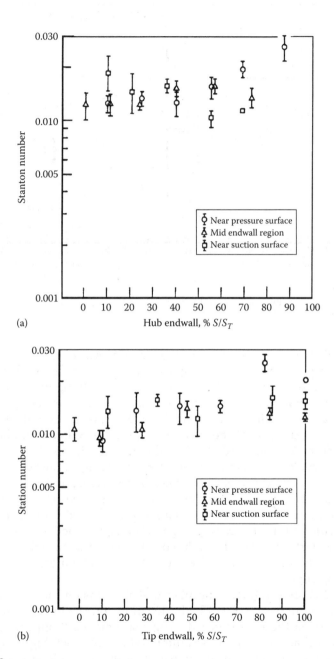

FIGURE 2.9
Stanton number distribution for LART NGV (a) hub endwall and (b) tip endwall. (From Dunn, M.G. et al., *ASME J. Turbomach.*, 108, 1986b.)

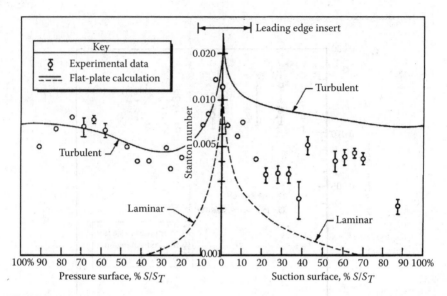

FIGURE 2.10
Stanton number distributions on the rotor blade. (From Dunn, M.G. et al., *ASME J. Turbomach.*, 108, 108, 1986b.)

point at 90% surface distance. On the blade pressure surface, the Stanton number drops from a maximum at 3.5% surface distance to a minimum at about 25% surface distance. This is a region with strong pressure gradient causing deceleration of flow on the pressure surface. Further downstream, the Stanton number increases again to a high value at around 70% surface distance as in the case of the suction surface. The Stanton number values decrease from 70% surface distance to the trailing edge on the pressure surface. The turbulent and laminar boundary layer predictions are also shown on the figure. The turbulent boundary layer prediction agrees well with the pressure surface data but is much higher than the suction surface data. Based on the results, we can say that the pressure-side boundary layer is fully turbulent from the leading edge, whereas the suction-side boundary layer may start laminar and undergoes transition to turbulent boundary layer along the surface and reaches fully turbulent boundary layer close to the trailing edge.

Figure 2.11 shows Stanton number distributions on the hub and tip of the blade. Dunn et al. (1986b) had 3 locations on the hub (platform) and 10 locations on the tip for heat flux measurements. It appears from the sparse data that the tip region Stanton data were much higher than that of the platform region. On comparison with the blade surface effects, the tip region Stanton numbers are of the order of blade leading edge, which indicates extremely high heat transfer. More will be discussed in Section 2.6 later in this chapter.

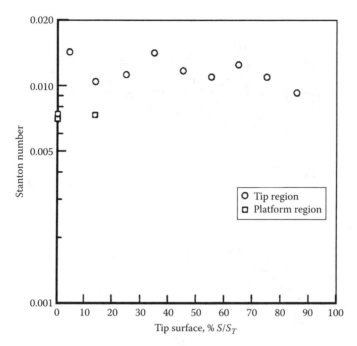

FIGURE 2.11

Stanton number distribution for the LART rotor blade tip and platform regions. (From Dunn, M.G. et al., *ASME J. Turbomach.*, 108, 108, 1986b.)

2.2.3 Simulated Turbine Stage

Blair et al. (1989) conducted experiments on a large-scale ambient tempera-ture, turbine-stage model. The turbine-stage model consisted of a stator, a rotor, and an additional stator behind the rotor. They also studied the effects of inlet turbulence, stator-rotor axial spacing, and relative circumferential spacing of the first and second stators on turbine airfoil heat transfer. A com-prehensive report of this experimental study can be obtained from Dring et al. (1981, 1982, 1986) at United Technologies Research Center (UTRC). Details on the experimental rig and instrumentation can also be found in the aforementioned studies. Figure 2.12 shows the large-scale rotating-rig geometry and airfoil flow angles. This test facility was designed for con-ducting detailed experimental investigations for flow around turbine and compressor blading. The assembly consists of 1½ stages for this turbine geometry study.

Figure 2.13 presents the measured static pressures on the airfoils. For the first stator, the flow was well behaved over the entire airfoil surface. On the pressure surface, the local flow continuously accelerated from the leading edge to the trailing edge gradually, with a large part of the increase after 60% surface distance. On the suction surface, the flow accelerated initially, then decelerated and accelerated strongly toward the throat region and then

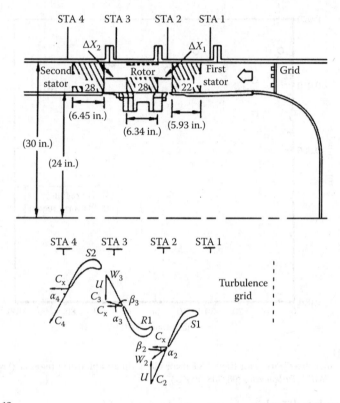

FIGURE 2.12
Large-scale rotating-rig geometry and velocity triangles. (From Blair, M.F. et al., *ASME J. Turbomach.*, 111, 87, 1989.)

smoothly decelerated toward the trailing edge without flow separation. For the rotor blade, flow deceleration occurred near the leading edge for both pressure and suction surface. On the pressure surface after the initial deceleration to about 3% surface distance, the flow continuously accelerated toward the trailing edge. On the suction surface, the flow accelerated from 5% surface distance to about 25% surface distance. The flow velocity was approximately constant from about 25% to 70% surface distance and then decelerated toward the trailing edge. For the second stator the, pressure distribution was similar to that for the first stator except in the immediate vicinity of the leading edge. On the pressure surface, there was a strong acceleration followed by a slight deceleration just downstream of the leading edge. On the suction surface, there was a continuous acceleration of flow toward the throat, and then the flow decelerated toward the trailing edge after achieving maximum velocity at the throat.

Figure 2.14 presents the Stanton number distributions for each airfoil based on the exit velocity and density for each airfoil. Midspan heat-transfer distributions are presented for the case where stator 1 to rotor and rotor to stator 2

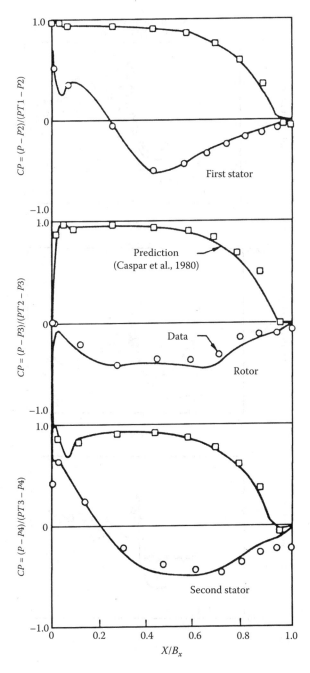

FIGURE 2.13
Measured static pressures on all three airfoils. (From Blair, M.F. et al., *ASME J. Turbomach.*, 111, 87, 1989.)

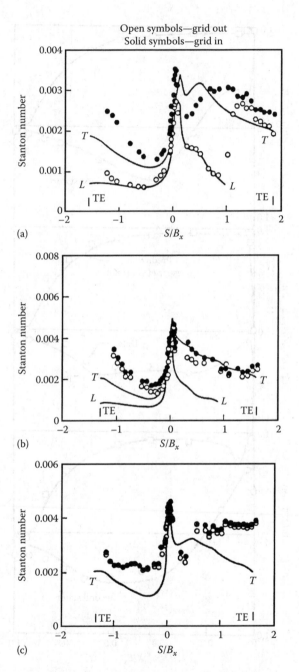

FIGURE 2.14

Stanton number distributions for each airfoil under two free-stream turbulence conditions: (a) first stator; (b) rotor; and (c) second stator. (From Blair, M.F. et al., *ASME J. Turbomach.*, 111, 87, 1989.)

spacing was 65% and 63% of the rotor blade chord, respectively. Results are compared for a case with a turbulence grid just upstream of stator 1 versus no-grid case. Grid turbulence generated was about 9.8% at the first stator inlet. Stator 1 shows almost laminar boundary layer over the entire pressure surface without a grid. However, with grid, Stanton numbers become higher, indicative of transition to turbulence immediately downstream of the leading edge. On the suction surface, transition occurs at $S/B_x = 1$ (surface distance to airfoil axial chord ratio) without grid. Transition moves to $S/B_x = 0.2$ with grid. This indicates that a highly turbulent oncoming flow advances boundary layer transition location on both suction and pressure surfaces and increases heat-transfer levels significantly. In the real engine, the combustor exit gases are highly turbulent, with turbulence levels up to 15%–20% at the first-stage stator inlet.

For the rotor, the effect of turbulence is not as strongly evident as in the case for the first stator. The low-turbulence case indicates that the boundary layer appears to be laminar only in the vicinity of the leading edge. Pressure surface heat-transfer levels show a strong turbulent boundary layer development after $S/B_x = 0.2$. The rotor heat transfer is not significantly affected by the grid-generated turbulence due to two reasons: (1) The rotor flow is already highly disturbed by the unsteady wakes generated by the upstream stator, and (2) the grid-generated turbulence decays downstream of the first stator exit due to flow acceleration. The levels of unsteadiness in the flow generated by the rotor relative to the stationary stator wakes are much higher than that generated by the grid. Only transition location on the suction surface is affected slightly. Unsteady wake produces periodic turbulence in the free-stream, which on averaging produces turbulence in the levels of 10%–15% over the already existing grid-generated turbulence (or combustor generated in the real engine). This may be the cause for higher Stanton numbers on the rotor blade than on the first stator.

For the second stator, the grid turbulence effect is nonexistent due to further acceleration of flows downstream of the first blade. The boundary layer is affected by upstream unsteady wakes and secondary flows generated by the rotor. Both pressure surface and suction surface heat-transfer levels are much higher than those for the first stator in the case of relatively low turbulence. The complexity of the flow increases with every row of airfoils. This is very much evident in the measured heat-transfer distributions. More information on the experiment and other variations can be obtained from Dring et al. (1981, 1982, 1986).

Figure 2.15 shows the effect of the mainstream Reynolds number on the first stator Stanton number distributions under high grid turbulence (~9.8%). On the suction surface, a decrease in Reynolds number (Re) moves the boundary layer transition location farther away from the leading edge. For $Re = 642,900$, the transition begins almost immediately downstream of the leading edge. For $Re = 242,800$, transition begins only at $S/B_x = 1.0$. With an increase in Reynolds number, the transition location moves closer to the leading edge, in turn reducing the near-laminar heat-transfer zone and expanding the fully turbulent heat-transfer zone on the airfoil surface. On the pressure surface, the effect of

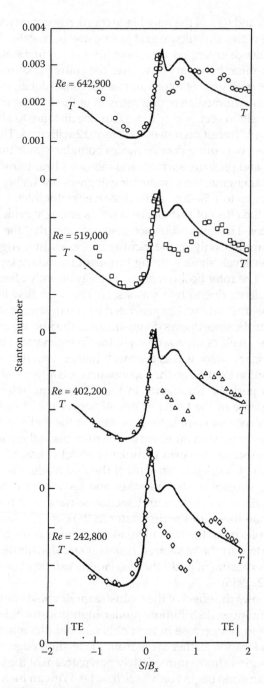

FIGURE 2.15
Effect of cascade exit flow Reynolds number on first stator Stanton number distributions for turbulence = 9.8%. (From Blair, M.F. et al., *ASME J. Turbomach.*, 111, 87, 1989.)

Reynolds number is not felt up to a surface $S/B_x = -0.4$. Further downstream, the effect can be separated into two groups. For $Re > 400,000$, the measured heat transfer exceeded fully turbulent predictions, whereas for lower Reynolds numbers, the measured heat-transfer levels matched the fully turbulent profile.

2.2.4 Time-Resolved Heat-Transfer Measurements on a Rotor Blade

Guenette et al. (1989) presented heat-transfer measurements for a fully scaled transonic turbine blade. The measurements were performed in the MIT blowdown turbine tunnel. The facility has been designed to simulate the flow Reynolds number, Mach number, Prandtl number, corrected speed and weight flow, and gas-to-metal temperature ratios associated with turbine fluid mechanics and heat transfer. They used thin-film heat flux gauges to make surface heat-transfer measurements. More details on the measurement technique can be found in the work by Guenette et al. (1989). Figure 2.16 shows the MIT blowdown turbine flow arrangement. The turbine-stage geometry is shown in the inset.

Figure 2.17 presents the calculated profile pressure distributions for the rotor blade. On the pressure surface, there is an initial flow deceleration to

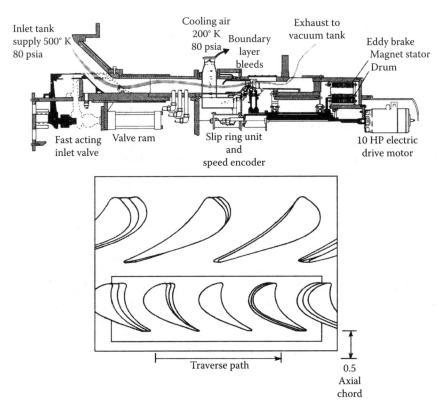

FIGURE 2.16
MIT blowdown turbine arrangement. (From Guenette, G.R. et al., *ASME J. Turbomach.*, 111, 1, 1989.)

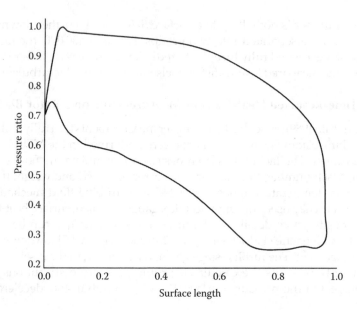

FIGURE 2.17
Cascade blade profile pressure distributions. (From Guenette, G.R. et al., *ASME J. Turbomach.*, 111, 1, 1989.)

about 5% surface distance, and then a gradual acceleration to about 60% surface distance and a strong acceleration toward the trailing edge. On the suction surface, there is a slight deceleration immediately downstream of the leading edge. The flow then accelerates strongly to about 75% surface distance, which is the throat location, and then decelerates toward the trailing edge. The blade is aft loaded as the highest local flow velocity (or lowest pressure location) occurs at a location closer to the trailing edge.

Figure 2.18 presents the ensemble average of the time-resolved heat flux measurements about the rotor blade midspan. The ensemble averaging is done over 360 consecutive blade passings. The figure shows measurements at 12 locations on the blade surface. The investigators observed a broad coherent disturbance on the pressure surface along the blade. The investigators indicate that the disturbance may be indicative of wake convection down the passage. The blade-passing period is indicated on one of the time profiles. The attenuation of the disturbance may be due to the acceleration of the passage flow from leading edge to trailing edge. On the suction surface, the blade-passing modulation is stronger at the leading edge (70%–90%) and attenuates toward the trailing edge (30%–40%). The wake effect of upstream NGV decreases as the flow enters the blade passages due to strong acceleration of flow from leading edge to trailing edge. The steep variations of the data on the suction surface indicate strong wake propagation toward the suction surface near the leading edge and moving toward the pressure surface near the trailing edge.

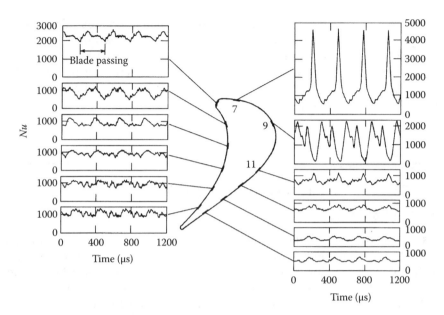

FIGURE 2.18
Ensemble average of time-resolved heat flux measurements about the rotor blade midspan.
(From Guenette, G.R. et al., *ASME J. Turbomach.*, 111, 1, 1989.)

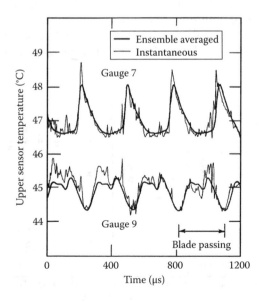

FIGURE 2.19
Comparison of measured instantaneous and ensemble averaged wall temperatures at two suction surface locations. (From Guenette, G.R. et al., *ASME J. Turbomach.*, 111, 1, 1989.)

Figure 2.19 presents a comparison of the ensemble-averaged and instantaneous temperatures measured by the top sensors at two heat flux gauge locations on the suction surface. The noise (small) fluctuations—like slip ring noise, electrical noise, and flow fluctuation—are filtered by the ensemble technique. As can be seen from the measured time profiles, the flow is periodic with blade passing.

2.3 Cascade Vane Heat-Transfer Experiments

2.3.1 Introduction

As indicated before, all the earlier studies conducted experiments relating to stator/rotor interactions. They studied the heat transfer for the entire rotor/stator. Studies focusing only on the first-stage NGV heat transfer did not consider the rotor to create any upstream effects. The presence of the rotor was not expected to affect the first-stage vane heat-transfer characteristics. Based on this assumption, there were several studies that focused only on all the parametric effects on NGV heat transfer.

First of all, we would like to confirm that the presence of a downstream rotor does not significantly affect the upstream first-stage vane heat-transfer characteristics. Dunn et al. (1984) studied the influence of the rotor on the Stanton number distributions for the upstream vane. Figure 2.20 shows the comparisons for a vane-only data and vane data with downstream rotor for a $T_w/T_0 = 0.53$. The filled circles are for the vane-only data, and the open circles

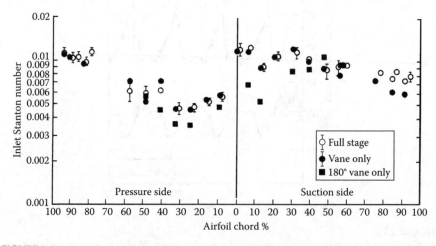

FIGURE 2.20
Influence of rotor on NGV Stanton number distributions. (From Dunn, M.G. et al., *ASME J. Eng. Gas Turbines Power*, 106, 229, 1984.)

are for the full-stage data. The additional filled squares represent data for vane only for $T_w/T_0 = 0.21$. Comparison of open to closed circles clearly shows that the rotor does not affect the vane Stanton number distribution. However, there is a small effect on the suction side near the trailing edge. The presence of the rotor enhances the Stanton number up to 25% near the trailing edge on the suction surface. Since this is a small region compared to the entire vane surface, it can be sufficiently stated that the vane heat-transfer characteristics are not significantly affected by the presence of a downstream rotor.

2.3.2 Effect of Exit Mach Number and Reynolds Number

Nealy et al. (1984) presented heat-transfer distributions on highly loaded turbine NGV in moderate-temperature, three-vane cascades under steady-state conditions. They varied parameters such as Mach number, Reynolds number, turbulence intensity, and wall-to-gas temperature ratio. The experimental data were obtained in the aerothermodynamic cascade facility at the Allison Engine Company. Nealy et al. (1984) also indicated the basic mechanisms that affected gas-to-airfoil heat transfer. They attributed boundary layer transitional behavior, free-stream turbulence, airfoil surface curvature, airfoil surface roughness, pressure gradient, coolant injection location, flow separation and reattachment, and shock/boundary layer interaction as the basic mechanics, affects of which need to be ascertained on the airfoil heat transfer. In this study, they focused their attention on the cascade exit Mach number, Reynolds number, and airfoil shape. Figure 2.21 presents the surface profiles for the two cascade vanes. The two vane designs, named Mark II and C3X, have significantly different suction surface geometries. The tests on these two designs provided insight into the effect of suction surface geometry on surface heat transfer.

Figure 2.22 presents the surface static pressure distributions on the two vanes for three different vane exit Mach numbers. The pressure distributions

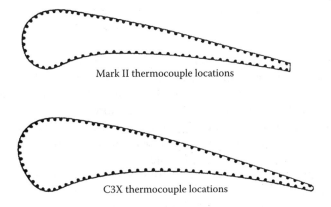

Mark II thermocouple locations

C3X thermocouple locations

FIGURE 2.21
Designs of Mark II and C3X vanes. (From Nealy, D.A. et al., *ASME J. Eng. Gas Turbines Power*, 106, 149, 1984.)

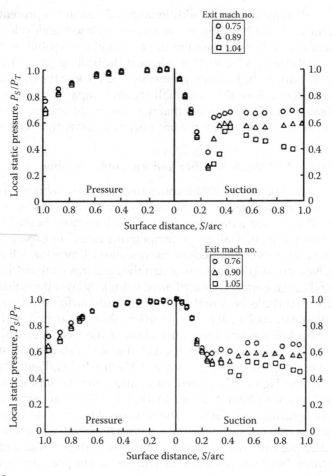

FIGURE 2.22
Surface static pressure distributions for both Mark II and C3X vanes. (From Nealy, D.A. et al., *ASME J. Eng. Gas Turbines Power*, 106, 149, 1984.)

on the two vanes are very similar on the pressure surface. However, the effect of the vane profile is significant on the suction surface. There is a strong adverse pressure gradient on the suction side of the Mark II vane at about 20% surface distance. The C3X vane shows a more gradual acceleration. The effect of the exit Mach number is also significant on the suction surface of both the vanes. A higher exit Mach number transforms into a higher velocity on the suction side near the trailing edge.

Indicative of the varied pressure distributions on the suction surface on the two vanes, the measured surface heat-transfer distributions also show different characteristics. Figure 2.23 presents the effect of exit Mach number on heat-transfer distributions for the Mark II vane. As from the pressure distributions, the heat-transfer distribution on the pressure surface is not affected by changing the

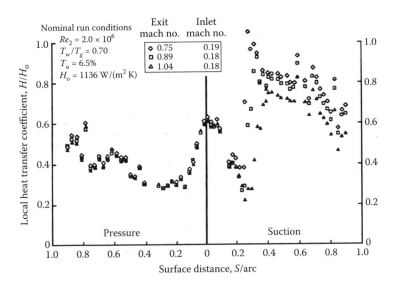

FIGURE 2.23
Effect of exit Mach number on heat-transfer distributions for Mark II vanes. (From Nealy, D.A. et al., *ASME J. Eng. Gas Turbines Power*, 106, 149, 1984.)

exit Mach number. Heat-transfer coefficients on the pressure surface decrease rapidly from the leading edge to about 20% surface distance and then gradually increase toward the trailing edge. On the suction surface, heat-transfer coefficient distributions show laminar boundary layer separation, transition, and turbulent reattachment at about 25% surface distance. The location of the laminar boundary layer separation seems to move upstream with decreasing exit Mach number. Also downstream of that location, heat-transfer coefficients are higher with decreasing exit Mach numbers. In regions where the boundary layer remained attached, there is no apparent effect of the exit Mach number.

Figure 2.24 shows the effect of exit Mach numbers on surface heat-transfer distributions for the C3X vane. Again, there is no effect of the exit Mach number on the pressure surface. The heat-transfer coefficient drops rapidly from the leading edge to about 20% surface distance and increases gradually toward the trailing edge. On the suction surface, the heat-transfer distributions are typical of a classic boundary layer transition. Heat-transfer coefficients decrease with increasing surface distance up to about 25% surface distance. Then the heat-transfer coefficients increase due to the start of transition from laminar–turbulent boundary layer. Transition is completed at about 50% surface distance downstream of which heat-transfer coefficients decrease with increasing fully turbulent boundary layer thickness. The transition location moves closer to the leading edge with a decrease in exit Mach number. This heat-transfer behavior on the C3X vane is the more typical distribution attributed to airfoils.

Figure 2.25 presents the effect of the Reynolds number on C3X vane heat transfer. The flow conditions are shown on the figure. The transition location

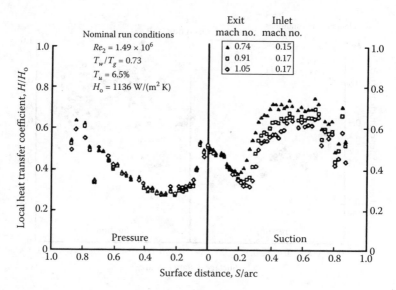

FIGURE 2.24
Effect of exit Mach number on heat-transfer distributions for C3X vanes. (From Nealy, D.A. et al., *ASME J. Eng. Gas Turbines Power*, 106, 149, 1984.)

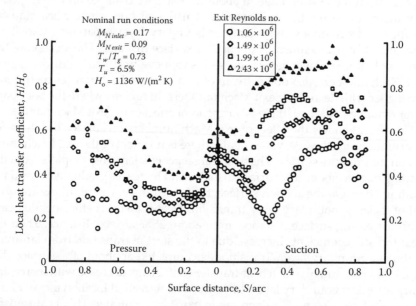

FIGURE 2.25
Effect of Reynolds number on C3X vane heat transfer. (From Nealy, D.A. et al., *ASME J. Eng. Gas Turbines Power*, 106, 149, 1984.)

on the suction surface moves closer to the leading edge with an increase in Reynolds number. The pressure surface heat-transfer distributions shows increased tendency toward a transitional behavior as observed on the suction surface. Overall, heat-transfer coefficients over the entire airfoil surface showed significant increases with an increase in Reynolds number.

This study was the first comprehensive study on NGV heat transfer. The heat-transfer characteristics are very much dependent on the vane geometry, as shown in the aforementioned study. The strong differences in the Mark II and C3X vane distributions and the effects of other parameters are clearly evident. More information on this study has been reported in the NASA report (Hylton et al., 1982).

2.3.3 Effect of Free-Stream Turbulence

One of the primary effects on vane heat transfer is the free-stream turbulence generated at the combustor exit. Combustor-generated turbulence contributes to significant heat-transfer enhancement. However, the effect of free-stream turbulence on vane heat transfer is not well documented. Turbulence can strongly affect laminar heat transfer to the stagnation region, pressure surface, transition, and turbulent boundary layer heat transfer. Ames (1997) investigated the influence of combustor-simulated turbulence on turbine vane heat transfer. More details on turbulence and flow characterization can be obtained from Ames (1997) and from Ames and Plesniak (1997). The four-vane cascade had airfoils scaled 4.5 times the C3X vanes. The vane profiles are a centerline slice of the first-stage nozzle of an Allison Engine Company helicopter engine. The vane geometry is similar to the C3X vane configuration used by Nealy et al. (1984). Ames studied the effects of both turbulence intensity and length scale on vane heat transfer. Ames (1997) developed a combustor turbulence generator where the turbulence levels and the length scale associated with it were identical to those produced by a real engine combustor. Figure 2.26 shows the schematic of the combustor turbulence generator. Ames simulated different levels of turbulence by moving the turbulence generator closer to the vanes for the high-turbulence case and moving it away for other low-turbulence cases.

Figure 2.27 shows the schematic of the four-vane cascade used by Ames (1997) for flow and heat-transfer measurements.

Figure 2.28 presents the measured heat-transfer data for a $Re = 790,000$ (based on axial chord) comparing four different turbulence conditions. The four cases for which data are presented are (1) a baseline case where $Tu = 1.1\%$ and $L_u = 6.6$ cm, (2) a combustor-simulated turbulence where $Tu = 12\%$ and $L_u = 3.36$ cm, (3) a combustor-simulated turbulence where $Tu = 8.3\%$ and $L_u = 4.26$ cm, and (4) a grid-generated turbulence case where $Tu = 7.8\%$ and $L_u = 1.36$ cm. The elevated turbulence data show significant augmented Stanton numbers over the low-turbulence case. The laminar region included the stagnation region, the entire pressure surface, and the favorable pressure gradient region of

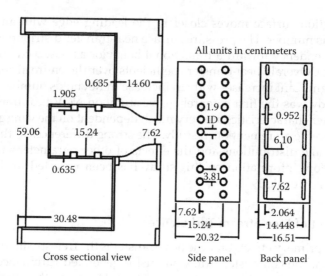

All units in centimeters

Cross sectional view   Side panel   Back panel

FIGURE 2.26
Schematic of a mock combustor turbulence generator. (From Ames, F.E., *ASME J. Turbomach.*, 119, 23, 1997b.)

the suction surface. Transition occurs on the suction surface during which the heat-transfer levels increase rapidly. Further downstream, the Stanton numbers decrease with development of the fully turbulent boundary layer. However, with increase in free-stream turbulence, the transition location moves closer to the leading edge. After transition, higher turbulence intensity does not appear to further enhance heat-transfer levels. Results also indicate a strong effect of turbulence length scale (L_u). The grid-generated turbulence is similar to the lower combustor-generated turbulence case, but there is a strong difference in the levels of the length scale. Large length scale turbulence seems to produce lower heat-transfer augmentation than small length scales over the baseline case at similar turbulence levels. This study provided a strong case for researchers to consider free-stream turbulence together with length scales in their future studies. The effect of turbulence in increasing heat-transfer levels and also causing earlier boundary layer transition on the suction surface is confirmed by the aforementioned study.

2.3.4 Effect of Surface Roughness

Another severe heat-transfer enhancement factor for NGV heat transfer is the surface roughness effect. In real engines, surface roughness becomes a parameter due to initial manufacturing finish and engine deposits after many hours of operation. Combustion deposits may make the vane surface rough after several hours of operation, and this roughness could be detrimental to the life of the vane due to increased heat-transfer levels that are much higher than design conditions. Abuaf et al. (1998) studied the effects of surface roughness on airfoil heat-transfer and aerodynamic performance.

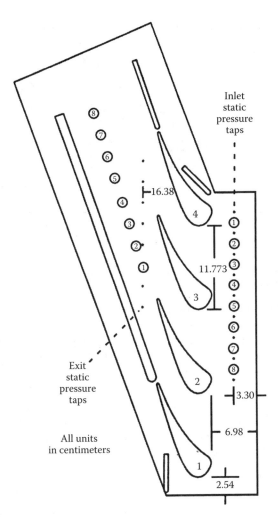

FIGURE 2.27
Schematic of the four-vane cascade used by Ames. (From Ames, F.E., *ASME J. Turbomach.*, 119, 23, 1997b.)

This was one of the first studies to focus on surface roughness effects to vane heat transfer. Prior studies on surface roughness focused on flat surfaces or rotor blades. (This will be discussed in a later section in this chapter.) Three different levels of roughness were studied by Abuaf et al. (1998). The roughness levels were characterized using a scanning interference microscope. Airfoil A was vapor grit blasted and coated with Codep, an oxidation-resistant coating. Airfoil B was vapor grit blasted, tumbled, and then coated with chemical vapor-deposited (CVD) aluminide. Airfoil C was vapor grit blasted, tumbled, polished, coated with CVD aluminide, and finally polished again. Selected regions of the airfoil surfaces were measured using the microscope

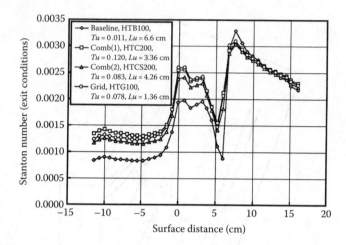

FIGURE 2.28
Heat-transfer data on the vane with free-stream turbulence effects. (From Ames, F.E., *ASME J. Turbomach.*, 119, 23, 1997b.)

for roughness levels. The centerline average roughness (R_a) values of 2.98, 0.94, and 0.77 μm were obtained for airfoils A, B, and C, respectively.

Figure 2.29 presents a comparison of the average heat-transfer coefficient distributions for airfoils A, B, and C. The effect of surface roughness on the pressure surface is small. There is a slight enhancement in the heat-transfer coefficient values for airfoil A with highest roughness. On the suction surface, the effect is strong, with the transition location moved to almost the leading edge for airfoil A. The effect of surface roughness is also evident in the fully turbulent region of the suction surface. It appears that surface roughness not only enhances heat-transfer coefficients but also hastens boundary

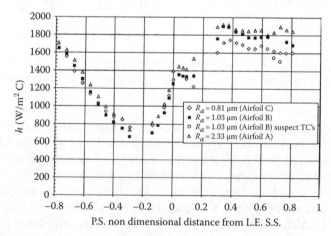

FIGURE 2.29
Effect of surface roughness on heat-transfer coefficients. (From Abuaf, N. et al., *ASME J. Turbomach.*, 120, 522, 1998.)

layer transition on the suction surface of the airfoil. Higher surface roughness causes even earlier transition, leading to the belief that surface roughness is detrimental to the life of the vane. Designers need to take surface finish of airfoils into consideration while studying cooling possibilities for NGVs.

There is another factor that needs consideration. The combination of high free-stream turbulence combined with surface roughness may nullify the single effect of roughness. If surface roughness effect on vane heat transfer already influenced by high free-stream turbulence is only incremental, then surface roughness as a factor may be only a secondary effect. Hoffs et al. (1996) made heat-transfer measurements on turbine airfoils under high free-stream turbulence and increased surface roughness conditions. The surface roughness effects were determined by the liquid crystal coating in their study. The naturally coated liquid crystal surface acted as the rough surface with surface roughness level $R_z = 25\,\mu m$. For the smooth surface, the liquid crystal-coated surface was polished with very fine sandpaper to obtain smooth and repeatable surface conditions. The surface roughness value R_z was equal to $15\,\mu m$. Higher turbulence levels were generated by using turbulence-generating grids upstream of the airfoil cascade.

Figure 2.30 presents the heat-transfer coefficients for the effect of free-stream turbulence and also the addition of surface roughness to an airfoil

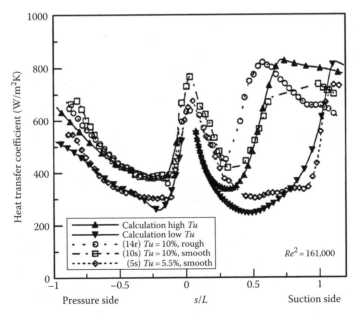

FIGURE 2.30
Effect of surface roughness and free-stream turbulence on airfoil heat-transfer coefficient distributions. (From Hoffs, A. et al., Heat transfer measurements on a turbine airfoil at various Reynolds numbers and turbulence intensities including effects of surface roughness, ASME Paper 96-GT-169, 1996.)

already affected by high free-stream turbulence by Hoffs et al. (1996). Three experimental cases are shown: The first case is for a smooth surface with Tu = 5.5%; the second case is for a smooth surface with Tu = 10%; the third case is for the rough surface at Tu = 10%. Comparing the first two cases for the smooth surface, it is evident that the heat-transfer coefficients are significantly enhanced on the pressure surface due to increased turbulence. The suction surface results show that the transition location has moved upstream due to increased free-stream turbulence from s/L = 1.0 to s/L = 0.25. This is typical of the results earlier discussed for a classic turbine vane. Now let's compare the smooth surface to the rough surface at Tu = 10%. The pressure surface is unaffected by the surface finish at high turbulence. The already enhanced heat-transfer coefficients due to high free-stream turbulence are unaffected by the surface roughness. However, the effect on suction surface is significant. The transition location does not seem to be affected by the rough surface. It is the length of the transition that is greatly reduced by addition of surface roughness. A combination of surface roughness with high free-stream turbulence causes the boundary layer to undergo transition more rapidly than for the high free-stream turbulence case only. Bunker (1997) also studied the combined effect of high free-stream turbulence and surface roughness on vanes. His study also shows similar conclusions.

2.3.5 Annular Cascade Vane Heat Transfer

Martinez-Botas et al. (1995) used a blowdown tunnel and an annular NGV cascade for heat-transfer measurements. The blowdown tunnel is presented in detail by Baines et al. (1982). A cold heat-transfer tunnel (CHTT) was first used by Martinez-Botas et al. (1993). In this tunnel, the principal fluid need not be heated. The NGV is heated and suddenly exposed into the blowdown tunnel. The heat-transfer coefficients are measured by recording the changing blade temperature using the transient liquid crystal technique.

To perform the transient heat-transfer test, the NGVs are preheated before the run by isolating the heat-transfer cassette (shown in Figure 2.31) consisting of four passages using a shutter mechanism. The shutter is removed during the run when the transient test is initiated. Figure 2.32 shows the isentropic Mach number distribution on the NGV at three different span locations of 10%, 50%, and 90%. The data show an inward (tip to hub) radial pressure gradient near the back of the airfoil. Figure 2.33 shows the experimental Nusselt number distribution at midspan of the airfoil. The highest levels of heat transfer occur at the trailing edge on the pressure surface. On the suction surface, the peak occurs at 8% surface distance and remains constant for a surface distance greater than 20%. On the pressure surface, the Nusselt number initially decreases and then continually increases beyond 15% of the surface distance toward the trailing edge. A 2-D boundary layer heat-transfer prediction is also included for comparisons.

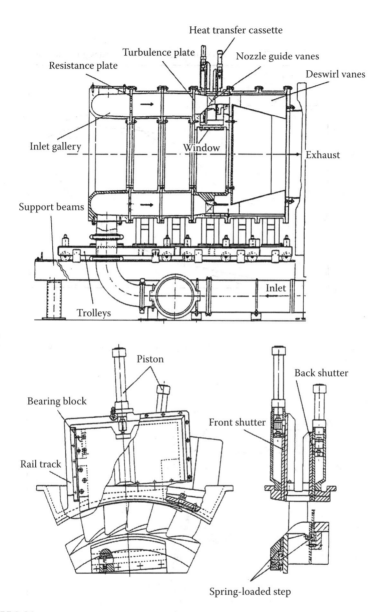

FIGURE 2.31
CHTT and the heat-transfer cassette. (From Martinez-Botas, R.F. et al., *ASME J. Turbomach.*, 117, 425, 1995.)

Figure 2.34 shows the surface heat-transfer coefficient distributions on both the pressure and suction surfaces. The pressure surface shows minimal spanwise variation. The heat-transfer coefficients increase toward the trailing edge, as seen in Figure 2.33. The suction surface shows strong spanwise variations. There are high heat-transfer regions near the

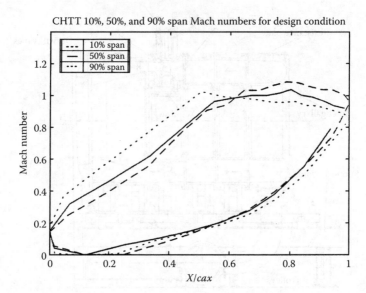

FIGURE 2.32

Isentropic Mach number distributions on the airfoil. (From Martinez-Botas, R.F. et al., *ASME J. Turbomach.*, 117, 425, 1995.)

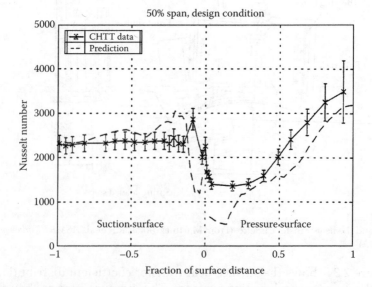

FIGURE 2.33

Midspan heat-transfer distributions at engine design conditions. (From Martinez-Botas, R.F. et al., *ASME J. Turbomach.*, 117, 425, 1995.)

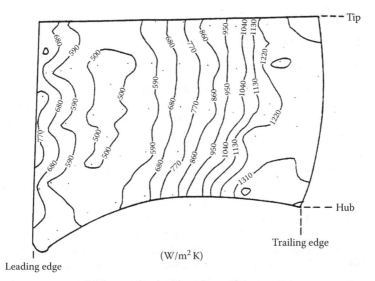

Pressure surface heat transfer coefficient contours

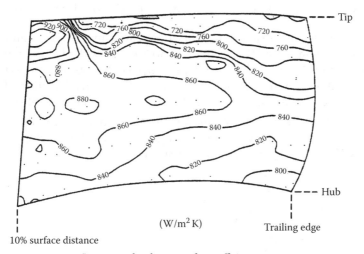

Suction surface heat transfer coefficient contours

FIGURE 2.34
Heat-transfer coefficient contours on the NGV. (From Martinez-Botas, R.F. et al., *ASME J. Turbomach.*, 117, 425, 1995.)

tip at the leading edge, followed by a low heat-transfer region where separation may be occurring. The endwall effects resulting in the passage and horseshoe vortices may cause the strong spanwise variations on the suction surface. More details on the endwall effects are presented in Section 2.5.

2.4 Cascade Blade Heat Transfer

2.4.1 Introduction

As the flow passes through the NGV passages, the free-stream turbulence intensity reduces due to acceleration of flow through the vane throat. The reduced free-stream turbulence effects on the rotor blade heat transfer are not as significant as that of the high free-stream turbulence effects on the vane heat transfer. Typically, combustor-generated, free-stream turbulence levels are around 15%–20% at the first-stage vane leading edge, and due to acceleration of flow in the vane passage, the turbulence intensity at the first-stage rotor blade leading edge is typically around 5%–10%. However, the rotor blade heat transfer is affected by another important parameter: the effect of unsteadiness in the flow. The unsteadiness of flow arises from the relative motion of the rotor blade rows with reference to alternate stationary vane rows. Figure 2.35 shows a conceptual view of the unsteady wake propagation through a rotor blade row. The shaded regions indicate where unsteadiness is caused by the upstream airfoils. For a first-stage blade, the principal components of unsteadiness as summarized by Doorly (1988) are as follows:

1. *Wake passing.* The exit flow from the upstream vane row is nonuniform in the circumferential direction due to the wakes shed at the trailing edge of upstream vanes. These wakes subject the blades to a periodic flow velocity and turbulence field, since the relative rotation of the rows causes downstream blades to "chop" through these wakes.

2. *Shock wave passing (for transonic turbines only).* Shock waves are generated by a transonic vane row impinging on the downstream blade row. These are in addition to the wake effect.

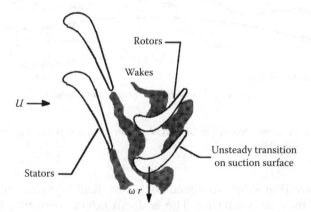

FIGURE 2.35
Conceptual view of the unsteady wake propagation through a rotor blade row. (From Mayle, R.E. and Dullenkopf, K., *ASME J. Turbomach.*, 113, 428, 1991.)

3. *Potential flow interactions.* Periodic variations in the potential field are caused by relative motion of vane and blade rows. Increasing the blade vane row spacing can reduce this type of effect.

4. *Additional high-energy turbulence.* This is the level of free-stream turbulence that may persist through the vane passage.

2.4.2 Unsteady Wake-Simulation Experiments

Wake-simulation experiments are a compromise between using rotating or stationary components. The upstream vane-generated unsteady wake impinging on the downstream rotor is simulated using a stationary blade cascade and an upstream wake 'generator. Wake-simulation experiments typically used either a rotating spoked-wheel wake generator (Ashworth et al., 1985; Doorly and Oldfield, 1985; O'Brien and Capp, 1989; Dullenkopf et al., 1991; Han et al., 1993) or a rotating squirrel cage wake generator (Priddy and Bayley, 1988; Schobeiri et al., 1992) upstream of the stationary blade cascade to simulate the relative motion of vane trailing edges. Figure 2.36 shows the schematic of the squirrel cage wake generator (Priddy and Bayley, 1988) and the rotating spoked-wheel wake generator (Doorly and Oldfield, 1985). The relative motion of the rods on the wake generator creates unsteady wakes that impinge on the downstream blade cascade.

Guenette et al. (1989) confirmed the validity of using rotating-bar simulation by comparing wake characteristics obtained with both stator–rotor interaction and upstream rotor bar simulations. They compared the time-resolved rotor data on the suction surface of a rotor blade with a 2-D bar-passing cascade data at two different locations of $x/s = 0.10$ and 0.31 on the blade surface. Figure 2.37 presents the comparison at $x/s = 0.31$. The unsteady interactions measured for the rotor and the 2-D cascade are similar. However, the double spikes are more prominent for the rotor than in the 2-D cascade. Guenette et al. (1989) imply that the unsteady interactions between the wake and blade are stronger for the rotor than in the 2-D cascade, although they are similar in nature. The unsteady data for the 2-D cascade were obtained from a study by Doorly and Oldfield (1985).

Doorly and Oldfield (1985) simulated the effects of shock wave and wake passing on a turbine rotor blade. They used a spoked-wheel wake generator to generate the shock wave and wake passing. Figure 2.38 shows the time history of an isolated wake and shock wave passing over a blade cascade. The positions of the shock wave and wake are traced over a number of successive instants. The heat-transfer traces for different gauge locations are also plotted. All the gauges are located on the suction surface of the blade from the leading edge to about throat location (1–11). The shock wave is indicated with solid lines. The propagation of the shock wave from leading edge to trailing edge of the passage is clearly shown. Also, the wake shed by the

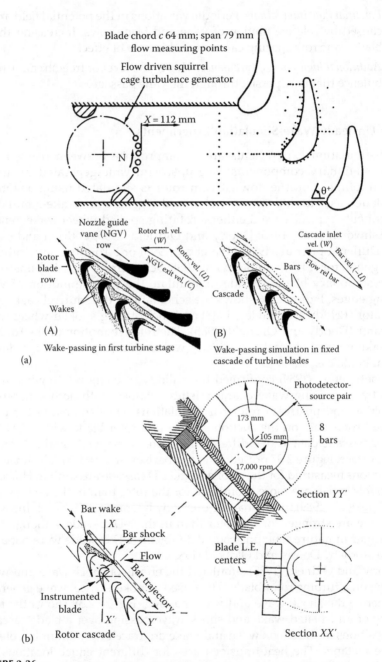

FIGURE 2.36
Schematic of the (a) squirrel cage wake generator. (From Priddy, W.J. and Bayley, F.J., *ASME J. Turbomach.*, 110, 73, 1988.); (b) rotating spoked-wheel generator. (From Doorly, D.J. and Oldfield, M.L.G., *ASME J. Eng. Gas Turbines Power*, 107, 998, 1985.)

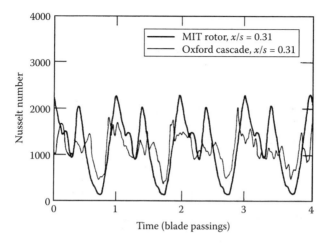

FIGURE 2.37
Comparison of time-resolved data from an actual rotor blade with 2-D bar-passing cascade data. (From Guenette, G.R. et al., *ASME J. Turbomach.*, 111, 1, 1989.)

rotating bar causes a periodic velocity deficit that envelops the blade passage as the bars pass in front of the cascade. The propagation of this velocity deficit region is clearly indicated in the time traces. According to Doorly and Oldfield, the main effect of periodic wakes is to influence the boundary layer transition on the suction surface of the downstream blade. Heat transfer on the blade surface is significantly affected by the wakes due to the increased turbulent-spot generations and also the transition. The impingement of the isolated shock wave on an otherwise laminar boundary layer on the suction surface produced a separation bubble. The very high transient heat transfer associated with the shock was caused by a turbulent patch (behind the bubble) as it swept down the boundary layer. The effect of shock waves acting simultaneously with wakes produces turbulent boundary layer patches that eventually merge, forming a continuously turbulent boundary layer. The combined shock wave and wake interaction on the blade boundary layer is very complicated, and Doorly and Oldfield (1985) have provided a detailed explanation of this phenomenon.

Ashworth et al. (1985) studied the effects of wake interactions from an upstream NGV on a rotor blade flow and heat transfer. They simulated a moving wake system simulated by a spoked-wheel wake generator. They measured local heat transfer using heat flux gauges at 22 locations along the blade surface. Figure 2.39 shows the heat-transfer distributions obtained for two free-stream turbulence conditions of 4% and 0.8% with unsteady wake interactions. The instantaneous signals of some of the local heat flux gauges are also indicated on the figure. The mean heat-transfer comparisons are shown as solid lines. It appears that the heat-transfer coefficients along the blade are greatly enhanced for $Tu = 0.8\%$ all along the blade surface. The highest increases are observed near the leading edge. Also interesting to note

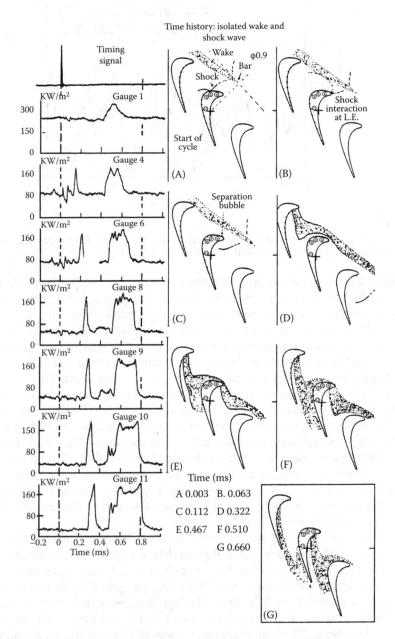

FIGURE 2.38
Time history of an isolated wake and shock-wave interaction over a blade cascade. (From Doorly, D.J. and Oldfield, M.L.G., *ASME J. Eng. Gas Turbines Power*, 107, 998, 1985.)

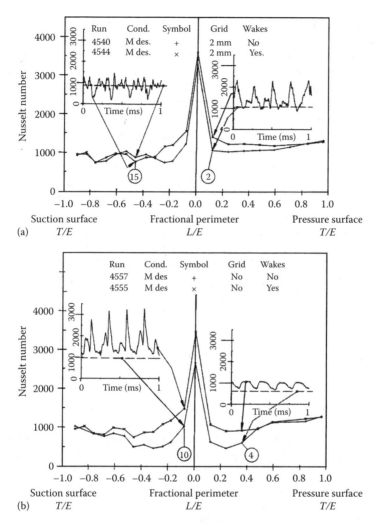

FIGURE 2.39
Effect of wake interaction on mean heat transfer to blade: (a) $Tu = 4\%$. Effect of wake interaction on mean heat transfer to blade with free-stream turbulence $\approx 4\%$, nominal "design" operating conditions. Insets are typical transient recorder signals at two x/s locations showing a 1 ms interval of the record together with the time-averaged value in the absence of wake interaction, (b) $Tu = 0.8\%$. Effect of wake interaction on mean heat transfer to blade with low ($\approx 0.8\%$ $u'/\bar{U}$) free-stream turbulence at nominal "design" operating conditions. Insets are typical transient recorder signals together with the time-averaged value in the absence of wake interaction. (From Ashworth, D.A. et al., *ASME J. Eng. Gas Turbines Power*, 107, 1022, 1985.)

is the shift of the transition location toward the leading edge for the passage with wake interactions. It is fairly evident that the presence of the unsteady wake causes early boundary layer transition. However, with an increase in free-stream turbulence, the effects of the unsteady wake are not as significant. Free-stream turbulence level without wake enhances heat transfer along the surface of the blade and also induces early transition on the pressure surface. The effect of the addition of unsteady wake is only incremental.

Dullenkopf et al. (1991) conducted time-mean heat-transfer measurements using a laboratory simulated spoked-wheel wake generator. The cascade consisting of five blades is incorporated in the hot-gas test facility. The hot gas generated by a combustion chamber passes through a turbulence grid and a rotating wake generator. More details on the test apparatus can be obtained from Wittig et al. (1987, 1988) and Dullenkopf et al. (1991). Dullenkopf et al. made hot-wire measurements and recorded instantaneous velocity signals with the wake generator rotating. The measurements were made at the entrance of the blade cascade. A typical hot-wire signal of a wake passing upstream of the cascade is illustrated in Figure 2.40. The signal clearly indicates the velocity deficit of each passing rod followed by the normal flow between the rods. The important parameters like the time-dependent velocity $v(t)$ composed of the periodically variable $\tilde{v}(t)$ and the stochastic velocity fluctuations $\tilde{v}'$ are shown in Figure 2.41. The ensemble-averaging analysis for obtaining the ensemble-averaged turbulence distributions is described later. The distribution of the turbulence-fluctuating component over one spacing is evaluated by means of the following equation:

$$\tilde{v}'^2(t_i) = \frac{1}{m-1}\sum_{j=1}^{m}\left[v_j(t_i) - \tilde{v}(t_i)\right]^2 \qquad (2.1)$$

Figure 2.42 clearly illustrates the evaluation of the ensemble-averaged turbulence distribution.

Figure 2.43 presents the heat-transfer coefficient on the whole blade surface for different inflow conditions. The Reynolds number based on blade chord was 380,000 for six different conditions: no bars, 4 bars, 7 bars, 14 bars, 28 bars, and the fully turbulent case with a grid-generated turbulence of 8%. The wake rotating frequency (f_w) and the relative duration of wake (d_w) are given for each case. The relative duration of wake is the ratio of the wake-duration time to the wake-passing period. Observing the different cases, it is instantly evident that the suction surface heat-transfer coefficients are higher for each case due to earlier boundary layer transition. The transition location with increasing wake frequency moves upstream closer to the leading edge. The transition location moves from a surface distance $s/l \sim 1.0$ to about 0.3 for the highest wake frequency. The grid-generated turbulence case has a transition location at around $s/l \sim 0.2$. Dullenkopf et al. (1991) indicated that the turbulent region and the free-stream wake outside the boundary layer act independently once transition is triggered by the wake at any location. The resultant time-mean heat-transfer coefficient distribution is composed of intermittent laminar and turbulent time fraction, in which the turbulent time fraction increases over the surface length. This explains the increased transitional length compared to the baseline case (no bars). The pressure

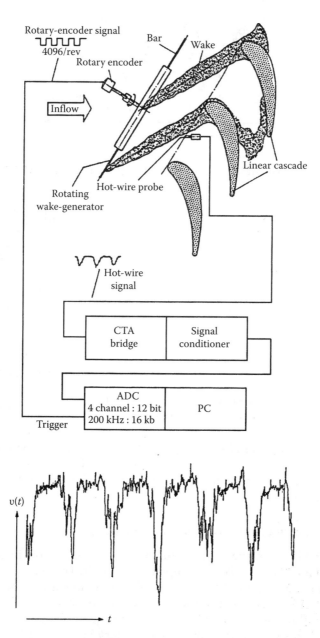

FIGURE 2.40
Instantaneous hot-wire signal of wake flow passing through a cascade. (From Dullenkopf, K. et al., *ASME J. Turbomach.*, 113, 412, 1991.)

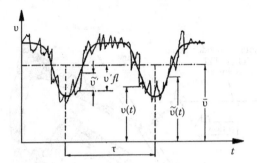

FIGURE 2.41
Parameters for hot-wire signal analysis. (From Dullenkopf, K. et al., *ASME J. Turbomach.*, 113, 412, 1991.)

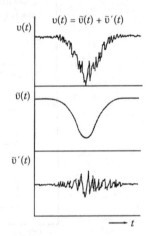

FIGURE 2.42
Ensemble averaged turbulence distributions. (From Dullenkopf, K. et al., *ASME J. Turbomach.*, 113, 412, 1991.)

surface shows a lesser effect compared to that on the suction surface. This may occur because the local intermittency caused by wake is approximately constant. More details on the effects of local intermittency will be discussed in the next section.

2.4.3 Wake-Affected Heat-Transfer Predictions

As described earlier, one of the main causes of unsteady flow in gas turbines is the propagation of wakes from upstream airfoils. These wakes impose the free stream with a periodic unsteady velocity, temperature, and turbulence intensity. The velocity deficit associated with the wake may also produce a convective flow either toward or away from the surface. The wakes cause an early unsteady laminar–turbulent boundary layer transition to occur on the suction side. The heat transfer associated with the unsteady flow effects clearly indicates the early boundary layer transition (Figure 2.43). In this section, we will focus on Mayle and his coinvestigators' transition theory to predict heat transfer on surfaces affected by unsteady wake passing.

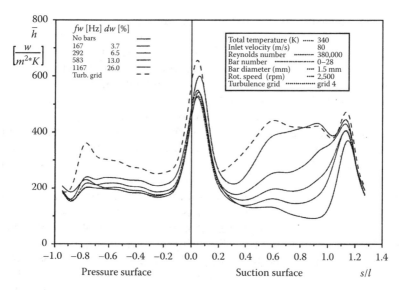

FIGURE 2.43

Heat-transfer coefficients on a blade under unsteady wake effects. (From Dullenkopf, K. et al., *ASME J. Turbomach.*, 113, 412, 1991.)

Mayle (1991) presented the impact of laminar–turbulent transition in gas turbine engine design and provided valuable suggestions for future studies. He presented a generalized description of transition and its various modes, and examined theoretical and experimental points of view for each mode of transition. There are three modes of transition: natural transition, bypass transition, and separated-flow transition: Natural transition begins with a weak instability in the laminar boundary layer and proceeds through various stages of amplified instability to fully turbulent flow. This was first described by Schlichting (1979). Bypass transition is caused by large disturbances in the external flow and completely bypasses the Tollmein–Schlichting mode of instability. This is typical of gas turbines. Separated-flow transition occurs in a separated laminar boundary layer and may or may not involve the Tollmein–Schlichting mode. This occurs mostly in compressors and low-pressure turbines. Mayle (1991) describes more details on the modes of transition. In this section, we will focus only on the bypass transition caused by unsteadiness introduced into the flow by periodic wakes and Mayle's analysis of this phenomenon.

Mayle also described periodic unsteady transition and reverse transition. Periodic unsteady transition occurs due to periodic passing of wakes from upstream airfoils or obstructions and is called wake-induced transition (Mayle and Dullenkopf, 1990). Transition induced by wakes or shocks (transonic turbines) appears to bypass the first stages of the natural transition process. Transition in gas turbines is sometimes considered multimoded, as different conditions can exist at different locations on the same surface at the same time. Mayle points out that an instantaneous snapshot of the flow over a gas turbine airfoil may show a laminar boundary layer near the leading edge, followed by a wake- or shock-induced transition to turbulent flow and another laminar flow region with a subsequent transition to turbulent flow by any other mode. This process is not completely understood and is not considered by designers. Reverse transition is the transition from turbulent to laminar flow. Reverse transition involves a balance between convection, production, and dissipation of turbulent kinetic energy within the boundary layer. It can be expected to occur at low turbulence levels if the acceleration parameter, $K = (v/U^2)(dU/dx)$, is greater than 3×10^{-6}. More details are discussed by Julien et al. (1969) and Jones and Launder (1972).

Mayle (1991) provides a detailed summary of the major works involving understanding of transition in gas turbine engines. Several experimental works of interest were presented by Hodson (1984), Ashworth et al. (1985), Doorly and Oldfield (1985), Dring et al. (1986), Dunn et al. (1979, 1984, 1986a,b), LaGraff et al. (1989), and Wittig et al. (1987, 1988). Models for unsteady wake-induced transition have also been proposed by Hodson (1984), Doorly (1988), Sharma et al. (1988), Addison and Hodson (1990a,b), Mayle and Dullenkopf (1990, 1991), and, recently, Chakka and Schobeiri (1999). Pfeil and Herbst (1979) and Pfeil et al. (1983) showed that the wakes impinge on the plate and cause the boundary layer to become turbulent. This phenomenon of wake impingement caused wake-induced transition zones propagating at lower velocities than the wake-passing velocity. They implied that the time-averaged condition of the boundary layer may be obtained from

$$\tilde{f} = (1 - \tilde{\gamma})f_L + \tilde{\gamma}f_T \tag{2.2}$$

where
 f is a boundary layer flow-related quantity
 f_L is its laminar value
 f_T is its turbulent value
 γ, the intermittency, is the fraction of time the flow is turbulent

Mayle and Dullenkopf (1990) developed their model based on Emmons's (1951) turbulent-spot theory. They extended Emmons's theory on the basis of two main points. First, they considered that two or more modes of transition can occur simultaneously on the same surface. Second, they considered the production term for turbulent spots to be a function of space and time. They also assumed that the production of turbulent spots due to wake impingement was so intense that the spots immediately coalesced into turbulent strips that then propagated and grew along the surface within the laminar boundary layer. Figure 2.44 illustrates the coalescence of turbulent spots into turbulent strips as a result of a passing highly turbulent wake.

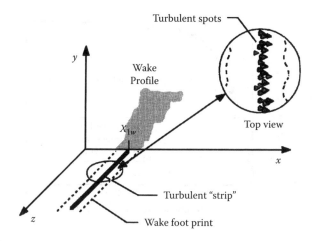

FIGURE 2.44

Coalescence of turbulent spots into turbulent strips as a result of a passing highly turbulent wake. (From Mayle, R.E. and Dullenkopf, K., *ASME J. Turbomach.*, 112, 188, 1990.)

Assuming that the turbulent spots produced normally or wake induced are independent of each other, Mayle and Dullenkopf defined the time-averaged intermittency as

$$\tilde{\gamma}(x) = 1 - [1 - \gamma_n(x)][1 - \tilde{\gamma}_w(x)] \tag{2.3}$$

where γ_n and $\tilde{\gamma}_w$ are the normal and wake-induced intermittencies, respectively. The tilde over the intermittency value indicates a time-averaged quantity over a wake-passing period. The normal mode intermittency is defined as

$$\gamma_n = 1 - \exp[-\hat{n}\sigma(Re_x - Re_{xt})^2] \tag{2.4}$$

where
$\hat{n} = nv^2/U^3$ is the dimensionless spot production parameter
σ is the turbulent spot propagation parameter
Re_x is the local flow Reynolds number
Re_{xt} is the Reynolds number at the location for onset of normal transition

Mayle and Dullenkopf assumed a square wave distribution for the turbulent strip production function and evaluated a production rate from experiments. They arrived at a simple expression for the time-averaged, wake-induced intermittency as

$$\tilde{\gamma}_w(x) = 1 - \exp\left[-1.9\left(\frac{x - x_{tw}}{U\tau}\right)\right] \tag{2.5}$$

where
U is the airfoil's incident velocity
τ is the wake-passing period

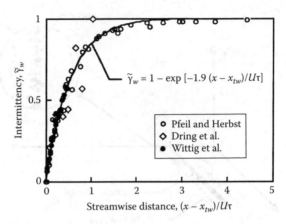

FIGURE 2.45
Measured intermittency results compared to calculated intermittency values. (From Mayle, R.E., *ASME J. Turbomach.*, 113, 509, 1991.)

Combining the expressions for natural-mode intermittency and wake-induced intermittency, Mayle and Dullenkopf arrived at a general expression for time-averaged intermittency for a surface affected by wakes as

$$\tilde{\gamma}(x) = 1 - e^{-0.412\left(x - x_{tn}/x_{75} - x_{25}\right)^2} e^{-b\left(\tau_w/\tau\right)\left(x - x_{tw}/U_s\right)} \tag{2.6}$$

where
x_{75} and x_{25} are locations on the surface where intermittency is 0.75 and 0.25
τ_w is the wake-duration time over location x_{tw}
τ is the wake-passing period

Figure 2.45 compares the measured intermittency results from Pfeil and Herbst, Dring et al., and Wittig et al. with the calculated intermittency values from the aforementioned theory. The agreement is very good with all three studies. Mayle (1991) shows some comparisons between experimental heat-transfer data and predictions based on the Mayle–Dullenkopf model in Figure 2.46. However, the beginning of transition was determined from experimental measurements to apply the transition model and calculate the heat-transfer coefficients. Han et al. (1993) also show some comparisons between the Mayle–Dullenkopf transition model and their heat-transfer experiments using a spoked-wheel wake generator.

2.4.4 Combined Effects of Unsteady Wake and Free-Stream Turbulence

Han et al., at Texas A&M University, studied the effects of unsteady wake combined with higher free-stream turbulence on heat-transfer coefficients on a rotor blade. They studied the primary effects of unsteady wake

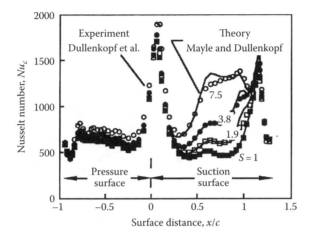

FIGURE 2.46
Nusselt number distributions on the suction side of a turbine blade: experiments versus predictions. (From Mayle, R.E., *ASME J. Turbomach.*, 113, 509, 1991.)

(Han et al., 1993) and free-stream turbulence (Zhang and Han, 1994) and then combined both effects (Zhang and Han, 1995) and compared that to the primary effects. Upstream wakes were generated using the spoked-wheel wake generator as shown in Figure 2.47. Zhang and Han (1995) defined a mean turbulence intensity ($\overline{Tu}$) for defining the turbulence generated by the combination of both free-stream turbulence and unsteady wakes. More details are given by Zhang and Han (1995).

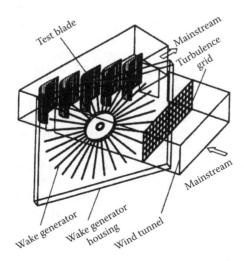

FIGURE 2.47
Spoked-wheel wake generator and turbulence grid used by Zhang and Han. (From Zhang, L. and Han, J.C., *ASME J. Heat Transfer*, 117, 296, 1995.)

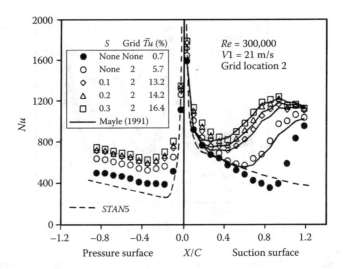

FIGURE 2.48
Effect of unsteady wake on local heat-transfer coefficient distributions for a given high oncoming free-stream turbulence condition. (From Zhang, L. and Han, J.C., *ASME J. Heat Transfer*, 117, 296, 1995.)

Figure 2.48 presents the effect of unsteady wake strength on blade surface heat-transfer coefficients from Zhang and Han's study. The effect of increasing wake strength is clearly evident. With increasing wake strength, the transition location on the suction surface moves closer to the leading edge. Two additional cases where the free-stream turbulence is 0.7% and a grid-generated turbulence of 5.7% are also included. All the wake cases had an upstream grid-generated turbulence of 5.7%. The solid lines on the suction surface are predictions based on the time-averaged intermittency factor method discussed earlier. The values of x_{tw}, the onset location for wake-induced transition, are 0.25C, 0.2C, and 0.15C, respectively for S = 0.1, 0.2, and 0.3 (C is the airfoil chord). The results show a good match in the region downstream of onset of transition. However, upstream of transition, the theory underpredicts heat-transfer levels. These results are presented to show the ability of Mayle's transition theory to predict time-averaged heat-transfer levels for blades under unsteady wake effects.

Figure 2.49 compares three cases where the turbulence is generated in different ways but the mean turbulence intensity of all three cases is similar at $\overline{Tu}$ = 13.5%. Each case has different grid-generated turbulence levels and wake strengths. However, the time-averaged heat-transfer coefficients are insensitive to the way the turbulence is generated. It is important to note that the turbulence characteristics for the three cases will be different, but their mean turbulence intensity is still the same. These results show that the characteristics of the turbulence may not be very important if the mean turbulence intensity can be simulated to be about the same as that in a gas

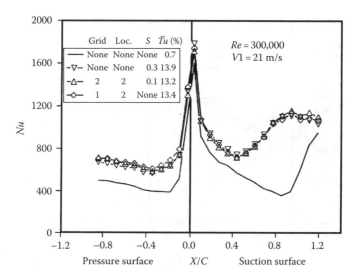

FIGURE 2.49
Effect of oncoming free-stream, unsteady wake, and combined conditions on local heat-transfer coefficient distributions. (From Zhang, L. and Han, J.C., *ASME J. Heat Transfer*, 117, 296, 1995.)

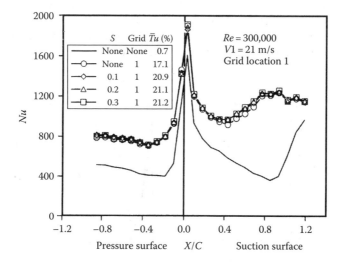

FIGURE 2.50
Effect of unsteady wake on local heat-transfer coefficient distributions for a given low oncoming free-stream turbulence condition. (From Zhang, L. and Han, J.C., *ASME J. Heat Transfer*, 117, 296, 1995.)

turbine engine blade passage. Figure 2.50 presents the effect of increasing upstream unsteady wake strength for a given oncoming high free-stream turbulence ($Tu = 17.2\%$). The superimposed wake Strouhal numbers for those cases are 0.1, 0.2, and 0.3. The calculated mean turbulence intensities $\overline{Tu}$ at the cascade inlet are 20.9%, 21.1%, and 21.2%, respectively. The Nusselt number distributions show little sensitivity to the change in Strouhal number. The results in the figure are fairly significant in that they indicate that the upstream wake-shedding frequency does not have any effect when the free-stream is highly turbulent ($Tu = 17.2\%$).

Du et al. (1997) used the same test setup of Han et al. (1993) and studied the additional effect of upstream vane trailing-edge ejection with unsteady wake on the downstream blade heat-transfer coefficient. Coolant ejection from upstream trailing ejection was achieved by injecting coolant from discrete holes on hollow-wake generator rods. They measured detailed heat-transfer coefficients on the blade midspan using a transient liquid crystal technique. The net effect was to increase heat-transfer coefficients on both pressure and suction surface around the leading-edge region. Figure 2.51 presents the combined effect of unsteady wake, free-stream turbulence, and coolant ejection on heat-transfer distributions of a turbine rotor blade. It compares cases for a clear wind tunnel (case 7, $\overline{Tu} = 0.7\%$), unsteady wake only (case 8, $\overline{Tu} = 10.4\%$), grid and wake (case 9, $\overline{Tu} = 13.7\%$), grid, wake, and ejection jet blowing rate $M = 0.25$ (case 10, $\overline{Tu} = 13.4\%$), and, finally, grid, wake, and ejection jet $M = 0.5$ (case 11, $\overline{Tu} = 13.0\%$). Adding wake effect itself makes the transition location on the suction side move from $X/SL = 0.5$ to $X/SL = 0.25$. With addition of free-stream grid turbulence, there is a further movement of transition location upstream. The jet effect on the suction surface is to increase heat transfer upstream of the transition location up to 25%, as compared to the case without jet ejection (case 9).

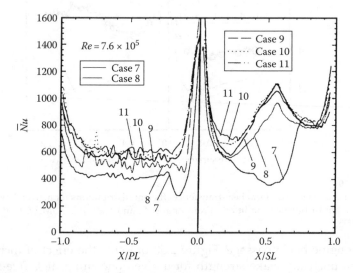

FIGURE 2.51
Combined effect of unsteady wake, free-stream turbulence and trailing-edge ejection on spanwise averaged Nusselt number distributions. (From Du, H. et al., *ASME J. Heat Transfer*, 119, 242, 1997.)

The jet effect diminishes in the transition and fully turbulent regions. The pressure surface heat-transfer does not show very significant increases as seen on the suction surface. However, there is up to a 15% increase in the heat-transfer levels in the region close to the leading edge. Dunn (1984) reported similar conclusions from their earlier rotor/stator engine model tests. It may be concluded that the addition of trailing-edge ejection is to compensate the velocity defect due to the wake, cause an increase in impinging mainstream velocity, and produce a more uniform turbulence intensity profile.

2.5 Airfoil Endwall Heat Transfer

2.5.1 Introduction

The airfoil endwall region is the most complicated region for analysis by gas turbine engineers. The endwall represents a large fraction of the surface to be cooled in a modern low-aspect ratio, low-solidity design. With increasing operating turbine inlet temperatures and pressures, the endwall-cooling problem has become significant. The endwall region is a complex region due to the presence of largely 3-D secondary flows. York et al. (1984) indicated that the essential characteristics of the endwall problem are as follows:

- The existence of pressure and temperature gradients in the inlet flow that induce secondary flows as the flow turns in the vane passages
- Roll-up of inlet boundary layer at the leading edge of each airfoil into a discrete horseshoe vortex that convects through the passage
- Three-dimensional boundary layers on the endwall with large cross flow components induced by strong cross stream pressure gradients
- Interaction of the horseshoe vortex in the suction surface–endwall corner

To understand these statements, one must understand the endwall flow field. A large number of studies have focused on the endwall flow field in an effort to understand the secondary flow structures and the associated heat transfer.

2.5.2 Description of the Flow Field

Wang et al. (1997) summarized the various secondary models published on the flow field near the endwall of a turbine vane. Langston (1980), Sharma and Butler (1987), and Goldstein and Spores (1988) presented three main representative models. Figure 2.52 presents the flow field as summarized by Langston (1980). The picture shows the evolution and development of two legs of a horseshoe vortex and a passage vortex. Figure 2.53 summarizes the findings of Sharma and Butler (1987). They indicate that the pressure-side

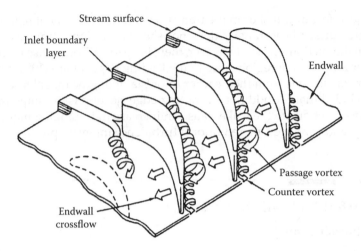

FIGURE 2.52
Endwall flow field as summarized by Langston. (From Langston, L.S., *ASME J. Eng. Power*, 102, 866, 1980.)

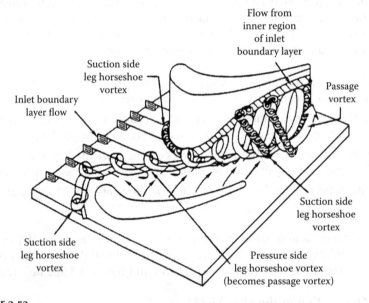

FIGURE 2.53
Flow field visualization based on the study by Sharma and Butler. (From Sharma, O.P. and Butler, T.L., *ASME J. Turbomach.*, 109, 229, 1987.)

leg horseshoe vortex becomes the passage vortex and moves closer to the suction-side leg horseshoe vortex. The suction-side leg horseshoe vortex then wraps itself around the passage vortex instead of adhering to the suction surface. The vortices also carry the inlet boundary layer flow toward the tip of the airfoil.

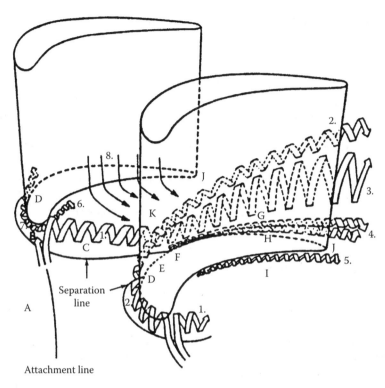

FIGURE 2.54
Endwall vortex pattern as described by Goldstein and Spores. (From Goldstein, R.J. and Spores, R.A., *ASME J. Heat Transfer*, 110, 862, 1988.)

Goldstein and Spores (1988) also presented their results of the endwall flow held. Figure 2.54 presents the vortex pattern as described by Goldstein and Spores. The figure indicates all the active vortices. The secondary flows of the endwall region are primarily the result of two main pressure gradients in the vane passage. The pressure variation in the passage is caused by the boundary layer velocity distribution and flow stagnation on the blade. These pressure variations force the flow toward the endwall and roll up into two legs of the leading-edge vortex. The turning angle of the flow between the airfoils causes a strong pressure gradient across the passage. This gradient affects the paths of the tow legs of the horseshoe vortex and also the low-momentum flow adjacent to the endwall. This variation causes a downflow on the pressure surface and an upflow on the suction surface. The pressure-side leg horseshoe vortex combines with the low-momentum flow near the endwall and forms the passage vortex. The passage vortex drifts from the pressure-side leading edge toward the suction-side trailing edge of the adjacent airfoil across the passage. This vortex (labeled 3) lifts off the endwall close to the suction side and adheres along the suction side as it moves downstream in the passage. The suction-side horseshoe vortex (labeled 2) remains close to the endwall until it reaches the separation line on the suction side and then lifts off the endwall and continues downstream along the suction side. It is interesting to note that the suction-side vortex wraps around the passage vortex; this point of wrapping occurs at different locations for different cascade geometries. There are other low-momentum vortices generated called the corner vortices. There also exists a downward component of velocity on the pressure side toward the endwall that may be pushing the passage vortex toward the suction side. These complex secondary flow distributions may result in complex heat-transfer behavior that may indicate strong variations across the endwall of the passage. Secondary flows can enhance heat-transfer significantly, and flat-plate models cannot predict heat transfer accurately for endwalls. Hence, it is important to obtain accurate heat-transfer measurements for endwall flows.

2.5.3 Endwall Heat Transfer

Graziani et al. (1980) explained that the streamline curvature of the mainstream flow establishes a strong pressure gradient, from pressure to suction surface, which is impressed on the endwall. The endwall secondary flow is made of low-momentum fluid that interacts with the mainstream fluid. Graziani et al. conducted experiments in a large-scale, low-speed, open-circuit cascade wind tunnel facility. They measured the blade surface and endwall surface heat-transfer distributions for two different inlet boundary layer thickness. Blade surface and endwall pressure distributions were obtained by using static pressure taps. Heat-transfer distributions were obtained using electrically heated, instrumented, uniform heat flux models. More details on the experimental setup and procedure can be obtained from Graziani et al. (1980).

Figure 2.55 presents the endwall static pressure coefficient contours on the blade endwall for two different boundary layer thicknesses. Immediate comparison shows that the thin boundary layer pattern is much less complex with fewer local distortions. The saddle point of separation on the endwall, the path of the passage vortex, and endwall separation line are recognizable only for the thicker boundary layer. Only near the suction surface at the minimum pressure location (or throat) is the passage vortex evident.

Figure 2.56 presents the endwall Stanton number distributions for a thick boundary layer thickness. Heat transfer is high near the airfoil leading edge due to the presence of a horseshoe vortex. The vortex path is clearly marked with a pattern of steep gradients originating in the saddle-point area and extending across the endwall to the suction surface. The Stanton numbers gradually decrease toward the suction surface. The lowest values of the heat transfer on the endwall are found near the pressure surface. The low Stanton number region forms a loop extending from the saddle-point region toward the pressure surface from out into the passage near the trailing edge.

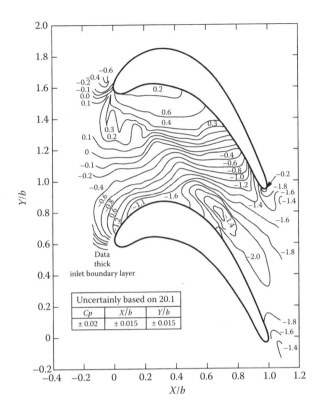

FIGURE 2.55
Endwall static pressure coefficient contours on blade endwall. (From Graziani, R.A. et al., *ASME J. Eng. Power*, 102, 257, 1980.)

A system of closed Stanton number contour regions indicates the highest regions of heat transfer in the airfoil wake region. It is clear that the endwall, 3-D boundary layer flows exist, for which conventional boundary layer assumptions may not be applicable.

Figure 2.57 presents the Stanton number contours on the blade and endwall surfaces for a thick inlet boundary layer. The existing vortex effects, secondary flows around the endwall blade hub region, are strongly evident in the distributions. Suction surface heat-transfer distributions are strongly 3-D similar to the endwall region. However, pressure surface heat-transfer distributions are lower and less varying over the surface. This was a premier study focusing on the airfoil endwall region and presenting a very complex view of the phenomenon. The study directed investigators to focus not only on the airfoil surface but also on the endwall and tip region of the airfoil. Heat-transfer levels are significantly high and necessitate cooling of endwalls. Other studies by Dunn and Stoddard (1979), Dunn and Hause (1979), York et al. (1984), Wedlake et al. (1989), Boyle and Russell (1989), and Blair (1994) have also focused on the endwall heat-transfer levels.

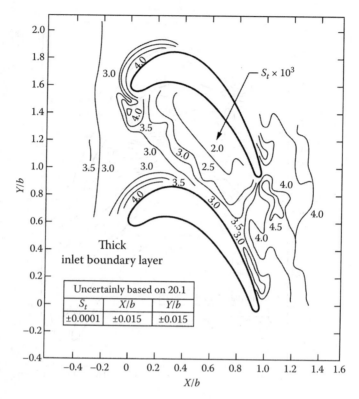

FIGURE 2.56
Endwall Stanton number distributions for a thick boundary layer. (From Graziani, R.A. et al., *ASME J. Eng. Power*, 102, 257, 1980.)

Chen and Goldstein (1992), Goldstein et al. (1995), and Wang et al. (1997) have provided detailed mass-transfer distributions on the endwall. They investigated mass-transfer characteristics on the endwall and on the suction and pressure surfaces close to the endwall. Their studies were based on low Reynolds number flows, and provided valuable insight into the endwall secondary flows and their effects on blade pressure and suction surfaces.

2.5.4 Near-Endwall Heat Transfer

Few studies have focused on the effects of endwall-generated vortices interactions with the blade surface and the effect on heat transfer. Graziani et al. (1980) and Sato et al. (1987) are among a few who studied the blade surface heat transfer in the near-endwall region. Chen and Goldstein (1992) studied heat (mass) transfer on the suction-side near-endwall region. They indicated that the heat-transfer distributions show a pattern of laminar separation, transition, and reattachment due to an adverse pressure gradient. Goldstein et al. (1995) studied the influence of secondary flows near-endwall and

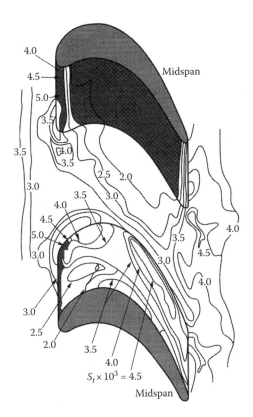

FIGURE 2.57
Endwall and blade surface Stanton number distributions for a thick boundary layer. (From Graziani, R.A. et al., *ASME J. Eng. Power*, 102, 257, 1980.)

boundary layer disturbance on the airfoil surface heat transfer. The near-endwall region is strongly 3-D due to the three-dimensionality of the flow in this region. The complicated flow pattern has already been explained.

Figure 2.58 shows the detailed Sherwood number distributions on the blade pressure and suction surfaces by Goldstein et al. (1995). As can be seen from the distributions, the horseshoe vortex does not influence the suction surface heat transfer near the leading edge. Around an axial distance of $S/C = 0.37$, the surface begins to be affected by the horseshoe vortex. The passage vortex influence originates at this point. The suction-side vortex is then pushed away from the endwall and moves above the passage vortex along the suction surface, creating the strong spanwise variations in heat-transfer distributions. The location of the peak heat transfer is along this vortex as it climbs up the blade suction side and is carried downstream by the passage vortex. On the pressure surface, the heat-transfer (mass-transfer) distribution is at most 2-D. There is not much effect on the pressure side due to the secondary flows as the passage vortex moves toward the suction side, and the pressure-side vortex is much smaller in magnitude causing large-scale

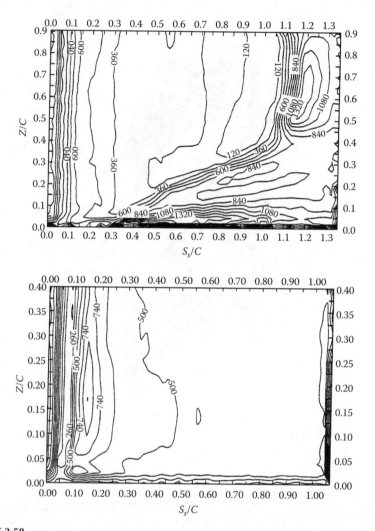

FIGURE 2.58

Detailed Sherwood numbers on blade suction and pressure surfaces near the endwall. (From Goldstein, R.J. et al., *ASME J. Turbomach.*, 117, 657, 1995.)

effects. The interaction between the pressure-side vortex and the different boundary layer regimes causes some flow separation, transition, and reattachment, causing some enhancement very close to endwall. The 3-D effects on the pressure surface are limited to a small region close to the endwall.

2.5.5 Engine Condition Experiments

Wedlake et al. (1989) ran a series of tests on an annular cascade of NGVs in an isentropic light piston cascade (ILPC). The ILPC is expected to produce flow conditions similar to that for a gas turbine engine. They simulated

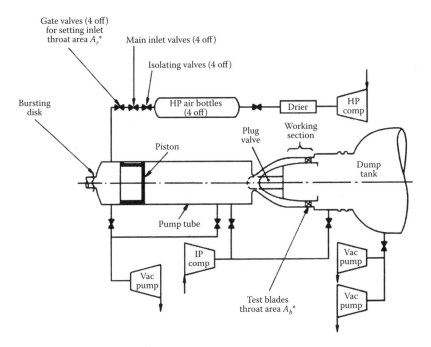

FIGURE 2.59
Configuration of ILPC facility. (From Wedlake, E.T. et al., *ASME J. Turbomach.*, 111, 36, 1989.)

gas-to-metal temperature ratios rather than using actual temperature levels that exist in a real gas turbine. Figure 2.59 shows the ILPC setup. Heat-transfer measurements were made by passing moderately heated air over test blades for a period just long enough to record the response of surface temperatures to the passage of the hot gas. Figure 2.60 shows the outer endwall Nusselt number contours for an exit Mach number of 1.4. The distributions are uniform across the vane inlet at midpassage, with high levels concentrated under the leading-edge horseshoe vortex regions. A saddle-point region of moderate heat transfer is evident early in the passage, with increasing heat transfer as the pressure-side vortex moves from the pressure side of one airfoil to the suction side of the adjacent airfoil. This is consistent with the low-speed results provided by Graziani et al. (1980). Giel et al. (1998) presented endwall heat-transfer measurements in a transonic turbine cascade. They investigated the effects of exit Mach number and free-stream turbulence on endwall heat transfer. They indicated that supersonic exit Mach numbers break up the smooth Stanton number distributions seen in subsonic exit Mach number cases. Also, they indicated that they obtained lower-peak heat-transfer levels with the addition of grid-generated turbulence. They mused that this may result because of the changes in secondary flows brought about by thinner boundary layer thickness. Recently, Harvey et al. (1999) also investigated endwall heat transfer. They compared their experimental results with computational simulations.

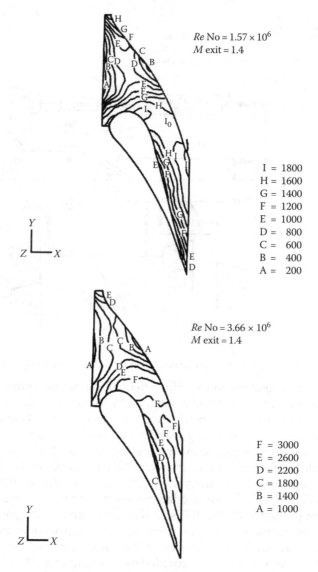

FIGURE 2.60
Outer endwall Nusselt number contours. (From Wedlake, E.T. et al., *ASME J. Turbomach.*, 111, 36, 1989.)

2.5.6 Effect of Surface Roughness

Blair (1994) studied the effect of roughness on the endwall and blade surface heat transfer. Figure 2.61 presents the airfoil surface and hub endwall Stanton number contours for smooth and rough walls. The smooth surface was a liquid crystal-coated surface, and the rough surface was said to be 100 times rougher than the liquid crystal coating. A screened grit of 660 μm-sized

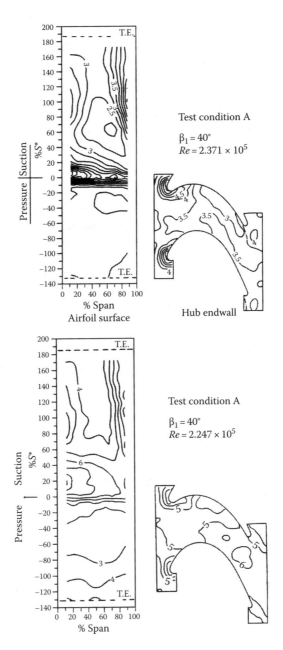

FIGURE 2.61
Airfoil surface and hub endwall Stanton number distributions for smooth and rough surfaces.
(From Blair, M.F., *ASME J. Turbomach.*, 116, 1, 1994.)

particles was applied uniformly over the entire rotor test model to simulate the rough surface. The results for the airfoil surface are presented here to show the 3-D effects that the endwall secondary flows have on the airfoil surface heat transfer near the endwall. It is clear from the smooth-surface results that the endwall secondary flows have a stronger effect on suction surface Stanton number distributions than on the pressure surface. The heat-transfer distributions clearly show that the endwall boundary layers are being swept across the passage toward the suction surface by the cross passage pressure gradient, as explained earlier. The boundary layers then roll up into a pair of vortices near the airfoil/hub/endwall corner and the airfoil tip/endwall corner. The effect on the suction surface is to produce a pattern of Stanton number contours that corresponds directly with the shapes of vortices and lines of separation deduced from the flow field visualization studies. The hub endwall heat-transfer map shows significantly high Stanton numbers near the airfoil leading edges. The leading-edge horseshoe vortices also significantly enhance the endwall heat transfer as they enhanced the airfoil surface heat transfer. The maximum heat transfer on the endwall occurred at the leading edges of the airfoils. The high heat-transfer region near the leading edge extends slightly onto the suction side. Further downstream into the passage, higher Stanton numbers are observed nearer the suction surface than the pressure surface. This may be due to the migration of the passage vortex toward the suction side.

Comparison of rough-surface heat-transfer map to the smooth-surface heat-transfer map clearly shows the significantly increased heat-transfer levels. The boundary layers are expected to be fully turbulent for the rough surfaces that may produce higher heat-transfer levels. However, it is important to note that the flow patterns are similar for both smooth and rough surfaces, both on the airfoil and the hub endwall. Heat-transfer levels are augmented between 20% and 40%, depending on location on the airfoil surface and endwall.

More recent studies have concentrated on using film cooling to cool the endwall region. These studies will be discussed in the next chapter on film cooling.

2.6 Turbine Rotor Blade Tip Heat Transfer

2.6.1 Introduction

In modern gas turbine engines, unshrouded blades rotate in close proximity to a stationary shroud or outer wall of the turbine housing. The tip gap is typically 1.5% of the blade span. Blade tip failures are primarily caused by the hot leakage flow through the tip gap. The pressure difference between the pressure and suction surfaces induces the flow through the gap between the rotating blade and the stationary shroud. Although undesirable, it is impractical to eliminate the gap entirely as the clearance accommodates for the centrifugal growth of the blade as well as for the differential thermal expansion between the shroud and blade. Metzger et al. (1991) pointed out that the blade tip leakage flow is normally associated with degradation in blade aerodynamic performance, and that the second aspect, that of increased convective heat-transfer load to the blade tip, is ignored. Heat-transfer rates at the tip can be the highest over the entire blade surface, leading to large temperature gradients and related durability problems. Studies by Dunn et al. (1984) in real rotor/stator environment have shown that the thermal load on the blade tip surfaces are so high that they can produce structural damage if the tip area heating is not adequately included in design considerations. The blade tip heat transfer is directly associated with the tip leakage flow characteristics. To fully understand the tip heat transfer, researchers need also to study the blade tip inlet flow and exit flow characteristics and its effects on pressure and suction surface heat transfer near the tip regions.

2.6.2 Blade Tip Region Flow Field and Heat Transfer

The blade tip leakage flow, being a pressure-driven flow, may affect the heat transfer on the pressure surface near the tip at entrance and also on the suction surface near the tip at the exit. Some studies have focused on the blade tip leakage flow for aerodynamic purposes and studied the effects of blade tip leakage flow on blade performance. Studies by Bindon and Morphus (1988) and Bindon (1989) have contributed to the general understanding of tip-gap flow patterns. Bindon made pressure and flow field measurements on an axial turbine blade tip in a linear cascade under low-speed conditions. Figure 2.62 presents Bindon's explanation of the tip-gap leakage flow. He suggested that the pressure-side near-tip flow rolls over the tip due to strong active pressure gradient across the tip. The leakage flow possesses an axial velocity component due to the secondary flows. The leakage flow then encounters the oncoming suction-side wall flow and tends then to separate and roll into a vortex. The growth of the vortex as the tip flow exits the gap causes an increase in the total amount of secondary flow within the turbine blade passage, thus resulting in large aerodynamic losses. The tip leakage flow has a significant effect on the blade tip heat transfer, the pressure-side near-tip region heat transfer, and the suction-side near-tip region heat transfer. The leakage flow has a sink-like character on the pressure side near the tip. The sink flow characteristics vary on the pressure side from leading edge to trailing edge as dictated by the local pressure difference across the tip. The superposed sink flow accelerates the flow through the small gap that leads to the tinning and possible relaminarization of the near-tip boundary layer (Metzger et al., 1991). The leakage flow on the suction side is also typical of a source flow entering an already existing main flow. The suction-side heat

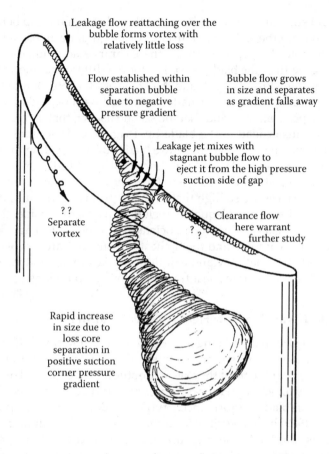

FIGURE 2.62
Tip leakage flow as explained by Bindon. (From Bindon, J.P., *ASME J. Turbomach.*, 111, 258, 1989.)

transfer is also affected by the vortex generation. The leakage flow consists of hot mainstream gases, and oxidation of the metal occurs. The oxidized gap becomes wider, which in turn increases leakage flow and leads to failure of the blade tip. The pressure-side tip corner from midchord to trailing edge suffers the greatest damage due to oxidation. Yaras et al. (1989) and Moore et al. (1989) also contributed to the understanding of flow over tip gaps.

Figure 2.63 illustrates flow measurements between adjacent blades in the passage using a five-hole pitot probe as presented by Yamamoto (1989). The results indicate two kind of vectors: (1) the velocity projected on to the end-wall (W) and the magnitude of the secondary flow vector based on mass-averaged flow direction ($V_s^{\square}$), and (2) the streaklines on a blade-to-blade plane and downstream total pressure loss contours. The relation between the endwall flow and the leakage flow to the cascade loss generation process is represented effectively. Leakage flows through the blade tip clearance interact near the suction surface with the endwall flow and roll down away

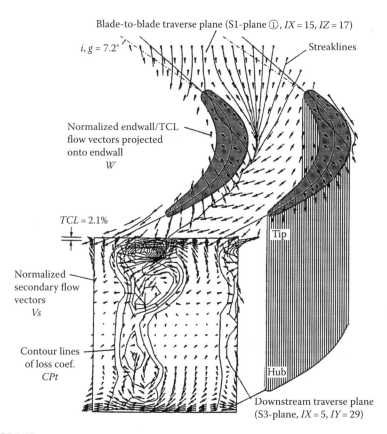

FIGURE 2.63
Overall view of leakage loss mechanisms between adjacent blades. (From Yamamoto, A., *ASME J. Turbomach.*, 111(3), 264, 1989.)

from the endwall along an interaction (or separation) line, and form a leakage vortex with significant loss generation. The vortex, as clearly seen in the figure, interacts intensely with the tip-side passage vortex, which is rotating in the opposite direction to the leakage vortex. The flow mechanisms are significantly affected by clearance gap sizes, incidence angles of the inlet guide plates, and the position of the blade-to-blade planes within the gap.

Metzger et al. (1991) measured blade tip and shroud heat flux for a Garrett 731-2 full-stage rotating turbine. This work was similar to the works published by Dunn et al. (1984, 1986a,b). Figure 2.64 shows the measured heat flux values at the blade tip along the chord from leading edge to about 30% blade chord for two tip/shroud clearance values. The dashed line is the estimated tip heat flux value based on a simple 1-D flow model as an entrance flow into a short stationary duct. The estimation is in good agreement with the data for clearance gap = 0.015 in., for which it is calculated. The heat flux values for a clearance gap of 0.025 in. are higher as a result of higher flow rate.

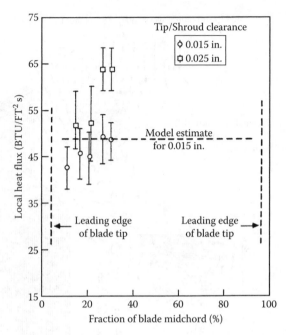

FIGURE 2.64
Measured and estimated local heat flux on blade tip. (From Metzger, D.E. et al., *ASME J. Turbomach.*, 113, 502, 1991.)

Yang and Diller (1995) presented a study of heat transfer and flow for a grooved turbine blade tip in a transonic cascade. They measured heat transfer on the grooved tip at a single location for different tip gap flows under high-speed transonic conditions. More detailed measurements under high-speed flow conditions on an actual turbine cascade are required to estimate better the heat load to the blade tips.

2.6.3 Flat-Blade Tip Heat Transfer

The earliest study on blade tip heat transfer was done by Mayle and Metzger (1982). They studied heat transfer to a simulated tip, with and without a rotating shroud. In their study, the blade tip heat-transfer model was held stationary. The adjacent wall, provided by the rim surface of a rotating disk, was moving. The heat-transfer surface was a heated strip representing a typical pressure-side to suction-side flow across the tip. This experiment was a simplistic model that provided heat-transfer data on a blade tip for a stationary or rotating shroud. Figure 2.65 presents the heat-transfer data with rotation (Nu) normalized by heat-transfer data without rotation (Nu_o) vs. a relative velocity parameter ($R\omega/u$). R is the radius of the shroud from center of rotation, ω is the angular speed, and u is the mean velocity of flow in the clearance gap. From the data, it is evident that the effect of moving wall is

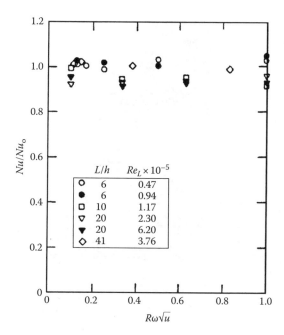

FIGURE 2.65

Nusselt number ratio versus the relative velocity parameter. (From Mayle, R.E. and Metzger, D.E., Heat transfer at the tip of an unshrouded turbine blade, *Proceedings of the 7th International Heat Transfer Conference*, Munich, Germany, Vol. 3, pp. 87–92, 1982.)

negligible over the entire range of parameters considered in the study. The study varied the L/h ratio that represents the length on the tip (L) to clearance gap height (h) ratio. The study also varied the flow length Reynolds number (UL/v). Mayle and Metzger (1982) then concluded that experiments that neglected the effect of blade rotation could be used to assess the blade tip heat transfer as long as the flow conditions are similar.

Metzger and Rued (1989) presented heat-transfer and flow results in the near-tip region of the pressure surface using a blade tip simulation and a sink flow. This model simulated a sink flow similar to the sink-like characteristics of the blade tip gap on the pressure side. The sink flow was monitored and simulated to be similar to actual blade tip leakage flows. Rued and Metzger (1989) presented heat-transfer and flow results in the near-tip region of the suction surface using a blade tip simulation and a source flow. This model simulated a blade suction surface with a gap feeding a source flow similar to the flow exiting the tip gap on the suction surface of an airfoil.

2.6.4 Squealer- or Grooved-Blade-Tip Heat Transfer

Typically, turbine blade tips are grooved chordwise to reduce tip flow and heat transfer. The groove acts as a labyrinth seal to increase flow resistance and thus reduce leakage flow that in turn reduces heat transfer.

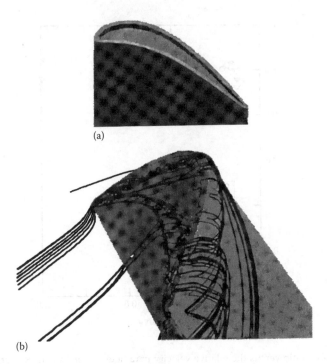

FIGURE 2.66
(a) Grooved tip blade and (b) blade tip flow patterns. (From Ameri, A.A. et al., Effect of squealer tip on rotor heat transfer and efficiency, ASME Paper 97-GT-128, 1997.)

Figure 2.66 shows the grooved tip of a blade and also the streamlines showing the tip flow patterns (Ameri et al., 1997). Two distinct vortices can be clearly seen within the groove. A separation vortex is generated as the incoming flow separates the inner edge of the pressure-side rim. This vortex hugs the pressure surface sidewall and spills out of the groove near the trailing edge of the blade. The second vortex, also a separation vortex, runs from stagnation region to the suction surface of the blade. Pressure-side rim, suction-side rim, and tip flow vortex are also generated in the process.

Chyu et al. (1989) and Metzger et al. (1989) studied heat transfer on grooved turbine tips extensively. They used naphthalene sublimation technique to obtain mass-transfer coefficients inside the grooved areas of the simulated tips. Figure 2.67 shows an illustration of their grooved-blade-tip model. In their model, the blade is stationary, and moving the shroud at a known constant speed creates relative motion. The direction of the leakage flow will be in the opposite direction from pressure to suction side. The figure also shows the parametric representations used in both the studies. Figure 2.68 shows the effect of gap clearance (solid symbols) on heat transfer for flow over the tip (Chyu et al., 1989). The cavity depth-to-width ratio (D/W), flow Reynolds number ($Re = U \times C/v$, C = clearance gap) are kept constant for all cases with

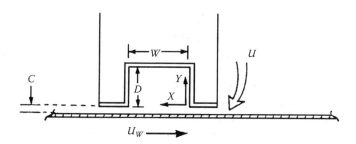

FIGURE 2.67
Shrouded rectangular cavity blade tip model. (From Metzger, D.E. et al., *ASME J. Heat Transfer*, 111, 73, 1989.)

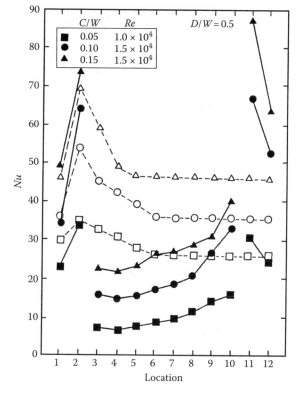

FIGURE 2.68
Effect of gap clearance on heat transfer for flow over a tip model. (From Chyu, M.K. et al., *ASME J. Turbomach.*, 111, 131, 1989.)

no shroud velocity. The local heat-transfer coefficient increases with increasing gap clearance-to-cavity width ratio (C/W). Additionally, the heat-transfer distribution inside the cavity varies with different clearance gaps. According to the authors, the larger gap clearance permits the separated shear layer to grow thicker, inducing higher turbulence levels in the gap mainstream.

The heat transfer on the cavity wall is influenced by the characteristics of this shear layer. The flat tip results (open symbols) are also shown for comparison. Heat transfer in the upstream end of the cavity is greatly reduced compared to the flat tip. Downstream of the cavity, heat-transfer levels for the grooved tip are higher due to flow reattachment inside the cavity. The heat-transfer levels in the downstream edge of the grooved tip are much higher than for the flat tip due to the accelerated flow from the cavity into the small clearance gap. Figure 2.69 summarizes the results with area-averaged heat transfer on the cavity floor with respect to increasing cavity depth. Computed results show minimal effect of the shroud velocity on the heat-transfer coefficient, as earlier stated. Reduced clearance gap can greatly reduce heat-transfer load to the blade tip (refer to results for $C/W = 0.05$). Increase in cavity depth also contributes to reduced heat load to the tip.

Most of the studies on blade tip have been for either low-speed flows or for turbine blade tip models. Such results, although helpful, do not provide a clear understanding of the complex blade tip leakage flow and heat transfer. The most recent study on turbine blade tip heat transfer and flow in a linear cascade was carried out by Bunker et al. (1999). They used a hue detection-based liquid crystal technique to obtain detailed heat-transfer coefficient

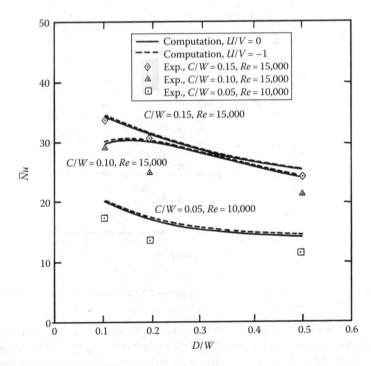

FIGURE 2.69
Summary of area-averaged heat-transfer results on the cavity floor. (From Chyu, M.K. et al., *ASME J. Turbomach.*, 111, 131, 1989.)

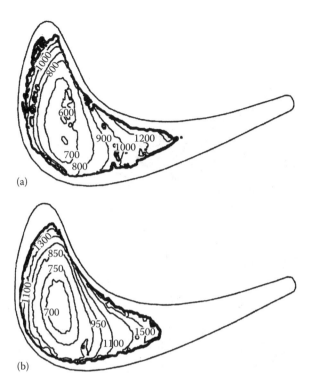

FIGURE 2.70
Effect of free-stream turbulence on radius-edge blade tip heat-transfer coefficient distributions. (a) Radius-edge blade tip heat transfer coefficients for 2.03 mm clearance and $Tu = 5\%$ (W/m²/K). (b) Radius-edge blade tip heat transfer coefficients for 2.03 mm clearance and $Tu = 9\%$ (W/m²/K). (From Bunker, R.S. et al., Heat transfer and flow on the first stage blade tip of a power generation gas turbine. Part 1: Experimental results, *ASME International Gas Turbine and Aero-Engine Congress*, Indianapolis, IN, ASME Paper No. 99-GT-169, June 7–10, 1999.)

distributions on the flat tip surface with both sharp and rounded edges for a large power generation turbine under engine representative flow conditions. Figures 2.70 and 2.71 show the blade tip heat-transfer coefficients for various tip clearances and free-stream turbulence levels. Results show that the blade tip heat-transfer coefficients can be varied up to three times and there exists a central sweet spot of low heat transfer that extends into the midchord region and toward the suction side. An increase in the free-stream turbulence intensity level from 5% to 9% raises the overall tip heat transfer by about 10%. Decreasing the tip clearance by 38% of the nominal value (2.03 mm) results in a decrease of about 10% in heat transfer, while a 38% increase in tip clearance results in a 10% increase in heat transfer. Azad et al. (2000a,b) reported flow and heat transfer on a blade tip surface of a GE–E³ aircraft rotor blade in a five-bladed linear cascade under engine representative flow conditions. They used a transient liquid crystal technique to obtain detailed heat-transfer coefficient distributions for both flat and squealer tips for various tip

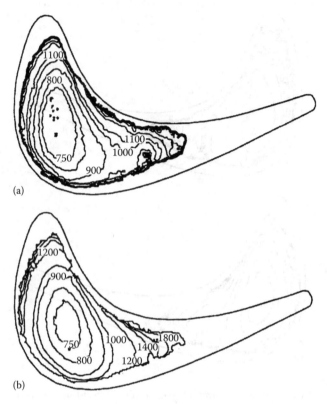

FIGURE 2.71
Effect of tip clearance on radius-edge blade tip heat-transfer coefficient distributions.
(a) Radius-edge blade tip heat transfer coefficients for 1.27 mm clearance and Tu = 9%
(W/m²/K). (b) Radius-edge blade tip heat transfer coefficients for 2.79 mm clearance and
Tu = 9% (W/m²/K). (From Bunker, R.S. et al., Heat transfer and flow on the first stage blade tip
of a power generation gas turbine. Part 1: Experimental results, *ASME International Gas Turbine
and Aero-Engine Congress*, Indianapolis, IN, ASME Paper No. 99-GT-169, June 7–10, 1999.)

clearances and free-stream turbulence levels. Figure 2.72 shows that various
regions of high and low heat-transfer coefficient on the tip surface, with high
heat transfer near the pressure side and low heat transfer near the suction
side as well as toward leading and trailing-edge regions. Figure 2.73 shows
that averaged tip heat-transfer coefficients increase with increasing either tip
clearance or free-stream turbulence intensity level. Figure 2.74 compares the
heat-transfer coefficient distributions between the flat tip and squealer tips
with various tip clearances. Results show that the lower heat-transfer region
shifts from suction-side leading edge of the flat tip to the squealer cavity
toward the trailing edge of the squealer tip blade. The squealer tip blade
provides a lower overall heat-transfer coefficient when compared to the flat
tip blade. Figure 2.75 shows that the average heat transfer on the squealer
rim is higher than that on the squealer cavity and increases with increasing
the tip clearance.

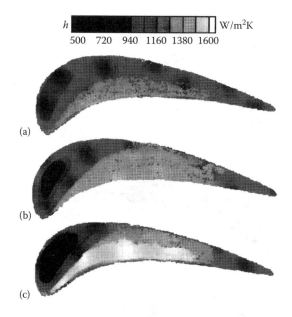

FIGURE 2.72
Effect of tip clearance on flat tip heat-transfer coefficient distributions at $Tu = 9.7\%$. (a) 1% tip gap; (b) 1.5% tip gap; and (c) 2.5% tip gap. (From Azad, G.M.S. et al., Heat transfer and pressure distributions on a gas turbine blade tip, *ASME International Gas Turbine and Aero-Engine Congress*, Munich, Germany, ASME Paper 2000-GT-194, May 8–11, 2000a.)

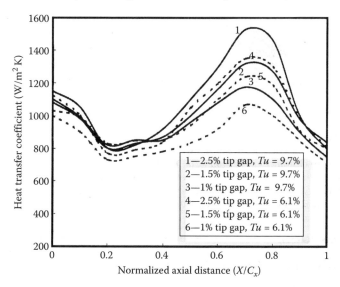

FIGURE 2.73
Averaged heat-transfer coefficient for various tip clearances and free-stream turbulence levels. (From Azad, G.M.S. et al., Heat transfer and pressure distributions on a gas turbine blade tip, *ASME International Gas Turbine and Aero-Engine Congress*, Munich, Germany, ASME Paper 2000-GT-194, May 8–11, 2000a.)

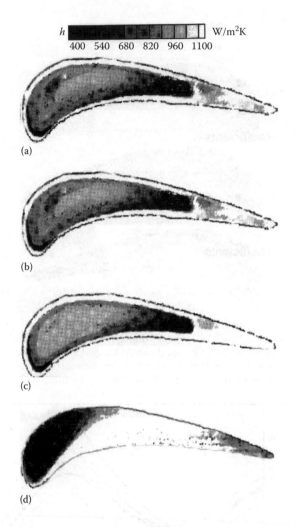

FIGURE 2.74

Heat-transfer coefficient comparison between the flat tip blade and the squealer tip blade with various tip clearances at $Tu = 9.7\%$. (a) 1% tip gap; (b) 1.5% tip gap; (c) 2.5% tip gap; and (d) 1.5% tip gap, flat tip. (From Azad, G.M.S. et al., Heat transfer and flow on the squealer tip of a gas turbine blade, *ASME International Gas Turbine and Aero-Engine Congress*, Munich, Germany, ASME Paper No. 2000-GT-195, May 8–11, 2000b.)

2.7 Leading-Edge Region Heat Transfer

2.7.1 Introduction

The leading-edge region of the airfoil is required to have a moderately small radius to ensure better aerodynamic design and blend smoothly into the rest of the airfoil shape. The leading edge is the most critical heat-transfer area on

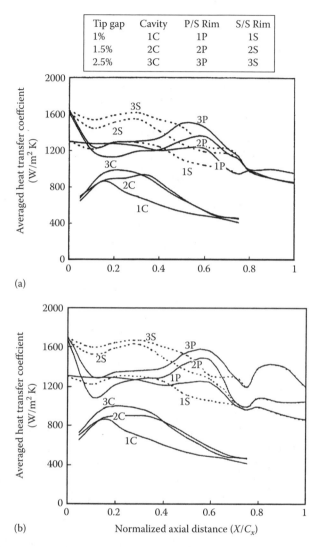

FIGURE 2.75
Averaged heat-transfer coefficients on both the squealer rim and cavity for various tip clearances and free-stream turbulence levels. (a) $Tu = 6.1\%$ and (b) $Tu = 9.7\%$. (From Azad, G.M.S. et al., Heat transfer and flow on the squealer tip of a gas turbine blade, *ASME International Gas Turbine and Aero-Engine Congress*, Munich, Germany, ASME Paper No. 2000-GT-195, May 8–11, 2000b.)

the turbine airfoil. In most cases, the highest heat-transfer rates on the airfoil occur on the stagnation region of the leading edge. Frossling (1958) modeled the stagnation region as a circular or elliptic cylinder in cross flow for a laminar free stream. Frossling's results are used for comparison in all later studies on leading-edge heat transfer. Several studies have focused only on the leading-edge region by modeling it as a cylinder in a cross flow or a blunt body with a semicircular or elliptic leading edge. Individual effects such as free-stream turbulence, unsteady wakes, surface roughness, and geometry have been the focus of several studies.

2.7.2 Effect of Free-Stream Turbulence

High turbulence is generated typically by using grids upstream of the test section in a wind tunnel. The grid generates turbulence by acting as a blockage to the flow. Turbulence decays behind a grid, necessitating proximity of the test section to the grid for very-high-turbulence-intensity effects. Jet grids are also used to generate high-turbulence intensities. A jet grid consists of an array of hollow rods or pipes through which air is forced and delivered to small holes on the rods that exit into the mainstream. The jets can be oriented to blow upstream or downstream with controlled flow rates. Studies by Smith and Keuthe (1966), Kestin and Wood (1971), and Lowery and Vachon (1975) have demonstrated the effect of free-stream turbulence on stagnation region heat transfer. Lowery and Vachon (1975) developed a correlation to predict heat transfer in the stagnation region under free-stream turbulence. They presented the following equation for stagnation point heat transfer:

$$\frac{Nu}{Re^{0.5}} = 1.01 + 2.624 \frac{Tu(Re)^{0.5}}{100} - 3.07 \left[\frac{Tu(Re)^{0.5}}{100} \right]^2 \qquad (2.7)$$

All the aforementioned studies reported a strong effect of turbulence intensity on stagnation point heat transfer. Studies indicate that vortex stretching causes the heat-transfer augmentation. Vortices with components of their axes normal to the cylinder stagnation line and normal to the free-stream flow direction are stretched and tilted due to divergence of streamlines and acceleration around the body. Conservation of angular momentum causes this vorticity to be amplified.

O'Brien and Van Fossen (1985) studied the influence of jet grid turbulence on stagnation region heat transfer for a cylinder in cross flow. Figure 2.76 presents the circumferentially local heat-transfer results for the front one-half of the cylinder from stagnation point to about 50° from stagnation. The Frossling number $Nu/Re^{1/2}$ is plotted as a function of angle from cylinder stagnation line. Data are provided for three turbulence conditions: no grid, square-bar grid, and jet grid. The no-grid case turbulence intensity was 0.6%, the square-bar-grid case was around 7.4%, and the jet grid was around 11.0%. The solid line represents the classical Frossling theoretical calculation for a laminar boundary layer around a circular cylinder with zero free-stream turbulence. The data for low-turbulence cases for all three Reynolds numbers agree well with the Frossling solution. The highest Frossling numbers are obtained for the case with the highest Reynolds number and highest

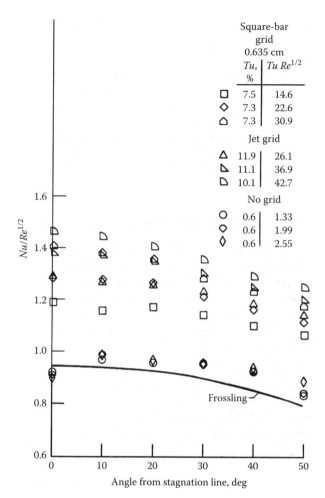

FIGURE 2.76
Circumferentially local heat-transfer results on a cylindrical leading-edge model. (From O'Brien, J.E. and Van Fossen, G.J., The influence of jet-grid turbulence on heat transfer from a stagnation region of a cylinder in crossflow, ASME Paper 85-HT-58, 1985.)

turbulence intensity ($Tu\ Re^{1/2} = 42.7$). The heat-transfer augmentation is a combination of both the free-stream turbulence intensity (Tu) and the free-stream Reynolds number (Re). In all cases, the heat-transfer augmentation is fairly uniform over the entire cylinder surface.

Stagnation region heat-transfer results from O'Brien and Van Fossen (1985) are compared with the correlations from Smith and Keuthe (1966), Kestin and Wood (1971), and Lowery and Vachon (1975) in Figure 2.77. Smith and Keuthe's correlation shows a monotonic increase in stagnation region Frossling number with increasing $Tu\ Re^{1/2}$. Kestin and Wood's and Lowery and Vachon's correlations show a leveling off in Frossling number for

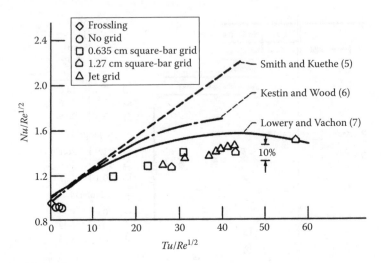

FIGURE 2.77
Stagnation point heat-transfer results. (From O'Brien, J.E. and Van Fossen, G.J., The influence of jet-grid turbulence on heat transfer from a stagnation region of a cylinder in crossflow, ASME Paper 85-HT-58, 1985.)

$Tu\,Re^{1/2}$ values greater than 30. Results from O'Brien and Van Fossen (1985) are lower than all three correlations.

Mehendale et al. (1991) studied the effect of free-stream turbulence on heat transfer for a blunt body with a semicircular leading edge and flat afterbody. High mainstream turbulence levels were obtained using jet grids. The tests were performed in a low-speed, open-circuit wind tunnel. Different turbulence levels (1%–15%) were studied for a range of Reynolds numbers (25,000–100,000). Heat-transfer measurements were made using instrumented strip heaters and embedded thermocouples.

Figure 2.78 presents Nusselt number distributions for different turbulence levels at $Re = 100,000$. As can be observed, the high free-stream turbulence ($Tu = 12.9\%$) augments heat-transfer levels up to 60% over the low-turbulence case ($Tu = 0.75\%$). The effect of the grid-generated turbulence is insignificant beyond the curved region on the flat afterbody. Mehendale et al. (1991) observed that their stagnation point heat-transfer data were in good agreement with the Lowery–Vachon correlation. However, they proposed a new correlation for the overall averaged heat transfer on the entire leading-edge region. The correlation they proposed was

$$\frac{Nu}{Re^{0.5}} = 0.902 + 2.14\frac{Tu(Re)^{0.5}}{100} - 2.89\left[\frac{Tu(Re)^{0.5}}{100}\right]^2 \qquad (2.8)$$

This correlation will help in predicting overall averaged heat-transfer data on the cylindrical leading-edge region for known free-stream turbulence conditions and Reynolds number based on leading-edge diameter. Figure 2.79 shows the

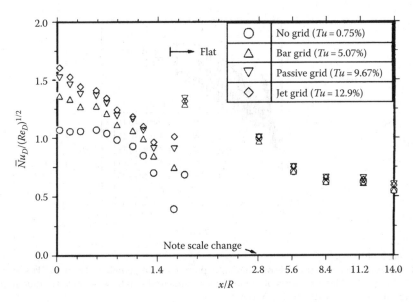

FIGURE 2.78
Effect of turbulence intensity on spanwise-averaged Nusselt number distributions for $Re_D =$ 100,000. (From Mehendale, A.B. et al., *ASME J. Heat Transfer*, 113, 843, 1991.)

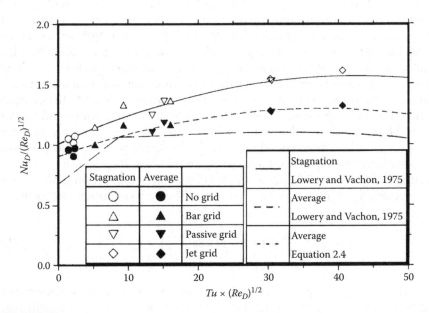

FIGURE 2.79
Comparison of leading-edge heat-transfer data under free-stream turbulence effects. (From Mehendale, A.B. et al., *ASME J. Heat Transfer*, 113, 843, 1991.)

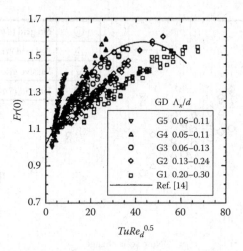

FIGURE 2.80
Effect of turbulence length scale on stagnation point Frossling number. (From Van Fossen, G.J. and Ching, C.Y., Measurements of the influence of integral length scale on stagnation region heat transfer, ASME Paper 94-HT-5, 1994.)

comparison of leading-edge heat-transfer data with high turbulence to the correlations developed by Lowery and Vachon and Mehendale et al.

Van Fossen and Ching (1994) studied the effect of length scale on heat-transfer augmentation in stagnation region for a given turbulence intensity. The heat-transfer model had an electric foil heater with embedded thermocouples to measure the surface temperature. They used five grids to generate turbulence levels up to 8% and different length scales. Figure 2.80 presents the effect of length scale on the stagnation point Frossling number. Grids G1–G5 generate from 8% high turbulence to low turbulence (Tu = 1.0%), respectively. The length scales (Λ) are also shown in the figure. To increase the turbulence intensity, the grid was moved closer to the cylinder. The range of length scale varied from 0.05 to 0.30. The Lowery and Vachon (Ref. 14 in figure) correlation predicts the heat-transfer data only in a narrow range of length scales. Above a $Tu\,Re^{0.5}$ value of 40, the Lowery–Vachon correlation turns downward against the trend of the data of Van Fossen and Ching (1994). Figure 2.81 presents the stagnation point Frossling number vs. a correlation parameter, $Tu\,Re^{0.5}\,(\Lambda/d)^{-0.574}$. The data collapse to the form

$$Fr(0) = A_{Fr}\sqrt{TuRe^{\xi}\left(\frac{\Lambda}{d}\right)^{\beta} + c} \qquad (2.9)$$

where c, A_{Fr}, ξ, and β are determined from least-square fit. Augmentation of heat transfer in the stagnation region increases as integral length scale decreases. As Ames (1997) pointed out in his study, the turbulence generated by the combustor in a gas turbine engine is typically large scale. Gas turbine

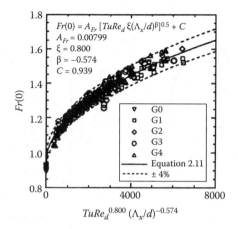

FIGURE 2.81
Stagnation point Frossling number vs. a new correlation parameter. (From Van Fossen, G.J. and Ching, C.Y., Measurements of the influence of integral length scale on stagnation region heat transfer, ASME Paper 94-HT-5, 1994.)

heat-transfer designers need to focus more on the large-scale turbulence effects on heat transfer specifically in the leading edge of the airfoil.

2.7.3 Effect of Leading-Edge Shape

Van Fossen and Simoneau (1994) studied the additional effect of leading-edge shape under the same measurement conditions as that of Van Fossen and Ching (1994). Figure 2.82 presents the four leading-edge shapes used in their study. The ratios of major to minor axes were 1:1, 1.5:1, 2.25:1,

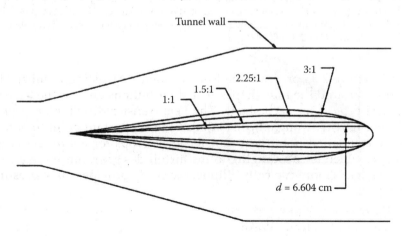

FIGURE 2.82
Leading-edge profiles studied by Van Fossen and Simoneau. (From Van Fossen, G.J. and Simoneau, R.J., Stagnation region heat transfer: The influence of turbulence parameters, Reynolds number and body shape, *ASME HTD*, 271, 177, 1994.)

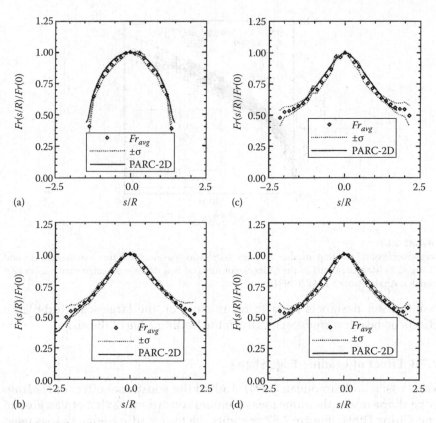

FIGURE 2.83
Low turbulence Frossling number distributions around four leading-edge shapes. (a) 1:1 ellipse; (b) 1.5:1 ellipse; (c) 2.25:1 ellipse; and (d) 3:1 ellipse. (From Van Fossen, G.J. and Simoneau, R.J., Stagnation region heat transfer: The influence of turbulence parameters, Reynolds number and body shape, *ASME HTD*, 271, 177, 1994.)

and 3:1. Figure 2.83 presents the low-turbulence Frossling number distributions around the four shapes. The distributions around the leading edges are typical for each shape. The heat-transfer distribution profile is thinner for the strong ellipse (3:1). Another important point to note is that the stagnation point heat-transfer level drops for a stronger elliptic shape, which is encouraging to the airfoil designer. Since the airfoil leading edge is more typically elliptic, the study provides an interesting comparison.

2.7.4 Effect of Unsteady Wake

The effect of unsteady wake on blade surface heat transfer and flow was discussed in detail in earlier sections. However, few studies have focused only

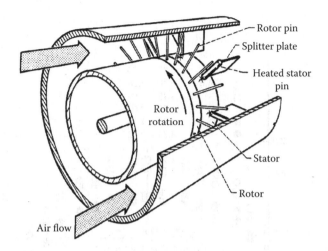

FIGURE 2.84
Concept of rotor wake heat-transfer experiment. (From O'Brien, J.E. et al., Unsteady heat transfer and direct comparison to steady-state measurements in a rotor-wake experiment, *Proceedings of the 8th International Heat Transfer Conference*, Vol. 3, pp. 1243–1248, San Francisco, CA, 1986.)

on the effect of unsteady wake on leading-edge heat transfer. The upstream wakes are simulated using rotating rods that shed wakes on stationary cylindrical models downstream. Most of the work relating to unsteady wake effect on leading-edge models was done by Simoneau et al. (1984), Morehouse and Simoneau (1986), O'Brien et al. (1986), and O'Brien and Capp (1989). They developed an experiment to simulate the wake dynamics of a turbine rotor/stator interaction. They used a spoked-wheel rotating in an annular flow to produce impinging wakes on a simulated stator leading edge composed of a circular cylinder with a splitter plate behind it. Figure 2.84 shows the concept of the rotor wake heat-transfer experiment conducted by the aforementioned group of researchers.

Morehouse and Simoneau (1986) measured typical instantaneous- and ensemble-averaged wake profiles produced in the rig. They used different pin sizes and mesh grid combinations to simulate different unsteady flow conditions. Figure 2.85 presents the effect of various types of unsteadiness on the circumferential heat-transfer distribution on the cylinder. The values of unsteady turbulence parameter Tu' are varied from 1.12% to 9.80%. Based on their results, the unsteady wake effect is not as significant as that of grid-generated turbulence. However, there is a need to understand the unsteady wake flow phenomena to account for the heat-transfer behavior.

O'Brien and Capp (1989) measured the axial and tangential components of the unsteady turbulent flow downstream of the wake generator used by

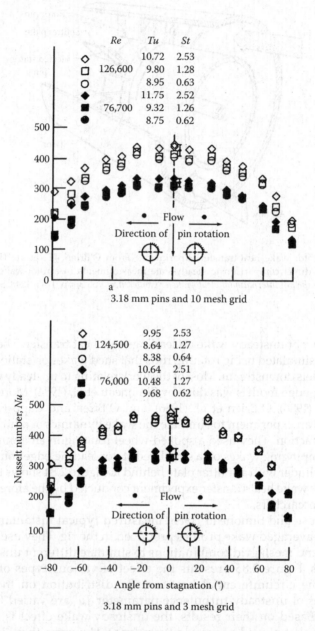

FIGURE 2.85
Effect of various types of unsteadiness on the circumferential distribution of heat transfer. (From Morehouse, K.A. and Simoneau, R.J., Effect of rotor wake on the local heat transfer on the forward half of a circular cylinder, *Proceedings of the 8th International Heat Transfer Conference*, Vol. 3, pp. 1249–1256, San Francisco, CA, 1986.)

the earlier studies. They also defined a new parameter called the rotor flow coefficient U/V_b, which can be related to the Strouhal number (S) as

$$\frac{\bar{U}}{V_b} = \frac{\bar{U}}{r_b(2\pi \cdot f_b/24)} = \frac{24d_b}{r_b S} \qquad (2.10)$$

where

d_b is the rotating bar diameter

r_b is the midspan radius of the wake-generating bars measured from rotor centerline

The flow measurements presented by this study help one to understand the wake characteristics and thus add to one's understanding their effects on heat-transfer augmentation.

Funazaki (1996) also presented time-averaged, heat-transfer distributions around the leading edge of a blunt body affected by periodic wakes. He observed that there is strong heat-transfer enhancement with increasing bar-passing wake Strouhal number. Figure 2.86 presents the wake-affected heat-transfer distributions on the test model for several unsteady flow conditions for a $Re = 67{,}000$. The profiles of wake-affected heat transfer are more asymmetric with increasing wake strength. A stationary wake condition is also shown that is not typical of any real case. The trend for the stationary rod case is shown just for comparison and flows of dissimilar trend compared to other cases. The data of Funazaki clearly show the strong wake effect on heat-transfer around the leading edge. The test model used by Funazaki (1996)

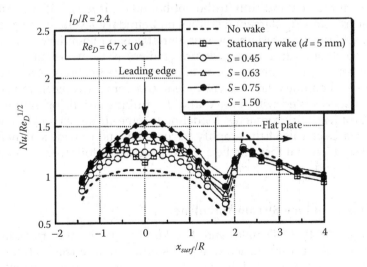

FIGURE 2.86
Wake-affected heat-transfer distributions on the leading-edge model. (From Funazaki, K., *ASME J. Turbomach.*, 118, 452, 1996.)

was similar to Mehendale et al. (1991), except that Funazaki had a wake generator upstream of the blunt body. The bars were rotated at different frequencies to obtain different Strouhal numbers. Funazaki (1996) also attempted to correlate his data with respect to Strouhal number. More attempts similar to this study will help designers account for the wake-affected heat transfer on the leading-edge region of airfoils.

The combined effect of free-stream turbulence and unsteady wake impingement is also important on the leading-edge-region heat transfer. However, there has been no study focusing on these effects on leading-edge heat transfer. Since a combination of free-stream turbulence and unsteady wake can cause a significantly different heat-transfer response compared to the singular effects, it is important for future research to focus on this aspect of leading-edge heat transfer. Also, there are no time-dependent data available for the leading-edge region as presented on the blade by Guenette et al. (1989).

2.8 Flat-Surface Heat Transfer

2.8.1 Introduction

Some of the early studies researching the effects of free-stream turbulence conducted measurements on flat, simple surfaces. The effect of free-stream turbulence, as seen in the earlier sections, is a major influence on laminar, laminar–turbulent, and fully turbulent boundary layers, and subsequently on heat transfer. Fundamental studies on boundary layers were conducted on flat surfaces to ascertain the singular effects on boundary layer transition. Free-stream turbulence, pressure gradient, streamwise curvature, and surface roughness were the focus of several investigations. Accurate predictions of boundary layer development and convective heat-transfer distributions associated with it are critical for turbine airfoils operating under high-turbulence environments. Modeling turbine airfoils as convex and concave surfaces can also help simplify the problem and provide a more fundamental solution that then can be applied to actual airfoils under real engine conditions.

2.8.2 Effect of Free-Stream Turbulence

Bradshaw (1974), McDonald and Kreskovsky (1979), and Hancock and Bradshaw (1983) presented aerodynamic studies on the effects of free-stream turbulence. They concluded that heat-transfer rates would be enhanced with increases in free-stream turbulence if some form of Reynolds analogy held for these types of flows. Bradshaw (1974) developed an intensity- and length

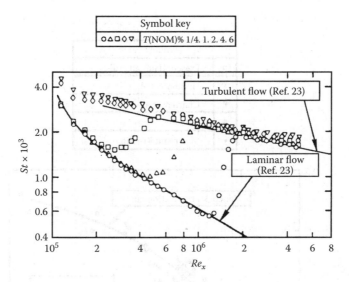

FIGURE 2.87
Heat-transfer distributions along the flat surface test wall for five free-stream turbulence levels. (From Blair, M.F., *ASME J. Heat Transfer*, 105, 33, 1983.)

scale-dependent correlation for skin friction data. Simonich and Bradshaw (1978) and Blair (1983) focused on the effects of free-stream turbulence on transitional boundary layer affected heat transfer on a flat surface.

Hancock and Bradshaw (1983) presented a new correlating parameter based on enhancement of skin friction due to grid-generated turbulence. The correlating parameter α is both a function of turbulence level and ratio of dissipation length scale to boundary layer thickness and is defined as

$$\alpha = \frac{Tu\%}{(L_u^\varepsilon/\delta) + 2} \tag{2.11}$$

Blair (1983) presented the heat-transfer distributions along the flat test wall at five increasing free-stream turbulence levels up to $Tu = 6\%$. As indicated earlier in the chapter, increase in free-stream turbulence induces earlier boundary layer transition. Figure 2.87 presents the heat-transfer distributions for five different turbulence levels. As seen clearly, transition location moves progressively upstream with increasing turbulence level. For higher levels of turbulence, the boundary layer is already fully turbulent. Blair indicated that heat transfer in the fully turbulent region increased up to 36%, for an increase of free-stream turbulence from 0.25% to 6%. Figure 2.88 presents the experimental data compared to the enhanced skin friction factor ($\Delta C_f/C_{fo}$) and the enhanced heat transfer ($\Delta St/St_0$) correlations from Hancock (1980). The turbulent intensity on the abscissa has been modified to include the empirical term β. Hancock's correlation

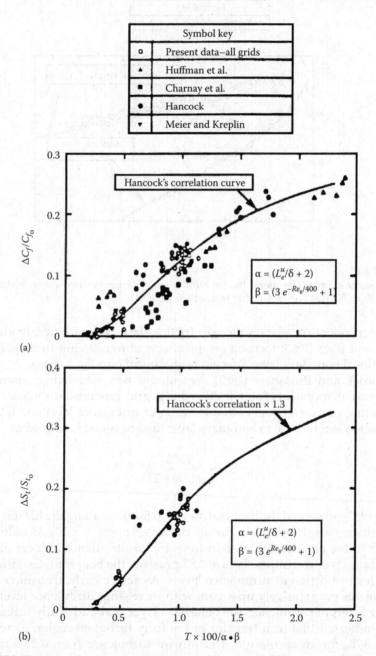

(a)

(b)

FIGURE 2.88
(a) Skin friction and (b) Stanton number free-stream turbulence correlations. (From Blair, M.F., *ASME J. Heat Transfer*, 105, 33, 1983.)

appears to provide a fairly good prediction of the effects of free-stream turbulence on turbulent boundary layer. However, the heat-transfer information from Blair's study and other studies indicated in the figure do not seem to provide enough validation of Hancock's correlation. Zhou and Wang (1995) studied the effect of elevated free-stream turbulence levels (up to $Tu = 6.4\%$) on flow and thermal behavior for transitional boundary layers. They provided statistical results of the streamwise and cross stream velocity fluctuations, temperature fluctuations, Reynolds stresses, and Reynolds heat fluxes.

Maciejewski and Moffat (1992a,b) presented a flat-plate boundary layer study where the free-stream turbulence levels are as high as 60% using a free-jet flow facility. This facility may not produce turbulence characteristics similar to that in a gas turbine airfoil passage. Figure 2.89 presents the data by Maciejewski and Moffat (1992a) in comparison to results from MacMullin et al. (1989), who used a free-jet facility for turbulence generation and from Blair (1983) who used grid turbulence generation. The results show that the Stanton number continually increases with increasing free-stream turbulence, independent of the mechanism used to generate the turbulence in the free stream. This conclusion is similar to the conclusions stated earlier in this chapter. The laminar and turbulent correlation (Kays and Crawford, 1980) are also shown in the figure. All the data presented are higher than the turbulent correlation. Maciejewski and Moffat (1992b) also proposed a new parameter St', which uses the maximum rms velocity found in the near-wall region. The new parameter will be discussed in detail later in this section. Ames and Moffat (1992) studied both skin friction and heat-transfer enhancement

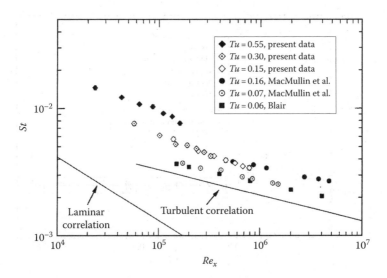

FIGURE 2.89
High-turbulence effects on flat-surface Stanton numbers. (From Maciejewski, P.K. and Moffat, R.J., *ASME J. Heat Transfer*, 114, 834, 1992b.)

using a combustor simulator (Figure 2.26). They proposed a new parameter *TLR*, which uses integral quantities rather than boundary layer thickness:

$$TLR = Tu\left(\frac{\Delta}{L_u^\varepsilon}\right)^{0.33}\left(\frac{Re_\Delta}{1000}\right)^{0.25} \tag{2.12}$$

They used $\Delta = \theta$ for shear stress enhancement and $\Delta = \Delta_2$ for heat-transfer enhancement.

Most of the studies are referenced by Thole and Bogard (1995) in a more recent study on surface heat transfer and skin friction enhancements for high free-stream turbulence levels. Data from several investigators are plotted in terms of the new parameter St', introduced by Maciejewski and Moffat (1992) in Figure 2.90. The new parameter is defined as $St' = h/\rho C_p u'_{max}$, where u'_{max} is maximum standard deviation in the streamwise component of velocity in the wall-affected region, which is the region near the surface that exhibits "law-of-the-wall" behavior in the mean velocity profile. Data from Maciejewski and Moffat (1992b), MacMullin et al. (1989), Ames and Moffat (1992), Thole and Bogard (1995), and Sahm and Moffat (1992) are presented for comparison purposes. The collapsing of the data to the correlation is very good. However, most of the data do not show the 46% increase around $Tu = 11\%$, as predicted by Maciejewski and Moffat.

It is quite evident that existing accepted turbulence correlations fall very short in predicting heat-transfer enhancements for high free-stream turbulence levels. Also, the physics of heat-transfer enhancement for higher free-stream turbulence levels is indicated by the previously mentioned studies to

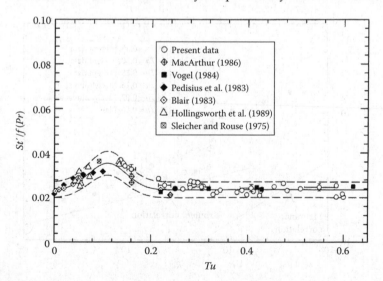

FIGURE 2.90

Comparison of present data in terms of Maciejewski and Moffat. (From Maciejewski, P.K. and Moffat, R.J., *ASME J. Heat Transfer*, 114, 834, 1992b.)

be quite different than that for low free-stream turbulence cases. Although fundamentally important, this study of such extremely high free-stream turbulence levels (Tu = 30%–55%) is not critical to gas turbine situations. Highest levels of turbulence for gas turbine situations are around 15%–20% at the NGV entrance.

2.8.3 Effect of Pressure Gradient

Also essential to modeling a turbine situation is the pressure gradient effect. Since both effects exist in the turbine stages, it is essential to understand the effects of pressure gradient on heat-transfer rates. Pressure gradient can influence boundary layer transition in the actual engine environment.

Blair (1982) conducted experiments on a flat wall with a zero-pressure gradient and for two levels of sink streamwise acceleration. He also varied free-stream turbulence levels from 0.7% to 5.0%. Pressure gradient effects are defined on the basis of the acceleration parameter $K = (v/U_e^2)(\partial U_e/\partial x)$. Blair tested two levels of accelerating flows with accelerating parameter $K = 0.2$ and 0.75×10^{-6}. The results for the zero-pressure gradient flow are shown in Figure 2.87. The results for the two accelerating flows are shown in Figure 2.91. The heat-transfer distributions demonstrate the interplay of the effects of acceleration and turbulence on the transition process. Compare the cases in Figure 2.91 to those in Figure 2.87; it is evident that acceleration suppresses the natural boundary layer transition phenomena. For lower-turbulence cases, the boundary layer remained laminar over the entire test surface length, unlike the zero-pressure gradient case. Two points to note are that (1) the increased acceleration pushes the boundary layer transition location downstream, and that (2) if transition begins to occur, the transition distance to the fully turbulent boundary layer is longer. Negative-accelerating parameter or decelerating flow field may be expected to provide exactly the opposite effect. However, flow separation is a big problem when the flow field is exposed to strongly adverse pressure gradients. This is quite evident on the blade suction surfaces. Designers have to be careful on the suction surface close to the trailing edge where the flow is decelerating due to adverse pressure gradient.

Rued and Wittig (1986) presented data that matched with the correlation presented by Kays and Crawford (1980) for flows with pressure gradient effects:

$$St = 0.0295(Pr)^{-0.4}Re^{-0.2}\left[1 - F_k\frac{K(x)}{St}\right]F(Tu) \qquad (2.13)$$

where
   K is the acceleration parameter
   $F(Tu)$ is a function of turbulent length scale and intensity
   F_k accounts for the effects of pressure gradient (value varies from 90 to 165)

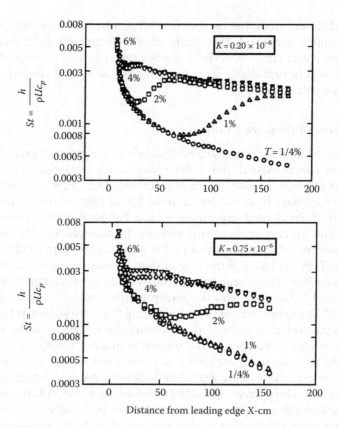

FIGURE 2.91
Heat-transfer distributions for accelerating flows. (From Blair, M.F., *ASME J. Eng. Power*, 104, 743, 1982.)

2.8.4 Effect of Streamwise Curvature

Mayle et al. (1979) presented heat-transfer measurements on convex and concave surfaces with a turbulent boundary layer. Figure 2.92 presents the heat-transfer distributions on curved surfaces compared to the flat-surface correlation. As curvature increases, the convex-surface, heat-transfer levels drop below the flat-surface correlation, and concave-surface heat transfer is higher. As indicated by Mayle et al. (1979), the deviation for convex surfaces may be caused by the 2-D nature of the flow that reduces the turbulent shear stress, and thus reduces heat transfer. However, for concave surfaces, the deviation is most likely caused by the steady Taylor–Gortler vortices within the boundary layer. The correlation for heat transfer in a turbulent boundary layer by Reynolds et al. (1958) is also plotted for comparison:

$$St\left(\frac{T_w}{T_\infty}\right)^{0.4} = 0.0296 Pr^{-0.4} Re_x^{-0.2} \qquad (2.14)$$

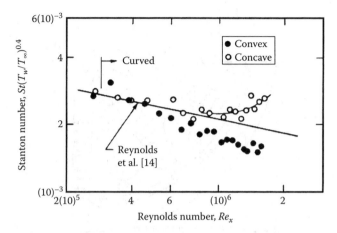

FIGURE 2.92
Stanton number measurements on curved surfaces and comparison with flat-plate correlation. (From Mayle, R.E. et al., *ASME J. Heat Transfer*, 101, 521, 1979.)

Simon and Moffat (1982) reported heat-transfer data for convexly curved surfaces. Other studies that also focused on boundary layers over a convex-curved surface were by Gillis and Johnston (1983) and You et al. (1986). You et al. (1986) documented the effects of free-stream turbulence intensity on heated convexly curved flows. They reported that (1) the curvature effect dominated the turbulence intensity effect, (2) the curvature effect was stronger on the skin friction coefficient than the Stanton number, and (3) similar streamwise turbulence intensity and shear stress profiles were measured at the end of the curve in spite of the different profiles upstream of the curves. Kim and Simon (1987) investigated the effects of streamwise convex curvature, recovery, and free-stream turbulence intensity (up to 2%) on the turbulent transport of heat and momentum in a mature turbulent boundary layer.

Data on the effects of elevated free-stream turbulence on concave-curved turbulent boundary layers are very limited. Kim et al. (1992) studied concave-curved boundary layers for turbulence intensities as high as 8.6%. They reported the cross transport of momentum by boundary work in the core of the flow over a concave wall when turbulence intensity was elevated to 8.6%. Kestoras and Simon (1995, 1997) studied the effects of turbulence intensity on curved wall boundary layer and heat transfer. Figure 2.93 illustrates the changes in turbulence structure effected by introduction of concave curvature with high free-stream turbulence conditions. High free-stream turbulence provides large-scale, high-momentum eddies in the streamwise direction from the core flow to the boundary layer. In the inner region, curvature effects dominate the free-stream turbulence effect. More insights into the concave-curved wall boundary layer can be obtained from Kim et al. (1992) and Kestoras and Simon (1995, 1997).

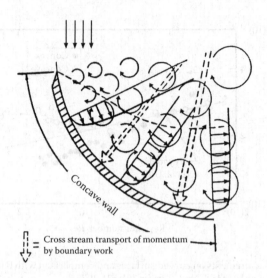

FIGURE 2.93
Changes in turbulence structure effected by the introduction of concave curvature. (From Kestoras, M.D. and Simon, T.W., *ASME J. Turbomach.*, 117, 206, 1995.)

2.8.5 Surface Roughness Effects

Roughness is typically characterized by the roughness Reynolds number, $k^+ = kU_\tau/v$, where U_τ is the friction velocity and k is the roughness element characteristic height. Blair (1994) presented several correlations for heat-transfer predictions on rough walls. Kadar and Yaglom (1972) presented their data based on flat-plate experiments as

$$St = \frac{\sqrt{C_f Pr^{-1}}}{4.3\ln(Re_x C_f) + 3.8} \tag{2.15}$$

Bogard et al. (1996) presented a study on characterization and laboratory simulation of turbine airfoil surface roughness. They simulated the physical characteristics of surface roughness observed on the first-stage, high-pressure vanes on a flat surface. Figure 2.94 presents a comparison of the sample vane and modeled rough-surface profilometer traces. Figure 2.95 presents the heat-transfer results for smooth and rough surfaces. The rough-surface data were compared with correlation from Kays and Crawford (1980) for fully rough flows (fully rough flow for $Re_{k_s} > 70$):

$$St = \frac{C_f}{2}\left(Pr_t + \frac{(C_f/2)^{0.5}}{St_{ks}}\right)^{-1} \tag{2.16}$$

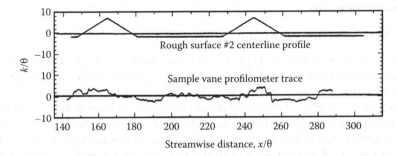

FIGURE 2.94
Comparison of sample vane and modeled rough surfaces. (From Bogard, D.G. et al., Characterization and laboratory simulation of turbine airfoil surface roughness and associated heat transfer, ASME Paper 96-GT-386, 1996.)

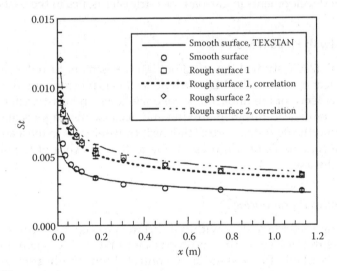

FIGURE 2.95
Stanton number versus distance on heat flux plate for smooth and rough surfaces. (From Bogard, D.G. et al., Characterization and laboratory simulation of turbine airfoil surface roughness and associated heat transfer, ASME Paper 96-GT-386, 1996.)

where roughness Stanton number (St_{ks}) is

$$St_{ks} = CRe_{ks}^{-0.2}Pr^{-0.4} \qquad (2.17)$$

Results show an increase of over 50% for rough surfaces. The correlation showed a clear increase from rough surface 1 to rough surface 2, whereas measurements did not show any further effect of the difference in roughness levels.

The results for surface roughness effects on flat surfaces indicate the importance for designers to consider surface roughness as a heat-transfer enhancement factor that may directly affect the operational life of the engine component.

2.9 New Information from 2000 to 2010

Since 2000, most of the focus of research has been on gas-path heat-transfer issues on extremities like endwalls and tips. Endwall contouring (EWC) to reduce secondary flows and thus reduce heat load to endwalls has been a focus of many studies. Turbine tip desensitization has moved away from simplistic models to more realistic loaded turbine airfoils simulating pressure-driven tip leakage flows. Some of the new areas of research have been in the area of effects of deposition on turbine surfaces and associated heat transfer. There have been more data available on engine-type configurations with results on turbine stages. Challenges still exist in getting detailed data on actual engine-type configurations. We are updating the community on some important developments in gas-path heat-transfer issues in the past decade.

2.9.1 Endwall Heat Transfer

Ames et al. (2003) studied the effect of high free-stream turbulence on vane endwall heat transfer. The high-turbulence heat-transfer contours showed weaker evidence of the impact of secondary flows on heat-transfer patterns due to the enhanced turbulent mixing and because the larger scales tend to push around the flow structures. High heat-transfer rates in the leading edge and wake regions were still present due to the influence of the horseshoe vortex and wake.

2.9.1.1 Endwall Contouring

Controlling secondary flow within a turbine passage has been an area of active research for some time with passage EWC being one out of many methods available for passage flow control. From the heat-transfer perspective, it is considered that reduction in secondary flow generation along the endwall will reduce heat transfer to the endwall, thus requiring less cooling of the endwall surface. In addition, reduction in secondary flow implies reduction in pressure losses and increment in turbine performance. EWC has become more complicated and 3-D with advanced computer-based optimization techniques. NGV and blade endwalls differ in the levels of Mach number resulting in different methodologies for contouring.

2.9.1.1.1 Endwall Contouring for Nozzle Guide Vane Passages

During one of the earliest studies, Morris and Hoare (1975) studied the effect of meridional EWC and demonstrated a reduction in overall secondary loss by 25%. The nonaxisymmetric endwall profile used in the study, however, did not show promising results. The study was conducted at a very low Mach number and Reynolds number as compared to those encountered in

modern HP turbines. Kopper et al. (1980) studied an axisymmetrically contoured vane passage at high exit Mach number of 0.85 and noted about 17%
reduction in mass-averaged total losses. The vane was a low turning angle
(70°) profile with a low aspect ratio of about 0.5. Mass contained in the secondary flow structures is a significant portion of the total mass flow in such
cases. The secondary losses were over half of the total losses.

Many other researchers tried such axisymmetric EWC, especially during
1990s, but none of the studies involving turbine blade row flow control showed
significant improvement. Burd and Simon (2000) and Lin et al. (2000) all presented results for contoured endwalls. Shih et al. simulated contoured endwalls
and showed that the secondary flow reduction and minimization is achieved
by the pressure gradient induced by the area contraction of the contouring.
They also indicated that flow reversals and crossflows are greatly minimized.
Nagel and Baier (2005) used combination of pressure and suction side shape
functions with a circumferentially varying decay function to generate endwall
profile for a turbine vane cascade. These functions were based on the passage
design parameters. The exit Mach number for the flow was reported to be 0.59.

Rose (1994) proposed a method of nonaxisymmetric EWC using a combination of two profiles, where these profiles specified axial and circumferential
shape variations for the endwall of an HP turbine NGV. He suggested a sinu-
soidal circumferential profile variation for subsonic flow fields and a Fourier
series based profile variation for supersonic flow field. However, there are
very little data on heat transfer for endwall contoured vanes and there is little
evidence of heat transfer reduction due to EWC. (Germain et al., 2010).

2.9.1.1.2 Endwall Contouring for Blades

The experiments conducted by Duden et al. (1998) with different types
of meridional endwall profiling for a highly loaded turbine cascade with
about 100° turning, however, did not show any significant overall reduction
in overall losses. Harvey et al. (2000) used a combination of Fourier series
perturbations in pitchwise direction and a b-spline fit in axial direction to
generate nonaxisymmetric endwall profile for a 100° turning turbine blade
cascade. Experimental analysis carried out at a very low Mach number (~0.1)
showed reduction in total loss by about 20%. CFD predictions used for the
design iterations had reported a total loss reduction of only about 0.5%. This
method has been used in many other cascade studies as well as real engine
experiments and has been found to produce improved results at similar low
Mach numbers. Hartland et al. (2000a) and Hartland and Gregory-Smith
(2012) reported 6% reduction in secondary losses for an endwall surface that
was defined using a half cosine wave in pitchwise direction and an axial
profile based on the blade camber line shape. The investigations were done
on the same Durham cascade used by Harvey et al. (2000). Mahmood et al.
(2005) Saha and Acharya (2006) followed the approach of generating a geometry using combination of streamwise and pitchwise height variation curves.
They numerically studied nine geometries with such nonaxisymmetric

contours. Reported reduction in mass-averaged total losses was about 3.2% through numerical computations for the geometry that was finally selected. During the low Mach number experimental investigations for this blade profile, Gustafson et al. (2007) reported 50% reduction in mass-averaged pressure losses. Praisner et al. (2007) used direct surface modification using control point heights to generate profiled endwall. Numerical computations showed 12% reduction in total row-loss for the optimized endwall. However, the experimental results showed 25% reduction for the same geometry. The exit Mach number was about 0.1, a very low value. Snedden et al. (2009) applied Durham cascade hub profile to the annular endwall of a 1½ stage rotating rig and observed about 0.4% improvement in rotor efficiency. However, the rotor exit relative velocities were very low in the range of about 50 m/s. Again, there are no published studies of measured heat transfer on highly contoured endwall surfaces for turbine blades. Lynch et al. (2009, 2011) presented experimental and computational measurements of flow and heat transfer with nonaxisymmetric EWC.

2.9.1.2 Leading-Edge Modifications to Reduce Secondary Flows

Using leading-edge fillets to eliminate horseshoe vortices was recently proposed by Bancalari and Nordlund (1999), and then further studied by Shih and Lin (2002). Zess and Thole (2001) provided a combined numerical and experimental study on leading-edge airfoil fillet design and found that the pointed/sharp type of fillet shown in Figure 2.96 can indeed diminish leading-edge-induced horseshoe vortices. Sauer et al. (2001) studied a different type of fillet that enhances instead of diminishes horseshoe vortices.

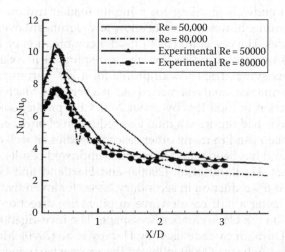

FIGURE 2.96
Nusselt number augmentation for Re = 50,000 and Re = 80,000 along the liner wall. (From Patil, S. et al., *ASME J. Turbomach.*, January, 011028, 1, 7, 2011.)

Their fillet, a bulb-type Figure 2.96, is intended to intensify horseshoe vortices on the suction side in order to reduce the pressure-side to suction-side secondary flow (passage vortex).

Lethander et al. (2003) presented optimization results that showed that a large fillet was needed for the maximum thermal benefit and that this benefit was most influenced by the height of the fillet. Analysis of the filleted vane indicated that three benefits were realized with a fillet. First, there is a reduction in the overall surface area that needs to be cooled when considering the combined fillet, endwall, and vane surfaces. Second, there is a reduction of the secondary flows resulting from the acceleration of the flow in the endwall region, which prevents the flow from separating and forming a leading-edge horseshoe vortex. Third, a reduction of the secondary flows for the filleted vane allows cooler near-wall fluid to remain along the endwall and fillet surfaces rather than being lifted into the hot mainstream, which occurs when a strong passage vortex is present.

2.9.1.3 Endwall Heat-Transfer Measurements

Radomsky and Thole (2000) studied the effect of high free-stream turbulence on the naturally occurring secondary flow vortices to determine the effects on the flow field and the endwall convective heat transfer. The endwall surface heat-transfer measurements showed an increase in the endwall heat transfer due to high free-stream turbulence which was similar to that seen on the suction and pressure sides of the airfoil. Lethander and Thole (2003) looked at secondary flow mitigation generated by leading-edge horseshoe vortices achieved through optimizing a fillet in the vane endwall juncture to minimize adiabatic wall temperatures. The optimization results from their study indicated that a large fillet was needed for the maximum thermal benefit and that this benefit was most influenced by the height of the fillet. While a thermal benefit was predicted for the endwall, the maximum predicted benefit occurred on the vane itself at the 20% span location. Goldstein and Han (2005) compared Sherwood number distributions on an endwall with a simple blade and a similar blade having modified leading-edge by adding a fillet. With low free-stream turbulence intensity, the fillet appeared to significantly reduce the horseshoe vortex and delayed formation of the passage vortex. With high-turbulence intensity, the passage vortex is not clearly observed with a simple blade nor with a similar blade with fillets. The fillet significantly reduces the horseshoe vortex with high-turbulence intensity. Lynch et al. (2009) presented heat-transfer measurements for a baseline flat endwall, as well as for a contoured endwall geometry, for three cascade exit Reynolds numbers. The contoured endwall heat-transfer results were compared with the flat endwall, indicating high augmentation levels near the saddle point, but low near the blade pressure side–endwall corner. Low augmentation levels were also seen near the suction side of the blade, around the region where the passage vortex impinges on the blade.

2.9.2 Turbine Tip and Casing Heat Transfer

Developments in computational methods and advanced experimental techniques resulted in a number of high-resolution results from the tip gap and tip clearance vortex region. There had been few studies on actual turbine blade tips under realistic flow conditions prior to 2000. However, several research groups have worked on blade tip heat transfer and cooling in the past 10 years. Bunker et al. (2000) obtained tip heat transfer and pressure measurements in a three-bladed linear cascade arrangement simulating the first-stage blade geometry on a large power-generating turbine. They used flat and smooth tip surfaces with sharp and rounded edges. They noticed a central sweet spot of low heat transfer extending into the midchord region and toward the suction side. They observed that the measured surface heat-transfer coefficients increase 10%–20% when the free-stream turbulence intensity level was increased from 5% to 9%. When the sharp tip edge was rounded (small edge radius), the tip heat transfer increased by about 10%, presumably due to higher allowed tip leakage flow. This was the main motivation to redesign blade tips to have a recessed center to allow for increased heat transfer in the so-called sweet spot in that area and reduce heat transfer toward the trailing edge. Ameri and Bunker (2000) also published the results of a study dealing with the numerical prediction of the tip gap heat transfer based on the measurements of Bunker et al. (2000), using the k-ω low Reynolds number turbulence model. Dunn and Haldeman (2000, 2003) presented Nusselt number values in the recessed tip region of a single-stage turbine. The Nusselt numbers measured at the floor of the recess near the leading edge of the blade and on the suction side lip were found to be in excess of the blade stagnation value. Azad et al. (2000a) produced heat-transfer results on the flat tip section of a turbine blade in a five-bladed linear cascade arrangement. Azad et al. (2000b) also investigated heat transfer and flow characteristics on the squealer tip of a gas turbine blade representative of E^3 design. They found a higher heat-transfer coefficient on the rim surface because of the entrance and exit effect. The measured heat-transfer coefficient was much lower in the midchord toward the pressure side and downstream end of the cavity when compared to the flat tip case. The heat-transfer coefficient in a squealer tip was higher near the central upstream end of the cavity and the trailing edge region. As a conclusion, they found that the squealer tip provided an overall lower heat-transfer coefficient when compared to the flat tip case. Urban and Vortmeyer (2000) tested different tip arrangements such as plain and squealer with and without cooling in a steam turbine rig and made measurements of leakage flow and thermal load. Bunker and Bailey (2001) studied the effect of squealer cavity depth and oxidation on turbine blade tip heat transfer for a similar blade studied in previous studies by same group. Blade tip heat transfer with chordwise sealing strips was studied by Bunker and Bailey (2000). A significant aerodynamic modification on the leakage flow behavior was observed in a linear cascade arrangement with a stationary outer casing. Bunker (2001) presented a review of all the work published prior to 2000 in his 2001 paper on the state of blade tip heat-transfer studies.

Chana and Jones (2002) performed detailed experimental investigations to measure the heat transfer and static pressure distributions on the rotor tip and rotor casing of a gas turbine stage with a shroudless rotor blade. The unsteady heat-transfer results showed little difference in form between the uniform and nonuniform cases. They saw that highest heat-transfer rate on the shroud was above the pressure side of the blade. A new tip desensitization method based on a pressure side tip extension was discussed in Dey and Camci (2001). This investigation presented time-accurate and phase-averaged total pressure measurements performed in a single-stage turbine. The goal was to quantify the aerodynamic structure of a tip leakage vortex so that the suggested passive tip treatments could be evaluated in real time. The study indicated the "momentum" defect in the tip vortex of an untreated turbine blade tip could be effectively eliminated by a suggested "pressure side tip platform extension." A thermodynamic explanation of the desensitization process was attempted. Pressure side tip extensions were found to be highly effective passive desensitization schemes. However, they caused increased heat transfer on the extensions resulting in durability issues. Acharya et al. (2003) presented computational predictions of heat transfer for a similar pressure side winglet as that by Dey and Camci (2001). They saw increased heat load on the extension but reduced heat load on the actual surface.

Saxena et al. (2003) investigated the effect of crosswise trip strips to reduce leakage flow and associated heat transfer with strips placed along the leakage flow direction, against the leakage flow and along the chord. Cylindrical pin fins and pitch variation of strips over the tip surface are also investigated. Results show that the trip strips placed in a direction orthogonal to the leakage flow direction produce the lowest leakage flow and heat-transfer coefficient over the tip. The other configurations produce similar levels to the plain tip or are higher than the plain tip. Azad et al. (2000) and Kwak and Han (2002) investigated the heat transfer on several different squealer geometries. They found that a suction side squealer tip gave the lowest heat transfer among all cases studied. Heat-transfer coefficient distributions for plane tip, squealer tip, and near-tip regions were presented by Kwak and Han (2002) in two papers. Results showed that, in general, the heat-transfer coefficient on the tip was higher than that on the shroud and on the near-tip region of the pressure and suction side. Results also showed that the heat-transfer coefficients on the tip and the shroud increased as the tip gap clearance increased. But the heat-transfer coefficients on the near-tip region of the pressure and suction sides were not sensitive to the tip gap clearance.

Squealer rims were located by Kwak et al. (2003) along (a) the camber line, (b) the pressure side, (c) the suction side, (d) the pressure and suction sides, (e) the camber line and the pressure side, and (f) the camber line and the suction

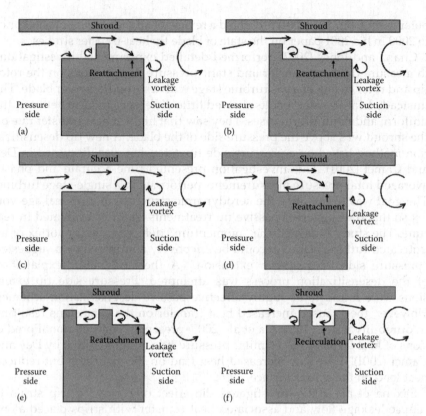

FIGURE 2.97
The conceptual view of the flow leakage near the tip region as shown by Kwak et al. (a) Squealer along CL, (b) Squealer along PS, (c) Squealer along SS, (d) Squealer along PS + SS, (e) Squealer along CL + PS, and (f) Squealer along CL + SS. (From Kwak, J.S. et al., Heat transfer coefficients on the squealer tip and near tip regions of a gas turbine blade with single or double squealer, ASME Paper GT2003-38907, 2003.)

side, as shown in Figure 2.97, respectively. Results showed that a different squealer geometry arrangement changed the leakage flow path and resulted in different heat-transfer coefficient distributions. As shown in Figure 2.98, the suction side squealer tip provided the lowest heat-transfer coefficient on the blade tip and near-tip regions compared to the other squealer geometry arrangements. By using a squealer tip, heat transfer was generally found to decrease on the tip and near-tip regions.

Newton et al. (2006), Whitney et al. (2005), Kanjirakkad et al. (2006), Krishnababu et al. (2008), and Lehmann et al. (2009) presented local measurements of the heat-transfer coefficient and pressure coefficient on the tip and near-tip region of a generic turbine blade in a five-blade linear cascade with two different tip clearance gaps. Three different tip geometries were investigated: A flat (plain) tip, a suction-side squealer, and a cavity squealer.

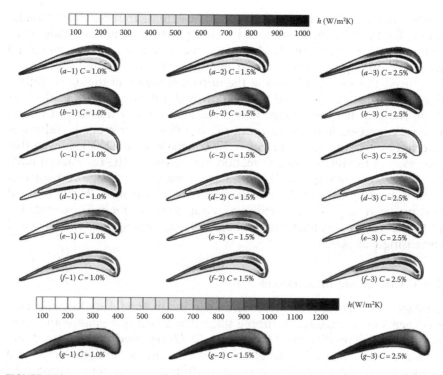

FIGURE 2.98

Heat-transfer coefficient on the tip as shown by Kwak et al. (a) squealer along CL; (b) squealer along PS; (c) squealer along SS; (d) squealer along PS + SS; (e) squealer along CL + PS; (f) squealer along CL + SS; (g) plane tip. (From Kwak, J.S. et al., Heat transfer coefficients on the squealer tip and near tip regions of a gas turbine blade with single or double squealer, ASME Paper GT2003-38908, 2003.)

Both squealer geometries eliminated the peak in heat transfer associated with flow reattachment from the pressure-side rim which dominated the case of the plain tip. However, high local heat transfer was observed at positions on the squealers which corresponded to areas of impingement. The cavity squealer also featured local spots of high heat transfer at the base of the squealer near the leading edge, possibly associated with vertical flow drawing fluid into the cavity. This was seen in all previous studies also. Molter et al. (2006) presented measurements of heat flux and pressure in the blade tip region of a modern one-and-one-half-stage high-pressure turbine operating at design-corrected conditions in a rotating rig. Both flat tip and recessed, or squealer, tip blades were used in the experiments. For the uncooled airfoils used in these experiments, the measurements showed that the tip has about 25% higher heat load than the blade at 90% span. The squealer tip increased the heat load even further showing a 30% increase over the flat tip blade in two localized regions just inside the leading edge and near the pressure side at midchord.

Rhee and Cho (2005a,b) presented local heat/mass transfer characteristics on the flat tip by using a low-speed rotating turbine annular cascade. They pointed out that when the blade rotates, the heat/mass transfer enhanced region on the tip is shifted toward the downstream side, and the level of heat/mass transfer coefficients at the upstream region of the tip is slightly decreased due to the reduced tip gap flow with rotation. Palafox et al. (2005, 2006) showed that the leakage flow structure and heat transfer on the tip in a very large-scale linear cascade with a stationary or moving endwall is affected by relative endwall motion. Krishnababu et al. (2007) showed that the effect of the relative casing motion was to reduce the tip leakage mass flow rate. Some changes in the tip leakage vortex and passage vortex structures were observed with the introduction of relative motion. It was observed that the strength of the tip vortex was reduced considerably. The circulation of the tip vortex was found to decrease with decreasing clearance size and increasing wall speed.

2.9.3 Vane-Blade Interactions

Bergholz et al. (2000) presented techniques and methodology to make measurements on a large-scale turbine stage using heat flux gauges and comparing to CFD predictions. Didier et al. (2002) reported time-resolved and time-averaged convective heat-transfer coefficients around the rotor of a transonic turbine stage. The tests are performed in the compression tube turbine test rig CT3 of the von Karman Institute, allowing a correct simulation of the operating conditions encountered in modern aero-engines. Their results showed that the time-averaged Nusselt number distribution had a strong dependence on both blade Mach number distribution and Reynolds number. However, the time-resolved heat-transfer rate was mostly dictated by the vane trailing edge shock impingement on the rotor boundary layer. Dunn and Haldemann (2004) presented heat-transfer measurements and predictions for the 20%, 50%, and 80% span locations on the vane, the vane inner and outer endwall, the 20% and 96% span location on the blade, the blade tip (flat tip), and the stationary blade shroud for the same geometry as used in Bergholz et al. (2000). Their results showed that the variation on the blade and vanes as a function of span indicated the 3-D characteristics of the flow at even modest span differences of 20%, 50%, and 80% on the vane. Haldemann et al. (2008b) presented measurements and computational fluid dynamics (CFD) predictions for a fully cooled, high-work single-stage HP turbine operating in a short-duration blowdown rig. Part I of this paper (Haldeman et al., 2008a, 130(2), p. 021015) presented the experimental approach. The measurements presented are both time-averaged and time-accurate results that include heat transfer at multiple spans on the vane, blade, and rotor shroud as well as flow path measurements of total temperature and total pressure. Tallman et al. (2006) presented heat-transfer measurements for an uncooled turbine stage at a range of operating conditions representative of the engine: in terms of corrected speed, flow function, stage pressure ratio, and gas-to-metal temperature ratio.

As controlled laboratory experiments using full-stage turbines are expanded to replicate more of the complicated flow features associated with real engines, it is imperative that experiments to understand the influence of the vane inlet temperature profile on the high-pressure vane and blade heat transfer as well as its interaction with film cooling need to be performed (Ainsworth et al., 2004; Thorpe and Ainsworth, 2006; Atkins et al., 2008). Mathison et al. (2010a,b) and Dunn et al. (2010) studied various effects of vane inlet temperature profiles and associated flow migration due to the temperature distribution. The migration of these temperature profiles through the turbine stage is examined with and without cooling flows. They indicate that the uniform and radial profiles both exhibit some smoothing in transmission from the vane inlet to the rotor inlet. The uniform flow temperature at inlet showed a nearly flat exit profile, and the radial case only had a small drop in temperature at the inner spans of the passage.

2.9.3.1 Cascade Studies

Giel et al. (2000) made measurements in a linear cascade with high-turning power generation turbine blades. They studied the effect of Reynolds number and angle of incidence of the flow on turbine blade surface heat transfer. The data quantified the effect of laminar–turbulent transition on the blade surface heat transfer. They compared their experimental results to CFD predictions. There were regions of good agreement and some regions of needed improvement. Giel et al. (2003) made similar measurements on a transonic turbine blade. The data from their experiments appeared to agree well with appropriate scaling laws and showed good agreement with a stagnation point heat-transfer correlation. They also made comparisons with CFD predictions. Carullo et al. (2007) experimentally investigated the effect of high free-stream turbulence intensity, turbulence length scale, and exit Reynolds number on the surface heat-transfer distribution of a turbine blade at realistic engine Mach numbers. From their results, increasing the exit Reynolds number was shown to increase the heat-transfer levels and cause earlier boundary layer transition. The boundary layer transition was observed to be influenced primarily by the Reynolds number of the flow. An increase in the exit Reynolds (Mach) number caused the boundary layer to transition closer to the leading edge. The local peak on the pressure side increased with increasing Reynolds number for the baseline case.

Olson et al. (2009) investigated the effect of wakes in the presence of varying levels of background free-stream turbulence on the heat (mass) transfer from gas turbine blades. The results from their study showed that increasing the wake-blade pitch relative to the turbine-blade pitch has some effect when the vane is directly ahead of the turbine blade. On the suction surface, the start of transition to a turbulent boundary layer was delayed, while on the pressure surface, the minimum mass transfer coefficient where separation sometimes occurs was increased.

2.9.4 Deposition and Roughness Effects

Bons (2002) presented experimental measurements of skin friction (c_f) and heat-transfer (St) augmentation for low-speed flow over turbine roughness models. The models were scaled from surface measurements taken on actual, in-service land-based turbine hardware. Both existing c_f and St correlations severely underpredicted the effect of roughness for $k^+ < 70$ (when k_s, as determined by the roughness shape/density parameter, is small). A new k_s correlation based on the rms surface slope angle was generated to overcome this limitation. From their study, comparison of data from real roughness and simulated (ordered cones or hemispheres) roughness suggested that simulated roughness was fundamentally different from real roughness. Boyle and Senyitko (2005) measured turbine vane surface temperatures coated with thermal barrier coating (TBC) in a linear cascade using an infrared non-contact thermal detector. The temperature drop across the relatively thick TBC was used to determine heat-transfer coefficients.

Recent experimental work in measuring the formation of deposits has been done under the UTSR program by Bons-Fletcher and coworkers (Bons and McClain, 2003; Fletcher et al., 2008; Smith et al., 2010). In a series of experiments in an accelerated test facility, Bons et al. have presented a comparative analysis of various fuel alternatives at actual engine condition. They found that gas temperature, gas velocity, and particle loading govern the deposition on the turbine blade. Wammack et al. (2006) investigated the physical characteristics and evolution of surface deposition on bare polished metal, polished TBC with bond coat (initial average roughness was less than 0.6 μm), and unpolished oxidation-resistant bond coat (initial average roughness ~16 μm). Based on these results, they inferred that the initial surface preparation has a significant effect on deposit growth, that thermal cycling combined with particle deposition caused extensive TBC spallation while thermal cycling alone caused none, and finally that the deposit penetration into the TBC is a significant contributor to spallation. Subsequently, Bons et al. (2006) made convective heat-transfer measurements using scaled models of the deposited roughness and found that the Stanton number was augmented by between 15% and 30% over a smooth surface. Crosby et al. characterized the effect of particle size, gas temperature, and backside cooling of the blade sample on deposition from coal-derived fuels. They concluded that deposition increased by a factor of two as the mass mean diameter of the particle was increased from 3 to 16 μm. Second, particle deposition decreased with decreasing gas temperature and with increased backside cooling. They found a threshold gas temperature for deposition to occur at 960°C.

Erickson et al. (2010) used a realistically rough-surface distribution generated by Brigham Young University's accelerated deposition facility for their vane heat-transfer study. The surface represented a TBC surface which has accumulated 7500 h of operation with particulate deposition due to a mainstream concentration of 0.02 ppmw. Their results indicated that realistic roughness appeared to have no influence on heat-transfer rates as long as boundary layers remained laminar. However, realistic roughness led to early transition in all cases and transition onset appeared to be independent of the inlet turbulence condition. This observation that roughness dominates transition was consistent with the findings of Boyle and Senyitko (2005).

Lorenz et al. (2010) presented a comprehensive heat-transfer analysis on a highly loaded low-pressure turbine blade and endwall with varying surface roughness. A set of different arrays of deterministic roughness was investigated, varying the height and eccentricity of the roughness elements, showing the combined influence of roughness height and anisotropy of the rough surfaces on laminar–turbulent transition and the turbulent boundary layer as well as boundary layer separation on the pressure and suction sides. In the case of streamwise stretched roughness (as in particle deposits), laminar–turbulent transition occurred considerably downstream compared to isotropic roughness. In contrast, spanwise stretched roughness elements caused the onset of transition to move considerably upstream. Higher turbulent heat transfer was observed for spanwise stretched and lower heat transfer for streamwise stretched roughness when compared to isotropic roughness elements.

2.9.5 Combustor–Turbine Effects

Van Fossen and Bunker (2000) made heat-transfer measurements in the stagnation region of a flat plate with a circular leading edge. The model was immersed in the flow field downstream of an approximately half-scale model of a can-type combustor from a low NOx, ground-based power-generating turbine. The tests were conducted with room temperature air with no reacting flow. Room air flowed into the combustor through six vane-type fuel/air swirlers. The combustor can contained no dilution holes. The fuel/air swirlers all swirled the incoming airflow in a counterclockwise direction (facing downstream). Although the measured turbulence generated appeared to be isotropic, Stagnation heat transfer in the turbulent exhaust of the DLN combustor liner was found to be higher than that for a laminar free-stream by a factor of 1.77 and stagnation heat-transfer augmentation was found to be 14% higher than predicted by a correlation developed from isotropic grid-generated turbulence. Van Fossen and Bunker (2002) did another similar study on an elliptical leading-edge model to simulate more engine-type vane geometry downstream of combustors. Ames et al. (2003b) documented heat-transfer rates on an aft-loaded vane subject to turbulence generated by mock combustion configurations representative of recently developed catalytic and dry low NOx (DLN) combustors. Ames et al. (2002a,b) also studied the effect of combustor-generated turbulence on endwall heat-transfer distributions of a linear vane cascade. Their results showed weaker evidence of the impact of secondary flows on heat-transfer patterns due to the enhanced turbulent mixing and because the larger scales tend to push around the flow structures. Barringer et al. (2006) investigated the effects of different profiles representative of those exiting aero-engine combustors on high-pressure turbine vane aerodynamics and heat transfer. The various profiles were produced using the non-reacting, inlet profile generator in the Turbine Research Facility (TRF) located at the Air Force Research Laboratory (AFRL). Their results showed that the baseline heat transfer was reduced 30%–40% over the majority of the vane surface near the ID endwall. They saw that certain inlet profiles could increase the baseline heat transfer by 20%–30% while other profiles resulted in a decrease of the baseline heat transfer by 30%–40% near the OD endwall. Barringer et al. (2007) and Barringer et al. (2007) also studied the effect of combustor exit profiles on vane endwall heat transfer.

Jenkins and Bogard (2003, 2004) discussed the effects of varying pitch position on the attenuation of a simulated hot streak in a vane cascade under conditions of low and high turbulence. Measurements describe the effect of both small and large changes in pitch away from the stagnation line as evidenced by hot streak variations in the wake and at the trailing edge. They indicated that a major factor in the vane influence on the hot streak attenuation was the interruption of the hot streak dispersion caused when a hot streak impacts the vane and is split into two parts. This effect was more pronounced when the peak of the hot streak passed to either side of the vane so that a lower temperature outer edge, or "tail," of the hot streak was split from the core. Basol et al. (2010) explored numerically the effect of the hot streak's orientation at the stator inlet on the rotor blade heat load and on the tip in particular. The inlet boundary conditions are taken from the hot streak experiment conducted in the axial turbine facility "LISA" at ETH Zurich. Their results indicated that hot streak clocking led to a heat load redistribution in the rotor row in the spanwise direction. They concluded that shifting the hot streak toward the stator pressure side reduced the rotor blade tip temperatures and increased the heat load around midspan. The hot streak migration pattern can help facilitate the design of the combustor temperature profiles for optimum heat load distribution on the rotor blade leading to an overall increased blade life.

Patil et al. (2010) presented a study measuring heat-transfer coefficients on the gas side wall of a can combustor shell with cold flow through a swirler nozzle. Previous studies had focused on the downstream effect of combustor swirl but this study measured heat transfer on the combustor liner wall itself. Predictions using CFD were also used for comparison and to help understand the swirling flow behavior inside the can combustor. Figure 2.96 shows the measured and predicted heat-transfer coefficients on the combustor liner wall indicating local hot spots where the swirling flow impinges on the liner. This information of heat transfer on the liner wall could be useful in designing targeted cooling for combustor liners. Sedalor et al. (2010) studied the effects on annular combustor models indicating variability between convex and concave wall behavior.

2.9.6 Transition-Induced Effects and Modeling

Roach and Brierley (2000) described a new method of determining boundary layer transition with zero mean pressure gradient. The approach examines the development of a laminar boundary layer to the start of transition, accounting for the influences of free-stream turbulence and test surface geometry. Their method was developed from results obtained in laboratory experiments where the free-stream turbulence intensity and length scale were well documented. Schultz and Volino (2001) carried out an experimental investigation on a transitional boundary layer subject to high (initially 9%) free-stream turbulence, strong acceleration as high as 9×10^6, and strong concave curvature (boundary layer thickness between 2% and 5% of the wall radius of curvature). They indicated that curvature caused higher momentum transport in the outer part of the boundary layer, resulting in a more rapid transition to turbulence and higher skin friction. Conditional sampling showed that the curvature effect is present in both the turbulent and nonturbulent zones of the transitional flow. Johnson (2002, 2003) presented a new procedure capable of predicting the development of the fluctuations in the laminar boundary layer from values of the free-stream turbulence intensity and length scale, and hence determining the start of transition without resorting to any empirical correlation. Roberts and Yaras (2003a) presented measurements of free-stream turbulence, streamwise pressure gradients, and flow Reynolds number effects on attached-flow transition. An improved spot production rate correlation is proposed that when used in conjunction with the intermittency model, yielded satisfactory prediction of the measured intermittency distributions. Roberts and Yaras (2003b) also presented measurements of the combined effects of free-stream turbulence and periodic streamwise velocity variations on separation-bubble transition. They presented results that demonstrated that the time-averaged locations of flow separation, transition inception, and reattachment are essentially insensitive to free-stream streamwise velocity oscillations over the ranges of frequency and amplitude considered in the study. Chong and Zhong (2003) presented a detailed measurement of turbulent spots propagating in a laminar boundary layer over a flat plate at a zero-pressure gradient and three favorable pressure gradients. Their experiment revealed some interesting aspects of the overall structure of the spots and the role that regions with negative velocity perturbations play in spanwise growth of turbulent spots. Lodefier et al. (2003) and Dick and Lodefier (2003) presented a transition model for describing bypass transition based on a two-equation k-ω model and a dynamic equation for intermittency factor. The intermittency factor was a multiplier of the turbulent viscosity computed by the turbulence model. The results for the suction side comparing with experiments were relatively good. The transition was detected at the correct position and also had the correct length. However, the heat-transfer levels after transition were too low. This was indicated as a deficiency of the turbulence model. Dris and Johnson (2004) made boundary layer measurements on the concave surfaces of two constant curvature blades using hot-wire anemometry. Their measurements showed that the Taylor–Goertler (T-G) vortices resulting from the concave curvature made the laminar and turbulent boundary layer profiles fuller and increased the skin friction coefficient by up to 40% compared to flat-plate values. Praisner et al. (2004a,b), Clark and Praisner (2004) and Grover et al. (2004) developed two correlations, one for the onset of attached-flow transition and the other for the length of a separation bubble prior to turbulent reattachment. Their models were based on local flow field parameters and appeared to have greater efficacy than a number of extant correlations. In particular, the model for attached-flow transition appeared to have a physical basis with respect to the fundamental mechanism of bypass transition in compressible flow. Their models provided a significant improvement in the accuracy of predicted profile losses for both cascades and multistage LPT rigs over the fully turbulent assumption. Stripf et al. (2008a,b) presented two extended models for the calculation of rough wall transitional boundary layers with heat transfer. Both their models comprised of a new transition onset correlation, which accounted for the effects of roughness height and density, turbulence intensity, and wall curvature. The computationally more expensive DEM-TLV-T model gave slightly better predictions than the KS-TLK-T model in the majority of test cases. The KS-TLK-T model, however, allowed reasonable predictions as well and is less complicated to implement in existing CFD codes. Test cases included flat-plate turbulent boundary layers on rough walls with and without a pressure gradient as well as heat-transfer measurements on two different turbine cascades.

There has been lot of focus on the separated-flow transition in LP turbines in the recent past. Volino (2002) studied the scales in a transitional boundary layer subject to high (initially 8%) free-stream turbulence and strong acceleration (K as high as 9×10^6) using wavelet spectral analysis and conditional sampling of experimental data. The results from their study suggested some steps for improving transition models, demonstrating the clear differences between the two zones of the intermittent flow and suggesting the benefit of modeling these zones separately. Separated-flow transition for an LP turbine blade has been documented by Volino (2002a,b) for cases with Reynolds numbers ranging from 25,000 to 300,000 at both high and low free-stream turbulence intensity. Stieger and Hodson (2003) presented a detailed experimental investigation into the interaction of a convected wake and a separation bubble on the rear suction surface of a highly loaded low-pressure (LP) turbine blade. Their measurements showed the separated shear layer associated with the inflexional profiles of the reestablishing separation bubble formed roll-up vortices formed by an inviscid Kelvin–Helmholtz mechanism, beneath the passing wake. The roll-up vortices were observed to break down into highly turbulent flow that convected along the blade surface. Simon and Jiang (2003a,b) presented a model for laminar–turbulent transition for an LP turbine due to passing wakes. They indicated that when the wakes are directly over the separation-to-transition region, the actual transition length is much shorter than (about 1/20th) the length predicted by the Mayle short bubble model if transition onset is taken to be the point of first perceivable rise in intermittency. Coton and Arts (2004a,b) provided a new test case for high lift LP turbine profiles. The high front loading led to the presence of a separation bubble downstream. Experiments were performed in close engine conditions of Reynolds number, Mach number, and turbulence level. The effect of periodically passing wakes was simulated and the correct wake flow coefficient reproduced. In steady conditions, the heat-transfer coefficient distributions confirmed that the profile loss generally decreased with the Reynolds number and the inlet turbulence level as they promoted the transition and the bubble reattachment. The results also showed the change of the dominant mode of the transition, from separation to bypass, with the upstream shift of the transition onset when Re and Tu_∞ are increased. Ozturk et al. (2005, 2007) experimentally studied the effects of periodic unsteady wake flow and Reynolds number on boundary layer development, separation, reattachment, and the intermittency behavior along the suction surface of a low-pressure turbine blade. Their intermittency analysis of the current boundary layer experimental data with the flow separation determined the minimum, maximum, and the relative intermittency functions. The calculated relative intermittency factor followed a Gaussian distribution confirming the universal character of the relative intermittency function.

2.10 Closure

Turbine airfoil designers need to account for each of the aforementioned effects in their design. Depending on the operating needs and requirements, airfoils operate at varied conditions that may cause a mixed combination of each of the effects. It is imperative that correlations consider all the effects for final design. It is difficult to simulate a combination of all the parameters and tests' under real conditions. There is a need to consider matching CFD correlations to reduce the burden on the experimentally derived correlations. However, CFD calculations need to match some existing experimental results. Accurate predictions by CFD codes are still far from matching the requirements of airfoil designers.

References

Abuaf, N., Bunker, R.S., and Lee, C.P., 1998. Effects of surface roughness on heat transfer and aerodynamic performance of turbine airfoils. *ASME Journal of Turbomachinery*, 120, 522–529.

Acharya, S., Saba, A., Prakash, C., and Bunker, R.S., 2003. Blade tip leakage flow and heat transfer with pressure-side winglet. ASME Paper GT2003-38620.

Addison, J.S. and Hodson, H.P., 1990a. Unsteady transition in an axial flow turbine. Part I: Measurements on the turbine rotor. *ASME Journal of Turbomachinery*, 112, 206–214.

Addison, J.S. and Hodson, H.P., 1990b. Unsteady transition in an axial flow turbine. Part II: Cascade measurements and modeling. *ASME Journal of Turbomachinery*, 112, 215–221.

Ainsworth, R., Thorpe, S., and Allan, W., 2004. Unsteady heat transfer measurements from transonic turbine blades at engine representative conditions in a transient facility. ASME Paper, GT2004-53835.

Ameri, A.A., Steinthorsson, E., and Rigby, D.L., 1997. Effect of squealer tip on rotor heat transfer and efficiency. ASME Paper 97-GT-128.

Ameri, A.A. and Bunker, R., 2000. Heat transfer and flow on the first-stage blade tip of a power generation gas turbine: Part 2—Simulation results. *ASME Journal of Turbomachinery*, 122(2), 272–277.

Ames, F.E., 1997a. Film cooling: A short course. Conducted by DOE-AGTSR, Clemson, SC.

Ames, F.E., 1997b. The influence of large-scale high-intensity turbulence on vane heat transfer. *ASME Journal of Turbomachinery*, 119, 23–30.

Ames, F.E., Barbot, P.A., and Wang, C., 2002a. Effects of aeroderivative combustor turbulence on endwall heat transfer distributions acquired in a linear vane cascade. ASME Paper, GT2002-30525.

Ames, F.E., Barbot, P., and Wang, C., 2003a. Effects of catalytic and dry low NOx combustor turbulence on endwall heat transfer distributions. ASME Paper, GT2003-38507.

Ames, F.E. and Moffat, R.J., 1992. Heat transfer with high intensity, large scale turbulence: The flat plate turbulent boundary layer and the cylindrical stagnation point. Stanford University Report No. HMT-44, Stanford, CA.

Ames, F.E. and Plesniak, M.W., 1997. The influence of large-scale high-intensity turbulence on vane aerodynamic losses, wake growth, and the exit turbulence parameters. *ASME Journal of Turbomachinery*, 119, 182–192.

Ames, F.E., Wang, C., and Argenziano, M., 2003b. Measurement and prediction of heat transfer distributions on an aft loaded vane subjected to the influence of catalytic and dry low NOx combustor turbulence. ASME Paper, GT2003-38509.

Ames, F.E., Wang, C., and Barbot, P.A., 2002b. Measurement and prediction of the influence of catalytic and dry low NOx combustor turbulence on vane surface heat transfer. ASME Paper, GT2002-30524.

Ashworth, D.A., LaGraff, J.E., Schultz, D.L., and Grindrod, K.J., 1985. Unsteady aerodynamic and heat transfer processes in a transonic turbine stage. *ASME Journal of Engineering for Gas Turbines and Power*, 107, 1022–1030.

Atkins, N., Thorpe, S., and Ainsworth, R., 2008. Unsteady effects on transonic turbine blade-tip heat transfer. ASME Paper, GT2008-51177.

Azad, G.S., Han, J.C., and Boyle, R.J., 2000a. Heat transfer and pressure distributions on a gas turbine blade tip. ASME Paper, 2000-GT-0194.

Azad, G.S., Han, J.C., and Boyle, R.J., 2000b. Heat transfer and flow on squealer tip of a gas turbine blade. ASME Paper, 2000-GT-0195.

Azad, G.M.S., Han, J.C., Teng, S., and Boyle, R.J., 2000a. Heat transfer and pressure distributions on a gas turbine blade tip. *ASME International Gas Turbine and Aero-Engine Congress*, May 8–11, 2000, Munich, Germany, ASME Paper 2000-GT-194.

Azad, G.M.S., Han, J.C., Teng, S., and Boyle, R.J., 2000b. Heat transfer and flow on the squealer tip of a gas turbine blade. *ASME International Gas Turbine and Aero-Engine Congress*, May 8–11, 2000, Munich, Germany, ASME Paper No. 2000-GT-195.

Baines, N.C., Oldfield, M.L.G., Jones, T.V., Shultz, D.L., King, P.I., and Daniels, L.C., 1982. A short-duration blowdown tunnel for aerodynamic studies on gas turbine blading. ASME Paper 82-GT-312.

Bancalari, E. and Nordlund, S., 1999. *DOE AGTSR Aero-Heat Transfer Workshop III*, February 10–12, 1999, Austin, TX.

Barringer, M., Thole, K., and Polanka, M., 2006. Effects of combustor exit profiles on high pressure turbine vane aerodynamics and heat transfer. ASME Paper, GT2006-90277.

Barringer, M., Thole, K., and Polanka, M., 2007. An experimental study of combustor exit profile shapes on endwall heat transfer in high pressure turbine vanes. ASME Paper, GT2007-27156.

Basol, A.M., Ibrahim, M., Kalfas, A.I., Jenny, P., and Abhari, R.S., 2010. Hot streak migration in a turbine stage: Integrated design to improve aero-thermal performance. ASME Paper, GT2010-23556.

Bergholz, R.F., Steuber, G.D., and Dunn, M.G., 2000. Rotor/stator heat transfer measurements and CFD predictions for short-duration turbine rig tests. ASME Paper, 2000-GT-0208.

Bindon, J.P., 1989. The measurement and formation of tip clearance loss. *ASME Journal of Turbomachinery*, 111, 258–263.

Bindon, J.P. and Morphus, G., 1988. The effect of relative motion, blade edge radius and gap size on the blade tip pressure distribution in an annular turbine cascade with clearance. ASME Paper 88-GT-256.

Blair, M.F., 1982. Influence of free-stream turbulence on boundary layer transition in favorable pressure gradients. *ASME Journal of Engineering for Power*, 104, 743–750.

Blair, M.F., 1983. Influence of free-stream turbulence on turbulent boundary layer heat transfer and mean profile development. Part I: Experimental data. Part II: Analysis of results. *ASME Journal of Heat Transfer*, 105, 33–47.

Blair, M.F., 1994. An experimental study of heat transfer in a large-scale turbine rotor passage. *ASME Journal of Turbomachinery*, 116, 1–13.

Blair, M.F., Dring, R.P., and Joslyn, H.D., 1989. The effects of turbulence and stator/rotor interactions on turbine heat transfer. Part I: Design operating conditions. Part II: Effects of Reynolds number and incidence. *ASME Journal of Turbomachinery*, 111, 87–103.

Boelter, L.M.K., Young, G., and Iversen, H.W., 1948. An investigation of aircraft heaters. XXVII: Distribution of heater transfer rate in the entrance region of a tube. NACA TN 1451.

Bogard, D.G., Schmidt, D.L., and Tabbita, M., 1996. Characterization and laboratory simulation of turbine airfoil surface roughness and associated heat transfer. ASME Paper 96-GT-386.

Bohn, D. and Krewinkel, R., 2008. Influence of a broken-away TBC on the flow structure and wall temperature of an effusion cooled multi-layer plate using the conjugate calculation method. ASME Paper, GT2008-50378.

Bons, J.P., 2002. St and Cf augmentation for real turbine roughness with elevated freestream turbulence. ASME Paper, GT2002-30198.

Bons, J.P. and McClain, S.T., 2003. The effect of real turbine roughness with pressure gradient on heat transfer. ASME Paper, GT2003-38738.

Bons, J., Wammack, J., Crosby, J., Fletcher, D., and Fletcher, T., 2006. Evolution of surface deposits on a high pressure turbine blade: Part II-Convective heat transfer. ASME Paper, GT2006-91246.

Boyle, R.J. and Russell, L.M., 1989. Experimental determination of stator endwall heat transfer. ASME Paper 89-GT-219.

Boyle, R. and Senyitko, R.G., 2005. Effects of surface roughness on turbine vane heat transfer. ASME Paper, GT2005-69133.

Bradshaw, P., 1974. Effect of free-stream turbulence on turbulent shear layers. Aeronautical Research Council, Paper 35468.

Bunker, R.S., 1997. Separate and combined effects of surface roughness and turbulence intensity on vane heat transfer. ASME Paper 97-GT-135.

Bunker, R., 2008. The effects of manufacturing tolerances on gas turbine cooling. ASME Paper, GT2008-50124.

Bunker, R.S. and Bailey, J.C., 2000. Blade tip heat transfer and flow with chordwise sealing strips, *International Symposium on Transport Phenomena and Dynamics of Rotating Machinery (ISROMAC)*, Honolulu, HI, pp. 548–555.

Bunker, R., Bailey, J.C., and Ameri, A.A., 2000. Heat transfer and flow on the first stage blade tip of a power generation gas turbine: Part 1: Experimental results. *ASME Journal of Turbomachinery*, 122(2), 263–271.

Bunker, R.S., 2001. A review of turbine blade tip heat transfer, heat transfer in gas turbine systems. *Annals of the New York Academy of Sciences*, 934, 64–79.

Bunker, R.S. and Bailey, J.C., 2001. Effect of squealer cavity depth and oxidation on turbine blade tip heat transfer. ASME Paper, 2001-GT-0155.

Bunker, R.S., Bailey, J.C., and Ameri, A., 1999. Heat transfer and flow on the first stage blade tip of a power generation gas turbine. Part 1: Experimental results, *ASME International Gas Turbine and Aero-Engine Congress*, June 7–10, 1999, Indianapolis, IN, ASME Paper No. 99-GT-169.

Burd, S.W. and Simon, T.W., 2000. Flow measurements in a nozzle guide vane passage with a low aspect ratio and endwall contouring. ASME Paper 2000-GT-0213.

Carullo, J., Nasir, S., Cress, R.D., Ng, W., Thole, K., Zhang, L., and Moon, H.K., 2007. The effects of freestream turbulence, turbulence length scale and exit Reynolds number on turbine blade heat transfer in a transonic cascade. ASME aper, GT2007-27859.

Chakka, P. and Schobeiri, M.T., 1999. Modeling of unsteady boundary layer transition on a curved plate under periodic unsteady flow condition: Aerodynamic and heat transfer investigations. *Journal of Turbomachinery—Transactions of the ASME*, 121, 88–97.

Chana, K.S. and Jones, T.V., 2002. An investigation on turbine tip and shroud heat transfer. ASME Paper, GT2002-30554.

Chen, P.H. and Goldstein, R.J., 1992. Convective transport phenomena on the suction surface of a turbine blade including the influence of secondary flows near the endwall. *ASME Journal of Turbomachinery*, 114, 776–787.

Cho, H.H. and Rhee, D.H., 2005a. Local heat/mass transfer characteristics on a rotating blade with flat tip in a low speed annual cascade: Part 1–Near-tip surface. ASME Paper, GT2005-68723.

Cho, H.H. and Rhee, D.H., 2005b. Local heat/mass transfer characteristics on a rotating blade with flat tip in a low speed annual cascade: Part 2–Tip and shroud. ASME Paper, GT2005-68724.

Chong, T.P. and Zhong, S., 2003. On the three-dimensional structure of turbulent spots. ASME Paper, GT2003-38435.

Chyu, M.K., Moon, H.K., and Metzger, D.E., 1989. Heat transfer in the tip region of grooved turbine blades. *ASME Journal of Turbomachinery*, 111, 131–138.

Clark, J.P. and Praisner, T., 2004. Predicting transition in turbomachinery, Part I—A review and new model development. ASME Paper, GT2004-54108.

Coton, T. and Arts, T., 2004a. Investigation of a high lift LP turbine blade submitted to passing wakes. Part 1: Profile loss and heat transfer. ASME Paper, GT2004-53768.

Coton, T. and Arts, T., 2004b. Investigation of a high lift LP turbine blade submitted to passing wakes. Part 2: Boundary layer transition. ASME Paper, GT2004-53781.

Dey, D. and Camci, C., 2001. Aerodynamic tip de-sensitization of an axial turbine rotor using tip platform extensions. ASME paper 2001-GT-484.

Dick, E.R. and Lodefier, B.K., 2003. Transition modelling with SST turbulence model and an intermittency transport equation. ASME Paper, GT2003-38282.

Didier, F., Denos, R., and Arts, T., 2002. Unsteady rotor heat transfer in a transonic turbine stage. ASME Paper, GT2002-30195.

Doorly, D.J., 1988. Modeling the unsteady flow in a turbine rotor passage. *ASME Journal of Turbomachinery*, 110, 27–37.

Doorly, D.J. and Oldfield, M.L.G., 1985. Simulation of the effects of shock-waves passing on a turbine rotor blade. *ASME Journal of Engineering for Gas Turbines and Power*, 107, 998–1006.

Doorly and Oldfield, M.J., 1985. Simulation of the effect of shock wave passing on a turbine rotor blade. *Journal of Engineering for Gas Turbines and Power*, 107, 998–1006.

Dring, R.P., Blair, M.F., and Joslyn, H.D., 1986. The effects of inlet turbulence and rotor–stator interactions on the aerodynamics and heat transfer of a large-scale rotating turbine model. NASA CR 4079, NASA CR 179467-69.

Dring, R.P. and Joslyn, H.D., 1981. Measurements of turbine rotor blade flows. *ASME Journal of Engineering for Power*, 103, 400–405.

Dring, R.P., Joslyn, H.D., Hardin, L.W., and Wagner, J.H., 1982. Turbine rotor–stator interactions. *ASME Journal of Engineering for Power*, 104, 729–742.

Dris, A. and Johnson, M.W., 2004. Transition on concave surfaces. ASME Paper, GT2004-53352.

Du, H., Ekkad, S., and Han, J.C., 1997. Effect of unsteady wake with trailing edge coolant ejection on detailed heat transfer coefficient distributions for a gas turbine blade. *ASME Journal of Heat Transfer*, 119, 242–248.

Duden, A., Raab, I., and Fottner, L., 1998. Controlling the secondary flow in a turbine cascade by 3D airfoil design and endwall contouring, *International Gas Turbine & Aeroengine Conference & Exhibition*, Stockholm, Sweden, 98-GT-72.

Dullenkopf, K., Schulz, A., and Wittig, S., 1991. The effect of incident wake conditions on the mean heat transfer of an airfoil. *ASME Journal of Turbomachinery*, 113, 412–418.

Dunn, M.G., 1984. Turbine heat flux measurements: Influence of slot injection on vane trailing edge heat transfer and influence of rotor on vane heat transfer. ASME Paper 84-GT-175.

Dunn, M.G., 1986. Heat flux measurements for a rotor of a full-stage turbine. Part I: Time averaged results. *ASME Journal of Turbomachinery*, 108, 90–97.

Dunn, M.G., George, W.K., Rae, W.J., Woodward, S.H., Moller, J.C., and Seymour, J.P., 1986a. Heat flux measurements for a rotor of a full-stage turbine. Part II: Description of analysis technique and typical time-resolved measurements. *ASME Journal of Turbomachinery*, 108, 98–107.

Dunn, M.G. and Haldeman, C.W., 2000. Time-averaged heat flux for a recessed tip, lip, and platform of a transonic turbine blade. ASME Paper, 2000-GT-0197.

Dunn, M.G. and Haldeman, C.W., 2003. Heat transfer measurements and predictions for the vane and blade of a rotating high-pressure turbine stage. ASME Paper, GT2003-38726.

Dunn, M.G. and Hause, A., 1979. Measurement of heat flux and pressure in a turbine stage. *ASME Journal of Engineering for Power*, 101.

Dunn, M.G., Martin, H.L., and Stanek, M.J., 1986b. Heat flux and pressure measurements and comparison with prediction for a low aspect ratio turbine stage. *ASME Journal of Turbomachinery*, 108, 108–115.

Dunn, M.G. and Haldeman, C.W., 2000. Time-averaged heat flux for a recessed tip, lip and platform of a transonic turbine blade. ASME Paper 2000-GT-0197.

Dunn, M., Mathison, R., and Haldeman, C., 2010. Aerodynamics and heat transfer for a cooled one and one-half stage high-pressure turbine: Part III: Impact of hot streak characteristics on blade row heat flux. ASME Paper GT2010-23855.

Dunn, M.G., Rae, W.J., and Holt, J.L., 1984. Measurement and analyses of heat flux data in a turbine stage. Part I: Description of experimental apparatus and data analysis. Part II: Discussion of results and comparison with predictions. *ASME Journal of Engineering for Gas Turbines and Power*, 106, 229–240.

Dunn, M.G. and Stoddard, F.J., 1979. Measurement of heat-transfer rate to a gas turbine stator. *ASME Journal of Engineering for Power*, 101, 275–280.

Ekkad, S.V., Nasir, H., and Saxena, V., 2003. Effect of blade tip geometry on tip flow and heat transfer for a blade in a low speed cascade. ASME Paper, GT2003-38176.

Emmons, H.W., 1951. The laminar-turbulent transition in boundary layer: Part I. *Journal of Aeronautical Science*, 18, 490–498.

Erickson, E., Ames, F., and Bons, J., 2010a. Effects of a realistically rough surface on vane aerodynamic losses including the influence of turbulence condition and Reynolds number. ASME Paper, GT2010-22173.

Erickson, E., Ames, F., and Bons, J., 2010b. Effects of a realistically rough surface on vane heat transfer including the influence of turbulence condition and Reynolds number. ASME Paper, GT2010-22174.

Epstein, A.H., Guenette, G.R., and Norton, R.J.G., 1984. The MIT blowdown turbine facility. ASME Paper 84-GT-116.

Fletcher, T.H., Ai, W., Murray, N., Harding, S., Lewis, S., and Bons, J., 2008. Deposition near film cooling holes on a high pressure turbine vane. ASME Paper, GT2008-50901.

Frossling, N., 1958. Evaporation heat transfer and velocity distribution in two-dimensional and rotationally-symmetric laminar boundary layer. NACA, TM-1432.

Funazaki, K., 1996. Studies on wake-affected heat transfer around the circular leading edge of blunt body. *ASME Journal of Turbomachinery*, 118, 452–460.

Germain, T., Nagel, M., Raab, I., Schüpbach, P., Abhari, R.S., and Rose, M., 2010. Improving efficiency of a high work turbine using nonaxisymmetric endwalls— Part I: Endwall design and performance. *ASME Journal of Turbomachinery*, 132(2), 021007, http://dx.doi.org/10.1115/1.3106706

Giel, P.W., Boyle, R.J., and Bunker, R.S., 2003. Measurements and predictions of heat transfer on rotor blades in a transonic turbine cascade. ASME Paper, GT2003-38839.

Giel, P.W., Bunker, R.S., Van Fossen, G.J., and Boyle, R.J., 2000. Heat transfer measurements and predictions on a power generation gas turbine blade. ASME paper, 2000-GT-0209.

Giel, P.W., Thurman, D.R., Van Fossen, G.J., Hippensteele, S.A., and Boyle, R.J., 1998. Endwall heat transfer measurements in a transonic turbine cascade. *ASME Journal of Turbomachinery*, 120, 305–313.

Gillis, J.C. and Johnston, J.P., 1983. Turbulent boundary layer flow and structure on a convex wall and its redevelopment on a flat wall. *Journal of Fluid Mechanics*, 135, 123–153.

Goebel, S.G., Abuaf, N., Lovett, J.A., and Lee, C.P., 1993. Measurements of combustor velocity and turbulence profiles. ASME Paper 93-GT-228.

Goldstein, R. and Han, S., 2005. Influence of blade leading edge geometry on turbine endwall heat (mass) transfer. ASME Paper, GT2005-68590.

Goldstein, R.J., Lau, K.Y., and Leung, C.C., 1983. Velocity and turbulence measurements in combustion systems. *Experiments in Fluids*, 1, 93–99.

Goldstein, R.J. and Spores, R.A., 1988. Turbulent transport on the endwall in the region between adjacent turbine blades. *ASME Journal of Heat Transfer*, 110, 862–869.

Goldstein, R.J., Wang, H.P., and Jabbari, M.Y., 1995. The influence of secondary flows near the endwall and boundary layer disturbance on convective transport from a turbine blade. *ASME Journal of Turbomachinery*, 117, 657–663.

Graziani, R.A., Blair, M.F., Taylor, J.R., and Mayle, R.E., 1980. An experimental study of endwall and airfoil surface heat transfer in a large scale turbine blade cascade. *ASME Journal of Engineering for Power*, 102, 257–267.

Grover, E.A., Clark, J.P., Rice, M.J., and Praisner, T., 2004. Predicting transition in turbomachinery: Part II—Model validation and benchmarking. ASME Paper, GT2004-54109.

Guenette, G.R., Epstein, A.H., Giles, M.B., Hanes, R., and Norton, R.J.G., 1989. Fully scaled transonic turbine rotor heat transfer measurements. *ASME Journal of Turbomachinery*, 111, 1–7.

Gustafson, R., Mahmood, G., and Acharya, S., 2007. Aerodynamic measurements in a linear turbine blade passage with three-dimensional endwall contouring, *Proceedings of GT2007, ASME Turbo Expo 2007: Power for Land, Sea and Air*, May 14–17, 2007, Montreal, Quebec, Canada, GT2007-28073.

Haldeman, C., Mathison, R., Dunn, M., Southworth, S., Harral, J., and Heitland, G., 2006a. Aerodynamic and heat flux measurements in a single stage fully cooled turbine—Part I: Experimental approach. ASME Paper, GT2006-90966.

Haldeman, C., Mathison, R., Dunn, M., Southworth, S., Harral, J., and Heitland, G., 2006b. Aerodynamic and heat flux measurements in a single stage fully cooled turbine—Part II: Experimental results. ASME Paper, GT2006-90968.

Haldeman, C.W., Mathison, R.M., Dunn, M.G., Southworth, S.A., Harral, J.W., and Heitland, G., 2008a. Aerodynamic and heat flux measurements in a single-stage fully cooled turbine—Part I: Experimental approach. *Journal of Turbomachinery*, 130, 021015 (10 pages).

Haldeman, C.W., Mathison, R.M., Dunn, M.G., Southworth, S.A., Harral, J.W., and Heitland, G., 2008b. Aerodynamic and heat flux measurements in a single-stage fully cooled turbine—Part II: Experimental results. *Journal of Turbomachinery*, 130, 021016 (10 pages).

Han, J.C., Zhang, L., and Ou, S., 1993. Influence of unsteady wake on heat transfer coefficients from a gas turbine blade. *ASME Journal of Heat Transfer*, 115, 904–911.

Hancock, P.E., 1980. Effect of free-stream turbulence on turbulent shear layers, PhD thesis, Imperial College, London, U.K.

Hancock, P.E. and Bradshaw, P., 1983. The effect of free-stream turbulence on turbulent boundary layers. *ASME Journal of Fluids Engineering*, 105, 284–289.

Hartland, J. and Gregory-Smith, D.G., 2002. A design method for the profiling of end walls in turbines, *Proceedings of ASME Turbo Expo 2002*, June 3–6, 2002, Amsterdam, the Netherlands, GT-2002-30433.

Hartland, J.C., Gregory-Smith, D.G., Harvey, N.W., and Rose, M.G., 2000. Nonaxisymmetric turbine end wall design: Part II—Experimental validation. *ASME Journal of Turbomachinery*, 122, 286–293.

Harvey, N.W., Rose, M.G., Coupland, J., and Jones, T.V., 1999. Measurement and calculation of nozzle guide vane endwall heat transfer. *ASME Journal of Turbomachinery*, 121, 184–190.

Harvey, N.W., Rose, M.G., Taylor, M.D., Shahpar, S., Hartland, J., and Gregory-Smith, D.G., 2000. Nonaxisymmetric turbine end wall design: Part I—Three-dimensional linear design system. *ASME Journal of Turbomachinery*, 122, 278–285.

Harvey, N.W., Rose, M.G., Taylor, M.D., Shahpar, S., Hartland, J., and Gregory-Smith, D.G., 2000. Nonaxisymmetric turbine end wall design: Part I—Three-dimensional linear design system. *Journal of Turbomachinery—Transactions of the ASME*, 122, 278–285.

Harvey, N., Yoshino, S., Miller, R., Thorpe, S.J., and Ainsworth, R.W., 2005. The effect of work processes on the casing heat transfer of a transonic turbine. ASME Paper, GT2005-68437.

Hodson, H.P., 1984. Boundary layer and loss measurements on the rotor of an axial flow turbine. *ASME Journal of Engineering for Gas Turbines and Power*, 106, 391–399.

Hoffs, A., Drost, U., and Boles, A., 1996. Heat transfer measurements on a turbine airfoil at various Reynolds numbers and turbulence intensities including effects of surface roughness. ASME Paper 96-GT-169.

Hylton, L.D., Mihelc, M.S., Turner, E.R., Nealy, D.A., and York, R.E., 1982. Analytical and experimental evaluation of the heat transfer distributions over the surfaces of turbine vanes. NASA CR-168015.

Jenkins, S. and Bogard, D.G., 2004. The effects of the vane and mainstream turbulence level on hot streak attenuation. ASME Paper, GT2004-54022.

Jenkins, S., Bogard, D.G., and Varadarajan, K., 2003. The effects of high mainstream turbulence and turbine vane film cooling on the dispersion of a simulated hot streak. ASME Paper, GT2003-38575.

Jin, P. and Goldstein, R.J., 2002. Local mass/heat transfer on turbine blade near-tip surfaces. ASME Paper, GT2002-30556.

Johnson, M.W., 2002. Predicting transition without empiricism or DNS. ASME Paper, GT2002-30238.

Johnson, M.W., 2003. A receptivity based transition model. ASME Paper, GT2003-38073.

Jones, W.P. and Launder, B.E., 1972. The prediction of laminarization with a two-equation model of turbulence. *International Journal of Heat and Mass Transfer*, 15, 301–314.

Julien, H.L., Kays, W.M., and Moffat, R.J., 1969. The turbulent boundary layer on a porous plate: Experimental study of the effects of a favorable pressure gradient. Stanford University, Thermo-Sciences Division, Report HMT-4, Stanford, CA.

Kadar, B.A. and Yaglom, A.M., 1972. Heat and mass transfer laws for fully turbulent wall flows. *International Journal of Heat and Mass Transfer*, 15, 2329–2351.

Kanjirakkad, V., Thomas, R.L., Hodson, H., Janke, E., Haselbach, F., and Whitney, C., 2006. Passive shroud cooling concepts for HP turbines: Experimental investigations. ASME Paper, GT2006-91250.

Kaszeta, R.W., Simon, T.W., and Ashpis, D.E., 2001. Experimental investigation of transition to turbulence as affected by passing wakes. ASME Paper, 2001-GT-0195.

Kays, W.M. and Crawford, M.E., 1980. *Convective Heat and Mass Transfer*, 2nd edn. McGraw-Hill, New York.

Kestin, J. and Wood, R.T., 1971. The influence of turbulence on mass transfer from cylinders. *ASME Journal of Heat Transfer*, 93, 321–327.

Kestoras, M.D. and Simon, T.W., 1995. Effects of free-stream turbulence intensity on a boundary layer recovering from concave curvature effects. *ASME Journal of Turbomachinery*, 117, 206–214.

Kestoras, M.D. and Simon, T.W., 1997. Turbulent transport measurements in a heated boundary layer: Combined effects of free-stream turbulence and removal of concave curvature. *ASME Journal of Turbomachinery*, 112, 206–214.

Kim, J. and Simon, T.W., 1987. Measurements of the turbulent transport of heat and momentum in convexly curved boundary layers: Effects of curvature, recovery, and free-stream turbulence. ASME Paper 87-GT-199.

Kim, J., Simon, T.W., and Russ, S.G., 1992. Free-stream turbulence and concave curvature effects on heated, transitional boundary layers. *ASME Journal of Heat Transfer*, 114, 338–347.

Kopper, F.C., Milano, R., and Vanco, M., 1980. An experimental investigation of endwalls profiling in a turbine vane cascade. AIAA Paper No. 80–1089.

Krishnababu, S., Hodson, H.P., Booth, G., Dawes, W., and Lock, G., 2008. Aero-thermal investigations of tip-leakage flow in a film-cooled industrial gas turbine rotor. ASME Paper, GT2008-50222.

Krishnababu, S.K., Newton, P.J., Dawes, W., Lock, G., Hodson, H., Hannis, J., and Whitney, C., 2007a. Aero-thermal investigations of tip leakage flow in axial flow turbines: Part I—Effect of tip geometry. ASME Paper, GT2007-27954.

Krishnababu, S.K., Newton, P.J., Dawes, W., Lock, G., Hodson, H., Hannis, J., and Whitney, C., 2007b. Aero-thermal investigations of tip leakage flow in axial flow turbines: Part II—Effect of relative casing motion. ASME Paper, GT2007-27957.

Kwak, J.S., Ahn, J., Han, J.C., Lee, C.P., Bunker, R.S., Boyle, R.J., and Gaugler, R.E., 2003. Heat transfer coefficients on the squealer tip and near tip regions of a gas turbine blade with single or double squealer. ASME Paper, GT2003-38907.

Kwak, J.S. and Han, J.C., 2002a. Heat transfer on a gas turbine blade tip and near-tip regions, *The 8th AIAA/ASME Joint Thermophysics and Heat Transfer Conference*, June 24–27, 2002, St. Louis, MI, AIAA-2002-3012.

Kwak, J.S. and Han, J.C., 2002b. Heat transfer coefficient on the squealer tip and near squealer tip regions of a gas turbine blade, *Proceedings of ASME IMECE'02*, November 17–22, 2002, New Orleans, LA, IMECE 2002-32109.

LaGraff, J.E., Ashworth, D.A., and Schultz, D.L., 1989. Measurement and modeling of the gas turbine blade transition process as disturbed by wakes. *ASME Journal of Turbomachinery*, 111, 315–322.

Langston, L.S., 1980. Crossflows in a turbine cascade passage. *ASME Journal of Engineering for Power*, 102, 866–874.

Lehmann, K., Thomas, R., Hodson, H., and Stefanis, V., 2009. Heat transfer and aerodynamics of over-shroud leakage flows in a high pressure turbine. ASME Paper, GT2009-59531.

Lethander, A.T., Thole, K.A., Zess, G., and Wagner, J., 2003. Optimizing the vane-endwall junction to reduce adiabatic wall temperatures in a turbine vane passage, *Proceedings of the ASME Turbo Expo*, Atlanta, GA, Paper No. GT-2003-38939.

Lethander, A.T. and Thole, K.A., 2003. Optimizing the vane-endwall junction to reduce adiabatic wall temperatures in a turbine vane passage. ASME Paper GT2003-38940.

Lin, Y.-L., Shih, T.I.-P., and Simon, T.W., 2000. Control of secondary flows in a turbine nozzle guide vane by endwall contouring. ASME Paper GT2000-556.

Lodefier, K., Merci, B., De Langhe, C., and Dick, E., 2003. Transition modeling with the SST turbulence model and an intermittency transport equation. ASME Paper GT-2003-38282.

Lorenz, M., Schulz, A., and Bauer, H.J., 2010a. Experimental study of surface roughness effects on a turbine airfoil in a linear cascade—Part I: External heat transfer. ASME Paper, GT2010-23800.

Lorenz, M., Schulz, A., and Bauer, H.J., 2010b. Experimental study of surface roughness effects on a turbine airfoil in a linear cascade—Part II: Aerodynamic losses. ASME Paper, GT2010-23801.

Lowery, G.W. and Vachon, R.I., 1975. The effect of turbulence on heat transfer from heated cylinders. *International Journal of Heat and Mass Transfer*, 18, 1229–1242.

Lynch, S.P., Sundaram, N., Thole, K.A., Kohli, A., and Lehane, C., 2009. Heat transfer for a turbine blade with non-axisymmetric endwall contouring. ASME Paper, GT2009-60185.

Lynch, S.P., Thole, K.A., Kohli, A., and Lehane, C., 2011. Computational predictions of heat transfer and film-cooling for a turbine blade with nonaxisymmetric endwall contouring. *ASME Journal of Turbomachinery*, 133(4), 041003, http://dx.doi.org/10.1115/1.4002951

Maciejewski, P.K. and Moffat, R.J., 1992a. Heat transfer with very high free-stream turbulence. Part I: Experimental data. *ASME Journal of Heat Transfer*, 114, 827–833.

Maciejewski, P.K. and Moffat, R.J., 1992b. Heat transfer with very high free-stream turbulence. Part II: Analysis of results. *ASME Journal of Heat Transfer*, 114, 834–839.

MacMullin, R., Elrod, W., and Rivir, R., 1989. Free-stream turbulence from a circular wall jet on a flat plate heat transfer and boundary layer. *ASME Journal of Turbomachinery*, 111, 78–86.

Mahmood, G.I., Gustafson, R., and Acharya, S., 2005. Experimental investigation of flow structure and Nusselt number in a low speed linear blade passage with and without leading edge fillets. *ASME Journal of Heat Transfer*, 127, 499–512.

Martinez-Botas, R.F., Lock, G.D., and Jones, T.V., 1995. Heat transfer measurements in an annular cascade of transonic gas turbine blades using the transient liquid crystal technique. *ASME Journal of Turbomachinery*, 117, 425–431.

Mathison, R., Haldeman, C., and Dunn, M., 2010a. Aerodynamics and heat transfer for a cooled one and one-half stage high-pressure turbine: Part I: Vane inlet temperature profile generation and migration. ASME Paper GT2010-22716.

Mathison, R., Haldeman, C., and Dunn, M., 2010b. Aerodynamics and heat transfer for a cooled one and one-half stage high-pressure turbine: Part II: Vane inlet temperature profile on blade row and shroud. ASME Paper GT2010-22718.

Mathison, R., Haldeman, C., and Dunn, M., 2010c. Heat transfer for the blade of a cooled stage and one-half high-pressure turbine: Part I: Influence of cooling variation. ASME Paper GT2010-22713.

Mathison, R., Haldeman, C., and Dunn, M., 2010d. Heat transfer for the blade of a cooled stage and one-half high-pressure turbine: Part II: Independent influence of vane trailing edge and purge cooling. ASME Paper GT2010-22715.

Martinez-Botas, R.F., Main, A.J., Lock, G.D., and Jones, T.V., 1993. Cold heat transfer tunnel for gas turbine research on an annular cascade. ASME Paper 93-GT-248.

Mayle, R.E., 1991. The role of laminar-turbulent transition in gas turbine engines. *ASME Journal of Turbomachinery*, 113, 509–537.

Mayle, R.E., Blair, M.F., and Kopper, F.C., 1979. Turbulent boundary layer heat transfer on curved surfaces. *ASME Journal of Heat Transfer*, 101, 521–525.

Mayle, R.E. and Dullenkopf, K., 1990. A theory for wake-induced transition. *ASME Journal of Turbomachinery*, 112, 188–915.

Mayle, R.E. and Dullenkopf, K., 1991. More on the turbulent strip theory for wake-induced transition. *ASME Journal of Turbomachinery*, 113, 428–432.

Mayle, R.E. and Metzger, D.E., 1982. Heat transfer at the tip of an unshrouded turbine blade, *Proceedings of the 7th International Heat Transfer Conference*, Munich, Germany, Vol. 3, pp. 87–92.

Mazzotta, D., Karaivanov, V., Slaughter, W.S., Chyu, M.K., and Alvin, M.A., 2008. Gas-side heat transfer in syngas hydrogen fired and oxy-fuel turbines. ASME Paper, GT2008-51474.

McDonald, H. and Kreskovsky, J.P., 1979. Effect of free-stream turbulence on the turbulent boundary layer. *International Journal of Heat and Mass Transfer*, 26, 31–36.

Mehendale, A.B., Han, J.C., and Ou, S., 1991. Influence of mainstream turbulence on leading edge heat transfer. *ASME Journal of Heat Transfer*, 113, 843–850.

Metzger, D.E., Bunker, R.S., and Chyu, M.K., 1989. Cavity heat transfer on a transversed grooved wall in a narrow flow channel. *ASME Journal of Heat Transfer*, 111, 73–79.

Metzger, D.E., Dunn, M.G., and Hah, C., 1991. Turbine tip and shroud heat transfer. *ASME Journal of Turbomachinery*, 113, 502–507.

Metzger, D.E. and Rued, K., 1989. The influence of turbine clearance gap leakage on passage velocity and heat transfer near blade tips. Part I: Sink flow effects on blade pressure side. *ASME Journal of Turbomachinery*, 111, 284–292.

Molter, S., Duun, M., Haldeman, C., Bergholz, R.F., and Vitt, P., 2006. Heat-flux measurements and predictions for the blade tip region of a high-pressure turbine. ASME Paper, GT2006-90048.

Moore, J., Moore, J.G., Henry, G.S., and Chaudhry, U., 1989. Flow and heat transfer in turbine tip gaps. *ASME Journal of Turbomachinery*, 111, 301–309.

Morehouse, K.A. and Simoneau, R.J., 1986. Effect of rotor wake on the local heat transfer on the forward half of a circular cylinder, *Proceedings of the 8th International Heat Transfer Conference*, San Francisco, CA, Vol. 3, pp. 1249–1256.

Morris, A.W.H. and Hoare, R.G., 1975. Secondary loss measurements in a cascade of turbine blades with meridional wall profiling. ASME Paper No. 75-WA/GT-13.

Moss, R.W. and Oldfield, M.L.G., 1991. Measurements of hot combustor turbulence spectra. ASME Paper 91-GT-351.

Nagel, M.G. and Baier, R.D., 2005. Experimentally verified numerical optimization of a three-dimensional parameterized turbine vane with nonaxisymmetric end walls. *ASME Journal of Turbomachinery*, 127, 380–387.

Nealy, D.A., Mihelc, M.S., Hylton, L.D., and Gladden, H.J., 1984. Measurements of heat transfer distribution over the surfaces of highly loaded turbine nozzle guide vanes. *ASME Journal of Engineering for Gas Turbines and Power*, 106, 149–158.

Newton, P.J., Lock, G.D., Krishnababu, S.K., Hodson, H.P., Dawes, W.N., Hannis, J., and Whitney, C., 2006. Heat transfer and aerodynamics of turbine blade tips in a linear cascade. *Journal of Turbomachinery—Transactions of the ASME*, 128(2), 300–309.

O'Brien, J.E. and Capp, S.E., 1989. Two-component phase-averaged turbulence statistics downstream of a rotating spoked-wheel wake generator. *ASME Journal of Turbomachinery*, 111, 475–482.

O'Brien, J.E., Simoneau, R.J., LaGraff, J.E., and Morehouse, K.A., 1986. Unsteady heat transfer and direct comparison to steady-state measurements in a rotor-wake experiment, *Proceedings of the 8th International Heat Transfer Conference*, San Francisco, CA, Vol. 3, pp. 1243–1248.

O'Brien, J.E. and Van Fossen, G.J., 1985. The influence of jet-grid turbulence on heat transfer from a stagnation region of a cylinder in crossflow. ASME Paper 85-HT-58.

O'Dowd, D., Zhang, Q., Usandizaga, I., He, L., and Ligrani, P., 2010. Transonic turbine blade tip aero-thermal performance with different tip gaps: Part II—Tip aerodynamic loss. ASME Paper GT2010-22780.

Olson, S.J., Sanitjai, S., Ghosh, K., and Goldstein, R.J., 2009. Effect of wake-disturbed flow on heat (mass) transfer to a turbine blade. ASME Paper, GT2009-60218.

Ozturk, B., Schobeiri, M., and Ashpis, D., 2005a. Effect of Reynolds number and periodic unsteady wake flow condition on boundary layer development, separation, and re-attachment along the suction surface of a low pressure turbine blade. ASME Paper, GT2005-68600.

Ozturk, B., Schobeiri, M., and Ashpis, D., 2005b. Intermittent behavior of the separated boundary layer along the suction surface of a low pressure turbine blade under periodic unsteady flow conditions. ASME Paper, GT2005-68603.

Ozturk, B. and Schobeiri, M.T., 2007, Effect of turbulence intensity and periodic unsteady wake flow condition on boundary layer development, separation, and re-attachment over the separation bubble along the suction surface of a low pressure turbine blade. *ASME Journal of Fluids Engineering*, 129, 747–763.

Palafox, P., Oldfield, M.L.G., and Lagraff, J.E., 2005. PIV maps of tip leakage and secondary flow fields on a low speed turbine blade cascade with moving endwall. ASME Paper No. GT-2005-68189.

Palafox, P., Oldfield, M., LaGraff, J.E., Jones, T.V., and Ireland, P.T., 2006. Blade tip heat transfer and aerodynamics in a large scale turbine cascade with moving endwall. ASME Paper, GT2006-90425.

Patil, S., Abraham, S., Tafti, D., Ekkad, S.V., Kim, Y., Dutta, P., Moon, H.K., and Srinivasan, R., 2011. Experimental and numerical investigation of convective heat transfer in a gas turbine can combustor. *ASME Journal of Turbomachinery*, January, 011028-1:7.

Pfeil, H. and Herbst, R., 1979. Transition procedure of instationary boundary layers. ASME Paper 79-GT-128.

Pfeil, H., Herbst, R., and Schroder, T., 1983. Investigation of the laminar-turbulent transition of boundary layers disturbed by wakes. *ASME Journal of Engineering for Power*, 105, 130–137.

Praisner, T.J. and Clark, J.P., 2004a. Predicting transition in turbomachinery, Part I—A review and new model development. ASME Paper GT-2004-54108.

Praisner, T.J. and Clark, J.P., 2004b. Predicting transition in turbomachinery, Part II—Model validation and benchmarking. ASME Paper GT-2004-54109.

Praisner, T.J., Allen-Bradley, E., Grover, E.A., Knezevici, D.C., and Sjolander, S.A., 2007. Application of non-axisymmetric endwall contouring to conventional and high-lift turbine airfoils, *Proceedings of GT2007, ASME Turbo Expo 2007: Power for Land, Sea and Air*, May 14–17, 2007, Montreal, Quebec, Canada, GT2007-27579.

Priddy, W.J. and Bayley, F.J., 1988. Turbulence measurements in turbine blade passages and implications for heat transfer. *ASME Journal of Turbomachinery*, 110, 73–79.

Qureshi, I., Povey, T., and Beretta, A., 2010. Effect of simulated combustor temperature nonunformity on HP vane and endwall heat transfer: An experimental and computational investigation. ASME Paper, GT2010-22702.

Radomsky, R.W. and Thole, K., 2000. High freestream turbulence effects on endwall heat transfer for a gas turbine stator vane. ASME Paper, 2000-GT-0201.

Reynolds, W.C., Kays, W.M., and Kline, S.J., 1958. Heat transfer in a turbulent incompressible boundary layer. I: Constant wall temperature. NASA Memo 12-1-58W.

Roach, P.E. and Brierley, D.H., 2000. Bypass transition modelling: A new method which accounts for free-stream turbulence intensity and length scale. ASME Paper, 2000-GT-0278.

Roberts, S.K. and Yaras, M.I., 2003a. Measurements and prediction of free-stream turbulence and pressure-gradient effects on attached-flow boundary-layer transition. ASME Paper, GT2003-38261.

Roberts, S.K. and Yaras, M.I., 2003b. Effects of periodic-unsteadiness, free-stream turbulence, and flow Reynolds number on separation-bullbe transition. ASME Paper, GT2003-38262.

Rose, M.G., 1994. Non-axisymmetric endwall profiling in the HP NGV's of an axial flow gas turbine, *International Gas Turbine and Aeroengine Congress and Exposition*, The Hague, the Netherlands, 94-GT-249.

Rued, K. and Metzger, D.E., 1989. The influence of turbine clearance gap leakage on passage velocity and heat transfer near blade tips. Part I: Source flow effects on blade suction side. *ASME Journal of Turbomachinery*, 111, 293–300.

Rued, K. and Wittig, S., 1986. Laminar and transitional boundary layer structures in accelerating flow with heat transfer. ASME Paper 86-GT-97.

Saha, A.K. and Acharya, S., 2006. Computations of turbulent flow and heat transfer through a three-dimensional non-axisymmetric blade passage, *Proceedings of GT2006, ASME Turbo Expo 2006: Power for Land, Sea and Air*, May 8–11, 2006, Barcelona, Spain, GT2006-90390.

Sahm, M.K. and Moffat, R.J., 1992. Turbulent boundary layers with high turbulence: Experimental heat transfer and structure on flat and convex walls. Stanford University Report No. HMT-45, Stanford, CA.

Sato, T., Aoki, S., Takeishi, K., and Matsuura, M., 1987. Effect of three-dimensional flow field on heat transfer problems of a low aspect ratio turbine nozzle. Presented at the 1987 *Tokyo International Gas Turbine Congress*, Paper No. 87-TOKYO-IGTC-59.

Sauer, H., Müller, R., and Vogeler, K., 2001. Reduction of secondary flow losses in turbine cascades by leading edge modifications at the endwall. *ASME Journal of Turbomachinery*, 123, 207–213.

Saxena, V., Nasir, H., and Ekkad, S.V., 2003. Effect of blade tip geometry on tip flow and heat transfer for a blade in a low speed cascade. *ASME Journal of Turbomachinery*, 126, 130–138.

Schlichting, H., 1979. *Boundary Layer Theory*. McGraw-Hill, New York.

Schobeiri, T. and Pardivala, D., 1992. Development of a subsonic flow research facility for simulating the turbomachinery flow and investigating its effects on boundary layer transition, wake development and heat transfer, *Fourth International Symposium on Transport Phenomena and Dynamics of Rotating Machinery, ISROMAC*, pp. 98–114.

Schultz, M.P. and Volino, R.J., 2001. Effects of concave curvature on boundary layer transition under high free-stream turbulence conditions. ASME Paper, 2001-GT-0191.

Scrittore, J., Thole, K., and Burd, S., 2006. Investigation of velocity profiles for effusion cooling of a combustor liner. ASME Paper, GT2006-90532.

Sedalor, T., Ekkad, S., Patil, S., Tafti, D.K., Moon, H.K., Kim, Y., and Srinivasan, R., 2010. Study of flow and convective heat transfer in a simulated scaled up low emission annual combustor. ASME Paper, GT2010-22986.

Sharma, O.P. and Butler, T.L., 1987. Predictions of endwall losses and secondary flows in axial flow turbine cascades. *ASME Journal of Turbomachinery*, 109, 229–236.

Sharma, O.P., Renaud, E., Butler, T.L., Milsaps, K., Jr., Dring, R.P., and Joslyn, H.D., 1988. Rotor–stator interaction in axial flow turbines. AIAA Paper 88-3013.

Shih, T.I-P. and Lin, Y.-L., 2002a. Controlling secondary-flow structure by leading-edge airfoil fillet and inlet swirl to reduce aerodynamic loss and surface heat transfer, *Proceedings of the ASME Turbo Expo*, Amsterdam, the Netherlands, Paper No. GT-2002-30529.

Shih, T.I.-P. and Lin, Y.-L., 2002b. Controlling secondary-flow structure by leading-edge airfoil fillet and inlet swirl to reduce aerodynamic loss and surface heat transfer. ASME Paper No. GT-2002-30529.

Simon, T. and Jiang, N., 2003a. Modeling laminar-to-turbulent transition in a low-pressure turbine flow which is unsteady due to passing wakes: Part I: Transition onset. ASME Paper GT2003-38787.

Simon, T. and Jiang, N., 2003b. Modeling laminar-to-turbulent transition in a low-pressure turbine flow which is unsteady due to passing wakes: Part I: Transition path. ASME Paper GT2003-38963.

Simon, T.W. and Moffat, R.J., 1982. Turbulent boundary layer heat transfer experiments: A separate effects study on a convexly curved wall. *ASME Journal of Heat Transfer*, 105, 835–840.

Simoneau, R.J., Morehouse, K.A., and Van Fossen, G.J., 1984. Effect of a rotor wake on heat transfer from a circular cylinder. ASME Paper 84-HT-25.

Simonich, J.C. and Bradshaw, P., 1978. Effect of free-stream turbulence on heal transfer through a turbulent boundary layer. *ASME Journal of Heat Transfer*, 100, 671–677.

Smith, C., Barker, B., Clum, C., and Bons, J., 2010. Deposition in a turbine cascade with combusting flow. ASME Paper GT2010-22855.

Smith, M.C. and Keuthe, A.M., 1966. Effects of turbulence on laminar skin friction and heat transfer. *Physics of Fluids*, 9(12), 2337–2344.

Snedden, G., Dunn, D., Ingram, G., and Gregory-Smith, D., 2009. The application of non-axisymmetric endwall contouring in a single-stage rotating turbine. ASME Paper No. GT2009-59169.

Stieger, R. and Hodson, H., 2003. The transition mechanism of highly-loaded LP turbine blades. ASME Paper, GT2003-38304.

Stripf, M., Schulz, A., Bauer, H-J., and Wittig, S., 2008a. Extended models for transitional rough wall boundary layers with heat transfer—Part I: Model formulations. ASME Paper, GT2008-50494.

Stripf, M., Schulz, A., Bauer, H-J., and Wittig, S., 2008b. Extended models for transitional rough wall boundary layers with heat transfer—Part II: Model validation and benchmarking. ASME Paper, GT2008-50495.

Tallman, J., Haldeman, C., Dunn, M., Tolpadi, A.K., and Bergholz, R.F., 2006. Heat transfer measurements and predictions for a modern high-pressure trasonic turbine including endwalls. ASME Paper, GT2006-90927.

Thole, K.A. and Bogard, D.G., 1995. Enhanced heat transfer and shear stress due to high free-stream turbulence. *ASME Journal of Turbomachinery*, 117, 418–423.

Thorpe, S.J. and Ainsworth, R.W., 2006. The effects of blade passing on the heat transfer coefficient of the over-tip casing in a transonic turbine stage. ASME Paper, GT2006-90534.

Urban, M.F. and Vortmeyer, N., 2000. Experimental investigations on the thermal load and leakage flow of a turbine blade tip section with different tip section geometries. ASME Paper, 2000-GT-0196.

Van Duikeren, B. and Heselhaus, A., 2008. Experimental investigations at the over-tip casing of a HP-turbine stage including off-design conditions and heat transfer correlations. ASME Paper, GT2008-50625.

Van Fossen, G.J. and Bunker, R.S., 2000. Augmentation of stagnation region heat transfer due to turbulence from a DLN can combustor. ASME Paper, 2000-GT-0215.

Van Fossen, G.J. and Bunker, R.S., 2002. Augmentation of stagnation region heat transfer due to turbulence from an advanced dual-annular combustor. ASME Paper, GT2002-30184.

Van Fossen, G.J. and Ching, C.Y., 1994. Measurements of the influence of integral length scale on stagnation region heat transfer. ASME Paper 94-HT-5.

Van Fossen, G.J., and Simoneau, R.J., 1994. Stagnation region heat transfer: The influence of turbulence parameters, Reynolds number and body shape. *ASME HTD*, 271, 177–191.

Volino, R.J., 2002a. An investigation of the scales in transitional boundary layers under high free-stream turbulence conditions. ASME Paper, GT2002-30233.

Volino, R.J., 2002b. Separated flow transition under simulated low-pressure turbine airfoil conditions: Part 1—Mean flow and turbulence statistics. ASME Paper, GT2002-30236.

Volino, R.J., 2002c. Separated flow transition under simulated low-pressure turbine airfoil conditions: Part 2—Turbulence spectra. ASME Paper, GT2002-30237.

Wammack, J., Crosby, J., Fletcher, D., Bons, J., and Fletcher, T., 2006. Evolution of surface deposits on a high pressure turbine blade: Part I—Physical characteristics. ASME Paper, GT2006-91246.

Wang, H., Olson, S.J., Goldstein, R.J., and Eckert, E.R.G., 1997. Flow visualization in a linear turbine cascade of high performance turbine blades. *ASME Journal of Turbomachinery*, 119, 1–8.

Wedlake, E.T., Brooks, A.J., and Harasagama, S.P., 1989. Aerodynamic and heat transfer measurements on a transonic nozzle guide vane. *ASME Journal of Turbomachinery*, 111, 36–42.

Wheeler, A.P., Atkins, N.R., and He, L., 2009. Turbine blade tip heat transfer in low speed and high speed flows. ASME Paper, GT2009-59404.

Whitney, C., Dawes, B., Hannis, J., Lock, G., Newton, P., Krishnababu, S., and Hodson, H., 2005. Heat transfer and aerodynamics of turbine blade tips in a linear cascade. ASME Paper, GT2005-69034.

Wittig, S., Dullenkopf, K., Schulz, A., and Hestermann, R., 1987. Laser-Doppler studies of the wake-effected flow field in a turbine cascade. *ASME Journal of Turbomachinery*, 109, 170–176.

Wittig, S., Schulz, A., Dullenkopf, K., and Fairbanks, J., 1988. Effects of free-stream turbulence and wake characteristics on the heat transfer along a cooled turbine blade. ASME Paper 88-GT-179.

Yamamoto, A., 1989. Endwall flow/loss mechanisms in a linear turbine cascade with blade tip clearance. *ASME Journal of Turbomachinery*, 111(3), 264 (12 pages).

Yang, T.T. and Diller, T.E., 1995. Heat transfer and flow for a grooved turbine blade tip in a transonic cascade. ASME Paper 95-WA/HT-29.

Yaras, M., Yinkang, Z., and Sjolander, S.A., 1989. Flow field in the tip gap of a planar cascade of turbine blades. *ASME Journal of Turbomachinery*, 111, 276–283.

York, R.E., Hylton, L.D., and Mihelc, M.S., 1984. An experimental investigation of endwall heat transfer and aerodynamics in a linear vane cascade. *ASME Journal of Engineering for Gas Turbines and Power*, 106, 159–167.

You, S.M., Simon, T.W., and Kim, J., 1986. Free-stream turbulence effects on convex-curved turbulent boundary layers. *ASME Journal of Heat Transfer*, 111, 66–72.

Zess, G.A. and Thole, K.A., 2001. Computational design and experimental evaluation of using a leading edge fillet on a gas turbine vane, *Proceedings of the ASME Turbo Expo*, New Orleans, LA, Paper No. GT-2001-0404.

Zhang, L. and Han, J.C., 1994. Influence of mainstream turbulence on heat transfer coefficients from a gas turbine blade. *ASME Journal of Heat Transfer*, 116, 896–903.

Zhang, L. and Han, J.C., 1995. Combined effect of free-stream turbulence and unsteady wake on heat transfer coefficients from a gas turbine blade. *ASME Journal of Heat Transfer*, 117, 296–302.

Zhang, Q., O'Dowd, D., He, L., Oldfield, M., and Ligrani, P., 2010. Transonic turbine blade tip aero-thermal performance with different tip gaps: Part I—Tip heat transfer. ASME Paper GT2010-22779.

Zhou, D. and Wang, T., 1995. Effects of elevated free-stream turbulence on flow and thermal structures in transitional boundary layers. *ASME Journal of Turbomachinery*, 117, 407–412.

3

Turbine Film Cooling

3.1 Introduction

Chapter 2 discussed gas-path heat-transfer fundamentals. In the next few chapters, the focus is on cooling the gas-path components to temperatures several hundred degrees below gas temperatures to allow continued operation of these components without failure. High metal temperatures lead to higher thermal stresses and lead to ultimate failure of the component. Typically, cooling is provided by extracting some of the compressed air and bypassing the combustor directly into the turbine. The extraction of the working fluid imposes a penalty to system efficiency, as work has been done on the fluid to compress it. Also, addition of the cooler coolant back into the hot mainstream imposes further thermodynamic penalties. Cooling of turbine components is achieved by cooling from inside and outside. Internal cooling typically involves impingement cooling, turbulated serpentine passages, and flow through pin-fin arrays. Internal cooling will be the focus of Chapters 4 and 5. External cooling involves allowing the coolant from inside the blade to eject out onto the hot-gas path-side surface through discrete holes. This method is called film cooling, which is a complicated process and the focus of this chapter.

3.1.1 Fundamentals of Film Cooling

Film Cooling is the introduction of a secondary fluid (coolant or injected fluid) at one or more discrete locations along a surface exposed to a high temperature environment to protect that surface not only in the immediate region of injection but also in the downstream region

Goldstein (1971)

Turbine airfoil surfaces, shrouds, blade tips, and endwalls are all cooled using discrete-hole film cooling. A typical cooled airfoil is shown in Figure 3.1. The figure shows the various locations where coolant is injected into the mainstream from inside the airfoil through discrete holes. Film cooling protects the airfoil surface directly, compared to internal cooling techniques that remove heat from the inside surface. Film cooling also removes

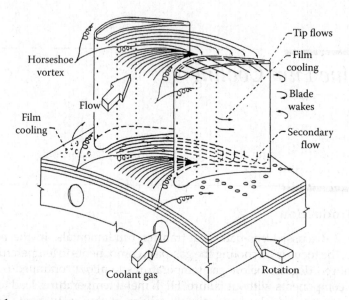

FIGURE 3.1
Typical cooled airfoil. (From Hyams, D. et al., Effects of geometry on slot-jet film cooling performance, ASME Paper No. 96-GT-187, 1996.)

heat from the blade surface through the film hole by internal convection. The thermal protection is expected to provide reduced heat load to the airfoil surface. Designers need to know the net heat load into the component surface when film is injected. Due to the complex nature of discrete hole injection, there is a need to know the local wall temperature (T_w) under the film and the gas-side heat-transfer coefficient with film injection. Both these components are required to estimate reduced heat load to the surface.

Figure 3.2 shows the film injection geometry and heat-transfer model associated with it. Typically, the heat load to surface without film cooling is represented as heat flux:

$$q_0'' = h_0(T_g - T_w) \tag{3.1}$$

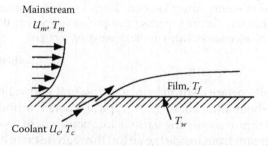

FIGURE 3.2
Schematic of film-cooling concept. (From Han, J.C. and Ekkad, S.V. Recent development in turbine blade film cooling. Invited paper for ISTP-II, Hsinchu, Taiwan, November, 1998.)

where h_0 is the heat-transfer coefficient on the surface, with wall temperature T_w and oncoming gas temperature (T_g). When film is injected on the surface, the driving temperature is T_f, film temperature, which is a mixture of gas (T_g) and coolant temperature (T_c):

$$q'' = h(T_f - T_w) \tag{3.2}$$

where h is the heat-transfer coefficient on the surface with film injection. Also, a new term, him effectiveness (η), is introduced, where $\eta = (T_g - T_f)/(T_g - T_c)$. To obtain any benefit from film cooling, the heat load ratio q''/q_0'' should be below 1.0:

$$\frac{q''}{q_0''} = \frac{h}{h_0}\frac{T_f - T_w}{T_g - T_w} = \frac{h}{h_0}\left\{1 - \eta\frac{T_g - T_c}{T_g - T_w}\right\} \tag{3.3}$$

Two methods have been used to determine heat load reduction to surfaces with film cooling. There are two approaches described in detail by Moffat (1985). The approaches are called the superposition approach and adiabatic wall effectiveness approach. Both approaches are well established and are conceptually equivalent. Data can be converted from one approach to another without any corrections.

In the superposition approach, actual surface temperatures and gas temperatures are measured to calculate the heat transfer. The effects of injection are all accounted for by the value of the heat-transfer coefficient. Experiments are run for cases where the surface has no film cooling and cases where film cooling is introduced onto the surface. The ratio of gas temperature to coolant temperature (T_g/T_c) is similar to values that exist in the real engine conditions. The heat-transfer coefficient depends on the injection temperature as well as on the injection rate. The heat-transfer coefficient obtained from this method can be directly applied to the real engine condition with the same injection rate without corrections. The heat-transfer coefficient with film cooling obtained for this approach is lower than the heat-transfer coefficient without film cooling. The heat-transfer coefficient is based on the local wall temperature and injection rate. A single test will provide all the heat-transfer information for this approach. Typically, the result is presented as a parameter called Stanton number reduction (SNR), defined as

$$\text{SNR} = 1 - \frac{St_{film}}{St_{no-film}} \tag{3.4}$$

For the adiabatic wall effectiveness approach, the adiabatic wall effectiveness and the heat-transfer coefficient are determined from separate experiments. The details of such experiments are described in Chapter 6. The adiabatic surface temperature is reported as a function of position and injection rate.

The heat-transfer coefficient is also reported for the same position and injection rate using a different experiment. Both the heat-transfer coefficient and the adiabatic wall temperature (or effectiveness) are used in Equation 3.3 to obtain the heat load reduction. In these experiments, the heat-transfer coefficient ratio is enhanced due to turbulent mixing of the jets with the mainstream and is normally greater than 1.0. The temperature ratio, which is related to the film effectiveness, should be much lower than 1.0 so that the heat load ratio is lower than 1.0.

Film-cooled airfoils have leading-edge film-cooling holes called showerhead and downstream pressure and suction-surface film cooling. Shower-head cooling has several film hole rows near the stagnation point on the blade leading edge. Pressure- and suction-surface cooling is achieved by single or multiple rows of injection holes at several locations. Figure 3.1 also indicates all the parameters affecting airfoil film cooling. The oncoming mainstream flow, blade wakes, rotation, secondary flow effects on endwalls, and tip flows are important external effects. The film hole geometry and configuration and coolant temperature and mass flow through the holes also affect film-cooling performance.

Figure 3.3 shows computed temperature distributions downstream of injection through an angled hole (Leylek and Zerkle, 1994). The temperature distributions clearly show the counterrotating vortex structure and kidney-shaped cross section of the coolant jet. Two coolant-to-mainstream blowing rates of 0.5 and 2.0 are shown. At a low blowing rate of 0.5, the jet velocity is lower than the mainstream flow. The stronger mainstream forces the jet down toward the surface. However, for a blowing rate of 2.0, the jet mainstream interaction region is larger. The jet appears to lift off farther from the surface. Also, more mainstream flow appears to entrain underneath the jet.

This chapter deals with individual effects and combined effects of the aforementioned parameters on film-cooling performance. Studies have focused on complicated problems such as rotational effects on airfoils to simple flat-surface stationary models to study film-cooling performance. The effects of each parameter on heat-transfer coefficient and film effectiveness are presented. Some correlations to predict heat-transfer coefficient and film effectiveness are also presented.

3.2 Film Cooling on Rotating Turbine Blades

Although designers expected to see actual heat-transfer data under real experimental conditions, it is very difficult to obtain results under real engine condition experiments. Due to this reason, there are very few results on real rotating blades. Dring et al. (1980) were among the earliest to study film-cooling performance in a low-speed rotating facility. Abhari and Epstein (1994)

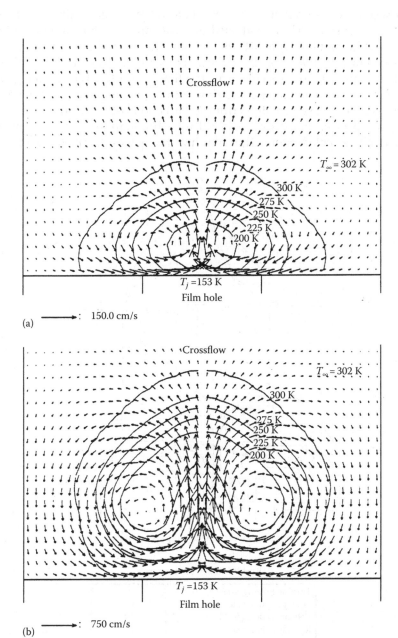

FIGURE 3.3
Computed temperature distributions downstream of injection through angled holes. (a) $M = 0.5$, $DR = 2.0$ (computed) and (b) $M = 2.0$, $DR = 2.0$ (computed). (From Leylek, J.H. and Zerkle, R.D., *ASME J. Turbomach.*, 116, 358, 1994.)

presented time-resolved measurements of heat transfer on a fully cooled transonic turbine stage. They conducted experiments in a short-duration blowdown turbine test facility at MIT, which simulated full engine scale Reynolds
number, Mach number, Prandtl number, temperature ratios of gas to wall and
coolant to mainstream, specific heat, and flow geometry. The turbine geometry and cooling arrangement are shown in Figure 3.4. More details on the
blade geometry can be obtained from Abhari and Epstein (1994).

Figure 3.5 presents the time-resolved rotor heat-transfer measurements for
cooled and uncooled rotors. Thin-film heat flux gages distributed across the

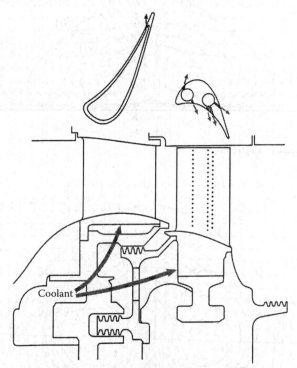

Turbine geometry and cooling arrangement

Turbine design parameters

Turbine loading, $\Delta H/U^2$	−2.3
Total pressure ratio	4.3
Velocity ratio, C_x/U	0.63
Rotor aspect ratio	1.5
NGV exit mach no.	1.18
Rotor coolant/main flow	6%
NGV coolant/main flow	3%

FIGURE 3.4
Turbine geometry and cooling arrangement. (From Abhari, R.S. and Epstein, A.H., *ASME J.
Turbomach.*, 116, 63, 1994.)

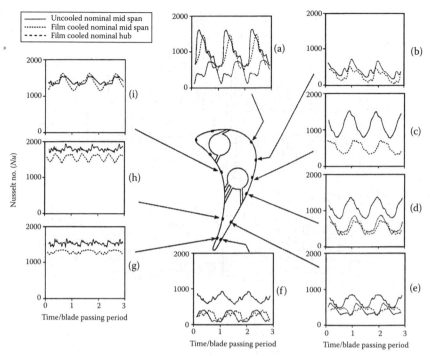

FIGURE 3.5
Time-resolved rotor heat-transfer measurements for cooled and uncooled rotors. (From Abhari, R.S. and Epstein, A.H., *ASME J. Turbomach.*, 116, 63, 1994.)

blade surface were used to measure the time-resolved heat flux distributions. Over the pressure surface, there appears to be very little effect of the film injection on heat-transfer levels. Comparing the midspan levels for uncooled and cooled rotors, a strong effect of film injection appears over the entire suction surface. The effect reduces toward the trailing edge. The influence of film injection is to produce a strong reduction in heat-transfer levels on the suction surface and a smaller effect on the pressure surface. Film injection, thus, can produce a significant reduction in heat load to the blade from the hot mainstream gases. The heat-transfer coefficient provided by this study is inclusive of all the sources of unsteadiness existing in turbine rotors. It is noted that interactions between blade rows and NGV wake impingement and shock waves on the rotor are primary sources for unsteadiness. The time-resolved data at all locations indicate clearly the perturbations in heat transfer due to this unsteadiness. The results obtained by Abhari and Epstein (1994) are based on the superposition approach.

Dring et al. (1980) were the first to measure film-cooling effectiveness on the rotating blade. Takeishi et al. (1992) also presented the film effectiveness distributions on a low-speed stator-rotor stage. The experiment was carried out in a rotating rig with a one-stage turbine model. A cross section of the rotating test equipment is shown in Figure 3.6. The stator stage had

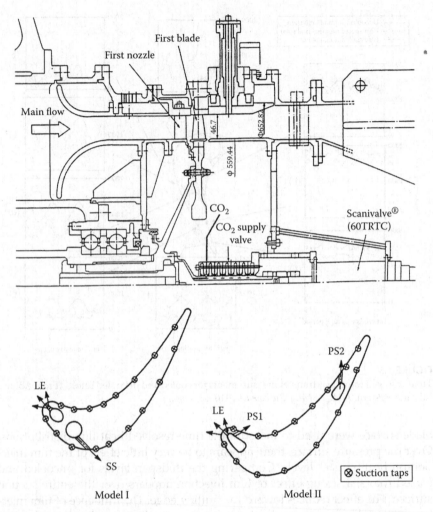

FIGURE 3.6
Cross section of the rotating test rig and model blades. (From Takeishi, K. et al., *ASME J. Turbomach.*, 114, 828, 1992.)

32 vanes; the rotor stage had 72 blades. Two model blades were tested for film effectiveness. The aerodynamically simulated conditions for the film-cooling tests were carried out for an actual heavy-duty gas turbine. They simulated actual flow velocity triangles and pressure ratios. The coolant-to-mainstream mass flux ratios were similar to engine conditions ($M = 0.6–1.0$). However, the coolant density was the same as that of the mainstream, whereas the coolant density is 1.5–2.5 times mainstream density in the real engine. Figure 3.6 also shows the blade models with different cooling configurations. A mass/heat-transfer analogy was applied to measure film effectiveness, as explained in Chapter 6.

Figure 3.7 presents the film effectiveness results on the suction and pressure surfaces due to LE hole row injections. The air turbine results are for the rotating turbine blades. Other results are from a two-dimensional (2-D) stationary cascade test. More data were measured for the stationary cascade, but few data were measured for the air turbine test. It is evident that there are

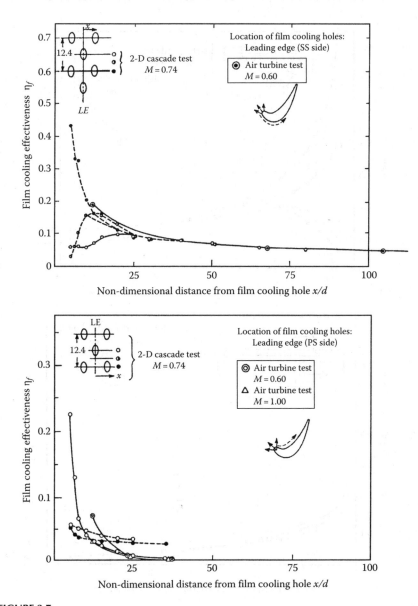

FIGURE 3.7
Film-cooling effectiveness of leading-edge injection on suction (SS) and pressure (PS) surfaces. (From Takeishi, K. et al., *ASME J. Turbomach.*, 114, 828, 1992.)

more difficulties involved in measurements on a rotating blade than on a stationary blade. The blowing ratio (*M*) values are also not similar but are close to similar conditions for both cases. It is evident from the results that the stationary cascade results match the rotating results on the suction surface very well. However, on the pressure surface, rotating results decrease more

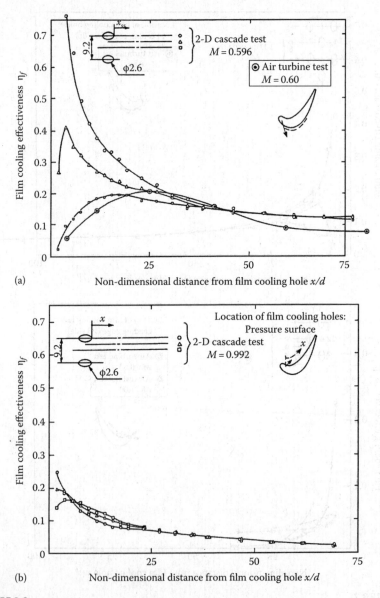

(a)

(b)

FIGURE 3.8
Film-cooling effectiveness of (a) suction-surface injection on suction surface and (b) pressure-surface injection on pressure surface. (From Takeishi, K. et al., *ASME J. Turbomach.*, 114, 828, 1992.)

rapidly than for a stationary cascade. Takeishi et al. explained this phenomenon as an effect of the radial flow and strong mixing on the pressure surface.

The suction-surface injection results are shown in Figure 3.8. Effectiveness results for the rotating turbine match well with the lowest reading on the stationary cascade. Effectiveness due to rotation could cause axial smearing on the suction surface caused by deflection of jet. The low film-cooling zone on the blade is in the 2-D film-cooling region and is estimated to be caused by the divergent flow on the suction surface. The pressure-surface injection results were presented only for the stationary cascade. The results show a continual decrease in effectiveness values downstream of injection. Also, the levels of film effectiveness are lower on the pressure surface compared to the suction surface. Takeishi et al. also attempted to correlate their results for the 2-D cascade. The correlated results on the suction and pressure surfaces are shown in Figure 3.9. The effectiveness results are correlated against x/MS, where x is the distance downstream of the film-cooling hole, M is the mass flux ratio, and S is the equivalent film-cooling slot width ($=\pi d^2/4P$ for SS, PS1, PS2, and 1.5 $\pi d^2/4P$ for LE). Based on their study, the authors indicate that the significant differences in the film effectiveness between the concave (pressure) and convex (suction) surfaces depend on strong lateral mixing that causes fundamentally different roles of concave and convex surfaces on boundary layer stability. The film effectiveness results obtained from this study will have to be supported for design purposes by heat-transfer coefficient data for similar flow conditions. The results are based on the adiabatic wall effectiveness approach as described earlier.

Although, there are few studies on rotating turbine blades, the results indicate a very small effect of rotation on external surface temperatures, except on the pressure-side near-tip region. Based on this ideology, several researchers ignored that effect of rotation for turbine blades and used stationary linear cascade data for designing blades with film cooling.

3.3 Film Cooling on Cascade Vane Simulations

3.3.1 Introduction

As explained earlier in Chapter 2, nozzle guide vanes, being just downstream of the combustor exit, experience the hottest gas-path temperatures. The vanes also experience high free-stream turbulence caused by combustor mixing flows. Depending on the requirements, vanes are cooled internally, and some coolant is ejected out as film cooling. It is important to understand the effects of high free-stream turbulence on vane surface heat transfer. Since vanes operate at the highest temperatures in the turbine section, there is an important need for adequate and efficient cooling systems to ensure that vanes survive for their estimated life.

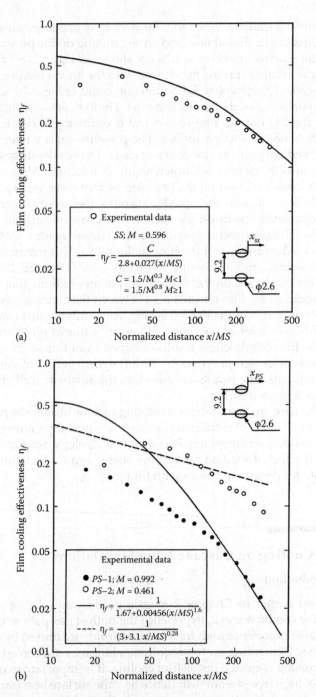

FIGURE 3.9
Film-cooling effectiveness correlations for blowing through (a) suction side and (b) pressure side. (From Takeishi, K. et al., *ASME J. Turbomach.*, 114, 828, 1992.)

3.3.2 Effect of Film Cooling

Hylton et al. (1988) and Nirmalan and Hylton (1990) studied the effects of film cooling on a turbine vane in an aerothermodynamic cascade. They maintained conditions similar to ranges existing in real engines. They performed experiments on a three-vane cascade similar to the C3X cascade used by Nealy et al. (1984), described in Chapter 2. A typical film-cooled vane is shown in Figure 3.10. The vane has an internal impingement tube through which the coolant enters the core of the vane and then impinges on to the inside surface of the outer wall. The coolant is ejected out of discrete film hole rows at various axial locations on the airfoil. They used the superposition approach for measuring heat transfer with film cooling.

Nirmalan and Hylton studied the effects of parameters such as Mach number, Reynolds number, turbulence intensity, coolant-to-gas temperature ratio, and coolant-to-gas pressure ratio. More details on the hardware used in this study can be obtained from their papers. The aerothermodynamic cascade facility consists of a burner, a convergent section, a free-stream section with instrumentation and optical access, a test section with instrumentation, a quench zone with back-pressure regulation, and an exhaust system. The film-cooling geometry consists of hole arrays on the leading edge, the suction surface, and the pressure surface. The leading edge has a shower-head array of five equally spaced rows, with the central row located at the aerodynamic stagnation point. Two rows each on the pressure and suction surfaces are located downstream. The hole rows are also staggered, with holes in the second row located midway between the holes in the first row.

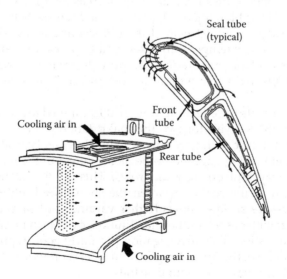

FIGURE 3.10
Typical film-cooled NGV. (From Nirmalan, N.V. and Hylton, L.D., *ASME J. Turbomach.*, 112, 477, 1990.)

More details on the film hole geometry are given in their papers. Typical results are presented for a cascade exit Mach number of 0.9 and exit Reynolds number of 2.0×10^6.

The results here are presented in the form of Stanton number reduction (SNR), which is mainly to isolate the differences between non-film-cooled and film-cooled heat transfer. The SNR is defined as $SNR = 1 - (St_{FC}/St_{NFC})$. When SNR is greater or less than zero, it implies reduced or increased heat-transfer levels for film-cooled airfoil over an uncooled airfoil, respectively. Figure 3.11 shows the effects of varying blowing strength (i.e., varying P_c/P_t ratio) at two thermal dilution levels (T_c/T_g). In this case, there is only downstream blowing. There is no film cooling from the leading-edge showerhead region. A positive SNR value is seen on both surfaces with film cooling. However, the effect of blowing strength is evident on the pressure surface only. With increasing blowing strength, the effect on the pressure surface increases farther downstream. However, the suction surface shows little effect of blowing for this reason: The film coolant flow on the suction surface is choked over this range of pressure ratios ($P_c/P_t = 1.02$–1.61), which makes the effect of blowing invariant. With higher $T_c/T_g = 0.85$, the effects are similar to that for lower $T_c/T_g = 0.65$ on the suction surface. However, the effect of blowing on the pressure surface is reversed as the SNR values are affected just downstream of hole rows. The levels of SNR values are lower on both surfaces.

Figure 3.12 presents the effects of thermal dilution (T_c/T_g) at two blowing strengths of 1.02 and 1.10. In both cases, there is a significant effect of T_c/T_g on both surfaces. Increasing blowing ratio amplifies the thermal dilution effect. SNR values increase with decreasing T_c/T_g values. At high $T_c/T_g = 0.84$ and high blowing of 1.1, SNR values decrease below zero on the pressure surface just downstream of the film-cooling holes. This indicates that the coolant-to-gas temperature effect is a significant parameter in film cooling. Lower coolant temperature will provide greater resistance for heat flow from hot-gas mainstream to the airfoil surface. This is consistent with the physics.

Figure 3.13 presents the effects of combined leading-edge and downstream blowing on the SNR levels for two thermal dilution levels. The trends are very similar to the case with only downstream blowing. Typically the effects of blowing from leading edge would be most evident in the region between showerhead array and downstream rows. This study did not have any instrumentation in the region between the showerhead and downstream rows. The pressure surface shows higher variance in heat-transfer reduction due to parameters such as thermal dilution, exit Mach number, and turbulence augmentation. This investigation is a fairly good representation of actual conditions existing in real engines. Such tests provide valuable insight into film-cooling behavior on actual airfoils.

Abuaf et al. (1997) presented heat-transfer coefficients and film effectiveness for a film-cooled vane. They used a linear cascade in a warm wind

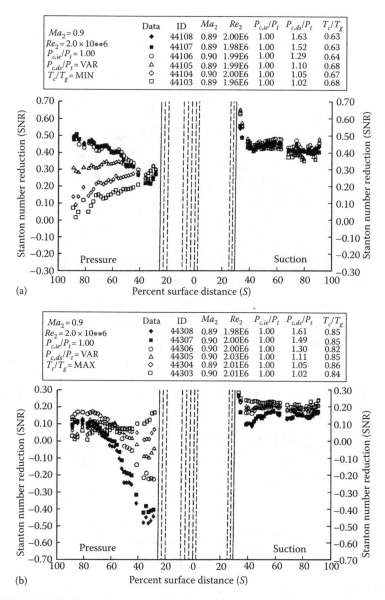

The tables within the figure:

	Data	ID	Ma_2	Re_2	$P_{c,ie}/P_t$	$P_{c,ds}/P_t$	T_c/T_g
$Ma_2 = 0.9$	◆	44108	0.89	2.00E6	1.00	1.63	0.63
$Re_2 = 2.0 \times 10{**}6$	■	44107	0.89	1.98E6	1.00	1.52	0.63
$P_{c,ie}/P_t = 1.00$	○	44106	0.90	1.99E6	1.00	1.29	0.64
$P_{c,ds}/P_t = $ VAR	△	44105	0.89	1.99E6	1.00	1.10	0.68
$T_c/T_g = $ MIN	◇	44104	0.90	2.00E6	1.00	1.05	0.67
	□	44103	0.89	1.96E6	1.00	1.02	0.68

	Data	ID	Ma_2	Re_2	$P_{c,ie}/P_t$	$P_{c,ds}/P_t$	T_c/T_g
$Ma_2 = 0.9$	◆	44308	0.89	1.98E6	1.00	1.61	0.85
$Re_2 = 2.0 \times 10{**}6$	■	44307	0.90	2.00E6	1.00	1.49	0.85
$P_{c,ie}/P_t = 1.00$	○	44306	0.90	2.00E6	1.00	1.30	0.82
$P_{c,ds}/P_t = $ VAR	△	44305	0.90	2.03E6	1.00	1.11	0.85
$T_c/T_g = $ MAX	◇	44304	0.89	2.01E6	1.00	1.05	0.86
	□	44303	0.90	2.01E6	1.00	1.02	0.84

FIGURE 3.11
Effect of downstream blowing on Stanton number reduction. (a) $T_c/T_g = $ MIN and (b) $T_c/T_g = $ MAX. (From Nirmalan, N.V. and Hylton, L.D., *ASME J. Turbomach.*, 112, 477, 1990.)

tunnel and a transient test to measure heat-transfer coefficients and film effectiveness simultaneously. Figure 3.14 presents the schematic of the test cascade and the test airfoil with film hole locations. They had 14 rows of film-cooling holes, 9 in the leading-edge showerhead region and 5 on the suction side. They used a grid upstream of the cascade to generate 14%

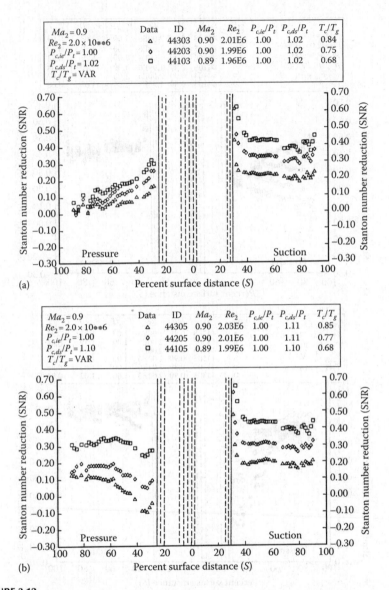

FIGURE 3.12
Effect of downstream film-cooling thermal dilution on Stanton number reduction. (a) $P_c/P_t = 1.02$ and (b) $P_c/P_t = 1.10$. (From Nirmalan, N.V. and Hylton, L.D., *ASME J. Turbomach.*, 112, 477, 1990.)

free-stream turbulence intensity. Figure 3.15 presents the heat-transfer coefficient and film effectiveness results for $Re = 517,000$. The results for the case with plugged film holes are also presented to compare the effect of film cooling. The coolant-to-mainstream density ratio was obtained to be around 2.0, and the coolant blowing ratio was as high as 2.7 in the showerhead region and as low as 1.2 on the suction surface. The heat-transfer

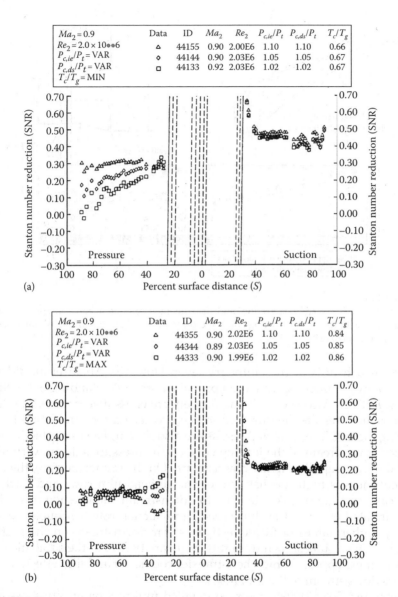

FIGURE 3.13

Effects of combined leading-edge and downstream blowing on Stanton number reduction. (a) $T_c/T_g = $ MIN and (b) $T_c/T_g = $ MAX. (From Nirmalan, N.V. and Hylton, L.D., *ASME J. Turbomach.*, 112, 477, 1990.)

coefficient on the pressure surface shows very little effect except for a small region near the leading edge where there is some interaction with the film. However, the suction-side heat-transfer coefficient shows significant increases due to film cooling. It is fairly evident that the heat-transfer coefficient is strongly enhanced on introduction of film cooling due to increased

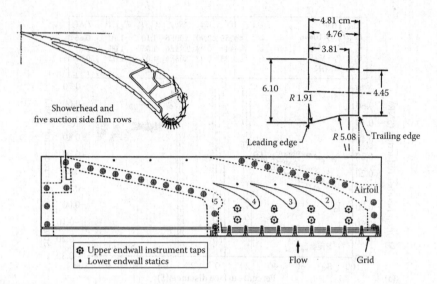

FIGURE 3.14
Schematic of the cascade test section, and instrumented cascade airfoil configuration with thermocouple locations, film-cooling holes, and three-film feed cavities. (From Abuaf, N. et al., *ASME J. Turbomach.*, 119, 302, 1997.)

local disturbances in the boundary layer. However, Abuaf et al. did not vary many parameters, which can provide more understanding of the film-cooling effect. Also shown is the film effectiveness distribution. The film effectiveness results measured have been corrected for conduction losses during the test. The corrected results are also shown. The film effectiveness values downstream of the leading edge on the pressure side are very high and decrease rapidly further downstream. On the suction surface, the film effectiveness is high downstream of showerhead film rows and starts to decrease and again spikes up due to downstream suction-side rows. The accumulated effect of the showerhead and suction rows produces significantly higher values of film effectiveness far downstream on the suction surface. The results shown from this study are typical of the adiabatic wall effectiveness approach: high heat-transfer coefficient and film effectiveness values downstream of the film holes.

Drost and Bolcs (1999) presented detailed film-cooling effectiveness and heat transfer on a turbine NGV airfoil for suction-side film cooling with single and double rows and pressure-side film cooling with a single row. The film holes were angled both in the streamwise and spanwise direction to increase the film-cooled area on the airfoil surface. They used a transient liquid crystal technique to measure heat-transfer coefficient and film effectiveness on the blade surface (see Chapter 6). Figure 3.16 shows the suction-side film-cooling effectiveness (η) and heat-transfer coefficient ratio (a_f/a_o) distributions downstream of a single row of holes. The heat flux ratio (Q_f/Q_o) combining the

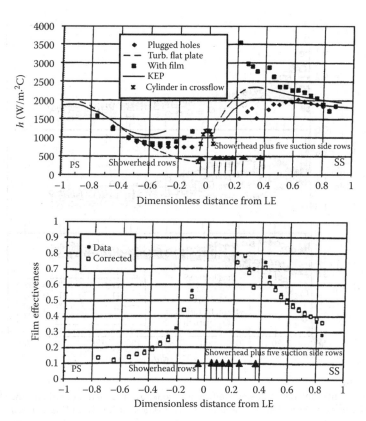

FIGURE 3.15
Heat-transfer coefficient and film effectiveness distributions at an air Reynolds number of 517,000, with film injection through all three cavities. (From Abuaf, N. et al., *ASME J. Turbomach.*, 119, 302, 1997.)

heat-transfer coefficient ratio and film effectiveness is also presented. Results are shown for different blowing ratios (*G*) for a free-stream turbulence case of 10% and $Re_2 = 1.45 \times 10^6$. The film effectiveness distributions show that the lower blowing ratio produces higher film effectiveness immediately downstream of the holes and decreases rapidly downstream. As blowing ratio increases, film effectiveness near the hole decreases, but increases downstream. As greater coolant mass is introduced into the boundary layer, the coolant jets lift off near the hole and attach downstream, producing higher effectiveness downstream. Lower blowing ratio jets are pushed toward the surface as they exit the hole and thus provide higher effectiveness immediately downstream of the holes. Heat-transfer coefficient ratio, however, continually increases with increasing blowing ratio. However, the peak values depend on where the coolant jets reattach to the surface. Heat flux reduction is the combined effect of both heat-transfer coefficient ratio and film

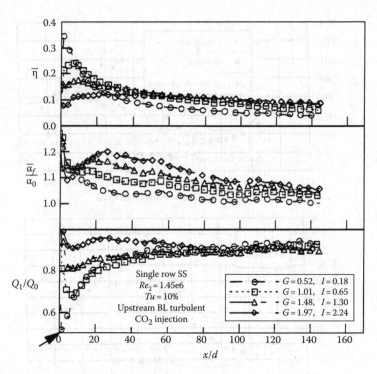

FIGURE 3.16
Film effectiveness (η), heat-transfer coefficient ratio ($\bar{a}_f/a_0$), and heat flux ratio for the SS single row configuration. (From Drost, U. and Bolcs, A., *ASME J. Turbomach.*, 121, 233, 1999.)

effectiveness. Heat flux reduction is greater near the hole for low blowing ratios due to high film effectiveness and low heat-transfer coefficient ratio. Further downstream, all blowing ratios produce similar heat flux reduction levels. The higher film effectiveness values offset the high heat-transfer coefficient ratios for high blowing ratios.

Similar data are shown for double-row film cooling on the suction surface in Figure 3.17. The variance of data is stronger for this case compared to the single-row case. The trends are similar to the single-row case for film effectiveness data. However, much higher effectiveness is observed for the high blowing ratios at downstream locations. This is due to coolant accumulation. Injecting large amount of coolant with subsequent rows causes a large accumulation of coolant downstream, causing higher effectiveness values. This also leads to higher heat-transfer coefficients. Heat-transfer coefficients are significantly higher for higher blowing ratios. The combined effects of film effectiveness and heat-transfer coefficient reveal interesting trends. At low blowing ratios, net heat flux reduction is significant near the holes and decreases (~1.0) rapidly downstream. However, high blowing ratios show a reversal of trends with lesser heat flux reduction values near the holes and increasing downstream,

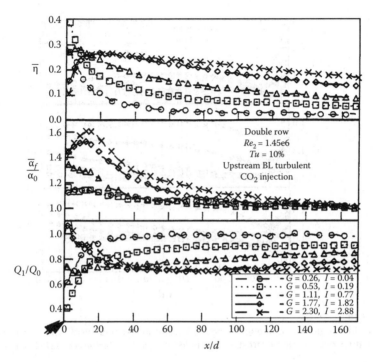

FIGURE 3.17
Film effectiveness (η), heat-transfer coefficient ratio ($\overline{a}_f/a_o$), and heat flux ratio for the SS double-row configuration. (From Drost, U. and Bolcs, A., *ASME J. Turbomach.*, 121, 233, 1999.)

because of the significantly high film effectiveness values far downstream at high blowing ratios due to jet separation and reattachment.

Figure 3.18 shows the pressure-side single-row injection results. The holes in this case are inclined streamwise and spanwise. Film effectiveness results show that highest effectiveness near the hole is observed for $G = 1.30$, which is not the lowest blowing ratio. Farther downstream, higher blowing ratio provides higher film effectiveness. The lowest blowing ratio of 0.58 provides lowest effectiveness over the entire surface. Heat-transfer coefficient trends are similar with higher blowing ratio providing higher values. The variance in values is significant far downstream than near the holes. This is due to reattachment of coolant jets and also the additional spanwise or lateral momentum of jets due to the spanwise angle. The heat flux ratio values are low at lower blowing ratios and increase with increasing blowing ratio. At lowest blowing ratio, the heat flux ratio is low although effectiveness is low, because the heat-transfer coefficient ratio appears to be around 1.0 over the entire surface. At low blowing ratios, the effect of film injection on heat-transfer coefficient is minimal. The results presented by Drost and Bolcs (1999) on a turbine airfoil indicate that the heat flux ratio is greater than 1.0 for blowing ratios $G > 2.54$.

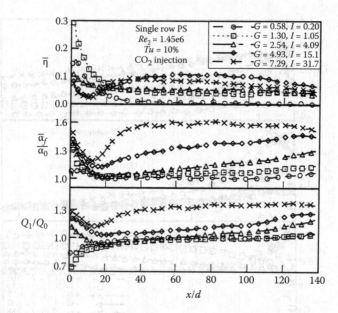

FIGURE 3.18
Pressure-side single-row, film effectiveness (η), heat-transfer coefficient ratio ($\bar{\alpha}_f/\alpha_0$), and heat flux ratio distributions. (From Drost, U. and Bolcs, A., *ASME J. Turbomach.*, 121, 233, 1999.)

3.3.3 Effect of Free-Stream Turbulence

Ames (1998a,b) studied film cooling on a turbine NGV airfoil (C3X vane) under high free-stream turbulence conditions. Figure 3.19 shows the effect of free-stream turbulence on pressure surface adiabatic effectiveness for different velocity ratios. The ratio of film effectiveness for the high turbulence case

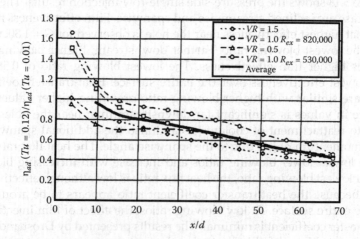

FIGURE 3.19
Influence of turbulence on relative level of pressure-surface adiabatic effectiveness, two rows, 30° holes, $DR = 0.94$, $P/D = 3.0$. (From Ames, F.E., *ASME J. Turbomach.*, 120, 777, 1998b.)

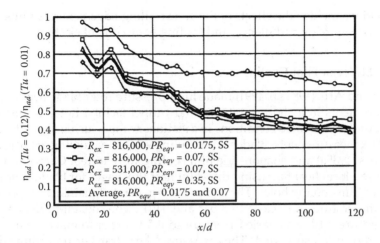

FIGURE 3.20
Influence of turbulence on relative level of suction-surface adiabatic effectiveness for shower-head array, $DR = 0.94$. (From Ames, F.E., *ASME J. Turbomach.*, 120, 777, 1998b.)

divided by value for the low turbulence case at same mainstream Reynolds number and velocity ratio is plotted versus normalized surface distance from edge of hole. Lower value of ratio indicates lower film effectiveness at high turbulence ($Tu \sim 12\%$). Turbulence has a significant effect on the pressure surface film effectiveness. At a velocity ratio of 1.0, the film effectiveness values for high turbulence are higher than for the low turbulence case near injection. However, effectiveness for higher and lower velocity ratios decreases with increase in free-stream turbulence levels. Far downstream, film effectiveness is significantly reduced due to free-stream turbulence. Figure 3.20 shows the effect of free-stream turbulence on suction-surface adiabatic effectiveness for showerhead cooling. Free-stream turbulence causes effectiveness to decrease for all cases at all locations. The higher blowing ratio (PR) case withstands high free-stream turbulence better than for low PR cases. Effectiveness drop far downstream of injection is significant.

3.4 Film Cooling on Cascade Blade Simulations

3.4.1 Introduction

Most experimental results for turbine blades were obtained on simulated cascades under simulated engine flow conditions. Most of these studies used the adiabatic wall effectiveness technique. However, a few studies have also reported results with the "superposition" approach. The cascade blade simulations include the effects of the vane trailing-edge-generated unsteady wakes, free-stream turbulence, and trailing-edge coolant ejection.

3.4.2 Effect of Film Cooling

Camci and Arts (1985a,b, 1991) made high-pressure rotor blade heat-transfer measurements in a short-duration wind tunnel facility. Rotor blades were tested under simulated conditions in a stationary mode using short-duration heat flux measurements. Blades with leading-edge film-cooling array (Camci and Arts, 1985a) and suction-side film cooling were tested. Camci and Arts measured local heat flux using thin-film heat flux gages for blades with and without film-cooling holes. The compressor-generated mainstream pressure and temperature were close to 3 bar and 410 K. The coolant-to-mainstream temperature ratio was varied from 0.4 to 0.7. Coolant-to-mainstream mass flow ratio was also varied. They measured heat transfer on a film-cooled blade with the "superposition" approach.

Figure 3.21 presents the data for a blade with leading-edge film holes. Heat-transfer distributions for $\dot{m}_c/\dot{m}_\infty = 0.6\%$ and $T_c/T_\infty = 0.52$ and 0.59. The figure also shows the film hole geometry. The three-film hole rows are designated middle (M), pressure (P), and suction (S). Analysis of blowing rates shows that the local coolant blowing ratio was 4.17 for row M, 1.26 for row P, and 0.85 for row S. The effect of introducing film cooling is the lowering of heat transfer into the turbine blade compared to levels obtained on the blade without film holes. The effect of film cooling in reducing heat-transfer levels is strongly evident in the figure. The high blowing ratio out of row M produces strong jets with high momentum compared to the mainstream. These jets penetrate the mainstream at the leading edge and create a separated flow region on suction surface upstream of row S. This causes a slight increase in the heat-transfer values on suction surface just upstream of row S. The coolant flow then reattaches downstream of row S and causes reduced heat-transfer levels. This is due to combination of the reattached jets from row M and the new coolant jets from row S. Heat-transfer levels increase gradually downstream and reach uncooled levels close to trailing edge. The cooling efficiency increases with decreasing T_c/T_∞, which is as expected. When cooler flow is injected, the blade receives more protection downstream of injection. In that case, lower heat-transfer levels are expected.

Figure 3.22 presents the effect of suction-side film cooling on heat transfer for different T_c/T_∞ levels at two different coolant-blowing rates. The figure also shows the airfoil suction-side cooling geometry. Significant heat-transfer reductions are observed in the region for $X/D < 40$ for both blowing ratios of 0.4 and 0.96. Further downstream, the effects are reduced, and the heat-transfer levels approach uncooled levels. However, the film-cooling effect is still significant at $X/D = 100$ for $M = 0.96$. Also, the effect of temperature ratio is also stronger at higher blowing ratio. The results indicate that at low blowing ratios, the effect of temperature ratio is smaller compared to that for $M = 0.96$.

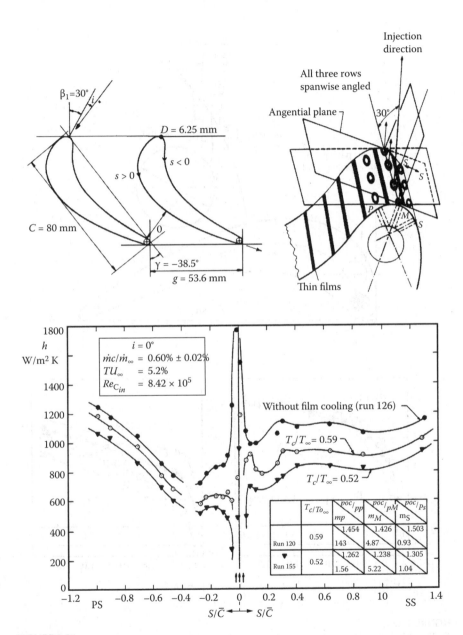

FIGURE 3.21
Cooling configuration and leading-edge heat transfer with film cooling, effect of mass weight ratio. (From Camci, C. and Arts, T., *ASME J. Eng. Gas Turbines Power*, 107, 991, 1985a.)

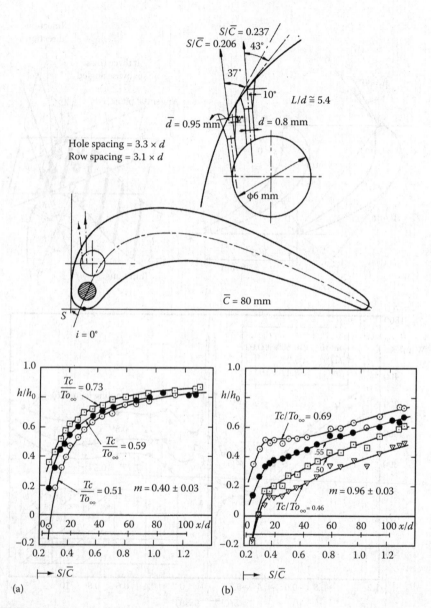

FIGURE 3.22
Turbine blade cooling geometry and effect of temperature ratio on suction-side heat transfer with film cooling, (a) $m = 0.4$ and (b) $m = 0.96$. (From Camci, C. and Arts, T., *ASME J. Eng. Gas Turbines Power*, 107, 1016, 1985a.)

3.4.3 Effect of Free-Stream Turbulence

Mehendale et al. (1994a) studied the effect of high free-stream turbulence on a turbine rotor blade. They used grids upstream of a five-blade linear cascade to generate high free-stream turbulence. Figure 2.47 (see Chapter 2) shows a schematic of the linear turbine blade cascade and geometry. Since there was no relative motion of airfoil simulated, the results may be applicable to NGV airfoil situations. Figure 3.23 shows the airfoil with film hole locations. The airfoil had three showerhead rows, and two rows each on the pressure and suction sides. The film hole configuration including the streamwise location, diameter, length, spanwise spacing, and compound angle were all typical of a modern film-cooled gas turbine engine blade. Some details of the film hole are shown in the figure. The blade cascade was scaled up five times for the study. Mehendale et al. used thermocouples to measure surface temperatures during a steady-state surface-heating test. They simulated four different turbulence levels of 0.7%, 5.0%, 13.4%, and 17% by using a combination of grid size and grid location from the cascade. They also used air and CO_2 to simulate different coolant densities. CO_2 was 1.5 times heavier than the mainstream air.

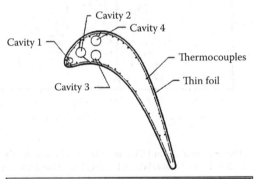

Film hole Row location	P/D	Axial angle	Radial angle	Tangential angle
Leading edge (all three rows)	7.31	90°	27°	–
Cavity 2 pressure side	6.79	–	32°	55°
Cavity 2 suction side	4.13	–	90°	45°
Cavity 3 pressure side	5.00	–	35°	50°
Cavity 4 suction side	5.71	–	90°	30°

FIGURE 3.23
Schematic of the turbine blade with film holes. (From Mehendale, A.B. et al., *Int. J. Heat Mass Transfer,* 37, 2707, 1994a.)

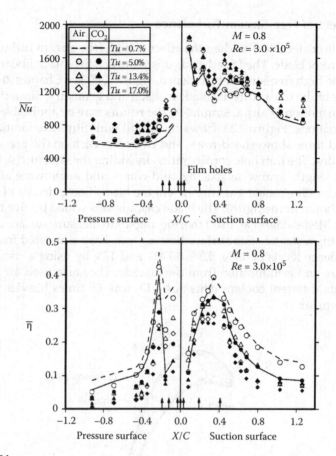

FIGURE 3.24

Effect of mainstream turbulence and air and CO_2 injection on Nusselt number and film effectiveness distributions for $M = 0.8$. (From Mehendale, A.B. et al., *Int. J. Heat Mass Transfer*, 37, 2707, 1994a.)

Figure 3.24 shows the effect of free-stream turbulence on Nusselt number (hC/k) and film effectiveness for a coolant-blowing ratio of 0.8. Results for low free-stream turbulence at $Tu = 0.7\%$ are shown as lines, and higher free-stream turbulence cases are shown as symbols. In general, Nusselt numbers increase and film effectiveness values decrease with increasing free-stream turbulence. The effect on Nusselt number is not as significant as the effect on film effectiveness. Film injection by itself significantly enhances Nusselt number on the blade without any free-stream turbulence. Further addition of free-stream turbulence on a film-cooled blade causes only slight enhancements of heat-transfer coefficient.

3.4.4 Effect of Unsteady Wake

As described in Chapter 2, downstream turbine blades are affected by unsteady wakes shed by upstream vane trailing edges. Wake simulation

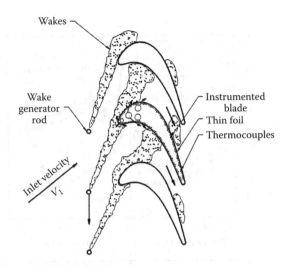

Wakes

Wake
generator
rod

Inlet velocity V_1

Instrumented
blade

Thin foil

Thermocouples

FIGURE 3.25
Conceptual view of unsteady wake effect on a model blade with film holes.
(From Mehendale, A.B. et al., *ASME J. Turbomach.*, 116, 730, 1994b.)

experiments focus on the effects of unsteady wakes on downstream film-cooled turbine blades. The blades are stationary, and upstream rotating rods simulate the unsteady wakes. The spoked-wheel-type wake generator is similar to the one described in Chapter 2. Ou et al. (1994) and Mehendale et al. (1994b) presented the effect of unsteady wake on downstream blade heat-transfer coefficient and film effectiveness, respectively. A typical advanced high-pressure turbine blade was chosen for the study. The selected blade had three rows of film cooling on the leading edge and two rows each on the pressure and suction surfaces. Figure 3.25 illustrates the film-cooled turbine blade in a cascade under unsteady wake effects. Wake Strouhal numbers $[S = 2\pi N dn/(60V_1)]$ were simulated to be consistent with real engine; values [wake generator speed (N), wake rod diameter (d), number of rods (n), and upstream mainstream velocity (V_1)]. Effects of coolant-to-mainstream blowing ratio and density ratio were also investigated. In the real engine, coolant is at higher pressure and lower temperature than the mainstream, and hence at a higher density. This makes it necessary to simulate coolant density effect in film-cooling experiments.

Figure 3.26 shows the heat-transfer coefficient, film effectiveness, and heat load ratio results for a low blowing ratio of $M = 0.4$. Two coolant density ratios of 1.0 and 1.5 were tested using air and CO_2 as the coolants. Four conditions were compared for an uncooled blade without wake, cooled blades without wake, and wake Strouhal numbers of 0.1 and 0.3. It is evident that film injection by itself significantly enhances heat-transfer coefficients on the blade surface. Film injection increases local turbulence and causes earlier boundary layer transition on the suction surface. Unsteady wake effect

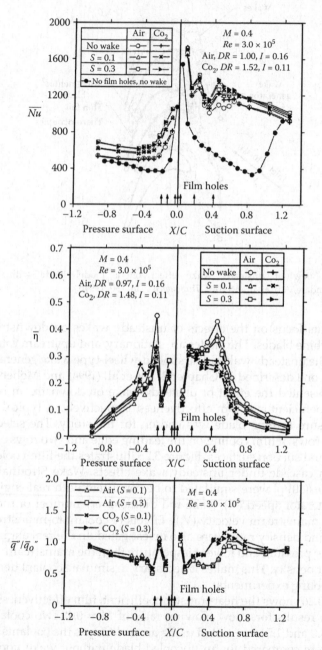

FIGURE 3.26
Effect of unsteady wake on heat-transfer coefficient, film effectiveness, and heat load ratio for
$M = 0.4$. (From Ou, S. et al., *ASME J. Turbomach.*, 116, 721, 1994; Mehendale, A.B. et al., *ASME J. Turbomach.*, 116, 730, 1994b.)

further enhances heat-transfer coefficients. However, film injection effect dominates the unsteady wake effect. The higher-density coolant causes higher heat-transfer coefficients on the suction surface. On the pressure surface, the lower-density coolant appears to provide higher heat-transfer coefficients. This dissimilarity has to be due to the curvature of each side. The pressure side is concave; the suction side is convex. We will discuss curvature effects later in this chapter. The film injection effect on the pressure side is not as strong as on the suction side. Addition of unsteady wake enhances heat-transfer coefficients on the pressure surface. Film effectiveness values are highest for both coolants without any wake. However, unsteady wake in the free stream causes reduction in film coverage. Unsteady wakes cause coolant jet structures to break down and reduce film coverage. Air as coolant provides higher effectiveness than CO_2 for $M = 0.4$ at all given free-stream conditions. The figure also presents the heat load ratio comparing heat flux with film cooling to that without film cooling. A value below 1.0 indicates reduced heat load to the blade. Film injection generally reduces heat load to the blade due to significantly high film effectiveness. Regions with high heat-transfer coefficients and low film effectiveness as on the suction surface show higher heat load ratio values than the entire surface. As film injection effects are not significant near the trailing edge of the blade, the heat load ratio is typically 1.0.

Figure 3.27 shows similar results for a high blowing ratio of 1.2. The heat-transfer coefficient distributions are similar at $M = 1.2$ as in the case of $M = 0.4$. Film effectiveness levels, however, do not seem to be affected by unsteady wake as was the case for $M = 0.4$. At higher blowing ratios, the coolant jets possess higher momentum and thus can sustain strong free-stream effects such as unsteadiness. The jet structures withstand the unsteady effects farther downstream and produce effectiveness levels similar to the no-wake case. The heat flux ratio distribution clearly shows that neither the unsteady wake nor the coolant density has a very significant effect on the heat load ratios.

Du et al. (1998) presented detailed surface heat-transfer distributions on a similar blade in the same cascade setup used by Mehendale et al. (1994) and Ou et al. (1994). They used a transient liquid crystal technique to provide detailed heat-transfer coefficient and film effectiveness distributions. Figure 3.28 shows the detailed Nusselt number distributions on the entire blade surface for an uncooled blade without wake, uncooled blade with $S = 0.1$, cooled blade with $M = 0.8$, and a cooled blade with $M = 0.8$ and $S = 0.1$. The detailed distributions on the suction side clearly show the advancement of transition location toward the leading edge due to unsteady wake and film injection as in cases 2 and 3. The combined effect of unsteady wake and film injection (case 4) does not show any further enhancement of heat-transfer coefficient over case 3. On the pressure surface, heat-transfer coefficients are significantly enhanced for cases 2 and 3 over case 1. The combined effect of unsteady wake and film injection further enhances heat-transfer coefficients on the pressure side.

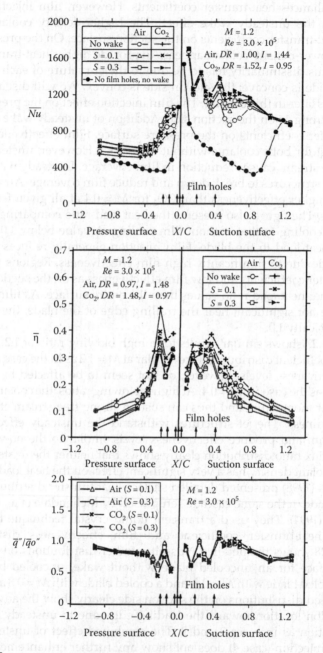

FIGURE 3.27
Effect of unsteady wake on heat-transfer coefficient, film effectiveness, and heat load ratio for $M = 1.2$. (From Ou, S. et al., *ASME J. Turbomach.*, 116, 721, 1994; Mehendale, A.B. et al., *ASME J. Turbomach.*, 116, 730, 1994b.)

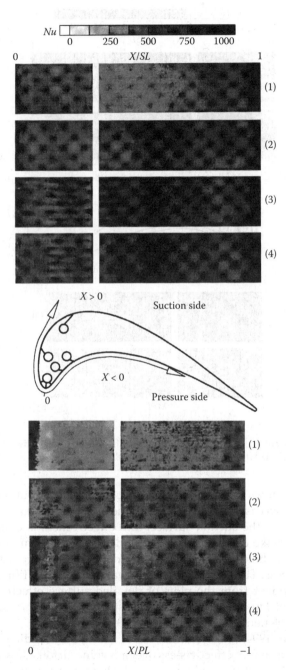

FIGURE 3.28
Detailed Nusselt number distributions showing the effect of unsteady wake for $M = 0.8$. (From Du, H. et al., *ASME J. Turbomach.*, 120, 808, 1998.)

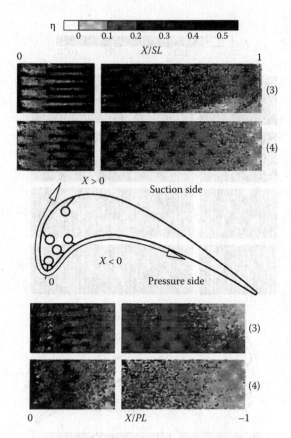

FIGURE 3.29
Detailed film effectiveness distributions showing the effect of unsteady wake for $M = 0.8$. (From Du, H. et al., *ASME J. Turbomach.*, 120, 808, 1998.)

Figure 3.29 shows the detailed film effectiveness distributions for cases 3 and 4. These two cases directly compare the effect of unsteady wake on film effectiveness. On the suction surface, film effectiveness is clearly lower for the unsteady wake case. This is most evident downstream of the first suction-side row of holes. On the pressure side, there is a smaller decrease in film effectiveness. The effect of unsteady wake to decrease film effectiveness is strongly evident from the plots of detailed surface effectiveness. These observations of Du et al. (1998) are consistent with the thermocouple data presented by Mehendale et al. (1994) and Ou et al. (1994).

Ou and Han (1994) and Heidmann et al. (1997) also studied the effects of unsteady wake on film effectiveness over a turbine airfoil. Ou and Han (1994) studied the effects of unsteady wake on the heat-transfer characteristics of a blade with only one row of film hole each on the pressure and suction sides. Heidmann et al. (1997) studied the effects of unsteady wake on blade surface heat transfer similar to the other studies.

3.4.5 Combined Effect of Free-Stream Turbulence and Unsteady Wakes

Ekkad et al. (1997a) studied the combined effect of unsteady wake and free-stream turbulence on blade surface heat transfer with film cooling. They concluded that superimposing grid turbulence on unsteady wakes in a free-stream produced higher surface heat-transfer coefficients and lower film effectiveness compared to the individual effects of free-stream turbulence or unsteady wake. The variance of effect depended on the strength of free-stream turbulence or unsteadiness in the oncoming mainstream and also on the coolant-blowing ratio. Du et al. (1999) studied the effect of the combination of trailing-edge ejection with unsteady wake on the downstream blade surface heat transfer. They concluded that additional effect of trailing-edge ejection was negligible on a film-cooled blade already affected by unsteady wakes and free-stream turbulence. Even if the effect existed, the increase was minor compared to other major effects such as unsteady wake or film injection or free-stream turbulence.

3.5 Film Cooling on Airfoil Endwalls

3.5.1 Introduction

Endwall film cooling has gained significant importance due to the usage of low-aspect ratio and low-solidity turbine designs. Film cooling and associated heat transfer are strongly influenced by the secondary flow effects. Figure 3.30 shows the conceptual view of the three-dimensional (3-D) flow field inside a first-stage vane passage (also explained in Chapter 2). The location of film holes and their cooling effectiveness is strongly influenced by the presence of these secondary flows. Locating film-cooling holes for cooling on endwalls requires better understanding of secondary-flow behavior and associated heat transfer. To that end, heat-transfer researchers have focused on several film-cooling configurations for endwall film cooling. Two of the most common configurations are discussed in this section. Several studies have focused on endwall flow behavior and heat transfer, as discussed in Chapter 2. Cooling on endwalls has gained importance with increasing turbine inlet temperature and flat inlet temperature profiles, making it a requirement.

3.5.2 Low-Speed Simulation Experiments

Studies by Georgiou et al. (1979) and Dunn and Stoddard (1979) focused on the turbine endwall heat transfer, coupled with some effects of cooling from the blade surface near the endwall. However, near-endwall cooling is not effective in the midpassage region where the hot gases may be most

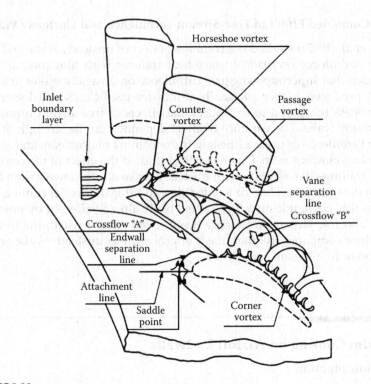

FIGURE 3.30
3-D endwall vortex pattern as described by Takeishi et al. (1990). (From Takeishi, K. et al., *ASME J. Turbomach.*, 112, 488, 1990.)

detrimental. Takeishi et al. (1990) studied the effect of endwall film cooling in a low-speed open circuit fully annular cascade with 13 vanes having an aspect ratio of $h/c = 0.5$. They had film-cooling holes on the model vanes and also on the inner and outer endwalls. Film-cooling holes were placed at three locations on the endwall. The first location was at the leading edge, and the other two locations are downstream near the throat of the nozzle. Figure 3.31 shows the film-cooling geometry on the outer endwall of the vane passage. Takeishi et al. (1990) measured film effectiveness distributions for all three locations at coolant blowing ratios of 0.5–2.5 using heater foils and surface-mounted thermocouples.

Figure 3.32 presents the film-cooling effectiveness distributions on the outer endwall for the leading-edge-row film cooling. Film effectiveness is highest immediately downstream of injection close to the pressure surface of the vane. Film effectiveness is low in the middle of the passage close to suction surface leading edge. This is consistent with the secondary flow structures. Effectiveness values are around 0.1 in the downstream region. However, it appears that the coolant flow migrates from the pressure-surface leading-edge region of one vane to the suction-side trailing-edge of the adjacent vane. This causes the typical film effectiveness distributions shown in

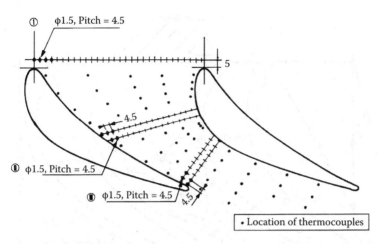

FIGURE 3.31
Film-cooling geometry on the outer endwall. (From Takeishi, K. et al., *ASME J. Turbomach.*, 112, 488, 1990.)

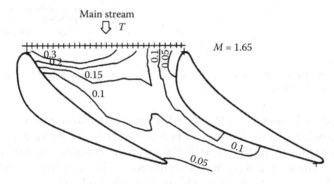

FIGURE 3.32
Film-cooling effectiveness distribution on the outer endwall. (From Takeishi, K. et al., *ASME J. Turbomach.*, 112, 488, 1990.)

the figure. Downstream film hole rows are added to further increase the film effectiveness and also reduce the secondary flow effects on the cooling jets.

Jabbari et al. (1996) chose a film-cooling hole configuration based on the flow behavior along the endwall. They designed a hole pattern using a combination of flow visualization results and heat-transfer measurements on uncooled endwalls. Figure 3.33 shows the endwall-cooling pattern with film hole ejection directions. Also shown is the sampling tap and injection hole row notation. A total of 14 discrete holes were used to cool the endwall. Effectiveness was measured at discrete points based on the sampling locations and presented in the study.

Jabbari et al. (1996) studied the effect of blowing rate, density ratio, and mainstream Reynolds numbers on film effectiveness distributions. Figure 3.34

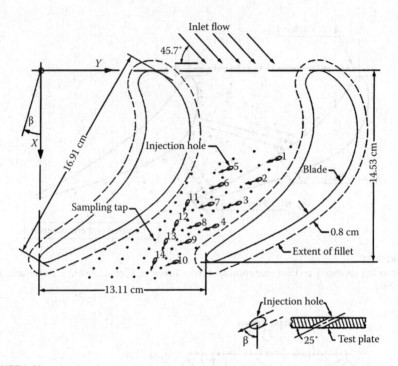

FIGURE 3.33
Endwall cooling pattern as used by Jabbari et al. (1996). (From Jabbari, M.Y. et al., *ASME J. Turbomach.*, 118, 278, 1996.)

shows the visualization of jet traces on the endwall for different coolant conditions. The mainstream velocity field is 3-D, and the velocity varies in magnitude and direction along the endwall. Each jet encounters different mainstream velocity, and the jet trajectory at exit depends on this mainstream velocity and direction. The traces clearly show the jet direction and, to an extent, the magnitude of the jet velocity component. The first two traces compare the effect of blowing ratio ($M = 1.0$ vs. $M = 2.0$) for the same coolant density ratio of $R = 2.0$. The third trace is for a blowing ratio of $M = 2.0$ and a lower density ratio of $R = 0.93$. The traces of the film-cooling jets were obtained using the ammonia-diazo visualization technique. The traces clearly indicate the direction and strength of the coolant jet. Comparing the first two cases for the effect of blowing ratio, it is evident that the lower blowing ratio is more effective in the middle of the passage and the higher blowing ratio jets are more effective in the region close to the suction surface. The jets all appear to be blowing from the pressure side toward the suction surface. The secondary flow structures clearly show the movement of the pressure-side vortex moving across the passage toward the suction side. This passage vortex deflects the jets toward the suction side. Further downstream, the passage vortex does not exist, and the jets appear to be directed along the mainstream. The third trace can be compared to the second trace for density ratio effects. The lower-density coolant

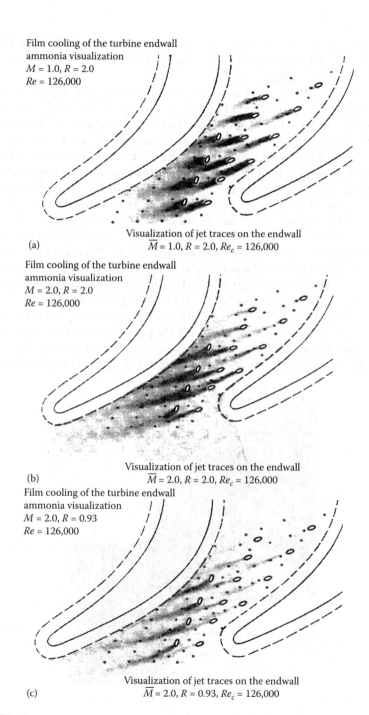

Film cooling of the turbine endwall
ammonia visualization
$M = 1.0, R = 2.0$
$Re = 126,000$

(a)

Visualization of jet traces on the endwall
$\overline{M} = 1.0, R = 2.0, Re_c = 126,000$

Film cooling of the turbine endwall
ammonia visualization
$M = 2.0, R = 2.0$
$Re = 126,000$

(b)

Visualization of jet traces on the endwall
$\overline{M} = 2.0, R = 2.0, Re_c = 126,000$

Film cooling of the turbine endwall
ammonia visualization
$M = 2.0, R = 0.93$
$Re = 126,000$

(c)

Visualization of jet traces on the endwall
$\overline{M} = 2.0, R = 0.93, Re_c = 126,000$

FIGURE 3.34
Visualization of jet traces on the endwall for $Re_c = 126,000$: (a) $M = 1.0$, $R = 2.0$; (b) $M = 2.0$, $R = 2.0$; and (c) $M = 2.0$, $R = 0.93$. (From Jabbari, M.Y. et al., *ASME J. Turbomach.*, 118, 278, 1996.)

($R = 0.93$) is not as effective as the higher-density coolant ($R = 2.0$). Higher-density coolant does not lift off and protects the surface more effectively than the lower-density coolant.

Friedrichs et al. (1996) presented adiabatic film-cooling effectiveness distributions on the endwall of a large-scale, low-speed linear turbine cascade using the ammonia-diazo technique. The endwall test surface is coated with a layer of diazo film. The cooling air is seeded with ammonia gas and water vapor. The ammonia gas reacts with the diazo coating, leaving traces of varying darkness on the test surface. An optical scanner was used to obtain a calibration curve between the darkness level and relative concentration. Thus, the adiabatic film-cooling effectiveness is quantified using the heat/mass transfer analogy (see Chapter 6). Figure 3.35 shows a typical trace of surface coating on a film-cooled endwall. The film hole locations are clearly evident, and the traces downstream of the endwall show different darkness levels. Figure 3.36 shows the endwall cooling effectiveness distribution measured from the traces in Figure 3.35. This figure shows that cooling occurs immediately downstream of holes and upstream of liftoff lines indicated by dark dashed lines. The levels of effectiveness increase downstream into the passage. The horseshoe vortex from the pressure side moving across the passage pushes most of the coolant from the middle of the passage toward the suction surface. The coolant jets closer to the suction surface show little

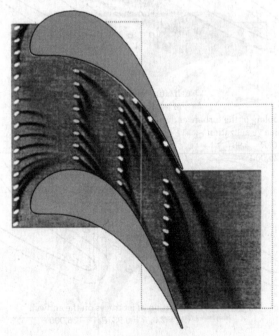

FIGURE 3.35
Traces on the diazo surface coating on the turbine endwall. (From Friedrichs, S. et al., *ASME J. Turbomach.*, 118, 613, 1996.)

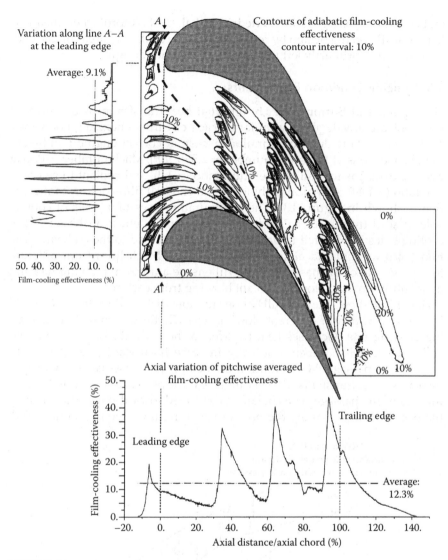

FIGURE 3.36
Endwall film-cooling effectiveness distributions as measured by Friedrichs et al. (1996). (From Friedrichs, S. et al., *ASME J. Turbomach.*, 118, 613, 1996.)

spreading and persist longer distances downstream. This may be, as indicated, due to accelerated flow in this region. A large region exists on the pressure surface immediately downstream of the leading edge where there is no cooling. This area as indicated in Chapter 2 is a high heat-transfer area. The addition of cooling in this area is assessed as a possible solution to this problem. Figure 3.36 also shows the local variation along the passage and across the passage. The integrated average along the passage is 12.3% and that across the passage is 9.1%. Friedrichs et al. (1998) improved on their earlier endwall-cooling design by redistributing the coolant to areas on the endwall that were primarily left uncooled in the earlier design.

3.5.3 Engine Condition Experiments

Harasgama and Burton (1992a,b) reported heat-transfer measurements on film-cooled endwalls of an annular cascade of turbine nozzle guide vanes. They used a short duration annular cascade at correct engine nondimensional conditions of Reynolds number (= 2.55×10^6), Mach number (= 0.93 at cascade exit), gas-to-wall temperature ratio (= 1.3), and coolant-to-gas density ratio (= 1.80). A typical turbine engine endwall film-cooling configuration is shown in Figure 3.37. The configuration uses four rows of cooling holes placed from blade to blade. The downstream effect of these rows of cooling holes is expected to provide reduced heat load to the endwalls. The film holes were placed along the measured iso-Mach lines from pressure to suction side of the passage. This will ensure a uniform blowing rate and momentum flux ratio for the coolant blowing from each row.

Figure 3.38 shows the endwall Nusselt number distributions for $\theta = (T_g - T_c)/(T_g - T_w) = 1.62$ at four different blowing rates (G). Results show that increasing blowing rate decreases heat transfer to the endwall. For the uncooled endwall, Nusselt numbers gradually increase from leading edge to trailing edge of passage. With addition of coolant, the cross-passage variation increases. Heat transfer is still very high near the pressure surface at the trailing end of the passage. The coolant is convected toward the suction side by the passage vortex leaving the pressure-side trailing-edge region uncooled.

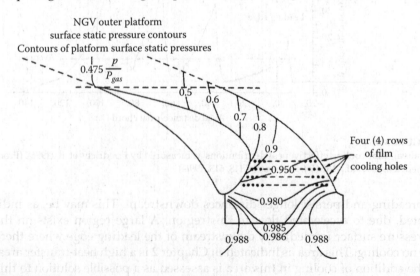

FIGURE 3.37
Typical engine endwall cooling. (From Harasgama, S.P. and Burton, C.D., *ASME J. Turbomach.*, 114, 734, 1992a.)

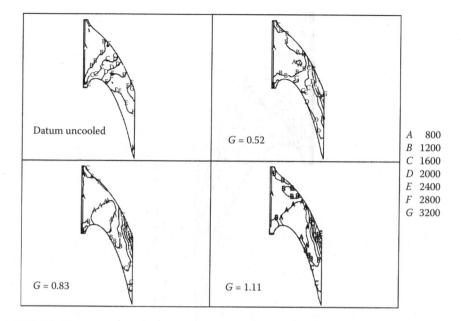

Datum uncooled	$G = 0.52$
$G = 0.83$	$G = 1.11$

A	800
B	1200
C	1600
D	2000
E	2400
F	2800
G	3200

FIGURE 3.38

Endwall Nusselt number distribution for θ = 1.62 for all blowing rates. (From Harasgama, S.P. and Burton, C.D., *ASME J. Turbomach.*, 114, 734, 1992a.)

Increased blowing rate provides slightly better heat-transfer rates. However, this region is significantly undercooled compared to the midpassage region and the suction-surface corners. The airfoil-to-airfoil variation is caused by the cross-passage pressure gradient, which produces the passage secondary flow vortex.

Figure 3.39 presents the ratio of cooled to uncooled Nusselt numbers for $G = 1.11$ and θ = 1.62. It is fairly evident that the entire endwall passage is cooled varyingly. There is no cooling effect near the pressure-side leading edge where the ratios are nearly 1.0. Most of the passage has reduced heat-transfer levels up to 60%. The pressure-side trailing edge is undercooled, with ratios of 80% indicating a cooling efficiency of only 20%.

3.5.4 Near-Endwall Film Cooling

The influence of the endwall on film cooling of gas turbine blades is important for holes on the blade close to the endwall. The complex flow near the endwall may have a significant impact on the film-cooling characteristics of film injection from holes very close to the endwall corner. Goldstein and Chen (1985) studied the effect of endwall on a single row of holes on a blade. Figure 3.40 shows the test blade and the sampling locations for film effectiveness measurements close to the endwall. The measurement regions are also shown for different locations along the span of the blade. The mass/heat-transfer analogy was used to determine the film-cooling effectiveness.

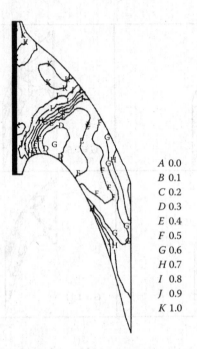

A	0.0
B	0.1
C	0.2
D	0.3
E	0.4
F	0.5
G	0.6
H	0.7
I	0.8
J	0.9
K	1.0

FIGURE 3.39
Ratio of cooled to uncooled Nusselt number for $G = 1.11$ and $\theta = 1.44$. (From Harasgama, S.P. and Burton, C.D., *ASME J. Turbomach.*, 114, 734, 1992a.)

The injected flow consists of helium as a tracer gas, and samples are measured near wall to measure the concentration of helium after the coolant mixes with the mainstream air. The ratio of local concentration to concentration of the original injected coolant provides the local film effectiveness (see Chapter 6).

Figure 3.41 presents the average film effectiveness on the convex surface (suction) at a blowing rate $M = 0.5$. The film effectiveness is plotted with normalized spanwise distance at difference normalized streamwise locations. A location $H/D = 0$ indicates the endwall corner with the blade surface. A small X/D indicates a streamwise location closer to the injection hole. It is evident from the figure that film effectiveness increases from endwall (small H/D) to midspan (larger H/D) of the airfoil for all stream-wise locations. Effectiveness near the endwall is significantly lower than the midspan value. This explains why the secondary flow generated by endwall: interactions reduces film effectiveness for holes close to the endwall corner. The downstream effect is even more significant, with the effectiveness almost nonexistent near the endwall. The film is either swept upward by the passage vortex or gets broken down by the complex interactions of the vortices near the suction surface.

Figure 3.42 presents the average film effectiveness on the concave surface (pressure) at $M = 0.8$. The trends are very different for the concave surface. At small X/D, film effectiveness is lowest near the endwall and immediately increases to a high value and decreases toward the midspan region.

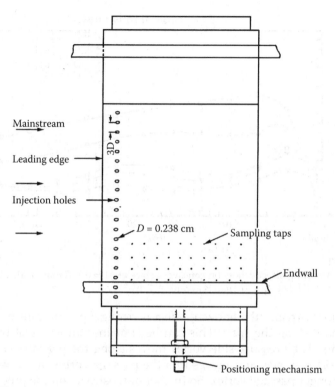

FIGURE 3.40
Test blade (suction-side view) used by Goldstein and Chen (1985). (From Goldstein, R.J. and Chen, H.P., *ASME J. Eng. Gas Turbines Power*, 107, 117, 1985.)

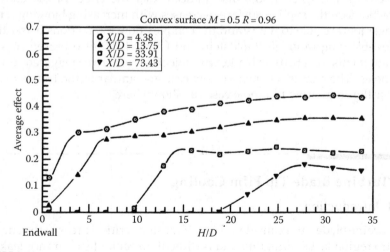

FIGURE 3.41
Average film-cooling effectiveness on convex surface at $M = 0.5$. (From Goldstein, R.J. and Chen, H.P., *ASME J. Eng. Gas Turbines Power*, 107, 117, 1985.)

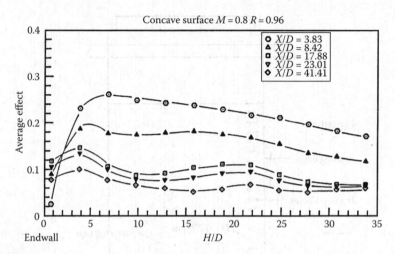

FIGURE 3.42
Average film-cooling effectiveness on concave surface at $M = 0.8$. (From Goldstein, R.J. and Chen, H.P., *ASME J. Eng. Gas Turbines Power*, 107, 117, 1985.)

Further downstream, the endwall effect is negligible, with almost uniform effectiveness along the span. This can be explained in terms of the vortex movements. The pressure-side vortex moves along the pressure surface for only a short distance, and moves into the passage vortex and combines to form a stronger passage vortex. So further downstream on the pressure surface, there is no vortex interaction with the film-cooling jets. The passage vortex then moves toward the suction surface and hugs the suction surface. So the pressure surface film cooling is much less affected by endwall interactions.

Goldstein and Chen (1985) also indicate that the effect of the endwall interactions on the suction surface decreases with increasing blowing ratios. Stronger jets are able to withstand the complex flow interactions better than the low-blowing ratio jets. Goldstein and Chen (1987) also presented such measurements for a turbine blade, with injection from two staggered rows of film holes. The results for the two-row case are similar to the trends exhibited for the single-row film-hole cases as shown here.

3.6 Turbine Blade Tip Film Cooling

3.6.1 Introduction

The turbine blade tip is considered one of the most critical areas of the engine. The tip region lacks durability and is difficult to cool. The clearance leakage flow across the tip from the pressure side to suction side is the cause of blade

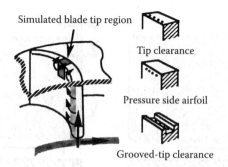

Simulated blade tip region

Tip clearance

Pressure side airfoil

Grooved-tip clearance

FIGURE 3.43
Simulated blade tip models used by Kim et al. (1995). (From Kim, Y.W. et al., *ASME J. Turbomach.*, 117, 1, 1995.)

tip failure. A detailed explanation of blade tip leakage flow and associated heat transfer was presented in Chapter 2. A summary of cooled-tip heat-transfer investigations by Dr. Metzger (Kim and Metzger, 1995; Kim et al., 1995) was published a few years ago. This study presents a comparison of various tip-cooling configurations and their effects on film effectiveness and heat-transfer coefficient.

Figure 3.43 presents the modeled situation used by Metzger et al. to simulate blade tip leakage flow with cooling. A forced flow through a thin clearance gap was used to simulate the blade tip leakage flow. Film coolant can be injected from the blade tip surface, from the airfoil pressure-side face near the tip, or from a grooved-tip cavity (squealer). The simulated blades were made of acrylic material and coated with a thin liquid crystal layer. A transient liquid crystal technique was used to measure the heat-transfer coefficient and film effectiveness distributions on the tips. Figure 3.44 shows four tip-cooling geometries used in the study. The first cooling configuration is called discrete slot, the second is called round hole, the third is called the pressure-side flared hole, and the last one is round hole on a grooved tip. They varied the mass flow of film-cooling flow with reference to the tip leakage flow for each cooling configuration.

3.6.2 Heat-Transfer Coefficient

Figure 3.45 shows the effect of mass flow ratio ($R = \dot{m}_f/\dot{m}_m$) on the Nusselt number for the discrete-slot, round-hole, and pressure-side flared-hole injection. All cases except for the highest mass flow ratio show a classical sharp-edged channel entrance distribution for the discrete slot injection. At the low mass flow ratio of 0.025, the effect of injection is negligible on the Nusselt number distribution. However, the values suddenly increase for $R = 0.074$. A further increase in mass flow ratio only causes the flow reattachment location to

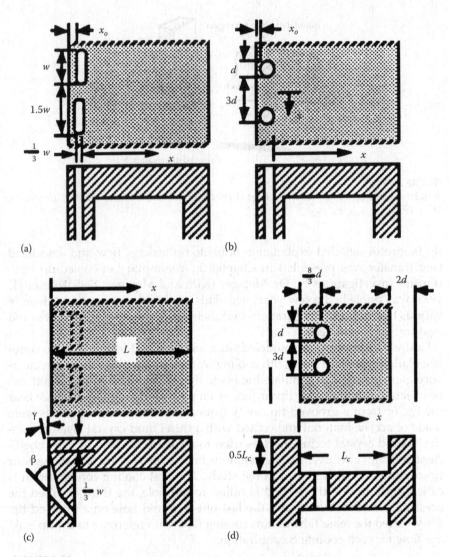

FIGURE 3.44

Tip-cooling geometries studied by Kim et al. (1995). (a) Discrete slot, (b) round hole, (c) pressure hole, and (d) grooved tip. (From Kim, Y.W. et al., *ASME J. Turbomach.*, 117, 1, 1995.)

move on to the test surface. For the round-hole injection, it is fairly evident that the coolant-to-leakage mass flow ratio has no effect on the heat-transfer coefficient on the blade tip compared to a blade tip with no leakage. Heat-transfer coefficient decreases with increasing distance from pressure side to suction side. The figure also shows the effect of mass flow ratio for the pressure-side flared-hole injection. Nusselt number variations are significant for the flared-hole injection. The Nusselt numbers increase with increasing

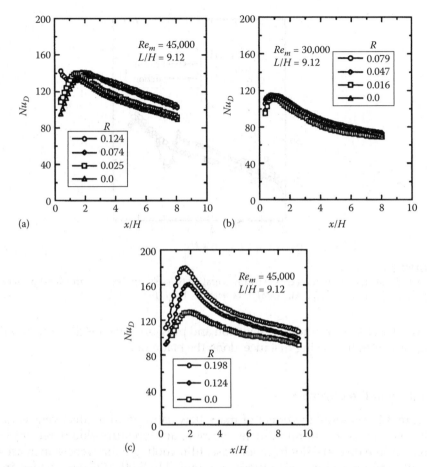

FIGURE 3.45
Effect of mass flow ratio on film-cooled Nusselt number for (a) discrete-slot, (b) round-hole, and (c) pressure-side flared-hole injection. (From Kim, Y.W. et al., *ASME J. Turbomach.*, 117, 1, 1995.)

mass flow ratio at all locations on the tip region. This may be due to increased local interaction between the coolant jets and the leakage flow. The leakage flow may be pushed away from the tip surface, causing complex flow fields compared to the round-hole injection case.

Figure 3.46 shows the grooved-tip results. In the grooved tip, the effect of varying injection hole row location was studied. The comparison was made with one case where the holes were closer to the pressure side and one where the holes were closer to the suction side. Nusselt number distributions show very small variations between the two cases, except in the immediate vicinity of injection hole location. The dips in the distribution for both cases indicate the hole location. In the grooved-tip case, the Nusselt numbers are

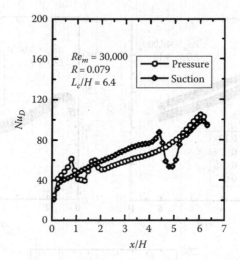

FIGURE 3.46
Effect of injection location on cross-stream averaged Nusselt numbers for grooved-tip cavity injection. (From Kim, Y.W. et al., *ASME J. Turbomach.*, 117, 1, 1995.)

lower than for the round-hole injection and pressure-side injection cases and increase with increasing distance along the blade tip.

3.6.3 Film Effectiveness

Figure 3.47 presents the effect of mass flow ratio on film effectiveness for discrete-slot injection, round-hole injection, and pressure-side flared injection. For the discrete-slot injection case, film-cooling effectiveness increases significantly with increasing mass flow ratio. High effectiveness values are obtained immediately downstream of the slots and decrease rapidly downstream. For the round-hole injection, low mass flow ratio produces low film effectiveness even immediately downstream of the injection hole. Slots cover more of the test section span, and this produces higher laterally averaged effectiveness for the slot injection case. However, a significant increase in film effectiveness is recorded for a mass flow ratio of 0.047 compared to 0.016. Further increase in mass flow ratio does not produce any appreciable increase in film effectiveness. This may be because at a low mass flow ratio of 0.016, the coolant flow is dominated by the leakage flow. However, with an increase in coolant flow for the same leakage flow, the coolant flow stays closer to the tip surface and pushes the leakage flow away, producing higher effectiveness. Further increase in coolant flow may not cause any further changes in the flow field to produce increases in film effectiveness. The right hand side of the figure shows the results for the pressure-side flared injection. The effect of mass flow ratio is insignificant for the two mass flow ratios shown here. The effect may be similar for lower mass flow ratios for this case

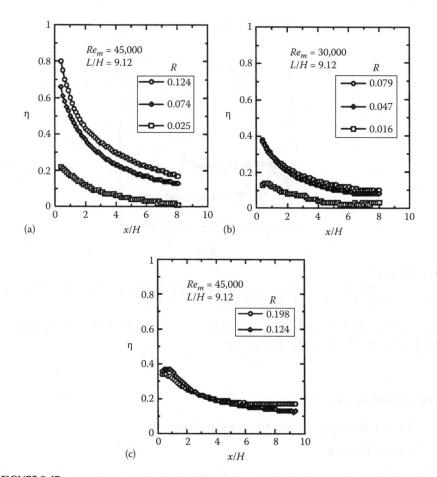

FIGURE 3.47
Effect of mass flow ratio on film-cooling effectiveness for (a) discrete-slot, (b) round-hole, and (c) pressure-side flared hole injection. (From Kim, Y.W. et al., *ASME J. Turbomach.*, 117, 1, 1995.)

as was in the round-hole injection. However, the authors did not study lower mass flow ratios for this case.

Figure 3.48 presents the effect of injection location on film effectiveness for the grooved-tip injection. The trend clearly shows that suction-side injection inside the grooved tip produces higher film effectiveness than the pressure-side injection. It may be due to the cavity flow effect where the coolant is pushed on to the tip surface by the reattaching leakage flow. The authors however recommend further flow and heat-transfer studies to confirm their understanding from the results.

The study by Kim et al. (1995) does not simulate the actual leakage flow characteristics required by engine designers. They used a 2-D rectangular model to simulate the leakage flow. The typical pressure-driven flow across the tip is more complex than the simple leakage flow modeled by this study.

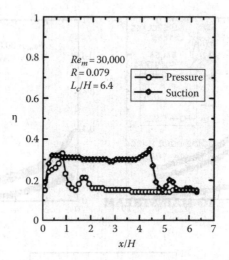

FIGURE 3.48
Effect of injection location on cross-stream averaged film effectiveness for grooved-tip cavity injection. (From Kim, Y.W. et al., *ASME J. Turbomach.*, 117, 1, 1995.)

More real leakage flow studies with film-cooling measurements are required for improved blade tip cooling design.

3.7 Leading-Edge Region Film Cooling

3.7.1 Introduction

Film cooling around the turbine airfoil leading-edge region not only protects the region from hot gases but also affects the aerodynamics and heat transfer over the entire airfoil. The leading-edge region of the airfoil has the highest heat-transfer rate over the entire airfoil, as observed in the early part of this chapter. Typically, coolant from within the airfoil is injected through several rows of discrete hole rows around the leading-edge region to protect the surface and reduce heat-transfer rates. The interaction between the mainstream gases and coolant jets varies from stagnation with increased interaction downstream on the suction and pressure surfaces. As local velocity on the airfoil increases and decreases depending on the curvature of the airfoil, coolant–mainstream interaction varies with other factors also affecting heat-transfer rates. This important region on the airfoil is affected by free-stream turbulence, unsteady wakes shed by upstream airfoil trailing edges, and mainstream Reynolds numbers.

Some of the early work on leading-edge film cooling was done by Luckey et al. (1977) at Purdue University. They simulated the airfoil leading edge as a cylinder in a cross-flow with film hole rows. They studied different hole angles of 20°, 30°, and 40° to the surface in the spanwise direction. They simulated a typical engine condition coolant-to-mainstream density ratio of 2.15. They also presented a correlation for the optimum blowing ratio in terms of coolant-to-mainstream velocity ratio and injection angle.

3.7.2 Effect of Coolant-to-Mainstream Blowing Ratio

Mick and Mayle (1988) studied the stagnation region as a semicircular leading edge with a flat afterbody. They presented heat-transfer coefficient and film effectiveness results for two rows of holes at ±15° and ±44° from stagnation. They varied the coolant blowing rate from $M = 0.38$ to 0.97. The cylinder had two rows of injection on one side and one row on the other side of the stagnation. Figure 3.49 shows the semicircular cylinder with the hole geometry. The holes in each row were spaced four-hole diameters apart and were angled 30° to the spanwise direction. The test model had a heater surface with thermocouples embedded to measure surface temperature during the steady-state heat-transfer tests. Heat-transfer and film effectiveness tests were run separately. The methodology of the steady-state thin-foil heater experiment is described in detail in Chapter 6.

Figure 3.50 shows the film effectiveness distributions for three injection rates. Effectiveness values downstream of the first row are lower than values downstream of the second row. The coolant from the first row combines with jets from the second row and produces a higher effectiveness downstream of the second row. Strong spanwise variations are clearly evident downstream of the holes. From the figure, it appears that effectiveness values decrease with increasing blowing rate. Also, the lower injection rate coolant stays closer to the surface from hole location to much farther distance downstream, producing higher effectiveness. Higher blowing-rate jets penetrate the mainstream boundary layer, and coolant mass protecting the surface is reduced thereby, producing lower film effectiveness. It is also interesting to note that jets at a higher injection rate show stronger jet tilt along the hole angle. The span-averaged film effectiveness distributions clearly indicate the trend discussed earlier, as seen in Figure 3.50. As described from the

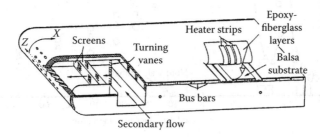

FIGURE 3.49
Heat-transfer model with leading-edge injection and injection hole geometry. (From Mick, W.J. and Mayle, R.E., *ASME J. Turbomach.*, 110, 66, 1988.)

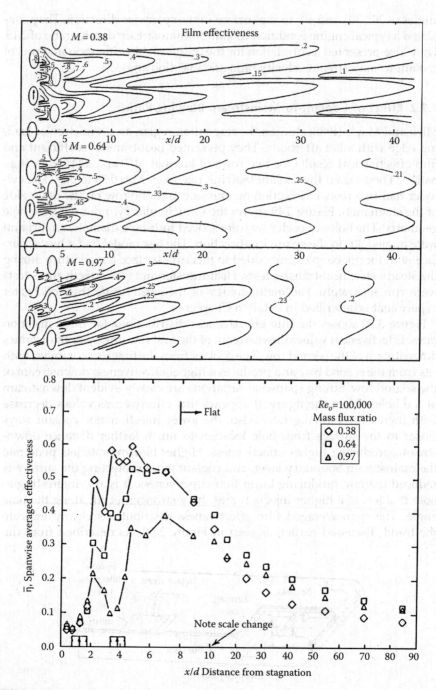

FIGURE 3.50

Detailed and spanwise-averaged film effectiveness distributions for three injection rates. (From Mick, W.J. and Mayle, R.E., *ASME J. Turbomach.*, 110, 66, 1988.)

detailed contours, the effect of blowing rate is clearly evident. Coolant protection at higher injection rates suffers from excessive mainstream boundary layer penetration, jet interference, and strong turbulent mixing.

Figure 3.51 shows the detailed Nusselt number distributions for the same conditions. The Nusselt number downstream of injection increases with increasing injection rate. Coolant interaction with mainstream increases with increasing injection, producing locally high turbulence, thus enhancing heat-transfer coefficients. Downstream of the second row, the combination of coolant from the first row and second row enhances Nusselt numbers even further. Far downstream of injection, on the flat surface, Nusselt numbers are unaffected by injection rate. Figure 3.51 also shows the span wise averaged Nusselt number distributions for all three blowing rates and the uncooled surface. It is clearly evident that film injection enhances Nusselt number up to two to three times over that of uncooled surface. Further increases of injection rate produce slight increases in the Nusselt numbers. Far downstream of injection, film-cooled Nusselt numbers approach values of uncooled surface.

Combining the film effectiveness and heat-transfer coefficient results, an assessment of film cooling toward heat-transfer rate reduction can be made. The methodology was described in the beginning of this chapter. Figure 3.52 shows the heat-transfer rate ratio of cooled leading-edge model to uncooled leading-edge model. The lowest blowing rate of $M = 0.38$ produces the best heat-transfer rate reduction. Heat-transfer rates increase with increasing blowing rate. At $M = 0.97$, some locations on the surface downstream of second-row injection show higher heat flux ratio values than uncooled surface. At higher blowing ratios, higher Nusselt numbers and lower film effectiveness contribute to increased heat-transfer rate.

This study clearly shows the effect of coolant injection rate on leading-edge heat transfer. However, the effect of free-stream turbulence has not been studied. In real engines, higher free-stream turbulence is an important parameter.

3.7.3 Effect of Free-Stream Turbulence

As indicated in Chapter 2, in real engines, higher free-stream turbulence is an important parameter. Higher free-stream turbulence can significantly reduce film protection and increase heat-transfer rates to the airfoil surface. Airfoil designers consider the free-stream turbulence effect as an important parameter in designing the film-cooled airfoil. Correlated results of film-cooling heat-transfer rate with increased free-stream turbulence would help designers consider the combustor-generated turbulence.

Mehendale and Han (1992) studied the effect of high free-stream turbulence on a film-cooled leading-edge model. Their test model was similar to the one used by Mick and Mayle (1988). Figure 3.53 shows the schematic of the test model with injection and thermocouple locations. The hole locations and geometry on the semicircular leading edge were also similar.

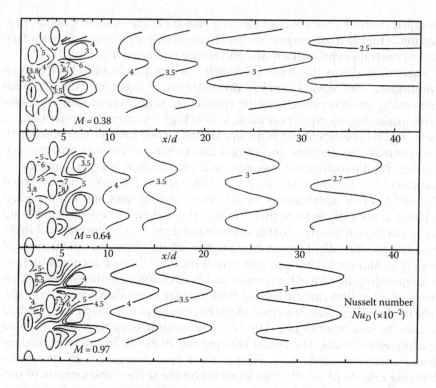

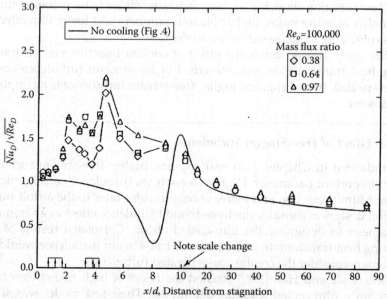

FIGURE 3.51

Detailed and spanwise-averaged Nusselt number distributions for three injection rates. (From Mick, W.J. and Mayle, R.E., *ASME J. Turbomach.*, 110, 66, 1988.)

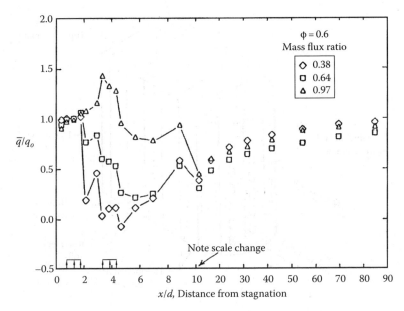

FIGURE 3.52
Effect of film cooling on the surface heat load ratio. (From Mick, W.J. and Mayle, R.E., *ASME J. Turbomach.*, 110, 66, 1988.)

Higher free-stream turbulence was generated using passive and jet grids. The passive grid generated turbulence of 9.67% and the jet grid generated 12.9%.

Mehendale and Han (1992) measured Nusselt numbers and film effectiveness on the semicircular leading edge with flat afterbody using heater foils and thermocouples. The coolant-blowing rate (*B*) was varied from 0.4 to 1.2 for all three free-stream turbulence levels. Figure 3.54 presents the effect of free-stream turbulence on spanwise averaged Nusselt number distributions for two blowing rates of 0.4 and 1.2. Nusselt number distributions on an uncooled leading edge are also shown for all three turbulence levels. As indicated earlier, film cooling by itself enhances Nusselt number significantly downstream of injection. Further addition of higher free-stream turbulence causes only slight increases in Nusselt numbers for both blowing rates. Far downstream of injection, the effect of free-stream turbulence decreases as the turbulence dissipates. Film injection locally produces high turbulence due to coolant-mainstream interactions. The levels of turbulence are as high as 15%. A free-stream with higher turbulence does not seem to increase the turbulence in the interaction; thus only a slight increase in heat-transfer coefficient is observed.

Figure 3.55 shows the effect of free-stream turbulence on film effectiveness for a blowing rate of 0.4. Free-stream turbulence causes reduced film effectiveness over the entire surface. Increased free-stream turbulence causes coolant jets to dissipate faster into the mainstream. Also, the unsteady free stream penetrates the protective layer of the film and reduces film effectiveness.

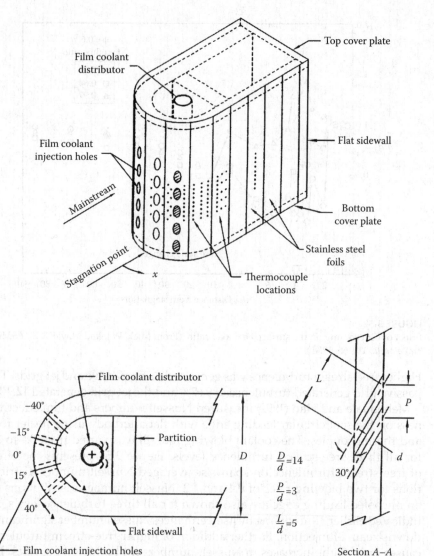

FIGURE 3.53
Schematic of the leading-edge model with film holes and thermocouples. (From Mehendale, A.B. and Han, J.C., *ASME J. Turbomach.*, 114, 707, 1992.)

At low free-stream turbulence, the coolant jet structures are maintained over a longer distance downstream of injection, producing higher effectiveness. With an increase in blowing ratio, the jet momentum is stronger and the diffusion of the jets in the mainstream is slower. The coolant jet structures are maintained over a longer distance, producing a very minimal effect on film effectiveness. Higher free-stream turbulence thus causes reduced film effectiveness and slightly increased Nusselt numbers at lower blowing ratios,

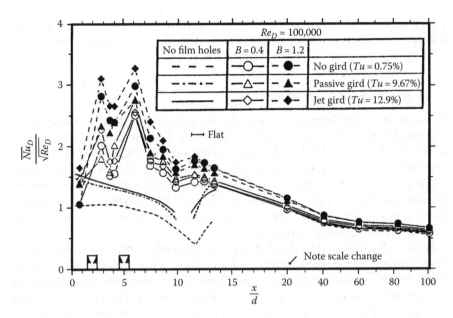

FIGURE 3.54
Effect of mainstream turbulence on spanwise-averaged Nusselt number distributions for
$B = 0.4$ and 1.2. (From Mehendale, A.B. and Han, J.C., *ASME J. Turbomach.*, 114, 707, 1992.)

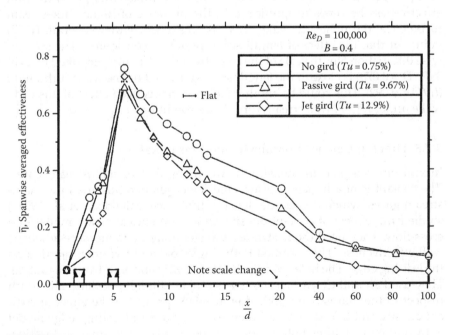

FIGURE 3.55
Effect of mainstream turbulence on spanwise-averaged film effectiveness distributions for
$B = 0.4$. (From Mehendale, A.B. and Han, J.C., *ASME J. Turbomach.*, 114, 707, 1992.)

which may further cause an increase in heat-transfer rates to the airfoil. This information is very useful to the airfoil designer designing efficient film cooling.

3.7.4 Effect of Unsteady Wake

Funazaki et al. (1997) presented the effect of periodic unsteady wake passing on film effectiveness for an airfoil leading edge. Their test model was similar to the blunt used by Mick and Mayle (1988) and Mehendale and Han (1992). The hole locations and the test model dimensions are also similar. However, Funazaki et al. (1997) used a spoked-wheel-type generator described earlier to simulate the periodic unsteady wake effect that exists for real airfoils. They superimposed three free-stream turbulence levels for the different unsteady wake passing strengths.

Figure 3.56 presents the effect of periodic unsteady wake on spanwise averaged film effectiveness distributions for a blowing rate (B) of 0.4 under three different free-stream turbulence conditions. Four unsteady wake strengths are presented for each free-stream turbulence case. As can be noticed, for a grid 1 ($Tu = 1\%$), an increase in periodic wake strength causes a slight reduction in film effectiveness. Unsteady wakes may cause mainstream to penetrate the film protection and reduce film effectiveness. With increase from grid 1 to grid 2 ($Tu = 3\%$), the effect of unsteady wake strength on the film effectiveness becomes insignificant. In the presence of higher free-stream turbulence, the unsteady wake effect is weakened. Funazaki et al. (1997) point out that the effect of length scale should be considered. The streamwise dissipation length scale of the free-stream turbulence generated by grid 3 ($L_e/D = 0.078$) is close to the length scale generated by the wake turbulence ($L_e/D = 0.092$). This could be another factor of the reduced effect of unsteady wake under higher free-stream turbulence conditions.

3.7.5 Effect of Coolant-to-Mainstream Density Ratio

In real engines, coolant density is two times that of mainstream gases. The coolant is at a higher pressure and lower temperature than the mainstream gases, which causes the density difference. Salcudean et al. (1994a) studied the effects of coolant density on a symmetrical airfoil model in a cross-flow. They used the heat/mass transfer analogy to measure adiabatic wall effectiveness. They studied both single-row and two-row injection on the leading edge. The hole rows are located at ±15° and ±44° from stagnation. They used air and CO_2 as injectants to simulate different density ratios with respect to the mainstream. Ekkad et al. (1998) presented the effect of coolant density and free-stream turbulence on a cylindrical leading-edge model with one row of film-cooling holes on both sides of stagnation location. They used a transient liquid crystal technique to obtain detailed heat-transfer coefficient and film effectiveness distributions. To obtain a higher

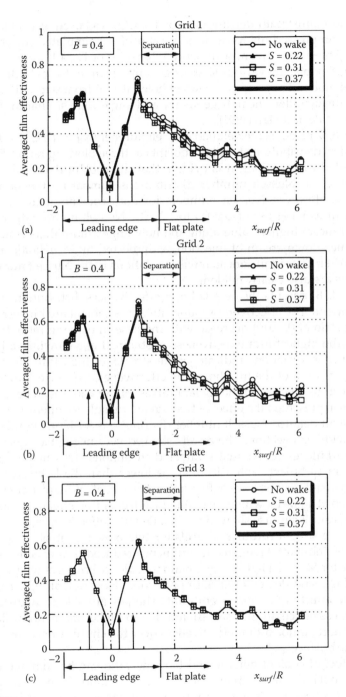

FIGURE 3.56
Average film effectiveness with wake and free-stream turbulence effect for $B = 0.4$: (a) grid 1, (b) grid 2, and (c) grid 3. (From Funazaki, K. et al., *ASME J. Turbomach.*, 119, 292, 1997.)

coolant-to-mainstream density ratio of 1.5, CO_2 was used as coolant, using regular compressed air for a density ratio of 1.0 simulated lower-density coolant. Three free-stream turbulence intensities of 1%, 4.1%, and 7.1% were generated by placing various sizes of grid upstream of the cylindrical leading-edge model. They had two rows of holes at ±15° from stagnation on both sides of the cylinder The hole geometry and angle were similar to the study by Mehendale and Han (1992).

Figure 3.57 presents the detailed Nusselt number ($Nu = hD/k$; D = diameter of cylinder) distributions for both coolants at three blowing ratios. Based on the detailed distributions, it is evident that the coolant density has a very small effect on Nusselt number distributions comparing the same blowing ratios. However, Nusselt numbers for both coolants increase with an increase in blowing ratio from 0.4 to 1.2. The Nusselt number distributions clearly indicate higher value along the hole angle and enhancement up to three times downstream of injection, as indicated by Mehendale and Han (1992) earlier. The significant spanwise variations in the Nusselt number distributions are also clearly evident.

Figure 3.58 presents the detailed film effectiveness distributions for similar conditions. The effectiveness values for air as coolant are highest at a low blowing ratio of 0.4 and decrease with increase in blowing ratio. However, for CO_2, the highest film effectiveness is obtained at a blowing ratio of 0.8. The spanwise variations indicate jetlike streaks of high film effectiveness along the bottom of the film hole. The effectiveness decreases downstream as jet effect dissipates and the coolant mixes totally with the mainstream.

They also presented the effect of free-stream turbulence intensity (1%, 4.1%, and 7.1%) on detailed Nusselt number distributions for blowing ratios of 0.4 and 1.2 with air injection. The results showed very negligible effect of free-stream turbulence on Nusselt number distributions. This may be because free-stream turbulence intensity may be lower than the large-scale turbulence generated downstream of film holes due to coolant-mainstream interactions. Even at a low blowing rate of 0.4, the effect of free-stream turbulence is not significant. They also showed the detailed film effectiveness distributions at three free-stream turbulence levels for two blowing ratios of 0.4 and 1.2 with air as injectant. Free-stream turbulence reduced film effectiveness significantly for $M = 0.4$. Higher free-stream turbulence breaks down the coolant jet structures and decreases surface protection. However, at higher blowing ratios, the effect is not very significant. The jets are able to withstand the unsteadiness effects in the mainstream due to higher momentum flux. The result is more uniform film effectiveness over the entire surface due to free-stream turbulence.

The effect of coolant density and blowing ratio can be combined as the effect of coolant-to-mainstream momentum flux ratio ($I = \rho_c U_c^2 / \rho_\infty U_\infty^2$). The film effectiveness and Nusselt number results can be explained in terms of momentum ratios. Figure 3.59 presents the area-averaged Nusselt number

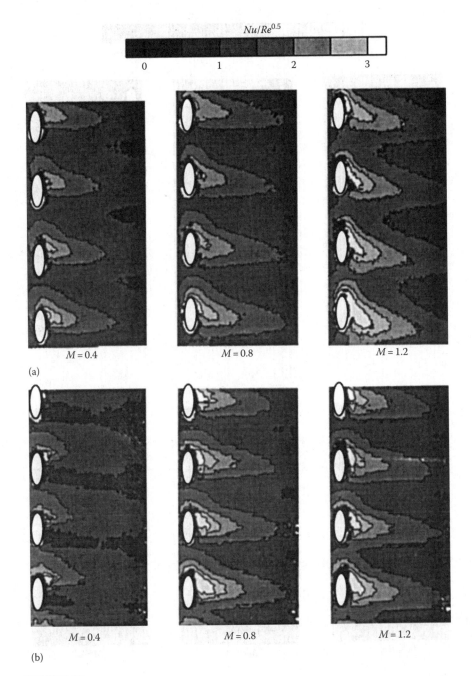

FIGURE 3.57
Effect of blowing ratio on detailed Nusselt number distributions for air and CO_2 injection.
(a) Air ($DR = 1.0$). (b) CO_2 ($DR = 1.5$). (From Ekkad, S.V. et al., *ASME J. Turbomach.*, 120, 799, 1998.)

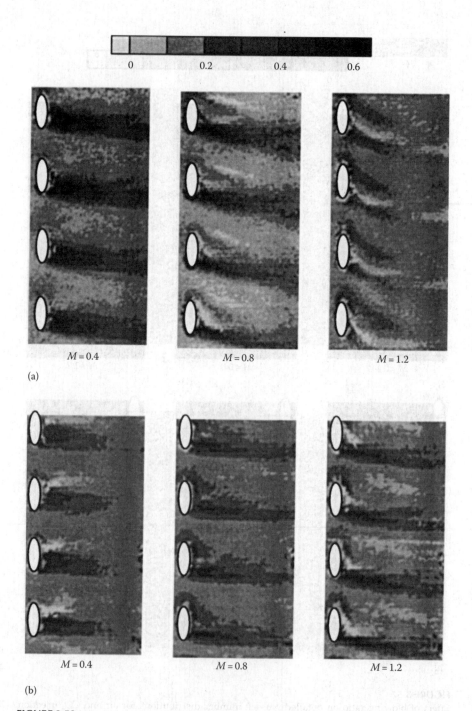

FIGURE 3.58
Effect of blowing ratio on detailed film effectiveness distributions for air and CO_2 injection.
(a) Air ($DR = 1.0$). (b) CO_2 ($DR = 1.5$). (From Ekkad, S.V. et al., *ASME J. Turbomach.*, 120, 799, 1998.)

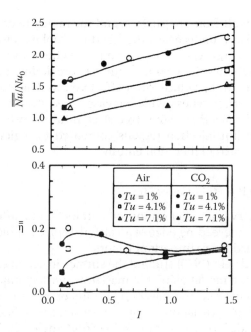

FIGURE 3.59
Variation of spatially averaged Nusselt in number ratio and film effectiveness with momentum flux ratio. (From Ekkad, S.V. et al., *ASME J. Turbomach.*, 120, 799, 1998.)

ratio $(\overline{Nu/Nu_0})$ and film effectiveness results as a function of momentum ratio (I). The Nusselt number values with film cooling (Nu) are normalized with the Nusselt number on an uncooled surface (Nu_0) to obtain the Nu/Nu_0 value. The area-averaged values are presented for three free-stream turbulence cases of 1%, 4.1%, and 7.1%. As seen from the figure, it is clear that the Nusselt number ratio increases with increasing coolant momentum. An increase in coolant momentum produces increased coolant-mainstream interaction, thus producing higher heat-transfer coefficients. The coolant CO_2 simulates lower momentum ratio for the same blowing ratio compared to air due to higher density. The results obtained for CO_2 are in line with the results obtained for air. Increase in free-stream turbulence causes a reduced Nusselt number ratio. This is because free-stream turbulence enhances heat-transfer coefficient on an uncooled surface but has negligible effect on a film-cooled surface. Thus, Nusselt number ratios decrease with increase in free-stream turbulence at the same I.

Film effectiveness results show different trends to that of Nusselt number distributions. At low free-stream turbulence, film effectiveness increases toward a high value at an optimum I value and decreases with increasing I value. At low I, the coolant possesses low momentum and cannot protect the surface effectively. As coolant momentum increases, the film stays closer to the surface and also has substantial momentum to deflect the mainstream away from the surface. Further increases in coolant momentum cause increased jet blowoff from the surface, and the mainstream comes into contact with the surface underneath the lifted-off coolant jets. An increase in free-stream turbulence causes significant reductions in film effectiveness for low-momentum jets. Low-momentum jets break down under high free-stream turbulence and produce lower effectiveness. Higher-momentum jets produce similar effectiveness with increased free-stream turbulence. This study clearly indicates that higher free-stream turbulence is detrimental to low-momentum film injection regions compared to regions on the airfoil where the jets possess higher momentum.

3.7.6 Effect of Film Hole Geometry

Film hole geometry is another important parameter that affects film-cooling performance. On the leading edge of an airfoil, typically hole location, shape, and angle of injection are important factors.

Karni and Goldstein (1990) studied the surface injection effects on a cylindrical leading-edge model using the mass-transfer technique. They studied the effect of blowing rate and injection location on mass-transfer coefficient. They presented the variation of spanwise averaged Sherwood number around the cylinder for a streamwise injection angle of 37°, and spanwise injection of 90° at three injection locations of 10°, 20°, and 30° from stagnation location. The effect of blowing ratio is shown for all three injection locations. Injection causes an enhancement of up to two to three times in the Sherwood number, with film cooling compared to a no-holes case. Downstream of injection, Sherwood numbers drop rapidly and increase again around 90° from stagnation due to boundary layer transition to turbulence. This transition occurs only for the film-cooling cases. For an injection location of 30° from stagnation, the injection-enhanced high Sherwood numbers overlap the transition location, causing a reduced drop downstream of injection. The other side of the cylinder without film cooling behaves similarly to the no-injection holes case, as expected. Film injection induces boundary layer transition to turbulence on the cylinder near about 90° from stagnation. This is due to the boundary layer disturbance caused by the jets, which causes the sudden transition to turbulence. This phenomenon was described earlier in the chapter.

Ou et al. (1992) presented the effect of film hole location on leading-edge film-cooling heat transfer. They used the test model used by Mehendale and Han (1992). The effect of hole location was studied for two locations of ±15° and ±40° from stagnation. Mehendale and Han (1992), as described earlier in this section, presented the cumulative effect of two-row injection on leading-edge film cooling. Figure 3.53 shows the leading-edge model, with hole locations marked. The ±40° holes were plugged for ±15° film-cooling tests, and vice versa.

Figure 3.60 shows the effect of film hole row location on spanwise-averaged film effectiveness distributions. The data for two-row injection from

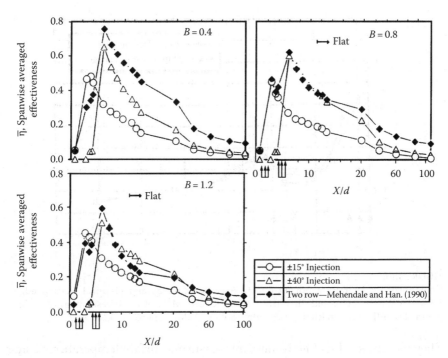

FIGURE 3.60
Effect of film hole row location on spanwise-averaged film effectiveness distributions and comparison with two-row data for $Tu = 9.67\%$. (From Ou, S. et al., *ASME J. Turbomach.*, 114, 716, 1992.)

Mehendale and Han (1992) are also shown for comparison. It is clearly evident that two-row injection performs better at low blowing ratio. However, the ±15° injection-only case performs better immediately downstream compared to combined row injection. At higher blowing ratio, two-row injection and one-row injection film effectiveness distributions overlap. The ±40° holes produce higher effectiveness than ±15° immediately downstream of injection. This may be because the apparent blowing ratio is much lower for ±15° injection than for ±40° injection. The local free-stream velocity causes the difference in blowing ratio for same mass flux injected. Salcudean et al. (1994b) also measured local film effectiveness distributions on a film-cooled leading-edge model similar to that in the study. They had two rows of holes on either side of the stagnation line. The hole rows are located at ±15° and ±44° from stagnation.

3.7.7 Effect of Leading-Edge Shape

Cruse et al. (1997) studied the effect of leading-edge shapes in a study that investigated several parametric influences on leading-edge film cooling. A circular leading edge was compared to a elliptic leading edge.

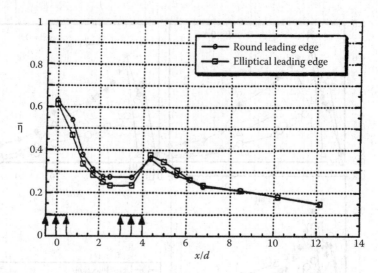

FIGURE 3.61
Comparison of laterally averaged effectiveness for circular and elliptical leading edges: $DR = 1.8\%$, $M = 2.0\%$, and $Tu = 0.5\%$. (From Cruse, M.W. et al., Investigation of various parametric influences on leading edge film cooling, ASME Paper 97-GT-296, 1997.)

They measured film effectiveness by measuring surface temperatures using an infrared camera. They injected air, cooled to 166 K, to simulate a density ratio of 1.8. They also measured the effect of free-stream turbulence up to 20% on film effectiveness distributions.

Figure 3.61 shows film effectiveness results on the two different leading-edge models. The film hole locations are shown on the figure. The span-wise-averaged film effectiveness distributions are very similar for both leading-edge geometries. This indicates that a circular leading edge can be used to simulate the elliptic leading edge of an airfoil. This is an important result for designers, as most of the results presented for leading-edge film cooling in the available literature have been for circular leading-edge models.

3.8 Flat-Surface Film Cooling

3.8.1 Introduction

Goldstein (1971) presented a review of all literature prior to 1971 on flat-surface film cooling. He presented the effects of various geometrical and flow parameters affecting film cooling. Many researchers have studied film cooling on flat surfaces. Flat-surface models can be used to study the effects of individual parameters with relative ease and are less expensive. Early studies have proved that the results obtained on simple flat-surface models can be applied to real engine design with slight corrections. The effects of geometrical parameters (hole geometry, shape, size, spacing) and flow parameters (coolant-to-mainstream mass flux, temperature ratio, mainstream Reynolds number, velocity, etc.) have been studied on flat surfaces. Also, the effects of pressure gradient and curved surface have also been studied. Some studies have focused only on the heat-transfer coefficient enhancement, and others have presented only film effectiveness results. The effect of each of the aforementioned parameters on flat-surface h and η will be discussed in the next section.

3.8.2 Film-Cooled, Heat-Transfer Coefficient

Heat-transfer coefficient downstream of film injection is enhanced due to increased turbulence produced by mixing of the coolant jets with the mainstream boundary layer. This increased turbulence locally enhances the heat-transfer coefficients. Figure 3.62 (Goldstein and Taylor, 1982) summarizes the effects of coolant injection through a discrete hole on local film-cooled heat-transfer coefficient. The coolant hole is inclined 35° along the mainstream flow direction (streamwise) and is referred to as simple angle injection. The figure clearly illustrates the local high and low heat-transfer regions around the hole due to injection. The results are illustrated in terms of the h/h_0 ratio. The heat-transfer coefficient h_0 is for a surface without film holes under the same mainstream conditions, whereas h is the heat-transfer coefficient with film injection. Region A corresponds to the unaffected region upstream of the hole where the heat transfer is related only to mainstream effects. At small blowing ratios ($M < 0.8$), the coolant jets do not hinder the forward movement of the mainstream at injection hole location, while at large M, the coolant jet starts to push the mainstream back and acts as a solid rod. In that case, the mainstream flow close to the wall seems to be similar to a flow around a solid cylinder. Since, typically, the film holes are inclined toward the mainstream direction, the effect would be significantly reduced. Region B is the

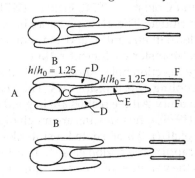

FIGURE 3.62
Regions of high and low heat/mass transfer around injection holes. (From Goldstein, R.J., and Taylor, J.R., *ASME J. Heat Transfer*, 104, 715, 1982.)

region between two adjacent holes. Injection again has little effect on the heat-transfer coefficient, except for high blowing ratios. This region will have enhanced heat-transfer coefficients only when jets coalesce due to small spacing or due to increased mainstream acceleration caused by blockage created by high blowing ratio jets. The region C is immediately downstream of the injection hole. The minimum heat-transfer ratio occurs in this region. This can be explained by the action of the jets in creating a stagnation region underneath the jet. The blowing ratio does not affect the presence of this stagnation region. However, one would infer that region C would become larger with increasing blowing ratio. Region D exists along the sides of the injection hole. The high heat-transfer region is caused by large shear stresses and eddies due to mainstream jet interaction. Region E is the region where the maximum heat-transfer coefficient occurs. The jets remain attached near the surface for small M, and the heat-transfer coefficient is low due to lack of strong mainstream jet interaction. As the blowing ratio increases, the region becomes larger, and increased mainstream jet interaction occurs. Further increase in blowing ratios ($M > 1.5$) can cause region E to move away from injection due to the jet lifting off at the hole and reattaching downstream. With separation, the mainstream penetrates underneath the jet and forms large eddies that further enhance heat-transfer coefficients. At high M, there exists another region F that occurs due to the partial reattachment of the jet to the surface.

3.8.2.1 *Effect of Blowing Ratio*

The effect of the coolant jet decreases downstream of injection as the jet structure dissipates and the mainstream dominates the coolant film completely. The high heat-transfer coefficient in the near injection region is due to the 3-D nature of the jet, and far downstream ($X/D > 15$), the jet structure is completely absent and is 2-D in nature. Ammari et al. (1990) presented a summary of results for film cooling on a flat surface with a single row of holes inclined at 35° along the mainstream direction. Results from Ammari et al. (1990), Leiss (1975), Eriksen and Goldstein (1974), Eriksen (1971), and Goldstein and Yoshida (1982) were included. The blowing ratio, boundary layer displacement thickness to hole diameter ratio, and mainstream Reynolds number were studied. The heat-transfer coefficient ratio decreases with increasing axial distance from the injection hole. About 15-hole diameters downstream of injection, the film-cooling effect disappears. The heat-transfer coefficient ratio is almost equal to unity. All the results are in good agreement with each other, except for Leiss (1975), whose results are much higher than those of the others. Only Ammari et al. (1990) presented results closer to injection. Other earlier studies do not capture the region close to injection hole.

Figure 3.63 presents the variation of the heat-transfer coefficient ratio (h/h_0) with the blowing ratio M for 35° injection (Hay et al., 1985). For all the studies, the effect of the blowing ratio is to increase the heat-transfer coefficient ratio. The ratio is closer to unity at lower blowing ratios but increases to

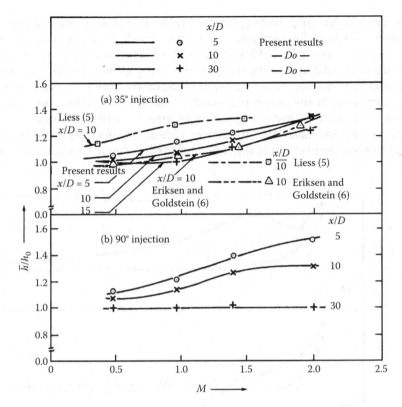

FIGURE 3.63
Variation of the heat-transfer coefficient ratio h/h_0 with the blowing parameter M. (From Hay, N. et al., *ASME J. Eng. Gas Turbines Power*, 107, 105, 1985.)

significantly high values at high blowing ratios. Also shown in the figure is the same M effect for 90° injection (normal to mainstream flow). The effect of the blowing ratio is similar to that for 35° injection. The effect of axial location is stronger for the normal injection compared to the 35° injection. The jet shoots into the mainstream at an angle closer to the surface for 35° injection. This reduces the possibility of jet liftoff at lower blowing ratios.

3.8.2.2 Effect of Coolant-to-Mainstream Density Ratio

Ammari et al. (1990) also presented the effect of density ratio on heat-transfer coefficient contours downstream of a film hole inclined 35° along the streamwise direction for two different coolant-to-mainstream density ratios of 1.0 and 1.52 for a constant blowing ratio of $M = 1.46$. Differences of 10% occurred when coolant densities were changed. It was observed that lower-density injectant provides higher heat-transfer coefficient at the same blowing ratio due to higher momentum. Higher-momentum injection causes jet liftoff immediately downstream of injection, causing the mainstream to flow beneath the jets, creating intense eddies and thus increasing the heat transfer. Further downstream, the reattachment of streamwise vortices enhanced heat-transfer coefficients along the hole location. Figure 3.64 presents the effect of density ratio on spanwise-averaged h/h_0 for 35° injection at four different blowing ratios. At low blowing ratios, the effect of density ratio is evident only in the region immediately downstream of injection. Lower-density injectant provides higher heat-transfer coefficient ratios. As the blowing ratio increases, the differences in heat-transfer coefficient increase significantly downstream of injection and further downstream. Effects of jet

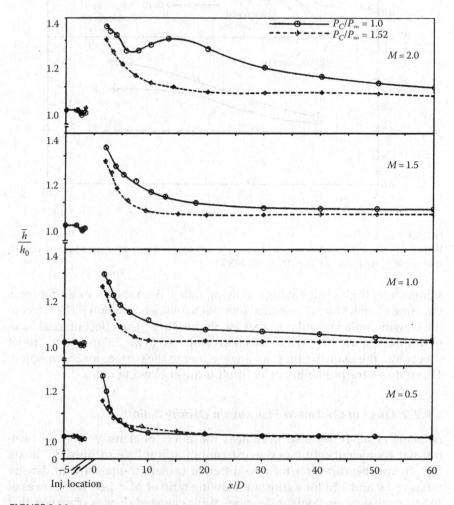

FIGURE 3.64
Effect of density ratio on heat-transfer coefficient ratio for 35° injection through a row of holes. (From Ammari, H.D. et al., *ASME J. Turbomach.*, 112, 444, 1990.)

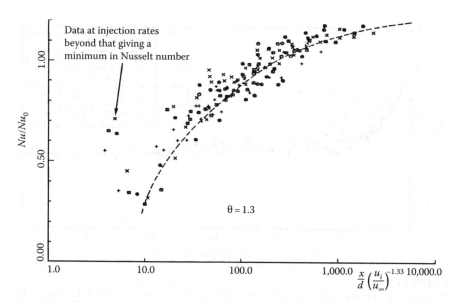

FIGURE 3.65

Correlation of single-row inclined-hole heat-transfer data. (From Forth, C.J.P. and Jones, T.V., Scaling parameters in film cooling, *Proceedings of the Eighth International Heat Transfer Conference,* San Francisco, CA, 1986.)

liftoff and reattachment are clearly evident for lower-density injectant at the highest blowing ratio of $M = 2.0$.

Forth and Jones (1986) have correlated some single-row, inclined-hole heat-transfer data using a parameter $(X/D)(u_c/u_\infty)^{-4/3}$ for the "strong injection" region, shown in Figure 3.65. They used the superposition approach for measuring heat transfer with film cooling. They noted that M is a not a good correlating parameter for angled-injection data. When X/D was combined with velocity ratio, it was found to give a better correlating parameter than M or I, which are very dependent on X/D. Figure 3.66 presents a correlation of h/h_0 data for angle 35° injection provided by Ammari et al. (1990). The figure includes data from Ammari et al. (1990), Goldstein and Yoshida (1982), and Eriksen and Goldstein (1974). From the figure, it is fairly clear that the correlation is valid only for the high density ratios, with significant deviations at low density ratios. This indicates that the momentum flux ratio instead of velocity ratio may be a better parameter to combine with the X/D location. This apparent deviation of low density ratio results closer to the hole location is due to jet liftoff and reattachment.

3.8.2.3 Effect of Mainstream Acceleration

Figure 3.66 shows that the heat-transfer data, at zero mainstream acceleration, including higher coolant density, correlated reasonably well with the

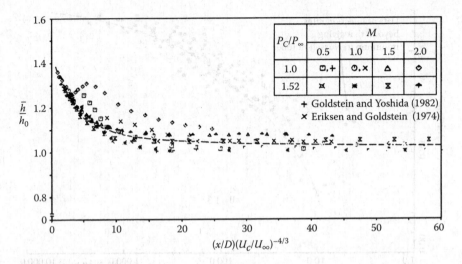

FIGURE 3.66
Correlation of heat-transfer coefficient ratio for 35° injection. (From Ammari, H.D. et al., *ASME J. Turbomach.*, 112, 444, 1990.)

parameter $(X/D)(u_c/u_\infty)^{-4/3}$ the strong injection regime. Ammari et al. (1991) correlated their higher-pressure gradient data using the same correlating parameter as described earlier. However, the effect of the pressure gradient acceleration parameter (K) needed to be included in the correlation. So this correlation incorporates the effect of K also. Figure 3.67 presents the

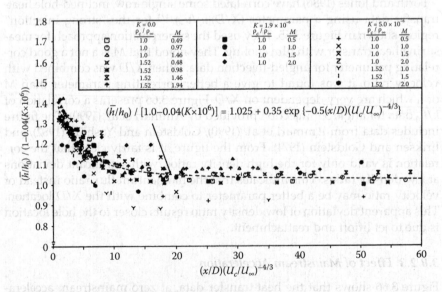

FIGURE 3.67
Correlation of heat-transfer coefficient ratio for 35° injection, including the effect of mainstream acceleration. (From Ammari, H.D. et al., *ASME J. Turbomach.*, 113, 464, 1991.)

correlated data for zero-pressure gradient and higher-pressure gradient. The figure shows all the conditions for each case studied by Ammari et al. (1991). The correlation is given as

$$\frac{\left(\bar{h}/h_0\right)}{[1.0-0.04(\kappa \times 10^6)]} = 1.025 + 0.35 \exp\left[-0.5\left(\frac{x}{D}\right)\left(\frac{u_c}{u_\infty}\right)^{-4/3}\right] \quad (3.5)$$

From the figure, it is evident that the scatter is stronger closer to injection. The effect of acceleration is to decrease the heat-transfer coefficient up to 25% depending on the blowing ratio. This correlation correlates both zero-pressure gradient and accelerated flows.

3.8.2.4 Effect of Hole Geometry

Ligrani et al. (1994) presented heat-transfer coefficient results for a single row of compound-angle holes with six-hole diameter spacing. They compared simple-angle holes with compound angle holes. They measured heat-transfer coefficients for three different film hole configurations. Figure 3.68 shows the three film-cooling configurations. Configurations 1 and 3 represent two different compound-angle hole configurations, and configuration 2 is the one with simple-angle holes. Configurations 1 and 3 differ in the hole spacing wherein the angle of injection is the same. So the difference in the two configurations is the hole–hole spacing.

Sometimes, cooling designers place two rows of film holes staggered or in line to provide better cooling efficiency. Ligrani et al. (1994a) presented results for two rows of holes with the same configurations as that shown in Figure 3.68. The hole-to-hole spacing in each row was three-hole diameters. Figure 3.69 compares the spanwise-averaged results for both single-row and two-row film cooling with simple and compound injection holes. Figure 3.69 shows the effect of blowing ratio. At low blowing ratios ($M < 1.0$), the effects of the number of film hole rows and compound angle are not very significant. However, with increasing blowing ratio, two-row injection shows higher heat-transfer coefficient enhancement compared to single-row injection for all three configurations. Also, compound-angle injection (1, 3) produces higher heat-transfer coefficient enhancement. At very high blowing ratios ($M > 3.0$), the trend appears to be reversing. However, there are few data available to corroborate the trends at high blowing ratios. Figure 3.69 also shows the effect of momentum flux ratio (I). The trends with I are similar to that observed for the blowing ratio except for the collapse of data at low I values.

Sen et al. (1996) studied the effect of compound-angle film-cooling on heat-transfer coefficients. They studied three different hole geometries and measured heat-transfer coefficient enhancements over a surface without film cooling. Compound-angle holes are holes that are inclined along the mainstream direction with an additional angle of inclination in the lateral or span-wise direction. Figure 3.70 illustrates the three-hole geometries studied

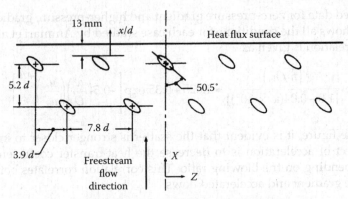

Test surface injection geometry for film cooling holes arranged
with configuration 1 compound angle holes

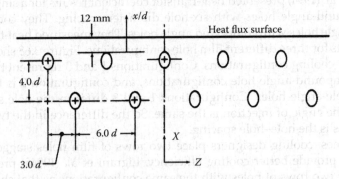

Test surface injection geometry for film cooling holes arranged
with configuration 2 simple angle holes

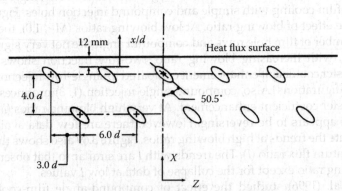

Test surface injection geometry for film cooling holes arranged
with configuration 3 compound angle holes

FIGURE 3.68
Three film-cooling hole configurations as studied by Ligrani et al. (1994). (From Ligrani, P.M.
et al., *ASME J. Heat Transfer*, 116, 341, 1994a.)

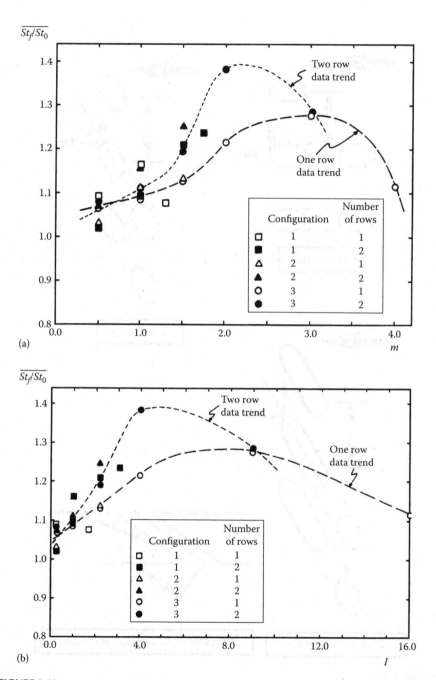

FIGURE 3.69
Effect of (a) blowing ratio and (b) momentum flux ratio on single-row and double-row injection through different hole configurations. (From Ligrani, P.M. et al., *ASME J. Turbomach.*, 114, 687, 1992.)

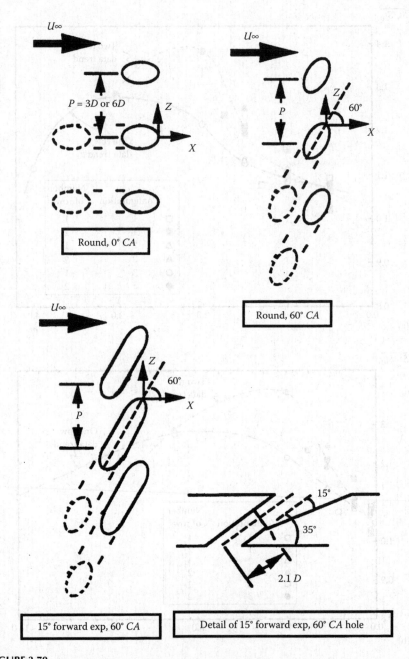

FIGURE 3.70
Illustration of the three-hole configuration studied by Sen et al. (1996). (From Sen, B. et al., *ASME J. Turbomach.*, 118, 800, 1996.)

by Sen et al. (1996). The first case was a round of holes indicated by "round, 0° *CA*." This kind of film hole is called the simple-angle hole, where the holes are inclined only along the mainstream, with no lateral angle. Most of the film-cooling studies discussed earlier in this section used simple-angle holes. The second case is the "round 60° *CA*" holes. These holes have an additional angle of 60° away from the mainstream in the lateral direction. The third case is called the "15° forward expansion with 60° *CA*" holes. As shown in the figure, these holes have the diffusing exit inclined further along the hole exit, at 15°.

Figure 3.71 presents the variation of spatially averaged heat-transfer coefficient with increasing momentum flux ratio (I). The heat-transfer coefficient ratio increases with increasing I for the round hole ($CA = 0$) and the round hole ($CA = 60$). The heat-transfer coefficient ratio is very high for the forward expansion hole. This may be due to increased lateral mixing of jets with the mainstream. However, the ratio decreases with increased forward-expansion hole at large I. The ratio values for the round hole, $CA = 60$, continually increase with increasing I.

Ekkad et al. (1997) compared two compound-angle holes with simple-angle injection. All the holes were inclined 35° in the streamwise direction. The compound-angle holes were inclined at 45° and 90° in the spanwise direction. The detailed heat-transfer coefficients were measured using a transient liquid crystal technique. Compound-angle injection provides higher heat-transfer coefficients than does simple-angle holes. Simple-angle injection causes limited interaction between mainstream and coolant jets. The jet structures move streamwise along the hole and dissipate at a slower rate compared to that for compound-angle injection. Due to the lateral momentum of the jet for compound-angle injection, the coolant jets interact with the mainstream, producing increased local turbulence and thus enhancing heat-transfer

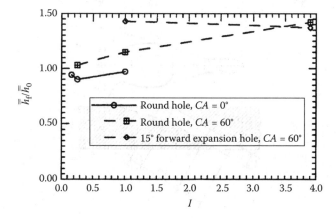

FIGURE 3.71
Variation of spatially averaged heat-transfer coefficient ratios with momentum flux ratio. (From Sen, B. et al., *ASME J. Turbomach.*, 118, 800, 1996.)

coefficients. This effect increases with higher blowing ratios. Higher blow-ing ratio jets possess higher lateral momentum. The detailed distributions clearly show the strong effect of compound-angle injection on heat-transfer coefficient enhancement.

Gritsch et al. (1998a) presented heat-transfer coefficient measurements for film holes with expanded exits. Such hole geometries are called shaped holes. Figure 3.72 shows the three-hole geometries studied by Gritsch et al. (1998a). The hole shapes are called cylindrical (*cyl*), fan-shaped (*fs*), and laid-back fan-shaped (*lb*) geometries. They presented the local heat-transfer coef-ficient ratio (h/h_0) distributions downstream of injection for all three-hole geometries at three different blowing ratios. The heat-transfer coefficient ratios are highest for the cylindrical hole. Heat-transfer coefficients for the fan-shaped holes are much lower due to the increased cross-sectional area at the hole exits. This decreases the momentum of the jet and reduces pen-etration into the mainstream. The laidback fan-shaped holes also provide low heat-transfer coefficients. However, the lowest heat-transfer coefficients are found at an off-center location. The jets spread out more laterally and

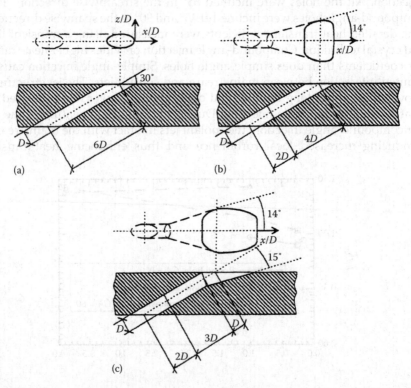

FIGURE 3.72
Shaped-hole configurations studied by Gritsch et al. (1998b). (a) Cylindrical hole, (b) flared hole, and (c) laidback fan shaped hole. (From Gritsch, M. et al., *ASME J. Turbomach.*, 120, 549, 1998b.)

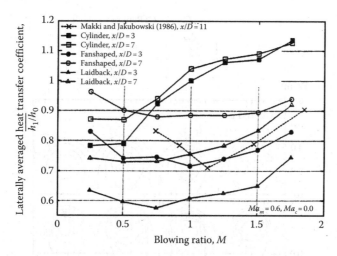

FIGURE 3.73
Laterally averaged heat-transfer coefficients for the three-hole shapes. (From Gritsch, M. et al., Heat transfer coefficient measurements of film cooling holes with expanded slots, ASME Paper 98-GT-28, 1998a.)

cause even more reduced heat-transfer coefficients. Figure 3.73 presents laterally averaged heat-transfer coefficients for all three-hole shapes versus the coolant-blowing ratio. For the cylindrical hole, the lowest blowing ratio provides the lowest heat-transfer enhancement. With increases in M, the heat-transfer coefficient ratio increases for both X/D locations. Minimum heat-transfer coefficient ratios are found at medium blowing ratios for the other two-hole shapes. The results are also compared with results for trapezoidal holes from Makki and Jakubowski (1986). The results strongly confirm the significant reduction of heat-transfer coefficients for shaped holes compared to the commonly used cylindrical holes. The reduction in jet momentum for the same blowing ratio reduces the interactions with mainstream and thus produces lower heat-transfer coefficients. The shaped holes may be better than compound-angle holes that produce much higher heat-transfer coefficients when compared to cylindrical simple-angle holes.

Byerley et al. (1988, 1992) presented detailed heat-transfer measurements near and within the entrance of film holes. They studied the effect of various velocity ratios on heat-transfer coefficients inside and outside the film hole.

3.8.3 Film-Cooling Effectiveness

The effects of all the aforementioned parameters are as important and may have a stronger effect on film-cooling effectiveness. It is important to measure film-cooling effectiveness distributions under various parametric effects and combine them with the heat-transfer coefficient results

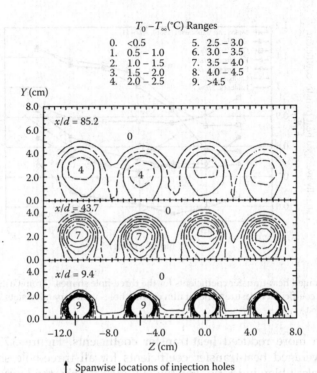

$T_0 - T_\infty$(°C) Ranges

0.	<0.5	5.	2.5 – 3.0
1.	0.5 – 1.0	6.	3.0 – 3.5
2.	1.0 – 1.5	7.	3.5 – 4.0
3.	1.5 – 2.0	8.	4.0 – 4.5
4.	2.0 – 2.5	9.	>4.5

↑ Spanwise locations of injection holes

FIGURE 3.74
Mean temperature field downstream of film injection for $M = 1.0$. (From Ligrani, P.M. et al., *ASME J. Heat Transfer*, 116, 353, 1994b.)

to obtain actual heat load reductions. Several researchers have presented the effects of various parameters on film-cooling effectiveness. Figure 3.74 presents the mean temperature field downstream of injection from an inclined row of holes at a blowing ratio of 1.0. In this experiment, Ligrani et al. (1994) injected hot coolant into an ambient mainstream and measured temperature contours. For small X/D location of 9.4, the coolant core is undiluted, and the hot coolant remains close to the surface. Further downstream, the hotter region moves away from the wall and is diluted by the cooler mainstream. This clearly indicates the decrease in coolant effectiveness downstream of injection.

Goldstein et al. (1974) were among the earliest to measure film effectiveness distributions downstream of various film hole configurations. They presented lateral variations of film-cooling effectiveness through a row of holes with three-diameter spacing for $M = 0.5$. The holes are inclined 35° in the direction of the mainstream. The film effectiveness distributions are shown at different X/D locations in Figure 3.75. At small X/D, the film effectiveness values along the hole are very high and drop sharply in the region between the holes. However, further downstream, the peak effectiveness decreases and the effectiveness in the region between the holes increases slightly.

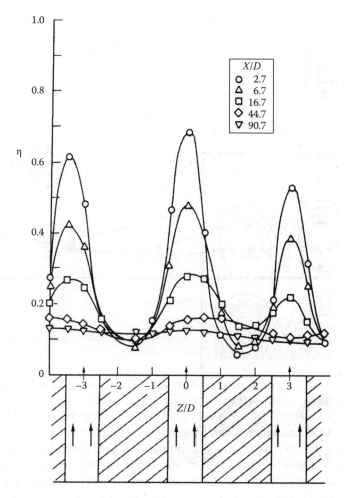

FIGURE 3.75

Spanwise film effectiveness distributions at different X/D locations. (From Goldstein, R.J. et al., *Int. J. Heat Mass Transfer*, 17, 595, 1974.)

The effectiveness decreases due to dilution of jet in the core region along the hole. However, some of the coolant spreads into the region between the holes and mixes to provide a higher effectiveness.

3.8.3.1 Effect of Blowing Ratio

Figure 3.76 presents the effect of blowing ratio on centerline film-cooling effectiveness for a row of simple injection holes inclined at 35° in the stream-wise direction (Goldstein et al., 1970). The centerline effectiveness values are plotted at different X/D locations. At near injection hole locations as in $X/D = 5.19$, the film effectiveness increases with increasing blowing rate initially and

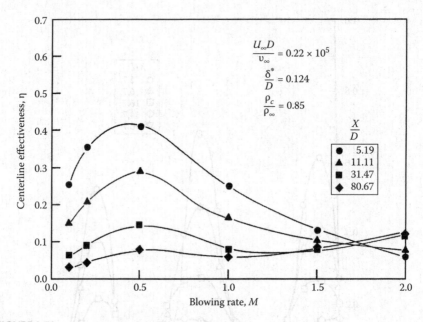

FIGURE 3.76
Effect of blowing ratio on centerline film effectiveness for simple angle holes inclined at 35°.
(From Goldstein, R.J., *Adv. Heat Transfer*, 7, 321, 1971.)

then reaches a peak and drops with further increases in blowing ratio. The
profiles become flatter farther away from the injection hole. At high blowing
ratios, distances far downstream of injection show similar effectiveness values as locations immediately downstream of injection. The jets possess higher
momentum than the mainstream and tend to lift off at the injection location
and reattach downstream. However, the effectiveness is not as significant for
downstream reattachment locations compared to a jet that stays closer to the
surface after injection, which is the case for low blowing ratios.

3.8.3.2 Effect of Coolant-to-Mainstream Density Ratio

Pedersen et al. (1977) presented the effect of coolant-to-mainstream density ratio on film-cooling effectiveness. They studied density ratios varying
from 0.75 to 4.17. Figure 3.77 presents the effect of blowing ratio on lateral
averaged film-cooling effectiveness as a function of coolant density for a single row of injection holes inclined 35° in the streamwise direction. Density
ratios below 1.0 show a decrease in effectiveness with increasing blowing
ratio. As density ratio increases, the peak effectiveness value moves toward
higher blowing ratio. Higher-density coolant tends to stay closer to the surface compared to the lower-density coolant at the same blowing ratio. The
jet momentum is higher for a lower-density coolant, and this may cause
jets to lift off and produce lower effectiveness. At very high-density ratios

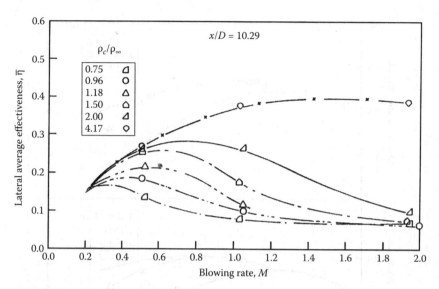

FIGURE 3.77

Effect of blowing ratio on laterally averaged film effectiveness for different coolant density ratios. (From Pedersen, D.R. et al., *ASME J. Heat Transfer*, 99, 620, 1977.)

of 4.17, effectiveness values are significantly higher and appear to increase even at high blowing ratios up to 2.0.

The effect of blowing ratio and density ratio on film effectiveness is clearly evident in Figure 3.78 (Sinha et al., 1991). For a constant-density ratio, the film effectiveness value increases with increasing blowing ratio and peaks and starts to drop off at very high blowing ratios. The blowing ratio at which peak effectiveness occurs is sometimes referred to as the optimum blowing ratio. It is typically in the range of 0.5–0.8. From the figure, it is evident that the effectiveness values for different coolant density ratios collapse into a single curve at low M values. However, the curves move away at higher blowing ratios with higher-density coolant providing higher effectiveness. This may be because higher-density coolant stays closer to the surface due to less mixing and lower momentum. It is important to note that there is no consistent trend in the way the film effectiveness values react to increasing blowing ratio for different coolant densities.

Figure 3.79 shows the variation of film effectiveness with momentum flux ratios (I) for the same coolant density ratios. It seems fairly evident that the film effectiveness is better scaled based on the momentum flux ratio than is the blowing ratio. The data for higher-density ratios of 1.6 and 2.0 collapse on a single line when plotted against I. Lower-density coolant values are slightly lower following the same trend. Since the momentum flux ratio combines the effects of both the blowing ratio and density ratio, it scales film effectiveness values better than the individual blowing ratios and density ratios ($I = M^2/DR$).

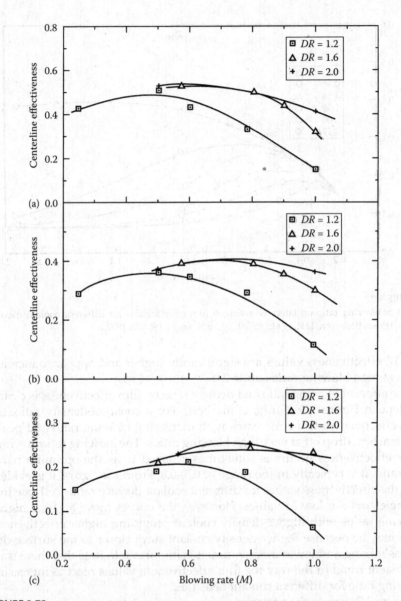

FIGURE 3.78
Effect of blowing ratio and density ratio on film effectiveness. (a) $X/D = 6$. (b) $X/D = 10$. (c) $X/D = 22$. (From Sinha, A.K. et al., *ASME J. Turbomach.*, 113, 442, 1991.)

3.8.3.3 Film Effectiveness Correlations

Goldstein (1971) presented an overview of the analysis of film cooling done by early researchers to determine correlating parameters. Figure 3.80 shows the control volume analysis for mass and energy balance analysis. Goldstein et al. (1966) determined that the blowing rate was not an effective parameter

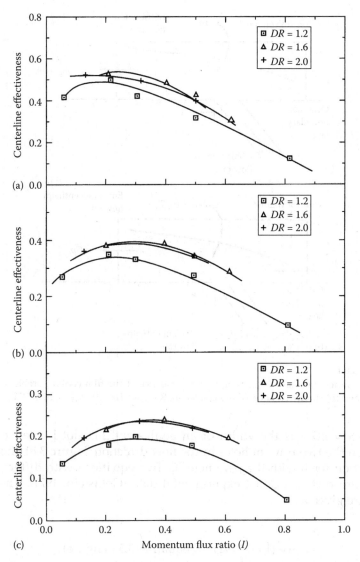

FIGURE 3.79
Effect of momentum flux ratio on film effectiveness. (a) $X/D = 6$. (b) $X/D = 10$. (c) $X/D = 22$.
(From Sinha, A.K. et al., *ASME J. Turbomach.*, 113, 442, 1991.)

to correlate film effectiveness as it is a function of downstream distance and momentum thickness Reynolds number.

Jabbari and Goldstein (1978) presented a correlation of laterally averaged film effectiveness versus a parameter ξ. The parameter ξ is defined as

$$\xi = \left[\frac{X}{MS}\right]\left(Re_2\,\frac{\mu_2}{\mu_\infty}\right)^{-0.25} \tag{3.6}$$

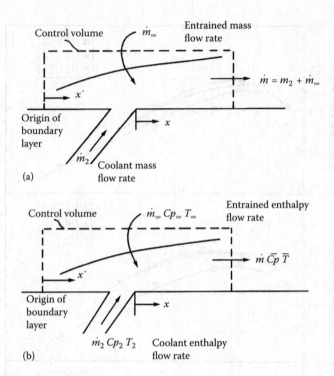

FIGURE 3.80
Control volume analysis for mass and energy balance of the film-cooling problem (a) mass
balance and (b) energy balance. (From Goldstein, R.J., *Adv. Heat Transfer*, 7, 321, 1971.)

where $S = \pi D/6$ is the width of an equivalent 2-D slot located midway
between the two rows of holes in the flow direction. Figure 3.81 shows the
correlation for $\bar{\eta}$ with the parameter ξ. Two equations with different constants are plotted against experimental data (Goldstein, 1971). The equations are given as

$$\bar{\eta} = (1 + 0.249\xi)^{-0.8} \text{ (Equation 3.5 in figure)} \tag{3.7}$$

$$\bar{\eta} = \frac{1.9Pr^{2/3}}{1 + 0.329\xi^{0.8}\beta} \text{ (Equation 3.6 in figure)} \tag{3.8}$$

For $\xi > 10$, the correlations work very well as the regime is more 2-D. For
$\xi < 10$, the regime is more 3-D, and the correlation grossly overpredicts the
film effectiveness. In this region, the blowing ratio effect is strong, and the
X/D effect is small.

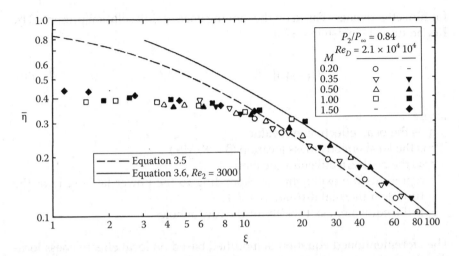

FIGURE 3.81

Correlation of laterally averaged film effectiveness with the parameter ξ. (From Goldstein, R.J., *Adv. Heat Transfer*, 7, 321, 1971.)

L'Ecuyer and Soechting (1985) examined the trends from data presented by Pedersen et al. (1977) and Afejuku (1977). They defined three regimes for characterizing the film effectiveness distribution on any surface:

Mass addition regime: Effectiveness increases with M due to increased thermal capacity of the coolant, but the effectiveness is independent of the density ratio and velocity ratio parameters.

Mixing regime: Effectiveness distribution depends on M, DR due to opposing influence of increased thermal capacity and increased coolant/free-stream mixing and penetration.

Penetration regime: Effectiveness distribution is completely dominated by a complex interaction of excessive coolant penetration and augmented turbulent diffusivity and turbulent diffusion of the coolants' thermal effect toward the surface.

Based on Pedersen's data, for angle of injection $\alpha_c = 35°$, $P/d = 3$, and for a single row of holes, we get

1. Mass addition regime: $V_c/V_g < 0.25$
2. Mixing regime: $0.25 < V_c/V_g < 0.8$
3. Penetration regime: $V_c/V_g > 0.8$

The major goal of L'Ecuyer and Soechting's study was to provide an effectiveness equation that was capable of representing the varied effectiveness distributions from all the aforementioned regimes. They arrived at a correlation for film effectiveness based on the moving energy sink solution presented by Kodatoni and Goldstein (1978):

$$\bar{\eta} = \eta_P (\beta/\beta_P)^{(a-1)/2} e^{1/2[1-(\beta/\beta_P)^{a-1}]} \tag{3.9}$$

where
    η_P is the peak effectiveness value
    β is the local effectiveness location ($\beta = X/MS_e$)
    β_P is the peak effectiveness location
    a represents the exponent corresponding to the power law decay of the turbulent thermal diffusivity $\varepsilon_H(x)$
    S_e is the equivalent slot width for the row of holes

The aforementioned equation is modified based on local effectiveness location in relation to peak effectiveness location, so that

$$\beta < \beta_P$$

$$\bar{\eta} = \eta_P (\beta/\beta_P)^{(a-1)/2} e^{\frac{1}{2}[1-(\beta/\beta_P)^{a-1}]} \tag{3.10}$$

$$\beta > \beta_P$$

$$\bar{\eta} = \eta_P (\beta/\beta_P)^{-\frac{1}{2}} e^{\frac{1}{2}[1-(\beta/\beta_P)^{-1}]} \tag{3.11}$$

$$\beta \gg \beta_P \tag{3.12}$$

$$\bar{\eta} = \eta_P \sqrt{\beta_P} e^{\frac{1}{2}} \beta^{\frac{1}{2}} \tag{3.13}$$

Three empirical constants ($\eta_P, \eta_P\sqrt{\beta_P}, a$) required for curve fit measured effectiveness data. The term a was selected to obtain the best fit for the data. Curve fitting the data of Pedersen et al. (1977), Goldstein et al. (1969), and Muska et al. (1976) resulted in the constants shown in their paper. Figure 3.82 shows the correlation of the peak effectiveness parameter, downstream effectiveness parameter, and the thermal diffusivity parameter using this correlation.

Figure 3.83 shows the adjustments required in correlating peak effectiveness (η_P) as a function of velocity ratio (V_c/V_g) for the effect of hole injection angle. A decrease of peak effectiveness is observed at low velocity ratios as the injection angle is decreased. Two important points to note are that at high velocity ratios, peak effectiveness is insensitive to injection angle indicative of a penetrative regime. Also, the downstream effectiveness constant

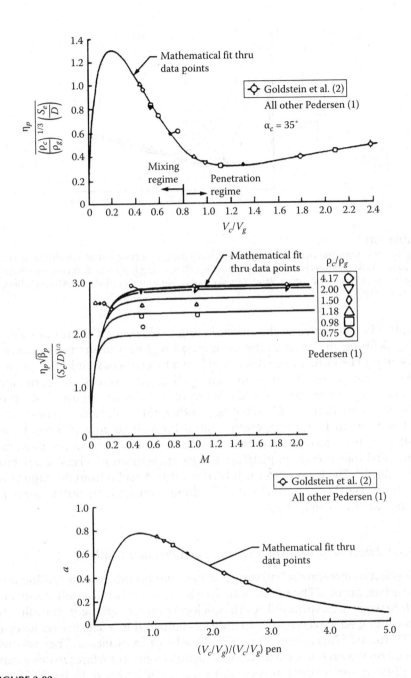

FIGURE 3.82
Correlation of the peak effectiveness, downstream effectiveness, and thermal diffusivity parameters using experimental data. (From L'Ecuyer, M.R. and Soechting, F.O., A model for correlating flat plate film cooling effectiveness for rows of round holes, *Heat Transfer and Cooling in Gas Turbines*, AGARD, Bergen, Norway, 1985.)

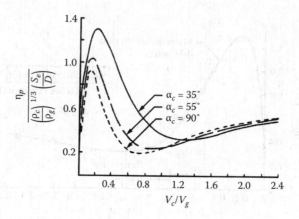

FIGURE 3.83
Adjustments required in correlating peak effectiveness as a function of velocity ratio. (From L'Ecuyer, M.R. and Soechting, F.O., A model for correlating flat plate film cooling effectiveness for rows of round holes, *Heat Transfer and Cooling in Gas Turbines*, AGARD, Bergen, Norway, 1985.)

$(\eta_P \sqrt{\beta_P})$ is insensitive to the injection angle. However, the critical velocity ratio, defining the onset of the penetration regime and the peak effectiveness, was determined to be directly affected by changes in injection angle.

Forth and Jones (1986) examined scaling parameters for correlating film-cooling results over a wide range of conditions. Figure 3.84a presents the variation of film-cooling isothermal wall effectiveness with coolant-to-mainstream temperature ratios for a single row of holes. Figure 3.84b presents a correlation for two rows of holes for all distances, mass flux ratios, and injection-to-mainstream temperature ratios in terms of a parameter $(x/s)I^{-2/3}$. The term $-B/A$ includes constants A and B from the superposition formula $q/K(T_\infty - T_\omega) = A + B\theta$. The dimensionless temperature term θ is defined as $(T_\infty - T_c)/(T_\infty - T_\omega)$.

3.8.3.4 Effect of Streamwise Curvature and Pressure Gradient

The effect of streamwise curvature is very important for film cooling over gas turbine airfoils. The pressure surface is a concave surface, and the suction surface is a convex surface. The effect of local pressure gradients can alter the film-cooling effectiveness of a cooling jet compared to a flat surface. Ito et al. (1978) studied film cooling on a gas turbine blade in a cascade. They injected coolant on the pressure (concave) and suction (convex) surfaces to determine the effects of surface curvature on film-cooling effectiveness. Figure 3.85 presents the lateral averaged film effectiveness results on a convex surface with a coolant-to-mainstream density ratio of 0.95. The effectiveness increases as M increases from 0.2 to 0.5 and peaks for an M value between 0.5 and 0.74. The effectiveness near injection decreases with increasing blowing ratio greater

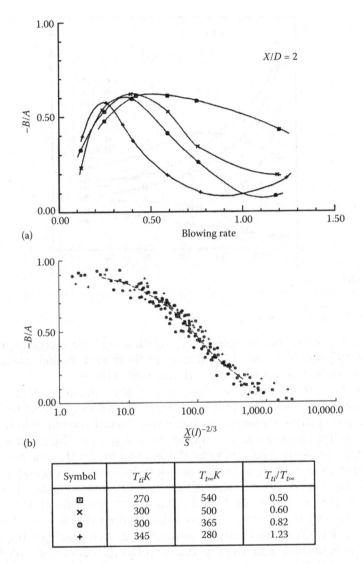

FIGURE 3.84
(a) Variation of B/A with blowing ratio at different injection-to-mainstream temperature ratios for a single row of holes; (b) correlation of film-cooling data for two rows of holes with x/s and I. (From Forth, C.J.P. and Jones, T.V., Scaling parameters in film cooling, *Proceedings of the Eighth International Heat Transfer Conference*, San Francisco, CA, 1986.)

than 0.7. For low blowing ratios, effectiveness is highest near the injection hole and decreases downstream. However, for blowing ratios greater than 0.7, the effectiveness is low immediately downstream of injection, reaches a peak farther downstream, and then starts decreasing gradually. This may be due to jet liftoff and reattachment downstream. The jets at high blowing ratio lift off immediately after injection and reattachment at a downstream

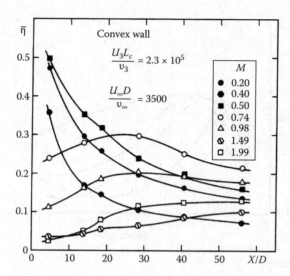

FIGURE 3.85
Laterally averaged film effectiveness results on a convex surface. (From Ito, S. et al., *ASME J. Eng. Power*, 100, 476, 1978.)

location, depending on jet momentum. As blowing ratio increases further, the jet reattachment is also reduced as jet is dissipated into the mainstream and the curvature is against the liftoff direction of the jet.

Figure 3.86 presents the laterally averaged film effectiveness results for a concave surface for a coolant-to-mainstream density ratio of 0.95. The trends are very different for concave surfaces. Overall, the film effectiveness is very low for concave surfaces compared to convex surfaces. This is due to the static pressure force acting on the jets that moves the jets away from the surface. Effectiveness near injection decreases with increasing blowing ratio. Unlike the convex surface, highest effectiveness immediately downstream of injection is obtained for the lowest blowing ratio of $M = 0.2$. As blowing ratio increases, effectiveness values farther downstream begin to increase as jets lift off and reattach downstream. At higher blowing ratios, the jets seem to show a stronger reattachment as the curvature is along the direction of liftoff.

Figure 3.87 presents the comparison of lateral averaged film effectiveness for different density ratios on convex, concave, and flat surfaces for a blowing ratio of $M = 0.5$. For any density ratio, the convex surface provided the highest film effectiveness. The lowest effectiveness is obtained on the concave surface. The trends with increasing density ratio are different for each surface. For the convex surface, the effectiveness decreases with increasing density ratio. For the flat surface, effectiveness increases significantly for an increase in density ratio from 0.75 to 1.5. Further increases in density ratio produce small increases in film effectiveness. For the concave surface, effectiveness increases with increasing density ratio.

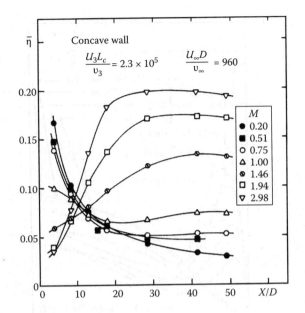

FIGURE 3.86
Laterally averaged film effectiveness results on a concave surface. (From Ito, S. et al., *ASME J. Eng. Power*, 100, 476, 1978.)

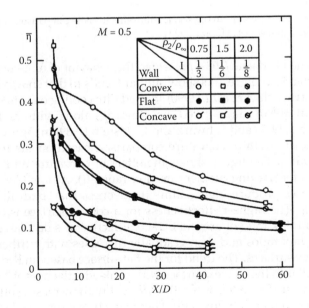

FIGURE 3.87
Comparison of lateral averaged film effectiveness for different density ratios and curvature effects. (From Ito, S. et al., *ASME J. Eng. Power*, 100, 476, 1978.)

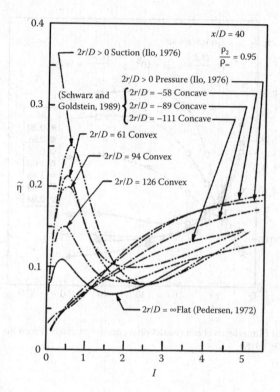

FIGURE 3.88
Behavior of film-cooling jets on curved surface–film effectiveness comparison. (From Schwarz, S.G. et al., *ASME J. Turbomach.*, 112, 472, 1990.)

Schwarz and Goldstein (1989) presented the behavior of film-cooling jets on concave surfaces. Their results show similar trends to that shown by Ito et al. (1978). Similarly, Schwarz et al. (1990) studied film cooling on convex surfaces. Schwarz et al. (1990) summarized their findings along with the findings of Pedersen et al. (1977) and Schwarz and Goldstein (1989) in Figure 3.88. They provided the strength of curvature comparison based on the results from all three studies. The figure shows the effect of curvature on film-cooling effectiveness for varying momentum flux ratios (I). At low blowing ratios, convex surfaces produce higher film effectiveness than both flat and concave surfaces. The highest effectiveness for a convex surface is obtained at a momentum flux ratio of around 0.5–1.0. Effectiveness decreases at higher momentum flux ratios and then gradually increases with further increases in I for convex surfaces. The trend for the flat surface based on Pedersen et al. (1977) is similar to the convex surface with peak effectiveness at $I = 0.5$–1.0.

The concave surface effect is totally different with increasing effectiveness by increasing momentum flux ratio. There may be a strong coolant accumulation effect as the I value increases for concave surfaces. The jets may lift off and reattachment occurs slightly downstream due to the concave nature of the surface. The results for convex and concave surfaces from the airfoil-cooling study by Ito et al. (1978) are also presented for comparison. The trends from their data are similar. It may be fairly reasonable to state that a convex surface enhances cooling effectiveness at low momentum flux ratios as does a concave surface at higher momentum flux ratios.

3.8.3.5 Effect of High Free-Stream Turbulence

As describe in Chapter 2 and earlier in this chapter, free-stream turbulence is a parameter that cannot be ignored in turbine passages. Film-cooling characteristics are significantly affected by presence of free-stream turbulence. Many researchers focused on the singular effect of free-stream turbulence on film effectiveness without other parametric effects and performed studies on flat surfaces with high free-stream turbulence.

Bons et al. (1996) studied the effect of high free-stream turbulence on film-cooling effectiveness. Film-cooling effectiveness data for Tu levels ranging from 0.9% to 17% at a constant free-stream Reynolds number and varying blowing ratios from 0.55 to 1.85 are presented. The most significant result from their study is shown in Figure 3.89. The hole centerline film effectiveness for two blowing ratios of $M = 0.75$ and 1.5 are presented at four different turbulence levels.

Let us focus on the data for $M = 0.75$. Film effectiveness at a low Tu level of 0.9% is very high near injection (0.65) and decreases steadily downstream. With increasing free-stream turbulence from $Tu = 0.9\%$ to 17%, the peak film effectiveness immediately downstream of injection decreases significantly. The effect of higher turbulence decreases far downstream, with lesser decreases in film effectiveness values. At low M, the coolant possesses low momentum. The low momentum jets stay closer to the surface and are weaker than the mainstream. With increased free-stream turbulence, large fluctuations in the mainstream break down the weak jets and reduce film coverage. The jet structure is not developed completely due to the strong fluctuations in the mainstream. This is the reason for the significant drop in film effectiveness at high Tu levels for low M jets.

At higher $M = 1.5$, the jets are stronger, possessing stronger momentum than the mainstream at the injection location. Since the jets lift off from the surface and weakly reattach downstream, effectiveness is low overall for high blowing ratios. The effect of free-stream turbulence is not as significant as that for $M = 0.75$. The strong coolant jet sustains under high free-stream turbulence and provides similar effectiveness as that for low Tu level for $X/D < 20$. Further downstream, effectiveness values decrease with increasing turbulence. After lifting off at injection, the jets are pushed back on to the surface farther downstream. On reattachment, the jets are dissipated faster under stronger free-stream turbulence. This may be the reason for lower effectiveness far downstream for high free-stream turbulence.

At high free-stream turbulence, heat-transfer coefficients with film cooling are not as significantly affected as the film effectiveness. Film injection by

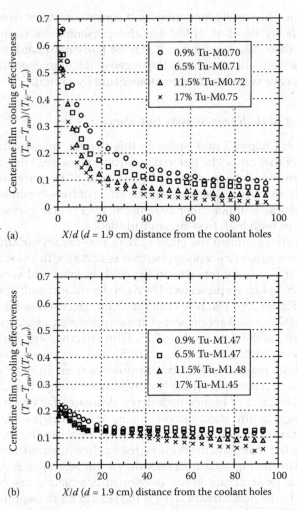

(a)

(b)

FIGURE 3.89
Effect of free-stream turbulence on film effectiveness. (a) Blowing ratio = 0.75 and (b) blowing ratio = 1.5. (From Bons, J.P. et al., *ASME J. Turbomach.*, 118, 814, 1996.)

itself produces high heat-transfer coefficient enhancement due to high turbulent mixing between jet mainstream. Since the heat-transfer coefficients are already significantly enhanced, higher free-stream turbulence does not seem to further enhance the heat-transfer coefficient. While heat-transfer coefficients are enhanced with film cooling, higher film effectiveness is a result of stable jet structure close to the surface. High free-stream turbulence does not affect the levels of turbulence produced in jet mainstream mixing, but it breaks down coolant jet structures. Therefore, high free-stream turbulence reduces film effectiveness significantly while affecting the heat-transfer coefficient slightly, thus causing a significant increase in heat flux into the film-cooled surface.

Burd et al. (1996) and Kohli and Bogard (1998) made flow field and temperature field measurements, respectively, to quantify the effect of free-stream turbulence on film-cooling jets.

3.8.3.6 Effect of Film Hole Geometry

Compound-angle injection provides higher film effectiveness than simple-angle injection for one row of holes. Ligrani et al. (1994) compared the effect of hole angle for two staggered rows. The hole geometry is shown in Figure 3.68. Figure 3.90 presents the laterally averaged film effectiveness distributions for configurations 2 and 3 at three blowing ratios. It is fairly evident that the simple-angle data are lower than the compound-angle data. Effectiveness for compound-angle injection is higher due to stronger lateral momentum and more spreading of jets, providing higher effectiveness over a larger spanwise region. The simple-angle film cooling is mostly along the hole in the downstream direction, with very little lateral spreading of jets. The effect is strongest in the near injection region ($X/D < 20$).

Ekkad et al. (1997) presented detailed film effectiveness distributions for three different compound-angle injection angles. The angle along the flow direction (α) was 35°. The angle lateral to the flow was varied from 0° (simple) to 45° to 90° (spanwise). For simple-angle injection ($\beta = 0°$), film effectiveness is high along the hole angle and decreases rapidly downstream. Effectiveness decreases with increasing blowing ratio. With a compound

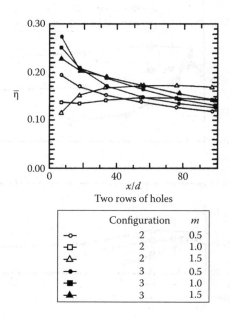

	Configuration	m
–o–	2	0.5
–□–	2	1.0
–△–	2	1.5
–•–	3	0.5
–■–	3	1.0
–▲–	3	1.5

FIGURE 3.90
Laterally averaged film effectiveness results for hole configurations shown in Figure 3.68. (From Ligrani et al., *ASME J. Heat Transfer*, 116, 353, 1994b.)

angle of $\beta = 45°$, film effectiveness is higher than for simple-angle injection. It is interesting to note that the high film effectiveness regions are along the film hole direction. The lateral smearing of high film effectiveness to the regions between the holes is the advantage of compound-angle injection. With lateral injection ($\beta = 90°$), film effectiveness between the hole is higher than for $\beta = 45°$. The spanwise injection is directly into the region between the holes, and effectiveness is higher due to increased film coverage. The higher the film effectiveness, the stronger the lateral smearing of the jets.

Figure 3.91 presents the film effectiveness results for the hole geometries presented in Figure 3.70. Schmidt et al. (1996) compared the spatially averaged film effectiveness for the three geometries against the momentum flux ratio (I). The results are shown at two hole-to-hole, pitch-to-diameter (P/D) ratios of 3.0 and 6.0. Again, it is evident that hole geometry does not provide significant effect at low momentum flux ratio (or blowing ratio). But with compound-angle hole injection ($CA = 60°$), the film effectiveness is significantly improved at higher momentum flux ratios. With forward expanded exits and $CA = 60°$, film effectiveness improves further at higher I values. The effect is reduced for the larger P/D ratio of 6.0. Overall effectiveness values are also lower for each case with an increased P/D ratio.

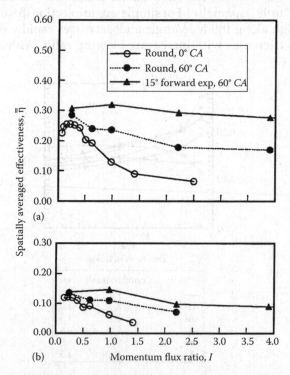

FIGURE 3.91
Spatially averaged film effectiveness results for hole configurations shown in Figure 3.70. (a) $P/D = 3$, (b) $P/D = 6$. (From Schmidt, D.L. et al., *ASME J. Turbomach.*, 118, 807, 1996a.)

Compared to cylindrical holes, expanded holes are expected to provide better thermal protection for the film-cooling surface at high blowing ratios. At low blowing ratios, the effect of film hole shape is not very significant. Some typical expanded hole geometries are shown in Figure 3.72. The first hole is the standard cylindrical hole; the second hole is called the fan-shaped hole where the inlet is cylindrical and the exit is flared. The third hole has an extra lip compared to the fan-shaped hole, making the exit larger. This hole is called a laidback fan-shaped hole. Gritsch et al. (1998a) studied all three hole shapes and compared film effectiveness results.

For the simple cylindrical hole, the film effectiveness decreases significantly with increasing blowing ratio. The jets penetrate the mainstream with increasing blowing ratio, resulting in lower film effectiveness. The distribution clearly shows jet separation at a higher blowing ratio. For the fan-shaped hole, the film effectiveness is not as significantly reduced with increasing blowing ratio. The jets spread better over a greater spanwise area due to the flared exit. However, at higher blowing ratio, the effect of jet separation is still evident just downstream of injection. For the laidback fan-shaped hole, the local film effectiveness downstream of the film hole is lower than that for the fan-shaped hole. However, the jet spreads more in the spanwise direction, providing higher film effectiveness laterally.

Figure 3.92 presents the laterally averaged film-cooling effectiveness versus the axial distance for the cases presented in Figure 3.72. For the simple cylindrical hole, a lower blowing ratio of $M = 0.5$ provides the highest effectiveness and decreasing effectiveness with increasing blowing ratio. For the fan-shaped hole, film effectiveness is highest at $M = 1.0$ and then decreases

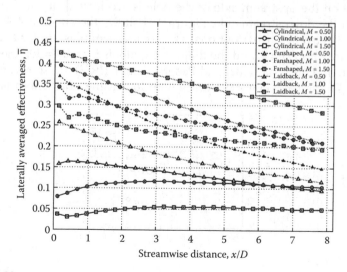

FIGURE 3.92
Laterally averaged film effectiveness results for shaped holes. (From Gritsch, M. et al., *ASME J. Turbomach.*, 120, 549, 1998b.)

with increasing blowing ratio. However, for the laidback fan-shaped hole, film effectiveness increases monotonically with increasing blowing ratio. It is interesting to note that the effectiveness for the laidback fan-shaped hole at $M = 0.5$ is comparable to the effectiveness obtained at $M = 1.0$ for the fan-shaped hole. The improved film protection characteristics of the expanded hole exit are clearly evident from this study.

3.8.3.7 Effect of Coolant Supply Geometry

Hole supply geometry typically indicates how the coolant enters the film hole and what happens to the coolant within the film hole. Pietrzyk et al. (1990) and Burd et al. (1996) indicated a strong effect of hole length-to-diameter ratio (L/D). Burd et al. (1996) made flow measurements on a flat surface with different hole lengths feeding the film coolant. Leylek and Zerkle (1994) indicated that short holes are subject to a "jetting" effect. Jetting occurs when the jet velocity profile is not uniformly distributed across the majority of the plane of exit into the mainstream. With short-hole injection, the coolant penetrates further into the mainstream in the wall-normal direction and spreads more in the spanwise direction. The effect of hole L/D is to skew the velocity profiles of the jets exiting the hole. Seo et al. (1998) recently studied the effect of hole L/D on heat transfer downstream of film-cooling holes. Figure 3.93 presents the injectant velocity profiles at the hole exit for three different hole L/D values of 1.6, 4, and 10. The distributions clearly show the nonuniform distribution of velocity inside the hole. The injectant velocity profiles become more uniform as the hole length increases. The skewing toward the upstream side of the hole decreases as the flow develops within the hole. The flow on the upstream side of the hole is accelerated due to flow separation on the downstream side. Figure 3.94 shows the effect of hole length on film effectiveness distributions. The circular symbols are for $L/D = 1.6$, squares are for $L/D = 4$, and the diamond symbols for $L/D = 10$. The data of Goldstein and Yoshida (1982) and Pedersen et al. (1977) are shown for comparison. That effectiveness increases with increasing hole length is clearly

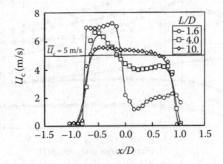

FIGURE 3.93
Injectant exit velocity profiles for holes with different length-to-diameter ratios. (From Seo, H.J. et al., *Int. J. Heat Mass Transfer*, 41(22), 3515, 1998.)

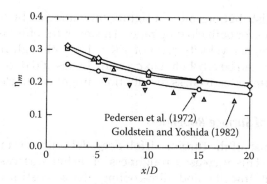

FIGURE 3.94
Effect of hole length on film effectiveness distributions. (From Seo, H.J. et al., *Int. J. Heat Mass Transfer*, 41(22), 3515, 1998.)

evident from the figure. A good exit condition for the film jet produces higher effectiveness.

In actual airfoils, the coolant is supplied for external cooling through internal channels. The coolant supply geometry is another important factor that may influence film effectiveness distributions downstream of injection. Burd and Simon (1997) studied the influence of coolant-supply geometry on film-cooling effectiveness. They studied four different coolant-delivery configurations. The first two configurations are for unrestricted coolant supply from a large plenum with short and long holes. Configuration (*c*) forces the coolant to enter the film-cooling holes flowing in the opposite direction of the mainstream and is called counterflow delivery. Configuration (*d*) delivers the coolant in the same flow direction as that of the mainstream and is called coflow delivery.

Figure 3.95 presents the effect of coolant supply on laterally averaged film effectiveness for short hole lengths. From the figure, it is evident that the film

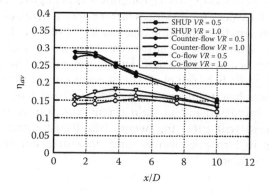

FIGURE 3.95
Effect of coolant supply geometry on laterally averaged film effectiveness. (From Burd, S.W. and Simon, T.W., The influence of coolant supply geometry on film coolant exit flow and surface adiabatic effectiveness, ASME Paper 97-GT-25, 1997.)

effectiveness distributions are not significantly affected by the coolant flow delivery systems for both blowing ratios. However, the effect is a significantly stronger at a higher velocity ratio of $VR = 1.0$. Although the coolant-flow delivery system can be attributed to pressure losses and different flow structures, overall film effectiveness distributions might be affected significantly.

3.8.3.8 Effect of Surface Roughness

The surface finish of the turbine airfoil degrades with continuous operation of the engine. This increased roughness of airfoil surfaces can adversely affect the heat transfer and film-cooling characteristics of the original design. It then becomes important to understand the extent of the effects of surface roughness on film-cooling performance. Two studies by Goldstein et al. (1985) and Barlow and Kim (1995) investigated the effects of surface roughness on film cooling on a flat surface. Goldstein et al. (1985) found that roughness decreased effectiveness by as much as 20% at low blowing ratios and increased it by 50% at high blowing ratios relative to that for a smooth wall. Barlow and Kim (1995) reported that film effectiveness is reduced by smaller roughness elements more than larger elements. They also presented heat-transfer coefficient enhancement results.

Schmidt et al. (1996) designed rough surfaces to match equivalent sand-grain roughness Reynolds number for typical engine vane surfaces after a number of hours of operation. Figure 3.96 presents the two roughness plate designs used in their study. They compared the results for the two rough surfaces with their results for the smooth surface. The roughness element dimensions are also shown in the figure. Figure 3.97 shows the effect of surface roughness on lateral averaged film effectiveness as a function of I. The effect is shown at different axial locations and is significant at $X/D = 3$, where there is a reduction of up to 20% in peak film effectiveness at low I values. At higher momentum flux ratios, the effect of surface roughness is negligible. The effect is also not evident at all I values for distances farther downstream ($X/D = 15, 90$).

Figure 3.98 present the centerline distribution of film-cooling heat-transfer coefficient enhancement for the smooth and rough surfaces. It appears that the film-cooling heat-transfer coefficient is not much affected by surface roughness. However, Schmidt et al. (1996) point out that surface roughness by itself enhances heat-transfer rates by 50% in the absence of film cooling and that these increased rates would occur with film cooling also. Thus, surface roughness will affect the overall heat-transfer rate even without a significant loss in film effectiveness and nonenhancement of film-cooled heat-transfer coefficient.

3.8.3.9 Effect of Gap Leakage

Yu and Chyu (1998) investigated the effect of gap leakage downstream of a row of film holes and how the presence of the gap affects the film-cooling

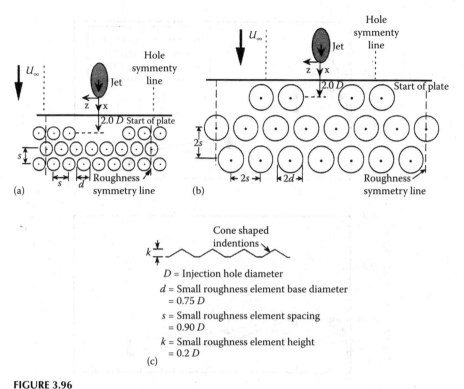

FIGURE 3.96
Roughness plate designs used by Schmidt et al. (1996). (a) Rough plate 1, (b) rough plate 2, and (c) side view of roughness. (From Schmidt, D.L. et al., *ASME J. Turbomach.*, 118, 807, 1996a.)

performance of the film hole row. The turbine structure on the endwall region of the assembly consists of several individual components instead of a single integral cast piece. Typically, assembly of several individual components presents such inevitables as gaps, which are present to accommodate for differential growth of these components. Depending on the pressure differences, the leak through the gap may flow either in or out of the gas path. Yu and Chyu (1998) investigated this aspect of turbine design. Figure 3.99 shows the conceptual view of a gap located downstream of film injection holes.

Yu and Chyu (1998) investigated the effect of positive and negative flows through the gap. Figure 3.100 shows the effect of coolant addition from the gap into the mainstream. The figure is divided into two parts. The first part shows the distribution of the first effectiveness (film hole row effect) for various positive gap leakage ratios. The ratio M2:M1 shows the ratio of gap leakage to film hole flowrate. It appears that a mild addition of coolant ejected out of the gap promotes better film protection. However, stronger gap injection eventually leads to lifting of the film boundary layer away from the wall, reducing film coverage immediately downstream of the gap. The second part of the figure shows the film effectiveness distribution due to gap injection.

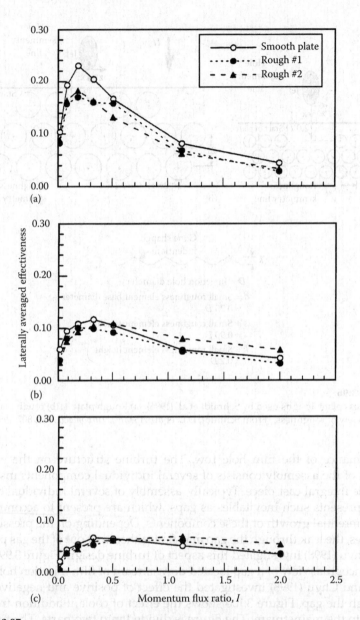

FIGURE 3.97
Effect of surface roughness on laterally averaged film effectiveness distributions. (a) $x/D = 3$, (b) $x/D = 15$, and (c) $x/D = 90$. (From Schmidt, D.L. et al., *ASME J. Turbomach.*, 118, 807, 1996a.)

The second injectant from the gap lacks sufficient momentum to penetrate the boundary layer, and it provided adequate film coverage when the film hole coolant flow is shut off. In the event that the coolant flow is added, film effectiveness downstream of the gap is substantially enhanced. A stronger blowing ratio (M) from the first injection further enhances effectiveness.

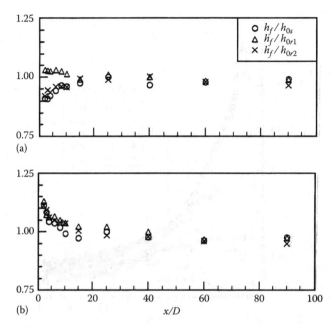

FIGURE 3.98
Effect of surface roughness on hole centerline film-cooled heat-transfer coefficient enhancement. (a) $I = 0.3$ and (b) $I = 1.1$. (From Schmidt, D.L. et al., *ASME J. Turbomach.*, 118, 807, 1996a.)

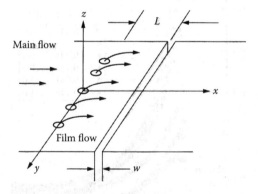

FIGURE 3.99
Illustration of typical gap leakage geometry. (From Yu, Y. and Chyu, M.K., *ASME J. Turbomach.*, 120, 541, 1998.)

Figure 3.101 presents the effect of negative gap leakage on film effectiveness. The leakage flow suction ratio is also provided in the figure. It appears that at low film hole blowing ratio ($M1$) of 0.5, the effect of negative gap leakage is small. However, the situation is different for $M1 = 1.0$. In this case, the effectiveness is enhanced when gap leakage increases from -0.042 to -0.084. However, further increase in $M2$ to -0.126 is detrimental to the

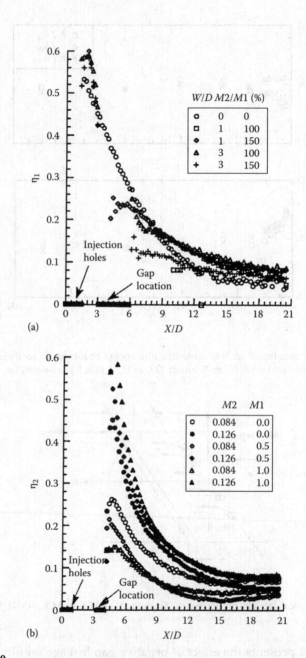

FIGURE 3.100
Effect of coolant addition from gap on film effectiveness: (a) effectiveness from injection hole and (b) effectiveness from gap flow. (From Yu, Y. and Chyu, M.K., *ASME J. Turbomach.*, 120, 541, 1998.)

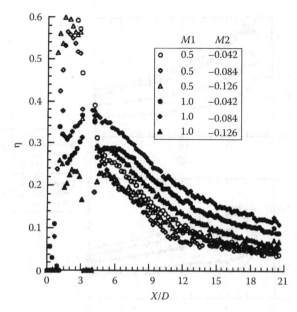

The chart legend:

	$M1$	$M2$
○	0.5	−0.042
◇	0.5	−0.084
△	0.5	−0.126
●	1.0	−0.042
◆	1.0	−0.084
▲	1.0	−0.126

FIGURE 3.101
Effect of negative gap leakage on film effectiveness. (From Yu, Y. and Chyu, M.K., *ASME J. Turbomach.*, 120, 541, 1998.)

film effectiveness. It appears that a moderate level of gap leakage drags the injection from the film hole toward the surface and enhances protection, but further increases deplete the coolant levels downstream of the gap, reducing film effectiveness.

3.8.3.10 Effect of Bulk Flow Pulsations

Recently, studies have focused on issues in film cooling that may have a serious impact on film-cooling performance of jets. Ligrani et al. (1997) and Seo et al. (1998) investigated the effects of bulk flow pulsations on film cooling through a row of holes. Periodically unsteady static pressure fields at the exits of injection holes result in pulsating coolant rates. In addition, passing potential flow disturbances and shock waves produce pulsating static pressure and velocity fields in the boundary layer just downstream from the injection holes. The pulsations produce important changes to the flow structure, especially because the pulsating coolant flowrates cause the unsteady static pressure fields. Figure 3.102 shows the effect of bulk flow pulsation on film effectiveness distributions. In general, the bulk flow pulsation is to decrease the film-cooling effectiveness.

3.8.3.11 Full-Coverage Film Cooling

Full-coverage discrete film cooling offers an effective cooling technique of the combustor wall and airfoils. The coolant air passes through multiple

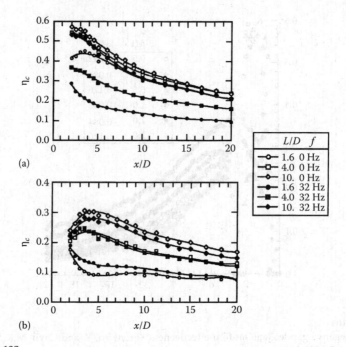

FIGURE 3.102

Effect of bulk flow pulsations on film effectiveness distributions (a) $M=0.5$ and (b) $M=1.0$. (From Seo, H.J. et al., Effects of bulk flow pulsations on film cooling from different length injection holes at different blowing ratios, ASME Paper 98-GT-192, 1998.)

rows of discrete holes cooling the inside wall and providing film to protecting the gas-side surface. It is also called effusion or transpiration cooling. Andrews et al. (1991) investigated the various effects of parameters on full-coverage film cooling. They presented the effects of hole length on film effectiveness to a surface covered by full-coverage film cooling. Martiny et al. (1995) also investigated film effectiveness on a plate with full-coverage film cooling. They used a test plate with a series of staggered film hole rows. The holes are all angled 17° along the streamwise direction. The spacing between holes in each row was $s/d = 4.48$, and the distance from row to row was $p/d = 7.46$.

Figure 3.103 shows the spanwise averaged film effectiveness along the test plate for the first four hole rows. The effect of blowing ratio is also shown. It appears that film effectiveness increases rapidly with increasing film attachment as coolant spreads both streamwise and spanwise. The coolant accumulation effect is significant at lower blowing ratios. Full-coverage film cooling is a very effective cooling technology used in cooling combustor walls. The limitation of coolant mass flow for providing cooling in airfoils is the main reason for the lack of using full-coverage film cooling in airfoils.

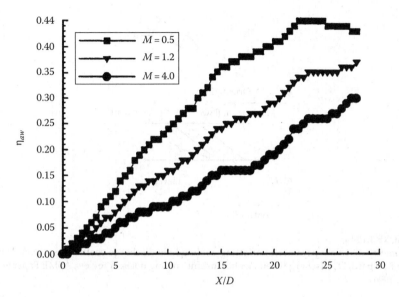

FIGURE 3.103
Spanwise-averaged film effectiveness along the test plate for first four-hole rows. (From Martiny, M. et al., Full coverage film cooling investigations: Adiabatic wall temperatures and flow visualizations, ASME Paper 95-WA/HT-4, 1995.)

3.9 Discharge Coefficients of Turbine Cooling Holes

Knowledge of the discharge coefficient C_d is vital in sizing film-cooling holes at the design stage to determine the requisite amount of coolant flow to produce the desired film effectiveness. The discharge coefficient depends on both the local geometry and the flow conditions upstream and downstream of the hole. Hay and Lampard (1996) presented a review of the published studies on discharge coefficients of turbine-cooling holes. The geometries and flow parameters that affected turbine-cooling hole discharge coefficients are illustrated in Figure 3.104. The various geometrical and flow parameters are listed by Hay and Lampard (1996). They indicated that the designer has three courses of action to estimate C_d for any hole:

1. Examine the flows in the immediate vicinity of the hole inlet and outlet separately, using empirical correlations, potential flow, or computational fluid dynamics (CFD) solutions as appropriate to get pressure loss coefficients for hole inlet and outlet. The complete hole C_d can be based on a simple loss coefficient analysis that included mixing within the hole.

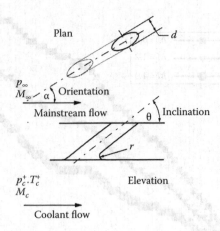

FIGURE 3.104
Geometry and flow parameters affecting cooling-hole discharge coefficients. (From Hay, N. and Lampard, D., Discharge coefficient of turbine cooling holes: A review, ASME Paper 96-GT-492, 1996.)

2. Modify the hole C_d using empirical correlations or other means to determine the changes that arise from the presence of cross-flows and from geometric variations such as inlet radiusing.

3. Measure or calculate C_d for the complete hole geometry under the specified flow conditions.

Presently, in the industry, methods 1 and 2 are commonly used. Although method 2 is the most popular technique for estimating C_d values, with complex hole shapes and orientations becoming common, the C_d values cannot be estimated empirically. There is a need for the first technique (1) that will take into effect most parameters that affect the discharge coefficient. Figure 3.105 shows the effect of velocity head ratio on discharge coefficients for normal cooling holes with an inlet cross-flow. A potential flow analysis solution predicts the measured data from Rohde et al. (1969), Hay et al. (1981), and Byerley (1989). The velocity head ratio $(p_c^+\text{-}p_\infty)/(p_c^+\text{-}p_c)$, has been shown to work over a wide range of inlet cross-flow Mach numbers for orifices with $L/d > 2$. This range is applicable for typical film-cooling holes in gas turbine blades and vanes. Figure 3.106 shows the effect of hole inclination and orientation on discharge coefficient from Byerley (1989). All holes show increasing C_d values for higher velocity head ratios. The best performance is achieved for a 30° inclined hole, which is the typical hole inclination in the streamwise direction for airfoil film holes.

Gritsch et al. (1998c) presented discharge coefficient measurements for film-cooling holes with expanded exits. They compared their expanded-hole exits with simple cylindrical-hole cases. The holes they used in their study are similar to those used by Gritsch et al. (1998a,b). The holes are illustrated in Figure 3.72. Figure 3.107 shows the effect of hole shape on the discharge coefficients for different pressure ratios. For zero internal and external

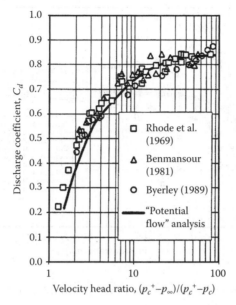

FIGURE 3.105
Effect of velocity head ratio on discharge coefficients through normal holes. (From Hay, N. and Lampard, D., Discharge coefficient of turbine cooling holes: A review, ASME Paper 96-GT-492, 1996.)

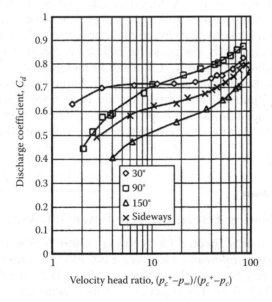

FIGURE 3.106
Effect of hole inclination and orientation on discharge coefficients. (From Byerley, A., Heat transfer near the entrance to a film cooling hole in a gas turbine blade, MS thesis, Air Force Institute of Technology, Wright-Patterson AFB, OH, 1989.)

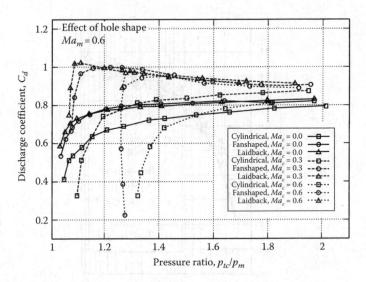

FIGURE 3.107
Effect of hole shape on discharge coefficients. (From Gritsch, M. et al., *ASME J. Turbomach.*, 120, 557, 1998c.)

cross-flow, the discharge coefficients with expanded exits were higher than the conventional cylindrical hole. With internal cross-flow applied, all holes showed an increase in discharge coefficients. At low-pressure ratios, there is a significant increase in C_d values over holes without internal cross-flow ($Ma_c = 0$). At higher-pressure ratios, the effect is reduced. This is due to the pressure recovery taking place in the diffuser section of the hole. This exists only for the cases with low-pressure ratios. At elevated pressure ratios, the pressure recovery is reduced, leading to moderate decrease of the discharge coefficient with pressure ratio. The discharge coefficients for both the expanded exits (laidback fan shaped vs. fan shaped) were similar. However, laidback fan-shaped holes offer improved thermal protection as compared to fan-shaped holes due to better lateral spreading of the cooling jets.

3.10 Film-Cooling Effects on Aerodynamic Losses

As discussed in this chapter, film cooling is an effective cooling solution for hot-gas-path components. However, film injection causes a decrease in thermal efficiency because the cooler injection gas is mixed with the hot gases in the main flow. The injection of this coolant into the mainstream also causes aerodynamic losses. Denton (1993) provides an extensive review of loss mechanisms in turbomachinery flows. Ito et al. (1980), Haller and Camus (1984), and Kollen and Koschel (1985) reported loss measurements

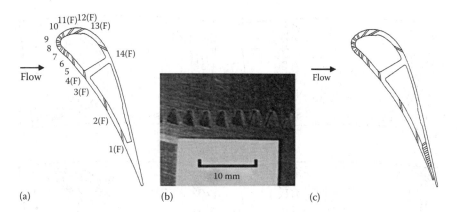

FIGURE 3.108
NGV with fanned-hole geometry studied by Day et al. (1999). (a) Cross-section of cylindrical hole cooled blade, (b) fan shaped hole, and (c) cross-section of cooled blade with TE ejection. (From Day, C.R.B. et al., *ASME J. Turbomach.*, 121, 145, 1999.)

for turbine airfoils with and without film cooling. They concluded that the loss mechanism depended on how and where the coolant was ejected into the mainstream. Day et al. (1999) presented aerodynamic loss measurements from film-cooled NGVs at engine representative Mach numbers and Reynolds numbers. However, they studied only cylindrical holes.

Day et al. (1998) provided efficiency measurements for the same bowed NGV model as Day et al. (1999) with different film-cooling geometries. The experiments were conducted in a cold heat-transfer tunnel (CTHT) with an annular transient blowdown cascade. Cylindrical hole injection was compared with fanned hole injection. Figure 3.108 shows the NGV with fanned-hole geometry used by Day et al. (1998).

Figure 3.109 shows the total pressure maps in the measurement plane normalized relative to the upstream total pressure. Figure 3.109a through e show different cooling configurations: Figure 3.109a is for no coolant ejection, Figure 3.109b is for cylindrical cooling geometry, Figure 3.109c is for fanned-hole geometry, Figure 3.109d is for trailing-edge ejection, and e is for trailing-edge ejection plus full cylindrical geometry. Figure 3.109a through c without trailing-edge ejection have similar wake shapes. The deepest point of the wake is along the midspan of the vane. The wakes become thinner and shallower toward the endwalls. The coolant injection basically broadens the wake slightly, and fan-shaped holes produce the broadest wake. It appears that coolant efflux velocities produce more aerodynamic losses. The passage with trailing-edge ejection produces a narrower and more bowed wake. Ejection produces a deeper wake (larger values in the core) without widening the wake shape (narrow or broaden) itself. It is clear from this result that wake shape and losses strongly depend on the type and location of coolant ejection.

Kubo et al. (1998) presented a CFD investigation on the total pressure loss variation for a linear NGV cascade due to individual film injections.

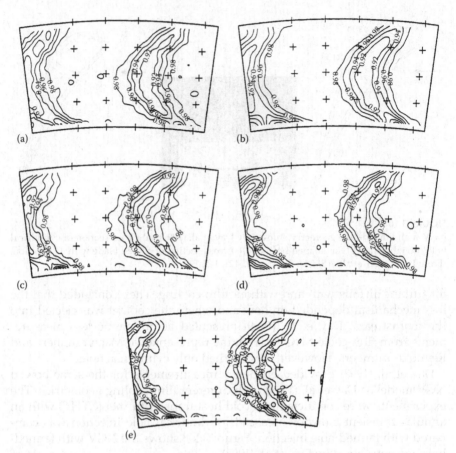

FIGURE 3.109
Total pressure maps normalized relative to upstream total pressure. (a) No coolant, (b) cylindrical hole, (c) fanned hole, (d) TE ejection, and (e) TE ejection with cylindrical hole. (From Day, C.R.B. et al., *ASME J. Turbomach.*, 121, 145, 1999.)

Figure 3.110 shows the vane and cooling hole configuration. They compared their computations to data generated by Otomo et al. (1997) on a low-speed wind tunnel facility. Total pressure loss coefficients c_{pt} between the inlet and outlet of the nozzle cascade were evaluated as $c_{pt} = (P_{t1} - P_{t2})/(P_{t1} - P_{s2})$ where P_{t1}, P_{t2}, and P_{s2} are inlet total pressure, the outlet total pressure, and the outlet static pressures, respectively.

Figure 3.111 indicates the experimentally measured loss coefficient profiles in the all-row injection case. This indicates that all the film hole rows are flowing. The mass flowrate ratio of each injection to total injection rate was maintained as 8.4% for SH rows, 14.4% for $S1$, 22.6% for $S2$, 11.3% for $P1$, 10.9% for $P2$, and 32.4% for TE, respectively, regardless of the total coolant-to-gas mass flow ratio (G_c/G_g). The results show a deviation on the suction-side slope of the wake and a simple increase of the peak value. Also the

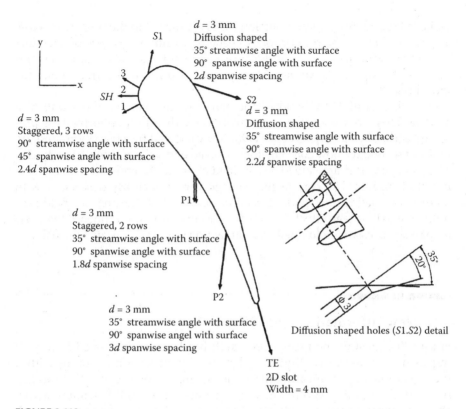

FIGURE 3.110
Vane and cooling-hole configuration used by Kubo et al. (1998). (From Kubo, R. et al., Aerodynamic loss increase due to individual film cooling injections from gas turbine nozzle surface, ASME Paper 98-GT-497, 1998.)

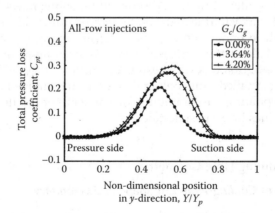

FIGURE 3.111
Experimentally determined loss coefficient profiles. (From Otomo, F. et al., Pressure loss characteristics of gas turbine nozzle blade row with individual film cooling injection rows, *Proceedings of the 25th Annual Gas Turbine Conference*, New Orleans, LA, JSME, 1997.)

wake is broader for coolant-injection cases compared to the no-coolant case. They also showed the proportions of each injection loss to type of injection. The ratios of the losses of $P1$ and $P2$ are significant at small G_c/G_g values and decrease at large G_c/G_g values. At large G_c/G_g values, TE ejection causes the most losses.

Friedrichs et al. (1997) studied the aerodynamic aspects of endwall film cooling. This work is related to the work on endwall film effectiveness presented earlier by Friedrichs et al. (1996). They showed the stagnation pressure loss and secondary flow contours for uncooled endwalls and cooled endwalls. The passage vortex appears to be confined closer to the endwall with increasing blowing rate. The core of the passage vortex is clearly descended with increasing blowing rate. Coolant ejection locally affects the pressure field near holes due to blockage effect. The coolant ejection will also influence the overall passage pressure field due to its interaction with the passage flow field.

3.11 New Information from 2000 to 2010

Since 2000, most of the focus of research has been on improved film cooling using aggressive cooling-hole designs. Improved film cooling with a focus on reduced coolant amount providing higher cooling effectiveness has resulted in innovative cooling-hole exit geometries. In addition, manufacturability issues and realistic effects of manufacturing have been considered. The effect of fuel-burn deposits on cooling-hole performance has been a focus of many studies. Turbine tip cooling has moved away from simplistic models to more realistic loaded turbine airfoils simulating pressure driven tip leakage flows interacting with discrete film-cooling flows. Endwall cooling has also been a focus due to the flatter temperature profiles and the presence of endwall secondary flows causing significant heat-transfer issues on the vane and blade platform areas. There have been many more studies on engine type configurations with results on turbine stages. Challenges still exist in getting detailed film-cooling data on actual engine type configurations. We are updating the community on some important developments in film-cooling issues in the past decade.

3.11.1 Film-Cooling-Hole Geometry

3.11.1.1 Effect of Cooling-Hole Exit Shape and Geometry

Sargison et al. (2001a,b) presented experimental measurements of the performance of a new film-cooling-hole geometry—the converging slot-hole or console. This novel, patented geometry has been designed to improve the heat transfer and aerodynamic loss performance of turbine vane and rotor blade cooling systems. The laterally averaged adiabatic effectiveness results demonstrated that the console cooling effectiveness approached that of the slot, as does the fan-shaped hole effectiveness. The console laterally averaged heat-transfer coefficient was similar to the slot, and higher than the cylindrical and fan-shaped hole results. The slot and console do not significantly change the boundary layer flow compared with the case of no film cooling, and hence the heat-transfer coefficient is similar. The aerodynamic performance of the console was also investigated, and it was shown that the aerodynamic loss due to a console is significantly less than for fan-shaped or cylindrical film-cooling holes. Saumweber et al. (2002) investigated the effect of elevated free-stream turbulence on film cooling performance of shaped holes. For cylindrical holes, elevated free-stream turbulence intensity lowers adiabatic film-cooling effectiveness at small to moderate blowing ratios. Reductions up to 40% for local effectiveness and 25% for laterally averaged effectiveness are found when the turbulence intensity is raised from 3.6% to 11%. They found that the effect of increased turbulence level is detrimental at all blowing ratios for the shaped holes as the lateral spreading of coolant could not be improved by elevated free-stream turbulence intensity. Bunker (2005) provided a review of shaped-hole technology both in terms of cooling effectiveness as well as manufacturing. Bunker quoted

> The current understanding of shaped film parametric effects and physics is relatively sparse, however, the acceptance of shaped film technology is based largely on the foundation of round film-hole data. The benefits of shaped-hole film cooling are real and substantial, so much so that these types of film holes are used whenever possible in the practice of cooling gas turbines. No single shaping of film hole stands out as an optimal geometry, as long as the general guidelines of diffusion angles and sufficient depth/length of transition are followed. As manufacturing methods progress further, additional unconventional shapes of film holes will become viable, possibly leading to even higher performance and the ability to cool components with less air

Bunker (2008) reported the effects of manufacturing tolerances on gas turbine cooling, and Bunker (2009a,b) presented breaking the limits of diffusion shaped holes for film cooling.

Colban et al. (2011) presented a study that extends the prediction capability for shaped holes through the development of a physics-based empirical correlation for predicting laterally averaged film-cooling effectiveness on a flat-plate downstream of a row of shaped film-cooling holes. A predictive correlation for $\bar{\eta}$ can be assumed of the form

$$\bar{\eta} = \frac{1}{P/t + C_1 M^{C_2} \xi^{C_3}}$$

where

$$\xi = \frac{4}{\pi} \frac{(X/D)(P/D)}{M \cdot AR}$$

P is the hole pitch
t is the width of the hole at exit
M is the blowing ratio at the metering location
D is the hole diameter
X is the downstream distance from the hole
AR is the area ratio between the inlet and outlet areas of the holes

The coefficients C_1, C_2, and C_3 are determined from existing data sets that incorporated variations in the parameters that have been included into the correlation. The correlation that was developed in this study was compared with existing forms that are used in industry, as well as to a cylindrical hole correlation. Based on those comparisons, the current model was considered as an improvement on the predictive capability for adiabatic film-cooling effectiveness downstream of a row of shaped holes. Mhetras et al. (2007, 2008) simulated the film-cooling hole arrangement of a typical film cooled blade design used in stage 1 rotor blades for gas turbines used for power generation. Six rows of compound angled shaped film-cooling holes were provided on the pressure side while four such rows were provided on the suction side of the blade. The holes had a laidback and fan-shaped diffusing cross section. Another three rows of holes are drilled on the leading edge to capture the effect of showerhead film coolant injection. However, they did not compare their shaped hole data with cylindrical hole measurements.

Saumweber and Schulz (2008a) studied the effects of free-stream Mach number and free-stream turbulence, including turbulence intensity, integral length scale, and periodic unsteady wake flow on a row of cylindrical and shaped film holes. The results clearly show that the performance gain with fan-shaped holes as compared to cylindrical holes is clearly overestimated at low laboratory type conditions. If considered at conditions with high pressure ratios across the hole, separation at the diffuser side wall caused significant performance degradation and elevated free stream turbulence levels and periodic unsteady wakes showed an increase of cooling performance for cylindrical holes at medium to high blowing rates. Therefore, an exact knowledge of all geometric and flow parameters are mandatory to conceive a reliable design since tendencies observed in the open literature for cylindrical holes cannot be transferred to shaped holes. Saumweber and Schulz (2008b) studied the effect of coolant flow direction prior to entrance into the film hole for both cylindrical and shaped holes. With a coolant supply channel orientation perpendicular to the cooling hole axis, they found that the counter rotating vortex pair inside the hole—typically found for plenum conditions at the inlet—is suppressed. Instead, they saw a single helical motion that is induced inside the cylindrical part of the hole. The resulting flow effects downstream of the ejection location were very different for cylindrical and fan-shaped holes. Saumweber and Schulz (2008c) studied the effect of the expansion angle of the diffuser, the inclination angle of the hole, and the length of the cylindrical part at the hole entrance on cooling performance of diffuser holes. They indicated that a combination of parameters affect overall cooling performance and different situations requires different hole geometries. Abraham et al. (2009) examined the external cooling effects using coolant holes that are a combination of both angled shaped holes as well as perpendicular holes. The inlet of the coolant hole was kept perpendicular to the direction of flow to enhance the internal side heat-transfer coefficient and the exit of the coolant hole was expanded and angled along the mainstream flow to prevent the coolant jet from lifting off from the blade external surface. Detailed film effectiveness measurements clearly showed that the new hole configurations performed better at high blowing ratios due to strong reattachment of jet downstream of the shaped exit. The effectiveness at lower blowing ratios was also comparable but was lower than the cylindrical angled hole case.

Gao et al. (2008, 2009) compared the effects of axial shaped hole and compound angled shaped hole on suction-side film-cooling distribution as well as on pressure-side film-cooling distribution by using pressure sensitive technique. Figure 3.112 shows that film traces were converged toward the suction-side midspan region, with no film effectiveness found near 25% of hub and 25% of tip, respectively. On the other hand, Figure 3.113 shows that film traces were diverged toward the pressure-side hub and tip, respectively. This distorted film-cooling distribution might be due to impact from passage vortex and tip vortex, respectively.

Wright et al. (2010) studied the effect of density ration on flat plate film cooling with shaped holes. They indicated that increasing the density ratio decreased the centerline film-cooling effectiveness. Although the centerline effectiveness was reduced, the increased density ratios enhanced the spreading of the jets, and thus the laterally averaged film-cooling effectiveness increased. Narzary et al. (2010) studied the effects of coolant blowing, density, and free stream turbulence intensity on adiabatic film-cooling effectiveness on a high pressure turbine rotor blade. Film effectiveness on the pressure side increases proportionally with density ratio. On the suction side, some improvement with density ratio is seen in the region $0.2 < x_{SS}/C_x < 0.45$ and a sizable improvement is seen in the region, $0.45 < x_{SS}/C_x < 0.75$. Similarly, free stream turbulence intensity has a detrimental effect on the pressure-side effectiveness throughout. On the suction side, however, film effectiveness deteriorates in the region, $x_{SS}/C_x < 0.45$, but improves dramatically in the region, $x_{SS}/C_x > 0.45$, with turbulence intensity. Figure 3.114 presents the detailed film effectiveness distributions on the blade surface with different coolant densities. As can be seen, coolant traces are significantly stronger for the higher density coolant especially on the pressure surface. The effect on the suction surface is not as significant. Figure 3.115 shows effects of coolant

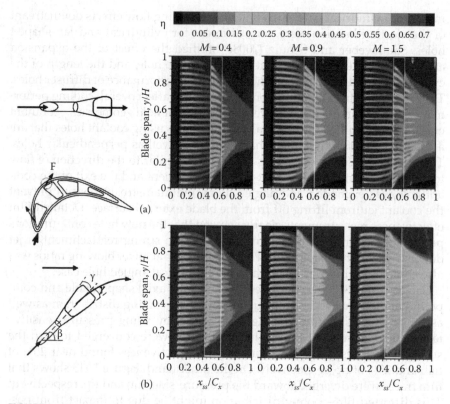

FIGURE 3.112

Measured film-cooling effectiveness on blade suction surface using axial and compound angled fan-shaped holes from Gao et al. (2008) and Gao et al. (2009). (a) Axial shaped holes and (b) compound angle shaped holes. (From Gao, Z. et al., *Int. J. Heat Mass Transfer*, 51(9–10), 2139, 2008.)

density and unsteady wake on blade film-cooling effectiveness (Rallabandi et al., 2010). Results show, as expected, film effectiveness increases with coolant density but decreases with unsteady wake.

Gao and Han (2009) studied leading-edge film cooling using pressure sensitive paint technique. Figure 3.116 compared film-cooling traces for (a) radial angled cylindrical holes, (b) radial angled shaped holes, (c) compound angled cylindrical holes, and (d) compound angled shaped holes. Results show shaped holes are better than cylindrical holes, and compound angled holes are better than radial angled holes.

Ahn et al. (2004, 2005b) studied effect of rotation on leading-edge film cooling using pressure sensitive paint technique. They used laser light through borescope to take pressure-sensitive paint (PSP)-emitted light from rotating blade leading edge for both two-row and three-row film-cooling model, respectively, as shown in Figure 3.117. Figure 3.118 shows three-row leading-edge film cooling model and the effect of blowing ratio on film-cooling traces under off-design condition (2400 rpm).

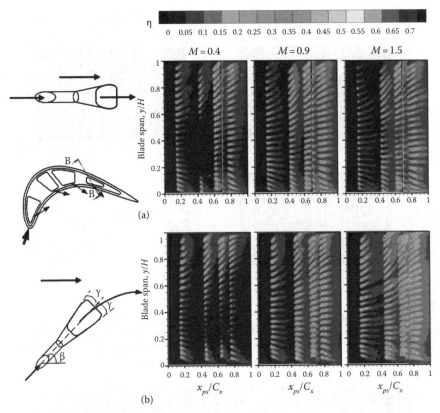

FIGURE 3.113
Measured film-cooling effectiveness on blade pressure surface using axial and compound angled fan-shaped holes Gao et al. (2008) and Gao et al. (2009). (a) Axial shaped holes and (b) compound angle shaped holes. (From Gao, Z. et al., *Int. J. Heat Mass Transfer*, 51(9–10), 2139, 2008.)

3.11.1.2 Trenching of Holes

Ekkad et al. (2005a,b) investigated the film-cooling performance for a row of cylindrical holes by embedding the row in transverse slots. The geometry of the transverse slot greatly affected the cooling performance downstream of injection. They investigated the effect of the slot exit area and edge shape. They indicated that embedding the holes in a slot reduced the jet momentum at exit and also spread the jets and provided 2-D slot jet coverage compared to the 3-D nature of individual jets. The wider spread of jets provided a better over film effectiveness with slight increases in heat-transfer coefficients. Waye and Bogard (2006a,b) investigated adiabatic film-cooling effectiveness of axial holes embedded within a transverse trench on the suction side of a turbine vane. The narrow trench configuration provided the best adiabatic effectiveness performance with increasing adiabatic effectiveness levels with increasing blowing due to the trench suppressed coolant jet separation.

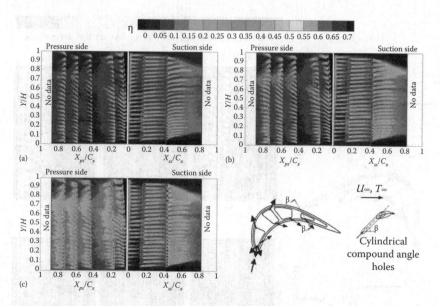

FIGURE 3.114
Coolant to mainstream density ratio on measured film-cooling effectiveness on surface using axial and compound angled cylindrical holes by Narzary et al. (2010). (a) $DR = 1.0$, (b) $DR = 1.5$, and (c) $DR = 2.5$. (From Narzary, D. et al., Influence of coolant density on turbine blade film-cooling using pressure sensitive paint technique, ASME Paper GT2010-22781, 2010.)

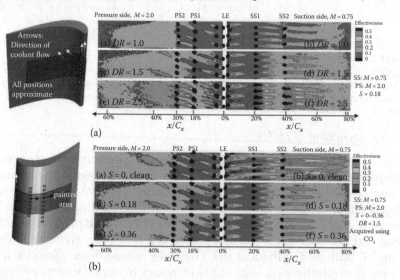

FIGURE 3.115
Results from Rallabandi et al. (2010) on blade surface using PSP measurement technique—(a) effect of coolant to mainstream density ratio and (b) effect of unsteady wake. (From Rallabandi, A.P. et al., Influence of unsteady wake on turbine blade film cooling using pressure sensitive paint technique, *Proceedings of the 14th IHTC (International Heat Transfer Conference)*, Washington, DC, August 8–13, 2010.)

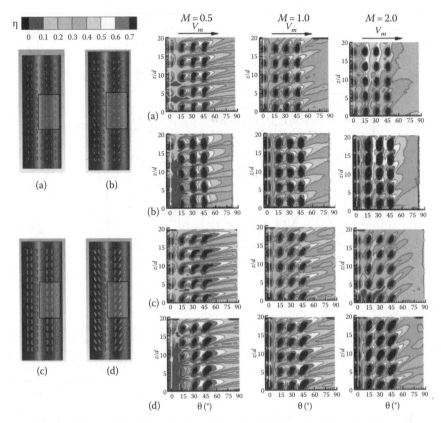

FIGURE 3.116

Results from Gao and Han (2009) on leading-edge film-cooling model: Models shown on left-hand side (a) radial angled cylindrical holes, (b) compound angled cylindrical holes, (c) radial angled shaped holes, and (d) compound angled shaped holes; contour plots shown on right-hand side (a) radial angled cylindrical holes, (b) radial angled shaped holes, (c) compound angled cylindrical holes, and (d) compound angled shaped holes. (From Gao, Z. and Han, J.C., *ASME J. Heat Transfer*, 131(6), 061701, 2009.)

Lu et al. (2007) presented an experimental investigation of film cooling from cylindrical holes embedded in transverse trenches where different trench depths were studied with two different trench widths. Figure 3.119 shows the various trench geometries studied. They presented overall heat flux ratio comparisons for different trenches and compared the performance of the trenched holes with a shaped hole configuration. Figure 3.120 shows the heat flux ratio comparison. Baseline and cases 1–6 are as shown in Figure 3.120. An optimum trench depth at 0.75*D* was identified as shallow and deep trenches showed worse performance. This was confirmed by the flow distributions obtained from CFD predictions. Shaped diffuser hole reduced the jet momentum at hole exit and outperformed the trench cases. The presence of a TBC layer on a shaped hole configuration is bound to change the dynamics

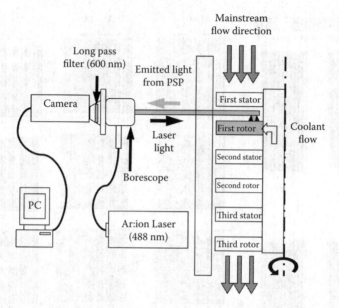

FIGURE 3.117
Schematic of optical components used for pressure sensitive paint measurements on the leading edge of a rotating blade from Ahn et al. (2004, 2005). (From Ahn, J.Y. et al., Film cooling effectiveness on the leading edge of a rotating turbine blade, *Proceedings of ASME-IMECE '04*, Anaheim, CA, November 13–19, Paper No. IMECE 2004-59852, 2004; Ahn, J.Y. et al., Film cooling effectiveness on the leading edge of a rotating film-cooled blade using pressure sensitive paint, *Proceedings of ASME Turbo-Expo 2005*, Reno, NV, June 6–9, Paper No. GT 2005-68344, 2005b.)

of interaction between coolant and mainstream and the results from this study provide understanding of this interaction.

Lu et al. (2007b) also presented results for cylindrical holes embedded in craters. Different crater geometries were considered for a typical crater depth, which was identified by Lu et al (2007a) as 0.75D. Figure 3.121 shows the different crater geometries investigated in this study. The location of the hole inside the crater will have significant effect on the downstream cooling behavior of these holes.

Figure 3.122 shows the overall heat flux ratio comparing heat flux with film cooling with heat flux on a surface without film holes. The cratered holes perform better at higher blowing ratios compared to the baseline. However, a 2-D trench is clearly superior to the craters.

Dorrington et al. (2007a,b) studied film-cooling adiabatic effectiveness on the suction side of a simulated vane for various configurations of coolant holes embedded in shallow transverse trenches or circular and elliptical shaped depressions. A shaped hole configuration was also tested. They compared the best trench configurations with a typical shaped hole configuration and showed similar film effectiveness performance over the range of blowing ratios. The trench configurations appeared to be continuing to improve with increasing blowing ratios, while the shaped holes reached a

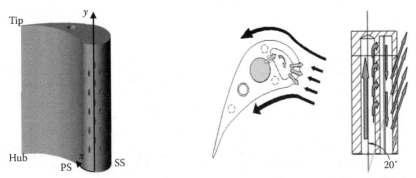

Experimental blade with leading edge film   Coolant path inside experimental film cooled blade

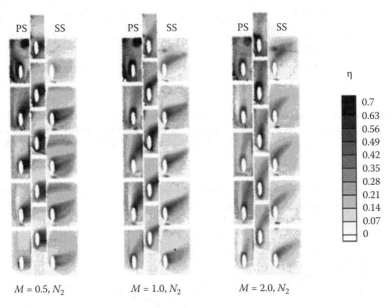

$M = 0.5, N_2$ $M = 1.0, N_2$ $M = 2.0, N_2$

FIGURE 3.118
Effect of rotation on the leading-edge film-cooling effectiveness (rotation speed = 2400 rpm, off-design condition) from Ahn et al. (2004, 2005). (From Ahn, J.Y. et al., Film cooling effectiveness on the leading edge of a rotating turbine blade, *Proceedings of ASME-IMECE '04*, Anaheim, CA, November 13–19, Paper No. IMECE 2004-59852, 2004; Ahn, J.Y. et al., Film cooling effectiveness on the leading edge of a rotating film-cooled blade using pressure sensitive paint, *Proceedings of ASME Turbo-Expo 2005*, Reno, NV, June 6–9, Paper No. GT 2005-68344, 2005b.)

maximum film effectiveness and showed decreasing film effectiveness with further increasing blowing ratio. Dorrington et al. (2007b) presented both film-cooling adiabatic effectiveness and heat-transfer coefficients for cylindrical holes embedded in a 1d transverse trench on the suction side of a simulated turbine vane to determine the net heat flux reduction. They indicated that the trench configuration had similar, if not greater augmentation

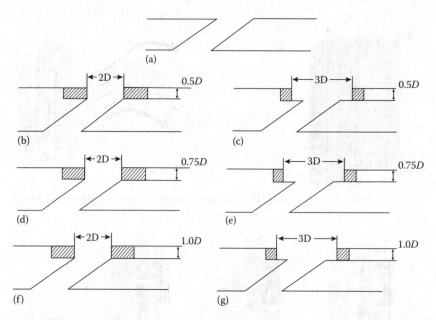

FIGURE 3.119
Trench geometries studied by Lu et al. (2007). (a) Baseline, (b) Case 1, (c) Case 2, (d) Case 3, (e) Case 4, (f) Case 5, and (g) Case 6. (From Lu, Y. et al., Effect of trench width and depth on film cooling from cylindrical holes embedded in trenches, ASME Paper GT2007-27388, 2007a.)

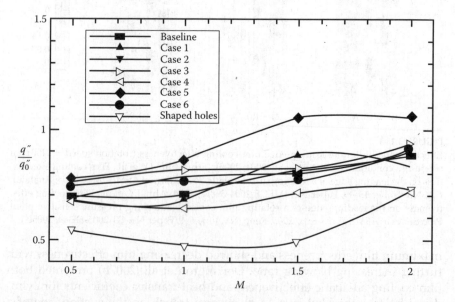

FIGURE 3.120
Overall heat flux ratio comparisons for different trench geometries. (From Lu, Y. et al., Film cooling measurements for cratered cylindrical inclined holes, ASME Paper GT2007-27386, 2007b.)

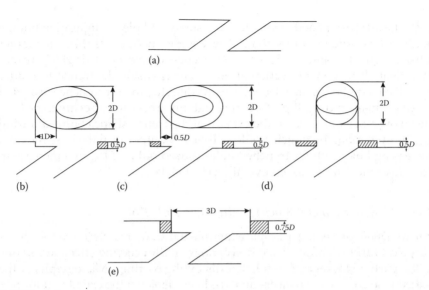

FIGURE 3.121
Crater geometries studied by Lu et al. (2007b). (a) Baseline, (b) Case 1, (c) Case 2, (d) Case 3, and (e) Case 4. (From Lu, Y. et al., Film cooling measurements for cratered cylindrical inclined holes, ASME Paper GT2007-27386, 2007b.)

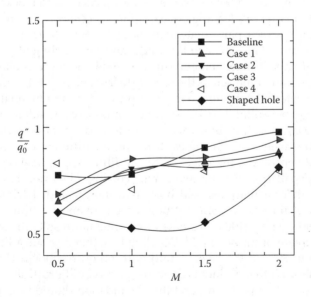

FIGURE 3.122
Comparison of crater geometries with baseline and trench for overall heat flux reduction. (From Lu, Y. et al., Film cooling measurements for cratered cylindrical inclined holes, ASME Paper GT2007–27386, 2007b; Zhang, L.J. and Jaiswal, R.S., Turbine nozzle endwall film cooling study using pressure sensitive paint, ASME Paper 2001-GT-0147, 2001.)

of the heat-transfer coefficients than the standard hole configuration but the net heat flux reduction was significantly higher because of the much larger adiabatic effectiveness produced by the trenched hole. Lu et al. (2008) further investigated the effect of different hole spacing inside the trench to reduce overall coolant flow usage. Downstream edge shaping for trenches provided small changes in overall performance but any trenching appears to be beneficial over baseline case. In the event that the trench shape changed during engine operation, the trench would still clearly outperform the baseline case. Increasing hole pitch for trenched holes caused only a 10%–15% increase in heat flux ratio but decreased overall performance.

3.11.1.3 Deposition and Blockage Effects on Hole Exits

Bunker (2000) presented a simple experimental study to determine the effects of typical turbine airfoil protective coatings on film-cooling effectiveness due to the partial blockage of film hole exits by the coatings. Micrographs of the coated hole cross sections were analyzed and depicted the extent and form of coating blockage during deposition. The measurements indicated significant degradation to film performance can result from coatings which are deposited in the hole exit regions, or inside the holes themselves, during the spray application process. Results also show that shaped film holes are generally very tolerant of coatings and do not show the degradation shown for cylindrical holes. Walsh et al. (2006) investigated the effects of sand blockage on film-cooling holes placed in a leading-edge coupon. Since most aero engines do not have filtration systems, particulates can entrain in both the main gas path and coolant streams and block coolant passages and film-cooling holes eventually leading to increased airfoil temperatures caused by reduced coolant available for a given pressure ratio across the cooling holes. The results presented in this paper represented a beginning to improve our understanding on the effects of sand blockage of cooling holes in turbine components. Sundaram and Thole (2006) used a large-scale turbine vane cascade to study the effect on film-cooling adiabatic effectiveness in the endwall region for different effects: the effect of near-hole deposition, the effect of partial film-cooling hole blockage, and the effect of spallation of a thermal barrier coating. Their hole blockage studies showed that partially blocked holes had the greatest detrimental effect on degrading film-cooling effectiveness downstream of a film-cooling row. Demling and Bogard (2006) presented experimental results for obstructions located upstream or inside of a film-cooling hole that degrade adiabatic effectiveness up to 80% of the levels found with no obstructions. The results of this study showed that coolant hole obstructions could drastically change the way the film coolant performs, and this degradation should be considered in the design of film-cooling configurations. Somawardhana and Bogard (2007) indicated that as much as 50% degradation occurred with upstream obstructions, but downstream obstructions actually enhanced film-cooling effectiveness. The transverse trench configuration performed significantly better than the traditional cylindrical holes, both with and without obstructions and almost eliminated the effects of both surface roughness and obstructions. Somawardhana and Bogard (2007) presented a study to determine the effects of varying surface roughness as well as near-hole obstructions on adiabatic effectiveness performance. They indicated an area of influence for film-cooling degradation due to obstructions could be narrowed to an area in the range of 1d upstream of the hole. Obstructions outside this area of influence cause negligible degradation to film-cooling performance.

Lewis et al. (2009) conducted experiments to determine the impact of synfuel deposits on film-cooling effectiveness and heat transfer. They used scaled up models made of synfuel deposits formed on film-cooled turbine blade coupons and were exposed to accelerated deposition. For roughness located primarily upstream of the film holes, area-averaged heat flux in the region up to $x/d = 7$ from the film holes is lower with the deposit than without. This was indicated as due to the "effective ramp" produced by the upstream deposit, providing a separation cavity for the film cooling to reside in. However, when roughness forms downstream of film holes, it was found primarily between the film holes. This provided an effective smooth channel or "furrow" for the coolant to reside in. Heat-transfer levels were correspondingly lower in this smooth channel and effectiveness levels were significantly higher. Due to the insulating effect of the deposits between the film-cooling "channels," the effective heat flux to the underlying smooth surface was significantly reduced even though the calculated rough-surface heat-transfer coefficients were higher. Ai et al. (2009) performed particulate deposition experiments in a turbine accelerated deposition facility to examine the effects of fly ash particle size and trench configuration on deposits near film-cooling holes. Trench tests conducted at an impingement angle of 30° suggested although the increased cooling effectiveness reduced deposition downstream, the capture efficiency was not improved due to the fact that the trench became a particulate collector. Deposits that accumulated on the downstream side of the trench between cooling holes eventually changed the geometry of the trench and clogging cooling holes.

3.11.2 Endwall Film Cooling

Endwall film cooling has been extensively studied in the past decade. There have been several research groups focused on various aspects of endwall film cooling. Majority of the work has focused on film cooling on vane endwalls. However, there are some studies that have also looked at film cooling of rotating blade platforms.

Zhang and Jaiswal (2001) studied endwall surface film-cooling effectiveness on a turbine vane endwall surface using the PSP technique. Two different geometries including a double staggered row of holes and a single row of discrete slots were used to supply film cooling in front of the nozzle cascade leading edges. Figure 3.123 shows the two geometries studied by Zhang and Jaiswal (2001).

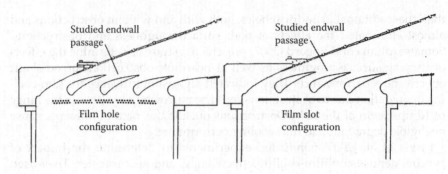

FIGURE 3.123

Film-cooling configurations studied by Zhang and Jaiswal (2001)—left sketch for double row of film hole injection geometry and right sketch for film slot injection geometry. (From Zhang, L.J. and Jaiswal, R.S., Turbine nozzle endwall film cooling study using pressure sensitive paint, ASME Paper 2001-GT-0147, 2001.)

Their results shows that the endwall secondary flow dominated the near endwall flow field at lower mass flow ratios where the cooling film was pushed to the suction side and decayed quickly for both hole and slot injections. However, the cooling film tended to dominate the near wall flow field and the secondary flow was suppressed when the jet momentum was significantly higher. The film moved to the pressure side and acceptable film effectiveness was obtained near the trailing edge. They also saw uniform effectiveness distribution with the discrete slot injection for higher mass flow ratios, while higher effectiveness near the trailing edge was obtained by double staggered rows of hole injection. Kost and Nicklas (2001) and Nicklas (2001) provided detailed aerodynamic and heat-transfer measurements on a NGV endwall under transonic conditions. They had a series of discrete holes near and around the leading edge of the passage and a slot upstream for providing the cooling for the endwall. They also indicated a weakening secondary flow structure due to coolant ejection but measured a significant increase in net secondary losses with coolant ejection. The coolant ejection caused a decrease of the passage vortex intensity near the endwall and thus a decrease in heat transfer at the pressure-side corner and downstream, next to the wake region. Coolant ejected from the holes caused a local sixfold increase in local turbulence levels directly downstream and boosts the heat-transfer coefficients in that area. There is a strong increase of the heat-transfer coefficients in the wake-region downstream of the trailing edge due to a rise in the downstream Mach number from transonic to supersonic speed range. From their results they indicated that the wake region at the endwall is a region where cooling is extremely difficult without additional trailing edge cooling.

Oke et al. (2001) and Oke and Simon (2002) studied film cooling from upstream slots for contoured endwalls for vanes. Film-cooling flow is introduced through two successive rows of slots, a single row of slots and slots that have particular area distributions in the pitch-wise direction.

Measurements are made by heating the film-cooling flow slightly above the main flow temperature and recording temperature distributions in the film-cooling flow and main flow mixing zone at various axial planes. The single and double slot injection cases represent base-line injection geometries. They concluded on the basis of results for different four configuration studies that specific slot configurations could be chosen to satisfy desired design criteria of uniform pitch-wise cooling, improved leading-edge cooling, etc. Colban et al. (2002a,b) examined the effect of varying the combustor liner film-cooling and junction slot flows on the adiabatic wall temperatures measured on the platform of the first vane and on the flow and thermal field downstream of injection inside the passage between the two vanes. Figure 3.124 shows the configurations studied by Colban et al. (2001a,b)

Figure 3.125 shows the detailed film effectiveness contours provided by Colban et al. (2002a) for cases (a–c). Different geometries provided

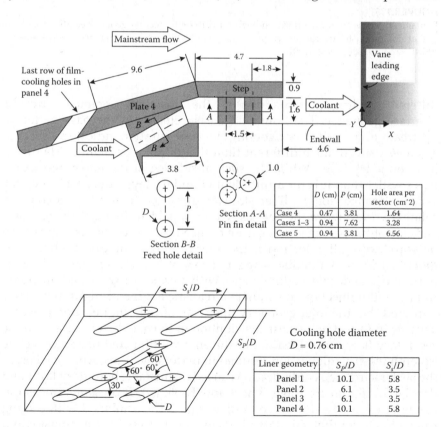

FIGURE 3.124
Slot and hole configurations studied by Colban et al. (2002). (From Colban, W.F. et al., Combustor turbine interface studies—Part 1: Endwall effectiveness measurements, ASME Paper GT2002-30526, 2002b.)

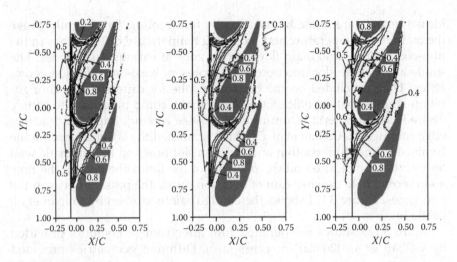

FIGURE 3.125
Film effectiveness contours on the endwall for different injection geometries. (From Colban, W.F. et al., Combustor turbine interface studies—Part 2: Flow and thermal field measurements, ASME Paper GT2002-30527, 2002a.)

different cooling patterns due to the upstream combustor–turbine interface geometry. The flow field measurements in the passage for an upstream film cooled liner/cooling slot configuration indicated a secondary flow pattern much different than that of an approaching 2-D turbulent boundary layer. While the suction-side leg of the horseshoe vortex was smaller than it is for a turbulent boundary layer, the passage vortex became stronger for the liner/slot configuration. In addition, a counter-rotating vortex above the passage vortex was identified and the strength and size of the passage vortex were reduced as a result of the higher total pressure exiting the slot and as the liner cooling is increased. Zhang and Moon (2003) used a double staggered row of holes to supply cooling air in front of the nozzle leading edges, with a back-facing located upstream from the film injection to simulate a realistic engine configuration. They indicated that the inlet geometry upstream of the film injection played a key role in film effectiveness distribution in the nozzle surface and in the secondary flow control feature of the endwall film cooling. The baseline (smooth inlet) configuration resulted in reasonable film coverage whereas the back-facing step configuration caused an unstable boundary layer and damaged the film coverage. Knost and Thole (2004) presented measurements of two endwall film-cooling hole patterns combined with cooling from a flush slot that simulated leakage flow between the combustor and turbine sections. Adiabatic effectiveness measurements showed the slot that the cooling flow adequately cooled portions of the endwall. Their measurements also showed two very difficult regions to cool, including the leading edge and pressure side–endwall junction. Colban et al. (2008) presented film-cooling effectiveness measurements comparing cylindrical and shaped holes on an endwall surface in a large-scale, low-speed, two-passage, linear vane cascade. Results showed that film-cooling effectiveness for the fan-shaped hole cooled passage showed increased film-cooling effectiveness with increasing blowing ratio and that they increased film-cooling effectiveness by an average of 75% over cylindrical holes for constant cooling flow.

Wright et al. (2008) used a five-blade, linear cascade to experimentally investigate turbine blade platform cooling. A 30° inclined slot upstream of the blades is used to model the seal between the stator and rotor, and 12 discrete film holes are located on the downstream half of the platform for additional cooling.

Combining upstream slot injection with downstream discrete film hole cooling showed the potential to further increase the endwall film-cooling effectiveness. At a mass flow ratio of 1.0%, the film-cooling effectiveness increased more than three times over the effectiveness resulting from only upstream slot injection. However, the levels were not sustained at higher injection conditions. The study showed that slot injection has the potential to provide adequate film-cooling protection on the platform surface and additional, limited protection on the suction surface of the blade. Cardwell et al. (2005) and Piggush and Simon (2006a,b) provided heat-transfer and film-effectiveness measurements on a contoured endwalls with additional features such as misalignment, leakage, and assembly features. Figure 3.126 shows the geometries studied by Piggush and Simon (2006a). They showed that a forward facing step at the transition section gap imposed a local acceleration on the flow, thinning boundary layers and caused slightly higher heat-transfer rates across the whole test section endwall, whereas the backward facing transition section step has the opposite effect.

Their results indicated that the addition of steps influenced the local boundary layer by either increasing or decreasing the mixing the leakage flow experiences. Steps also affected the distribution of leakage coming from the gaps. The leakage flow generally provided minimal cooling, except for the case of the 1.0% MFR at the transition section gap, which did provide some coverage at the inlet of the passage.

Suryanarayanan et al. (2006) carried out an experimental investigation in a three-stage turbine facility, designed and taken into operation at the Turbomachinery Performance and Flow Research Laboratory (TPFL) of Texas A&M University. This turbine rotor was designed to facilitate coolant injection through this stator-rotor gap upstream of the first stage rotor blade. The gap was inclined at 25° to mainstream flow to allow the injected coolant to form a film along the passage platform. The effects of turbine rotating conditions on the blade platform film-cooling effectiveness were investigated at three speeds of 2550, 2000, and 1500 rpm with corresponding incidence angles of 23.2°, 43.4°, and 54.8° respectively.

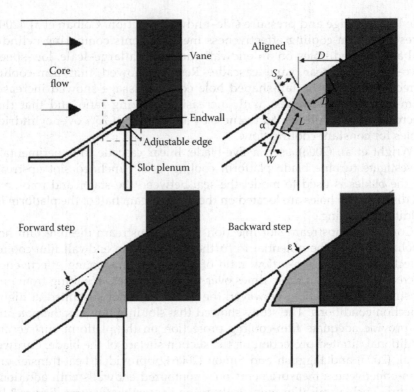

FIGURE 3.126
Different leakage flow configurations studied by Piggush and Simon (2006a). (From Piggush, J. and Simon, T., Heat transfer measurements in a first stage nozzle cascade having endwall contouring: Misalignment and leakage studies, ASME Paper GT2006-90575, 2006a.)

Figure 3.127 shows the actual three stage rotors form the test rig with clear indications of the purge flow locations. They used a PSP technique to measure local film effectiveness contours on a rotating platform.

Figure 3.128 shows the detailed film effectiveness contours for different coolant flow conditions on the rotating platform. It is clear that cooling effectiveness increases with increasing mass flow of coolant. It should be kept in mind that the coolant ejection is nonrotating whereas the platform is rotating. Figure 3.129 shows the averaged effectiveness for the same conditions. The cooling effect decreases even for the highest mass flow rate case due to the presence of the passage vortex and the secondary flows generated by the leading edge of the rotor and the trailing edge wakes of the upstream stator.

Gao et al. (2008) studied the effect of film-hole configurations on platform film cooling (Figure 3.130). The platform was cooled by purge flow from a simulated stator-rotor seal combined with discrete-hole film cooling within the blade passage (Figure 3.131). Two different hole exit shapes, cylindrical holes and laidback fan-shaped holes, are assessed in terms of film-cooling effectiveness and total pressure loss. The film holes were arranged on the

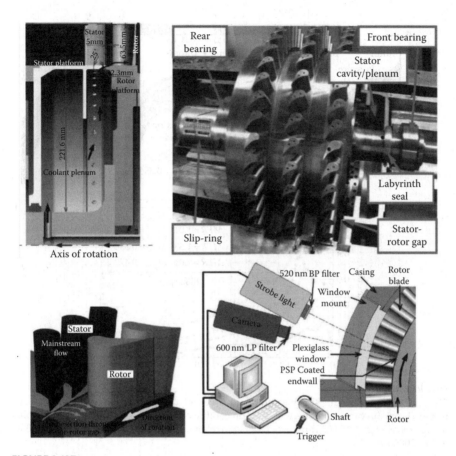

FIGURE 3.127
Details on the test rig used by Suryanaryanan et al. (2006). (From Suryanarayanan, A. et al., Film cooling effectiveness on a rotating blade platform, *Proceedings of ASME Turbo-Expo 2006*, Barcelona, Spain, May 8–11, Paper No. GT2006-90034, 2006.)

platform with two different layouts to obtain a total of four configurations. In one layout, the film-cooling holes were divided into two rows and more concentrated on the pressure side of the passage. In the other layout, the film-cooling holes were divided into four rows and loosely distributed on the platform. The purge flow through the seal is fixed at $MFR = 0.75\%$. Although shaped holes provided overall higher effectiveness, hole arrangement seemed to have little effect.

Calban et al. (2006) and Lynch and Thole (2007) presented measurements of adiabatic cooling effectiveness and heat-transfer coefficients on the endwall of a first vane, with the presence of leakage flow through a flush slot upstream of the vane. They found that contracting the slot width while maintaining the slot mass flow resulted in a larger coolant coverage area and higher effectiveness values, as well as slightly lower heat transfer coefficients. Wright et al. (2009) studied the effect of unsteady wake with vortex

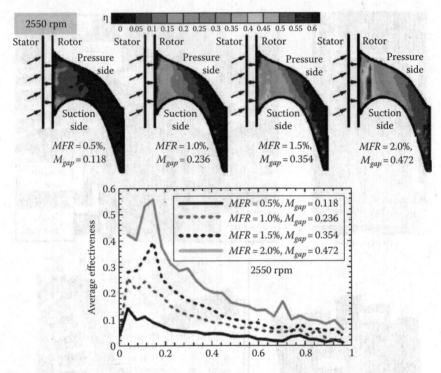

FIGURE 3.128
Film effectiveness results for one rotational speed of the rotor by Suryanaryanan et al. (2006). (From Suryanarayanan, A. et al., Film cooling effectiveness on a rotating blade platform, *Proceedings of ASME Turbo-Expo 2006*, Barcelona, Spain, May 8–11, Paper No. GT2006-90034, 2006.)

generators attached to a spoked wheel wake generator on endwall film cooling with blade upstream purge flow. They determined that the upstream wake had only a negligible effect on the platform film-cooling effectiveness. With the rod placed at four locations upstream of the blades, the film-cooling effectiveness was not significantly altered for any seal flow rate or rod position. However, the film-cooling effectiveness could be significantly reduced with the type of vortex generation upstream of the blade passage. They also indicated that it was clear that the secondary flow induced in the upstream vane passage had a profound effect on the rotor platform film-cooling effectiveness. The need for additional cooling on the latter half of the passage is obvious from the detailed film-cooling distributions. Suryanarayanan et al. (2006, 2007) presented more results on turbine blade platform with effects of rotation in the same rig as shown in earlier Figures 3.126 through 3.128. In this case, they had upstream purge flow and local discrete film-cooling injection inside the blade passage. They concluded that complete film-cooling protection on a rotating platform can be provided with combined upstream stator-rotor gap ejection and discrete-hole ejection. Positioning the holes with

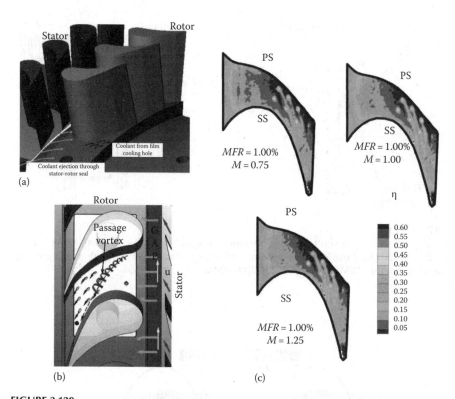

FIGURE 3.129
Film-cooling effectiveness on rotor endwall under rotating conditions. (a) Rotor-stator configuration with inclined slot, (b) schematic of expected fluid mechanics in endwall film coolings, and (c) measured film-cooling effectiveness values on endwall under rotation. (From Suryanarayanan, A. et al., Film-cooling effectiveness on a rotating turbine platform using pressure sensitive paint technique, *Proceedings of ASME Turbo-Expo 2007*, Montreal, Quebec, Canada, May 14–17, Paper No. GT2007-27122, 2007.)

angles oriented more toward the pressure surface will help reduce the effects of passage vortex and cross-flow. To optimize coolant usage, they suggested maintaining the stator-rotor gap injection close to $MFR = 1\%$ and increasing the number of holes on the platform without compromising the structural integrity is the way forward in providing proper film protection on the platform. Sundaram and Thole (2008) studied the effect of trenching a row of holes upstream of leading edge of a vane to determine whether trenching helps in increasing cooling effectiveness on the endwall. Film-cooling holes in a trench resulted in enhancing the adiabatic effectiveness levels at all blowing ratios compared with film-cooling holes without a trench even at a high blowing ratio of $M = 2.5$. Their flow field measurements indicated some clear differences in the flow field characteristics at the leading edge. In the presence of a trench, the coolant jets did not separate as severely from the endwall surface compared with the untrenched holes.

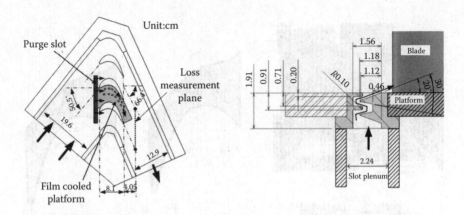

FIGURE 3.130
Configuration of upstream labyrinth-like stator rotor seal by Gao et al. (2008). (From Gao, Z. et al., Turbine blade platform film cooling with typical stator-rotor purge flow and discrete-hole film cooling, ASME Paper GT2008-50286, 2008a.)

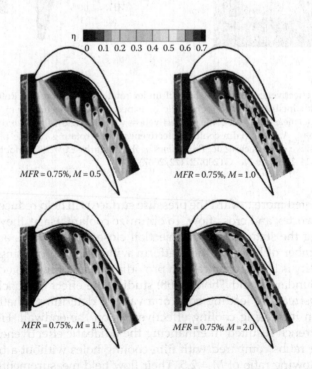

FIGURE 3.131
Film-cooling effectiveness in the hub region, from Gao et al. (2008). (From Kwak, J.S. and Han, J.C., Heat transfer coefficient and film cooling effectiveness on the squealer tip of a gas turbine blade, ASME Paper GT2002-30555, 2002b.)

Mahmood et al. (2009) measured flow field and temperature field near a 3-D asymmetric contour endwall in a linear blade cascade with and without film-cooling flow on the endwall. They indicated that the direction of the vortex flow axes changes as the flow yaw angle and pitch-wise velocity changes when the film-cooling flow is introduced. The influence on the directions of vortex axes was higher with the higher blowing ratio as the coolant flow strengthens the boundary layer when injected under the boundary layer. The vortices in the passage get pushed higher above the endwall with increasing blowing ratio. Film-cooling effectiveness data indicated that the coolant holes far upstream of the passage inlet provide insignificant cover on the passage wall. More coolant flow was required to provide adequate coolant cover on the endwall near the leading-edge and pressure-side regions. Thrift et al. (2010a,b) presented the effect of axisymmetric endwall contouring on the cooling performance of a film cooled endwall. They measured both adiabatic effectiveness and heat-transfer coefficients on the endwall with and without contours. Axisymmetric endwall contouring was shown to have a significant impact on cooling performance, both from an upstream slot and film-cooling holes. The contoured passage introduced a unique situation where each endwall was subject to increased acceleration relative to the planar case. In addition, the contoured endwall was under the influence of a streamline curvature. Increased acceleration was shown to have a limited impact on the cooling performance as indicated by a comparison between the flat endwalls of the planar and contoured passages. However, streamline curvature and mainstream impingement were shown to be detrimental to film-cooling performance on the contoured endwall. With and without leakage flow, the endwalls of the contoured passage were measured to have lower heat-transfer levels than the corresponding flat endwall of the planar passage. Post and Acharya (2010) presented temperature and heat flux measurements obtained in a film-cooled, heated fixed-vane cascade. They presented endwall and vane surface normalized metal temperatures (NMT), and surface heat fluxes, q'', in both local and averaged forms. Higher density coolant film appeared to provide greater cooling effectiveness along the endwall at higher M, with local increases in NMT of as much as 0.2, but the opposite trend was observed for lower M along the endwall, with potential lowering in NMT of 0.1 in the early-passage regions.

3.11.3 Turbine Blade Tip Film Cooling

In the past decade, the turbine airfoil tip and trailing edge have been identified as a critical region for failure. There has been significant in-road into understanding the physics of tip leakage flow and designing cooling for tips and also the difficulties of cooling the trailing edge with conventional cooling techniques.

Kwak and Han (2002a,b) and Mhetras et al. (2005) presented detailed distributions of heat-transfer coefficient and film-cooling effectiveness on a

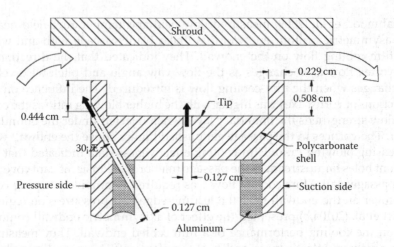

FIGURE 3.132
Blade tip cooling geometry. (From Kwak, J.S. and Han, J.C., Heat transfer coefficient and film cooling effectiveness on the squealer tip of a gas turbine blade, ASME Paper GT2002-30555, 2002b.)

gas turbine blade tip using a hue-detection-based transient liquid crystal technique. Tests were performed on a five-bladed linear cascade with blow down facility where the blade was a 2-D model of a first stage gas turbine rotor blade with a profile of the GE-E^3 aircraft gas turbine engine rotor blade tip section. Figure 3.132 shows the configuration tested on a real blade for squealer tip heat transfer with film cooling.

Figure 3.133 through 3.137 show the detailed heat-transfer coefficient and film effectiveness distributions for a tip clearance of 1.5% of total span with (a) tip injection only and (b) tip and pressure-side injection. The plane tip cases are included for comparison. The no-injection cases are also shown ($M = 0$). The results show that heat-transfer coefficient decreases with coolant injection but increases further with pressure injection added to tip injection. Film effectiveness is clearly enhanced due to addition of pressure injection, which acts as barrier to the leakage flow and enhances coverage from tip cooling holes. Increases coolant mass flow does not cause a dramatic increase in film effectiveness. Film-cooling traces follow tip leakage flow direction for the flat tip (plane tip) case. However, film-cooling traces follow cavity flow direction (recirculation) for the squealer tip case as shown in Figure 3.133.

Christophel et al. (2004a–c) presented film effectiveness and heat-transfer coefficient measurements over a tip in a low speed cascade experiment. They studied both pressure-side injection and on tip injection. Bailey et al. (2004) presented an experimental and computational investigation focused on the detailed distribution of heat-transfer effectiveness and pressure on an attached tip-shroud of a turbine blade. Temperatures and pressures were measured on the airfoil side and gap-side surfaces of the shrouded tip in a three-airfoil stationary cascade at simulated engine conditions.

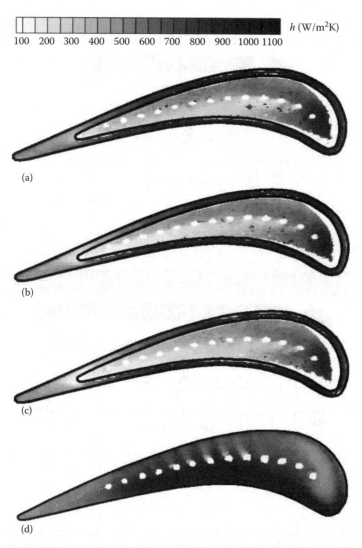

FIGURE 3.133
Heat-transfer coefficient for $C = 1.5\%$ and coolant injection from tip holes only. (a) Squealer tip, $M = 0$, (b) squealer tip, $M = 1$, (c) squealer tip, $M = 2$, and (d) plane tip, $M = 0$. (From Kwak, J.S. and Han, J.C., Heat transfer coefficient and film cooling effectiveness on the squealer tip of a gas turbine blade, ASME Paper GT2002-30556, 2002b.)

Results indicated that the shrouded tip would be at a higher temperature as gap increased. Results also indicated that the tip shroud near the radial exit holes was cooler with higher coolant flow but the extremities of the tip shroud did not see the effect of the higher coolant flow.

Nasir et al. (2004) presented detailed heat-transfer measurements on plain and squealer tips with film cooling into the tip gap through discrete holes.

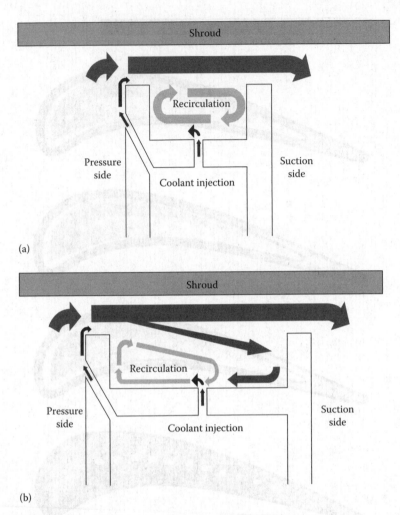

FIGURE 3.134
Conceptual view of flow in the cavity. (a) Cavity closer to the trailing edge and (b) cavity closer to the leading edge. (From Kwak, J.S. and Han, J.C., Heat transfer coefficient and film cooling effectiveness on the squealer tip of a gas turbine blade, ASME Paper GT2002-30557, 2002b.)

Figure 3.138 shows the detailed heat-transfer coefficients for both tips. The pattern of heat-transfer coefficients for the plain and squealer tips is very different with higher heat-transfer coefficient regions at different sides of the blade profile. Film cooling seems to significantly affect the tip leakage flow and associated heat-transfer pattern on the plain tip bit the squealer tip shows rapid reduction with increased coolant compared to no film-cooling case.

Figure 3.139 shows the film effectiveness distributions on the plain and squealer tips. The plain tips show localized streaks as the jets are bent by the oncoming leakage flow accelerating through the gap. However, there is

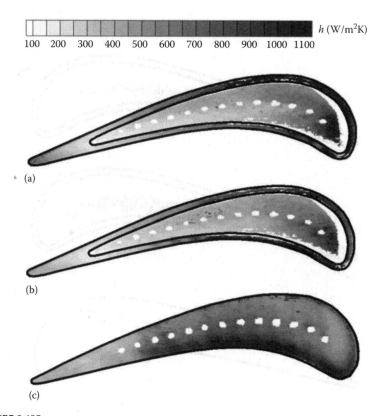

FIGURE 3.135
Heat-transfer coefficient for $C = 1.5\%$ and coolant injection from both tip and pressure-side holes. (a) Squealer tip, $M = 1$, (b) squealer tip, $M = 2$, and (c) plane tip, $M = 2$. (From Kwak, J.S. and Han, J.C., Heat transfer coefficient and film cooling effectiveness on the squealer tip of a gas turbine blade, ASME Paper GT2002-30558, 2002b.)

no cooling effectiveness for the squealer tip as the jets blow away from the surface and interact with the leakage flow.

Arts et al. (2004) and Kuwabara et al. (2004) presented a numerical and experimental program on the study of film-cooling effectiveness on a flat blade tip as a function of tip gap and mass flux ratios. They reported increased film effectiveness obtained with tighter clearances but no film-cooling effectiveness was observed in leading-edge area. Nasir et al. (2004) investigated the effects of coolant injection on adiabatic film effectiveness and heat-transfer coefficients from a plane and recessed tip of a HPT first stage rotor blade. Three cases where coolant is injected from (1) five orthogonal holes located along the camber line, (2) seven angled holes located near the blade tip along the pressure side, and (3) combination cases when coolant is injected from both tip and pressure-side holes were studied. Overall, it appeared from their results that the pressure-side injection reduced leakage flow into the tip gap and helped improve protection of the tip surface

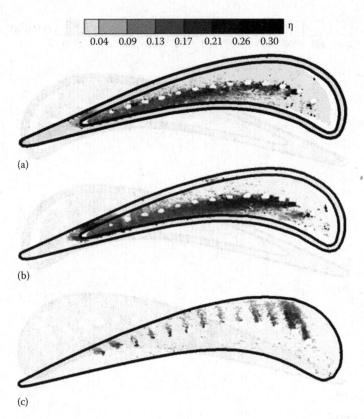

FIGURE 3.136
Film-cooling effectiveness for $C = 1.5\%$ and coolant injection from tip holes only. (a) Squealer tip, $M = 1$, (b) squealer tip, $M = 2$, and (c) plane tip, $M = 2$. (From Kwak, J.S. and Han, J.C., Heat transfer coefficient and film cooling effectiveness on the squealer tip of a gas turbine blade, ASME Paper GT2002-30559, 2002b.)

with the tip injection. However, pressure-side injection may also protect the pressure-side edge of the tip surface, which is typically prone to erosion.

Ahn et al. (2005) studied the blade tip film cooling with pressure-side injection using pressure sensitive paint technique. Figure 3.140 shows sample results on detailed film-cooling effectiveness behavior for a typical squealer/ recessed tip. Mhetras et al. (2006) investigated the effects of shaped holes on the tip pressure side, coolant jet impingement on the pressure-side squealer rim from tip holes, and varying blowing ratios for a squealer blade tip on film-cooling effectiveness. Effectiveness was measured on the cavity floor, rim, cavity rim walls, and near tip pressure side. Figure 3.141 shows a sample result presented by Mhetras et al. (2006) on a recessed tip and along its rim.

Gao et al. (2007) investigated the influence of incidence angle on film-cooling effectiveness for a cutback squealer blade tip. Three incidence angles are investigated—0° at design condition and ±5° at off-design conditions. The pressure-side squealer rim wall was cut near the trailing edge to allow the

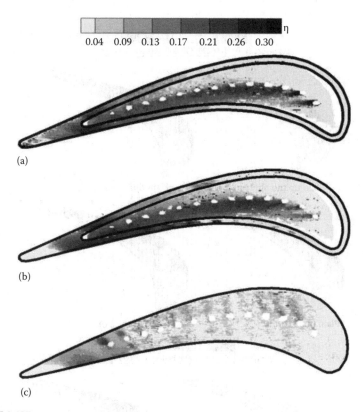

FIGURE 3.137
Film-cooling effectiveness for $C = 1.5\%$ and coolant injection from both tip and pressure-side holes. (a) Squealer tip, $M = 1$, (b) squealer tip, $M = 2$, and (c) plane tip, $M = 2$. (From Kwak, J.S. and Han, J.C., Heat transfer coefficient and film cooling effectiveness on the squealer tip of a gas turbine blade, ASME Paper GT2002-30560, 2002b.)

accumulated coolant in the cavity to escape and cool the tip trailing edge. The internal coolant-supply passages of the squealer tipped blade were modeled similar to those in the GE-E3 rotor blade. As the incidence angle varies from +5° to 0° and −5°, the coolant jet deflection was increased slightly. This caused the peak of laterally averaged effectiveness to shift; however, the area averaged cooling effectiveness on the pressure side was not affected much by the jet deflection and peak shifting. Park et al. (2010) presented detailed heat/mass transfer coefficients and film-cooling effectiveness on the tip and inner rim surfaces of a rotor blade with a squealer rim. They showed that the heat-transfer coefficients and high heat-transfer region on the tip surface decreased when the rim height was increased. However, the heat-transfer patterns on both inner rim surfaces were similar, irrespective of rim heights, because the amount of tip leakage flow was the same due to the same tip clearance. They concluded that the film-cooling effectiveness at the middle of the blade tip was high and was dependent on the film-cooling hole arrays.

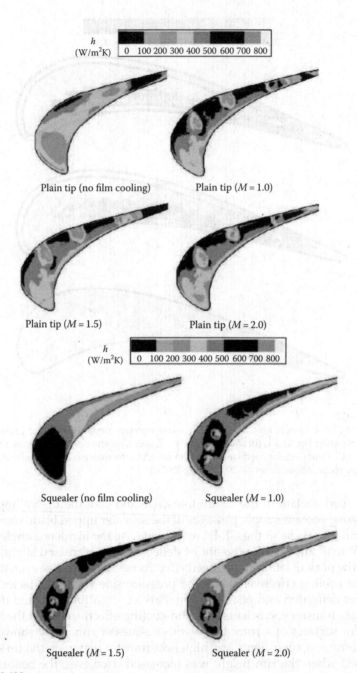

FIGURE 3.138
Detailed heat-transfer coefficient measurements on cooled and uncooled tips. (From Nasir, H. et al., Effects of tip gap film injection from plain and squealer blade tips, ASME Paper GT2004-53455, 2004.)

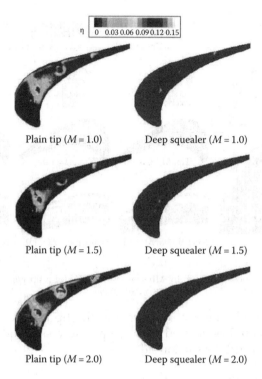

FIGURE 3.139
Detailed film effectiveness for plain and squealer tips. (From Nasir, H. et al., Effects of tip gap film injection from plain and squealer blade tips, ASME Paper GT2004-53455, 2004.)

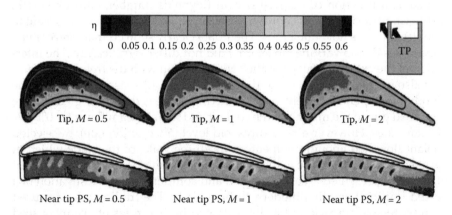

FIGURE 3.140
Sample results on detailed film effectiveness behavior for a recessed tip. (From Ahn, J.Y. et al., Film cooling effectiveness on the leading edge of a rotating film-cooled blade using pressure sensitive paint, *Proceedings of ASME Turbo-Expo 2005*, Reno, NV, June 6–9, Paper No. GT 2005-68344, 2005b.)

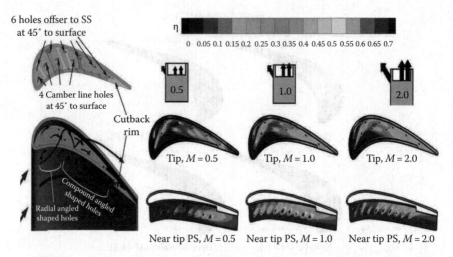

FIGURE 3.141

Film-cooling effectiveness recorded by Mhetras et al. (2006) for a tip-gap ratio of 2.1%. (From Mhetras, S.P. et al., *AIAA J. Prop. Power*, 22(4), 889, 2006.)

To protect efficiently the rim surfaces and the tip surface, a part of film-cooling holes should be located near the pressure side and others will be inclined holes toward the suction side.

3.11.4 Turbine Trailing Edge Film Cooling

Uzol et al. (2000) and Uzol and Cancia (2000) presented an experimental study of the internal fluid mechanic loss characteristics of a turbine blade with trailing edge coolant ejection. The discharge coefficient C_d was presented as a function of the free stream Reynolds number, cutback length, spanwise rib spacing, and chordwise rib length for a wide range of coolant to free stream mass flow rate ratios. Their data clearly showed that internal viscous losses due to varying rib lengths did not differ significantly. The interaction of the external wall jet in the cutback region with the free stream fluid was also a strong contributor to the losses. The results also showed that the aerodynamic penalty levels in the wake region near the trailing edge were increased due to the mixing of the coolant and main stream flows for 0%–3% ejection rates. However after a threshold level (5% ejection rate), the ejected coolant flow had enough momentum to fill the wake of the blade which in turn resulted in a decrease in the overall aerodynamic penalty levels. Martini et al. (2005) presented an experimental and computational investigation of a cooled trailing edge in a modern turbine blade. The trailing edge featured a pressure-side cutback and a slot, stiffened by two rows of evenly spaced ribs in an inline configuration. Cooling air was ejected through the slot and forms a cooling film on the trailing edge cutback region. They emphasized that the investigated cooling configuration was not desirable for film cooling of the trailing edge cutback as the attachment of coolant jets resulted in severe temperature gradients on the film cooled surface. Furthermore, they observed the phenomenon of re-grouping of attached coolant jets which led to additional thermal fatigue of the film cooled trailing edge. Martini et al. (2005) provided a study that dealt with trailing edge film cooling on the pressure-side cutback of gas turbine airfoils. In this case, before being ejected tangentially onto the inclined cutback surface, the coolant air passed a partly converging passage that was equipped with turbulators such as pin fins and ribs. The experimental results indicated that the extension of the core region, where η_{aw} was close to unity, was strongly influenced by the internal cooling design. However, the decay of η_{aw} downstream of the core region was similar for all trailing edge cooling slots. They concluded that the mixing generated by the lip ($t/H = 1$ for all slots) dominated in that region. The decay of η_{aw} was almost the same as it was observed for a clean thin lipped slot in the turbulent boundary layer region which was considered an indication for an upstream shift of the turbulent boundary layer region closer to the slot exit. Coon et al. (2005) measured detailed local heat transfer and film-cooling effectiveness from a 10× scale trailing-edge model of an industrial gas turbine airfoil in a low speed wind tunnel. A linear velocity gradient was imposed using an adjustable top wall to simulate the mainstream flow acceleration in vane and blade row passages. In an accelerating mainstream, the heat-transfer enhancement appeared a little bit more sensitive to the blowing ratio. The film effectiveness results displayed much stronger dependency on the blowing ratio with an improving trend with increasing blowing ratio. Telisinghe et al. (2006) investigated the aero performance differences between a conventional turbine blade trailing edge and a trailing edge with a sharp cutback. Both geometries included trailing edge film cooling with a scaled model of a conventional turbine blade trailing edge and the trailing edge with a sharp cutback. The scaled film-cooling-hole geometries were incorporated into flat plates. A loss coefficient for the two plates was determined by evaluating the kinetic energy at a mixed out plane far downstream from the traverse plane using momentum considerations. The losses indicated little difference between the two cases. Fiala et al. (2008a,b) presented the effects of blowing rate, Reynolds numbers and external turbulence on heat transfer and film-cooling distributions and aerodynamic losses for a vane trailing edge with letterbox partitions. Heat-transfer results on partition side correlated with both internal and external flow Reynolds number but turbulence condition was found to have no influence. Heat-transfer results on partition top were primarily a function of external Reynolds number although flow rate had a small influence at high flows. Partition side adiabatic effectiveness levels were seen to be high at low blowing rates but decreased with increasing blowing rate. The letterbox vane produced an incremental increase in loss over the base vane for all conditions and generally produced incremental losses of about 0.2% less than the gill slot vane at the same conditions. Rehder (2009) studied the aerodynamic and thermodynamic performance of a high pressure turbine cascade with different trailing edge cooling configurations in a wind tunnel with linear cascades. A transonic rotor profile with a rela-tive thick trailing edge was chosen for the experiments with three different trailing edge cooling configurations: a central trailing edge ejection, a trail-ing edge shape with a pressure-side cutback and slot equipped with a dif-fuser rib array, and a pressure-side film cooling through a row of cylindrical holes. The calculated separated contributions of loss indicated that the influ-ence of coolant ejection on the trailing edge and mixing loss was tremen-dous. Shock and boundary layer losses made major contributions to the total loss but were less affected by the coolant flow. Increased coolant mixing loss and hence higher total losses were achieved when a real temperature ratio between coolant and main flow was simulated. Central trailing edge ejec-tion was always superior compared to the other configurations. The coolant directly filled up and re-energized the blades wakes resulting in the low-est total loss level. Dannhauer (2009) investigated two different trailing edge geometries with coolant ejection. The first configuration was equipped with a pressure-side cutback while for the second configuration the pressure-side film cooling was realized by a row of cylindrical holes. The contour plots of the film-cooling efficiency for the cutback configuration showed a homo-geneous distribution of the coolant. The refilling of the separation region behind the cutback led to higher coolant concentrations. For pressure-side bleed, a continuous increase of the film-cooling efficiency with increasing mass flow rate was shown whereby the investigated outlet Mach numbers seemed to have only minor effects. Horbach et al. (2010) presented an experi-mental study on trailing edge film cooling of modern high-pressure turbine blades using coolant ejection through planar slots on a pressure-side cut-back. Variation of the ejection lip thickness had a pronounced effect on the mixing process of the cooling film as well as on the discharge coefficient. An increase in lip thickness intensified unsteady vortex shedding from the blunt lip, which further enhanced the mixing of the cooling film. The insertion of land extensions had an enhancing effect on film-cooling effectiveness and reduced vortex shedding at high blowing ratios. Furthermore, higher block-age of the ejection slot slightly reduced the discharge coefficient.

3.11.5 Airfoil Film Cooling

3.11.5.1 Vane Film Cooling

Cutbirth and Bogard (2002a,b) provided detailed measurements of the inter-action of the showerhead coolant injection with the pressure side film-cooling performance for a simulated turbine vane. They studied showerhead cool-ing with and without turbulence measurements. Cutbirth and Bogard (2003) studied the effect of density ratio on film-cooling performance where adia-batic effectiveness was measured in the showerhead region of the vane, and following the first row of coolant holes on the pressure side with density ratios up to 1.8, Schneider et al. (2003) investigated the effect of showerhead injection on the superposition of film cooling on the pressure side of a vane. Aerodynamic and thermal inlet conditions were measured to assure uniform approach conditions to the central test vane. Comparing their individual row results without showerhead blowing to multirow measurements by using the superposition approach, good agreement was found. Using the same data for superposition calculations with showerhead blowing, the reduction in film-cooling effectiveness for the first pressure side row was clearly observed.

Varadarajan and Bogard (2006) provided measurements of the hot streak temperatures in the showerhead region. When impacting the vane in regions of film cooling, the hot streak caused significant changes in adiabatic effectiveness. However, they indicated that the effect of the hot streak could be accounted for by using an "adjusted" mainstream temperature. By adjusting the mainstream temperature used in the definition of adiabatic effectiveness to equal the hot streak temperature near the wall, the adiabatic effectiveness was found to be the same as the no hot streak case.

3.11.5.2 Blade Film Cooling

Zhang and Pudupatty (2000) studied film cooling on turbine nozzle surface, including a showerhead. Later, Zhang and Moon (2007, 2009) studied turbine nozzle film cooling with showerhead injection and film hole location on the pressure side. Detailed film effectiveness was measured in a warm cascade simulating realistic engine operation conditions on both the pressure and suction surfaces of blade models using the PSP technique. Narzary et al. (2007) performed experimental tests on a fully cooled high-pressure turbine rotor blade with fan-shaped, laid-back compound angled holes. The shaped holes featured a 10° expansion in the lateral direction and an additional 10° in the forward direction. The effect of four different blowing ratios and the presence of stationary, upstream wakes are examined on blade film coverage.

3.11.5.3 Effect of Shocks

Ligrani et al. (2001) investigated interactions between shock waves and film cooling as they affect magnitudes of local and spanwise-averaged adiabatic film-cooling effectiveness distributions. A row of three cylindrical holes was studied at free-stream Mach numbers of 0.8 and 1.10–1.12 with coolant to free-stream density ratios of 1.5–1.6. Effectiveness values measured with supersonic approaching free-stream and shock waves decreased as the injection cross-flow Mach number increased. Such changes were due to altered flow separation regions in film holes, different injection velocity distributions at hole exits, and alterations of static pressures at film hole exits produced by different types of shock-wave events. Smith et al. (2003) measured and compared the unsteady heat flux created by the passing shock on the pressure surface of a turbine with similar measurements on the suction surface. Measurements were taken with and without film cooling and at varying blowing ratios. Because the shock is much weaker by the time it reaches the pressure surface, the uncooled shock-heating effect on the pressure surface was roughly 50% of the magnitude observed on the suction surface. With film cooling, however, the shock affected the flow and heat transfer on the pressure surface because it redirected cooling from the suction side through the internal film-cooling channel. This additional coolant actually lowers the heat transfer after the initial shock reaches the pressure side. Ochs et al. (2007) performed a detailed experiment to investigate the impact of a trailing edge shock wave on the film-cooling performance of a gas turbine airfoil suction side. Adiabatic film-cooling effectiveness as well as heat transfer coefficients were presented within the shock–film cooling interaction zone at high local resolution. Bow-shocks forming in front of the cooling holes varied in strength as the blowing ratio was increased, leading to an altered hot gas flow field. The presence of the main oblique shock, however, did not have a significant effect on adiabatic film-cooling effectiveness.

Liu et al. (2011) studied the effect of shock wave on turbine vane suction side film cooling using a conduction-free PSP technique. Tests were performed in a five-vane annular cascade with a blowdown flow loop facility. The exit Mach numbers were controlled to be 0.7, 1.1, and 1.3, from subsonic to transonic flow conditions. Two foreign gases N_2 and CO_2 were selected to study the effects of coolant density. Results showed that at lower blowing ratio, film-cooling effectiveness decreased with increasing exit Mach number. On the other hand, an opposite trend was observed at high blowing ratio. In transonic flow, the rapid rise in pressure caused by shock might benefit film-cooling by deflecting the coolant jet toward the vane surface at higher blowing ratio. Denser coolant performed better, typically at higher blowing ratio in transonic flow.

3.11.5.4 Effect of Superposition on Film Effectiveness

Figure 3.142 shows the spanwise averaged film-cooling effectiveness from superposition of individual cylindrical-hole rows and with all film holes blowing for all the blowing ratios by Mhetras and Han (2006). The film-cooling effectiveness from superposition as discussed by Sellers (1963) is calculated from the following equation:

$$\eta_s = \sum_{i=1}^{n} \eta_i \prod_{j=0}^{i-1} (1 - \eta_j) \quad \text{where } \eta_0 = 0$$

This equation is valid when the ejecting coolant temperature is the same for all film rows. Effectiveness from superposition compares reasonably well with full coverage film cooling on both pressure and suction surfaces. Results from superposition indicate a smaller drop in film-cooling effectiveness downstream of the holes. However, the peak difference between the overall

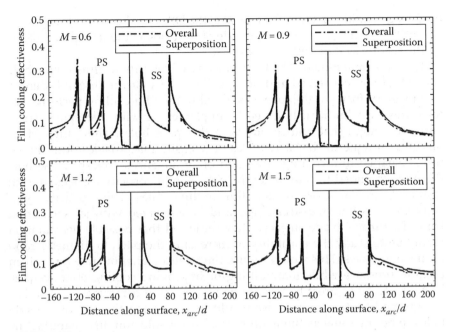

FIGURE 3.142
Effect of superposition on spanwise averaged film cooling effectiveness distribution. (From Mhetras, S.P. and Han, J.C., *2006 ASME IMECE*, Chicago, IL, November 5–10, 2006, ASME Paper IMECE2006-15084.)

and superposition results is small with a magnitude of only about 0.03 units. This indicates that the superposition principle can be successfully applied to predict film cooling in the absence of showerhead film cooling. Similar observations were made by Schneider et al. (2003) who investigated this effect with and without showerhead cooling on the pressure side. Good agreement was found between individual row and multiple row measurements using the superposition approach without showerhead injection. However, showerhead injection caused degradation in full coverage effectiveness in the downstream rows resulting in over prediction of the superposition effectiveness magnitudes. Similar observations were made by Mhetras et al. (2008) with compound angle shaped-hole film cooling on both pressure and suction surface, respectively, with and without showerhead film injection.

3.11.6 Novel Film-Cooling Designs

Kusterer et al. (2006) introduced the double-jet film-cooling (DJFC) as an alternative film-cooling technology to conventional film-cooling design. They showed improvement of the film-cooling effectiveness by application of the DJFC. They also showed basic applicability of the DJFC to a realistic blade cooling configuration and presented first test results under machine operating conditions. From their results, the DJFC has been able to replace a row of shaped holes on the suction side of the blade without negative effects on the suction side thermal load level.

Heidmann and Ekkad (2007) introduced a novel turbine film-cooling hole shape that was conceived and designed at NASA Glenn Research Center. The primary focus of this paper is to study the film-cooling performance for a row of cylindrical holes each supplemented with two symmetrical antivortex holes that branch out from the main holes. This "antivortex" design is unique in that it requires only easily machinable round holes, unlike shaped film-cooling holes and other advanced concepts. The hole design is intended to counteract the detrimental vorticity associated with standard circular cross section film-cooling holes. This vorticity typically entrains hot free-stream gas and is associated with jet separation from the turbine blade surface. They indicated that possible permutation would be to study the feasibility of increasing the pitch to diameter ratio of the main holes. This would reduce the total coolant flow rate for a given plenum pressure while maintaining or improving cooling effectiveness. This would be of great benefit to the engine cycle performance. Another design modification to consider would be increasing the size of the side holes to be the same as the main holes. This would simplify manufacturing and ease concerns about drilling very small holes. Dhungel et al. (2007)

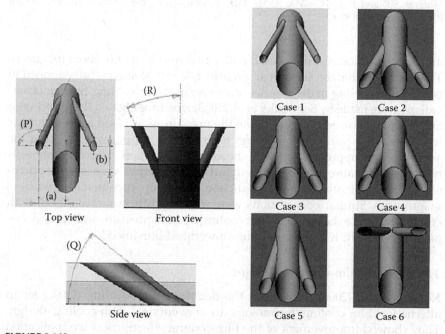

FIGURE 3.143
Design of anti-vortex holes and different configurations. (From Dhungel, A. et al., ASME Paper GT2007-27419, 2007.)

presented a detailed study on the effects of geometry, and orientations of antivortex holes branching out from the primary holes on film cooling have been presented. For all the cases, both heat transfer coefficients and film effectiveness were measured using a transient infrared thermography technique. It appears that the presence of antivortex holes kills the kidney pair vortices and also reduces the momentum of the main jet hence bettering the film coverage in both the downstream and the lateral direction. When the antivortex holes are nearer to the primary film-cooling holes and are developing from the base of the primary holes, better film cooling was accomplished as compared to other antivortex hole orientations. Leblanc et al. (2011) presented a new antivortex design geometry where the two side holes, also of the same diameter, branch out from the root at 15° angle, based on Dhungel et al. (2007). The pitch-to-diameter ratio was increased to 6.0 between the main holes. Results indicated significant improvement in film effectiveness with antivortex holes compared to cylindrical holes at all the blowing ratios studied. At any given blowing ratio, the antivortex hole design used 50% less coolant and provided at least 30%–40% higher cooling effectiveness. Figure 3.143 shows all the antivortex hole designs studied by Dhungel et al. (2007).

3.12 Closure

Turbine airfoil designers need to account for each of the aforementioned effects in their design. Depending on the operating needs and requirements, airfoils operate at varied conditions, which may cause a combination of each of these effects. It is imperative that correlations consider all the effects for final design. It is difficult to simulate a combination of all the parameters and test under real conditions. There is a need to consider matching CFD correlations to reduce the burden on the experimentally derived correlations. However, CFD calculations need to match some existing baseline experimental results. Accurate predictions by CFD codes are still far from matching the requirements of airfoil designers.

References

Abhari, R.S. and Epstein, A.H., 1994. An experimental study of film cooling in a rotating transonic turbine. *ASME Journal of Turbomachinery*, 116, 63–70.

Abraham, S., Navin, A.R., and Ekkad, S.V., 2009. Film cooling study of novel orthogonal entrance and shaped exit holes. ASME Paper GT2009-60003.

Abuaf, N., Bunker, R., and Lee, C.P., 1997. Heat transfer and film cooling effectiveness in a linear airfoil cascade. *ASME Journal of Turbomachinery*, 119, 302–309.

Afejuku, W.O., 1977. Superposition of cooling films. PhD thesis, University of Nottingham, Nottingham, U.K.

Ahn, J.Y., Mhetras, S.P., and Han, J.C., 2005a. Film-cooling effectiveness on a gas turbine blade tip using pressure sensitive paint. *ASME Journal of Heat Transfer*, May, 521–530.

Ahn, J.Y., Schobeiri, M.T., Han, J.C., and Moon, H.K., 2004. Film cooling effectiveness on the leading edge of a rotating turbine blade, *Proceedings of ASME-IMECE'04*, November 13–19, 2004, Anaheim, CA, Paper No. IMECE 2004-59852.

Ahn, J.Y., Schobeiri, M.T., Han, J.C., and Moon, H.K., 2005b. Film cooling effectiveness on the leading edge of a rotating film-cooled blade using pressure sensitive paint, *Proceedings of ASME Turbo-Expo 2005*, June 6–9, 2005, Reno, NV, Paper No. GT 2005-68344.

Ai, W., Laycock, R.G., Rappleye, D.S., Fletcher, T.H., and Bons, J.P., 2009. Effect of particle size and trench configuration on deposition from fine coal flyash near film cooling holes. ASME Paper GT2009-59571.

Ames, F.E., 1998a. Aspects of vane film cooling with high turbulence. Part I: Heat transfer. *ASME Journal of Turbomachinery*, 120, 768–776.

Ames, F.E., 1998b. Aspects of vane film cooling with high turbulence. Part II: Adiabatic effectiveness. *ASME Journal of Turbomachinery*, 120, 777–784.

Ammari, H.D., Hay, N., and Lampard, D., 1990. The effect of density ratio on the heat transfer coefficient from a film cooled flat plate. *ASME Journal of Turbomachinery*, 112, 444–450.

Ammari, H.D., Hay, N., and Lampard, D., 1991. Effect of acceleration on the heat transfer coefficient on a film cooled surface. *ASME Journal of Turbomachinery*, 113, 464–470.

Andrews, G.E., Bazdidi-Tehrani, F., Hussain, C.I., and Pearson, J.P., 1991. Small diameter film cooling hole heat transfer: The influence of hole length. ASME Paper 91-GT-344.

Arts, T., Soechting, F., and Tsukagoshi, K., 2004. High coverage blade tip film cooling. ASME Paper GT2004-53226.

Bailey, J.C., Nirmalan, N., and Braaten, M.E., 2004. Experimental and computational investigation of heat transfer effectiveness and pressure distribution of a shrouded blade tip section. ASME Paper GT2004-53279.

Barlow, D.N. and Kim, Y.W., 1995. Effect of Surface roughness on local heat transfer and film cooling effectiveness. ASME Paper 95-GT-14.

Bogard, D.G. and Cutbrith, J.M., 2003. Effects of coolant density ratio on film cooling performance on a vane. ASME Paper GT2003-38582.

Bons, J.P., MacArthur, C.D., and Rivir, R.B., 1996. The effect of high free-stream turbulence on film cooling effectiveness. *ASME Journal of Turbomachinery*, 118, 814–825.

Buck, F.A., Walters, D.K., Ferguson, J.D., McGrath, E.L., and Leylek, J.H., 2002. Film cooling on a modern HP turbine blade. Part I: Experimental and computational methodology and validation. ASME Paper GT2002-30470.

Bunker, R.S., 2000. Effect of partial coating blockage on film cooling effectiveness. ASME Paper 2000-GT-0244.

Bunker, R.S., 2005. A review of turbine shaped film cooling technology. *Journal of Heat Transfer*, 127, 441–453.

Bunker, R.S., 2008. The effects of manufacturing tolerances on gas turbine cooling. ASME Paper GT2008-50124.

Bunker, R.S., 2009a. Film cooling: Breaking the limits of diffusion shaped holes, *Turbine09 Symposium on Heat Transfer in Gas Turbine Systems*, August 9–14, 2009, Antalya, Turkey.

Bunker, R.S., 2009b. A study of mesh-fed slot film cooling. ASME Paper GT2009-59338.

Bunker, R.S. and Bailey, J.C., 2001. Film cooling discharge coefficient measurements in a turbulated passage with internal cross flow. ASME Paper 2001-GT-0135.

Burd, S.W., Kaszeta, R.W., and Simon, T.W., 1996. Measurements in film cooling flows: Hole L/D and turbulence intensity effects. ASME Paper 96-WA/HT-7.

Burd, S.W. and Simon, T.W., 1997. The influence of coolant supply geometry on film coolant exit flow and surface adiabatic effectiveness. ASME Paper 97-GT-25.

Byerley, A.R., Ireland, P.T., Jones, T.V., and Ashton, S.A., 1988. Detailed heat transfer measurements near and within the entrance of a film cooling hole, *ASME, Gas Turbine and Aeroengine Congress and Exposition*, Amsterdam, the Netherlands, June 6–9, 1988.

Byerley, A.R., Jones, T.V., and Ireland, P.T., 1992. Internal cooling passage heat transfer near the entrance to a film cooling hole—Experimental and computational results, *ASME, International Gas Turbine and Aeroengine Congress and Exposition, 37th*, Cologne, Germany, June 1–4, 1992, 12pp.

Byerley, A.R., 1989. Heat transfer near the entrance to a film cooling hole in a gas turbine blade, D Phil thesis, Department of Engineering Science, University of Oxford, Oxford, U.K.

Camci, C. and Arts, T., 1985a. Short-duration measurements and numerical simulation of heat transfer along the suction side of a gas turbine blade. *ASME Journal of Engineering for Gas Turbines and Power*, 107, 991–997.

Camci, C. and Arts, T., 1985b. Experimental heat transfer investigation around the film cooled leading edge of a high pressure gas turbine rotor blade. *ASME Journal of Engineering for Gas Turbines and Power*, 107, 1016–1021.

Camci, C. and Arts, T., 1991. Effect of incidence on wall heating rates and aerodynamics on a film cooled transonic turbine blade. *ASME Journal of Turbomachinery*. 113, 493–500.

Cardwell, N., Sundaram, N., and Thole, K., 2005. Effects of mid-passage gap, endwall misalignment and roughness on endwall film-cooling. ASME Paper GT2005-68900.

Christophel, J., Couch, E., Thole, K., and Cunha, F., 2004a. Measured adiabatic effectiveness and heat transfer for blowing from the tip of a turbine blade. ASME Paper GT2004-53250.

Christophel, J., Thole, K., and Cunha, F., 2004b. Cooling the tip of a turbine blade using pressure side holes: Part 1—Adiabatic effectiveness measurements. ASME Paper GT2004-53251.

Christophel, J., Thole, K., and Cunha, F., 2004c. Cooling the tip of a turbine blade using pressure side holes: Part 2—Heat transfer measurements. ASME Paper GT2004-53254.

Colban, W.F., Bogard, D., and Thole, K., 2008. A film-cooling correlation for shaped holes on a flat-plate surface. ASME Paper GT2008-50121.

Colban, W.F., Lethander, A.T., Thole, K., and Zess, G., 2002a. Combustor turbine interface studies—Part 2: Flow and thermal field measurements. ASME Paper GT2002-30527.

Colban, W., Thole, K., and Haendler, M., 2006. A comparison of cylindrical and fan-shaped film-cooling holes on a vane endwall at low and high freestream turbulence levels. ASME Paper GT2006-90021.

Colban, W.F., Thole, K.A., and Bogard, D., 2011. A film cooling correlation for shaped holes on a flat-plate surface. *ASME Journal of Turbomachinery*, January 2011, 133.

Colban, W.F., Thole, K., and Zess, G., 2002b. Combustor turbine interface studies—Part 1: Endwall effectiveness measurements. ASME Paper GT2002-30526.

Coon, C., Kim, Y., and Moon, H.K., 2005. Film-cooling characteristics of pressure-side discharge slots in an accelerating mainstream flow. ASME Paper GT2005-69061.

Cruse, M.W., Yuki, U.M., and Bogard, D.G., 1997. Investigation of various parametric influences on leading edge film cooling. ASME Paper 97-GT-296.

Cutbirth, J.M. and Bogard, D.G., 2002a. Evaluation of pressure side film cooling with flow and thermal field measurements, Part I: Showerhead effects. ASME Paper GT2002-30174.

Cutbirth, J.M. and Bogard, D.G., 2002b. Evaluation of pressure side film cooling with flow and thermal field measurements, Part I: Turbulence effects. ASME Paper GT2002-30175.

Dannhauer, A., 2009. Investigation of trailing edge cooling concepts in a high pressure turbine cascade-analysis of the adiabatic film cooling effectiveness. ASME Paper GT2009-59343.

Day, C.R.B., Oldfield, M.L.G., and Lock, G.D., 1999. The influence of film cooling on the efficiency of an annular nozzle guide vane cascade. *ASME Journal of Turbomachinery*. 121, 145–151.

Day, C.R.B., Oldfield, M.L.G., Lock, G.D., and Dancer, S.N., 1998. Efficiency measurements of an annular nozzle guide vane cascade with different film cooling geometries. ASME Paper 98-GT-538.

Demling, P. and Bogard, D., 2006. The effects of obstructions on film cooling effectiveness on the suction side of a gas turbine vane. ASME Paper GT2006-90577.

Denton, J.D., 1993. Loss mechanisms in turbomachines. *ASME Journal of Turbomachinery*, 115, 621–656.

Dhungel, A., Lu, Y., Phillips, A., Ekkad, S., and Heidmann, J., 2007. Film cooling from a row of holes supplemented with anti vortex holes. ASME Paper GT2007-27419.

Dorrington, R., Bogard, D.G., and Bunker, R., 2007. Film effectiveness performance for coolant holes imbedded in various shallow trench and crater depressions. ASME Paper GT2007-27992.

Dorrington, R., Bogard, D.G., and Bunker, R., 2007a. Film effectiveness performance for coolant holes imbedded in various shallow trench and crater depressions. ASME Paper GT2007-27992.

Dorrington, R., Bogard, D.G., Bunker, R., and Dees, J., 2007. Turbine airfoil net heat flux reduction with cylindrical holes embedded in a transverse trench. ASME Paper GT2007-27996.

Dorrington, R., Bogard, D.G., Bunker, R., and Dees, J., 2007b. Turbine airfoil net heat flux reduction with cylindrical holes embedded in a transverse trench. ASME Paper GT2007-27996.

Dring, R.P., Blair, M.F., and Joslyn, H.D., 1980. An experimental investigation of film cooling on a turbine rotor blade. *ASME Journal of Engineering for Power*, 102, 81–87.

Drost, U. and Bolcs, A., 1999. Investigation of detailed film cooling effectiveness and heat transfer distributions on a gas turbine airfoil. *ASME Journal of Turbomachinery*, 121, 233–242.

Du, H., Ekkad, S.V., and Han, J.C., 1999. Effect of unsteady wake with trailing edge coolant ejection on film cooling performance for a gas turbine blade. *ASME Journal of Turbomachinery*, 121, 448–455.

Du, H., Han, J.C., and Ekkad, S.V., 1998. Effect of unsteady wake on detailed heat transfer coefficient and film effectiveness distributions for a gas turbine blade. *ASME Journal of Turbomachinery*, 120, 808–817.

Dunn, M.G. and Stoddard, F.J., 1979. Measurement of heat-transfer rate to a gas turbine blade. *ASME Journal of Engineering for Power*, 101, 275–280.

Ekkad, S.V., Han, J.C., and Du, H., 1998. Detailed film cooling measurements on a cylindrical leading edge model: Effect of free-stream turbulence and coolant density. *ASME Journal of Turbomachinery*, 120, 799–807.

Ekkad, S., Lu, Y., and Nasir, H., 2005a. Film cooling from a row of holes embedded in transverse slots. ASME Paper GT2005-68598.

Ekkad, S.V., Mehendale, A.B., Han, J.C., and Lee, C.P., 1997a. Combined effect of grid turbulence and unsteady wake on film effectiveness and heat transfer coefficient of a gas turbine blade with air and CO_2 film injection. *ASME Journal of Turbomachinery*, 119, 594–600.

Ekkad, S., Nasir, H., and Bunker, R., 2005b. Effects of tip and pressure side coolant injection on heat transfer distributions for a plane and recessed tip. ASME Paper GT2005-68595.

Ekkad, S.V., Zapata, D., and Han, J.C., 1997b. Heat transfer coefficients over a flat surface with air and CO_2 film injection through compound angle holes using a transient liquid crystal image method. *ASME Journal of Turbomachinery*, 119, 580–586.

Ekkad, S.V., Zapata, D., and Han, J.C., 1997c. Film effectiveness over a flat surface with air and CO_2 film injection through compound angle holes using a transient liquid crystal image method. *ASME Journal of Turbomachinery*, 119, 587–593.

Eriksen, V.L., 1971. Film cooling effectiveness and heat transfer with injection through holes. NASA CR 72991.

Eriksen, V.L. and Goldstein, R.J., 1974. Heat transfer and film cooling following injection through inclined circular holes. *ASME Journal of Heat Transfer*, 96, 239–245.

Fiala, N., Jaswal, I., and Ames, F., 2008a. Letterbox trailing edge heat transfer—Effects of blowing rate, Reynolds number, and external turbulence on heat transfer and film cooling effectiveness. ASME Paper GT2008-50474.

Fiala, N., Johnson, J.D., and Ames, F., 2008b. Aerodynamics of a letterbox trailing edge heat transfer—Effects of blowing rate, Reynolds number, and external turbulence on aerodynamics losses and pressure distribution. ASME Paper GT2008-50475.

Forth, C.J.P. and Jones, T.V., 1986. Scaling parameters in film cooling, *Proceedings of the Eighth International Heat Transfer Conference*, San Francisco, CA.

Friedrichs, S., Hodson, H.P., and Dawes, W.N., 1996. Distribution of film-cooling effectiveness on a turbine endwall measured using the ammonia and diazo technique. *ASME Journal of Turbomachinery*, 118, 613–621.

Friedrichs, S., Hodson, H.P., and Dawes, W.N., 1997. Aerodynamics aspects of endwall film-cooling. *ASME Journal of Turbomachinery*, 119, 786–793.

Friedrichs, S., Hodson, H.P., and Dawes, W.N., 1998. The design of an improved end-wall film-cooling configuration. ASME Paper 98-GT-483.

Funazaki, K., Yokota, M., and Yamawaki, K., 1997. The effect of periodic wake passing on film effectiveness of discrete holes around the leading edge of a blunt body. *ASME Journal of Turbomachinery*, 119, 292–301.

Gao, Z. and Han, J.C., 2009. Influence of hole shape and angle on showerhead film cooling using PSP technique. *ASME Journal of Heat Transfer*, 131(6), 061701.

Gao, Z., Narzary, D., and Han, J.C., 2008a. Turbine blade platform film cooling with typical stator-rotor purge flow and discrete-hole film cooling. ASME Paper GT2008-50286.

Gao, Z., Narzary, D.P., and Han, J.C., 2008b. Film cooling on a gas turbine blade pressure side or suction side with axial shaped holes. *International Journal of Heat and Mass Transfer*, 51(9–10), 2139–2152.

Gao, Z., Narzary, D.P., and Han, J.C., 2009. Film cooling on a gas turbine blade pressure side or suction side with compound angle shaped holes. *ASME Journal of Turbomachinery*, 131(1), 011019.

Gao, Z., Narzary, D.P., Mhetras, S., and Han, J.C., 2007. Effect of inlet flow angle on gas turbine blade tip film cooling and heat transfer. ASME Paper GT2007-27066.

Georgiou, D.P., Godard, M., and Richards, B.E., 1979. Experimental study of iso-heat-transfer-rate lines on the endwall of a turbine cascade. ASME Paper 79-GT-20.

Goldstein, R.J., Ecket-t, E.R.G., Eriksen, V.L., and Ramsey, J.C.L., 1970. Film cooling following injection through inclined circular tubes. *Israel Journal of Technology*, 8, 135.

Goldstein, R.J., 1971. Film cooling. *Advances in Heat Transfer*, 7, 321–379.

Goldstein, R.J. and Chen, H.P., 1985. Film cooling on a gas turbine near the endwall. *ASME Journal of Engineering for Gas Turbines and Power*, 107, 117–122.

Goldstein, R.J. and Chen, H.P., 1987. Film cooling of a turbine blade with injection through two rows of holes in the near endwall region. *ASME Journal of Turbomachinery*, 109, 588–593.

Goldstein, R.J., Eckert, E.R.G., and Burgraff, F., 1974. Effects of hole geometry and density on three-dimensional film cooling. *International Journal of Heat and Mass Transfer*, 17, 595–605.

Goldstein, R.J., Eckert, E.R.G., Chiang, H.D., and Elovic, E., 1985. Effect of surface roughness on film cooling performance. *ASME Journal of Engineering for Gas Turbines and Power*, 107, 111–116.

Goldstein, R.J., Eckert, E.R.G., Tsou, F.K., and Haji-Sheikh, A., 1966. *AIAA Journal*, 4, 981.

Goldstein, R.J. and Taylor, J.R., 1982. Mass transfer in the neighborhood of jets entering a crossflow. *ASME Journal of Heat Transfer*, 104, 715–721.

Goldstein, R.J. and Yoshida, T., 1982. The influence of a laminar boundary layer and laminar injection on film cooling performance. *ASME Journal of Heat Transfer*, 104, 355–362.

Gritsch, M., Schulz, A., and Wittig, S., 1998a. Heat transfer coefficient measurements of film cooling holes with expanded slots. ASME Paper 98-GT-28.

Gritsch, M., Schulz, A., and Wittig, S., 1998b. Adiabatic wall effectiveness measurements of film cooling holes with expanded exits. *ASME Journal of Turbomachinery*, 120, 549–556.

Gritsch, M., Schulz, A., and Wittig, S., 1998c. Discharge coefficient measurements of film-cooling holes with expanded exits. *ASME Journal of Turbomachinery*, 120, 557–563.

Haller, B.R. and Camus, J.J., 1984. Aerodynamic loss penalty produced by film cooling transonic turbine blades. *ASME Journal of Engineering for Gas Turbines and Power*, 106, 198–205.

Han, J.C. and Ekkad, S.V., 1998. Recent development in turbine blade film cooling, Invited paper for *ISTP-11*, Hsinchu, Taiwan, November 1998.

Harasgama, S.P. and Burton, C.D., 1992a. Film cooling research on the endwall of a turbine nozzle guide vane in a short-duration annular cascade. Part I: Experimental technique and results. *ASME Journal of Turbomachinery*, 114, 734–740.

Harasgama, S.P. and Burton, C.D., 1992b. Film cooling research on the endwall of a turbine nozzle guide vane in a short-duration annular cascade. Part II: Analysis and correlation of results. *ASME Journal of Turbomachinery*, 114, 741–746.

Hay, N., Lampard, D., and Benmansour, S., 1983. Effect of cross-flow on the discharge coefficient of film cooling holes. *Journal of Engineering for Power*, 105, 243–248.

Hay, N. and Lampard, D., 1996. Discharge coefficient of turbine cooling holes: A review. ASME Paper 96-GT-492.

Hay, N., Lampard, D., and Saluja, C.L., 1985. Effect of cooling films on the heat transfer coefficient on a flat plate with zero mainstream pressure gradient. *ASME Journal of Engineering for Gas Turbines and Power*, 107, 105–110.

Heidmann, J. and Ekkad, S., 2007. A novel anti-vortex turbine film cooling hole concept. ASME Paper GT2007-27528.

Heidmann, J.D., Lucci, B.L., and Reshotko, E., 1997. An experimental study of the effect of wake passing on turbine blade film cooling. ASME Paper 97-GT-255.

Horbach, T., Schulz, A., and Bauer, H.-J., 2010. Trailing edge film cooling of gas turbine airfoils-external cooling performance of various internal pin fin configurations. ASME Paper GT2010-23578.

Hylton, L.D., Nirmalan, V., Sultanian, B.K., and Kaufmann, R.M., 1988. The effects of leading edge and downstream film cooling on turbine vane heat transfer. NASA CR-182133.

Ito, S., Eckert, E.R.G., and Goldstein, R.J., 1980. Aerodynamic loss in gas turbine stage with film cooling. *ASME Journal of Engineering for Power*, 102, 964–970.

Ito, S., Goldstein, R.J., and Eckert, E.R.G., 1978. Film cooling of a gas turbine blade. *ASME Journal of Engineering for Power*, 100, 476–481.

Jabbari, M.Y. and Goldstein, R.J., 1978. Adiabatic wall temperature and heat transfer downstream of injection through two rows of holes. *ASME Journal of Engineering for Power*, 100, 303–307.

Jabbari, M.Y., Marston, K.C., Eckert, E.R.G., and Goldstein, R.J., 1996. Film cooling of the gas turbine endwall by discrete hole injection. *ASME Journal of Turbomachinery*, 118, 278–284.

Kadotani, K. and Goldstein, R.J., 1978. Effect of mainstream variables on jets issuing from a row of inclined round holes, *American Society of Mechanical Engineers, Gas Turbine Conference and Products Show*, London, England, April 9–13, 1978, 7pp.

Karni, J. and Goldstein, R.J., 1990. Surface injection effect on mass transfer from a cylinder in crossflow: A simulation of film cooling in the leading edge region of a turbine blade. *ASME Journal of Turbomachinery*, 112, 418–427.

Kim, Y.W., Downs, J.P., Soechting, F.O., Abdel-Messeh, W., Steuber, G.D., and Tanrikut, S., 1995. A summary of the cooled turbine blade tip heat transfer and film effectiveness investigations performed by Dr. D.E. Metzger. *ASME Journal of Turbomachinery*, 117, 1–11.

Kim, Y.W. and Metzger, D.E., 1995. Heat transfer and effectiveness on film cooled turbine blade tip models. *ASME Journal of Turbomachinery*, 117, 12–18.

Knost, D. and Thole, K., 2004. Adiabatic effectiveness measurements of endwall film-cooling for a first stage vane. ASME Paper GT2004-53326.

Kohli, A. and Bogard, D.G., 1998. Effects of very high free-stream turbulence on the jet-mainstream interaction in a film cooling flow. *ASME Journal of Turbomachinery*, 120, 785–790.

Kost, F. and Nicklas, M., 2001. Film-cooled turbine endwall in a transonic flow field: Part I—Aerodynamic measurements. ASME Paper 2001-GT-0145.

Kubo, R., Otomo, F., Fukuyama, Y., and Nakata, Y., 1998. Aerodynamic loss increase due to individual film cooling injections from gas turbine nozzle surface. ASME Paper 98-GT-497.

Kusterer, K., Bohn, D., Sugimoto, T., and Tanaka, R., 2006. Double-jet ejection of cooling air for improved film-cooling. ASME Paper GT2006-90854.

Kuwabara, M., Tsukagoshi, K., and Arts, T., 2004. High coverage blade tip film cooling, *ASME Turbo Expo 2004: Power for Land, Sea, and Air (GT2004)*, ASME Paper GT2004-53226, pp. 189–199.

Kwak, J.S. and Han, J.C., 2002a. Heat transfer coefficient and film cooling effectiveness on a gas turbine blade tip. ASME Paper GT2002-30194.

Kwak, J.S. and Han, J.C., 2002b. Heat transfer coefficient and film cooling effectiveness on the squealer tip of a gas turbine blade. ASME Paper GT2002-30555.

L'Ecuyer, M.R. and Soechting, F.O., 1985. A model for correlating flat plate film cooling effectiveness for rows of round holes. *Heat Transfer and Cooling in Gas Turbines*, AGARD, Bergen, Norway.

Leblanc, C., Narzary, D., and Ekkad, S.V., 2011. Film cooling performance of an anti-vortex hole on a flat plate, *AJTEC2011-44161, 2011 AJTEC*, Honolulu, HI, March 2011.

Leiss, C., 1975. Experimental investigation of film cooling with injection from a row of holes for the application to gas turbine blades. *ASME Journal of Engineering for Power*, 97, 21–27.

Lewis, S., Barker, B., Bons, J.P., Ai, W., and Fletcher, T.H., 2009. Film cooling effectiveness and heat transfer near deposit-laden film holes. ASME Paper GT2009-59567.

Leylek, J.H. and Zerkle, R.D., 1994. Discrete-jet film cooling: A comparison of computational results with experiments. *ASME Journal of Turbomachinery*, 116, 358–368.

Ligrani, P.M., Gong, R., and Cuthrell, J.M., 1997. Bulk flow pulsations and film cooling: Flow structure just downstream of the holes. *ASME Journal of Turbomachinery*, 119, 568–573.

Ligrani, P.M., Saumweber, C., Schulz, A., and Wittig, S., 2001. Shock wave-film cooling interactions in transonic flows. ASME Paper 2001-GT-0133.

Ligrani, P.M., Wigle, J.M., Ciriello, S., and Jackson, S.W., 1994a. Film-cooling from holes with compound angle orientations. Part I: Results downstream of two staggered rows of holes with 3d spanwise spacing. *ASME Journal of Heat Transfer*, 116, 341–352.

Ligrani, P.M., Wigle, J.M., and Jackson, S.W., 1994b. Film-cooling from holes with compound angle orientations. Part 2: Results downstream of a single row of holes with 6d spanwise spacing. *ASME Journal of Heat Transfer*, 116, 353–362.

Liu, K., Narzary, D.P., Han, J.C., Mirzamoghadam, A.V., and Riahi, A., 2011. Influence of shock wave on turbine vane suction side film cooling with compound-angle shaped holes, *Proceedings of ASME Turbo-Expo 2011*, Vancouver, British Columbia, Canada, June 6–10, 2011, ASME Paper GT-2011-45927.

Turbine Film Cooling 323

Lu, Y., Dhungel, A., Ekkad, S, and Bunker, R., 2007a. Effect of trench width and depth on film cooling from cylindrical holes embedded in trenches. ASME Paper GT2007-27388.
Lu, Y., Dhungel, A., Ekkad, S., and Bunker, R., 2007b. Film cooling measurements for cratered cylindrical inclined holes. ASME Paper GT2007-27386.
Lu, Y., Ekkad, S., Bunker, R., 2008. Trench film cooling effect of trench downstream edge and hole spacing. ASME Paper GT2008-51207.
Luckey, D.W., Winstanley, D.K., Hames, G.J., and L'Ecuyer, M.R., 1977. Stagnation region gas film cooling for turbine blade leading-edge applications. AIAA Journal of Aircraft, 14, 494–501.
Lynch, S. and Thole, K., 2007. The effect of combustor-turbine interface gap leakage on the endwall heat transfer for a nozzle guide vane. ASME Paper GT2007-27867.
Mahmood, G.I., Gustafson, R., and Acharya, S., 2009. Flow dynamics and film cooling effectiveness on a non-axisymmetric contour endwall in a two dimensional cascade passage. ASME Paper GT2009-60236.
Makki, Y.H. and Jakubowski, G., 1986. An experimental study of film cooling from diffused trapezoidal shaped holes. AIAA Paper 86-1326.
Martini, P. and Schulz, A., 2003. Experimental and numerical investigation of trailing edge film cooling by circular coolant wall jets ejected from a slot with internal rib arrays. ASME Paper GT2003-38157.
Martini, P., Schulz, A., and Bauer, H.J., 2005. Film cooling effectiveness and heat transfer on the trailing edge cut-back of gas turbine airfoils with various internal cooling designs. ASME Paper GT2005-68083.
Martiny, M., Schulz, A., and Wittig, S., 1995. Full coverage film cooling investigations: Adiabatic wall temperatures and flow visualizations. ASME Paper 95-WA/HT-4.
Mehendale, A.B., Ekkad, S.V., and Han, J.C., 1994a. Mainstream turbulence effect on film effect on film effectiveness and heat transfer coefficient of a gas turbine blade with air and CO_2 film injection. International Journal of Heat and Mass Transfer, 37, 2707–2714.
Mehendale, A.B. and Han, J.C., 1992. Influence of high mainstream turbulence on leading edge film cooling heat transfer. ASME Journal of Turbomachinery, 114, 707–715.
Mehendale, A.B., Han, J.C., Ou, S., and Lee, C.P., 1994b. Unsteady wake over a linear turbine blade cascade with air and CO_2 film injection. Part II: Effect on film effectiveness and heat transfer distributions. ASME Journal of Turbomachinery, 116, 730–737.
Mhetras, S., Gao, Z., Yang, H., and Han, J.C., 2005. Film-cooling effectiveness on squealer rim walls and squealer cavity floor of a gas turbine blade tip using pressure sensitive paint. ASME Paper GT2005-68387.
Mhetras, S.P. and Han, J.C., 2006. Effect of superposition on spanwise film-cooling effectiveness distribution on a gas turbine blade, 2006 ASME IMECE, Chicago, IL, November 5–10, 2006, ASME Paper IMECE2006-15084.
Mhetras, S.P., Han, J.C., and Rudolph, R., 2007. Effect of flow parameter variations on full coverage film-cooling effectiveness for a gas turbine blade, Proceedings of ASME Turbo-Expo 2007, May 14–17, 2007, Montreal, Quebec, Canada, Paper No. GT2007-27071.
Mhetras, S.P., Han, J.C., and Rudolph, R., 2008. Film-cooling effectiveness from shaped film cooling holes for a gas turbine blade, Proceedings of ASME Turbo-Expo 2008, Berlin, Germany, June 9–13, 2008, ASME Paper GT-2008-50916.

Mhetras, S.P., Han, J.C., and Rudolph, R., 2008. Film-cooling effectiveness from shaped film cooling holes for a gas turbine blade, *Proceedings of ASME Turbo-Expo 2008*, June 9–13, 2008, Berlin, Germany, Paper No. GT-2008-50916.

Mhetras, S.P, Yang, H., Gao, Z., and Han, J.C., 2006. Film-cooling effectiveness on squealer cavity and rim walls of gas-turbine blade tip. *AIAA Journal of Propulsion and Power*, 22(4), 889–899.

Mick, W.J. and Mayle, R.E., 1988. Stagnation film cooling and heat transfer including its effect within the hole pattern. *ASME Journal of Turbomachinery*, 110, 66–72.

Moffat, R.J., 1985. Turbine blade cooling, *Symposium on Heat Transfer in Rotating Machinery*, Honolulu, HI.

Muska, J.F., Fish, R.W., and Suo, M., 1976. The additive nature of film cooling from rows of holes. *ASME Journal of Engineering for Power*, 457–464, 75-WA/GT-17, 1975.

Narzary, D.P., Gao, Z., Mhetras, S., and Han, J.C., 2007. Effect of unsteady wake on film-cooling effectiveness distribution on a gas turbine blade with compound shaped holes. ASME Paper GT2007-27070.

Narzary, D., Liu, K.C., Rallabandi, A., and Han, J.C., 2010. Influence of coolant density on turbine blade film-cooling using pressure sensitive paint technique. ASME Paper GT2010-22781.

Nasir, H., Ekkad, S.V., Bunker, R., and Prakash, C., 2004. Effects of tip gap film injection from plain and squealer blade tips. ASME Paper GT2004-53455.

Nealy, D.A., Mihelc, M.S., Hylton, L.D., and Gladden, H.J., 1984. Measurements of heat transfer distribution over the surfaces of highly loaded turbine nozzle guide vanes. *ASME Journal of Engineering for Gas Turbines and Power*, 106, 149–158.

Nicklas, M., 2001. Film-cooled turbine endwall in a transonic flow field: Part II—Heat transfer and film cooling effectiveness. ASME Paper 2001-GT-0146.

Nirmalan, N.V. and Hylton, L.D., 1990. An experimental study of turbine vane heat transfer with leading edge and downstream film cooling. *ASME Journal of Turbomachinery*, 112, 477–487.

Nix, A., Ng, W.F., Smith, A.C., and Diller, T., 2003. The unsteady effect of passing shocks on pressure surface versus suction surface heat transfer in film-cooled transonic turbine blades. ASME Paper GT2003-38530.

Ochs, M., Schulz, A, and Bauer, H.J., 2007. Investigation of the influence of trailing edge shock waves on film cooling performance of gas turbine airfoils. ASME Paper GT2007-27482.

Oke, R.A. and Simon, T.W., 2002. Film cooling experiments with flow introduced upstream of a first stage nozzle guide vane through slots of various geometries. ASME Paper GT2002-30169.

Oke, R., Simon, T., Shih, T., Zhu, B., Lin, Y.L., and Chyu, M.K., 2001. Measurements over a film-cooled contoured endwall with various coolant injection rates. ASME Paper 2001-GT-0140.

Otomo, F., Nakata, Y., Kubo, R., and Suga, T., 1997. Pressure loss characteristics of gas turbine nozzle blade row with individual film cooling injection rows, *Proceedings of the 25th Annual Gas Turbine Conference*, New Orleans, LA, JSME.

Ou, S. and Han, J.C., 1994. Unsteady wake effect on film effectiveness and heat transfer coefficient from a turbine blade with one row of air and CO_2 film injection. *ASME Journal of Heat Transfer*, 116, 921–928.

Ou, S., Han, J.C., Mehendale, A.B., and Lee, C.P., 1994. Unsteady wake over a linear turbine blade cascade with air and CO_2 film injection. Part I: Effect on heat transfer coefficients. *ASME Journal of Turbomachinery*, 116, 721–729.

Ou, S., Mehendale, A.B., and Han, J.C., 1992. Influence of high mainstream turbulence on leading edge film cooling heat transfer: Effect of film hole row location. *ASME Journal of Turbomachinery*, 114, 716–723.

Park, J.S., Lee, D.H., Rhee, D.H., Cho, H.H., and Kang, S.H., 2010. Heat transfer and effectiveness on the film cooled tip and inner rim surfaces of a turbine blade. ASME Paper GT2010-23203.

Pedersen, D.R., Eckert, E.R.G., and Goldstein, R.J., 1977. Film-cooling with large density differences between the mainstream and secondary fluid measured by the heat-mass transfer analogy. *ASME Journal of Heat Transfer*, 99, 620–627.

Pietrzyk, J.R., Bogard, D.G., and Crawford, M.E., 1990. Effects of density ratio on the hydrodynamics of film cooling. *ASME Journal of Turbomachinery*, 112, 437–443.

Piggush, J. and Simon, T., 2006a. Heat transfer measurements in a first stage nozzle cascade having endwall contouring: Misalignment and leakage studies. ASME Paper GT2006-90575.

Piggush, J. and Simon, T., 2006b. Adiabatic effectiveness measurements in a first stage nozzle cascade having endwall contouring, leakage and assembly features. ASME Paper GT2006-90576.

Polanka, M.D., Ethridge, M.I., Cutbirth, J.M., and Bogard, D.G., 2000. Effects of showerhead injection on film cooling effectiveness for a downstream row of holes. ASME Paper 2000-GT-0240.

Post, J. and Acharya, S., 2010. The role of density ratio and blowing ratio on film cooling in a vane passage. ASME Paper GT 2010-23680.

Rallabandi, A.P., Li, S.J., and Han, J.C., 2010. Influence of unsteady wake on turbine blade film cooling using pressure sensitive paint technique, *Proceedings of the 14th IHTC (International Heat Transfer Conference)*, August 8–13, 2010, Washington, DC.

Rehder, H.J., 2009. Investigation of trailing edge cooling concepts in a high pressure turbine cascade-aerodynamic experiments and loss analysis. ASME Paper GT2009-59303.

Rohde, J.E., Richards, H.T., and Metger, G.W., 1969. Discharge coefficients for thick plate orifices with approach flow perpendicular and inclined to the orifice axis, NASA TN D-5467.

Salcudean, M., Gartshore, I., Zhang, K., and Barnea, Y., 1994a. Leading edge film cooling of a turbine blade model through single and double row injection: Effects of coolant density. ASME Paper 94-GT-2.

Salcudean, M., Gartshore, I., Zhang, K., and McLean, I., 1994b. An experimental study of film cooling effectiveness near the leading edge of a turbine blade. *ASME Journal of Turbomachinery*, 116, 71–79.

Sargison, J.E., Guo, S.M., Oldfield, M.L.G., Lock, G.D., and Rawlinson, A.J., 2001a. A converging slot-hole film-cooling geometry. Part 1: Low-speed flat-plate heat transfer and loss. ASME Paper 2001-GT-0126.

Sargison, J.E., Guo, S.M., Oldfield, M.L.G., Lock, G.D., and Rawlinson, A.J., 2001b. A converging slot-hole film-cooling geometry. Part 2: Transonic nozzle guide vane heat transfer and loss. ASME Paper 2001-GT-0127.

Saumweber, C.J. and Schulz, A., 2003. Interactions of film cooling rows: Effects of hole geometry and row spacing on the cooling performance dowstream of the second row of holes. ASME Paper GT2003-38195.

Saumweber, C. and Schulz, A., 2008a. Free-stream effects on the cooling performance of cylindrical and fan-shaped cooling holes. ASME Paper GT2008-51030.

Saumweber, C. and Schulz, A., 2008b. Comparison the cooling performance of cylindrical and fan-shaped cooling holes with special emphasis on the effect of internal coolant cross-flow. ASME Paper GT2008-51036.

Saumweber, C. and Schulz, A., 2008c. Effect of geometry variations on the cooling
performance of fan-shaped cooling holes. ASME Paper GT2008-51038.

Saumweber, C., Schulz, A., and Wittig, S., 2002. Free-stream turbulence effects on film
cooling with shaped holes. ASME Paper GT2002-30170.

Schmidt, D.L., Sen, B., and Bogard, D.G., 1996a. Film cooling with compound angle
holes: Adiabatic effectiveness. *ASME Journal of Turbomachinery*, 118, 807–813.

Schmidt, D.L., Sen, B., and Bogard, D.G., 1996b. Effects of surface roughness on film
cooling. ASME Paper 96-GT-299.

Schneider, M., Parneix, S., and Wolfersdorf, J., 2003. Effect of showerhead injection on
superposition of multi-row pressure side film cooling with fan-shaped holes.
ASME Paper GT-2003-38693.

Schnieder, M., von Wolfersdorf, J., and Parneix, S., 2003. Effect of showerhead injection on superposition of multi-row pressure side film cooling with fan shaped
holes. ASME Paper GT2003-38693.

Schwarz, S.G. and Goldstein, R.J., 1989. The two-dimensional behavior of film cooling
jets on concave surfaces. *ASME Journal of Turbomachinery*, 111, 124–130.

Schwarz, S.G., Goldstein, R.J., and Eckert, E.R.G., 1990. The influence of curvature on
film cooling performance. *ASME Journal of Turbomachinery*, 112, 472–478.

Sellers, J.P., 1963. Gaseous film cooling with multiple ejection stations. *AIAA Journal*,
1(9), 2154–2156.

Sen, B., Schmidt, D.L., and Bogard, D.G., 1996. Film cooling with compound angle
holes: Heat transfer. *ASME Journal of Turbomachinery*, 118, 800–806.

Seo, H.J., Lee, J.S., and Ligrani, P., 1998. Effects of bulk flow pulsations on film cooling from
different length injection holes at different blowing ratios. ASME Paper 98-GT-192.

Sinha, A.K., Bogard, D.G., and Crawford, M.E., 1991. Film cooling effectiveness
downstream of a single row of holes with variable density ratio. *ASME Journal
of Turbomachinery*, 113, 442–449.

Somawardhana, R. and Borgard, D.G., 2007a. Effects of obstructions and surface
roughness on film cooling effectiveness with and without a transverse trench.
ASME Paper GT2007-28003.

Somawardhana, R. and Borgard, D.G., 2007b. Effects of surface roughness and near
hole obstructions on film cooling effectiveness. ASME Paper GT2007-28004.

Sundaram, N. and Thole, K., 2006. Effects of surface deposition, hole blockage, and
TBC spallation on vane endwall film-cooling. ASME Paper GT2006-90379.

Sundaram, N. and Thole, K., 2008. Film-cooling flow fields with trenched holes on an
endwall. ASME Paper GT2008-50149.

Suryanarayanan, A., Mhetras, S.P., Schobeiri, M.T., and Han, J.C., 2006. Film cooling
effectiveness on a rotating blade platform, *Proceedings of ASME Turbo-Expo 2006*,
May 8–11, 2006, Barcelona, Spain, Paper No. GT2006-90034.

Suryanarayanan, A., Ozturk, B., Schobeiri, M.T., and Han, J.C., 2007. Film-cooling
effectiveness on a rotating turbine platform using pressure sensitive paint
technique, *Proceedings of ASME Turbo-Expo 2007*, May 14–17, 2007, Montreal,
Canada, Paper No. GT2007-27122.

Takeishi, K., Aoki, S., Sato, T., and Tsukagoshi, K., 1992. Film cooling on a gas turbine
rotor blade. *ASME Journal of Turbomachinery*, 114, 828–834.

Takeishi, K., Matsuura, M., Aoki, S., and Sato, T., 1990. An experimental study of heat transfer and film cooling on low aspect ratio turbine nozzles. *ASME Journal of Turbomachinery*, 112, 488–496.

Telisinghe, J., Ireland, P.T., Jones, T.V., Son, C., and Barrett, D.W., 2006. Comparative study between a cut-back and conventional trailing edge film cooling system. ASME Paper GT2006-91207.

Thrift, A., Thole, K.A., and Hada, S., 2010a. Effects of an axisymmetric contoured endwall on a nozzle guide vane: Adiabatic effectiveness measurements. ASME Paper GT-2010-22968.

Thrift, A., Thole, K.A., and Hada, S., 2010b. Effects of an axisymmetric contoured endwall on a nozzle guide vane: Adiabatic heat transfer measurements. ASME Paper GT-2010-22970.

Uzol, O., Camci, C, and Glezer, B., 2000. Aerodynamic loss characteristics of a turbine blade with trailing edge coolant ejection: Part 1—Effect of cut-back length, spanwise rib spacing, free stream Reynolds number and chordwise rib length on discharge coefficients. ASME Paper 2000-GT-0557.

Uzol, O. and Canci, C., 2000. Aerodynamic loss characteristics of a turbine blade with trailing edge coolant ejection: Part 2—External aerodynamics, total pressure losses and predictions. ASME Paper 2000-GT-0258.

Varadarajan, K. and Bogard, D.G., 2004. Effects of hot streaks on adiabatic effectiveness for a film cooled turbine vane. ASME Paper GT2004-54016.

Walsh, W.S., Thole, K., and Joe, C., 2006. Effects of sand ingestion on the blockage of film-cooling holes. ASME Paper GT2006-90067.

Waye, S. and Bogard, D., 2006a. High resolution film cooling effectiveness measurements of axial holes embedded in a transverse trench with various trench configurations. ASME Paper GT2006-90226.

Waye, S. and Bogard, D., 2006b. High resolution film cooling effectiveness comparison of axial and compound angle holes on the suction side of a turbine vane. ASME Paper GT2006-90225.

Wright, L., Blake, S.A., Rhee, D.H., and Han, J.C., 2007. Effect of upstream wake with vortex on turbine blade platform film cooling with simulated stator-rotor purge flow. ASME Paper GT2007-27092.

Wright, L., Gao, Z., Yang, H., and Han, J.C., 2006. Film cooling effectiveness distribution on a gas turbine blade platform with inclined slot leakage and discrete film hole flows. ASME Paper GT2006-90375.

Wright, L.M., Blake, S.A., Rhee, D.H., and Han, J.C., 2009. Effect of upstream wake with vortex on turbine blade platform film cooling and simulated stator-rotor purge flow. *ASME Journal of Turbomachinery*, 131(2), 021017.

Wright, L.M., Gao, Z., Yang, H., and Han, J.-C., 2008. Film cooling effectiveness distribution on a gas turbine blade platform with inclined slot leakage and discrete film hole flows. *Journal of Heat Transfer*, 130, 071702 (11 pages)

Wright, L., McClain, S.T., and Clemenson, M., 2010. Effect of density ratio on flat plate film cooling with shaped holes using PSP. ASME Paper GT2010-23053.

Yu, Y. and Chyu, M.K., 1998. Influence of gap leakage downstream of the injection holes on film cooling performance. *ASME Journal of Turbomachinery*, 120, 541–548.

Zhang, L.J. and Jaiswal, R.S., 2001. Turbine nozzle endwall film cooling study using pressure sensitive paint. ASME Paper 2001-GT-0147.

Zhang, L. and Moon, H.K., 2003. Turbine nozzle endwall inlet film cooling: The effect of a back-facing step. ASME Paper GT2003-38319.

Zhang, L. and Moon, H.K., 2006. Turbine blade film cooling study: The effects of showerhead geometry. ASME Paper GT2006-90367.

Zhang, L. and Moon, H.K., 2007. Turbine blade film cooling study-the effect of film hole location on the pressure side. ASME Paper GT2007-27546.

Zhang, L. and Pudupatty, R., 2000. The effects of injection angle and hole exit shape on turbine nozzle pressure side film cooling. ASME Paper 2000-GT-0247.

4

Turbine Internal Cooling

4.1 Jet Impingement Cooling

4.1.1 Introduction

Among all heat-transfer enhancement techniques, jet impingement has the most significant potential to increase the local heat-transfer coefficient. However, the construction of this flow arrangement weakens the structural integrity, and therefore impingement cooling is used in locations where thermal loads are excessively high. Jet impingement heat transfer is most suitable for the leading edge of a rotor airfoil (bucket), where the thermal load is highest and a thicker cross section of this portion of the airfoil can suitably accommodate impingement cooling. The structural strength required in a stator airfoil (vane) is less than that required in a rotor airfoil; therefore, jet impingement in the midchord region is used for stator airfoils. There are several arrangements possible with cooling jets, and different aspects need to be considered before optimizing an efficient heat-transfer design. Figure 4.1 shows a schematic configuration of cooling jets in an inlet guide vane (Taylor, 1980). The shape of the jet nozzle, the layout of jet holes, the shape of confinement chambers, and the shape of target surfaces have a significant effect on the heat-transfer coefficient distribution. The following sections discuss these features in details.

4.1.2 Heat-Transfer Enhancement by a Single Jet

The boundary layer is thin and highly turbulent at and near the impingement region, and, therefore, heat transfer can be significantly augmented by jet impingement. A cooling jet can be defined as a high-velocity coolant mass ejected from a hole or slot that impinges on the heat-transfer surface. Figure 4.2 shows a typical surface impingement caused by a jet (Viskanta, 1993). There exists a stagnation region at the impingement location, and this stagnation region is surrounded by developing boundary layers. Though the convection effects are negligible in the stagnation region (because of zero velocity), the stagnation point is inherently unstable and moves around

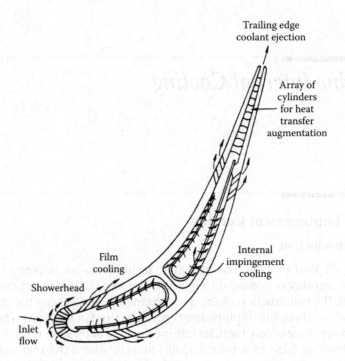

FIGURE 4.1
Schematic of impingement cooling arrangement in a first-stage turbine inlet guide vane. (From Taylor, J.R., Heat transfer phenomena in gas turbines, ASME Paper 80-GT-172, 1980.)

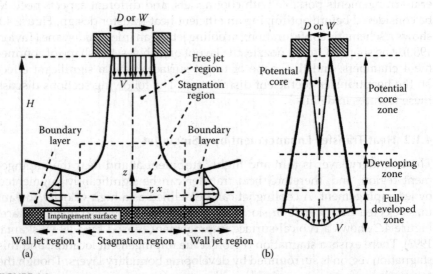

FIGURE 4.2
Comparison of flow regions in an impinging jet with flow regions of a free jet. (a) Typical jet structure and (b) jet core. (From Viskanta, R., *Exp. Therm. Fluid Sci.*, 6, 111, 1993.)

within a bound. Moreover, a cooling jet has a significantly higher momentum than its surroundings, and the spent jet after impingement creates a highly turbulent flow. Figure 4.2 shows that the free jet creates a stagnation zone right under the nozzle exit, and the jet retains the nozzle exit velocity in its potential core. Flow characteristics of impingement jets before and after striking the target plate have fundamental differences. After impingement, the spent jet creates a flow similar to a wall jet. Whereas, before striking the target plate, an impingement jet acts as a free jet. Several papers on fundamental fluid mechanics are available in the open literature that discuss the entrainment in free jets. Bradbury (1965) presented the typical velocity distribution and the jet axial velocity decay of a single free jet. The axial velocity in a developed jet is the highest at the center and gradually decreases toward its periphery. Velocity distribution of a free jet is characteristically different from velocity distribution in a viscous boundary layer. Unlike a viscous boundary layer, the velocity profile has a point of inflection. The nondimensional axial velocity of the jet center attenuates as the distance of the measured point from the jet nozzle exit increases. Initial decay is more rapid than the decay at the far region. The turbulence intensity at the core can be significantly high (as much as 25% turbulence). The high velocity coupled with high-velocity fluctuation, i.e., increased turbulence mixing, increases the heat-transfer enhancement capability of a jet significantly. However, there are design constraints and structural integrity problems with jet impingement. To create a jet impingement effect, a surface needs to be perforated, thus weakening the strength of that component.

Many more research publications are available on fundamentals of jet flow and related heat transfer in the published literature (see Incropera and DeWitt, 1996). Bradbury (1965) has discussed details of fluid mechanics in a free jet, and an interested reader can get substantial information on a single free jet from this article. However, this book is on gas turbine heat transfer, and therefore emphasis given in this chapter is on jet impingement related to gas turbine application. The following sections describe the key heat-transfer features of single and multiple jet impingement.

In a gas turbine application, mostly multiple jets are used instead of an isolated single jet, and velocity profiles in multiple jet impingement are laborious to obtain. Moreover, detailed velocity profile may be redundant in multiple jet impingement heat-transfer studies; instead, the static pressure distribution that provides overall pressure drop is more useful. There are typically two compartments or chambers in a jet impingement configuration. These two chambers are separated by a perforated surface. One of these chambers is pressurized, and the other experiences the impingement effects. The pressure inside a jet-creating pressurized chamber is higher than its outlet, and pressure gradient increases along the bulk flow. And this gradient also increases with an increase in the jet discharge. This pressure distribution is nonlinear and characteristically similar to an exponential distribution. The other chamber that contains the impingement target surface shows a different pressure distribution. The pressure change is more rapid near the exit of the impingement chamber, and for most of the channel, core pressure remains relatively flat.

4.1.2.1 Effect of Jet-to-Target-Plate Spacing

Figure 4.3 shows the heat-transfer results from a single round impingement jet. The Nusselt number for a jet Reynolds number of 23,000 for a jet nozzle diameter of 7.8 mm for different nozzle-to-target-plate distance H/D are plotted. In general, the heat-transfer coefficient in the stagnation point is higher compared to a location far from stagnation. However, depending on the jet Reynolds number and the jet-to-target-plate spacing, a secondary peak is observed in the heat-transfer distribution for circular jets. The inner peak is located at approximately $r/D = 0.5$, and the secondary peak is observed at $r/D = 1.75$. For a very close spacing of jet to the target plate ($H/D = 0.1$), the secondary peak dominates the heat-transfer pattern. For a typical gas turbine, the H/D ratio varies from 1 to 3. Sparrow and Wong (1975) studied the mass-transfer distribution for a single impingement jet. The jet diameter and the gap of the jet from target heat-transfer surface were constant. Their jet discharge was varied, and the mass transfer remained almost uniform at a high value near the impingement region. Mass transfer decreased with an increase in the radial distance measured from the center of impingement.

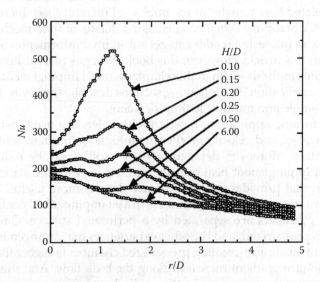

FIGURE 4.3
Effect of jet-to-target-plate spacing on the Nusselt number distribution. (From Viskanta, R., *Exp. Therm. Fluid Sci.*, 6, 111, 1993.)

4.1.2.2 Correlation for Single Jet Impingement Heat Transfer

Goldstein et al. (1986) developed an average heat-transfer coefficient correlation for single impingement jets. The Nusselt number is primarily a function of jet Reynolds number, the gap between the jet nozzle and impingement plate, and the jet nozzle exit area as

$$\frac{\overline{Nu}}{Re^{0.76}} = \frac{A - |L/D - 7.75|}{B + C(R/D)^{1.285}} \quad \text{for constant surface heat flux}$$

$$\frac{\overline{Nu}}{Re^{0.76}} = \frac{A - |L/D - 7.75|}{B + C\ (R/D)^{1.394}} \quad \text{for constant surface temperature}$$

(4.1)

The parameters A, B, and C are 24, 533, and 44, respectively, and L and D are jet-to-target-plate spacing and jet diameter, respectively. R is the radial distance from the geometrical center of the jet. Note that the effect of jet-to-target-plate spacing on Nusselt number is linear, and there is an optimum L/D value of 7.75 for a maximum average heat-transfer coefficient. The constant surface heat flux shows a higher heat-transfer coefficient. For atypical flow condition of $Re = 25{,}000$, $R/D = 5$, and $L/D = 7.75$, the average Nu is 60 for constant heat flux and 56 for the constant surface temperature heating conditions. The following year, Downs and James (1987) compiled available correlations for radial and slot jets to that date. Most correlations related to impingement heat transfer had a nonlinear relationship with exit jet Reynolds number as shown in the equations presented here.

The jet spreads by the mass entrainment as it develops from the nozzle exit. The stagnant coolant mass surrounding the jet slows the jet core velocity, and this loss of momentum in the jet core during flow development reduces the effectiveness of the jet on its impingement cooling capacity. Goldstein et al. (1986) have shown the effect of jet-to-impingement-plate spacing on heat transfer. Their results were represented by a recovery factor r, which is a function of the local adiabatic wall temperature and is given as

$$r = \frac{T_r - T_j}{u_j^2 / 2c_p}$$

(4.2)

The temperatures T_r and T_j are the recovery and jet temperatures, respectively. The velocity scale u_j is the mean jet velocity, and c_p is the specific heat of the coolant. The recovery temperature is the temperature of the surface without any addition or removal of heat flux from the heat-transfer surface (insulated), and the jet temperature is different from the mainstream temperature. Therefore, recovery factor essentially represents the ratio of stagnation to static temperatures. Jet-to-wall spacing can significantly alter the temperature distribution on the impingement plate. At small-jet-to-plate spacing, where the plate can be within the potential core of the issuing jet, the cooling jet is surrounded by a vortex ring around its free shear layer. Goldstein et al. (1986) observed a sharp drop in the adiabatic wall temperature and attributed the observed characteristics to that vortex ring. The effect of the surrounding vortex ring was diminished with an increase in the jet-to-plate spacing. They also observed that at a large L/D ratio ($L/D > 5$), the heat transfer directly beneath the jet was high, and the vortex ring surrounding the jet got weak. This reduced vortex structure decreased heat-transfer coefficient at the jet core periphery.

Goldstein and Seol (1991) have shown the effect of jet-to-wall spacing on the Nusselt number. Jets placed closer to the impingement wall showed higher Nusselt numbers. Interestingly, their results showed that the peak in stagnation heat-transfer coefficient was increased by a closer jet-to-jet spacing. Local minima were observed at the midway region of two adjacent jets, but the overall heat-transfer coefficient was higher for closely packed jets. The peak Nusselt number occurred directly below the jet, and the lowest Nusselt number was observed at 4 and 2 jet diameter distances (i.e., middle of two jets) for 8 and 4 jet diameter jet-to-jet spacings, respectively. The overall Nusselt number profile did not change with the wall-to-jet spacing, only the magnitude was higher for a smaller spacing.

4.1.2.3 Effectiveness of Impinging Jets

Effectiveness of impinging jets can be defined based on the analysis of film cooling. A higher effectiveness signifies reduced adiabatic wall temperature and therefore better cooling. A smaller gap between the jet to wall significantly improves the cooling effectiveness of a jet. Goldstein et al. (1990) observed that the spread of the effective cooling region did not significantly change with the wall-to-jet spacing. They also noted that jet effectiveness beyond 5 jet diameters was mostly unaffected by the wall-to-jet spacing. The effectiveness can be defined as

$$\eta = \frac{T_{aw} - T_r}{T_j - T_\infty} \tag{4.3}$$

where
  T_{aw} is the adiabatic wall temperature
  T_r is the recovery temperature
  T_j is the jet temperature
  T_∞ is the ambient temperature

To get jet impingement effectiveness, the experiment should have a different jet temperature than the ambient temperature.

The single jet impingement effectiveness is shown in Figure 4.4. Results indicate that the effectiveness at the stagnation point is independent of

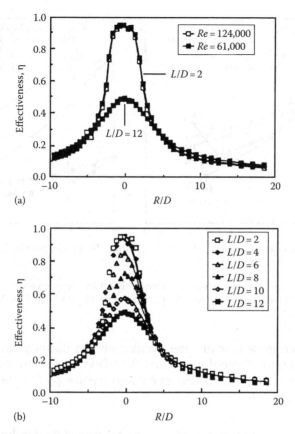

FIGURE 4.4
Impingement effectiveness distribution for (a) different jet Reynolds number and (b) different jet-to-target-plate spacing. (From Goldstein, R.J. et al., *ASME J. Heat Transfer*, 112, 608, 1990.)

Reynolds number but strongly dependent on the jet-to-target-plate distance. The local stagnation point effectiveness is significantly higher for the closely spaced jet. Results also indicate that the spreading of jet due to higher jet-to-target-plate spacing does not increase the effectiveness at locations far from the stagnation location.

4.1.2.4 Comparison of Circular to Slot Jets

Goldstein and Seol (1991) have compared the spanwise averaged Nusselt numbers and mass-transfer coefficients for two nozzle configurations: round circular jets and slot jets. Note that due to symmetry in the flow, slot jets are mostly two-dimensional, and multiple circular jets are three-dimensional in nature. Figure 4.5 shows the average Nusselt number and Sherwood number comparison of circular jets with slot jets. For a given jet Reynolds number, Sherwood numbers with slot jets show higher mass-transfer coefficients,

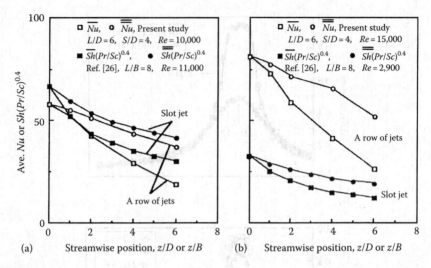

FIGURE 4.5
Comparison of span-averaged Nusslet numbers with slot jet impingement. (a) Same Reynolds number and (b) Same mass flow rate. (From Goldstein, R.J. and Seol, W.S., *Int. J. Heat Mass Transfer*, 34, 2133, 1991.)

but for a given mass flow rate the circular jets significantly outperform slot jets. Therefore, the three-dimensional flow field created by circular jets is more effective in transfer coefficient enhancement, especially near the direct impingement region.

4.1.3 Impingement Heat Transfer in the Midchord Region by Jet Array

Figure 4.6 shows a jet impingement configuration in the midchord region. This figure shows that the impingement target surface has a curvature, and impingement jets are nonuniformly distributed in the airfoil. The leading edge of the airfoil receives maximum thermal load and therefore has a concentrated jet impingement configuration. Though the airfoil has a curvature in the midchord region, the radius of curvature of the target surface to the

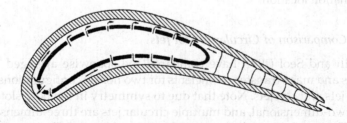

FIGURE 4.6
Impingement cooling arrangement of a typical gas turbine airfoil. Midchord jets are affected by cross-flow. (From Florschuetz, L.W. et al., *ASME J. Heat Transfer*, 106, 34, 1984.)

jet diameter ratio is large. For most midchord analysis, the target surface is assumed to be flat and perpendicular to the impingement jet direction. The trailing edge of the airfoil is too narrow to accommodate any impingement cooling. This narrow trailing edge is cooled by pin fins, as discussed in Section 4.5.3.

A group of jets performs differently from a single jet. The primary contributor to the difference is the cross-flow developed by spent jets. Florschuetz et al. (1980) and Koopman and Sparrow (1976) showed the effect of cross-flow on jet arrays. Mass flowing out of a jet, after impingement, moves in the cross-jet flow direction and can alter the performance of neighboring jets. The multiple jet arrangement can have two jet array configurations (inline and staggered; these configurations are similar to that discussed in the pin-fin section), and the impingement surface can be curved or flat. Though the midchord region of an airfoil can be considered flat for all practical purposes, the leading edge is significantly curved, and the effect of surface curvature cannot be neglected. This section discusses impingement heat transfer on a flat surface, followed by a discussion on curved surface impingement in the leading edge.

4.1.3.1 Jets with Large Jet-to-Jet Spacing

The effect of surrounding jets for large jet-to-jet spacing was presented by Hollworth and Berry (1978). Their results showed a significant dependency on the wall-to-jet spacing, and for a small wall-to-jet spacing the Nusselt number increased toward a secondary peak. On the other hand, larger wall-to-jet spacing showed that the Nusselt number profile of a particular jet was mostly unaffected by the neighboring jets. Figure 4.7 shows detailed distribution of jet impingement for multiple round jets and indicates that the circular structure of round jets is preserved in multiple jet formations. These detailed heat-transfer patterns obtained by Goldstein and Timmers (1982) showed that a round jet surrounded by a group of jets preserved the circular shape of the jet instead of forming a hexagon distribution. Smaller secondary circular regions in the interjet space were observed for higher wall-to-jet spacing; that resulted in lower heat-transfer coefficients in those regions. This figure also shows that closeness of the target plate from the jet has an impact on the local heat-transfer coefficient distribution.

4.1.3.2 Effect of Wall-to-Jet-Array Spacing

Figure 4.8 compares mean Nusselt numbers for different jet-to-target plate distances in an inline array with jet-to-jet spacing of 10 jet diameters. The left-side plots are for 6 jet diameters spacing in the other direction, and the plots on the right are for 8 jet diameters in the other direction. Results indicate that the mean Nusselt number does not change significantly for a narrow range of jet-to-target-plate spacing ($Z/d = 1-3$). Discussed earlier in the

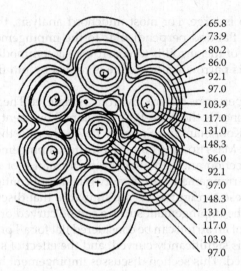

FIGURE 4.7
Detailed mass transfer of multiple jets shows that the circular structure of jets is preserved. (From Goldstein, R.J. and Timmers, J.F., *Int. J. Heat Mass Transfer*, 25, 1857, 1982.)

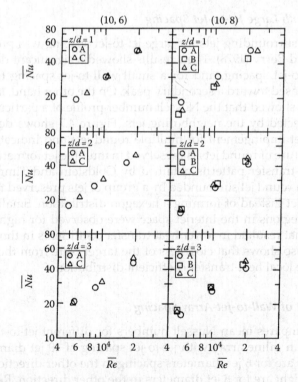

FIGURE 4.8
Mean Nusselt numbers for inline hole pattern. (From Metzger, D.E. et al., *ASME J. Heat Transfer*, 101, 526, 1979.)

single jet analysis is that impingement heat transfer from individual jets is significantly dependent on this distance. A closer spacing may increase the local peak, but the average value is similar to that of a larger spacing.

4.1.3.3 Cross-Flow Effect and Heat-Transfer Correlation

The flow in impingement chamber, i.e., the space enclosing the jet nozzle and impingement plate, differentiates the performance of multiple jets from single jets. The cross-flow in an impingement chamber is mostly developed by the spent jets. This flow is perpendicular to the bulk flow direction of a jet and is, therefore, called cross-flow. The cross-flow tries to deflect a jet away from its impinging location. If the cross-flow is strong and the jet is sufficiently far from the impingement plate, the cross-flow can deflect the jet away from the impingement surface, thus reducing the effectiveness of the impingement cooling. Florschuetz and Su (1987) have shown the effect of cross-flow on the Nusselt number. The cross-flow contributes in enhancing convection heat transfer, but the cooling by jet impingement decreases due to the jet deflection. Since the heat-transfer coefficient by jet impingement is significantly higher than cross-flow convection cooling, the average Nusselt number decreases in the presence of a cross-flow. However, for a smaller cross-flow, i.e., for cross-flow velocity less than 10% of the jet flow velocity, the cooling is better overall, and the Nusselt number increases.

Goldstein and Behbahani (1982) have shown the details of the Nusselt number distribution in a circular jet with and without cross-flow. The Nusselt number increased with an increase in the jet Reynolds number. A higher jet Reynolds number signified a higher exit flow velocity at the nozzle that could penetrate the cross-flow better. The distribution was not symmetric about the peak Nusselt number, and the peak Nusselt number shifted at a downstream location due to the jet deflection. In a separate study, Bouchez and Goldstein (1975) have shown the effectiveness (based on the film cooling concept) of jets in a cross-flow. The cooling effectiveness was significantly asymmetric, and the location of the most effective point was shifted by the cross-flow.

In a fundamental study to show the cross-flow effects, Sparrow et al. (1975) studied the recovery factor of a single jet in cross-flow. A mass velocity ratio M given by the ratio of mass flow rate of jet to that of the free stream was used to define relative strength of the cross-flow. A lower M represented a stronger cross-flow with respect to the jet exit velocity, and the asymmetric distribution of the recovery factor showed effects of jet deflection in cross-flow. For a lower M the deflection was stronger. The peak in recovery factor shifted significantly in $M = 4$, and for higher Ms the jet was strong enough to penetrate the cross-flow. The asymmetry in the recovery factors for higher Ms was due to the cross-flow convection effects. The recovery factor r qualitatively represented the heat-transfer coefficient distribution.

Bouchez and Goldstein (1975) have shown the effect of cross-flow for different jet hole diameters. The jet discharge distribution was dependent on the jet diameter, and a smaller jet hole diameter showed an increase in discharge at downstream locations. And though the cross-flow intensified at downstream locations, smaller diameter jets were more effective in cross-flow penetration. A larger jet hole diameter showed higher heat-transfer coefficients at the inlet where cross-flow intensities were small; but the heat-transfer coefficient was reduced at downstream locations due to stronger cross-flows and weaker jets. Spatial variations in the heat-transfer coefficient distribution increased with an increase in the jet hole diameter.

Womac et al. (1994) developed a correlation for small heat sources with multiple jet impingement. However, in a gas turbine application, the heat-transfer surface is comparatively large, and there are several studies available on this topic. Kercher and Tabakoff (1970) studied the impingement heat transfer in the perspective of turbine blade designers. Figures 4.9 and 4.10 show their correlation for Nusselt number distributions for multiple jet

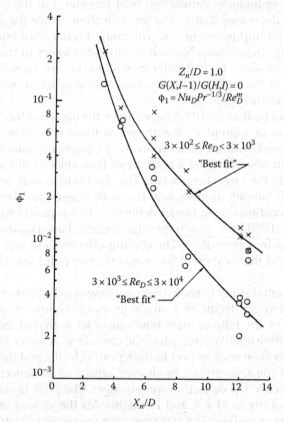

FIGURE 4.9

Heat-transfer coefficient correlation with a constant defined as ϕ_1. (From Kercher, D.M. and Tabakoff, W., *ASME J. Eng. Power*, 92, 73, 1970.)

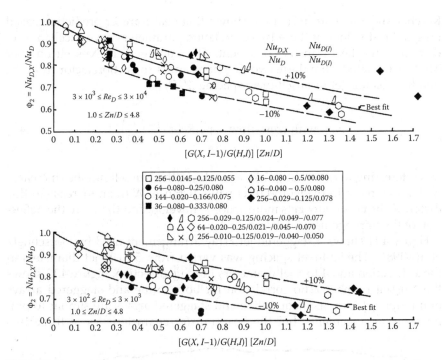

FIGURE 4.10
Impingement heat-transfer coefficient correlation based on degradation coefficient ϕ_2. (From Kercher, D.M. and Tabakoff, W., *ASME J. Eng. Power*, 92, 73, 1970.)

impingement on a flat plate. Two correlating parameters that are plotted are developed to correlate the Nusselt number distribution in the presence of cross-flow. The two parameters, ϕ_1 and ϕ_2, are defined as

$$\phi_1 = Nu_D Re_D^{-m} Pr^{-1/3} \tag{4.4}$$

where

Nu_D is the average Nusselt number based on the jet diameter in the absence of cross-flow

Re_D is the jet Reynolds number based on the jet diameter and average jet discharge velocity

This parameter ϕ_1 is evaluated at jet-to-target-plate spacing of 1 jet diameter. The other parameter, ϕ_2, is defined as

$$\phi_2 = \frac{Nu_{\text{with cross-flow}}}{Nu_{\text{without cross-flow}}}$$

$$= f\left(\frac{\text{Cross-flow mass flow}}{\text{Jet ass flow}} \frac{\text{Jet to plate distance}}{\text{Jet diameter}}, Re_D \right) \tag{4.5}$$

Kercher and Tabakoff (1970) have shown that heat-transfer prediction based on graphical values of these two correlation parameters gave excellent prediction of the heat-transfer coefficient. The impingement Nusselt number can be calculated from these two functions and a correction factor from the nondimensional jet-to-target-plate spacing (Z/D) as

$$Nu_D \text{ with cross-flow} = \phi_1\phi_2 Re_D^m Pr^{1/3}\left(\frac{Z}{D}\right)^{0.091} \qquad (4.6)$$

It is interesting to note that unlike a single jet, the Nusselt number increases with an increase in the jet-to-target-plate distance. With an increase in this distance, the cross-flow gets more space to develop, and therefore the deflection of the impingement jet is less.

Figure 4.11 shows a staggered jet array arrangement used by Florschuetz et al. (1981). The jet-to-jet spacing was varied from 5 to 15 jet diameters in the x direction and 4 to 8 jet diameters in the y direction. Figure 4.12 shows the Nusselt number ratio correlation for both inline and staggered arrays of multiple jet impingement. Inline and staggered arrays are similar to pin-fin array arrangements, but the heat-transfer characteristics of these two

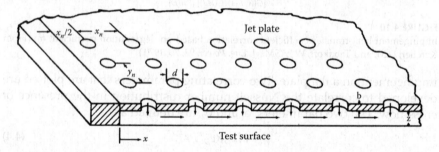

FIGURE 4.11
Test model used by Florschuetz et al. (From Florschuetz, L.W. et al., *ASME J. Heat Transfer*, 103, 337, 1981.)

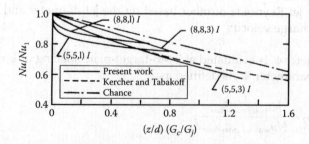

FIGURE 4.12
Effect of cross-flow on streamwise Nusselt numbers. (From Florschuetz, L.W. et al., *ASME J. Heat Transfer*, 103, 337, 1981.)

arrangements are entirely different. The fundamental difference in cross-flow effects between pin-fin heat transfer and that with impingement is this: In a pin-fin arrangement, cross-flow can increase heat transfer from the pins by impinging on them, whereas in a jet impingement arrangement, jets get deflected away from the target surface by the cross-flow, and therefore cross-flow effects are mostly detrimental for impingement cooling. The impingement heat-transfer correlation for both inline and staggered arrays is given as

$$\frac{Nu}{Nu_1} = 1 - C\left(\frac{x_n}{d}\right)^{n_x}\left(\frac{y_n}{d}\right)^{n_y}\left(\frac{z}{d}\right)^{n_z}\left(\frac{G_c}{G_j}\right)^{n} \tag{4.7}$$

where the normalizing Nusselt number is given as

$$Nu_1 = 0.363\left(\frac{x_n}{d}\right)^{-0.554}\left(\frac{y_n}{d}\right)^{-0.422}\left(\frac{z}{d}\right)^{0.068} Re_j^{0.727} Pr^{1/3} \tag{4.8}$$

The values of constants in the equation are dependent on the array configuration and are given as

Jet Array	C	n_x	n_y	n_z	n
Inline	0.596	−0.103	−0.38	0.803	0.561
Staggered	1.07	−0.198	−0.406	0.788	0.660

The confidence level for this correlation is 95%, and the advantage is that it is easy to compute. An inline arrangement shows a higher heat-transfer coefficient than that with a staggered arrangement. Inline array is less sensitive to jet-to-jet spacing (n_x and n_y are smaller) and more sensitive to jet-to-target-plate spacing (n_z is greater). A more detailed correlation developed earlier by Florschuetz et al. (1981) was

$$Nu = A\,Re_j^m\left\{1 - B\left[\left(\frac{z}{d}\right)\left(\frac{G_c}{G_j}\right)\right]^n\right\}Pr^{1/3} \tag{4.9}$$

where the constants are defined by geometrical parameters as A, m, B, or $n = C(x_n/d)^{nx}(y_n/d)^{ny}(z/d)^{nz}$. The constants to be used for this correlation are given in Table 4.1.

This correlation is based on the results presented in Figure 4.13, which shows the ratio of Nusselt numbers with staggered jet pattern to that with inline jet pattern. As the z/d ratio increases, the staggered jet pattern shows a reduction in the heat-transfer coefficient relative to the inline hole pattern. This is due to the associated spanwise distribution of cross-flow.

TABLE 4.1

Correlation Constants for Different Jet Arrangements

	Inline Pattern				Staggered Pattern			
	C	n_x	n_y	n_z	C	n_x	n_y	n_z
A	1.18	−0.944	−0.642	0.169	1.87	−0.771	−0.999	−0.257
m	0.612	0.059	0.032	−0.022	0.571	0.028	0.092	0.039
B	0.437	−0.095	−0.219	0.275	1.03	−0.243	−0.307	0.059
N	0.092	−0.005	0.599	1.04	0.442	0.098	−0.003	0.304

	W/H	E	e/D	P/e	α	$Re \times 10^{-3}$
Square channel	1	0.24	0.047	10	90°, 60°,	10
					45°, 30°	30
						60
Rectangular channel I	2	0.32	0.047	10	90°, 60°	10
					45°, 30°	30
						60
Rectangular channel II	2	0.32	0.078	10	90°, 60°	10
					45°, 30°	30
						60
Rectangular channel IA	2/4	0.32	0.047	10	90°, 60°	10
					45°, 30°	30
						60
Rectangular channel IIA	¼	0.32	0.078	10	90°, 60°	10
					45°	30
						60

Source: Florschuetz, L.W. et al., *ASME J. Heat Transfer*, 103, 337, 1981. With permission.

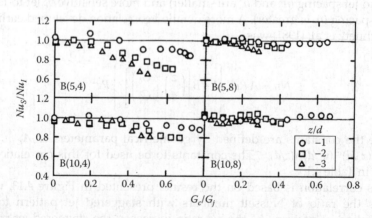

FIGURE 4.13

Nusselt number ratio of staggered arrangement and inline arrangement for different cross-flow conditions. (From Florschuetz, L.W. et al., *ASME J. Heat Transfer*, 103, 337, 1981.)

The tendency of the cross-flow to be channeled between adjacent stream-wise rows in the inline array reduces the impact of the cross-flow on the impinging jets. In contrast, the cross-flow distribution in the staggered array is more uniformly distributed and affects impinging jets more significantly. Note that a pin-fin array behaves exactly the opposite way. The heat-transfer coefficient in a pin-fin array improves in the staggered configuration compared to an inline arrangement. Since the cross-flow is detrimental in a jet impingement, an inline array that has less jet deflection from cross-flow performs better.

4.1.3.4 Effect of Initial Cross-Flow

In the previous section, we discussed the effect of cross-flow developed by the spent jets. Therefore, the first row of jets did not have any cross-flow effects. However, the cross-flow may develop from the upstream cooling conditions. For example, the cross-flow developed by impingement cooling in the leading edge can affect the midchord jet impingement. Presence of initial cross-flow can significantly alter the aerodynamic boundary layer and mass distribution in jet rows, and therefore heat-transfer patterns are different from those with spent jet cross-flows. The test configuration used by Florschuetz et al. (1984) is different from their previous study (Florschuetz et al., 1981) due to the initial cross-flow provided in the left side of the configuration. They studied the effect of initial cross-flow on mass distribution in a jet array. Results indicate that jet-to-jet variations in mass distribution are affected by the initial cross-flow. Mass distribution is more uniform for no cross-flow; and in the presence of a cross-flow, the mass issued from upstream jets is less compared to that issued from downstream jets.

Figure 4.14 shows the Nusselt number and jet impingement effectiveness for different initial cross-flow conditions. The effectiveness is defined as

$$\eta = \frac{T_{adibatic\,wall} - T_{jet}}{T_{cross\text{-}flow} - T_{jet}} \tag{4.10}$$

This effectiveness definition is analogous to the effectiveness used for film-cooling analysis. However, unlike for film cooling, for jet impingement the jet flow is the primary flow and the cross-flow is the secondary flow. Since the dominance of jet flow is desirable in jet impingement, lower effectiveness indicates a better impingement performance.

The cooling effectiveness of jets decreases with the axial distance measured in the direction of the cross-flow. The effect of initial cross-flow can clearly be seen in the heat-transfer pattern. A higher mass flow from a jet hole increases the related heat transfer. Thus, Figure 4.14 compares heat-transfer coefficients for staggered and inline jet array configurations. The Nusselt number is significantly reduced in the presence of cross-flow for both inline

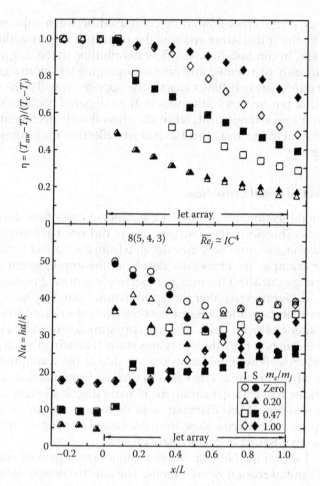

FIGURE 4.14

Effect of initial cross-flow on impingement effectiveness and Nusselt number for inline (I) and staggered (S) jet hole arrangements. (From Florschuetz, L.W. et al., *ASME J. Heat Transfer*, 106, 34, 1984.)

and staggered formations. The cooling effectiveness of inline formation is lower and the Nusselt number is higher than the corresponding staggered jet array formation.

4.1.3.5 Effect of Cross-Flow Direction on Impingement Heat Transfer

Huang et al. (1998) studied the effect of spent flow direction on impingement heat transfer. Since the spent jets develop a cross-flow, the exit and entrance directions play an important role in deciding the heat-transfer coefficient distribution. Figure 4.15 shows the experimental configuration used by Huang et al. (1998) to do the exit condition study. Three different configurations

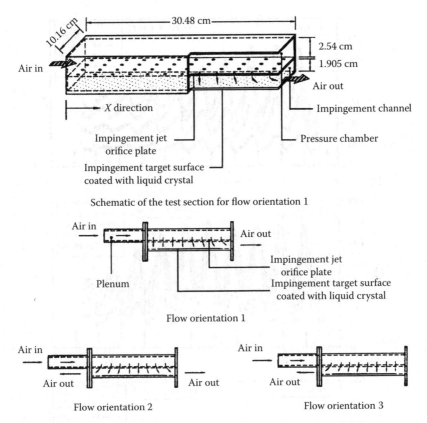

FIGURE 4.15
Schematic of the test setup and three exit flow orientations used by Huang et al. (From Huang, Y. et al., *AIAA J. Thermophys. Heat Transfer*, 12(1), 73, 1998.)

were studied. In configuration 1, the spent air went out from the furthest end relative to the entrance. In configuration 2, both exits were left open; in configuration 3, air went out from the nearest exit to the inlet. They presented local heat-transfer coefficient distributions measured with a transient liquid crystal technique. Configuration 1 shows that the impingement heat-transfer coefficient is nonuniform, and it is higher near the flow inlet (left side of the figure) and gradually reduces toward the flow outlet (toward the right). The impingement heat-transfer coefficient distribution is more uniform for configuration 2, where both outlets were left open and, therefore, the cross-flow effects are minimum. Configuration 3 shows more uniform heat-transfer coefficient distribution than configuration 1, but due to a stronger cross-flow, the peak Nusselt number is less than that in configuration 2.

Figure 4.16 shows span-averaged Nusselt number distributions for three different flow exit directions in different flow Reynolds numbers. Reynolds numbers were calculated based on average jet discharge velocity and jet diameter. Results show that the peak in Nusselt number is significantly reduced

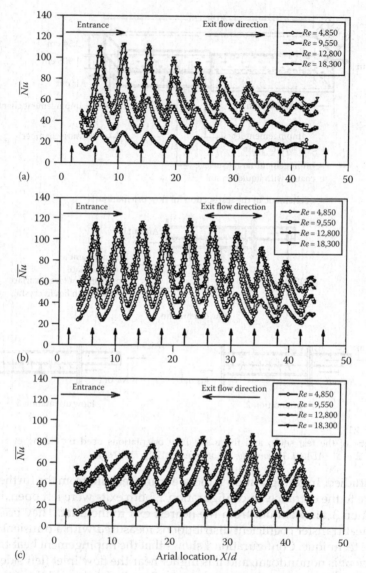

FIGURE 4.16
Effect of Reynolds number on span-averaged Nusselt number distribution for different exit flow conditions. (a) Orientation 1, (b) orientation 2, and (c) orientation 3. (From Huang, Y. et al., *AIAA J. Thermophys. Heat Transfer*, 12(1), 73, 1998.)

if the entrance and exit are in the same direction, i.e., configuration 3. Out of the three cases considered, the case with two ends open (configuration 2) has less cross-flow momentum and shows higher Nusselt numbers due to less jet deflection by cross-flow. Besides the peak Nusselt numbers, the effect of cross-flow on jet deflection is clearly visible in the impingement heat-transfer coefficient distribution. For convenience, jet locations are shown in these plots. An exit flow toward the right deflects impingement jets to the right as the peak Nusslet numbers shift to the right, whereas the flow exit toward the left deflects jets to the left. However, in configuration 2, the jet deflection is not symmetrical. Though both ends are open in this configuration, the jets are deflected to the right for most locations. The reason behind this asymmetry is that the relative strength of jets depends on the location of the flow inlet to the pressure chamber. Since jets near an inlet, i.e., left side of the plot, are stronger, they are less deflected by the cross-flow, whereas jets on the right are weaker and are more deflected by cross-flow.

Figure 4.17 shows the effect of cross-flow on the detailed Nusselt number distributions for the configuration 1 by using the transient liquid crystal technique. It is clear that the cross-flow is to shift the jet impingement location and reduce Nusselt number values downstream.

Figure 4.18 shows the channel-averaged heat-transfer results for different Reynolds numbers and different exit conditions. Correlation equations developed by other researchers are compared in this figure. Kercher and Tabakoff (1970) involved a cross-flow effect that is similar to configuration 1, whereas the correlation of Florschuetz et al. (1981) had smaller cross-flow effects. Correlation by Van Treuren et al. (1994) is for stagnation heat-transfer coefficient and, therefore, is highest in this group.

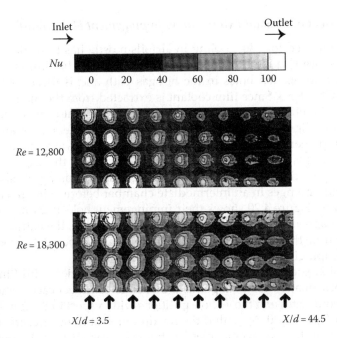

FIGURE 4.17
Effect of cross-flow on detailed Nusselt number distribution for exit flow configuration 1. (From Huang, Y. et al., *AIAA J. Thermophys. Heat Transfer*, 12(1), 73, 1998.)

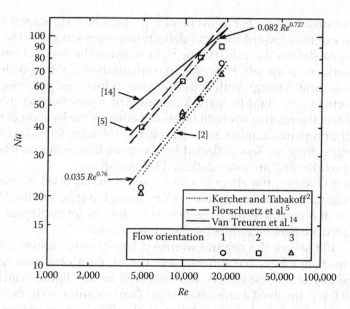

FIGURE 4.18
Overall Nusselt number averages obtained for different exit flow conditions of Huang et al. (From Huang, Y. et al., *AIAA J. Thermophys. Heat Transfer*, 12(1), 73, 1998.)

4.1.3.6 Effect of Coolant Extraction on Impingement Heat Transfer

The coolant used for film cooling in airfoils mostly has to work as internal coolant before it is bled to the main gas flow path. The film is developed from discrete ejection holes in the hot-gas path and is discussed in more detail in Chapter 3. Since film coolant is extracted from the internal coolant passages, the effect of this extraction is felt in the heat-transfer coefficient distribution. Figure 4.19 shows a typical flow arrangement in an airfoil with coolant impingement and extraction. The coolant side is at a higher pressure than the hot-gas side, and therefore the coolant passes through the coolant channels from the coolant side to the hot-gas side. This figure shows that the coolant impinges in an intermediate chamber before being released as a film coolant. Figure 4.20 shows their impingement hole and film extraction holes as mapped in an *x–y* plane. Note that, in general, the extraction holes are offset from the impingement holes to avoid the escapement of cooling jets without impinging the heat-transfer surfaces.

Figure 4.21 shows the local Nusselt number distribution with film extraction. Approximately 200 data points are used to construct each contour plot; *x* and *y* coordinates and the jet hole positions with respect to the film holes are shown in Figure 4.20. Note that the maximum heat-transfer coefficients are not exactly at the impingement jet locations. Instead they are shifted by the cross-flow toward the film-cooling holes. They showed the surface-averaged Nusselt numbers for different jet Reynolds numbers. Three different surfaces

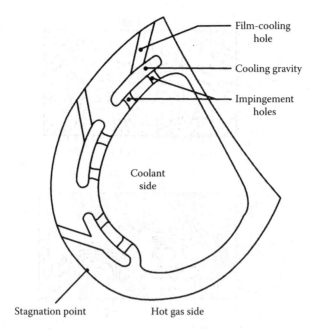

FIGURE 4.19
Cooling passages in the leading edge of a turbine airfoil. The spent jets are extracted for film cooling of the outer surface. (From Gillespie, D.R.H. et al., *ASME J. Turbomach.*, 120, 92, 1998.)

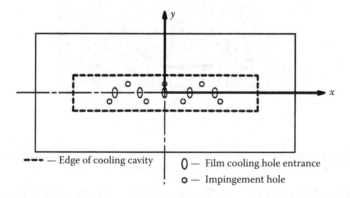

FIGURE 4.20
Arrangement of impingement holes with respect to the film extraction holes. (From Gillespie, D.R.H. et al., *ASME J. Turbomach.*, 120, 92, 1998.)

are considered. The first is the surface containing jet holes (upstream impingement surface), the second is the target surface for impingement, and the third is the surface that is film cooled (downstream impingement surface). As expected, the target surface Nusselt numbers are much higher than the other two surfaces, and downstream impingement surface has the least Nusselt number in the group. The impingement Nusselt number on the target plate more than doubles, for an increase in the jet Reynolds number from 21,000

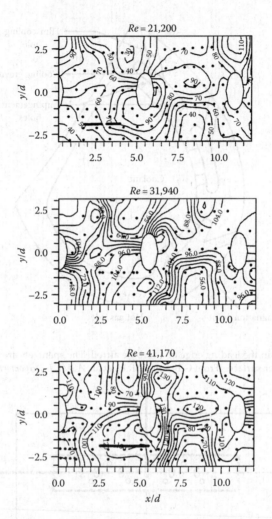

FIGURE 4.21
Nusselt number distribution on the target plate containing film extraction holes. (From Gillespie, D.R.H. et al., *ASME J. Turbomach.*, 120, 92, 1998.)

to 41,000. But the increase in the downstream surface shows insignificant Nusselt number enhancement by the same increase in jet Reynolds number.

Hollworth and Dagan (1980) studied heat transfer from arrays of impinging jets with spent fluid removal through inclined holes on the impingement target surface. Figure 4.22 shows their cooling arrangement. The target plate has inclined holes drilled through it, and therefore some of the coolant is extracted through these holes, decreasing the effect of cross-flow. Both inline and staggered vent hole configurations with respect to the impinging jets are considered. Figure 4.23 shows the two cooling arrangements. In the inline arrangement, the coolant extraction holes are directly under the jet issuing holes, and in the staggered arrangement they are offset from the

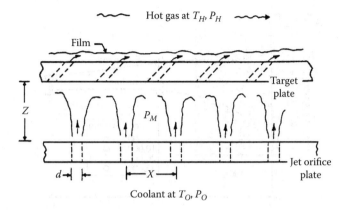

FIGURE 4.22
Impingement arrangement with inclined him extraction holes. (From Hollworth, B.R. and Dagan, L., *J. Eng. Power*, 102, 994, 1980.)

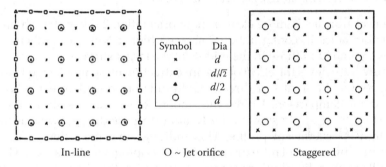

FIGURE 4.23
Different vent hole configurations with respect to the jet hole arrangement used by Hollworth and Dagan. (From Hollworth, B.R. and Dagan, L., *J. Eng. Power*, 102, 994, 1980.)

jet issuing holes. Figure 4.24 shows the impingement heat transfer for both staggered and inline orientations for different jet-to-target-plate spacings. The staggered arrangement shows that the heat-transfer coefficient is higher for a closer spacing. However, for the inline arrangement, the closer spacing deteriorates the heat-transfer performance. For a close spacing, most coolant escapes through the extraction hole without cooling the target plate, and therefore cooling performance decreases.

Ekkad et al. (1999) studied the heat-transfer enhancement in the presence of coolant extraction for their three different exit configurations. These exit configurations are similar to those discussed before in this chapter. The target impingement surface has 11 rows of 3 coolant holes, and the area ratio of the total coolant extraction holes to the jet holes is 33:48. They showed the detailed Nusselt numbers for three different exit configurations. The Nusselt numbers decrease significantly at the edges due to film extraction. The decrease in this

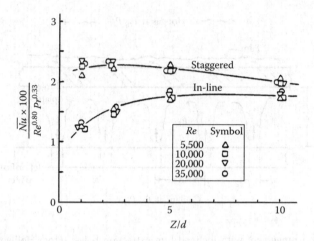

FIGURE 4.24
Average Nusselt number variations with different jet and extraction hole arrangements. (From Hollworth, B.R. and Dagan, L., *J. Eng. Power*, 102, 994, 1980.)

edge heat transfer is more clear in flow orientation 2. Note that absence of extraction holes in configuration 2 showed higher Nusselt numbers at the edges compared to those in the middle of the target surface.

Figure 4.25 shows the effect of coolant extraction on span-averaged Nusselt numbers for different exit directions. Results of no coolant extraction are plotted for comparison. Unlike the smooth target surface study, the film extraction holes study shows a fairly uniform periodic spanwise averaged heat-transfer coefficient profile. This indicates that cross-flow effects are weakened by the coolant extraction in the impingement channel. Though Nusselt number distribution is more uniform with film extraction, the span-averaged Nusselt number peaks are less with coolant extraction.

4.1.3.7 Effect of Inclined Jets on Heat Transfer

Huang et al. (1996) experimentally studied the heat-transfer coefficient distribution in an impingement heat transfer with an array of inclined jets using a transient liquid crystal technique. Figure 4.26 shows jet configurations used by them. The jets are oriented ±45° with respect to the anticipated cross-flow direction. Jet holes are 0.635 cm in diameter, and the orifice plate thickness is 1 jet diameter. Twelve rows of 4 inline holes are used. Six different cases, as shown in the figure, are studied. Figure 4.27 shows the span-averaged Nusselt numbers for different inclined impingement conditions. The span-averaged Nusselt number are higher for case 1 (+45°) and case 4 (−45°) inclined jets. These cases have both exits open for spent air to go out. However, this study shows that inclined jets have less heat-transfer coefficient than that of straight jets. It can be argued that instead of straightening the inclined jets, the cross-flow tends to diffuse the impingement effect, and therefore there is a net decrease in the impingement heat-transfer coefficient.

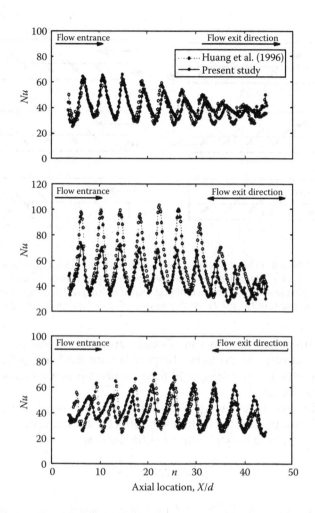

FIGURE 4.25
Effect of coolant extraction on span-averaged Nusselt number distribution at $Re = 9550$. (From Ekkad, S.V. et al., *AIAA J. Thermophys. Heat Transfer*, 13, 522, 1999.)

4.1.4 Impingement Cooling of the Leading Edge

4.1.4.1 Impingement on a Curved Surface

The leading edge of a turbine blade is significantly curved. The impingement characteristics of round jets are different on a curved impingement surface. Hrycak (1981) has measured the average Nusselt numbers for different Reynolds numbers at different curvatures. Experimental average Nusselt numbers are plotted in Figure 4.28. A stronger curvature (smaller radius of curvature) had a higher average Nusselt number, and the effect of curvature decreased with an increase in the jet Reynolds number. Results obtained with a row of air jets impinging on an electrically heated surface

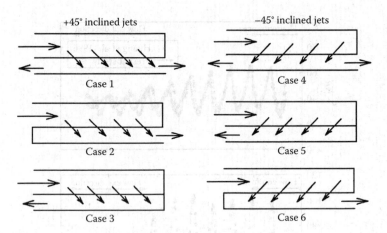

FIGURE 4.26
Schematic of different inclined jets used by Huang et al. with different exit arrangements. (From Huang, Y. et al., Detailed heat transfer coefficient distributions under an array of inclined impinging jets using a transient liquid crystal technique, *9th International Symposium on Transport Phenomena in Thermal Fluids Engineering (ISTP-9)*, Singapore, June 25–28, 1996.)

(constant surface heat flux) in a small-scale setup characterized a typical turbine blade and were found to be compatible with the average heat transfer from a geometrically similar but 10 times scaled-up steam-heated (constant temperature) surface.

Figure 4.28 shows the average Nusselt number on several curved surfaces. The Nusselt number plotted includes the effect of the curvature of the target surface. The resultant Nusselt number is defined as

$$Nu^* = 0.72Re_D^{0.63}Pr^{0.33} \tag{4.11}$$

This correlation is valid for $2 \leq Z_n/D \leq 8$ and is based on jet Reynolds number. The smaller power of Re_D compared to a flat-plate impingement is due to the restricted air entrainment from the surroundings in the presence of target plate curvature. Results plotted in Figure 4.28 show that the effect of curvature can be effectively included in the correlation by modifying the constant of correlation and the power of the Reynolds number. However, not enough information is available to create a correlation to include the target plate curvature effects in this correlation. The results shown in the inset of Figure 4.28 are based on the wall jet boundary layer thickness and show that the average Nusselt numbers for different jet nozzle diameters are scattered.

4.1.4.2 Impingement Heat Transfer in the Leading Edge

Figure 4.29 shows schematics of the impingement cooling at the leading edge. The leading edge has a sharp curvature, and the impingement flow turning

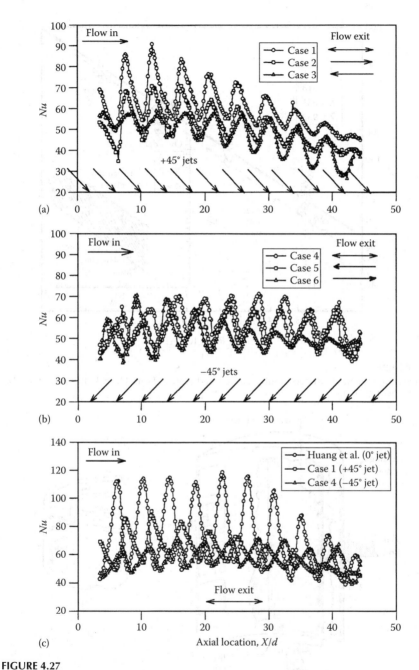

FIGURE 4.27
Span-averaged Nusselt number distributions show the effect of impingement jet angle. (a) Effect of orientation on +45° jets, (b) effect of orientation on −45° jets, and (c) effect of jet angle for orientation 2. (From Huang, Y. et al., Detailed heat transfer coefficient distributions under an array of inclined impinging jets using a transient liquid crystal technique, *9th International Symposium on Transport Phenomena in Thermal Fluids Engineering (ISTP-9)*, Singapore, June 25–28, 1996.)

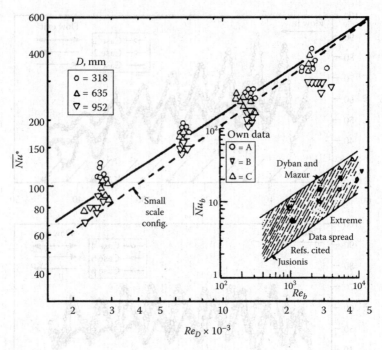

FIGURE 4.28
Average impingement Nusselt numbers on curved plates. (From Hrycak, P., *Int. J. Heat Mass Transfer*, 24, 407, 1981.)

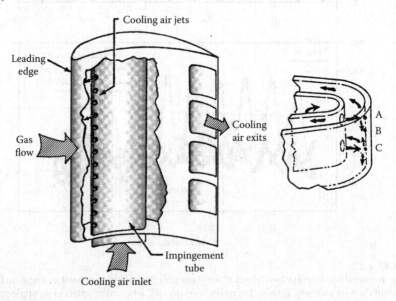

FIGURE 4.29
Schematic of the impingement cooling at the airfoil's leading edge. (From Chupp, R.E. et al., *AIAA J. Aircr.*, 6, 203, 1969.)

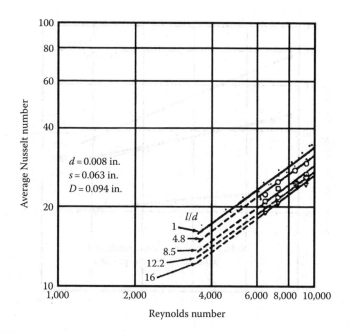

FIGURE 4.30
Average Nusselt number distribution in the leading-edge impingement experiment of Chupp et al. (From Chupp, R.E. et al., *AIAA J. Aircr.*, 6, 203, 1969.)

is significant. The curvature in the flow path creates a different cross-flow pattern, and therefore the resultant heat-transfer coefficients are different in the leading edge from those obtained for flat-plate impingement. Figure 4.30 shows the variation of average Nusselt numbers with Reynolds numbers at different jet-to-target-plate separation distances. For the range of parameters tested, the best-fit equation as developed by Chupp et al. (1969) is

$$Nu_{avg} = 0.63Re^{0.7}\left(\frac{d}{s}\right)^{0.5}\left(\frac{d}{D}\right)^{0.6}\exp\left[-1.27\left(\frac{l}{d}\right)\left(\frac{d}{s}\right)^{0.5}\left(\frac{d}{D}\right)^{1.2}\right] \quad (4.12)$$

and the stagnation Nusselt number is

$$Nu_{stag} = 0.44Re^{0.7}\left(\frac{d}{s}\right)^{0.8}\exp\left[-0.85\left(\frac{l}{d}\right)\left(\frac{d}{s}\right)\left(\frac{d}{D}\right)^{0.4}\right] \quad (4.13)$$

where
  l is the jet-to-target-plate separation distance
  d is the jet diameter
  s is the spacing between jet holes
  D is the diameter of the target plate

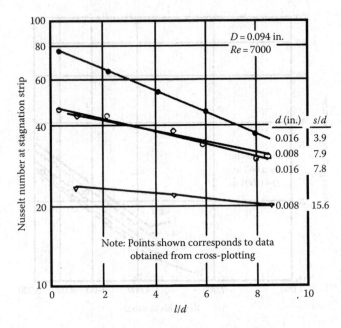

FIGURE 4.31
Nusselt number at the stagnation region for different hole arrangements. (From Chupp, R.E. et al., *AIAA J. Aircr.*, 6, 203, 1969.)

A single row of jets is used for this experiment. Results show that the average Nusselt number ratio decreases with an increase in the jet-to-target-plate distance, and an increase in the target plate curvature increases heat-transfer coefficient.

Figure 4.31 shows the Nusselt number variation along the stagnation line of the single row of impinging jets. An increase in the jet-to-target-plate distance decreases the Nusselt number, and an increase in the jet spacing also decreases the stagnation strip Nusselt numbers. The reduction in stagnation strip Nusselt number with increasing separation distance is more prominent for closer jet-to-jet spacing. Figure 4.32 shows the Nusselt number distribution away from the stagnation point for different jet-to-target-plate spacing. The normalized Nusselt number is only a weak function of s/d and practically independent of the jet Reynolds number. As expected, the Nusselt number is highest at the stagnation region and decays with the distance away from that location. The Nusselt number decreases to about 40% of the stagnation value at 4 jet diameters away from the stagnation point.

The leading edge of turbine airfoil can have different shapes. Bunker and Metzger (1990) studied the effect of target plate shape, and their test configuration is shown in Figure 4.33. The nozzle is located at a fixed place, and the fillet location of the target plate is changed. Figure 4.34 shows the Nusselt number difference in the impingement heat transfer for different target plate configurations. Location of the nose radius has a significant effect on

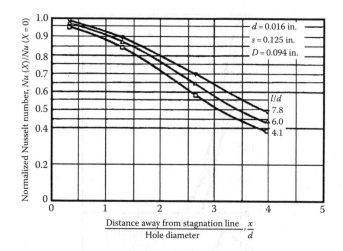

FIGURE 4.32
Normalized Nusselt number distribution for a typical impingement experiment of Chupp et al. (From Chupp, R.E. et al., *AIAA J. Aircr.*, 6, 203, 1969.)

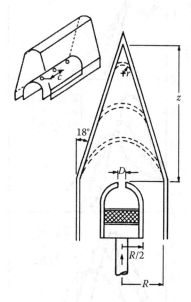

FIGURE 4.33
Leading-edge impingement test setup used by Bunker and Metzger. (From Bunker, R.S. and Metzger, D.E., *ASME J. Turbomach.*, 112, 451, 1990.)

the Nusselt number gradient distribution. The difference in peak Nusselt number and minimum Nusselt number shows the variation in the Nusselt number profile that contributes to the surface thermal stress. Results show that a sharper nose radius has more uniform Nusselt number distribution compared to a smooth-nosed leading-edge chamber.

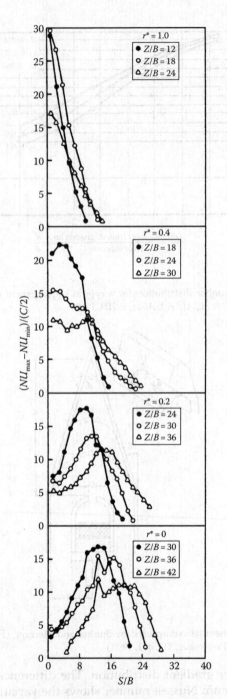

FIGURE 4.34
Span-averaged Nusselt number distributions for the leading-edge impingement configuration of Bunker and Metzger. (From Bunker, R.S. and Metzger, D.E., *ASME J. Turbomach.*, 112, 451, 1990.)

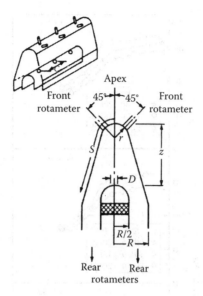

FIGURE 4.35

Leading-edge impingement with film extraction. (From Bunker, R.S. and Metzger, D.E., *ASME J. Turbomach.*, 112, 451, 1990.)

In another experiment done by Metzger and Bunker (1990), the effect of coolant bleeding was investigated. A discussion of coolant extraction from flat-plate jet impingement has been discussed earlier in this chapter. Figure 4.35 shows their experimental configuration. Note that two coolant extraction holes are symmetrically located about the stagnation location. Figure 4.36 shows the average Nusselt number in the leading edge with coolant bled for film cooling. In the plot, F represents front bleed, R represents rear flow, P is the pressure side bleed, and S is the suction-side bleed distribution. Equal bleed from suction and pressure sides show maximum reduction in the average Nusselt numbers.

4.2 Rib-Turbulated Cooling

4.2.1 Introduction

Based on the cooling technique used, there are three major internal cooling zones in a turbine rotor blade. The leading edge is cooled by jet impingement, the trailing edge is cooled by pin fins, and the middle portion is cooled by serpentine rib-roughened coolant passages. In advanced gas turbine blades, repeated rib turbulence promoters are cast on two opposite walls of internal cooling passages to enhance heat transfer. Thermal energy conducts

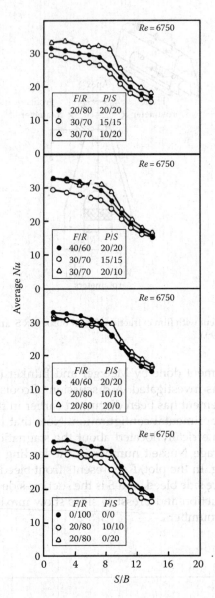

FIGURE 4.36
Average Nusselt numbers for the leading-edge impingement with bleed holes for film extraction. (From Bunker, R.S. and Metzger, D.E., *ASME J. Turbomach.*, 112, 451, 1990.)

from the external pressure and suction surfaces of turbine blades to the inner zones, and that heat is extracted by internal cooling. In laboratories, the internal cooling passages are mostly modeled as short rectangular or square channels with different aspect ratios. Han (1984) identified that the heat-transfer performance in a stationary ribbed channel primarily depends on the channel aspect ratio, the rib configuration, and the flow Reynolds number. There have been many fundamental studies to understand the heat-transfer enhancement phenomena by the flow separation caused by ribs. In general, ribs used for experimental studies are square in cross section, with a typical relative rib height of 5%–10% of channel hydraulic diameter and a *p/e* ratio varying from 7 to 15. However, today's airfoils have more complicated rib *shapes*, and smaller gas turbines have high-blockage ribs at closer spacing.

Specific configurations that characterize a ribbed channel include geometrical features such as rib height, pitch, and angle of attack. Figures 4.37 and 4.38 show the nomenclature commonly used for the geometrical features of ribs. Rib height is denoted as *e* and rib-to-rib spacing is referred to as *P*. This figure also shows a schematic of the flow past surface-mounted ribs, which indicates that a boundary layer separates upstream and downstream of ribs. Several studies have shown that another separation region may exist on top of the rib as well. These flow separations reattach the boundary layer to the heat-transfer surface, thus increasing the heat-transfer coefficient. Moreover, the separated boundary layer enhances turbulent mixing, and therefore the heat from the near-surface fluid can more effectively get dissipated to the main flow, thus increasing the heat-transfer coefficient. Ribs mostly disturb only the near-wall flow, and consequently the pressure drop penalty by ribs is acceptable for blade cooling designs.

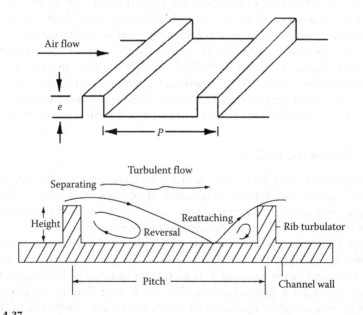

FIGURE 4.37
Schematic of flow separation and rib orientations in heat-transfer coefficient enhancement. (From Han, J.C. and Dutta, S., Internal convection heat transfer and cooling: An experimental approach, *Lecture Series 1995-05 on Heat Transfer and Cooling in Gas Turbines,* von Karman Institute for Fluid Dynamics, Belgium, Europe, May 8–12, 1995.)

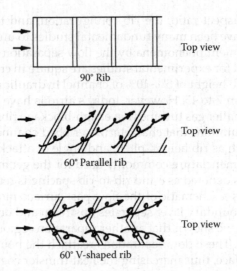

FIGURE 4.38
Schematic of secondary flow developed along angled-rib orientations. (From Han, J.C. and Dutta, S., Internal convection heat transfer and cooling: An experimental approach, *Lecture Series 1995-05 on Heat Transfer and Cooling in Gas Turbines*, von Karman Institute for Fluid Dynamics, Belgium, Europe, May 8–12, 1995.)

Figure 4.39 shows the orientation of a typical coolant channel in a turbine blade. Due to the curved asymmetric shape of a turbine blade, cooling channels near the trailing edge have broad aspect ratios, and those near the leading edge have narrow aspect ratios. Normally, the suction and pressure sides are ribbed. Angle of attack α is 90° for orthogonal or transverse ribs, and an angle of attack other than 90° is called skewed or angled ribs that develop secondary flows in the cross-stream direction. Most experimental ribs are square in rib cross section, but other cross sections are also discussed in this section.

4.2.1.1 Typical Test Facility

It is better for the reader to have an understanding about the experimental setups that are used for the measurement of presented results. Note that experiments provide ideal conditions, and measured data may need some extrapolation before they can be applied to the real design.

Figure 4.40 shows the cross section of a typical foil-heated test channel and rib geometry used by Park et al. (1992). Five straight rectangular channels of different aspect ratios were used. A plenum connected to the inlet of the test channel provided a hydrodynamically developing flow condition (sudden entrance). Since ribbed channels in gas turbines are short in length, most likely the flow and heat transfer in these channels are not fully developed. The spent air from the test channel was exhausted into the atmosphere. The test channels were heated by passing electrical current through 0.025 mm thick stainless-steel foils. Each wall of the rectangular duct had individual

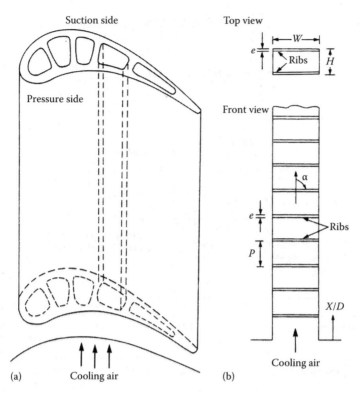

FIGURE 4.39
Typical coolant channels in turbine airfoil and internal rib arrangement. (a) Airfoil cutaway and (b) typical ribbed channel. (From Han, J.C., *ASME J. Heat Transfer*, 110, 321, 1988.)

heaters. These heaters were the electrical resistive type that provided uniform heat flux conditions. Brass ribs of square cross section were glued on the top and bottom walls of the foil-heated test channels. The ribs on the opposite walls were laid parallel to each other. The ribs were placed with a given spatial periodicity. The entire heated test channel was insulated from outside by fiberglass-insulating material to minimize ambient heat-transfer loss.

Han (1988) and Han et al. (1989) showed typical thermocouple locations in ribbed test channels. Each channel of their test facility had 180 36-gage copper-constantan thermocouples in strategic locations to measure the local surface temperature. Ninety thermocouples were placed on the ribbed wall and the other 90 on the smooth wall. Sixty thermocouples of each surface were placed along the centerline of the corresponding wall. Twenty of the remaining 30 thermocouples were distributed in the fully developed region, along the spanwise direction of each wall. A computerized data logger and a PC were used for temperature readings and data storage. Note that there was a higher concentration of thermocouples near the inlet to capture the developing heat-transfer distribution. The region $8.51 < X/D < 9.35$ was also well equipped to measure the periodically fully developed heat-transfer distribution.

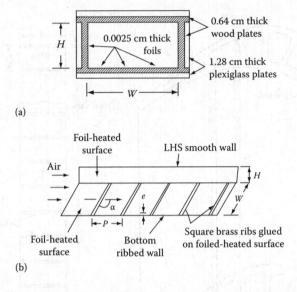

FIGURE 4.40
Typical foil-heated experimental setup and rib arrangements used by Park et al. (a) Cross-section of test channel and (b) test surface wall geometry. (From Park, J.S. et al., *Int. J. Heat Mass Transfer*, 35(11), 2891, 1992.)

	W/H	e	e/D	P/e	α	$Re \times 10^{-3}$
Square channel	1	0.24	0.047	10	90°, 60° 45°, 30°	10 30 60
Rectangular channel I	2	0.32	0.047	10	90°, 60° 45°, 30°	10 30 60
Rectangular channel II	4	0.32	0.078	10	90°, 60° 45°, 30°	10 30 60
Rectangular channel IA	2/4	0.32	0.047	10	90°, 60° 45°, 30°	10 30 60
Rectangular channel IIA	1/4	0.32	0.078	10	90°, 60° 45°	10 30 60

4.2.2 Effects of Rib Layouts and Flow Parameters on Ribbed-Channel Heat Transfer

Rib heat-transfer performance is significantly dependent on geometrical features and flow conditions. Effect of flow is characterized by the flow Reynolds number. That is followed by fundamental geometrical dependencies like rib spacing, angled ribs, and channel aspect ratios, which are discussed in this section. Han (1988) studied the effect of Reynolds number on the centerline heat-transfer coefficient of a square ($W/H = 1$) channel and two rectangular channels ($W/H = 2, 4$) for two rib spacings ($P/e = 10$ and 20). The heat-transfer distribution was presented by a Nusselt number ratio with several Reynolds numbers, and they showed similar trends except that the Nusselt number ratios decreased slightly with increasing Reynolds numbers.

4.2.2.1 Effect of Rib Spacing on the Ribbed and Adjacent Smooth Sidewalls

Figure 4.41 shows the effect of rib spacing on the heat-transfer coefficient distribution along the channel centerline. The ribs are placed orthogonal to the flow, and two rib spacings of $P/e = 10$ and 20 are used. Plotted results show that the local Nusselt number ratios on the smooth wall of the ribbed channels decrease significantly with increasing distance from the inlet for both P/e ratios. These are similar to the four-sided smooth-channel results, except that the ribbed-channel data have fluctuations and are about 20%–60% higher than the all-smooth-wall data. The adjacent ribbed walls cause these fluctuations. The local Nusselt number ratios on the ribbed wall have larger fluctuations. The magnitude of the heat-transfer coefficient decreases slowly with distance measured from inlet and settles into a periodic pattern

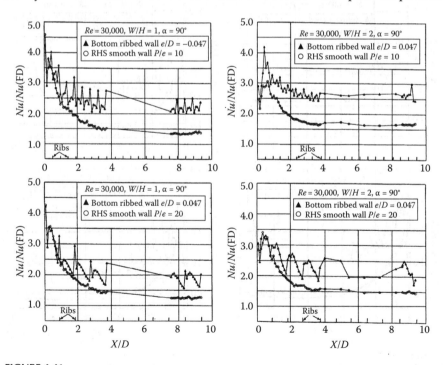

FIGURE 4.41
Effect of rib spacing on the channel centerline heat-transfer coefficient distributions for rectangular channels for $\alpha = 90°$ and $Re = 30,000$. (From Han, J.C., *ASME J. Heat Transfer*, 110, 321, 1988.)

just after the first few ribs ($X/D > 3.0$). For $P/e = 10$, heat-transfer coefficients increase near flow reattachment—i.e., downstream of each rib—and increase again as the flow approaches the next rib. Similar heat-transfer patterns are observed for $P/e = 20$, except that the heat-transfer coefficients are 10%–20% lower than those of $P/e = 10$. This reduction in heat-transfer coefficient is because of wider rib spacing, which develops a thicker boundary layer after flow reattachment between ribs. At $X/D < 1.0$, the Nusselt number ratios on the ribbed and smooth walls are about the same for both $P/e = 10$ and 20. The ribbed-wall Nusselt number ratios gradually differ from the smooth-wall values downstream of $X/D = 1.0$ and are up to two times higher at a downstream periodically fully developed region.

4.2.2.2 Angled Ribs

Ribs can also be placed at an angle to the bulk flow direction. Han and Park (1988) studied heat transfer with these nonorthogonal ribs. The thermocouple distributions were the same as those of the orthogonal rib arrangement of Han (1988), and therefore channel centerline results were obtained. Figure 4.42 shows the effects of the rib angle of attack on the centerline heat-transfer coefficients

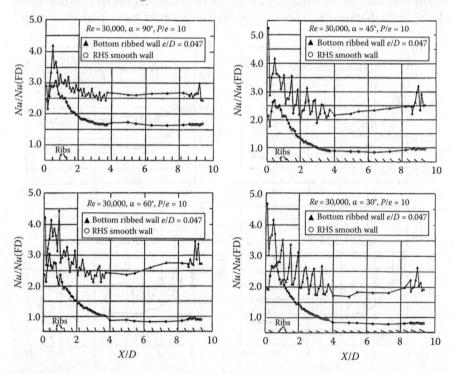

FIGURE 4.42
Effect of rib angle on the channel centerline heat-transfer coefficient distribution in a square channel for $p/e = 10$ and $Re = 30,000$. (From Han, J.C. and Park, J.S., *Int. J. Heat Mass Transfer*, 31(1), 183, 1988.)

for the case of *P/e* = 10 and *Re* = 30,000. The local heat-transfer distributions on the ribbed wall are not similar with respect to each other for different rib angles of attack. In a square channel (*W/H* = 1), the centerline Nusselt number ratios for α = 90° reach the fully developed periodic distribution after *X/D* > 3. This periodic distribution of the local Nusselt number ratio is due to the spatial periodicity of flow separation from ribs and flow reattachment between ribs. In contrast, the centerline Nusselt number ratios after *X/D* > 3 for α = 60°, 45°, and 30° are higher than those for α = 90° in a square channel, and the periodic Nusselt number ratios for α = 60°, 45°, and 30° increase after *X/D* > 3.

It is postulated that secondary flow induced by the angled ribs affects heat transfer significantly. Han and Park (1988) also discussed the effect of rib angle on the streamwise-averaged Nusselt numbers. The streamwise-averaged Nusselt number ratio was the average of the periodic Nusselt number ratios along the centerline, the middle line, and the edge line, respectively, in each of the test channels at the downstream regions. In the square channel (*W/H* = 1), the streamwise-averaged Nusselt number ratio on the ribbed side was mostly uniform in the lateral position for α = 90°. For other angles of attack—α = 60°, 45°, and 30°—the streamwise-averaged Nusselt number ratio on the ribbed side varied in the lateral direction. The Nusselt number ratios along the edge line were 30%–50% higher than corresponding values along the centerline. This was attributed to the secondary flow induced by the rib that moved from the right-hand swirl to the left-hand swirl, as depicted in Figure 4.43. Therefore, the Nusselt number ratio on the leading-edge side (right-hand side) was higher than that in the centerline and subsequently estimated to be higher than that on the left-hand side. Because of this secondary flow effect, the centerline Nusselt number ratios on both the ribbed sidewall and the smooth sidewall with angled ribs were higher than those with transverse ribs (α = 90°).

Figure 4.43 also shows the Nusselt number ratios for a rectangular channel (*W/H* = 2) with rib angle of attacks varying from 30° to 90°. In rectangular channels, the streamwise averaged Nusselt number ratio is similarly distributed to that in the square channel for α = 90°. In angled ribs—i.e., α = 60°, 45°, and 30°—the ribs are oriented to create a secondary flow that moves along the rib axes from the left to the right side. Similar results are obtained for *W/H* = 4. It is seen that in a wider rectangular channel, the centerline Nusselt number ratios on both the ribbed side and the smooth side walls for angled ribs are lower than that for α = 90°, i.e., orthogonal or transverse ribs.

4.2.2.3 Effect of Channel Aspect Ratio with Angled Ribs

Figure 4.44 shows the effect of channel aspect ratio on the heat-transfer enhancement and pressure drop for given rib angles at a Reynolds number of 30,000. For 60° ribs, the ribbed-side heat-transfer augmentations do not vary significantly, but the pressure drop penalties increase dramatically from 2- to 18-fold when the channel aspect ratio changes from narrow to wide (¼–4). Similar results are observed for 45° and 90° ribs. It is clear that

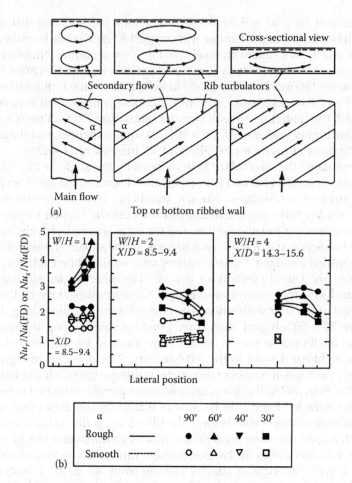

FIGURE 4.43

Effect of rib angle on secondary flow and lateral Nusselt number ratio at $p/e = 10$ and $Re = 30,000$. (a) Schematic of secondary flows and (b) comparison of lateral Nusselt number ratios. (From Han, J.C. and Park, J.S., *Int. J. Heat Mass Transfer*, 31(1), 183, 1988.)

a narrow-aspect ratio channel provides a better heat-transfer performance than a broad-aspect ratio channel. Both heat-transfer enhancement and pressure drop increment are relatively low for 30° ribs in narrow-aspect ratios. Whereas, compared to other ribs, the friction factor is significantly low for 30° ribs in a broad-aspect ratio duct ($W/H = 4$), and the heat-transfer coefficient is comparable to other angled ribs.

4.2.2.4 Comparison of Different Angled Ribs

Figure 4.45 shows a comparison of heat-transfer performance with different rib angles in different-aspect ratio channels, with aspect ratio varying from ¼ to 4. The results show that the ribbed-side heat-transfer enhancements are

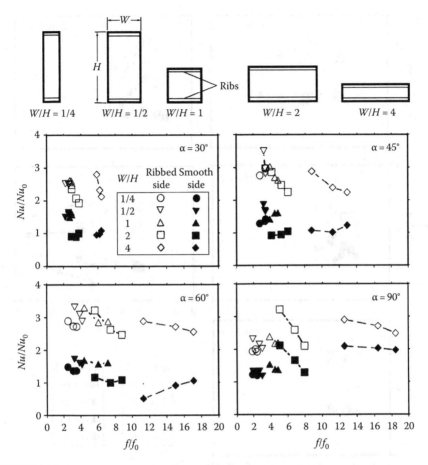

FIGURE 4.44
Effect of channel aspect ratio on heat-transfer performance. (From Park, J.S. et al., *Int. J. Heat Mass Transfer*, 35(11), 2891, 1992.)

about three times, and the pressure drop penalties are about four to eight times the values for 45° and 60° ribs compared to a smooth channel. The pressure drop penalties are only two to four times for the 45° and 60° angled ribs with the same level of heat-transfer enhancement for the narrow-aspect ratio channels ($W/H = \frac{1}{4}$). However, for the same level of heat-transfer enhancement in a broad-aspect ratio channel ($W/H = 4$), the pressure drop penalties are as high as 8–16 times the friction factor in a smooth channel for angled ribs. It is concluded by Park et al. (1992) that the narrow-aspect ratio channel performs better than a broad-aspect ratio channel with angled ribs. Figure 4.45 also shows that the 90° orthogonal ribs give the lowest heat-transfer and pressure drop augmentations in narrow-aspect ratio channels, whereas the 60° ribs provide the highest heat-transfer and pressure drop enhancements for narrow-aspect ratio ducts ($W/H = \frac{1}{2}$ and $\frac{1}{4}$). Results also indicate that the

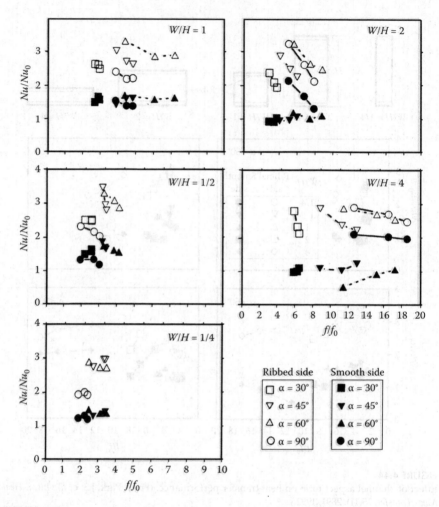

FIGURE 4.45
Effect of rib angle on heat-transfer performance. (From Park, J.S. et al., *Int. J. Heat Mass Transfer*, 35(11), 2891, 1992.)

30° angled ribs provide the worst, and both 60° angled and 90° orthogonal ribs provide better heat-transfer enhancements for W/H = 4 and 2.

The smooth-wall results in Figure 4.45 show a 10%–50% enhancement in heat transfer in square and narrow-aspect ratio ducts with the addition of ribs. The heat-transfer enhancement in smooth surfaces is more for nonorthogonal angled ribs than for the orthogonal ribs. Angled ribs develop a secondary flow that increases mixing in the entire duct. The smooth-side results for broad-aspect ratio ducts (W/H = 2 and 4) are higher compared to that for the narrow-aspect ratio ducts. Addition of 90° orthogonal ribs increases the smooth-side heat-transfer coefficient by 100% for W/H = 4. This increase is associated with the turbulence enhancement in the bulk flow-by-flow separation and reattachment processes. Unlike a 90° rib, 60° angled rib orientation causes a drop in the heat-transfer coefficient in the smooth side compared to that with all smooth walls for $W/H = 4$. It is possible that angled ribs in broad-aspect ratio ducts act as flutes in the wall to create mostly a swirl in the flow. The boundary layer separation effect is weaker than the swirl generated by the angled ribs in broad ducts. The swirl carries the warmer fluid from the ribbed surfaces toward the smooth surfaces. Thus the proximity of warmer fluid at smooth sides reduces the heat-transfer coefficients from them.

4.2.3 Heat-Transfer Coefficient and Friction Factor Correlation

Analytical methods for predicting the friction factor and heat-transfer coefficient for turbulent flow over rib-roughened surfaces are not available because of the complex flow created by periodic rib-roughness elements such as separation, reattachment, and recirculation. Thus heat-transfer designers depend on the semiempirical correlation over a wide range of rib geometry for their friction and heat-transfer calculations. These semiempirical correlations for the friction factors and the heat-transfer coefficients are derived from the laws of the wall similarity for flow over rough surfaces. Details of this correlation development are given in Han (1988). Some of the conceptual ideas behind those correlations are included here.

According to the concept of Nikuradse (1950) and Dipprey and Sabersky (1963), the laws of the wall in fully developed turbulent flow can be expressed by the velocity and temperature profiles normal to the rough surfaces as

$$u^+ = 2.5\ln\left(\frac{y}{e}\right) + R(e^+)$$

$$(4.14)$$

$$T^+ = 2.5\ln\left(\frac{y}{e}\right) + G(e^+, Pr)$$

where
   u^+ and T^+ are the dimensionless velocity and temperature at a distance y from the rough wall,
   e is the height of the ribs
   $R(e^+)$ and $G(e^+, Pr)$ are the dimensionless velocity and temperature at the tip of the ribs, i.e., at $7 = e$

The dimensionless average velocity and temperature are given as $\bar{u}^+ = (2/f)^{1/2}$ and $\bar{T}^+ = (f/2)^{1/2}/St$. Han (1988) developed a correlation to predict the performance of two-sided orthogonal ribbed rectangular channels. The roughness function R was given by

$$R(e^+) = \left(\frac{2}{f}\right)^{1/2} + 2.5\ln\left(\frac{2e}{D}\frac{2W}{W+H}\right) + 2.5 \qquad (4.15)$$

The heat-transfer roughness function G was given by

$$G(e^+, Pr) = R(e^+) + \frac{(f/2St_r) - 1}{(f/2)^{1/2}} \tag{4.16}$$

The roughness Reynolds number e^+ is given by $e^+ = (e/D)\,Re(f/2)^{1/2}$. The four-sided ribbed channel friction factor f is given by $f = \bar{f} + (H/W)(\bar{f} - f_s)$. $\bar{f}$ is the average friction factor in a channel with two opposite ribbed walls, f_s is the friction factor for four smooth-sided channels, H is the flow channel height, W is the flow channel width, and St_r is the ribbed sidewall center-line average Stanton number for flow in a channel with two opposite ribbed walls. R and G are correlated with experimentally measured $\bar{f}$ and St_r for fully developed turbulent flow in rectangular channels. These channels have two opposite ribbed walls for varying e/D ratio and Re. Therefore, $\bar{f}$ and St_r for a desired operating condition (given W/H, e/D, and Re) can be predicted from the experimentally obtained R and G correlation. Since $R(e^+)$ is inversely related to the friction factor f, it is better to have a larger R. A larger R means less pumping power. Whereas $G(e^+, Pr)$ is inversely proportional to the Stanton number St_r. Since a larger Stanton number is desirable, a smaller value of G is a better performing rib.

Figure 4.46 shows the friction roughness function R and heat-transfer roughness function G correlation developed by Han (1988). According to the friction similarity law, the measured average friction factor $\bar{f}$, the channel aspect ratio W/H, the rib height-to-hydraulic diameter ratio e/D, and the Reynolds number Re could be correlated with the friction roughness function $R(e^+)$. The geometrical parameter, P/e ratio, could also be taken into account. A plot of $R/(P/e/10)^{0.35}$ versus roughness Reynolds number e^+ is shown in Figure 4.46. The correlation of friction roughness function R is $R = 3.2(P/e)/10)^{0.35}$ for e^+ greater than or equal to 50. The equation correlates 95% of the experimental data within 6% deviation. Note that R is independent of the roughness Reynolds number e^+. This implies that the average friction factor is independent of roughness Reynolds number. Figure 4.46 also shows the variation of the heat-transfer roughness function G with respect to e^+. The effect of the P/e ratio on G is not significant. For a Prandtl number of 0.703, the correlation of G is presented by Han (1988) as $G = 3.7(e^+)^{0.28}$ for e^+ greater than or equal to 50. This equation correlates 95% of the experimental data within 8% deviation.

Figure 4.47 shows the correlations for friction factor and heat transfer in broad-aspect ratio rectangular ducts with angled ribs. The correlation for R with rib angle α is given as

$$\frac{R}{(P/e/10)^{0.35}(W/H)^m} = 12.3 - 27.07\left(\frac{\alpha}{90°}\right) + 17.86\left(\frac{\alpha}{90°}\right)^2 \tag{4.17}$$

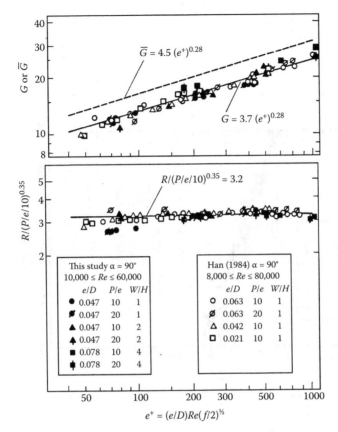

FIGURE 4.46
Friction and heat-transfer correlation for orthogonal 90° ribs. (From Han, J.C., *ASME J. Heat Transfer*, 110, 321, 1988.)

where $m = 0$ for $\alpha = 90°$ and $m = 0.35$ for $\alpha < 90°$. Another restriction is if $W/H > 2$, then set W/H equal to 2. This correlation is valid for the operating range of $P/e = 10$–20, rib blockage ratio $e/D = 0.047$–0.078, rib angle $\alpha = 90°$–30°, channel-aspect ratio $W/H = 1$–4, and Reynolds number of 10,000–60,000. The G correlation with e^+ is given as

$$G = 2.24 \left(\frac{W}{H} \right)^{0.1} \left(\frac{\alpha}{90°} \right)^{m} \left(\frac{p/e}{10} \right)^{n} (e^+)^{0.35} \tag{4.18}$$

where $m = 0.35$ and $n = 0.1$ for a square channel; $m = n = 0$ for rectangular channel. Therefore, the effect of rib angle α and rib spacing P/e is not significant in a rectangular channel for the G function. The e^+ is defined as $e^+ = (e/D) \, Re(f/2)^{1/2}$.

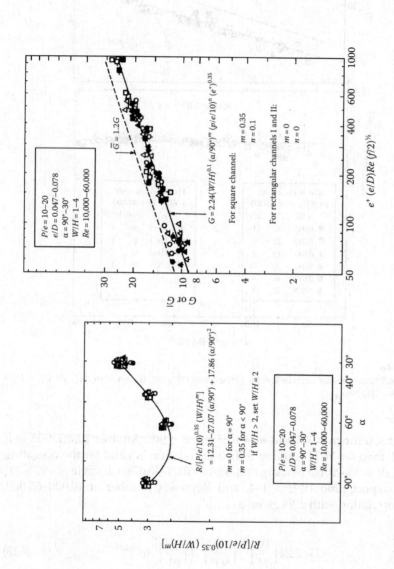

FIGURE 4.47

Friction factor and heat-transfer correlation in rectangular ribbed channels. (From Han, J.C. and Park, J.S., *Int. J. Heat Mass Transfer*, 31(1), 183, 1988.)

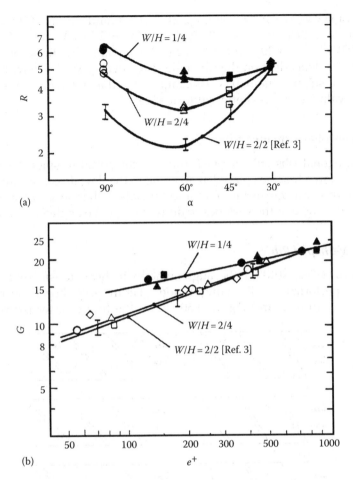

FIGURE 4.48
Friction factor and heat-transfer correlation in narrow-aspect ratio rectangular ribbed channels. (a) Correlation constant R behavior and (b) correlation constant G behavior. (From Han, J.C. et al., *Int. J. Heat Mass Transfer*, 31(1), 183, 1988.)

The dependence of R on α and W/H for narrow-aspect ratio channels is shown in Figure 4.48. The R function increases with decreasing W/H ratio. It is interesting to note that all three aspect ratios show the same value at $\alpha = 30°$. In rectangular channels of narrower-aspect ratios ($W/H = \frac{1}{2}$ and $\frac{1}{4}$), the friction factor, the friction roughness function, and the ribbed side-wall Stanton numbers are correlated with previously defined heat-transfer roughness function G, and their correlation of the heat-transfer data is also shown in Figure 4.48. For a Prandtl number of 0.7, the dependency of G on α, W/H, and e^+ is presented by Han et al. (1989) as

$$G = C(e^+)^n \tag{4.19}$$

where for $\frac{1}{2} < W/H < 1$: $n = 0.35$, $C = 2.24$ if $\alpha = 90°$, and $C = 1.80$ if $30° < \alpha < 90°$. And for $\frac{1}{4} < W/H < \frac{1}{2}$, $n = 0.35(W/H)^{0.44}$, $C = 2.24 (W/H)^{-0.76}$ if $\alpha = 90°$, $C = 1.80(W/H)^{-0.76}$ for $30° < \alpha < 90°$. The earlier equation has limits on the W/H ratio and is dependent on the rib angle. The effect of $W/H = \frac{1}{2}$ with respect to $W/H = 1$ (square channel) is less compared to that of $W/H = \frac{1}{4}$. Note that a lower G value refers to a better performance.

4.2.4 High-Performance Ribs

The orthogonal ribs enhance heat transfer compared to smooth channels, and it is shown that ribs placed at an angle to the flow direction can further enhance the heat-transfer performance of ribbed channels. In this section, we consider some of the varieties or derivatives of the angled-rib concept.

4.2.4.1 V-Shaped Rib

Han et al. (1991) studied the augmentation in heat transfer with different high-performance rib configurations. Figure 4.49 shows the rib configurations studied, and Figure 4.50 shows the heat-transfer performance

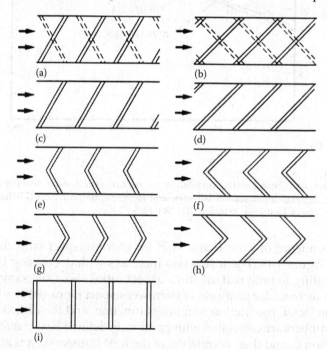

FIGURE 4.49
Different rib configurations studied by Han et al. (a) 60° crossed rib, (b) 45° crossed rib, (c) 60° parallel rib, (d) 45° parallel rib, (e) 60° V-shaped rib, (f) 45° V-shaped rib, (g) 60° Λ-shaped rib, (h) 45° Λ-shaped rib, and (i) 90° rib. (From Han, J.C. et al., *ASME J. Heat Transfer*, 113, 590, 1991.)

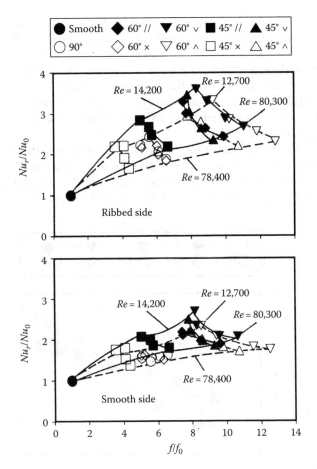

FIGURE 4.50
Heat-transfer performance for different rib configurations used by Han et al. (From Han, J.C. et al., *ASME J. Heat Transfer*, 113, 590, 1991.)

comparison of different rib configurations. The Nusselt number ratios of both ribbed and smooth sides versus the friction factor ratios for 90°, 60°, and 45° ribs over a range of studied Reynolds numbers are plotted. Results show that the Nusselt number ratio decreases, but the friction factor increases with increasing Reynolds number. This means that the improvement in heat-transfer performance decreases with increasing Reynolds number. These results also show that, in general, the Nusselt number ratio increase by a rib configuration is accompanied with increasing friction factor ratio. The 60° forward V-shaped rib provides 2.7–3.5 times the ribbed-side heat-transfer enhancement with 8–11 times the pressure drop penalty. Whereas the 60° crossed rib gives 1.8–2.2 times the ribbed-side heat-transfer augmentation with 5–7 times the pressure drop increment. The 90° rib gives about the same heat-transfer augmentation as the 60° and 45° crossed rib, but the friction factor with 45° crossed rib is less than that for 90° and 60° crossed ribs. The 60° inverted V rib provides slightly lower heat transfer but much higher pressure drop than the 60° V rib. The parallel ribs show higher heat-transfer coefficient compared to the corresponding crossed ribs. The 60° parallel rib gives slightly lower heat transfer than the 60° V rib with about the same amount of pressure-drop penalty. The performance of the 45° V-shaped rib is about the same as the 45° parallel rib, which is better than those of 45° crossed rib, 45° inverted V rib, and 90° rib.

Figure 4.51 shows the heat-transfer roughness function G versus the roughness Reynolds number e^+ for 90°, 60°, and 45° ribs for a range of Reynolds numbers. Similar plots for the average heat-transfer roughness function based on the average value of the ribbed-side and smooth-side heat-transfer

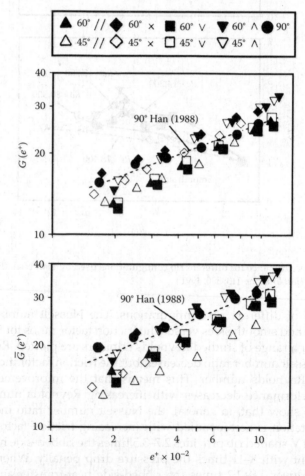

FIGURE 4.51
Heat-transfer correlation for different rib configurations. (From Han, J.C. et al., *ASME J. Heat Transfer*, 113, 590, 1991.)

coefficients are also plotted in Figure 4.51. The heat-transfer roughness function increases with increasing roughness Reynolds number for all studied rib configurations. The heat-transfer roughness function for the 90° rib is the same as the previous correlation that was developed for the 90° rib by Han (1988). The 60° V-shaped rib has the lowest G value, which means the highest heat-transfer coefficient. The 45° V-shaped rib and the 45° parallel rib have the same lower G values, which means that the heat-transfer coefficients are comparable for these two rib configurations. The slopes of heat-transfer roughness functions versus e^+ for 60° and 45° crossed, parallel, and V-shaped ribs are higher than the 90° rib. That means that the angled rib performances get progressively worse, and they become similar to each other at higher Reynolds numbers.

4.2.4.2 V-Shaped Broken Rib

Han and Zhang (1992) studied the high-performance broken parallel and V-shaped ribs. Figure 4.52 shows the different rib configurations studied by them. Figure 4.53 presents the Nusselt number ratio of ribbed and smooth-side versus the friction factor ratio (heat-transfer performance curve) for parallel broken and V-shaped broken ribs over a range of Reynolds numbers varying between 15,000 and 80,000. The data for parallel continuous and V-shaped continuous ribs are included for direct comparison. The results show that as observed before, for all ribs, the Nusselt number ratio decreases and friction factor ratio increases with increasing Reynolds number. The ribbed-side Nusselt number ratios for 60° and 45° parallel broken ribs or V-shaped broken ribs are much higher than the corresponding 60° and 45° parallel continuous ribs or V-shaped continuous ribs. However, the corresponding friction factor ratios are comparable with each other for broken and continuous rib configurations. The 60° and 45° parallel broken ribs or V-shaped broken ribs represent the similar characteristics of heat-transfer augmentation and pressure drop increment in which the 60° V-shaped broken rib is the best. On the other hand, the 60° and 45° parallel continuous ribs, V-shaped ribs, and the 90° broken ribs have the same heat-transfer characteristics in which the 60° V-shaped continuous ribs are the highest.

Figure 4.54 shows the heat-transfer roughness function G versus the roughness Reynolds number e^+ for 90°, 60°, and 45° broken ribs for a range of Reynolds numbers. Similar plots for the average heat-transfer roughness function, based on the average value of the ribbed-side and smooth-side heat-transfer coefficients, are also shown in Figure 4.54. The heat-transfer roughness function increases with an increase in the roughness Reynolds number for all studied rib configurations. The 60° and 45° parallel broken ribs or 60° V-shaped broken rib have lower G values, and that means they have higher heat-transfer coefficients. The slopes of G function with respect to e^+ for broken ribs are higher than those for the 90° transverse rib, which indicates that the broken ribs lose their heat-transfer benefits at higher Reynolds numbers.

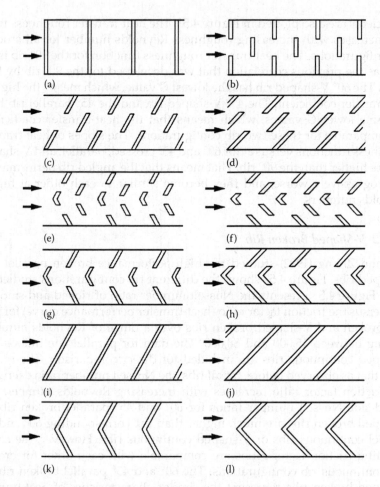

FIGURE 4.52
Rib configurations used by Han and Zhang. (a) 90° continuous rib, (b) 90° broken rib, (c) 60° parallel broken rib, (d) 45° parallel broken rib, (e) 60° V-shaped broken rib, (f) 45° V-shaped broken rib, (g) 60° V-shaped broken rib—A, (h) 45° V-shaped broken rib—A (i) 60° parallel continuous rib, (j) 45° parallel continuous rib, (k) 60° V-shaped continuous rib, and (l) 45° V-shaped continuous rib. (From Han, J.C. and Zhang, Y.M., *Int. J. Heat Mass Transfer*, 35(2), 513, 1992.)

However, the slopes for average G functions with respect to e^+ for broken ribs are about the same as that with the 90° transverse rib.

4.2.4.3 Wedge- and Delta-Shaped Rib

Two derivatives of the broken ribs are wedge-shaped and delta-shaped ribs. Figure 4.55 shows the continuous and broken wedge-shaped ribs along with delta-shaped ribs. These ribs can be forward or backward aligned; Han et al. (1993) have studied these special ribs. These heat-transfer enhancement

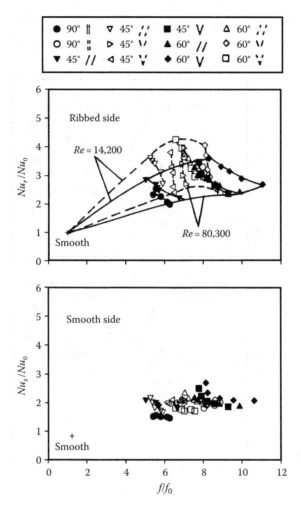

FIGURE 4.53
Comparison of heat-transfer performance between broken and continuous ribs. (From Han, J.C. and Zhang, Y.M., *Int. J. Heat Mass Transfer*, 35(2), 513, 1992.)

mechanisms combine the benefits of ribbed channel and the pin-finned channel. The isolated three-dimensional projections called broken ribs disturb the boundary layer and, like pins protruding in the flow, create a wake. Figure 4.56 shows the Nusselt number ratio of ribbed-side and smooth-side versus the friction factor ratio for eight rib configurations. It is clearly shown that the delta-shaped ribs enhance heat transfer more than the wedge-shaped ribs, as well as produce less pressure drop penalty. Highest Nusselt number ratios are obtained with backward-aligned delta-shaped ribs, whereas backward-offset delta ribs show a lower Nusselt number ratio but a friction factor ratio significantly lower than other high-performance ribs.

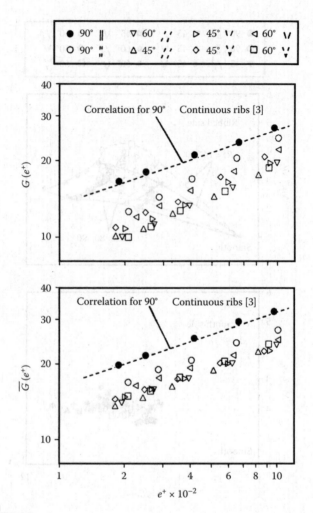

FIGURE 4.54
Heat-transfer correlation for broken ribs. (From Han, J.C. and Zhang, Y.M., *Int. J. Heat Mass Transfer*, 35(2), 513, 1992.)

Figure 4.57 shows the *R* function that characterizes pressure drop for given rib configurations. Results from Han (1988) for orthogonal continuous ribs are added for comparison. Figure 4.58 shows the heat-transfer roughness function versus the roughness Reynolds number for the three-dimensional ribs. The roughness function of continuous ribs is also included for comparison. The results show that the roughness functions for the three-dimensional ribs follow patterns similar (i.e., linear with a positive slope) to those of the continuous ribs. However, the *G* function is significantly lower for delta-shaped ribs than for continuous ribs.

Han et al. (1993) studied local heat-transfer and overall pressure drop measurements in a square channel with these specialty ribs and showed

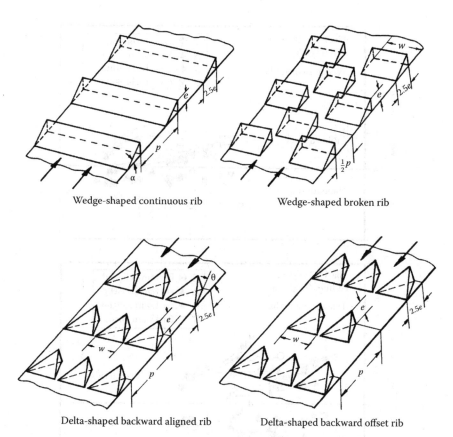

Wedge-shaped continuous rib   Wedge-shaped broken rib

Delta-shaped backward aligned rib   Delta-shaped backward offset rib

FIGURE 4.55
Wedge- and delta-shaped ribs used by Han et al. (From Han, J.C. et al., *Enhanced Heat Transfer*, 1(1), 37, 1993.)

that the delta-shaped ribs produce higher heat-transfer augmentation and lower pressure drop than those with wedge-shaped ribs. For delta-shaped ribs, the backward alignment shows better performance than the forward direction. The wedge-shaped ribs also showed that the broken ribs have better enhancing capabilities. The backward delta-shaped ribs, the overall best performer in the group, produced three to four times the heat-transfer augmentation over a smooth surface result; the pressure drop was seven to nine times higher.

4.2.5 Effect of Surface-Heating Condition

Han et al. (1992) studied the influence of surface heat flux ratio on heat transfer. Different parallel, crossed, and V-shaped ribs were used. Six different wall heat flux ratios were studied, as shown in Table 4.2.

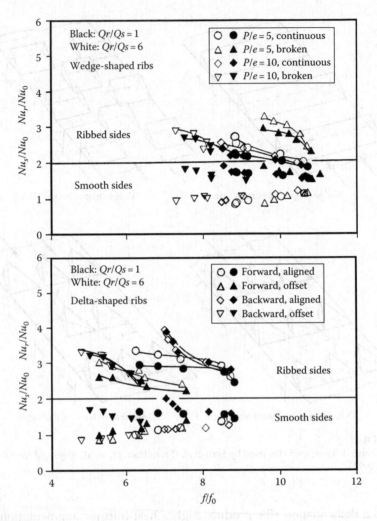

FIGURE 4.56
Heat-transfer performance for different rib arrangements used by Han et al. (From Han, J.C. et al., *Enhanced Heat Transfer*, 1(1), 37, 1993.)

The bulk temperature of the fluid differed by different heat input. The near-wall temperature of the coolant, along with the asymmetric temperature profile in the channel, created different heat-transfer patterns for different cases.

Figure 4.59 shows the heat-transfer performance comparison for three different wall heat flux ratios over a range of Reynolds numbers. The three different wall heat flux ratios are case 1, uniform wall heat flux; case 3, two ribbed wall heat flux is six times two smooth wall; and case 5, two ribbed walls heated but two smooth walls unheated. Plotted results show that the ribbed-side Nusselt number ratio with two ribbed walls heated (case 5) is

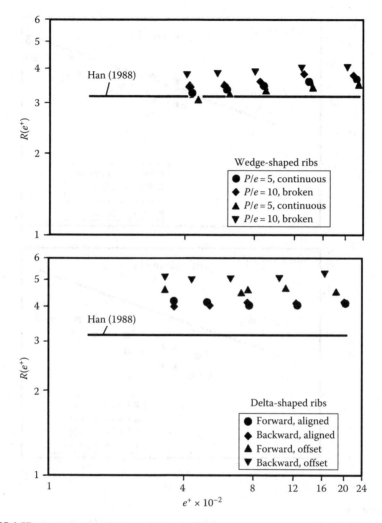

FIGURE 4.57

Friction factor correlation for wedge- and delta-shaped ribs. (From Han, J.C. et al., *Enhanced Heat Transfer*, 1(1), 37, 1993.)

higher than that with four uniform wall heating (case 1), whereas, the smooth-side Nusselt number ratio is higher with four-wall uniform heating than that with four-wall nonuniform heating (case 3). As observed in our earlier ribbed channels, the Nusselt number ratio decreases, but the friction factor increases with increasing Reynolds numbers. This means that the heat-transfer performance decreases (more pumping power but less proportionate heat-transfer augmentation) with increasing Reynolds numbers. The effect of wall heat flux on the heat-transfer performance decreases with increasing Reynolds numbers. The heat-transfer performance for the 60° V-shaped rib is higher than the 60° parallel rib, 60° crossed rib, and 90° rib,

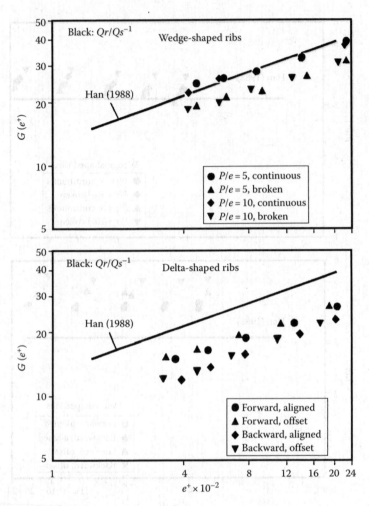

FIGURE 4.58
Heat-transfer correlation for wedge- and delta-shaped ribs. (From Han, J.C. et al., *Enhanced Heat Transfer*, 1(1), 37, 1993.)

regardless of the wall heat flux ratio. The effect of wall heat flux ratio on the four-side smooth channel heat transfer is also included for comparison. Note that the changes in Nusselt number ratio are only 3%–5% for different heating conditions in a smooth channel. This shows that the wall heat flux has an effect on the ribbed channel, but the effect is less significant on the smooth channel.

4.2.6 Nonrectangular Cross-Section Channels

Webb et al. (1971) and Gee and Webb (1980) studied the repeated-rib heat transfer in circular cross-sectioned tubes. Figure 4.60 shows the characteristic

TABLE 4.2

Different Wall Heating Conditions Used by Han et al.

Case No.	Wall Heat Condition	q_{rib1}/q_{smooth}	q_{rib2}/q_{smooth}
1	Four walls, uniform heat flux	1	1
2	Two ribbed walls greater than two smooth walls	3	3
3	Two ribbed walls greater than two smooth walls	6	6
4	Two ribbed walls different but greater than two smooth walls	6	4
5	Two ribbed walls heated and two smooth walls unheated	∞	∞
6	One ribbed wall heated and other three walls unheated	∞	0

Source: Han, J.C. et al., *ASME J. Turbomach.*, 114, 872, 1992.

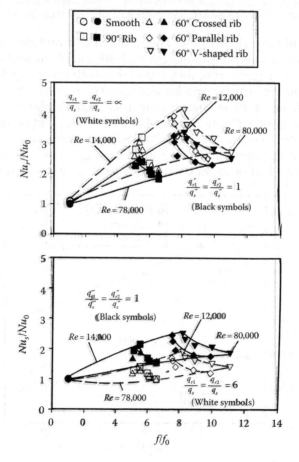

FIGURE 4.59

Heat-transfer performance for different rib orientations used by Han et al. (From Han, J.C. et al., *ASME J. Turbomach.*, 114, 872, 1992.)

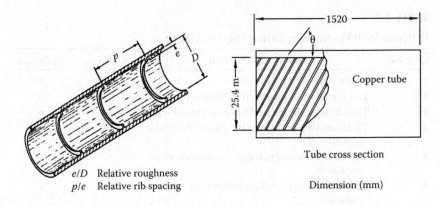

e/D   Relative roughness
p/e   Relative rib spacing

Tube cross section

Dimension (mm)

FIGURE 4.60
Repeated rib roughness arrangement in a circular pipe. (From Webb, R.L. et al., *Int. J. Heat Mass Transfer*, 14(4), 601, 1971; From Gee, D.L. and Webb, R.L., *Int. J. Heat Mass Transfer*, 23, 1127, 1980.)

dimensions of their ribbed tube. They studied different rib heights (e/D = 0.01 – 0.04) for various rib spacings (p/e = 10–40) and rib angles (θ = 90° – 30°).

For the perpendicular ribs, they found that both friction factor and heat-transfer coefficient increase with increasing e/D ratio for a given p/e ratio; however, both friction factor and heat-transfer coefficient decrease with increasing p/e ratio for a given e/D ratio. Figure 4.61 shows the effect of helix angle on Stanton number and friction factor. Both Stanton number and friction factor decrease with a decrease in the helix angle for a given p/e, e/D, and Reynolds number. But the friction factor decreases faster than the Stanton number.

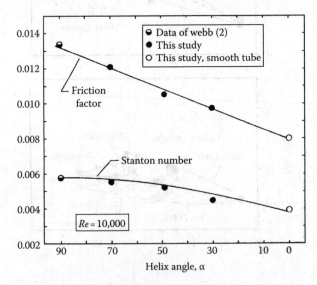

FIGURE 4.61
Effect of helix angle on Stanton number and friction factor. (From Gee, D.L. and Webb, R.L., *Int. J. Heat Mass Transfer*, 23, 1127, 1980.)

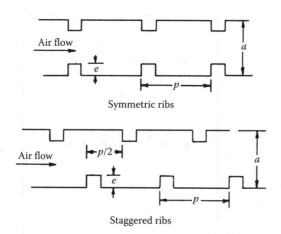

FIGURE 4.62

Symmetric and staggered rib arrangements in opposite walls. (From Han, J.C. et al., *Int. J. Heat Mass Transfer*, 21, 1143, 1978.)

Han et al. (1978) studied the rib-roughened heat-transfer patterns from a parallel-plate geometry. Parallel-plate configuration was simulated by a very high aspect ratio rectangular channel (W/H = 12). Figure 4.62 shows the two different rib orientations used by them. The symmetric ribs show exact alignment of ribs in the opposite parallel plates, and the staggered arrangement shows the misalignment in the ribs in opposite plates. They showed the effect on friction factor and Stanton number by different relative rib heights for p/e = 5. An increase in relative rib height indicates a higher Stanton number accompanied by a higher friction factor. However, e/D = 0.056 and 0.046 shows comparable Stanton numbers, but the friction factor is higher with the higher relative rib height of these two rib cases.

Figure 4.63 shows that the effect of rib arrangement (staggered and inline) on friction factor and that Stanton number is not so significant for the cases studied by Han et al. (1978). Figure 4.64 shows the effect of relative rib spacing on friction factor and Stanton number. Results indicate that the Stanton number is highest for the p/e = 10. This rib spacing also shows the highest friction factor among the group. Figure 4.65 shows the effect of rib angle of attack on friction factor and Stanton number. The friction factor shows a decrease in its value with decreasing angle of attacks. The Stanton number distribution, as shown in Figure 4.65, shows that the heat-transfer coefficient is mostly unaffected at angle of attacks 90° and 45°. The 30° angle of attack shows a sharp decrease in the heat-transfer coefficient.

The ribbed channel can be of nonrectangular shape. Coolant flow channels in the leading edge or near the trailing edge of an airfoil can be triangular in cross section. Metzger and Vedula (1987) studied triangular channels with angular ribs. Figure 4.66 shows three different ribs used by them. Ribs placed at an angle develop secondary flow by their orientation. Figure 4.66 also shows the schematic secondary flows developed by the three rib

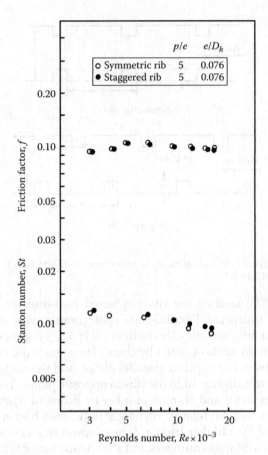

FIGURE 4.63
Friction factor and Stanton number distribution for symmetric and staggered rib arrangements. (From Han, J.C. et al., *Int. J. Heat Mass Transfer*, 21, 1143, 1978.)

orientations used. The upstream angled ribs and downstream angled ribs develop double vortex structures in the channel, whereas the spiral ribs create a single vortex structure. The first rib orientation in Figure 4.66 placed ribs on both walls that had a vertex angle pointing upstream. The second rib configuration shows the vertex angle to be pointing downstream. Both of these rib orientations develop symmetric secondary flow vortices. The third arrangement is the spiral arrangement, and as shown in Figure 4.66, a large single secondary flow vortex is envisioned.

Figure 4.67 shows the span-averaged Nusselt numbers for ribbed sides and smooth side of the triangular channel. The downstream angled ribs show higher heat-transfer coefficient in the ribbed side, and upstream angled ribs show higher heat-transfer coefficient for the smooth side. The secondary flows observed in Figure 4.66 explain the corresponding heat-transfer coefficients. The downstream angled ribs carry the core flow toward the ribbed

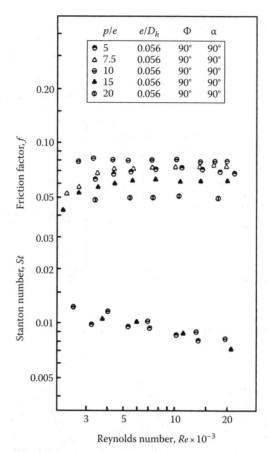

FIGURE 4.64

Effect of rib spacing on friction factor and Stanton number. (From Han, J.C. et al., *Int. J. Heat Mass Transfer*, 21, 1143, 1978.)

sides, and the upstream angled ribs carry the core flow toward the smooth side; that secondary flow pattern enhances heat-transfer coefficients from respective surfaces. Figure 4.68 shows the span-averaged Nusselt numbers for the three sides. It is interesting to note that the three walls approach a nearly equal fully developed value in this case. The ribbed sides for this rib configuration show lower heat-transfer coefficients than the other two rib configurations.

Zhang et al. (1994) studied heat transfer in triangular ducts with full and partial ribbed walls. Figure 4.69 shows the test channel cross section and rib arrangements for different cases. The objective of the work was to investigate the effect of vertex corners and full and partial ribbed walls on heat transfer and friction in triangular channels. Six triangular channels were tested. Figure 4.69 shows cross sections of the channels. Two of the channels had

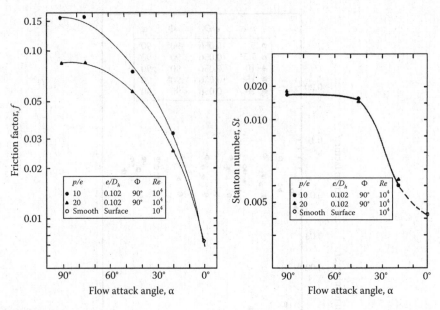

FIGURE 4.65
Effect of rib angle of attack on friction factor and Stanton number. (From Han, J.C. et al., *Int. J. Heat Mass Transfer*, 21, 1143, 1978.)

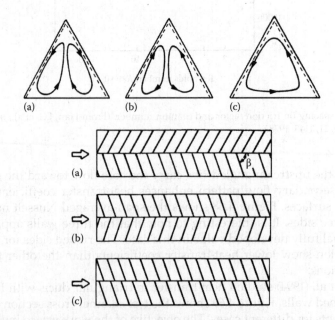

FIGURE 4.66
Schematic of secondary flows developed from different rib configurations used by Metzger and Vedula. (a) Downstream angled ribs, (b) upstream angled ribs, and (c) spiral ribs. (From Metzger, D.E. and Vedula, R.P., *Exp. Heat Transfer*, 1, 31, 1987.)

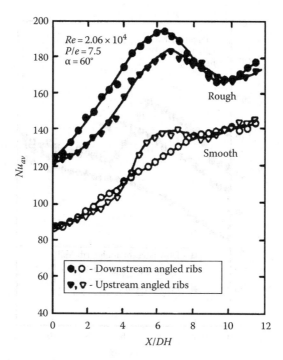

FIGURE 4.67
Effect of rib orientation on average Nusselt number. (From Metzger, D.E. and Vedula, R.P., *Exp. Heat Transfer*, 1, 31, 1987.)

transverse (90°) rib roughness on the walls (case 1, full ribs on three flat walls and on three vertex corners; case 2, ribs were discontinued on the vertex corners). The remaining three channels were roughened by 45° crossed ribs on the walls (case 3, full ribs on three flat walls and on three vertex corners; case 4, ribs were discontinued on three vertex corners; case 5, ribs were discontinued on three corners and removed from one flat wall). Case 6 was a smooth triangular channel that had the same shape and size as the other five ribbed channels. The heat-transfer coefficients on the ribbed flat walls and vertex corners of the triangular channels were obtained, and the heat-transfer coefficients with corresponding pressure drop performances were compared.

Figure 4.70 shows the heat-transfer performance comparison for the ribbed triangular channels over a range of Reynolds numbers. Results show that, in general, the Stanton number ratio (heat-transfer augmentation) increases while increasing the friction factor ratio (pressure drop increment). The Stanton number ratio decreases only slightly with increasing Reynolds number. However, the friction factor ratio is significantly enhanced with the increasing Reynolds numbers. The heat-transfer augmentation for the three-wall partial ribs is about 10% higher than that with the three-wall full ribs. The 90° transverse rib roughness provides 2.0–2.3 times the heat-transfer augmentation on flat walls with 90° corners and 3.6–6.6 times the pressure

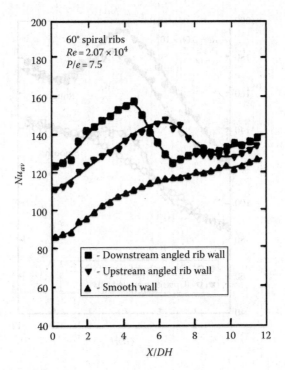

FIGURE 4.68
Average Nusselt number distribution for 60° spiral ribs. (From Metzger, D.E. and Vedula, R.P., *Exp. Heat Transfer*, 1, 31, 1987.)

drop penalty, whereas the 45° crossed rib roughness provides 1.6–1.8 times the heat-transfer augmentation on flat walls with 90° corners and 1.6–2.5 times the pressure drop penalty. The Stanton number ratio on 55° corners for the 90° transverse rib ducts is 1.4–1.7 times, whereas the heat-transfer augmentation on 55° corners for the 45° crossed ribbed ducts is 1.2–1.5 times. The Stanton number ratio on 35° corners is low for both the 90° transverse rib ducts and 45° crossed rib ducts.

Figure 4.71 shows the friction and heat-transfer correlation in ribbed triangular ducts as well as the R function distribution with respect to e^+ for ribbed triangular channels over a range of Reynolds numbers. Based on the rib-roughened channel analysis, the wall similarity laws correlate the friction and heat-transfer data for fully developed turbulent flow in the triangular channels. The friction roughness function for 90° full ribs (case 1) is very close to the previous correlation for 90° transverse ribs in the circular tubes of Webb et al. (1971). The values of R for the three-wall partial ribs (cases 2 and 4) are lower than those for the three full-walled ribs (cases 1 and 3). This lower value of R is because cases 2 and 4 have higher friction factors than those of cases 1 and 3. The effect of the rib angle on R is expressed by Zhang et al. (1994) as $R = 3.2(\alpha/90°)^{-0.95}$ for three-wall full ribs.

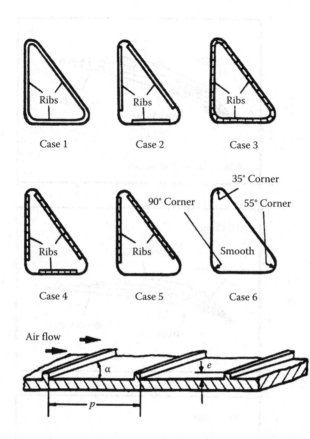

FIGURE 4.69
Schematic of triangular channels with full- and partial-ribbed walls used by Zhang et al. (From Zhang, Y.M. et al., *AIAA J. Thermophys. Heat Transfer*, 8(3), 574, 1994.)

Case No.	Wall Conditions	p/e	e/D	α
1	Three-wall full ribs	9.2	0.012	90°
2	Three-wall partial ribs	9.2	0.012	90°
3	Three-wall full ribs	9.2	0.012	45°
4	Three-wall partial ribs	9.2	0.012	45°
5	Two-wall partial ribs	9.2	0.012	45°
6	Smooth	—	—	—

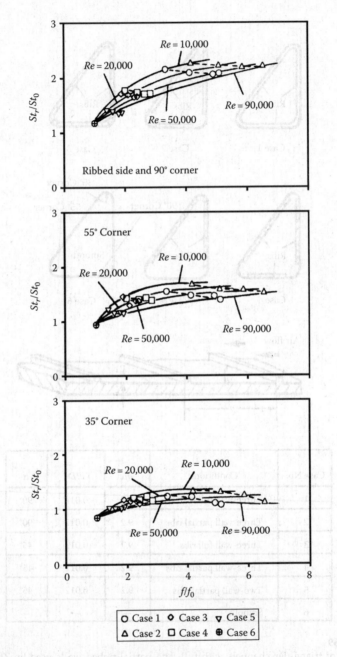

FIGURE 4.70
Heat-transfer performance curves for different ribbed triangular channels. (From Zhang, Y.M. et al., *AIAA J. Thermophys. Heat Transfer*, 8(3), 574, 1994.)

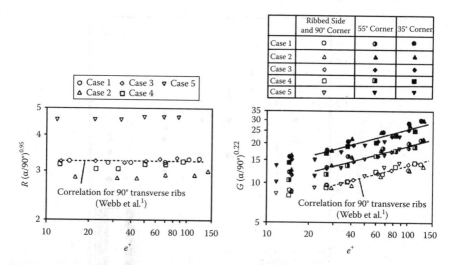

FIGURE 4.71
Friction factor and heat-transfer correlation in triangular ribbed channel. (From Zhang, Y.M. et al., *AIAA J. Thermophys. Heat Transfer*, 8(3), 574, 1994.)

Figure 4.71 also shows the averaged $G(e^+)$ distribution with respect to e^+ for rib-roughened walls over a range of Reynolds numbers. $G(e^+)$ increases with e^+ for all rib configurations. The heat-transfer roughness function for the ribbed side with 90° corner is very close to the previous correlation developed for 90° transverse ribs in circular tubes by Webb et al. (1971). The values of G with 55° and 35° corners are much higher than the ribbed side with 90° corner, which implies that the ribbed side with 90° corner has a higher heat-transfer coefficient than the 55° and 35° corners. The specific correlation equations are provided in Zhang et al. (1994).

Taslim et al. (1997) studied the coolant channel shapes that are relevant for the leading edge of an airfoil. The external contour of an airfoil dictates the shape of the internal coolant channel. Figure 4.72 shows the performance curves for a rounded-side rectangular channel. The narrow side of the channel is 38.1 mm wide, and the ribs are 11.18 mm high, resulting in a blockage ratio of 41.3%. Two types of ribs are used in this study. The J ribs are more streamlined than the I ribs, and J ribs are applied to the left side of the rectangular channel. Two rib configurations—90° (orthogonal) and 60° (tilted)—are presented. The streamlined ribs of the J ribs perform better than the I ribs. Moreover, a tilt in the rib alignment (i.e., 60°) increases heat-transfer performance for the straight sides. The curved side representing the leading edge shows a lower heat-transfer coefficient in the presence of tilt. The performance curve shows a decrease in the Nusselt number to friction factor ratio (plotted on the y axis) with an increase in the flow Reynolds number.

Figure 4.73 shows the performance comparison of triangular ribbed channels. Two test cases are presented: in one, the short side is not ribbed; in the other, the short side is ribbed. Geometry a (rib height of 8.53–8.89)

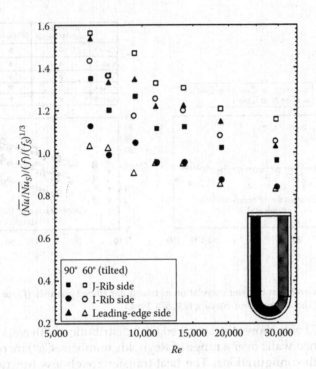

FIGURE 4.72
Thermal performance comparison of 90° and 60° rib arrangements. (Taslim, M.E. et al., *ASME J. Turbomach.*, 119, 601, 1997.)

refers to a higher blockage compared to geometry *b* (rib height of 7.87–8.18). Results indicate that the effect of higher blockage by ribs does not affect the straight sides significantly, but a higher blockage shows a higher heat-transfer coefficient for the leading edge (i.e., tip of the triangular channel) at low Reynolds numbers. Addition of ribs to the short side does not change the heat-transfer coefficients of the sidewalls but increases the leading-edge heat-transfer coefficient significantly at low Reynolds numbers. It can be argued that the addition of ribs in the short side increases flow resistance near that surface and more coolant flows closer to the leading edge. At higher Reynolds numbers, i.e., *Re* > 15,000, effects of geometry (i.e., *a* and *b*) are not reflected in the results.

Spence and Lau (1997) studied turbulated half-circle channels. Their test channel had ribs on the curved surface; the flat surface is left smooth. Figure 4.74 shows the Nusselt number profiles and thermal performance [$TP = Nu/Nu_{ref} \times (f/f_{ref})^{-1/3}$] for a semicircular channel. This thermal performance *TP* compares the heat transfer per unit pumping power for the ribbed segmental channel with that of a tube. Different heat flux ratios Q_r are used, defined as the ratio of heat flux applied to the ribbed side to that of the smooth side. The smooth side performance increases with a decrease in the Q_r, but no

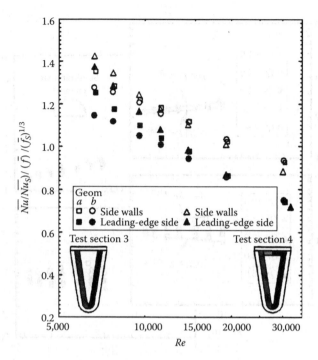

FIGURE 4.73
Thermal performance comparison for test configurations 3 and 4 used by Taslim et al. (Taslim, M.E. et al., *ASME J. Turbomach.*, 119, 601, 1997.)

significant effect is detected on the ribbed curved side. The ribbed curved side shows better thermal performance than that of the flat smooth side.

4.2.7 Effect of High Blockage-Ratio Ribs

Most experiments with ribs use about 10% blockage of the channel by ribs and the pitch-to-rib-height ratio of 10. However, in small gas turbine engines, the rib height can be significantly higher, and more blockages can occur. Moreover, orienting the ribs in an angle-to-flow direction makes it feasible to pack more ribs in the cooling channel with smaller p/e ratios. Metzger et al. (1988) showed that as more ribs are packed per unit surface area, the importance of the heat transfer occurring on the top surface of the rib increases. Figure 4.75 shows the Nusselt number ratio of the rib-top Nusselt number to that of the interrib Nusselt number. It also shows that the Nusselt number ratio increases with increasing p/e ratio for smaller ribs, whereas the Nusselt number ratio is less affected for large blockages ($e/D = 0.25$) and for angled ribs. Metzger et al. also showed the relative contribution of the rib-top Nusselt number for the total average Nusselt number. The contribution of the rib-top Nusselt number increases with a decrease in the p/e ratio.

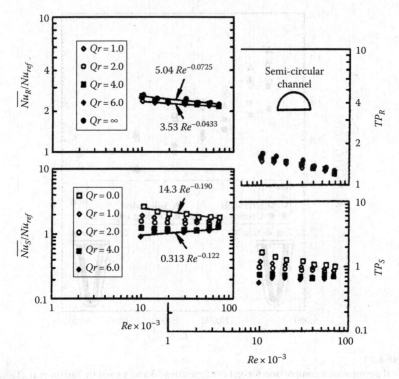

FIGURE 4.74
Effects of wall heat flux ratio on overall Nusselt number and heat-transfer performance in a semicircular channel. (From Spence, R.B. and Lau, S.C., *AIAA J. Thermophys. Heat Transfer*, 11(3), 486, 1997.)

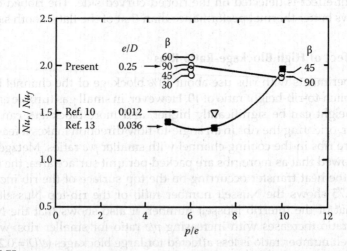

FIGURE 4.75
Effect of different rib arrangements on the average Nusselt number ratio. (From Metzger, D.E. et al., The contribution of on-rib heat transfer coefficients to total heat transfer from rib-roughened surfaces, Personal communication, 1988.)

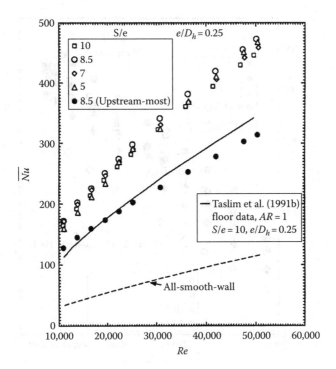

FIGURE 4.76
Effect of rib spacing on average Nusselt number. (From Taslim, M.E. and Wadsworth, C.M., *ASME J. Turbomach.*, 119, 381, 1997.)

Results indicate that the rib-top heat transfer can exceed 50% of the total heat transfer at smaller p/e ratios ($p/e < 6$).

The research group at Northeastern University has done several high-blockage rib studies. Taslim and Wadsworth (1997) studied three rib configurations corresponding to blockage ratios (e/D) of 0.133, 0.167, and 0.25 for pitch-to-height ratios (p/e) of 5, 7, 8.5, and 10. Figure 4.76 shows channel-averaged Nusselt numbers for the e/D ratio of 0.25. Note that the experimental setup had a 3.81 cm × 3.81 cm² cross-sectioned channel. Two opposite surfaces of the channel are rib roughened with a staggered rib arrangement, which means that ribs are not exactly on top of each other. Results for this high-blockage rib study indicate that variation of the average Nusselt number is not strongly dependent on the rib spacing. The upstream ribs have a lower heat-transfer coefficient that matched the nonstaggered rib data. The higher values of the present study indicate that the staggered rib arrangement enhances flow mixing; therefore heat transfer is increased.

Taslim and Lengkong (1998a) studied the heat-transfer coefficient in a channel with high-blockage staggered ribs placed at a 45° angle. Figure 4.77 shows the channel-averaged Nusselt number for different rib spacing with an e/D ratio of 0.25. Results show that the effect of rib spacing is more pronounced

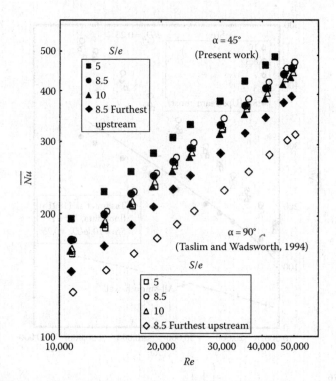

FIGURE 4.77
Effect of rib angle and rib spacing on average Nusselt number. (From Taslim, M.E. and Lengkong, A., 45° round-corner rib heat transfer coefficient measurements in a square channel, *International Gas Turbine and Aeroengine Congress and Exhibition*, Stockholm, Sweden, ASME Paper 98-GT-176, June 2–5, 1998a.)

in the 45° orientation. A closer rib spacing of 5 rib heights shows the highest heat-transfer coefficient distribution in the group. Figure 4.78 shows the channel-averaged friction factors for this rib configuration. A closer spacing increases the friction factor in this tilted rib, but for larger spacing (*S/e* = 8.5 and 10), the friction factor is less for the angled-rib orientation. The variation of the friction factor with Reynolds number is not significant with these ribs.

Taslim and Lengkong (1998b) studied the staggered 45° ribs with smaller blockage of *e/D* = 0.167. In general, the heat-transfer coefficients are higher than 90° ribs of similar arrangement. The friction factors of these ribs are presented by Taslim and Lengkong (1998b), and these friction factors for both *S/e* of 8.5 and 10 stayed close to 0.4, whereas the friction factor for *S/e* of 5 was highest in the group, with a value of 0.7.

4.2.8 Effect of Rib Profile

Figure 4.79 shows different rib shapes used by Han et al. (1978). The sharpness of the rib edges is eliminated by adding clay on the sides of the square

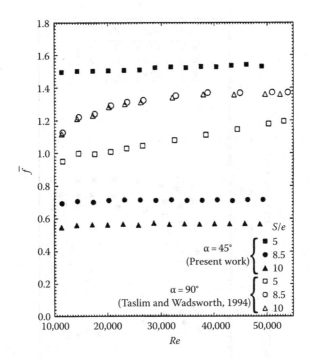

FIGURE 4.78
Effect of rib angle and rib spacing on friction factor. (From Taslim, M.E. and Lengkong, A., 45° round-corner rib heat transfer coefficient measurements in a square channel, *International Gas Turbine and Aeroengine Congress and Exhibition*, Stockholm, Sweden, ASME Paper 98-GT-176, June 2–5, 1998a.)

cross-sectioned ribs. Figure 4.79 also shows the friction factor and Stanton number resulting from different rib shapes arranged orthogonal to the flow. Results indicate that the rib geometry effect is more prominent at low Reynolds numbers. At lower Reynolds numbers, the Stanton number decreases by decreasing the sharpness of the rib profile. At higher Reynolds numbers, the Stanton numbers are the same for sharp and smooth ribs. However, the friction factor can be reduced up to 30% for smooth ribs at the studied Reynolds numbers.

Figure 4.80 shows the temperature distributions observed by Liou and Hwang (1993) in different shaped ribs. The mean flow direction is from left to right. The thermal boundary layer upstream of the triangular rib is observed to be thinner than the other two configurations. A real-time holographic interferometer is used to measure the temperature distribution in the air flow. A 3 W argon laser (514.5 nm wavelength) provided the coherent source of light required for interference. Air temperature was also checked by probing thermocouples in the air flow. A total of 30 copper-constantan thermocouples provided the ribbed-wall temperature. Holograms are taken

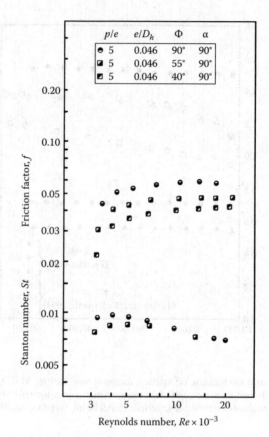

FIGURE 4.79
Different rib shapes and their effect on Stanton number and friction factor used by Han et al.
(From Han, J.C. et al., *Int. J. Heat Mass Transfer*, 21, 1143, 1978.)

at a location where both hydrodynamic and thermal fully developed conditions are achieved. The ribbed walls are heated with uniform wall heat flux.

Comparison of Nusselt number ratios for these rib geometries is shown in Figure 4.81. The local Nusselt number increases rapidly in the upstream face of ribs. The square cross-sectioned rib shows the highest Nusselt number upstream of the rib, followed by the lowest Nusselt number ratio at the downstream location. The semicircular cross section has the most uniform Nusselt number ratio distribution. The maximum local heat-transfer coefficients for all ribs are at the points of flow separation. The highest average Nusselt number ratio was obtained with the square cross section, and the lowest average Nusselt number ratio was obtained with the semicircular cross section.

Liou and Hwang (1993) studied the performance comparison of different shaped ribs. Figure 4.82 shows the performance comparison of the triangular, semicircular, and square-shaped ribs. Since pumping power is

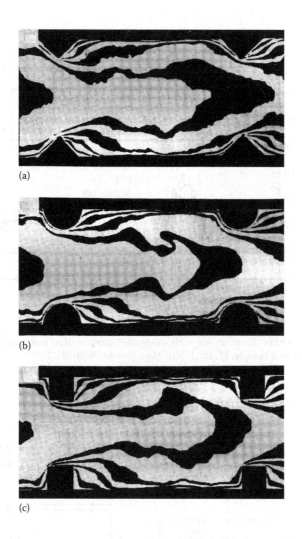

FIGURE 4.80
Temperature distributions in different shaped ribs using holographic interferograms. (a) Triangular rib, and (b) semicircular rib, and (c) square rib. (From Liou, T.M. and Hwang, J.J., *Int. J. Heat Mass Transfer*, 36(4), 931, 1993.)

proportional to $f \, Re^3$, this figure plots $(f \, Re^3)^{1/3}$ on the x axis. It shows the Nusselt number enhancement ratio for a given pumping power, i.e., performance. The plot shows that the rib shape does not have much effect on the performance. In general, the performance distribution stays above the Nusselt number ratio of 1. Figure 4.82 also shows the friction factor and Nusselt number correlation based on the method proposed by Han (1988). The technique of using the G and R functions has been discussed earlier in this section. A semicircular shape shows highest G and R functions, and a square shape shows the lowest.

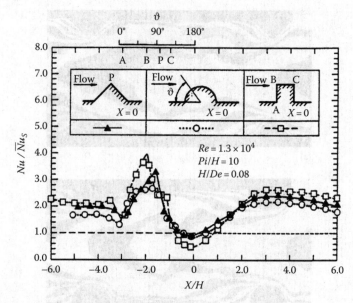

FIGURE 4.81
Local Nusselt number distributions with triangular, semicircular, and square-shaped ribs. (From Liou, T.M. and Hwang, J.J., *Int. J. Heat Mass Transfer*, 36(4), 931, 1993.)

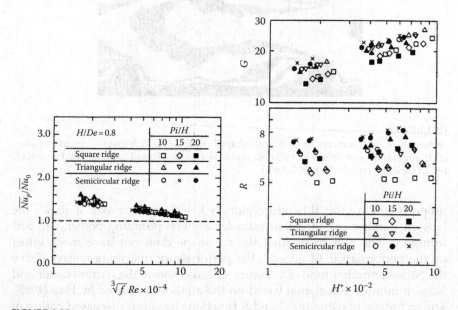

FIGURE 4.82
Friction factor and heat-transfer correlation with various rib shapes. (From Liou, T.M. and Hwang, J.J., *Int. J. Heat Mass Transfer*, 36(4), 931, 1993.)

TABLE 4.3

Rib Types and Test Configurations Used by Taslim and Spring

Test No.	e	w	p/e	e/D	AR	e/w	Remarks
1	0.5	0.5	5	0.235	0.55	1	Sharp corners
2	0.5	0.25	5	0.235	0.55	2	Sharp corners
3	0.44	0.44	5	0.22	0.5	1	Sharp corners
4	0.44	0.44	7.5	0.22	0.5	1	Sharp corners
5	0.44	0.44	10	0.22	0.5	1	Sharp corners
6	0.5	0.5	5	0.235	0.55	1	Round corners
7	0.5	0.25	5	0.235	0.55	2	Round corners
8	0.5	0.375	5	0.235	0.55	1.33	Trapezoidal
9	0.57	0.57	7.7	0.285	0.5	1	Maximum tolerance
10	0.31	0.57	14.2	0.155	0.5	0.55	Minimum tolerance
11	0.585	0.585	4.27	0.275	0.55	1	Maximum tolerance
12	0.44	0.585	5.68	0.207	0.55	0.77	Minimum tolerance
13	0.36	0.25	6.94	0.17	0.55	1.45	Round corners
14	0.3	0.3	10	0.15	0.5	1	Round top corners
15	0.3	0.48	10	0.15	0.5	0.625	Round top corners

Source: Taslim, M.E. and Spring, S.D., *AIAA J. Thermophys. Heat Transfer*, 8(3), 555, 1994.

The rib profiles used in experiments are mostly sharp edged and have nice smooth profiles. But that type of perfect ribs is difficult to fabricate. The casting of coolant channel ribs introduces edge roundness and base fillets. Taslim and Spring (1994) studied the effects of rib profile and spacing on heat transfer and friction. Their rib specifications are shown in Table 4.3.

Figure 4.83 shows the effects of fillets and corner radii on heat transfer and friction. This figure shows the effect on Nusselt number for different tested configurations. The effect of rounding decreases *Nu* for both configurations plotted. The higher-aspect ratio turbulator is more sensitive to the rounding effects, with a 17% reduction, whereas the lower-aspect ratio turbulator (*AR* = 1) shows only 5% reduction in Nusselt number with corner radii and fillets. The effect on the friction factor by edge smoothness for high-aspect ratio ribs is also shown in Figure 4.83. The friction factor decreases with rib roundness. It is argued by Taslim and Spring (1994) that the higher rib aspect ratio provides more space between turbulators for the flow to reattach, thus allowing more sensitivity to the rounded turbulator profile, whereas the lower rib aspect ratio does not provide sufficient space between turbulators for the flow to reattach, resulting in reduced sensitivity to rib profile.

Figure 4.84 shows the average Nusselt number and friction factors for different blockage ribs with sharp and rounded edges. The Nusselt numbers are comparable with two different corner shapes, but the friction factor is significantly higher with sharp edges. Especially for a high blockage rib (*e/D* = 0.25), the friction factor increases to more than double for sharp-edged

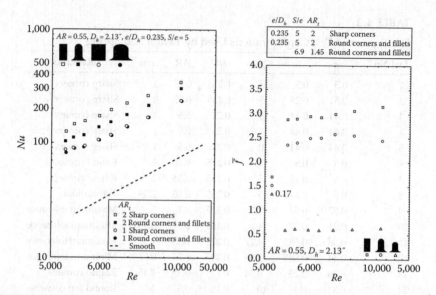

FIGURE 4.83
Effects of fillets and corner radii on heat-transfer coefficient and friction factor. (From Taslim, M.E. and Spring, S.D., *AIAA J. Thermophys. Heat Transfer*, 8(3), 555, 1994.)

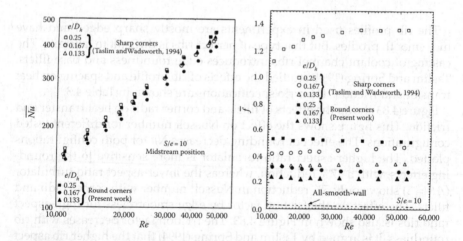

FIGURE 4.84
Effects of blockage ratios on average Nusselt number and friction factor with sharp-edged and rounded-cornered ribs. (From Korotky, G.J. and Taslim, M.E., *ASME J. Turbomach.*, 120, 376, 1998.)

ribs compared to rounded ribs. Results from other rib spacing indicate that the Nusselt number distribution is less sensitive to the edge condition than the friction factor, and the friction factor for the high blockage rib is most affected by the edge conditions. Sharp-edged ribs show a higher friction factor than round-edged ribs. Figure 4.85 summarizes the average Nusselt

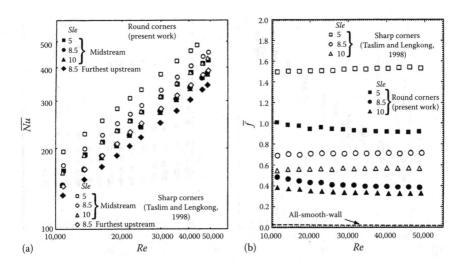

FIGURE 4.85
Effect of rib spacing on average Nusselt number and friction factor with sharp-edged and rounded ribs. (a) Round corners (present work) and (b) sharp corners. (From Taslim, M.E. and Lengkong, A., 45° round-corner rib heat transfer coefficient measurements in a square channel, *International Gas Turbine and Aeroengine Congress and Exhibition*, Stockholm, Sweden, ASME Paper 98-GT-176, June 2–5, 1998a.)

number and friction factors for the high-blockage rib. Results indicate that the sharp-edged ribs show increased Nusselt number and friction factors compared to corresponding round-edged ribs. Figure 4.86 summarizes the results presented in Taslim and Lengkong (1998a) for all blockage and pitch-to-height ratios. The highest friction factor is observed to be with $e/D = 0.25$ and $s/e = 5$, and the lowest friction factor is obtained with $e/D = 0.133$ and $s/e = 10$.

Chandra et al. (1997b) studied different rib shapes in a two-sided square channel. Figure 4.87 shows different rib profiles used by them and the Nusselt number ratios for the ribbed and smooth sides with friction factor ratios. Sharp-edged square ribs show the best performance in the group; circular-sectioned ribs show the worst performance. The rib heat-transfer correlation and friction correlation based on G and R functions are determined. Results show that the correlation lines have slopes similar to those observed for square-sectioned ribs. However, the variation with rib cross-section is less detectable in these plots than raw heat transfer and friction factors plotted in the earlier figure.

4.2.9 Effect of Number of Ribbed Walls

Since ribs alter the flow pattern, the overall flow distribution and turbulence characteristics may be affected in the presence of ribs. It is possible to influence the heat-transfer coefficient of a smooth wall by adding ribs to a neighboring

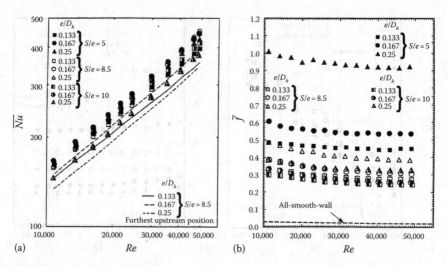

(a)   *Re*   (b)   *Re*

FIGURE 4.86
Average Nusselt numbers and friction factor for different rib arrangements tested by Taslim and Lengkong. (a) Average Nusselt number and (b) friction factor. (From Taslim, M.E. and Lengkong, A., 45° round-corner rib heat transfer coefficient measurements in a square channel, *International Gas Turbine and Aeroengine Congress and Exhibition*, Stockholm, Sweden, ASME Paper 98-GT-176, June 2–5, 1998a.)

wall. Chandra and Cook (1994) studied the effects of the number of ribbed walls on heat transfer and friction characteristics in a square channel. Figure 4.88 shows the different rib configurations used by them. The objective of this work was to investigate the effect of the number of ribbed walls on wall center-line average values of the heat-transfer coefficient. Since addition of a rib to one surface can change the entire flow distribution and turbulence level, smoothness or roughness of the neighboring walls can influence the heat-transfer pattern of a surface. Figure 4.88 also shows that case A was all smooth walls in a square channel, case B was ribs on one wall, case C was two opposite ribbed walls, case D was three ribbed walls, and case E was all four ribbed walls.

Figure 4.88 also shows the Nusselt number ratio variations with respect to the friction factor ratio for smooth and ribbed sides of the channel. As indicated before, this type of plot is also referred to as the heat-transfer performance curve. Results show that the Nusselt number ratio decreases, while the friction factor ratio increases with increasing Reynolds numbers. Case C, with transverse ribs on two opposite walls, provides 2.64–1.92 times the ribbed-side heat-transfer enhancement and 4.35–6.29 times the pressure drop increment, respectively, with increasing Reynolds number; whereas case E, with four ribbed surfaces, enhances the ribbed surface heat transfer by 2.99–2.12 times and increases the pressure drop by 7.96–11.45 times. The effect of ribbed surfaces on heat transfer decreases with increasing Reynolds number. Unlike heat transfer, the friction factor is higher for higher Reynolds numbers. It is interesting to note that the heat-transfer coefficient and the pressure

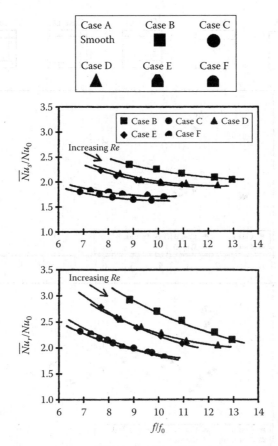

FIGURE 4.87

Heat-transfer performance comparison for different rib shapes studied by Chandra et al. (From Chandra, P.R. et al., *J. Thermophys. Heat Transfer*, 12(1), 116, 1997a.)

drop increase with an increase in the number of ribbed walls. This increase in heat-transfer coefficient is noted in both the smooth and ribbed walls. The smooth wall shows an increase of about 13% in heat-transfer coefficient for each additional ribbed surface, whereas the ribbed surface shows about a 10% increase in the heat-transfer coefficient with each additional ribbed surface.

Chandra and Cook (1994) also showed the friction factor and heat-transfer correlation for different ribbed-surface configurations. The wall similarity laws were used to correlate the friction and heat-transfer data for fully developed turbulent flow in a square channel with one, two, three, and four ribbed walls. The results show that the heat-transfer roughness function $G(e^+)$ increases with increasing roughness Reynolds number e^+ for all rib configurations. It can also be observed that the value of the heat-transfer roughness function decreases with each additional ribbed wall. Note that a decrease in $G(e^+)$ means an increase in heat transfer. T_r and T_i in the correlated equations

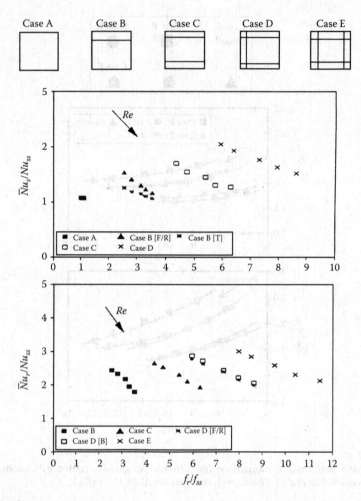

FIGURE 4.88
Effect of the number of ribbed walls on heat-transfer performance. (From Chandra, P.R. and Cook, M.M., Effect of the number of channel ribbed walls on heat transfer and friction characteristics of turbulent flows, General papers in heat and mass transfer, ASME HTD-Vol. 271, pp. 201–209, 1994.)

are the total width of the ribbed wall and total combined width of the channel walls, respectively.

Chandra et al. (1997a) studied the effect of ribbed walls on a rectangular channel. Similar to their square-channel configuration, they varied the number of ribbed walls. Figure 4.89 shows a schematic of the different cases studied by them. Figure 4.90 shows the ribbed-side and smooth-side Nusselt number ratios versus friction factor ratios. Both Nusselt number and friction factor increase with an increasing number of ribbed walls. It is interesting to note that the ribbed-side heat transfer shows a better performance when only one of the wider sides is rib roughened (case B) compared to when both

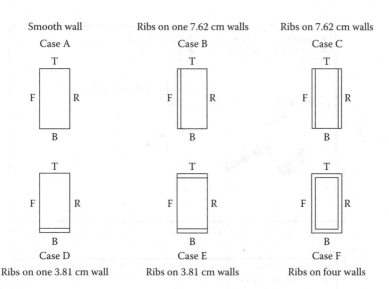

FIGURE 4.89
Different ribbed wall conditions studied by Chandra et al. (From Chandra, P.R. et al., *ASME J. Turbomach.*, 119, 374, 1997b.)

short sides are rib roughened (case E). The aspect ratio of the channel is 1:2, which means that the length of the ribbed sides is the same for cases E and B. Chandra et al. also presented the *G* and *R* correlation for these partially ribbed channels.

Taslim et al. (1998) measured heat-transfer coefficient and friction factors in partially ribbed passages. A liquid crystal technique was used, and channel cross sections were selected to represent the midchord section of an airfoil. Full-length ribs are used on the principal heat-transfer walls (i.e., suction and pressure sides). The divider or partitioning walls are roughened with half-length ribs. Several tests are run as listed in Table 4.4.

Figure 4.91 shows the average Nusselt numbers for different test cases. An increase in the number of ribbed walls results in an increase in the heat-transfer coefficient. This figure combines the results of five rib arrangements in the square channel. Rib profiles used for these test conditions simulate the effect of die wear-out during the manufacturing process. Rib shapes created with a new die differ from the shapes created by a worn die. The rib corners get rounded due to the gradual wearing of the core die. A decrease in the Nusselt number ranging from 1% to 7% is noted due to the shape change of ribs. A liquid crystal technique was used to get detailed Nusselt number profiles. The Nusselt number contours showed a significant change with the addition of half ribs on the partition walls. Note that two different staggered arrangements were studied. Results indicated that the addition of partial ribs significantly improved the heat transfer from the major heat-transfer surfaces (that have full-length ribs). Addition of partial ribs increased average wall heat-transfer coefficients by about 40%.

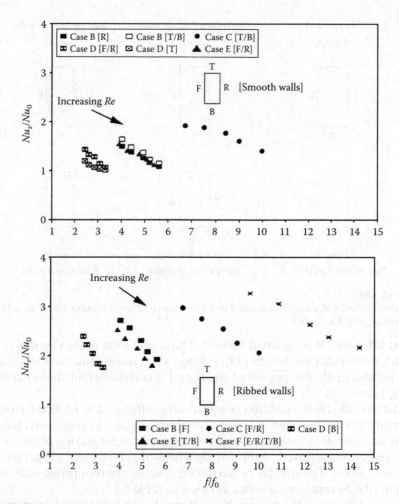

FIGURE 4.90

Heat-transfer performance variations with number of ribbed walls. (From Chandra, P.R. et al., *ASME J. Turbomach.*, 119, 374, 1997b.)

Figure 4.92 shows the heat-transfer coefficients and friction factors for a trapezoidal channel. Results from test 9, i.e., the trapezoidal test section with full-length ribs on two opposite primary walls, serve as the base line. Similar to that deserved in square channels, a nearly 20% increase in the primary wall heat-transfer coefficient is observed with the addition of half-length ribs on the bases of the trapezoidal channel. For this channel, roughening both the partition walls does not influence the heat-transfer coefficient significantly, compared to the results with one rib roughened partition wall. Figure 4.92 also shows the friction factors for the trapezoidal channel; results indicate that friction factors are more affected by the addition of ribs to the partition walls of the trapezoidal channel. Figure 4.93 shows the performance

TABLE 4.4

Test Configurations Used by Taslim et al.

Test No.	e/D	S/e	AR of Channel	AR of Rib	Fillet Radius/e	No. of Walls Ribbed
1	0.218	8.5	Square	1	0.0	2
2	0.218	8.5	Square	1	0.191	2
3	0.218	8.62	Square	1.47	0.191	2
4	0.218	8.5	Square	1	0.191	4
5	0.218	8.5	Square	1	0.191	4
6	0.203	9.11	Square	0.87	0.256	4
7	0.233	7.97	Square	1	0.224	4
8	0.218	8.5	Square	1	0.191	4
9	0.218	8.5	Trapezoid	1	0.164	2
10	0.218	8.5	Trapezoid	1	0.164	3
11	0.218	8.5	Trapezoid	1	0.164	4
12	0.202	9.16	Trapezoid	0.87	0.212	4

Source: Taslim, M.E. et al., *ASME J. Turbomach.*, 120, 564, 1998.

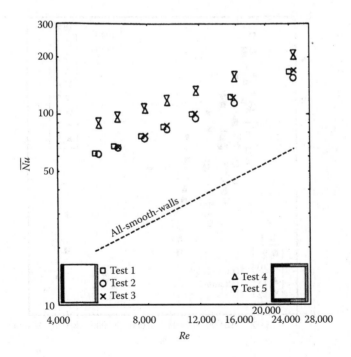

FIGURE 4.91

Comparison of average Nusselt numbers with different rib configurations used by Taslim et al.
(From Taslim, M.E. et al., *ASME J. Turbomach.*, 120, 564, 1998.)

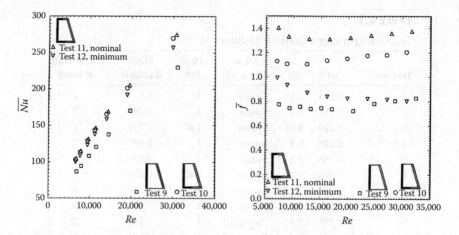

FIGURE 4.92

Comparison of Nusselt numbers and friction factors for four different ribbed channels used by Taslim et al. (From Taslim, M.E. et al., *ASME J. Turbomach.*, 120, 564, 1998.)

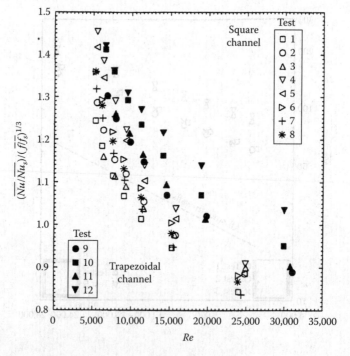

FIGURE 4.93

Heat-transfer performance comparison for all ribbed channels used by Taslim et al. (From Taslim, M.E. et al., *ASME J. Turbomach.*, 120, 564, 1998.)

comparison of the tests performed by Taslim et al. (1998). Test 12 shows a consistently higher performance for the Reynolds number range tested. Tests 4, 5, 10, and 11 have half-length ribs on partition walls. These geometries not only produce higher heat-transfer coefficients compared to their corresponding baseline cases without ribs on partition walls (tests 1 and 9), but they also have a superior thermal performance.

4.2.10 Effect of a 180° Sharp Turn

So far the discussion has been about the effects of ribs on heat transfer in straight channels. In a turbine blade, there are sharp 180° turns in the internal coolant passage. Some discussion on the effect of turn on heat transfer is provided here. Schabacker et al. (1998) used stereoscopic particle image velocimetry (PIV) to capture the time mean velocity and turbulence in a sharp bend. Figure 4.94 shows the time-averaged velocity field in the bend at the midplane. The separated flow is clearly seen in the downturn region. They also presented the small vortices created at corners. The midplane vortices are different from those observed near sidewalls. Figure 4.95 shows the secondary flow pattern developed by the bend. A double vortex structure, similar to a curved-channel flow, is recorded in the bend. Figure 4.96 shows the decay of the secondary flow downstream of the bend. This secondary flow promotes flow mixing, resulting in a heat-transfer enhancement.

In another turbulent flow analysis, Metzger et al. (1984) measured the pressure drop in a sharp turn. They varied the channel widths W_1 and W_2, as shown in Figure 4.97. The corner radius R is also varied to get pressure drop results that are shown by the pressure loss coefficient, $K = 2\Delta P/pV^2$. The test facility is designed to be assembled from interchangeable components to get a large range of geometrical combinations. The total width—i.e., the combined width of the two channels—is 7 cm. Dried air from the compressor is used

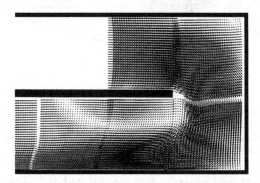

FIGURE 4.94
Flow distribution in a bend region. (From Schabacker, J. et al., PIV investigation of the flow characteristics in an internal coolant passage with two ducts connected by a sharp 180° bend, *International Gas Turbine & Aeroengine Congress & Exposition*, Stockholm, Sweden, ASME Paper 98-GT-544, June 2–5, 1998.)

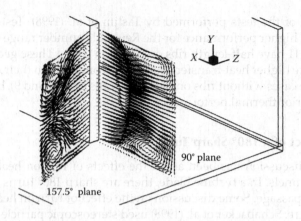

FIGURE 4.95
Secondary flow in a turn region. (From Schabacker, J. et al., PIV investigation of the flow characteristics in an internal coolant passage with two ducts connected by a sharp 180° bend, *International Gas Turbine & Aeroengine Congress & Exposition*, Stockholm, Sweden, ASME Paper 98-GT-544, June 2–5, 1998.)

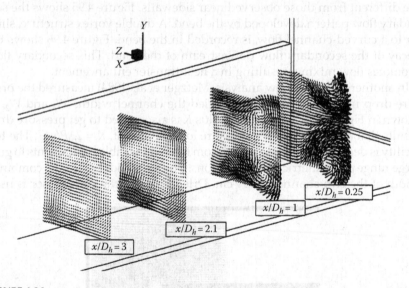

FIGURE 4.96
Secondary flow downstream of a bend. (From Schabacker, J. et al., PIV investigation of the flow characteristics in an internal coolant passage with two ducts connected by a sharp 180° bend, *International Gas Turbine & Aeroengine Congress & Exposition*, Stockholm, Sweden, ASME Paper 98-GT-544, June 2–5, 1998.)

for the experiment, and pressure taps are located at 2.54 cm intervals along both upstream and downstream of the turn region. They showed the typical normalized pressure distribution. There is significant activity in the bend, as characterized by the variations in the pressure measurement. Results indicate that the effect of $D^* = D/(W_1 + W_2)$ does not have a significant influence

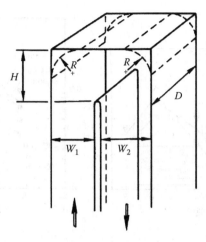

FIGURE 4.97
Schematic of a two-pass channel used by Metzger et al. (From Metzger, D.E. et al., Pressure loss through sharp 180° turns in smooth rectangular channels, ASME Paper 84-GT-154, 1984.)

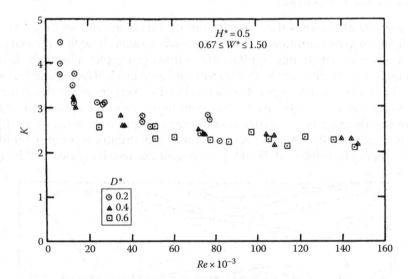

FIGURE 4.98
Loss coefficient K distribution in a bend. (From Metzger, D.E. et al., Pressure loss through sharp 180° turns in smooth rectangular channels, ASME Paper 84-GT-154, 1984.)

on the pressure drop. Figure 4.98 shows the pressure drop coefficient variations with the flow Reynolds number. The pressure drop coefficient K is higher at smaller D^*, but the effect is not significant.

Fan and Metzger (1987) studied heat transfer in a turn region. Figure 4.99 shows the Nusselt numbers for a two-pass channel. Region 3 is the turn region, region 1 is the inlet, and region 5 is the outlet of the two-pass channel. Results show that the peak Nusselt number occurs downstream of the turn.

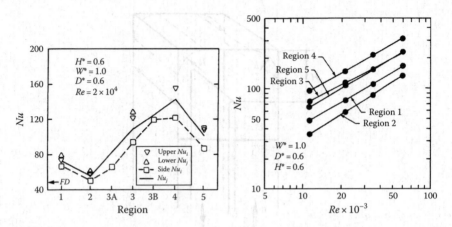

FIGURE 4.99
Nusselt number distribution in a turn region with different flow Reynolds numbers. (From Fan, C.S. and Metzger, D.E., Effects of channel aspect ratio on heat transfer in rectangular passage sharp 180° turns, *Gas Turbine Conference & Exhibition*, Anaheim, CA, ASME Paper 87-GT-113, May 31–June 4, 1987.)

Figure 4.99 also shows the Nusselt number variations in these regions for different Reynolds numbers. The heat-transfer results show that the slope of the lines does not change significantly in this log–log plot. However, different regions have different heat-transfer coefficient levels. The highest Nusselt number is obtained at region 4, which is the downstream of the sharp turn.

Ekkad and Han (1995) used a transient liquid crystal technique to measure the detailed heat-transfer coefficient profile in a smooth, sharp 180° turn. Figure 4.100 shows the Nusselt number distribution contour profiles for a Reynolds number of 25,000. The Nusselt number increases as the flow

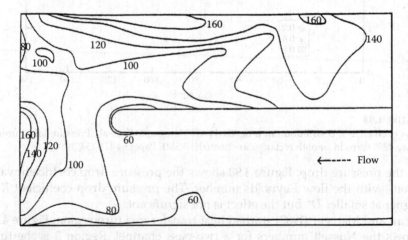

FIGURE 4.100
Detail Nusselt number distribution in the turn region for *Re* = 25,000. (From Ekkad, S.V. and Han, J.C., *J. Flow Vis. Image Process.*, 2, 285, 1995.)

approaches the turn. The highest Nusselt numbers are observed near the outer wall away from the divider. There exists a low heat-transfer region in the upper top-left corner of the turn region. This low heat-transfer region is perhaps due to corner flow entrapment. Nusselt numbers are higher further downstream in the middle of the channel due to reattachment of flow after a turn. A higher Reynolds number flow was also investigated. Since higher flow rate has higher convection effects, the turn effect was convected further by this higher flow rate. In higher Reynolds number flow, the fully developed profile in the second pass (after turn) was reached at a longer distance compared to that with lower Reynolds number flow.

Han et al. (1988) studied the mass transfer in a two-pass channel. Figure 4.101 shows their Sherwood number ratios in a smooth two-pass channel. As seen in earlier plots, the mass-transfer coefficient is highest just after the turn. It is interesting to note that the results from the three measurement

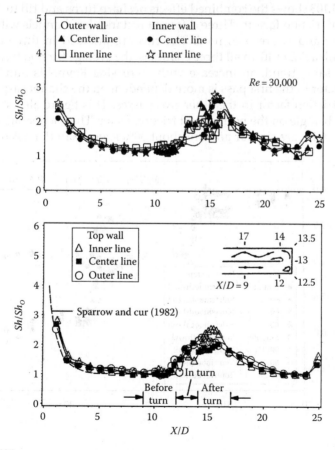

FIGURE 4.101

Local Sherwood number ratios for $Re = 30,000$ in a sharp-turn configuration studied by Han et al. (From Han, J.C. et al., *ASME J. Heat Transfer*, 110, 91, 1988.)

locations along the channel axis show very similar Sherwood number distribution in the straight channels. The three-dimensionality of the flow in the turn region is apparent from the differences in the Sherwood numbers at the same x/D locations.

Han and Zhang (1989) presented the friction factor and pressure drop in a three-pass channel. Their channel consisted of three straight sections connected by two sharp turns. They showed the pressure drop in the three-pass channel with and without rib-roughened wails. In the first pass, the pressure drop is almost linear until the turn, and a major pressure drop is noted in the turn region for both ribbed and smooth channels. Some pressure recovery is observed just downstream of the turn. The major pressure drop contribution in a smooth channel is the turn effect, whereas the ribbed channel shows the pressure drops along the channel due to turbulence enhancement by the ribs. Figure 4.102 shows the effect of rib angle on the pressure drop. The pressure drop in the ribbed channel is significantly greater than the smooth channel.

Figure 4.103 shows the combined effects of sharp turns and rib turbulators on average friction factors. Three different rectangular channels with aspect ratios 1, ½, and ¼ are presented. The ribs are oriented in three different angles, with a P/E of 10. In all three channels, the average friction factor ratios show the same trend: an increase with increasing Reynolds number. The friction factor in the first pass is more dependent on the channel aspect ratio than the friction factor in the other two passes. This figure also shows the effect of rib angle on the normalized friction factor. The friction factors with $\alpha = 90°$ and $45°$ are similar and are about 30% lower than that with $\alpha = 60°$.

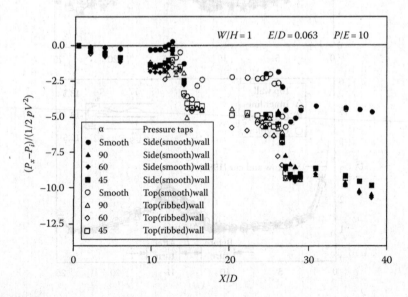

FIGURE 4.102
Effect of rib angle on friction factor at $Re = 30,000$. (From Han, J.C. and Zhang, P., *ASME J. Turbomach.*, 111, 515, 1989.)

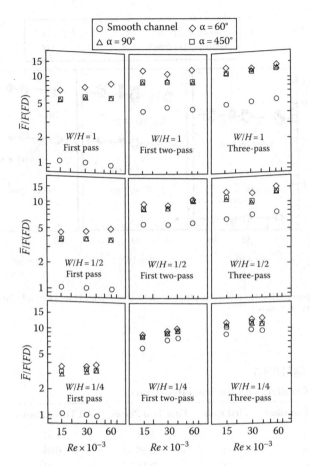

FIGURE 4.103
Average friction factors in multipass channels. (From Han, J.C. and Zhang, P., *ASME J. Turbomach.*, 111, 515, 1989.)

The effect of the rib orientation is more prominent in the first pass. In the second and third passes, the rib orientation effect is diluted by the mixing caused by two turns. The difference of the friction factor ratios between the angled ribs and transverse ribs is gradually diminished when the channel aspect ratio is changed from 1 to ¼ by keeping the same rib pitch and height in each of the channels. Figure 4.104 shows the friction factor correlation, which can be expressed in the general form as

$$\frac{\overline{F}/F(FD)}{(W/H)^{c}[(E/D)/0.063]^{d}[(P/E)/10]^{m}(\alpha/60°)^{n}} = aRe^{b} \qquad (4.20)$$

Empirically determined constants are given in Tables 4.5 and 4.6 for the rib-roughened and smooth channels.

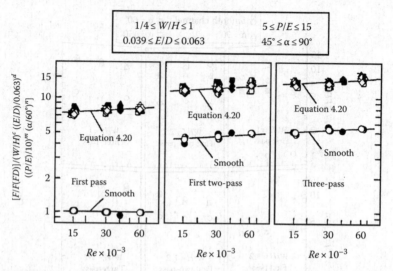

FIGURE 4.104

Friction factor correlation for ribbed multipass channels. (From Han, J.C. and Zhang, P., *ASME J. Turbomach.*, 111, 515, 1989.)

TABLE 4.5

Constants for the Smooth Channel

Constant	First Pass	First Two Passes	All Three Passes
a	1	2.02	2.32
b	0	0.08	0.08
c	0	−0.31	−0.41

TABLE 4.6

Constants for the Rib-Roughened Channel

Constant	First Pass	First Two Passes	All Three Passes
a	3.38	5.48	5.92
b	0.08	0.08	0.08
c	0.15	0.05	0.09
d	1.05	0.35	0.05
m	−0.45, $p/e \geq 10$	−0.27	−0.2
	0.21, $p/e < 10$	0.12	0.12
n	−0.7, $\alpha \geq 60°$	−0.5	−0.35
	0.8, $\alpha < 60°$	0.65	0.3

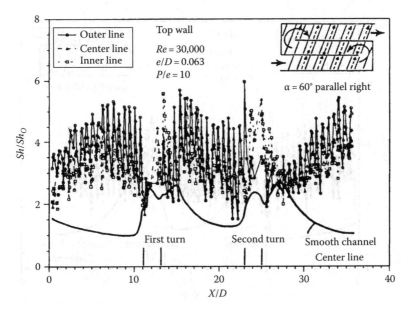

FIGURE 4.105
Local Sherwood number ratio for a 60° inclined ribbed channel with parallel ribs toward the left direction. (From Han, J.C. and Zhang, P., *ASME J. Turbomach.*, 113, 123, 1991.)

Han and Zhang (1991) studied the effect of rib angle orientation on local mass-transfer distribution in a three-pass rib-roughened channel. It was observed that the rib angle, rib orientation, and the sharp 180° turn significantly affected the local mass-transfer distributions. The combined effects of these parameters increased or decreased the mass-transfer coefficients after the sharp 180° turns. The angled ribs, in general, provided higher mass-transfer coefficients than the transverse ribs, and parallel ribs gave higher mass transfer than the crossed ribs.

Figures 4.105 and 4.106 show that the local Sherwood number ratios vary with different rib orientations. Figure 4.105 shows the distributions of the ribbed wall Sherwood number ratios along three different axial lines with $\alpha = 60°$ parallel ribs aligned toward the left. In the first pass, the ribbed wall Sherwood number ratios in different spanwise locations are comparable to those of the periodic Sherwood number distribution. However, the Sherwood number ratios along the outer line are always higher than the corresponding values along the inner line. It is conjectured by Han and Zhang (1991) that these spanwise variations are caused by the secondary flow developed along the rib axes from the outer line toward the inner line. The Sherwood number ratios immediately after the second 180° turn are much lower than those in the second pass of the channel. This may be caused by the main flow getting aligned with the rib axis. That means the ribs get aligned as longitudinal not orthogonal to the flow, and longitudinal ribs can no longer disturb the boundary layer. The Sherwood number ratios increase gradually

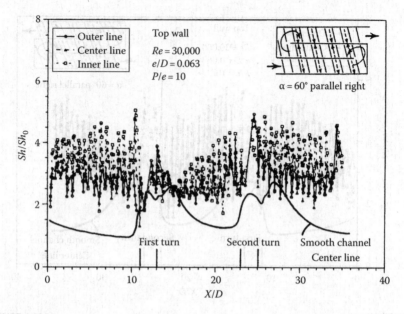

FIGURE 4.106
Local Sherwood number ratio for a 60° inclined ribbed channel with parallel ribs toward the right direction. (From Han, J.C. and Zhang, P., *ASME J. Turbomach.*, 113, 123, 1991.)

when the flow redevelops at the end of the third pass. Figure 4.106 shows the results for $\alpha = 60°$ parallel ribs aligned toward the right-hand direction. With this fib orientation, low Sherwood number ratios are observed after the first 180° turn. This decrease in Sherwood number is also due to the longitudinal alignment of ribs in the turn region. Therefore, the turn direction and rib orientation can cause a reduction in mass transfer. This study shows that care needs to be taken in rib alignment in the turn regions; guidance in that respect is provided in the discussed results.

4.2.11 Detailed Heat-Transfer Coefficient Measurements in a Ribbed Channel

So far, the results presented were more of the application type, and correlations were developed to aid heat-transfer analysts. More detailed measurements in the interrib region are also available in the literature and are presented here. These results provide an insight into the interaction of flow and heat-transfer process. Detailed results help to locate hot spots and provide guidance for effective rib arrangement. Kukreja et al. (1993) published detailed measurements in the interrib region of adjacent ribs to analyze the process of separation and reattachment in detail. They used a naphthalene sublimation technique and used both 90° transverse ribs and parallel and V-shaped continuous ribs. The reattachment of flow past ribs was clearly identified by this work. Though there are several fundamental studies on rib-turbulated surfaces (see Acharya et al., 1993, 1994), we shall address results that are related to the turbine applications. Acharya et al. (1997) reported detailed measurement of the velocity, temperature, and heat transfer in both developing and periodically developed regions of a ribbed duct. A transitional flow of $Re = 3,400$ and a fully turbulent flow at $Re = 24,000$ were studied. Among important observations, a sudden decrease in turbulence levels was noted in the vicinity of reattachment.

Abuaf and Kercher (1992) studied heat transfer and turbulence in a turbulated cooling channel. They used a model built with Plexiglas magnified 10 times to measure the detailed heat-transfer coefficients. Average axial turbulence intensities were also measured with a hot-wire anemometer. A CFD analysis predicted the details of the flow in the complicated coolant channel. Figure 4.107 shows their measurement and prediction comparisons of axial

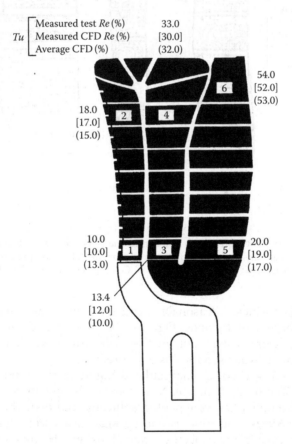

FIGURE 4.107
Axial turbulence intensities at different locations of a serpentine coolant passage. (From Abuaf, N. and Kercher, D.M., Heat transfer and turbulence in a turbulated blade cooling circuit, *International Gas Turbine & Aeroengine Congress & Exposition*, Cologne, Germany, ASME Paper 92-GT-187, June 1–4, 1992.)

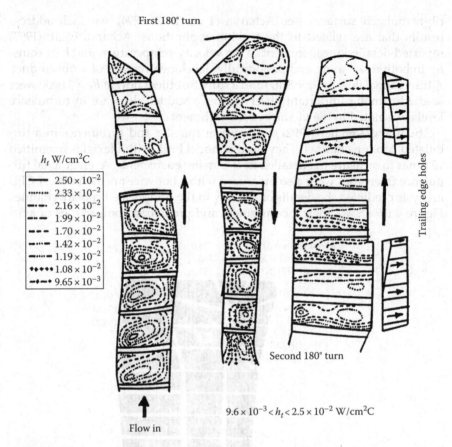

FIGURE 4.108
Heat-transfer coefficient distribution in a serpentine ribbed passage. (From Abuaf, N. and Kercher, D.M., Heat transfer and turbulence in a turbulated blade cooling circuit, *International Gas Turbine & Aeroengine Congress & Exposition*, Cologne, Germany, ASME Paper 92-GT-187, June 1–4, 1992.)

turbulence intensities. A reasonably good agreement was observed between measurement and predictions. Figure 4.108 shows the details of the heat-transfer coefficient distribution in their rib-turbulated channel. A liquid crystal technique was used for this measurement.

The test facility used an encapsulated liquid crystal layer between the heater and rib-mounted surface. Nine constant temperature lines that represented constant h lines were photographically recorded. The channel flow rate was 0.016 kg/s, and nine ascending heat fluxes were used. Note that the rib sizes vary in the coolant channel. Smaller ribs are used in the narrow regions; larger ribs are used in the wider regions. Heat-transfer results in Figure 4.108 show that the distribution is significantly two-dimensional, indicating the presence of three-dimensional flow. The span-averaged Nusselt number in each channel shows a mixed distribution resulting from

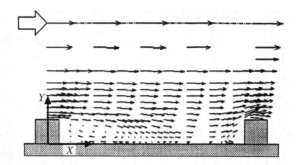

FIGURE 4.109
Flow vectors in a periodic rib arrangement. (From Rau, G. et al., *ASME J. Turbomach.*, 120, 368, 1998.)

ribs, change of cross-sectional area, and bends. The highest span-averaged Nusselt number is recorded just downstream of the first turn.

Rau et al. (1998) used a two-dimensional LDV system to measure the flow field in a ribbed channel. Figure 4.109 shows the velocity vectors in the channel centerline. The vectors show three separated regions: One is downstream of the upstream rib, the second is upstream of the downstream rib, and the third is on top of the downstream rib. The heat-transfer enhancement is caused by the boundary layer separation and boundary layer reattachment. More detailed turbulence and flow measurements and numerical predictions are discussed in Acharya et al. (1993, 1994).

Figure 4.110 shows the secondary flow in the ribbed channel. This one-sided ribbed channel shows a large vortex in the mainstream above the rib. However, due to expected symmetry in the flow, this single vortex structure is somewhat surprising. The vortex may be caused by an instability of the core flow. Whatever the cause, this vortex is partly responsible for increasing the heat-transfer coefficient of the smooth sides by the addition of ribs in one surface.

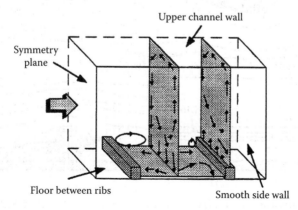

FIGURE 4.110
Secondary flow in a ribbed channel. (From Rau, G. et al., *ASME J. Turbomach.*, 120, 368, 1998.)

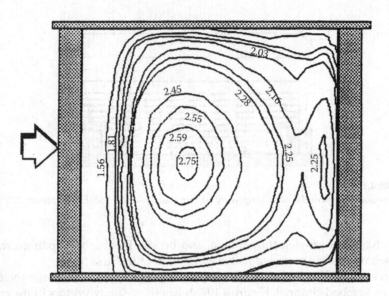

FIGURE 4.111
Heat-transfer enhancement factors in a ribbed channel. (From Rau, G. et al., *ASME J. Turbomach.*, 120, 368, 1998.)

Rau et al. (1998) also presented detailed heat-transfer coefficients in the interrib region and smooth side walls. Figure 4.111 shows the heat-transfer enhancement factors for a $p/e = 12$ ribbed surface. The zone with maximum heat-transfer coefficient is the region of flow reattachment. The pattern shows two-dimensionality, that is, the effect of sidewalls on the heat-transfer coefficient. Figure 4.112 shows the heat-transfer coefficient factor contours for

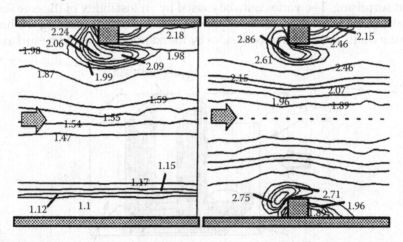

FIGURE 4.112
Heat-transfer coefficient enhancement on sidewalls. (From Rau, G. et al., *ASME J. Turbomach.*, 120, 368, 1998.)

the one- and two-sided ribbed channels. The heat-transfer enhancement for the two ribbed sides is higher than that for the single ribbed side. The flow is symmetric in the two-rib configuration.

Wang et al. (1998) presented detailed heat-transfer coefficient distribution in an angled rib configuration. Ekkad and Han (1997) studied different rib configurations in a two-pass channel. Figure 4.113 shows the span-averaged Nusselt number distributions at Reynolds numbers of 12,000 and 60,000. The 60° V shows the best performance in the group. Figure 4.114 shows the detailed Nusselt number distributions in a two-pass ribbed channel with smooth, orthogonal ribs, and high performance ribs. In the first pass of the smooth channel, there is no significant variation in the Nusselt number ratio. High heat transfer is observed away from the divider wall due to the flow impingement. The turbulent effects developed in the turn region are sustained in the inlet to the second pass. The addition of ribs in a 90° ribbed channel shows greater spanwise and axial variations in the Nusselt number distribution. Heat transfer is high in the middle of two ribs and comparatively low immediately before and after each rib. In general, high-performance ribs show higher Nusselt number ratios, and the turn effect is reduced compared to a smooth channel.

Figure 4.114c presents detailed Nusselt number ratios with 60° parallel ribs. The ribs in the first pass are angled away from the divider wall; the ribs in the second pass are angled toward the divider wall. This angled-rib orientation shows higher Nusselt number ratios than orthogonal 90° ribs (see Figure 4.114b). Highest local Nusselt numbers are observed on top of the ribs. In the turn region, the Nusselt number ratio is very high due to the combination of the 180° sharp turn and the ribs. The centrifugal effect on the flow is reduced by the presence of the ribs. In the second pass, Nusselt number ratios are higher near the outer wall and lower toward the divider wall. Nusselt number ratios gradually decrease in the second pass with increasing distance from the turn, and the distribution becomes periodic between ribs.

Figure 4.114d presents the Nusselt number ratio with the high performance 60° V ribs. A high-heat-transfer region is observed along the center-line of the channel. Nusselt number ratios are lower immediately upstream and downstream of each rib in the first pass due to the flow separation. Though this rib enhances heat transfer, it is noted that the rib-enhanced heat transfer is less than the heat-transfer coefficient enhancement by the combination of turn and ribs. Figure 4.114e shows the detailed Nusselt number ratio with 60° broken V ribs. The Nusselt number ratio near the turn region of the first pass is low compared to that of other rib configurations. The enhancement in the after-turn region is much higher than the enhancement observed in the first pass. The broken V rib develops a serpentine secondary flow that develops highest Nusselt numbers immediately downstream of ribs.

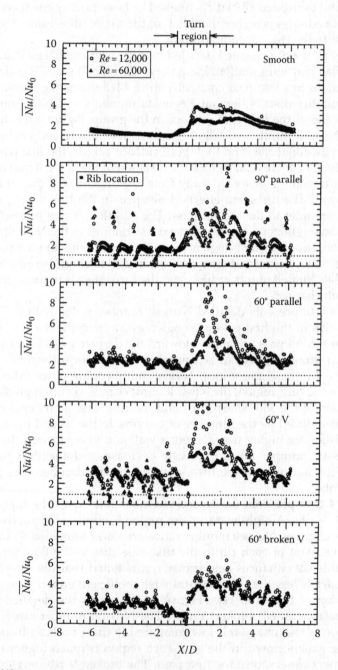

FIGURE 4.113

Span-averaged Nusselt number distributions for different configurations used by Ekkad and Han. (From Ekkad, S.V. and Han, J.C., *Int. J. Heat Mass Transfer*, 40(11), 2525, 1997.)

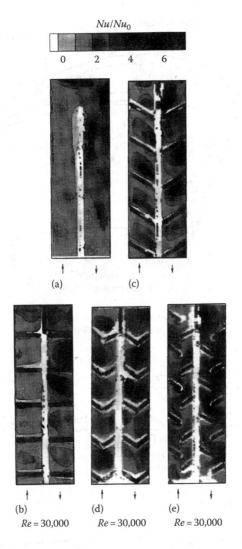

FIGURE 4.114
Detailed heat-transfer coefficient distributions for different rib arrangements studied by Ekkad and Han. (From Ekkad, S.V. and Han, J.C., *Int. J. Heat Mass Transfer*, 40(11), 2525, 1997.) (a) Smooth, (b) 90° parallel, (c) 60° parallel, (d) 60° V, and (e) 60° broken V.

4.2.12 Effect of Film-Cooling Hole on Ribbed-Channel Heat Transfer

Most modern turbine airfoils have ribs in the internal coolant channel and film cooling for the outside surface. Therefore, some of the cooling air is bled through the film-cooling holes. The presence of periodic ribs and bleed holes creates strong axial and spanwise variations in the heat-transfer distributions on the passage surface. Byerley et al. (1992) studied the internal-cooling passage heat transfer near the entrance to a film-cooling hole. Figure 4.115 shows the different orientations of holes studied. Figure 4.116

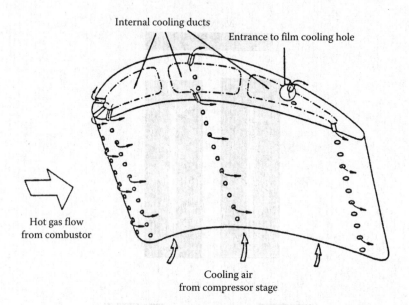

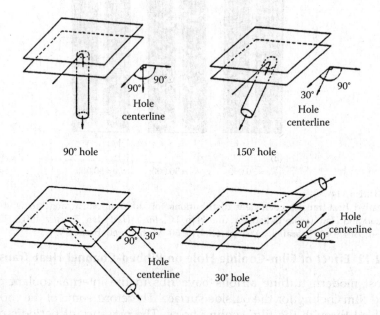

FIGURE 4.115
Different hole inclination arrangements used for film extraction by Byerley et al. (From Byerley, A.R. et al., Internal cooling passage heat transfer near the entrance to a film cooling hole: Experimental and computational results, *International Gas Turbine & Aeroengine Congress & Exposition*, Cologne, Germany, June 1–4, 1992.)

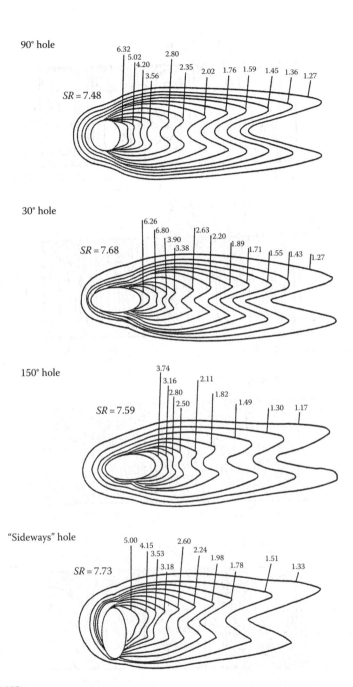

FIGURE 4.116
Heat-transfer enhancement factors for different film extraction holes. (From Byerley, A.R. et al., Internal cooling passage heat transfer near the entrance to a film cooling hole: Experimental and computational results, *International Gas Turbine & Aeroengine Congress & Exposition*, Cologne, Germany, June 1–4, 1992.)

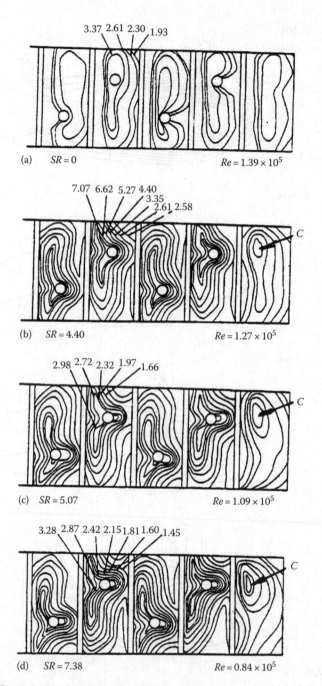

FIGURE 4.117
Heat-transfer coefficient enhancement factors in a ribbed channel with film extraction holes. (a) $SR=0$, (b) $SR=4.40$, (c) $SR=5.07$, and (d) $SR=7.38$ (SR=suction ratio). (From Shen, J.R. et al., *ASME J. Turbomach.*, 118, 428, 1996.)

shows the heat-transfer enhancement factors around the holes. The orientation and angle of the ejection hole have significant effect on the heat-transfer pattern. A 30° hole has more heat-transfer enhancement than a 150° hole arrangement.

Shen et al. (1996) studied the heat-transfer enhancement by ribs in the presence of coolant extraction. Several suction ratios *SR*, defined as the ratio of mean hole discharge velocity to mean channel velocity, are studied. Figure 4.117 shows the heat-transfer enhancement factors for several suction velocities. Figure 4.117a shows that, even in the absence of suction, the presence of holes acts as surface disturbances and affects the heat-transfer pattern. With increasing discharge through the film-cooling holes, the heat transfer initially enhances (see *SR* = 4.4) and then decreases with further increase in the coolant extraction.

Ekkad et al. (1998) studied the detailed heat-transfer coefficient distributions with different rib orientations in a two-pass channel with film-cooling bleed holes. Figure 4.118 shows the detailed Nusselt number distributions. Figure 4.118a shows details of heat-transfer coefficient distribution with bleed holes. The effect of turn and ribs has been discussed earlier in a test facility without bleed holes. Let us limit our discussion of the effects of bleed holes on heat-transfer coefficient distribution. In the first pass, the spanwise variation is not significant except around and downstream of

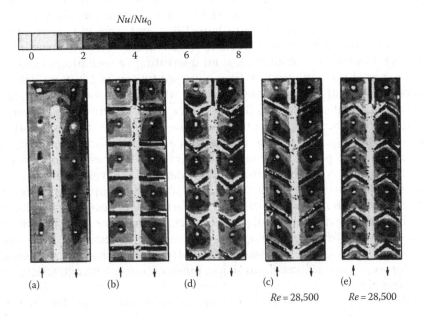

FIGURE 4.118
Detailed heat-transfer coefficient distributions in different rib arrangements with film extraction holes. (a) Smooth, (b) 90° parallel ribs, (c) 60° parallel ribs, (d) 60° V ribs, and (e) 60° inverted V ribs. (From Ekkad, S.V. et al., *Int. J. Heat Mass Transfer*, 41(13), 3781, 1998.)

each bleed hole. Since the flow is highly turbulent just downstream of the turn, the effect of bleed holes is not apparent in the Nusselt number ratio distribution. The heat-transfer coefficient distributions with the ribs (Figure 4.118b through e) show that the bleed holes increase heat-transfer coefficient in the near-hole regions, but no broader impact by these holes is noticeable in these results.

Figure 4.119 shows the span-averaged Nusselt number ratios at $Re = 28,500$. Figure 4.120 shows the regional-averaged Nusselt number ratios for different rib orientations. In this plot, the Nusselt numbers are averaged in selected regions. The values are almost identical with and without bleed hole extraction. This indicates that 20%–25% reduction of the main flow can be used for film cooling without significantly affecting the ribbed-channel cooling performance.

4.3 Pin-Fin Cooling

4.3.1 Introduction

Pin fins are mostly round projections protruding from the heat-transfer surface to the coolant flow path. As the name implies, this heat-transfer enhancement technique uses pin-shaped fins, and these pins are oriented perpendicular to the flow direction to maximize forced convected fin cooling. Pin-fin cooling analysis is an interesting combination of internal and external flow analyses. A flow bounded by top and bottom flat surfaces is a typical internal flow situation, whereas cross-flow over a tube bundle is a typical external flow situation. In a pin-fin-cooled channel, similar to tube bundles, pins protrude in the flow and the flow is contained in a flow passage. Thus, both internal and external cooling characteristics are present in a pin-finned channel. The wake shed by each pin increases the free-stream turbulence, and the boundary layer development over the pin-mounted surface gets disturbed. The wakes from upstream pins also affect both the flow and heat-transfer performance of downstream pins. In addition to flow disturbances, pins conduct thermal energy away from the heat-transfer surface, and long pins can increase the effective heat-transfer area. This section discusses the heat-transfer mechanism of a single pin, array of pins, effects of pin shape, pin-fin cooling with extraction, and missing pins.

Like cylinders in a cross-flow, pins shed wake at downstream flow. Besides this wake shedding, a horseshoe vortex originates just upstream of the base of the pin and wraps the pin around, causing more flow disturbances. Boundary layer also separates if the pin is placed as a three-dimensional

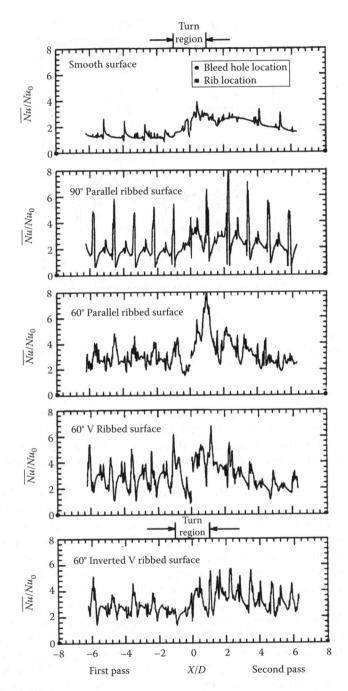

FIGURE 4.119
Span-averaged Nusselt number ratio for different rib arrangements studied by Ekkad et al.
(From Ekkad, S.V. et al., *Int. J. Heat Mass Transfer*, 41(13), 3781, 1998.) at *Re* = 28,500.

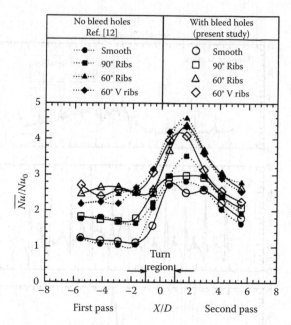

FIGURE 4.120

Effect of bleed holes on regional-averaged Nusselt number ratio at $Re = 28,500$. (From Ekkad, S.V. et al., *Int. J. Heat Mass Transfer*, 40(11), 2525, 1997.)

protrusion. These partial pins or three-dimensional protrusions do not extend to the top surface, and heat-transfer enhancement occurs on the pin-mounted surface. Interactions of all these flow disturbances (wakes, horseshoe vortex, boundary layer separation) increase heat transfer from the pin-mounted surface.

Figure 4.121 shows typical pin locations in a turbine airfoil. Pins are mostly used in the narrow trailing edge of an airfoil where impingement and ribbed channels cannot be accommodated due to manufacturing constraint. Pin fins perform better in low-aspect-ratio channels; Lau et al. (1985) observed that pin fins commonly used in turbine cooling have a pin height-to-diameter ratio between ½ and 4. Analysis of long cylinder arrays ($H/D > 10$) is done in the heat-exchanger industry. Heat transfer in these long cylinders is dominated by cylinder surfaces, and the effects from endwalls are not so significant. At the other extreme, very short pins of $H/D < ½$ are used in plate-fin heat exchangers where the endwall effect is dominating, and heat transfer from the pin's cylindrical surface is constrained by the endwall effect. Because of an intermediate H/D ratio of pins used in turbine airfoils, both endwall and cylinder effects are present. Armstrong and Winstanley (1988) observed that interpolation between the two extreme cases did not solve the case of intermediate-sized pin fins used in turbine blades and vanes.

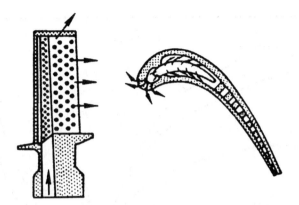

FIGURE 4.121
Trailing edge of an airfoil is cooled by pin fins. (From Metzger, D.E. et al., Pressure loss through sharp 180° turns in smooth rectangular channels, ASME Paper 84-GT-154, 1984.)

Therefore, heat-transfer analysis of pin fins related to turbine airfoil has its own performance characteristics.

Heat transfer in a turbine pin-fin array combines the cylinder heat transfer and endwall heat transfer. Due to the turbulence enhancement caused by pins, heat transfer from endwalls is higher than the smooth-wall condition; however, mounting of pins may cover a considerable surface area, and that area needs to be compensated by the pin surface area. Since the cylindrical surface area of short pins is less than the long pins, Armstrong and Winstanley (1988) observed that the average pin heat-transfer coefficient in short pins used in turbines is lower than long-cylinder heat-transfer coefficient. Zukauskas (1972) classified the flow in three domains for long pins. For the low Reynolds numbers, i.e., $Re < 10^3$, the flow is mostly laminar; for the high Reynolds numbers, i.e., $Re > 2 \times 10^5$, the flow is highly turbulent. The intermediate flow regime contains a mixture of laminar and turbulent characteristics in the interpin region.

Before we go into details of the pin-fin heat transfer, let us look at some of the fundamental aspects of the pin fin. Figure 4.122 shows the peripheral variation of Nusselt number in the first row of pins. There are no upstream pin effects in this row, and results indicate that the heat-transfer coefficient varies significantly over the circumference of the cylindrical pin. At downstream rows, local heat-transfer coefficient variation is dependent on the array arrangement and pitch of pin layout.

Besides the array arrangement, the material of the pin—i.e., conducting or nonconducting pin—can influence the heat-transfer result. Figure 4.123 shows the average Nusselt number variation for conducting and nonconducting pins. Though no significant variation can be observed in the average Nusselt number, the local Nusselt numbers can be significantly

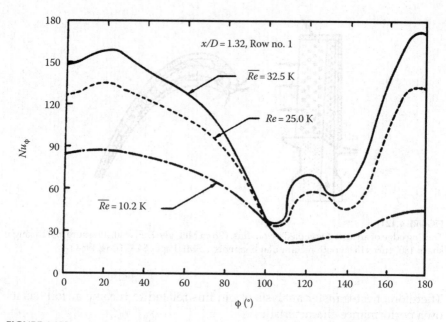

FIGURE 4.122
Circumferential variations in the local Nusselt number around a pin located in the first row. (From Metzger, D.E. and Haley, S.W., Heat transfer experiments and flow visualization for arrays of short pin fins, ASME Paper 82-GT-138, 1982.)

dependent on the conductivity of pins. At higher Reynolds numbers, the conducting pins show significantly higher Nusselt numbers than the nonconducting pins. This effect is less significant in the lower Reynolds number tested.

4.3.2 Flow and Heat-Transfer Analysis with Single Pin

Chyu and Natarajan (1996) have used experimental near-surface flow visualization to study the flow behavior near a pin and its mounting surface boundary layer. Figure 4.124 shows flow observations for a cylindrical standalone flow obstruction. Oil-graphite streak line patterns for the flow past the stand-alone pin are shown in this figure. Primary and secondary horseshoe vortices are evident from the cylindrical pin configuration. A dark line upstream of the cylinder signifies boundary layer separation, called a secondary vortex. The globular structures attached to the cylinder at the downstream location show strong rotating motion; and further downstream shear layer separation by the cylinder, similar to a rib surface roughness, dominates the flow along the channel centerline. It can be seen that the upstream horseshoe vortex cannot penetrate the core flow of high turbulence caused by this separated flow. The square cross-sectioned pin

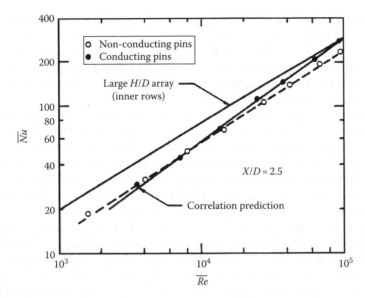

FIGURE 4.123
Effect of pin conductivity on average Nusselt number at different Reynolds numbers. (From Metzger, D.E. and Haley, S.W., Heat transfer experiments and flow visualization for arrays of short pin fins, ASME Paper 82-GT-138, 1982.)

FIGURE 4.124
Graphite streakline patterns near the endwall of a circular pin. (From Chyu, M.K. and Natarajan, V., *Int. J. Heat Mass Transfer*, 39(14), 2925, 1996.)

shows an accumulation of oil–graphite mixture near the sidewall. Low-frequency, periodic bursting is observed in this separated bubble. The necklace-shaped streakline formation, observed in a circular cylinder, is less prominent than that in a cube. It is obvious that a significant boundary layer disturbance is caused by these wall protrusions; therefore heat transfer is enhanced by pin fins.

Figure 4.125 shows the Stanton number ratio around a short cylinder in cross-flow. The cylinder height used by Goldstein et al. (1985) is the same as the diameter of the cylinder, i.e., aspect ratio of the cylinder is 1. The mass-transfer distributions are normalized by the measurements made on the flat naphthalene plate, without any protruding cylinder in the same tunnel operating with the same flow parameters. Results show that the mass transfer can be more than three times the smooth-plate condition. Manufacturing processes in pin-finned channels leave a fillet at the base of the pins, and heat-transfer pattern may change by the presence of those fillets. Figure 4.126 shows a pin with a large fillet radii, and details of the heat-transfer enhancement factor for such a fin. Contours shown are for local heat-transfer coefficients divided by local heat-transfer coefficients without the pin. Compared to a smooth straight pin, i.e., without a fillet, local heat-transfer coefficient enhancement ratios are lower. Moreover, the high heat-transfer coefficient enhancement by a smooth pin at $x/D = 2$ is not observed for a fillet pin.

Pins can be thick or thin with respect to pin height. The span-averaged endwall Stanton number ratios for two different aspect ratio pins

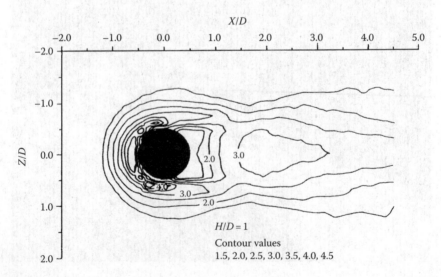

FIGURE 4.125
Stanton number ratio (St/St_0) distribution for a single pin. (From Goldstein, R.J. et al., *Int. J. Heat Mass Transfer*, 28(5), 977, 1985.)

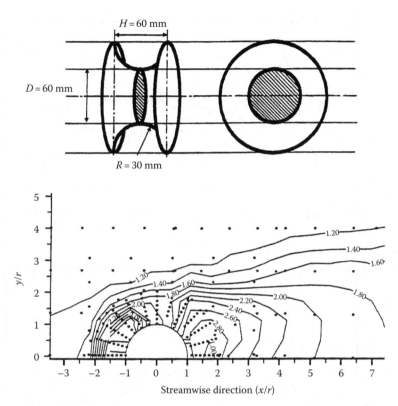

FIGURE 4.126
Pin with base fillet radii and heat-transfer enhancement factors as used by Wang et al. (From Wang, Z. et al., *ASME J. Turbomach.*, 117, 290, 1995.)

are plotted in Figure 4.127 (upper plot). Results show that upstream of the pin, the mass-transfer ratio is higher than that for a downstream region. Moreover, the aspect ratio does not have a significant effect on the upstream mass transfer, but in downstream locations, a shorter-aspect-ratio pin performs better than a taller pin. This span average includes the zero mass transfer at the under surface of the pin. A peak upstream of the pin fin signifies the formation of a horseshoe vortex that wraps around the pin. About one hydraulic diameter upstream of the pin, the Stanton number ratio is close to 1; therefore the effect of the pin is not felt upstream. The intense mixing near the mounting surface by the formation of the horseshoe vortex gives the highest mass transfer. At downstream locations, the mass transfer is increased by the flow separation. Experimental results show that a shorter pin with a diameter height of 1 has a three-dimensional vortex structure in the flow and significant curvature in the streamlines. A shorter pin indicates higher average mass transfer compared to that of a longer pin.

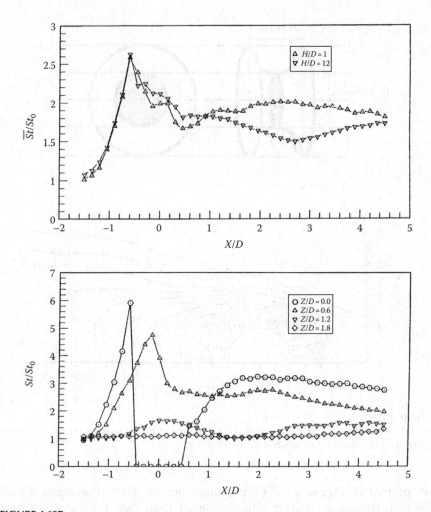

FIGURE 4.127

Span-averaged Stanton numbers for two different pin heights (upper plot) and local Stanton numbers for a short-aspect-ratio ($H/D = 1$) cylindrical pin (lower plot). (From Goldstein, R.J. et al., *Int. J. Heat Mass Transfer,* 28(5), 977, 1985.)

Figure 4.127 also shows the local Stanton number for a short cylindrical pin (lower plot). Immediately downstream of the cylinder along the centerline, the mass transfer is significantly low due to low air speed in the recirculating region. There is a clear reattachment region at the downstream location of the centerline. Moreover, as the flow bypasses the cylindrical pin, the mass transfer increases at the side of the cylinder. However, the effect of pins reduces with increasing distance from the pin, and at $Z/D = 1.8$, the mass transfer is nearly unaffected by the presence of this pin.

4.3.3 Pin Array and Correlation

A single pin can enhance mass and heat transfer in its neighborhood, but most cooling applications require a large area to participate in heat-transfer enhancement. Therefore, several pin fins are used to cover an area by distributing the pins in an array formation. There are two common array structures mostly used. Figure 4.128 shows two array configurations. One is the inline array (Figure 4.128a), and the other is the staggered array (Figure 4.128b). Figure 4.129 shows the pin surface and endwall heat transfers for two array configurations. Presented results are plotted as enhancement factors relative to the corresponding fully developed smooth-channel values. Results are row averaged, and except for the first two rows, the heat-transfer coefficient on the pin surface for both arrays is consistently higher than that of the endwall. The difference between the pin surface heat transfer and endwall Sherwood number ratios widens with a decrease in the flow Reynolds number. The pin surface heat transfer is observed to be 10%–20% higher for

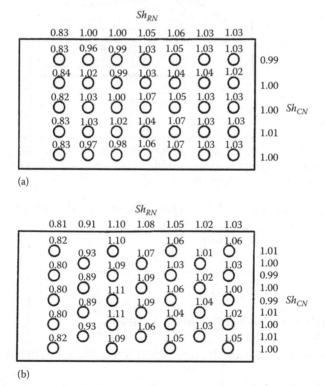

FIGURE 4.128
Relative heat-transfer coefficients on pin elements at $Re = 16,800$. (a) Inline array and (b) staggered array. (From Chyu, M.K. et al., Heat transfer contributions of pins and endwall in pin-fin arrays: Effects of thermal boundary condition modeling, *International Gas Turbine & Aeroengine Congress & Exhibition*, Stockholm, Sweden, ASME Paper 98-GT-l75, June 2–5, 1998b.)

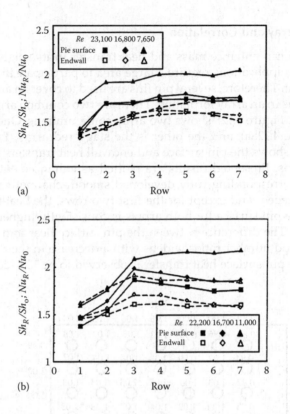

(a)

(b)

FIGURE 4.129
Heat-transfer coefficients on pin and endwall surfaces. (a) Inline array and (b) staggered array.
(From Chyu, M.K. et al., Heat transfer contributions of pins and endwall in pin-fin arrays:
Effects of thermal boundary condition modeling, *International Gas Turbine & Aeroengine
Congress & Exhibition*, Stockholm, Sweden, ASME Paper 98-GT-l75, June 2–5, 1998b.)

the presented case. Figure 4.130 shows that Sherwood numbers can be correlated for different Reynolds numbers as

$$\frac{Sh_A}{Sc^{0.4}} = \frac{Nu_A}{Pr^{0.4}} = aRe^b \tag{4.21}$$

where
Sh_A is the array-averaged Sherwood number
Sc is the Schmidt number
The constants a and b are the correlation constants, given in Table 4.7

Table 4.7 shows the value of constants used in the power correlation
between the average Nusselt number and the flow Reynolds number. The
inline array shows a stronger Reynolds number dependence compared to
the staggered arrangement. However, compared to previous observations of
Metzger et al. (1982), the present coefficients show a lower dependency on

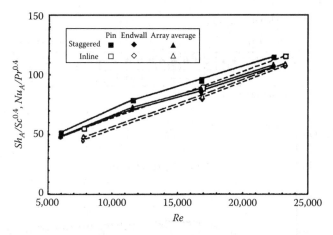

FIGURE 4.130
Effect of pin arrangement on array-averaged heat-transfer coefficient. (From Chyu, M.K. et al., Heat transfer contributions of pins and endwall in pin-fin arrays: Effects of thermal boundary condition modeling, *International Gas Turbine & Aeroengine Congress & Exhibition*, Stockholm, Sweden, ASME Paper 98-GT-l75, June 2–5, 1998b.)

TABLE 4.7

Correlation Constants for Heat-Transfer Coefficient

Array Configuration	Surface	a	b
Inline	Pin surface	0.155	0.658
	Endwall surface	0.052	0.759
	Average	0.068	0.733
Staggered	Pin surface	0.337	0.585
	Endwall surface	0.315	0.582
	Average	0.320	0.583

Source: Chyu, M.K. et al., Heat transfer contributions of pins and endwall in pin-fin arrays: Effects of thermal boundary condition modeling, *International Gas Turbine & Aeroengine Congress & Exhibition*, Stockholm, Sweden, ASME Paper 98-GT-l75, June 2–5, 1998b.

the flow Reynolds number. Metzger et al. (1982) showed $Nu/Pr^{04} = 0.08Re^{0.728}$ for their staggered arrangement. In other tests, VanFossen (1982) reported a 35% difference, and Metzger et al. (1984) observed nearly 100% difference between the two ratios.

Figure 4.131 shows the schematic of the pin-fin arrangement used by Metzger et al. (1982). They used a staggered arrangement with unheated entrance and exit channels. Figure 4.131 also shows the Nusselt number development in a staggered pin-fin array. One observation that is typical of the development region of the pin-fin array is the gradual increase of the Nusselt number near the inlet of a cooling channel. This is unlike common

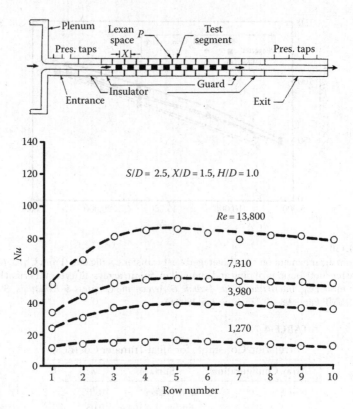

FIGURE 4.131

Typical test section for studying pin-fin heat transfer and development of the Nusselt number profile. (From Metzger, D.E. et al., *ASME J. Heat Transfer*, 104, 700, 1982.)

observations made in a flat-plate heat-transfer analysis. In a flat plate or a smooth channel, the Nusselt number is highest at the start of the thermal boundary layer and gradually decays to its fully developed value. Whereas in pin-fin arrays, the Nusselt number gradually increases before attaining a stable fully developed value. The increase in the development region is due to the interaction between wakes created by upstream pins and obstructions created by downstream pins. The first row of pins does not have any upstream wakes, and the heat transfer is mostly due to the boundary layer developed at the upstream location. The wake generated by the first row affects the second row of pin fins, and the heat-transfer coefficient is increased. The third row is affected by the first two rows, and heat-transfer coefficient augmentation is maximum. At downstream locations from the third row, the flow behavior becomes self-similar, and the heat-transfer coefficient stabilizes to its fully developed value. The amount of the increase of Nusselt number due to pin fins is strongly dependent on the Reynolds number. The amount of the increase is significantly higher for higher Reynolds numbers.

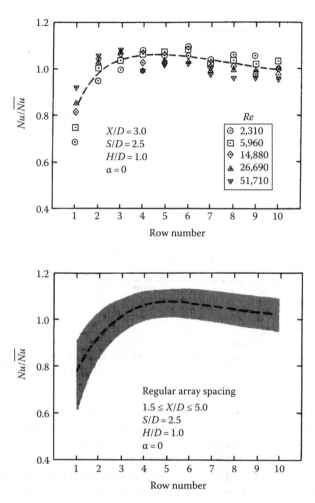

FIGURE 4.132

Row-resolved Nusselt numbers. (From Metzger, D.E. et al., Row resolved heat transfer variations in pin-fin arrays including effects of non-uniform arrays and flow convergence, *International Gas Turbine Conference and Exhibit*, Dusseldorf, West Germany, ASME Paper 86-GT-132, June 8–12, 1986.)

Metzger et al. (1986) used staggered arrays of circular pins with 1.5–5 pin diameter spacing. Figure 4.132 shows their span-averaged Nusselt number distribution along the channel. As discussed earlier, the Nusselt number initially increases, with rows reaching a peak at the third row, followed by a slow decrease with an increase in the row number. Figure 4.132 also shows the bandwidth of data dispersion for different Reynolds numbers. There is about ±10% variation in the observed Nusselt numbers. Figure 4.133 shows the average Nusselt number variations with Reynolds numbers. Results show a linear dependency with the log of Reynolds number. A correlation based

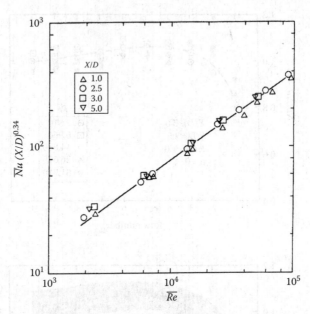

FIGURE 4.133
Average heat-transfer coefficients from a pin-mounted surface. (From Metzger, D.E. et al., Row resolved heat transfer variations in pin-fin arrays including effects of non-uniform arrays and flow convergence, *International Gas Turbine Conference and Exhibit*, Dusseldorf, West Germany, ASME Paper 86-GT-132, June 8–12, 1986.)

on spacing x/D is effectively developed in this figure. A closer spaced array (smaller x/D) shows a higher heat-transfer coefficient, and this heat-transfer coefficient pattern is validated by turbulent fluctuation measurements made by Metzger and Haley (1982). They found that the peak in turbulent fluctuation is much higher for $x/D = 1.32$ than coarsely spaced fins of $x/D = 2.19$.

Brigham and VanFossen (1984) studied the effect of pin-fin length on the heat-transfer coefficient. Figure 4.134 shows the Nusselt number distribution with different Reynolds numbers for different fin lengths. The results for different lengths collapse nicely on a single line in this log–log plot. This linear distribution has been possible by defining a length scale by $D = 4x$ channel volume/heat-transfer surface area. This length scale is different from classical hydraulic diameter, which is defined based on cross-sectional area and perimeter. Figure 4.134 also shows the percentage change in Nusselt number with different fin lengths. The Nusselt number variation is mostly due to the interaction of the pins with the endwalls. It is seen that the heat-transfer coefficient increases with an increase in the fin length. At larger fin lengths, results approximate those with the long cylinder.

Chyu and Goldstein (1991) observed that the peaks in mass transfer occurred near the cylinders' side surfaces. The staggered array performed better in the plateau region between two mass-transfer peaks. Peaks in mass transfer were observed at the side of the pins, and mass transfer decreased in the interpin

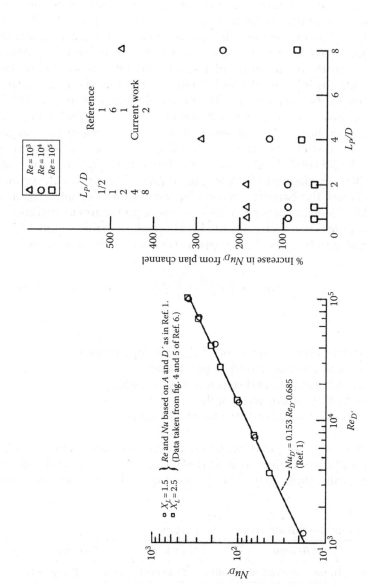

FIGURE 4.134

Array-averaged heat-transfer data and effect of pin length for pin-mounted surface. (From Brigham, B.A. and VanFossen, G.J., *ASME J. Eng. Gas Turbines Power*, 104(2), 268, 1982.)

region and again rose to a peak value at the side of the neighboring pin. Results showed that the staggered pin arrangement performed better in the interpin region. Their observations clearly indicated that addition of pin fins significantly enhanced the heat-transfer coefficient. However, addition of pins also increased the pressure drop in the flow channel. Goldstein et al. (1994) showed that the entrance and exit sections that did not have pins had a significantly lower pressure drop in these regions compared to the test section with pin fins.

Since heat-transfer coefficient is enhanced by the interaction of flow disturbances created by individual pins, it is obvious that there is an effect from pin spacing on the heat-transfer coefficient distribution. Jubran et al. (1993) developed a correlation for the Nusselt number with the Reynolds number and the spacing of the pin. They observed that staggered and inline arrangements have two distinct distributions. Based on a correlation equation of Sparrow and Molki (1982), Jubran et al. (1993) found the equation to the best-fit curve for the correlation to both staggered and inline arrays. According to this correlation, the Nusselt number depends on the Reynolds number and the packing densities of pins in the streamwise and spanwise directions. These pin-packing densities are included to give an idea of the amount of disturbances created in the flow. The general correlation can be represented as

$$Nu = a(Re)^b \left(\frac{S_z}{w} \right)^c \left(\frac{S_x}{L} \right)^d \tag{4.22}$$

where
  Nu is the Nusselt number in the thermally developed region
  Re is the Reynolds number based on pin diameter
  S_z and S_x are the spanwise and streamwise pin spacing
  w and L are channel width and length
  The constants a, b, c, and d are correlation constants

Note that aspect ratio of a pin is not taken into consideration in this correlation. With a best-fit curve, Jubran et al. (1993) found the values of these correlation coefficients; these constants are listed in Table 4.8. Three different

TABLE 4.8

Constants for the Nusselt Number Correlation

	C/H = 0.0		C/H = 0.5		C/H = 1.0	
Constant	Inline	Staggered	Inline	Staggered	Inline	Staggered
a	0.45	0.3	0.36	0.21	0.58	0.31
b	0.71	0.98	0.56	0.68	0.51	0.92
c	0.4	0.35	0.47	0.06	0.18	0.2
d	0.51	0.24	0.13	0.08	0.21	0.23

Source: Jubran, B.A. et al., ASME J. Heat Transfer, 115, 576, 1993.

cases are tabulated for both inline and staggered arrays. These three cases are (1) no clearance on top of pin ($C/H = 0.0$), (2) a clearance of half pin height ($C/H = 0.5$), and (3) a clearance of full pin height ($C/H = 1.0$). This correlation is based on ideal straight pins without base fillets; as mentioned earlier, heat transfer may be reduced in the presence of a fillet.

4.3.4 Effect of Pin Shape on Heat Transfer

In the introduction to this section on pin fins, it was discussed that a fillet radius at the base of the pin significantly affects the heat-transfer enhancement distribution. The pin shapes mostly studied are straight cylinders. However, the casting or other manufacturing processes cannot make perfect cylinders, and these manufacturing imperfections may affect the heat-transfer performance. Chyu (1990) studied the effect of a fillet at the base of the cylindrical pin. Inline and staggered arrays with and without base fillet are studied. Staggered straight cylinders perform best among the group, and addition of fillet reduces the heat- and mass-transfer performance in a staggered array. The inline array is less sensitive to this fillet. Figure 4.135 shows the total mass-transfer augmentation by pin fins. Total mass transfer means that all enclosure surfaces as well as the pin surfaces are included in this mass-transfer calculation. Results for both inline and staggered arrays and straight and fillet cylinders are presented. Straight cylinders in staggered array formation perform best, followed by fillet cylinders in the staggered formation. It is interesting to note that the fillet cylinder inline formation performs better than the straight cylinders in inline formation. However, the overall percentage augmentation of heat-transfer decreases with an increase in the Reynolds number. This increase in heat-transfer coefficients in fillet cylinder inline array from the straight cylinders may be due to the increase in area by the addition of fillets at the base of the cylindrical pins. Figure 4.135 also shows the performance curves of four different test cases. The performance is evaluated by taking the ratio of mass-transfer enhancement to the friction factor enhancement. The higher performance ratio means that the mass-transfer enhancement is associated with a lower friction factor enhancement. Though staggered array gives higher heat-transfer coefficient, performance of the inline straight cylinders is best among the group, and the fillet cylinders in staggered formation are worst. The fillet at the base of the pins affects the performance differently for the two array configurations.

In a different experimental work, Goldstein et al. (1994) studied the effect of stepped diameters on mass transfer. The diameter of the pin was axially varied. The base diameter was greater than the center diameter, and no fillet radius was provided. Figure 4.136 shows three different pins used by Goldstein et al. (1994), and the comparative mass-transfer enhancement and friction factor ratio for the double-stepped diameter pins. The array configuration was staggered. Results show that the mass transfer increases or remains the same compared to a straight cylinder pin array

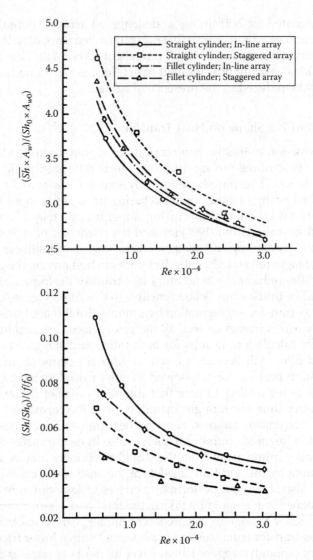

FIGURE 4.135
Effect of Reynolds number on mass-transfer augmentation and performance curves in straight pins and pins with fillet. (From Chyu, M.K., *ASME J. Heat Transfer*, 112, 926, 1990.)

when the radius is varied, but the pressure drop is reduced significantly for the stepped-diameter cylindrical pins. The mass-transfer enhancement increases with increasing Reynolds number. However, the pressure drop is significantly lower for lower Reynolds numbers. In a separate study, Olson (1992) presented an analysis of pin-fin cooling in microchannels. The pins were tapered, and heat-transfer results showed that the heat-transfer coefficient was lower than the large-channel heat-transfer analysis.

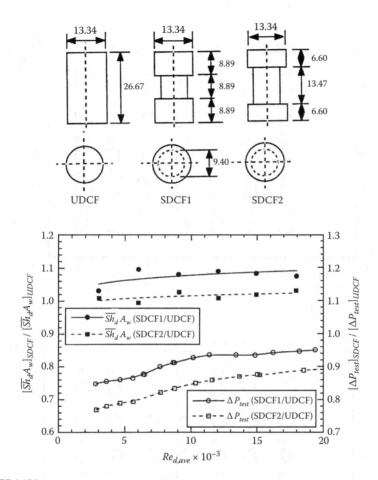

FIGURE 4.136
Comparison of mass transfer and pressure loss for different pin shapes. (From Goldstein, R.J. et al., *Int. J. Heat Mass Transfer*, 37(Suppl. 1), 149, 1994.)

Recently, Chyu et al. (1998a) used cube- and diamond-shaped pins to enhance heat-transfer coefficient from a surface. Figure 4.137 shows their mass-transfer results. The general trend of mass-transfer enhancement does not change by changing the shape of the pins. There is an initial increase in mass-transfer coefficient with increasing row number, and then the mass-transfer coefficient subsides to its fully developed value. In general, cube-shaped pins show higher mass-transfer coefficients near the inlet than that with diamond pins. Figure 4.138 show the array-averaged mass-transfer and pressure loss coefficients for cube- and diamond-shaped pins. It also shows that the cube-shaped pins have the highest mass-transfer coefficients among the shapes considered; round pins have the lowest mass-transfer coefficients. Corresponding pressure loss coefficients (lower plot) are higher for the cube- and diamond-shaped pins relative to the circular pins.

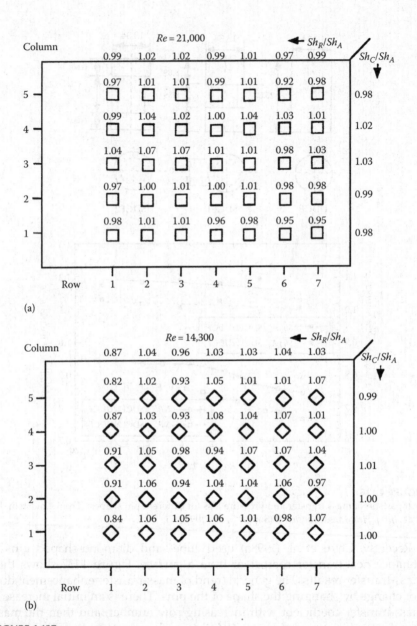

(a)

(b)

FIGURE 4.137

Relative heat-transfer (based on mass-transfer experiment) enhancement for different pin orientations of square cross-sectioned pins. (a) Inline cube array, (b) Inline diamond array, (From Chyu, M.K. et al., *ASME J. Turbomach.*, 120, 362, 1998a.)

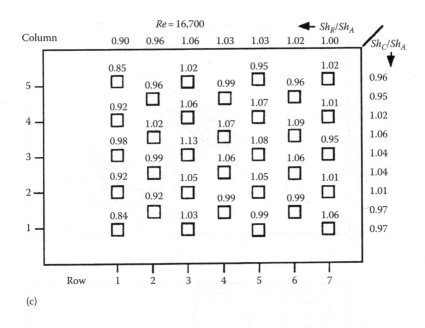

(c)

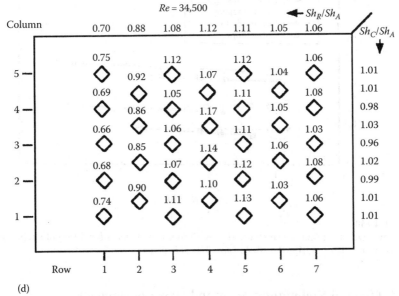

(d)

FIGURE 4.137 (continued)
Relative heat-transfer (based on mass-transfer experiment) enhancement for different pin orientations of square cross-sectioned pins. (c) Staggered cube array, and (d) Staggered diamond array. (From Chyu, M.K. et al., *ASME J. Turbomach.*, 120, 362, 1998a.)

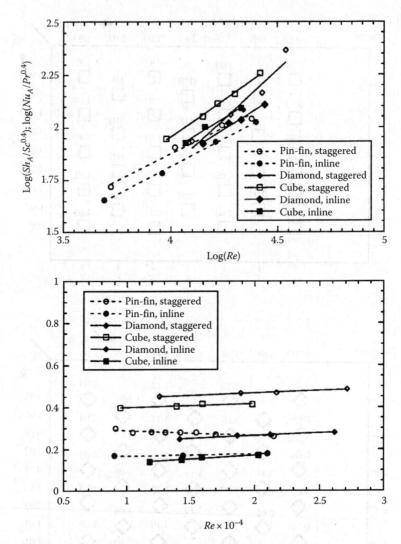

FIGURE 4.138

Array-averaged heat (mass)-transfer and pressure loss coefficients for different shaped pins. (From Chyu, M.K. et al., *ASME J. Turbomach.*, 120, 362, 1998a.)

4.3.5 Effect of Nonuniform Array and Flow Convergence

The flow channel in the trailing edge of an airfoil has a reducing cross section; therefore, the flow in the channel accelerates. Figure 4.139 shows the test geometries used by Metzger et al. (1986) to study the effect of this flow acceleration. Figure 4.140 shows the effect of flow convergence and a step change in the pin diameter. The results are row averaged, and the accelerating flow shows an increase in the heat-transfer coefficient. The

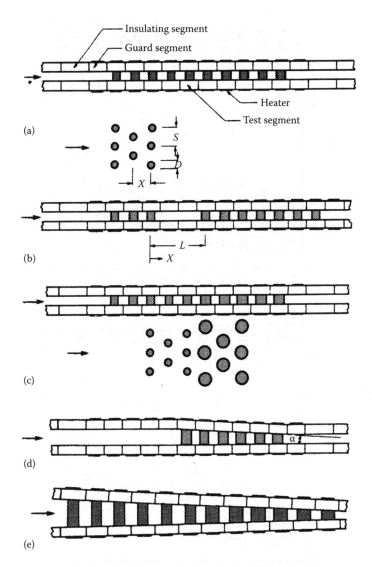

FIGURE 4.139
Nonuniform pin distributions and converging test section as used by Metzger et al. (a) Uniform cross-section, (b) staggered uniform pin spacing, (c) staggered varying pin sizes, (d) uniform inlet with converging channel, and (e) fully converging channel. (From Metzger, D.E. et al., Row resolved heat transfer variations in pin-fin arrays including effects of non-uniform arrays and flow convergence, *International Gas Turbine Conference and Exhibit*, Dusseldorf, West Germany, ASME Paper 86-GT-132, June 8–12, 1986.)

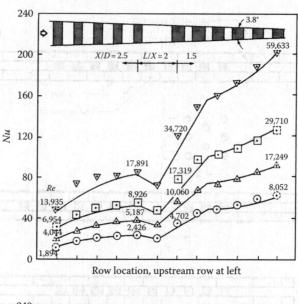

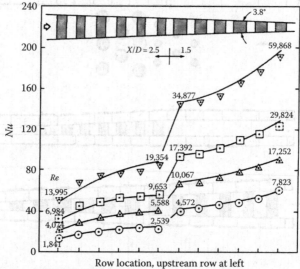

FIGURE 4.140
Measured and predicted Nusselt numbers with changing pin diameters, channel convergence, and row interruptions. (From Metzger, D.E. et al., Row resolved heat transfer variations in pin-fin arrays including effects of non-uniform arrays and flow convergence, *International Gas Turbine Conference and Exhibit*, Dusseldorf, West Germany, ASME Paper 86-GT-132, June 8–12, 1986.)

step increase in the pin diameter has a significant enhancement effect on the heat transfer. Since the flow area changes in the pin-finned channel, the flow velocity is different at different locations. The local Reynolds numbers are printed in the plot. Figure 4.140 also shows the effect of flow convergence, step change in pin diameter, and row interruption. The convergence angle is 3.8°. The missing pin row creates a drop in the Nusselt number profile, and the Nusselt number profile rapidly gains the values observed in Figure 4.140.

4.3.6 Effect of Skewed Pin Array

There are two pin configurations favored by heat-transfer designers, namely, staggered and inline. The staggered array has a higher heat-transfer coefficient compared to that with an inline array. However, the effect of skewness or the incident flow angle also has an effect on the heat-transfer pattern. Metzger et al. (1984) varied the incident angle from 0° to 40°. Figure 4.141 shows the flow incident angle on the pin array. They also considered oblong-shaped pins, as shown in Figure 4.141. The incident angles for oblong pins are denoted as γ and that for circular pins are denoted as α. Figure 4.142 shows the effect of incident angle α on circular pin arrays. The Nusselt number plotted is based on the overall heat-transfer coefficient, i.e., combined effect of pin surface and endwall. The $\alpha = 40°$ is essentially an inline pin array. Results indicate that at lower Reynolds number the inline array has a lower overall Nusselt number, and the staggered-array performance can be improved with small incident angles. At higher Reynolds numbers, the staggered performance shows lower Nusselt numbers compared to other formations, and the inline array shows highest overall Nusselt numbers. Most previous observations have shown that inline arrangement has a lower heat-transfer coefficient than staggered arrangement. Metzger et al. (1984) did not provide any explanation on this

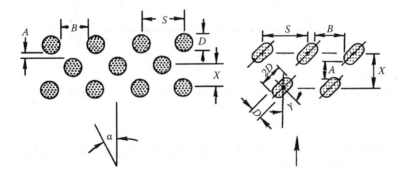

FIGURE 4.141
Pin arrangement with a skewed bulk flow (left) and oblong pin arrays (right) as used by Metzger et al. (From Metzger, D.E. et al., *ASME J. Eng. Gas Turbines Power*, 106, 252, 1984.)

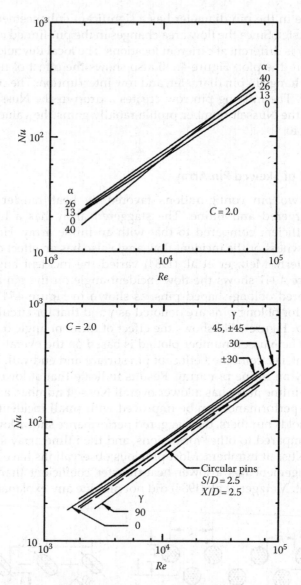

FIGURE 4.142
Effect of angle of attack on the heat transfer from both pin arrays. (From Metzger, D.E. et al., *ASME J. Eng. Gas Turbines Power*, 106, 252, 1984.)

result. One possibility could be that this experiment used low-conductivity pins that are different from prior experiments.

Figure 4.142 also shows the effect of flow incident angle on oblong pins. All incident angles except 90° yield higher Nusselt numbers than circular pins. The $\gamma = 90°$ array yields significantly lower Nusselt numbers, especially toward the lower end of the Reynolds number range. The $\gamma = 30°$ array has the highest Nusselt numbers, about 20% higher than the circular pin array on average.

Figure 4.143 shows the effect of incident flow on the friction factor. Figure 4.143 shows that the staggered circular pins with 0° incident angle have a decreasing friction factor with an increase in the Reynolds number. The nonzero incident angles show less dependency on the Reynolds number, and the inline array shows that the friction factor increases with an increase in the flow Reynolds number. Figure 4.143 also shows the friction factors for oblong pins at different flow incidence. Except for $\gamma = 90°$, the pressure drop for oblong pins is significantly higher than that for circular pins. This increase in the friction factor is associated with the flow turning caused by oblong pins.

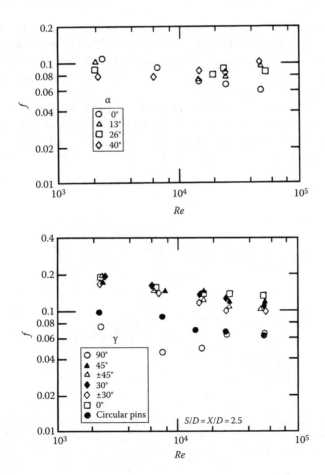

FIGURE 4.143
Effect of angle of attack on the friction factor. (From Metzger, D.E. et al., *ASME J. Eng. Gas Turbines Power*, 106, 252, 1984.)

4.3.7 Partial Pin Arrangements

Arora and Abdel-Messeh (1989) studied the effect of half pins on the heat-transfer coefficient. Figure 4.144 shows their schematics of half-pin arrangement. The figure shows that partial pins do not span the entire cross section of the flow channel, and there are tip clearances at one side. Steuber and Metzger (1986) have also reported the heat transfer from partial-length pins. Figure 4.145 compares the heat-transfer coefficients for full-length pins and partial pins. The partial pins used have a 15% clearance gap. The data for other Reynolds numbers and gaps showed similar Nusselt number patterns. Except for the first row, the Nusselt number for the pin-containing wall is comparable to the full-length pin. The other surface that does not have pins shows a lower Nusselt number. The local Nusselt numbers with the partial pins are slightly higher due to a higher Reynolds number than that used in full-length pins.

Figure 4.146 shows the average Nusselt number variation with flow Reynolds number with different tip clearances. The surface-containing pins do not have any effect on the pin-tip clearance, whereas the other surface that does not have pins shows a decrease in heat-transfer coefficient with an increase in the pin-tip clearance.

Figure 4.147 compares friction factors for different geometries of partial pins. The friction factor is lower for partial pins compared to full-length pins. Figure 4.148 compares the overall surface Nusselt numbers for different configurations. In general, heat-transfer coefficient decreases in partial pins.

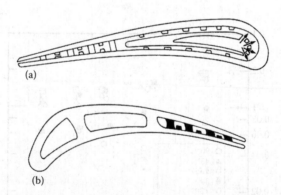

(a)

(b)

FIGURE 4.144
Partial pins for augmented cooling of high-pressure turbine airfoils. (a) Stator airfoil cross section and (b) rotor airfoil cross section. (From Arora, S.C. and Abdel-Messeh, W., Characteristics of partial length circular pin fins as heat transfer augmentors for airfoil internal cooling passages, *Gas Turbine & Aeroengine Congress & Exposition*, Toronto, Ontario, Canada, ASME Paper 89-GT-87, June 4–8, 1989.)

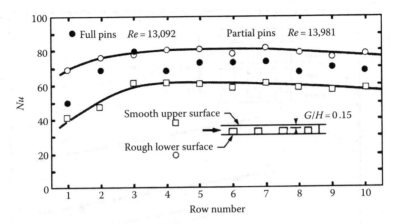

FIGURE 4.145
Local Nusselt number distribution for partial pins. (From Arora, S.C. and Abdel-Messeh, W., Characteristics of partial length circular pin fins as heat transfer augmentors for airfoil internal cooling passages, *Gas Turbine & Aeroengine Congress & Exposition*, Toronto, Ontario, Canada, ASME Paper 89-GT-87, June 4–8, 1989.)

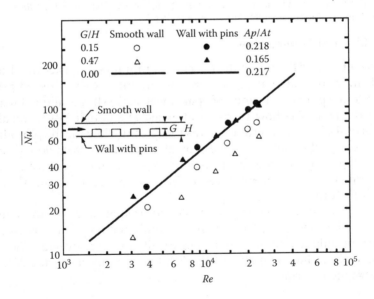

FIGURE 4.146
Average Nusselt number variations with flow Reynolds number for partial pins. (From Arora, S.C. and Abdel-Messeh, W., Characteristics of partial length circular pin fins as heat transfer augmentors for airfoil internal cooling passages, *Gas Turbine & Aeroengine Congress & Exposition*, Toronto, Ontario, Canada, ASME Paper 89-GT-87, June 4–8, 1989.)

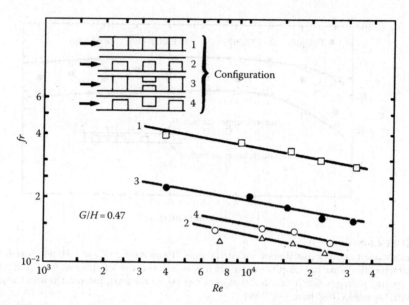

FIGURE 4.147
Comparison of friction factors for different configurations. (From Arora, S.C. and Abdel-Messeh, W., Characteristics of partial length circular pin fins as heat transfer augmentors for airfoil internal cooling passages, *Gas Turbine & Aeroengine Congress & Exposition*, Toronto, Ontario, Canada, ASME Paper 89-GT-87, June 4–8, 1989.)

4.3.8 Effect of Turning Flow

Sparrow et al. (1984) studied the effect of pin-fin mass transfer in a turning flow. Unlike conventional experiments, where inlets were directed perpendicular to the pin axis, the inlet of Sparrow et al. (1984) was aligned with the axis of the pins. The flow turned and exited at an angle perpendicular to the axis of the pins. Mass transfer was significantly low for the pins aligned parallel to the flow direction; the pins aligned in the cross-flow direction showed higher mass transfer. In real airfoils, the internal coolant path is often serpentine, and the pin-fin array is often located immediately downstream of a turn.

Chyu et al. (1992) used mass-transfer technique to study the effect of perpendicular flow entry in two pin-fin configurations. They showed that the turning inlet configuration always results in lower average Sherwood numbers. The reduction is about 40%–50% for the inline array and 20%–30% for the staggered array.

4.3.9 Pin-Fin Cooling with Ejection

The pin-fin chamber normally has ejection holes through which the Spent coolant exhausts to the mainstream flow. Lau et al. (1989a,b) tested several coolant ejection arrangements. Figure 4.149 shows selected configurations used by them. A total of nine cases were studied, as shown in Table 4.9.

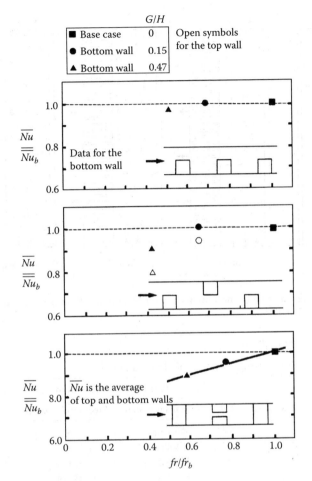

FIGURE 4.148

Relative performance of partial-length pins for Re = 20,000. (From Arora, S.C. and Abdel-Messeh, W., Characteristics of partial length circular pin fins as heat transfer augmentors for airfoil internal cooling passages, *Gas Turbine & Aeroengine Congress & Exposition*, Toronto, Ontario, Canada, ASME Paper 89-GT-87, June 4–8, 1989.)

Figure 4.150 shows the Nusselt number distribution with different Reynolds numbers. The log–log plot shows linear variations in Nusselt number distribution. The results show parallel lines for different cases, with an exception of case 1. The average Nusselt numbers can be correlated as $Nu = a - Re^b$. Values of a and b are tabulated in Table 4.10.

Heat-transfer coefficient enhancement is generally accompanied by a higher friction factor. Figure 4.151 shows the friction factor variations of different cases with changing Reynolds number. Like the Nusselt number variation, the friction factor varies linearly with Reynolds number in this log–log plot.

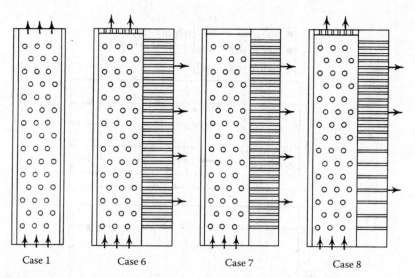

FIGURE 4.149
Four typical ejection hole configurations used by Lau et al. (From Lau, S.C. et al., *ASME J. Turbomach.*, 111, 116, 1989a.)

TABLE 4.9

Test Configurations Used by Lau et al.

Test Case	L/d	Exit-Hole Configuration	
		Radial-Flow Exit	**Ejection-Flow Exit**
1	2.0	Wide open	Block
2	2.0	6 holes	30 holes
3	2.0	Blocked	30 holes
4	2.0	6 holes	23 holes
5	2.0	Blocked	23 holes
6	12.5	6 holes	30 holes
7	12.5	Blocked	30 holes
8	12.5	6 holes	23 holes
9	12.5	Blocked	23 holes

Source: Lau, S.C. et al., *ASME J. Turbomach.*, 111, 116, 1989a.

Lau et al. (1992) used different ejection hole configurations and showed that an early extraction of coolant reduced the heat-transfer coefficient. Moreover, the Nusselt number was reduced with increasing ejection ratio McMillin and Lau (1994) also investigated the effect of trailing-edge ejection on local heat (mass) transfer in pin-array cooling. Both local heat-transfer distribution and pressure drop in a pin-finned channel that modeled an internal coolant passage were reported. McMillin and Lau (1994) noted an initial increase in the average Sherwood number, due to the increase in turbulence by wakes

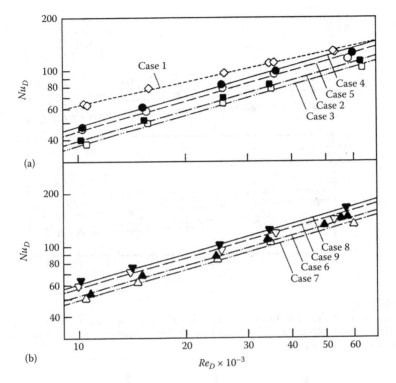

FIGURE 4.150
Overall heat-transfer coefficients for long and short pins. (From Lau, S.C. et al., *ASME J. Turbomach.*, 111, 116, 1989a.) (a) $L^*/d = 2$ and (b) $L^*/d = 12.5$.

TABLE 4.10

Coefficients Used in Heat-Transfer Correlation by Lau et al.

Test Case	a	b
1	1.1586	0.4335
2	0.2184	0.5642
3	0.2056	0.5642
4	0.2643	0.5642
5	0.2497	0.5642
6	0.2908	0.5642
7	0.2761	0.5642
8	0.3378	0.5642
9	0.3201	0.5642

Source: Lau, S.C. et al., *ASME J. Turbomach.*, 111, 116, 1989a.

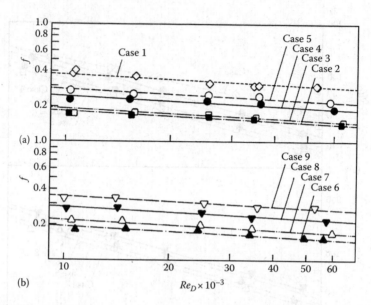

FIGURE 4.151
Overall friction factors for different extraction cases used by Lau et al. (a) $L^*/d = 2.0$ and (b) $L^*/d = 12.5$ (L^*/d = length of TE/diameter of TE holes). (From Lau, S.C. et al., *ASME J. Turbomach.*, 111, 116, 1989a.)

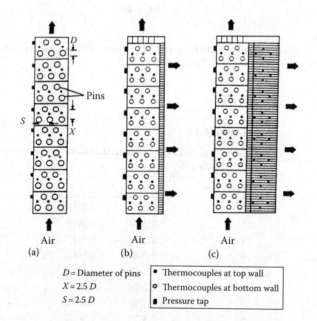

FIGURE 4.152
Schematic of test configurations used by Kumaran et al. (From Kumaran, T.K. et al., *Int. J. Heat Mass Transfer*, 34(10), 2617, 1991.): (a) straight flow (case 1), (b) short ejection holes (case 2), and (c) long ejection holes (case 3).

shed by pins. The decrease in Sherwood number with ejection was due to the lower mass flow at downstream locations by mass extraction in the ejection process. Coolant ejection also showed a lower pressure drop in comparison to instances without coolant ejection.

Kumaran et al. (1991) investigated the effects of the length of coolant ejection holes on the heat-transfer coefficient in pin fins. The length of the ejection hole can significantly alter the discharge rate of coolant. Figure 4.152 shows three cases used by them. Figure 4.153 compares the measured heat-transfer coefficients with the previous pin-fin correlation ($Nu = 0.298\ Re_{local}^{0.571}$). Results indicate that the correlation based on the local Reynolds number *can* satisfactorily predict the heat-transfer coefficient distribution except

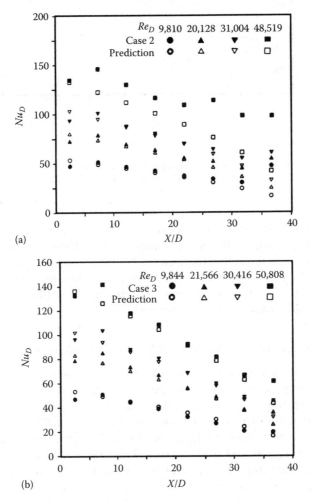

FIGURE 4.153
Comparison of measured and predicted local Nusselt numbers. (From Kumaran, T.K. et al., *Int. J. Heat Mass Transfer*, 34(10), 2617, 1991.): (a) case 2 and (b) case 3.

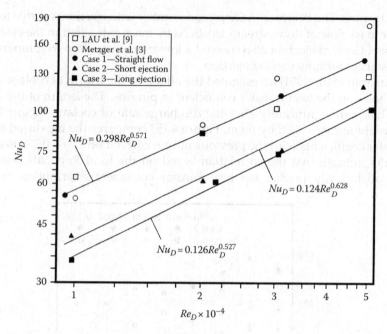

FIGURE 4.154
Variation of overall Nusselt number with Reynolds number for different extraction methods.
(From Kumaran, T.K. et al., *Int. J. Heat Mass Transfer*, 34(10), 2617, 1991.)

for higher Reynolds numbers. Figure 4.154 shows the variation of overall
Nusselt number, with a Reynolds number with different ejection arrangements. More coolant ejection reduces the Nusselt number significantly from
instances of no ejection. This decrease in the heat-transfer coefficient can be
explained by the fact that coolant mass is extracted from the coolant channel
before its cooling capacity is fully utilized.

4.3.10 Effect of Missing Pin on Heat-Transfer Coefficient

Since several pins are used simultaneously in an array formation, there are
possibilities of missing pins. A pin may be missing due to a problem during manufacturing, or a pin may structurally fail in operation; this missing
pin can disturb the harmony of a periodic flow pattern that develops in an
array. Sparrow and Molki (1982) showed two flow patterns in missing-pin
regions for staggered and inline arrays. Their figures showed that a missing pin has more influence in a staggered array. In a different study, Jubran
et al. (1993) provided heat flux data for missing-pin analysis. The reduction in
heat-transfer coefficient due to a missing pin was marginal and more noticeable in a staggered array.

4.4 Compound and New Cooling Techniques

4.4.1 Introduction

Several internal heat-transfer enhancement techniques are discussed in previous sections. The most common methods of heat-transfer augmentation in gas turbine airfoils are ribs, pins, jet impingement, and flow-disturbing inserts. It is shown that these enhancement techniques increase heat-transfer coefficients. But can combining these techniques increase the heat-transfer coefficient more? Several researchers have combined these heat-transfer enhancement techniques to improve the heat-transfer coefficient. However, as discussed in this section, the combination of more than one heat-transfer augmentation technique is not always recommended.

Besides compounding more than one heat-transfer enhancement technique, there are attempts to incorporate new concepts, e.g., heat pipes in turbomachinery cooling. Several studies are available on heat pipe applications; introductory concepts in that regard are also discussed here.

4.4.2 Impingement on Ribbed Walls

Studies on impingement on ribbed walls can be broadly classified as single-jet impingement and multiple-jet impingement. Haiping et al. (1997) intended to simulate the midchord region of a turbine airfoil. A single jet is used on a repeated rib-roughened surface. Figure 4.155 shows different jet impingement locations with respect to the rib location. A schematic of the flow and jet spreading is also shown in this figure. Effects of the relative position of the impinging jet on the local Nusselt numbers are also shown in Figure 4.155. Placement of the jet in between two ribs shows a higher Nusselt number compared to that in other positions. Gau and Lee (1992) also combined single-jet impingement on square cross-sectioned ribs. Figure 4.156 shows their Nusselt number distribution with and without ribs. Single-jet impingement shows that the heat-transfer coefficient is increased by the surface roughness, except for relatively small slot-width-to-rib-height ratios. The jet impinges on the centerline of a rib. This experiment used a smoke visualization technique to understand the flow related to the heat-transfer pattern. A wall jet effect was noticed after the jet impinged on the target surface. Results show that heat-transfer coefficients increase with an increase in the jet-to-target-plate spacing. It is argued that the jet becomes turbulent with an increase in the jet-to-target-plate spacing. Figure 4.157 shows the correlation of the stagnation point Nusselt number. Two different correlations are developed for two different jet-to-target-plate spacing. For larger spacing the correlation is given as

$$Nu_0 = 0.639 \left(\frac{b}{e}\right)^{0.33} \left(\frac{p}{e}\right)^{0.53} (Re)^{0.5} \left(\frac{z}{b}\right)^{-0.32} \quad \text{for } 10 \le \frac{z}{b} \le 16 \qquad (4.23)$$

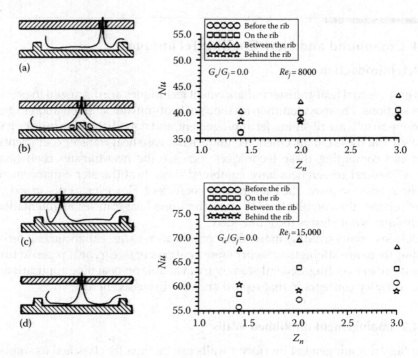

FIGURE 4.155
Effect of jet-to-target-plate spacing on impingement heat transfer for a rib-roughened surface.
(a) Downstream location, (b) on top of rib, (c) upstream location, and (d) center of pitch. (From
Haiping, C., et al., Impingement heat transfer from rib roughened surface within arrays of circu-
lar jet: The effect of the relative position of the jet hole to the ribs, Presented at the *International Gas
Turbine & Aeroengine Congress & Exhibition*, Orlando, FL, ASME Paper 97-GT-331, June 2–5, 1997.)

Correlations for smaller jet-to-target-plate spacing are given as

$$Nu_0 = 0.154\left(\frac{b}{e}\right)^{0.33}\left(\frac{p}{e}\right)^{0.22}(Re)^{0.5}\left(\frac{z}{b}\right)^{0.37} \quad \text{for } 2 \le \frac{z}{b} \le 8$$

$$Nu_0 = 0.335\left(\frac{b}{e}\right)^{0.31}\left(\frac{p}{e}\right)^{0.2}(Re)^{0.51}\left(\frac{z}{b}\right)^{-0.11} \quad \text{for } 8 \le \frac{z}{b} \le 16$$

(4.24)

Trabolt and Obot (1987) investigated the effects of cross-flow on multiple-jet
impingement heat transfer from rough surfaces. Figure 4.158 shows the layout of their rib-roughened surface. Two cross-flow schemes were tested: discharge of the spent jets through two sides and discharge through one side.
The rib height was 0.813 mm, and the pitch-to-rib-height ratio was varied as
6, 8, and 10. Figure 4.159 shows the Nusselt numbers for different Reynolds
numbers. The Nusselt numbers for smooth (without ribs) and ribbed channels at a Reynolds number of 11,000 show that addition of ribs reduces the
effective enhancement capability of jet impingement in the presence of

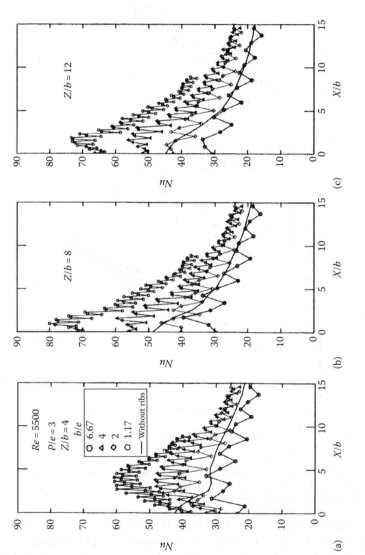

FIGURE 4.156

Local Nusselt number distribution for a single jet impinging on small ribs a different slot-width-to-rib-height ratios for $Re = 5500$. (a) $Z/b = 4$, (b) $Z/b = 8$, and (c) $Z/b = 12$. (From Gau, C. and Lee, C.C., *Int. J. Heat Mass Transfer*, 35(11), 3009, 1992.)

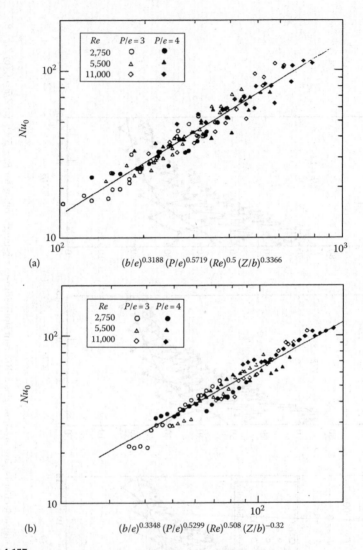

FIGURE 4.157
Stagnation point Nusselt number correlation for single-jet impingement on ribbed surface. (a) $2 \leq Z/b \leq 10$ and (b) $10 \leq Z/b \leq 16$ (Z = slot to plate distance; b = slot width). (From Gau, C. and Lee, C.C., *Int. J. Heat Mass Transfer*, 35(11), 3009, 1992.)

surface roughness. In this test condition, the exit at $x/L = 0$ is closed; the other side is open. The smooth-channel results show that impingement heat-transfer coefficient is high near the closed end due to lower cross-flow deflection. Addition of ribs reduces the heat-transfer coefficient in this region. The heat-transfer coefficient near the open side increases by the addition of ribs due to boundary layer separation and reattachment in the cross-flow.

Taslim et al. (1998) used ribs oriented in the direction of jet flows to simulate heat-transfer enhancement in the trailing edge of an airfoil. Several slots in

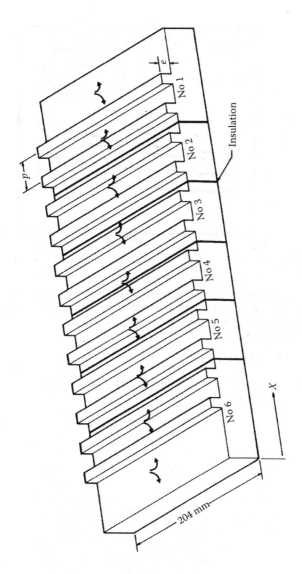

FIGURE 4.158
Multiple-jet impingement on a ribbed surface as used by Trabolt and Obot. (From Trabolt, T.A. and Obot, N.T., Impingement heat transfer within arrays of circular jets. Part II: Effects of cross-flow in the presence of roughness elements, Presented at the *Gas Turbine Conference & Exhibition*, Anaheim, CA, May 31–June 4, 1987.)

the high-pressure supply channel create impingement jets in the heat-transfer chamber. Spent jets are exhausted through exit slots. The impingement jet has an angle β and only one surface is used for heat-transfer analysis. Different test conditions can be summarized as shown in Table 4.11. The rib angle is 30° and ribs are 5.26 mm wide. Figure 4.160 shows their area-averaged Nusselt number variations with Reynolds number. A jet angle of $\beta = 6°$

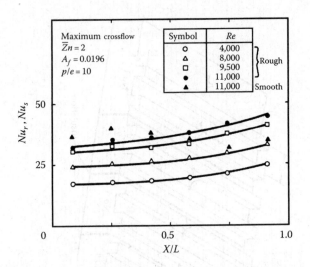

FIGURE 4.159

Nusselt number distribution in the rib-roughened channel of Trabolt and Obot. (From Trabolt, T.A. and Obot, N.T., Impingement heat transfer within arrays of circular jets. Part II: Effects of cross-flow in the presence of roughness elements, Presented at the *Gas Turbine Conference & Exhibition*, Anaheim, CA, May 31–June 4, 1987.)

TABLE 4.11

Test Configurations Used by Taslim et al.

Test	e (mm)	e/D	S/e
1	No ribs		
2	3.15	0.147	29.84
3	4.72	0.221	19.89
4	3.15	0.147	14.92
5	3.94	0.184	11.94
6	3.94	0.184	11.94

Source: Taslim, M.E. et al., Measurements of heat transfer coefficients in rib-roughened trailing-edge cavities with crossover jets, Presented at the *Gas Turbine & Aeroengine Congress & Exhibition*, Stockholm, Sweden, ASME Paper 98-GT-435, June 2–5, 1998.

shows a heat-transfer enhancement. In general, addition of ribs increases the heat-transfer coefficient for this configuration.

4.4.3 Impingement on Pinned and Dimpled Walls

Surface roughness can also be created with three-dimensional protrusions, called pins. Chakroun et al. (1997) studied heat transfer from a single round jet impinging on a surface containing three-dimensional cubes of

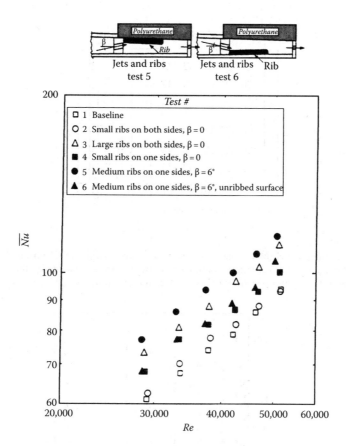

FIGURE 4.160
Average Nusselt number variations for different test configurations used by Taslim et al. (From Taslim, M.E. et al., Measurements of heat transfer coefficients in rib-roughened trailing-edge cavities with crossover jets, Presented at the *Gas Turbine & Aeroengine Congress & Exhibition*, Stockholm, Sweden, ASME Paper 98-GT-435, June 2–5, 1998.)

1 mm sides. The nozzle-to-impingement-plate distances ranged from 0.05 to 15 jet diameters to cover both the potential core of the jet and its far region. Figure 4.161 shows the schematics of the rough target plate used by them. The jet impinged at the center of the plate, and velocity measurements were taken at 1 and 2.5 jet diameters from the center of the plate.

Figure 4.162 shows the average Nusselt numbers for smooth and rough plates as a function of nozzle exit to plate distance. The heat-transfer coefficient of the rough plate increases from 8% to 28% compared to the smooth case. Note that the surface area increase from the addition of pins is only 7.5%. Therefore, heat-transfer enhancement is mostly due to the enhanced turbulent mixing caused by pins. The amount of enhancement is a function of Reynolds number and nozzle-to-target-plate separation distance. The effect of surface roughness is less significant for low

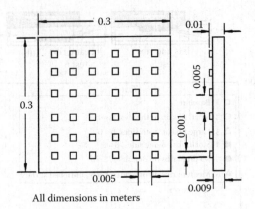

FIGURE 4.161
Schematic of the pin-roughened plate used by Chakroun et al. (From Chakroun, W.M. et al., Heat transfer augmentation for air jet impinged on rough surface, Presented at the *International Gas Turbine & Aeroengine Congress & Exhibition*, Orlando, FL, June 2–5, 1997.)

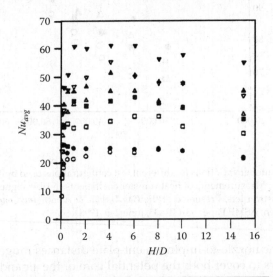

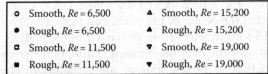

FIGURE 4.162
Average Nusselt numbers for smooth and rough plates as a function of nozzle-to-target-plate distance. (From Chakroun, W.M. et al., Heat transfer augmentation for air jet impinged on rough surface, Presented at the *International Gas Turbine & Aeroengine Congress & Exhibition*, Orlando, FL, June 2–5, 1997.)

Reynolds numbers. It can be argued that the pins are buried under the boundary layer for lower Reynolds numbers and therefore have less significant effect.

Azad et al. (2000a) extended their heat-transfer studies on jet impingement to surfaces containing pin fins. Pins had the same diameter as the impingement holes. The pins were 0.635 cm in diameter and 0.3175 cm in height, and had two different distributions. The sparse distribution used 11 rows of pins with 5 pins in each row; the dense distribution has 23 rows with 9 pins in each row. The impingement configuration had 12 rows of jets with 4 jets in each row. Both jets and pins were arranged in inline fashion. Figure 4.163 shows their impingement target surface with different exit flow directions. These exit flow configurations are similar to the conditions used for smooth-surface impingement tests by this research group. A transient liquid crystal method was used to measure the heat-transfer coefficient.

Figure 4.164 shows the average Nusselt number for different exit conditions. This plot also compares the heat-transfer results from the smooth-surface test conditions. The exit conditions significantly affect the discharge of jets and cross-flow effects. The smooth surface results show that flow orientation 2 provides the highest heat-transfer coefficient; flow orientation 3 provides the lowest. The Nusselt numbers for flow orientation 3 show a decrease in the overall heat-transfer coefficient with the addition of pins, whereas flow orientation 1 shows that the addition of pins significantly improves the heat-transfer coefficient. Flow condition 2 is less sensitive to the addition of pins.

Figure 4.165 compares average Nusselt numbers with other rough-surface impingement data. Pin fins show higher heat-transfer coefficients than the ribbed channel used by Trabolt and Obot (1987). Gau and Lee (1992) used a single-slot jet impingement and showed higher local heat-transfer coefficients at the stagnation point. This plot demonstrates that discontinuous pins may produce a higher heat-transfer coefficient in impingement heat transfer than continuous ribs. However, the cross-flow conditions are important for both pinned and ribbed surfaces.

Azad et al. (2000b) studied the impingement effect on a dimpled surface. Dimples are circular depressions on the target surface. Locations of dimples and jets are identical with the configurations used by Azad et al. (2000a). Different exit conditions similar to Azad et al. (2000a) are used for developing different cross-flow conditions. Figure 4.166 shows their test configuration. Figure 4.167 shows the average Nusselt numbers for pinned and dimpled-surface impingement test conditions. The heat-transfer results for the pinned surface and dimpled surface are comparable. At lower Reynolds numbers the pinned surface performed better than the dimpled surface. At higher Reynolds numbers, the dimpled surface performed better than the pinned surface for flow orientation 3.

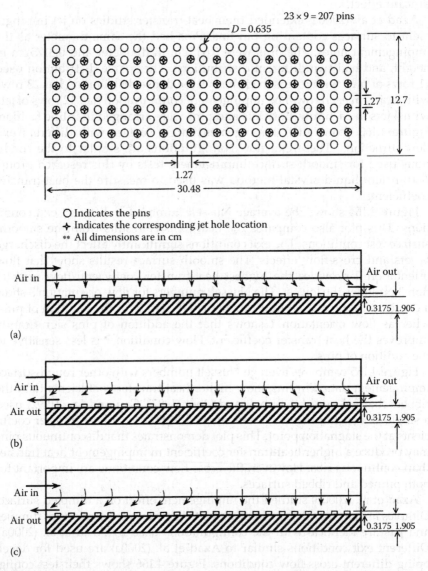

FIGURE 4.163
Schematic of the pinned surface of Azad et al. (From Azad, G.M. et al., Jet impingement heat transfer on pinned surfaces using a transient liquid crystal technique, *Proceedings of the 8th International Symposium on Transport Phenomena and Dynamics of Rotating Machinery*, Vol. II, pp. 731–738, March 2000a.) and different flow configurations. (a) Cross-flow exit direction 1, (b) cross-flow exit direction 2, and (c) cross-flow exit direction 3.

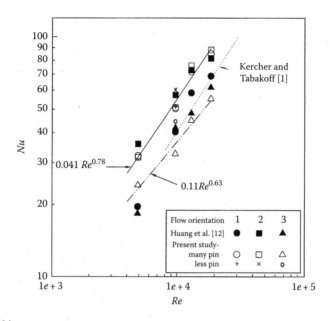

FIGURE 4.164
Average Nusselt number distributions for impingement on pin-roughened surface. (From Azad, G.M. et al., Jet impingement heat transfer on pinned surfaces using a transient liquid crystal technique, *Proceedings of the 8th International Symposium on Transport Phenomena and Dynamics of Rotating Machinery*, Vol. II, pp. 731–738, March 2000a.)

4.4.4 Combined Effect of Ribbed Wall with Grooves

Zhang et al. (1994) studied heat transfer and friction in rectangular channels with rib-groove combination. Figure 4.168 shows the conceptual flow pattern in a channel with rib-groove arrangement. The rib-groove combination is also referred to as compound roughness. Several cases were studied by Zhang et al. (1994), as shown in Table 4.12.

Figure 4.169 shows the heat-transfer performance comparison for the ribbed and ribbed-grooved ducts over a range of Reynolds numbers between 12,000 and 48,000. The results show that, in general, the Stanton number ratio (heat-transfer augmentation) increases with an increase in the friction factor ratio (pressure drop increment). The Stanton numbers for the ribbed-grooved walls (cases 1 and 2) are higher than that for the only ribbed walls (cases 7 and 8) at similar p/e values. The heat-transfer coefficient enhancement by the ribbed-grooved arrangement decreases with an increase in the p/e ratio. The ribbed-grooved duct with $p/e = 8$ (case 1) provides the highest heat-transfer augmentation. The Stanton number ratio is not sensitive to Reynolds numbers in the rib-groove arrangement, but the friction factor ratio increases with an increase in Reynolds numbers.

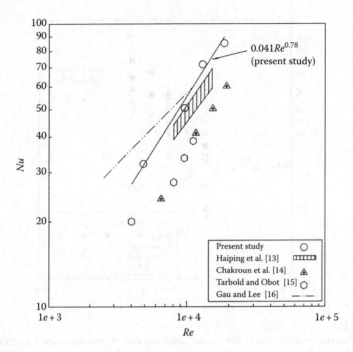

FIGURE 4.165
Comparison of pin-roughened impingement with other impingement data. (From Azad, G.M. et al., Jet impingement heat transfer on pinned surfaces using a transient liquid crystal technique, *Proceedings of the 8th International Symposium on Transport Phenomena and Dynamics of Rotating Machinery*, Vol. II, pp. 731–738, March 2000a.)

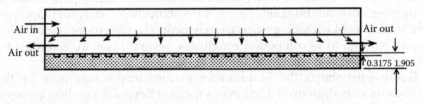

FIGURE 4.166
Schematic of the impingement on dimpled surface of Azad et al. for exit configuration 2. Locations of dimples are the same as pin locations used by them. (From Azad, G.M. et al., *AIAA J. Thermophys. Heat Transfer*, 14(2), 186, 2000b.)

Figure 4.170 shows the friction and heat-transfer correlation for the ribbed-grooved ducts. The effect of p/e ratio on the friction roughness function R for the ribbed-grooved duct was given by Zhang et al. (1994) as $R = 2.8(p/e/10)^{0.35}$. The heat-transfer roughness function for all rib and rib-groove configurations studied by Zhang et al. (1994) increases with an increase in the roughness Reynolds number.

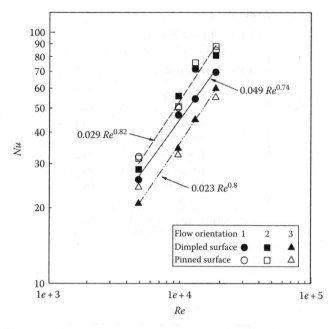

FIGURE 4.167
Comparison of average Nusselt numbers between pin-roughened and dimple-roughened target surfaces. (From Azad, G.M. et al., *AIAA J. Thermophys. Heat Transfer*, 14(2), 186, 2000b.)

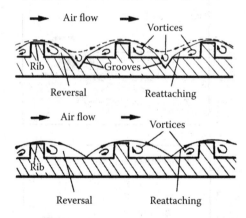

FIGURE 4.168
Conceptual flow patterns in rib-groove combination. (From Zhang, Y.M. et al., *ASME J. Heat Transfer*, 116, 58, 1994.)

4.4.5 Combined Effect of Ribbed Wall with Pins and Impingement Inlet Conditions

Metzger and Fan (1992) used a combination of pin fins with ribs and modified the inlet to have impingement effects. Figure 4.171 shows the schematic of their test configuration. A staggered configuration is used on the pin

TABLE 4.12

Different Surface Roughness
Conditions Used by Zhang et al.

Case No	Wall Condition	p/e	e/D
1	Rib-groove	8	0.028
2	Rib-groove	10	0.028
3	Rib-groove	15	0.028
4	Rib-groove	20	0.028
5	Rib-groove	25	0.028
6	Rib-groove	30	0.028
7	Rib only	8.5	0.02
8	Rib only	11.5	0.02

Source: Zhang, Y.M. et al., *ASME J. Heat Transfer*, 116, 58, 1994.

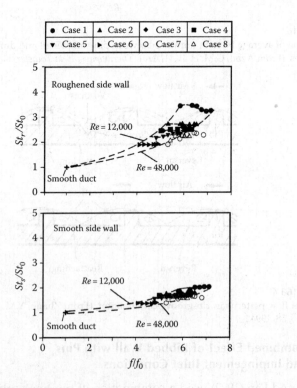

FIGURE 4.169

Heat-transfer performance curves for different rib and groove combinations. (From Zhang, Y.M. et al., *ASME J. Heat Transfer*, 116, 58, 1994.)

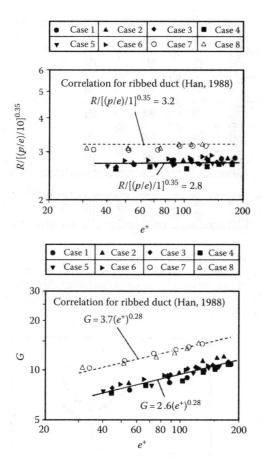

FIGURE 4.170
Friction factor and heat-transfer correlations in rib-groove combination. (From Zhang, Y.M. et al., *ASME J. Heat Transfer*, 116, 58, 1994.)

array. Roughened ribs with a square cross section on each side depend on the pin diameter used. Roughness ribs are mounted on the surface opposite to the pin-mounted surface. This figure shows that the inlet has a row of inclined wall jets, and exit is a combination of inclined jets and through flow. The inlet jets are oriented in alternating upward and downward directions. Results indicate a high heat transfer near the inlet due to these jet impingement, and it was argued that these significantly high heat-transfer rates near the inlet may be detrimental to the airfoil. That is because a highly nonuniform heat-transfer rate will develop a nonuniform surface temperature, which will create thermal stress in the airfoil.

Figure 4.172 shows the averaged Nusselt number variations with Reynolds numbers at the third row of pins. The dashed line in this plot indicates the pin-fin heat transfer without the ribs. Results indicate that the addition of ribs

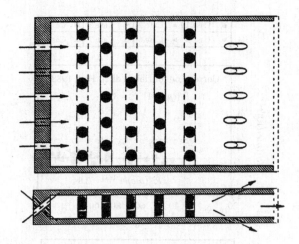

FIGURE 4.171
Jet impingement in a pin channel with rib-roughened walls. (From Metzger, D.E. and Fan, C.S., Heat transfer in pin-fin arrays with jet supply and large alternating wall roughness ribs, in *Fundamental and Applied Heat Transfer Research for Gas Turbine Engines*, HTD-Vol. 226, ASME, New York, pp. 23–30, 1992.)

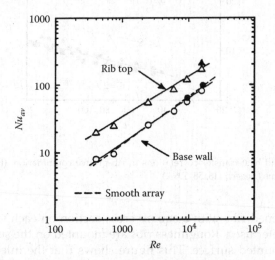

FIGURE 4.172
Average Nusselt numbers at the third row. (From Metzger, D.E. and Fan, C.S., Heat transfer in pin-fin arrays with jet supply and large alternating wall roughness ribs, in *Fundamental and Applied Heat Transfer Research for Gas Turbine Engines*, HTD-Vol. 226, ASME, New York, pp. 23–30, 1992.)

does not change the heat-transfer coefficient from the pin-mounted surface. However, the heat-transfer coefficient on the rib surface is significantly higher than that for the pin-mounted surface. Both pin-mounted surface and rib-top heat-transfer coefficients follow a power-law distribution with respect to the flow Reynolds number.

4.4.6 Combined Effect of Swirl Flow and Ribs

Kieda et al. (1984) experimentally investigated the single-phase water flow and heat transfer in a rectangular cross-sectioned twisted channel. Figure 4.173 shows their twisted-tube configuration. Several aspect ratios and twist pitches were used. Table 4.13 shows their test conditions.

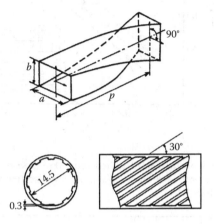

FIGURE 4.173
Twisted tube and spiral ribs used by Kieda et al. (From Kieda, S. et al., Heat transfer enhancement in a twisted tube having a rectangular cross-section with or without internal ribs, ASME Paper 84-HT-75, 1984.)

TABLE 4.13

Test Configurations Used by Kieda et al. in Their Twisted-Tube Experiment

Test No.	Aspect Ratio a/b	Twist Pitch p (mm)	Equivalent Dia. d_e (mm)	Pitch Ratio P/d_e
1	1	∞	5.01	∞
2	1	25.1	4.98	5.04
3	1	16.0	4.67	3.43
4	1	11.1	4.57	2.43
5	1	9.0	4.27	2.10
11	1.5	∞	4.71	∞
12	1.5	48.7	4.63	10.51
13	1.5	29.6	4.42	6.70
21	2.6	∞	5.52	∞
22	2.6	64.3	5.05	12.75
31	1	90.0	10.31	8.73
32	3	∞	8.67	∞
33	3	90	8.15	11

Source: Kieda, S. et al., Heat transfer enhancement in a twisted tube having a rectangular cross-section with or without internal ribs, ASME Paper 84-HT-75, 1984.

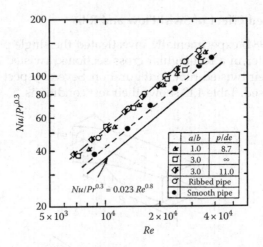

FIGURE 4.174
Nusselt number distribution for cooled ribbed tube. (From Kieda, S. et al., Heat transfer enhancement in a twisted tube having a rectangular cross-section with or without internal ribs, ASME Paper 84-HT-75, 1984.)

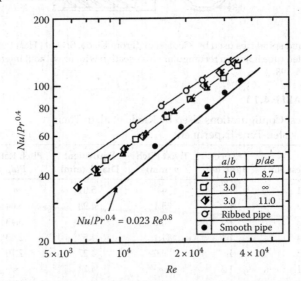

FIGURE 4.175
Nusselt number distribution for heated ribbed tube. (From Kieda, S. et al., Heat transfer enhancement in a twisted tube having a rectangular cross-section with or without internal ribs, ASME Paper 84-HT-75, 1984.)

Since the fluid used by Kieda et al. (1984) was water and since the flow was influenced by centrifugal effects, the cooling and heating applications had different heat-transfer performances. Figures 4.174 and 4.175 show the cooling and heating performances, respectively. Results indicate that in cooling application, this twisted channel performs similarly to a ribbed pipe. However, for

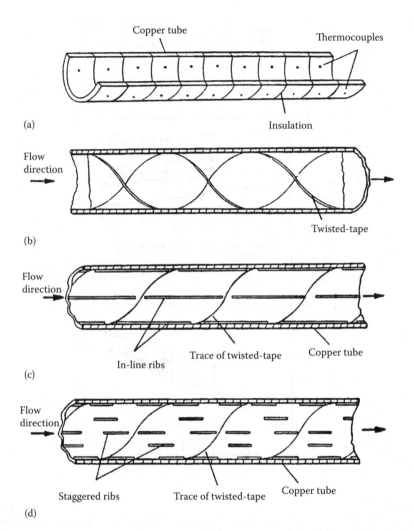

FIGURE 4.176
Schematic of the rib-roughened tube with twisted tape as used by Zhang et al. (a) Unenhanced test surface, (b) 2 turns of twisted tape, (c) single turn twisted tape with in-line ribs and (d) single turn twisted tape with staggered ribs. (From Zhang, Y.M. et al., *Enhanced Heat Transfer,* 4, 297, 1997.)

the heated test, twisted channels show a decrease in performance compared to the ribbed pipe. The centrifugal buoyancy plays a strong role in the twisted channels; therefore the heating and cooling heat-transfer patterns are different.

Zhang et al. (1997) used twisted-tape inserts and ribs in a circular tube. Figure 4.176 shows the schematics of their test configurations. Reynolds number was varied from 17,000 to 82,000. The ratio of rib height to pipe diameter was 0.125. Friction factors and Nusselt numbers obtained by their experiment are shown in Figure 4.177. Obtained results are compared with

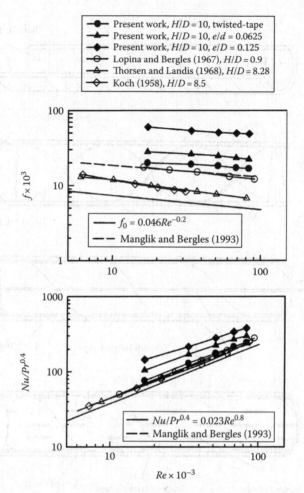

FIGURE 4.177

Friction factor and heat-transfer results for rib-roughened tube with twisted tape insert. (From Zhang, Y.M. et al., *Enhanced Heat Transfer*, 4, 297, 1997.)

tubes with twisted-tape inserts. Results indicate that both heat transfer and friction factors are higher with the addition of ribs in the pipe.

Later Zhang et al. (2000) used different types of inserts to study the combined rib and twisted-tape inserts in square ducts. Figure 4.178 shows different tape inserts used by them. Four test configurations are used: twisted tape, twisted tape with interrupted ribs, hemicircular wavy tape, and hemi-triangular wavy tape. Figure 4.179 shows the performance curves for these configurations as a function of the flow Reynolds numbers. This plot reflects the overall heat-transfer performance that includes the frictional losses. Two wall-heating conditions are tested. In one case, two opposite walls are heated; in the other, all four sides are heated. Twisted tape with interrupted ribs provides a higher overall heat-transfer performance over twisted tape

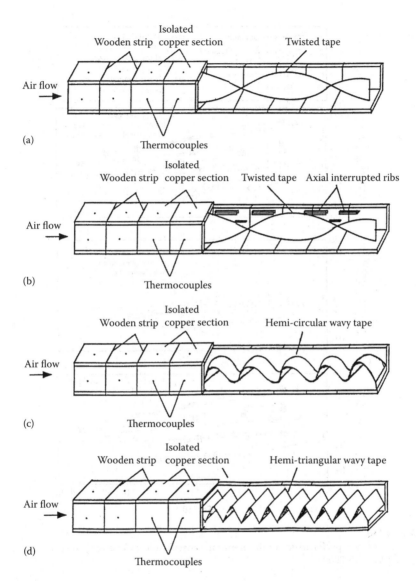

FIGURE 4.178
Schematic of the square channel with different types of tape inserts used by Zhang et al. (a) Twisted tape insert (b) twisted tape with axial interrupted ribs, (c) hemi-circular wavy tape, and (d) hemi-triangular wavy tape. (From Zhang, Y.M. et al., *Enhanced Heat Transfer*, 7, 35, 2000.)

without ribs and hemicircular wavy tape. The larger ribs (e/D = 0.33) performs better than smaller ribs (e/D = 0.25). The performance of the hemi-triangular wavy tape is comparable to the twisted tape plus interrupted ribs of e/D = 0.25 at lower Reynolds numbers (Re < 30,000), but performance is lower than twisted-tape performance at higher Reynolds numbers (Re > 30,000). Hemicircular wavy tapes show the lowest heat-transfer performance

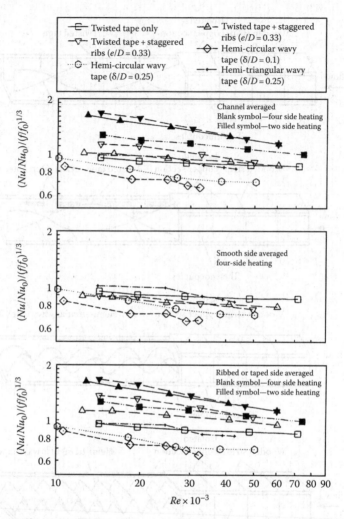

FIGURE 4.179
Channel-averaged performance of different tape inserts. (From Zhang, Y.M. et al., *Enhanced Heat Transfer*, 7, 35, 2000.)

in this group. Zhang et al. (2000) noted that a twisted-tape insert in a square cross-sectioned channel produced less pressure drop penalty compared to a similar case with circular pipe. The twisted-tape insert with ribs showed highest heat transfer and pressure drop in the group of test conditions considered.

4.4.7 Impingement Heat Transfer with Perforated Baffles

Dutta et al. (1998a) attempted to combine three popular techniques to enhance heat-transfer coefficient in a rectangular channel. Three enhancement

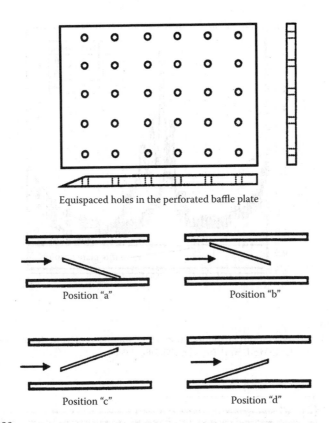

Equispaced holes in the perforated baffle plate

Position "a" Position "b"

Position "c" Position "d"

FIGURE 4.180
Perforated baffle used by Dutta et al. and different baffle positions in a rectangular channel.
(From Dutta, S. et al., *ASME J. Heat Transfer*, 120, 795, 1998a.)

techniques—boundary layer disturbance, impingement, and internal flow inserts—are combined to develop an inclined perforated baffle. Figure 4.180 shows the schematic of the perforated baffle and different locations of the baffle in the rectangular channel. In their experiment, only the top surface is heated with foil heaters, and the centerline temperature of the top surface is measured. The perforations create a jetlike flow. Out of these four configurations, configuration b showed the highest heat-transfer coefficients from the top heated surface. Figure 4.181 shows the Nusselt number distribution for configuration b. Two types of baffles are used, one with perforations and the other without. The solid baffle, i.e., without perforations, shows lower heat-transfer coefficients in the region covered by the baffle. The high Nusselt number near the tip of the baffle is due to wall-jet flow created by a small gap between the baffle and the heated surface. This gap was created to enhance the heat transfer. The perforated baffle shows clear impingement spikes in the Nusselt number profile. Downstream, the flow separation and reattachment in the main flow is stronger for the solid baffle; therefore, Nusselt numbers are higher with this baffle near the reattachment region.

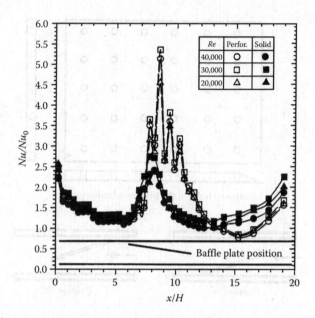

FIGURE 4.181
Local Nusselt number ratio distribution along the channel centerline for baffle position *b*. (From Dutta, S. et al., *ASME J. Heat Transfer*, 120, 795, 1998a.)

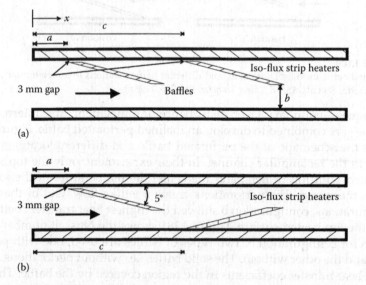

FIGURE 4.182
Two baffles used by Dutta et al. and their positions in a rectangular channel. (a) Position of baffles in the test section for arrangements A, B, and C. Both baffles perforated. (b) Position of baffles in the test section for arrangements D, E, and F. Both baffles perforated. (From Dutta, P. et al., Internal heat transfer enhancement by two perforated baffles in a rectangular channel, Presented at the *International Gas Turbine & Aeroengine Congress & Exhibition*, Stockholm, Sweden, ASME Paper 98-GT-55, June 2–5, 1998b.)

After successfully completing studies of the single baffle, one studies heat transfer with multiple baffles. Dutta et al. (1998b) studied eight different configurations of two baffles. Both solid and perforated baffles were used. Figure 4.182 shows schematics of the different configurations used. Configuration a showed the best heat-transfer enhancement from the top heated surface. Figure 4.183 shows the centerline heat-transfer coefficient distribution. The impingement indeed significantly enhances heat transfer in the regions covered by these baffles. The divergent formation of the baffles provides sufficient space for the cross-flow to pass without significantly deflecting the jets.

Figure 4.184 shows the performance of the tested perforated baffle configurations. Friction factor ratios are plotted with average Nusselt number ratios. Results indicate that the friction factors can significantly vary with the arrangements, but the average Nusselt number ratio is bounded by closer upper and lower limits. The Nusselt number ratio is higher with two baffles, but the friction factor can be higher or lower than the single-baffle arrangement. It was concluded that the heat transfer in a two-baffle case is due to more jet impingement, and friction factor is comparable with the single baffle due to a weaker reattachment region downstream of the first baffle.

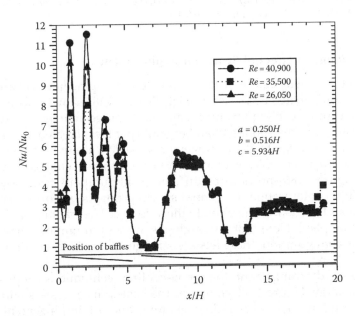

FIGURE 4.183
Local Nusselt number distribution along the channel centerline for configuration A of Dutta et al. (From Dutta, P. et al., Internal heat transfer enhancement by two perforated baffles in a rectangular channel, Presented at the *International Gas Turbine & Aeroengine Congress & Exhibition*, Stockholm, Sweden, ASME Paper 98-GT-55, June 2–5, 1998b.)

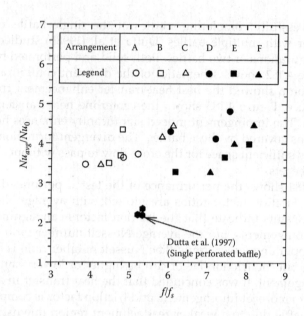

FIGURE 4.184
Performance comparison of double baffle arrangements used by Dutta et al. (From Dutta, P. et al., Internal heat transfer enhancement by two perforated baffles in a rectangular channel, Presented at the *International Gas Turbine & Aeroengine Congress & Exhibition*, Stockholm, Sweden, ASME Paper 98-GT-55, June 2–5, 1998b.)

4.4.8 Combined Effect of Swirl and Impingement

Glezer et al. (1996) modified the leading-edge impingement chamber to provide swirl created by the issuing jets. Figure 4.185 shows a schematic of their conceptual cooling arrangement. The impingement jets are released like a wall jet in the cooling chamber, and spent jets created a swirl flow. This flow is similar to the arrangements used in centrifugal separators or vortex chambers. Glezer et al. (1996) called this a screw-cooling technique. Figure 4.186 shows three different configurations tested by them. In the first configuration, tangential swirling flow is uniformly supplied throughout the cooling chamber. In the second configuration, partial swirl jets are supplied. Note that although jets are less in number than in the previous configuration, these jets are stronger than those in the previous configuration.

In the third arrangement, the screw-cooling technique is coupled with film discharge. Figure 4.187 compares the heat-transfer results with these three configurations with ribbed channel and jet impingement channels. Results indicate that the film extraction significantly lowers the performance of the screw cooling. However, the second configuration has a very high heat-transfer coefficient; it is higher than ribbed and regular impingement heat-transfer methods. The second configuration

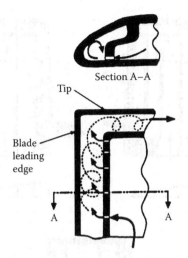

Section A–A

Tip

Blade
leading
edge

A

A

FIGURE 4.185
Schematic of the screw-cooling arrangement used by Glezer et al. (From Glezer, B. et al.,
A novel technique for the internal blade cooling, ASME Paper 96-GT-l 81, 1996.)

is comparable to the impingement with film extraction. Recall that film
extraction in impingement cooling reduces the cross-flow effect; therefore,
heat-transfer coefficients are higher in impingement with film extraction
than impingement without film extraction. Since the proposed screw-
cooling technique can match the heat-transfer coefficient of impingement
with film extraction, the internal cooling may be possible without using a
film. Note that though film cooling reduces the thermal load on the airfoil,
it has significant undesirable side effects, like thermal stress and aerodynamic imperfections.

4.4.9 Concept of Heat Pipe for Turbine Cooling

Heat pipes have very high effective thermal performance (Zuo et al., 1997,
1998). Therefore, they can transfer heat from the high-temperature to the low-
temperature regions. This concept may be used in airfoil cooling. Anderson
et al. (1993) investigated the possible use of heat pipes in the reduction of
cooling air. Figure 4.188 shows the schematic of a typical heat pipe, which
constitutes a closed container that is filled with a liquid–vapor mixture.
A wick is used to transport liquid by surface tension effects, and the vapor
moves by pressure difference. The liquid evaporates at the high-temperature
end and creates a high pressure by expansion, whereas the vapor condenses
in the lower-temperature end and, therefore, creates a lower-pressure region.
The vapor transport is pressure driven, and the vapor is transported from the
high-temperature end to the low-temperature end. The wick helps to transfer
the liquid from the low-pressure end to the high-pressure end. Therefore, the
surface tension effects should be higher than the pressure gradient effect to

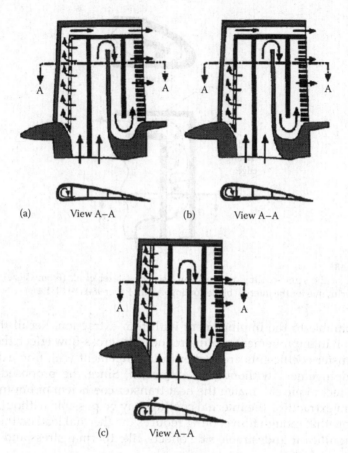

FIGURE 4.186
Schematics of different screw-cooling arrangements by Glezer et al. (a) Tangential swirl flow, (b) partial swirl jets, and (c) screw cooling with film dischage. (From Glezer, B. et al., A novel technique for the internal blade cooling, ASME Paper 96-GT-l 81, 1996.)

transfer liquid from the low-temperature end to the high-temperature end. This concept has several important attributes: First, it is capable of accepting very high heat fluxes; second, since heat transfer is done by phase change, it can maintain nearly constant temperature of the cooling surface; and third, since coolant flow inside the heat pipe is driven by the thermal load, it is self-controlling. That means the coolant and heat flow increase with an increase in the temperature difference.

Figure 4.189 shows a schematic arrangement of a heat pipe for effective cooling and recovery of heat in a multistage configuration. Heat is removed from the initial stage stator airfoils and delivered at a later stage to heat up the main flow. This way the heat extracted can be recycled to the main flow. Another cooling concept developed by Yamawaki et al. (1997) is shown in Figure 4.190. In this concept, the heat is conducted

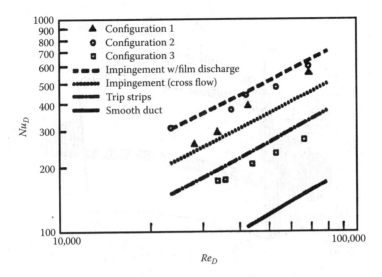

FIGURE 4.187
Comparison of different cooling techniques. (From Glezer, B. et al., A novel technique for the internal blade cooling, ASME Paper 96-GT-l 81, 1996.)

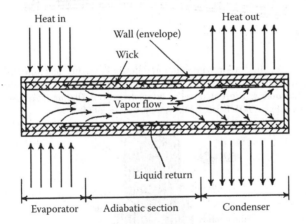

FIGURE 4.188
Schematic of a heat pipe. (From Anderson, W.G. et al., Heat pipe cooling of turboshaft engines, Presented at the *International Gas Turbine & Aeroengine Congress & Exposition*, Cincinnati, OH, May 24–27, 1993.)

away from the hot air-foil to the fin assembly. This passive heat extraction reduces the required cooling air.

Most heat pipe applications are designed for the stator airfoils, where it is easier to mount the connecting pipes or fins. Recently, Kerrebrock and Stickler (1998) proposed a design to incorporate a heat pipe in the rotor. Figure 4.191 shows a schematic of their concept. This heat pipe relies on the

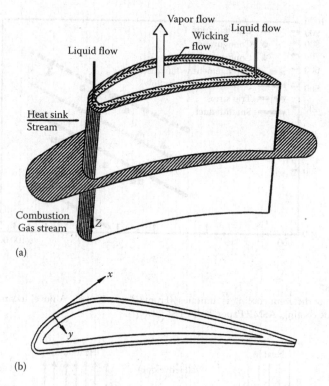

(a)

(b)

FIGURE 4.189
Schematic of a vane configuration with vapor and liquid flows. (a) Heat pipe turbine vane cooling and (b) Representative vane cross-section. (From Zuo, Z.J. et al., A parametric study of heat pipe turbine vane cooling, Presented at the *International Gas Turbine & Aeroengine Congress & Exhibition*, Orlando, FL, ASME Paper 97-GT-443, June 2–5, 1997.)

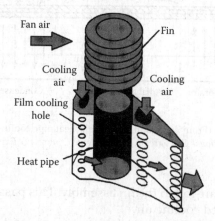

FIGURE 4.190
Schematic of a heat pipe application for vane cooling. (From Yamawaki, S. et al., Fundamental heat transfer experiments of heat pipes for turbine cooling, Presented at the *International Gas Turbine & Aeroengine Congress & Exhibition*, Orlando, FL, ASME Paper 97-GT-438, June 2–5, 1997.)

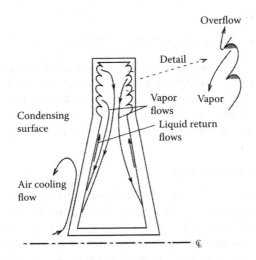

FIGURE 4.191
Schematic of an evaporative cooling system for rotor. (From Kerrebrock, J.L. and Stickler, D.B., Vaporization cooling for gas turbines, the return-flow cascade, Presented at the *Gas Turbine & Aeroengine Congress & Exhibition*, Stockholm, Sweden, ASME Paper 98-GT-177, June 2–5, 1998.)

centrifugal force to transfer liquid from the cold end to the hot end. The narrow top end of the cooling system is inserted in the rotor airfoil, and the wide bottom is extended to the rotor disk. The cooling air cools the vapor to liquid in the disk, and the condensed liquid moves radially outward to the rotating airfoil by centrifugal force. A special surface is prepared on the evaporating side to effectively evaporate the cooling fluid.

4.4.10 New Cooling Concepts

Commonly a closed loop steam cooled nozzle with thermal barrier coatings (TBC) is used in order to reduce the hot gas temperature drop through the first stage nozzle (Corman and Paul, 1995). A closed looped mist/steam cooling was reported by Guo et al. (1999). Results show that an average heat transfer enhancement of 100% can be achieved with 5% mist (fine water droplets) compared to the steam cooling.

The concept of cooled cooling air systems, through a heat exchanger, for turbine thermal management was reported by Bruening and Chang (1999). Results show that the use of a cooled cooling air system can make a positive impact on overall engine performance for land-based turbines.

The concept of using concavities (dimples) for heat-transfer enhancement in turbine blade cooling channels was reported by Chyu et al. (1997) and Moon et al. (1999). Results show that the use of dimples can enhance reasonable amount of heat transfer with relatively low pressure drop penalty.

4.5 New Information from 2000 to 2010

Since 2000, there have been lots of studies focused on pushing heat-transfer enhancement above conventional levels of 2.5 and above. Most of these studies have looked at very focused improvements based on flow physics and turbulence generation and also toward combining two or more enhancement techniques to promote increased heat transfer.

4.5.1 Rib Turbulated Cooling

Casarsa and Arts (2002) provided a detailed aero/thermal investigation of the turbulent flow inside a rib-roughened turbine blade cooling channel with blockage as high as 30% of the channel height by means of PIV and liquid crystal thermometry (LCT). Enhancement levels were of the order of 2.2 with a 90° rib. No mention was made regarding increased pressure drop. Chanteloup and Boelcs (2003) studied the effect of bend on heat transfer in a two-pass channel with and without film extraction. Three-dimensional velocity measurements and heat-transfer measurements were obtained in two-legged blade coolant passage models with ribs orientated 45° to the passage. Cho et al. (2003) investigated the effects of rib arrangements and aspect ratios of a rectangular duct simulating the cooling passage of a gas turbine blade. Two different V-shaped rib configurations were tested in a rectangular duct with the aspect ratios (W/H) varying from 3 to 6.82. The major conclusion from their study was when the thermal performances of two rib configurations were compared, the 45° discrete V-shaped rib configurations showed lower heat-transfer performance compared to the 60° continuous V-shaped rib configurations due to the weaker strength of the generated secondary flow. However, as duct aspect ratio increased, the discrepancy of thermal performances decreased, and for the highest aspect ratio of $AR = 6.82$, the thermal performances for both their rib configurations were almost the same. Bunker and Bailey (2003) measured heat transfer and friction coefficients within a rectangular passage of aspect ratio 0.4 containing 45° staggered turbulators of very high blockage. Using a constant pitch-to-height ratio of 10 for all geometries, turbulator height-to-channel hydraulic diameter ratios from 0.193 to 0.333 were investigated with a range of e/D creating actual channel blockage ratios e/H from 0.275 to 0.475, presenting significant flow area restrictions. They found that high-blockage turbulators with 45° orientation provide significantly lower friction factors when compared to 90° turbulators. In contrast to low-blockage turbulated channels, however, the high-blockage 45° turbulated Nusselt numbers are found to be lower than that those at 90° orientation, given very similar e/D and e/H values.

Bunker (2004) employed a simple radial vortex channel design throughout the channel with subchannel aspect ratios near unity, and Reynolds numbers from 20,000 to 100,000. Figure 4.192 shows the concept proposed by Bunker (2004).

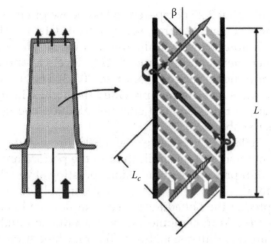

FIGURE 4.192
Latticework design proposed by Bunker. (From Bunker, R., Latticework (vortex) cooling effectiveness, Part 1: Stationary channel experiments, ASME Paper GT2004-54157, 2004.)

Two test methods are used to determine the local and overall heat-transfer coefficients for a vortex channel with crossing angle of 45°. Both liquid crystal and infrared thermography methods were used on acrylic and metallic models to discern the heat-transfer coefficients without and with the effects of internal rib fin effectiveness. Tests with insulating ribs determined the heat transfer on the primary surfaces representing the pressure and suction side walls of an airfoil. Tests with integral metal ribs determined the additional impact of the fin effectiveness provided by the ribs. Overall vortex channel heat-transfer coefficient enhancement levels are shown to be 2.5–3. This rivals the heat transfer of commonly used turbulated channels. In narrow vortex channels, overall enhancements are about 3, while in the wider vortex channels, turn effects are less of the total effect producing overall enhancements levels of about 2.8. Bunker et al. (2004) investigated a new class of in-wall mesh cooling techniques intended to produce significant wetted surface area heat-transfer enhancement products, $h_A A_{wetted}$, with low to moderate pressure losses. Cooling networks (meshes) are simulated using round pins and rounded diamond shaped pins with H/D_p ratios of 0.2 and S/D_p ratios of 1.5, as well as less dense rounded diamond pins of smaller diameter with H/D_p of 0.3 and S/D_p of 2.14. This investigation was the first of its kind to demonstrate feasible in-wall cooling methods that could be realized in practice for investment cast turbine blades. This study aims to be the first step toward a more comprehensive thermal-mechanical design solution for in-wall cooling that includes weight and stress considerations. Their specific conclusions are (1) dense pin meshes are more capable of producing about 3× average channel heat-transfer coefficients over those of a smooth channel, with

"hA" increases of about 2.25 compared to less dense pin meshes that agree with conventional staggered array, short pin-fin data. (2) The addition of turbulators or dimple arrays within the less dense pin mesh recovered heat-transfer coefficients and enhanced "hA" levels to about 2.5. Friction coefficient enhancements ranged from 6 to 20. (3) The use of pin meshes with both turbulators and dimples provided the highest heat transfer, with "hA" of 3 and more times smooth channel. Friction coefficient enhancement was higher at about 25.

Amaral et al. (2008) and Verstraete et al. (2008) presented a conjugate heat transfer (CHT) method and its application to the performance and lifetime prediction of a high-pressure turbine blade operating at very high inlet temperature. Amaral et al. (2008) used three separate solvers: a Navier–Stokes (NS) solver to predict the nonadiabatic external flow and heat flux, a finite element analysis (FEA) to compute the heat conduction and stress within the solid, and an one-dimensional aero-thermal model based on friction and heat-transfer correlations for smooth and rib-roughened cooling channels. Using the method developed by Amaral et al. (2008), Verstraete et al. (2008) optimized the channel and rib geometry for a cooled blade. They indicated that optimizing the geometry and position of the cooling channels resulted in a considerable increase of lifetime for the cooled blade. The optimization system was shown to be cost-effective and is suited for industrial applications.

Zehnder et al. (2009) investigated the influence of different turning vane configurations in the bend region of internal channels on pressure loss and local heat-transfer distribution. Results showed that the influence of investigated turning vane configurations was generally to increase heat transfer in the bend region: about 5% for the inner turning vane and outer turning vane configurations, for the configuration using both turning vanes the increase was only marginal. Their study showed that by the application of turning vanes in the bend region the pressure loss could be significantly reduced while keeping heat-transfer levels reasonably high.

More recently, Rallabandi et al. (2009b) performed systematic experiments to measure heat transfer and pressure losses in a stationary square channel with 45° square/sharp edged ribs at a wide range of Reynolds numbers ranging from 30 K to very high flows of $Re = 400$ K. These high Reynolds are typical of land-based turbines. Later, Rallabandi et al. (2009a) included 45° round-edged ribs as well to account for manufacturing effects. Figure 4.193a and b shows the rib profiles that were studied in the respective publications. In both studies, an array of blockage ratios and rib spacing ratios were considered. The correlations of Han (1988), Han et al. (1988), and Han and Park (1988) were modified to fit into the new extended parameter range. The parameters G and R (heat-transfer roughness and friction roughness functions, respectively) have been used in literature to absorb the effect of rib height (e/D) and Reynolds number (Re) into one variable. The heat-transfer roughness (G) is an indicator of thermal performance. For large values of G,

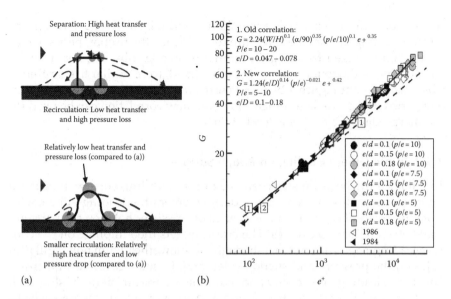

FIGURE 4.193
(a and b) Profile of ribs studied and (c) heat-transfer roughness variation with Reynolds number roughness for sharp ribs. (From Rallabandi, A.P. et al., *ASME J. Heat Transfer*, 131(7), 071703, 2009a; From Rallabandi, A.P. et al., Heat transfer and pressure drop measurements for a square channel with 45° round edged ribs at high Reynolds numbers, *Proceedings of ASME Turbo-Expo 2009*, Orlando, FL, ASME Paper GT2009-59546, June 8–12, 2009b.)

thermal performance will be low. Thus, it is desirable to achieve as low a value for G as possible. These parameters (G and R) have been correlated with e^+ (a nondimensional roughness Reynolds number) and P/e. Expressions of the form $R = C_1(P/e)^m$ and $G = C_2(P/e)^m(e^+)^n$ have typically been utilized to accurately correlate experimental data. In their studies, however, the blockage ratio (e/D) had to be explicitly included in the correlations for R and G. Figure 4.193c presents the results from the authors work using sharp ribs. Notable is the very well correlating of the heat-transfer roughness (G) with the roughness Reynolds number (e^+), which is defined as $e^+ = (e/D)(Re)(f/2)^{1/2}$. This work has extended the e^+ range of previous work from $e^+ = 1,000$ ($Re = 70$K $e/D = 0.078$) to $e^+ = 18,000$ ($Re = 400$K $e/D = 0.18$). Figure 4.193c indicates that the correlations for R and G do not agree with the earlier published correlations in the extended range. The authors attribute this to the parameter range differences, specifically the e/D ratio that was considerably larger than in prior work. The authors further explain that the turbulent boundary layer universal logarithmic velocity profile assumption is valid when the surface roughness is relatively small. Larger rib thickness could however invalidate this assumption owing to greater form drag caused by flow separating and reattaching due to the rib in comparison with the skin friction. This causes a dependence of R and G on e/D, as well as on e^+. However, round-edged ribs results coincidently showed agreement between prior correlations and the new data in the extended range. With round-edged ribs, the friction was lower, resulting in a smaller pressure drop. In fact, the friction performance for the round-edge ribs was quite similar to smaller ribs. The heat-transfer coefficients for the round ribs, on the other hand, were similar to sharp-edge ribs. Therefore, in Figure 4.193c, old correlation can be used for the round-edged ribs in the extended range. However, new correlation can be used for the sharp-edged ribs in the extended range.

4.5.2 Impingement Cooling on Rough Surface

To enhance the heat-transfer coefficient on the leading-edge wall of these cavities, the cooling flow in some designs enters the leading edge cavity from the adjacent cavity through a series of crossover holes on the partition wall between the two cavities. The crossover jets then impinge on the concave leading-edge wall and exit through the showerhead film holes, gill film holes on the pressure and suction sides, and, in some cases, form a cross-flow in the leading-edge cavity and move toward the airfoil tip. Taslim et al. (2002) studied the effects that racetrack crossover jets, in the presence of film holes on the target surface, have on the impingement heat-transfer coefficient. Their results indicated that the racetrack crossover holes produce higher impingement heat-transfer coefficients when compared with the round crossover jets. Without the inclusion of the heat-transfer area increase, the addition of surface features such as smaller conical bumps, longitudinal ribs produced comparable results to the smooth surface with crossover holes while bigger cones on the impingement surface produced higher impingement heat-transfer coefficients by about 30%. Bailey and Bunker (2002) studied the effect of the impingement jet array with a fixed target surface to cover a wide range of conditions, extending beyond the currently available literature data. They looked at axial and lateral jet spacing values of x/D and y/D of 3, 6, and 9, all with square orientation and in-line jets. The jet plate-to-target surface distance z/D was also varied from 1.25 to 5.5 with jet Reynolds numbers ranging from 14,000 to 65,000. They compared their results with the correlations of Florschuetz et al. and Kercher and Tabakoff indicating deviations in the streamwise row heat transfer behavior, which were sometimes minor and in several instances major. The detailed surface impingement heat-transfer distributions provided more insight into the character of such dense arrays in confined geometries, shedding light on the transitional nature of the flows between impingement dominated regions and channel-like flows.

Gaddis et al. (2004) experimentally investigated the mist/steam heat transfer of three rows of circular jet impingement in a confined channel. Fine water droplets with an average diameter of 3 mm were generated by atomizing water through small nozzles under high pressure, thus producing the mist. The circular jets had a uniform diameter of 8 mm, and the distance between adjacent jets in a row was 3 diameters. Jets in different rows were staggered and the distance between rows was 1.5 diameters with a nozzle-to-target spacing of 2.8 diameters. Experiments were conducted with Reynolds numbers at 7,500 and 15,000 and heat fluxes ranging from 3,350 to 13,400 W/m². The experiment results indicated that the wall temperature decreased significantly because of mist injection. A region of high cooling enhancement was observed and more extensive than those observed for one row of circular jets or slot jet. The overall performance of the mist/steam cooling showed potential for future gas turbine applications Kanokjaruvijit and Martinez-Botas (2004) experimentally investigated jet impingement on a dimpled surface for Reynolds numbers in the range 5,000–11,500, and jet-to-plate spacing from 2 to 12 jet-diameters. Two different dimple geometries were considered: hemispherical dimples and double or cusp elliptical dimples. Jet-to-plate spacing had a significant impact on the heat transfer of the dimple impingement. The narrow spacing seemed to cause recirculation, which diminished the heat transfer. Concurrently, at wide spacing, a lesser enhancement was measured. The depth of the dimples significantly influenced the heat transfer. The shallower dimples (d/Dd = 0.15) enhanced higher heat transfer than the deeper ones (d/Dd = 0.25) plausibly because the vortices formed inside the dimples could be shed more easily than the deeper dimples.

Bailey and Bunker (2005) presented detailed local distributions of heat-transfer coefficients for a platform cooling solution that utilizes a single impingement jet directed to the underside of the platform from the forward shank face. The cooled platform region is formed by the cavity between the pressure side and suction side of two adjacent blade castings. The effects of jet impingement distance to the surface from Z/D of 1.5 to 8, and jet Reynolds number from 65,000 to 155,000 was investigated with added surface roughness to further enhance heat transfer. The roughness benefit was not discernible if the impingement cooling jet was not relatively close to the platform surface. Kanokjaruvijit and Martinez-Botas (2005) presented heat transfer and pressure results for an inline array of round jets impinging on a staggered array of dimples with the consideration of various geometric and parametric effects. Dimpled impingement results were normalized against flat-plate data. The hemispherical dimples resulted in higher heat transfer compared to cusped dimples. This was thought due to the fact that two hemispherical dimples that formed a cusped elliptical dimple caused higher degree of recirculation than a normal hemispherical dimple. They also indicted that the shallow dimples led to better heat-transfer improvement than the deep ones, because they formed less severe recirculation, and shed the vortices faster. Nakamata et al. (2005) conducted experimental and numerical studies for the development of the integrated impingement and pin-fin cooling configuration. They focused on the spatial arrangements of impingement hole, pin-fin and film cooling (discharge) hole. They found that a configuration where there is a film hole in the center of four impingement jets provides the best overall thermal performance compared to various other configurations investigated in the study.

Son et al. (2005) presented shear pattern visualization on the impingement target surface, pressure loss measurements, and heat-transfer coefficient measurements for an impingement cooling system with simply shaped
roughness elements—specifically cylindrical and diamond pimples. They
found that the largest contribution to heat transfer is the impingement stagnation point and the developing wall jet regions. However, the research also
showed that the low heat-transfer coefficient region could be made to contribute more by using strategically located roughness elements. They designed
a patented hexagonal rim to cover the complete low heat-transfer coefficient
region midway between neighboring jets. Park et al. (2006) provided results
on the separate effects of Reynolds number and Mach number for an array of
impinging jets in the form of discharge coefficients, local and spatially averaged Nusselt numbers, and local and spatially averaged recovery factors.
They presented a new correlation equation for spatially averaged Nusselt
numbers for Mach numbers beyond 0.2 up to 0.6 as the existing correlations
under-predicted heat transfer for the same Reynolds numbers. Chambers
et al. (2006) presented a computational and experimental investigation into
the use of shaped elliptical or elongated circular impingement holes designed
to improve the penetration of the impinging jet across the coolant passage.
Experimental tests of an elongated hole with semicircular ends and aspect
ratio 1.35 confirmed the changes in behavior predicted by the CFD and provided reliable measurements of the Nusselt number enhancement. From the
tests performed it may be concluded that a low aspect ratio hole, of axes ratio
1.2–1.5, was required to obtain heat-transfer enhancement; average Nusselt
number enhancement in low cross-flow region of the channel was between
28% and 77% and a weak function of the jet Reynolds number, being driven
by a drop in the hole discharge coefficient; average Nusselt number enhancement in the cross-flow dominated region was 16% and independent of Re_{jet} at
the limited number of Reynolds numbers tested.

Bergholz (2008) described a process for conceptual and preliminary turbine airfoil cooling design based on the assembly of primitive cooling elements to define an overall "realizable" three-dimensional impingement
cooled airfoil cooling structure. The results from this study indicated that,
for sparse impingement arrays or low-conductivity airfoil materials, the use
of correlations based on spatially averaged data resulted in significant errors
in the prediction of airfoil temperatures and thermal gradients.

Xing and Weigand (2010) investigated heat-transfer behavior on a 9 × 9
staggered jet array impinging on a flat or rib roughened plate at Reynolds
numbers from 15,000 to 35,000. The heat-transfer performance on the roughened plate was better for both maximum and minimum cross-flow. The heat-
transfer enhancement ratio increased with increasing Reynolds numbers.
The narrow jet-to-plate spacing $H/d = 3$ resulted in the highest heat-transfer
enhancement for different cross-flow schemes with the highest enhancement ratio up to 7.5%. Segura and Acharya (2010) presented heat-transfer
results for a given slot-shaped channel with a 3:1 aspect ratio using various

methods to enhance swirl in the channel including helical shaped-trip-strips and swirl-jets issuing from the side walls. Their results indicate that the jet enhancement configurations show a substantial improvement in mean heat-transfer enhancement values when compared to published results that utilize trip strips, but did not provide a uniform distribution of enhancement. The trapezoid-shaped double helix trip strips provided a more uniform distribution of heat transfer, similar to trip strips used in rectangular channels. The overall heat-transfer enhancement values ranged from 2.36 at $Re = 10\,K$ to 3.02 at $Re = 50\,K$.

4.5.3 Trailing Edge Cooling

Lau et al. (2006) used naphthalene sublimation experiments to study heat transfer for flow through blockages with holes in an internal cooling passage near the trailing edge of a gas turbine airfoil. They investigated two configurations and indicated that neither the average heat (mass) transfer nor the distribution of the local heat (mass) transfer was significantly affected by the geometries of the entrance channel and the exit slots considered in this study. Chen et al. (2006) presented an experimental study of heat transfer over a trailing edge configuration preceded with an internal cooling channel of pedestal array that consisted of both circular pedestals and oblong-shaped blocks. The overall heat-transfer enhancement for the entire pedestal array, relative to the corresponding fully developed condition in a smooth channel, ranged from 2.3 to 2.8 for $3500 < Re < 9500$. Data also showed that the end-wall had about 30%–40% higher heat-transfer coefficient than the pedestals. The pedestal-induced mixing and turbulence upstream to the slot injection appeared to have significant impact on the film cooling over the cutback land region. Saha et al. (2008) presented an experimental study of the heat-transfer distribution and pressure drop through a converging lattice-matrix structure representing a gas turbine blade trailing-edge cooling passage for a scaled up model under three Reynolds numbers ($24,000 < Re < 60,000$). They reported a relatively higher value of average heat-transfer coefficient compared to the pin fins, which are widely used in trailing edge regions of gas turbine blades. Pin fins had suggested average Nu enhancement factor in the range of 1.8–3.6 for different orientations of the pin fins, whereas this lattice-matrix geometry showed an average Nu enhancement factor of 3–4.

Armellini et al. (2008) and Coletti et al. (2008) presented a comprehensive aero-thermal experimental and computational study of a trapezoidal cross-section model simulating a trailing edge cooling cavity with one rib roughened wall. The flow to the cavity was fed through tilted slots on one side wall and exited through straight slots on the opposite side wall. A three-dimensional reconstruction of detailed two-dimensional PIV measurements shows that the insertion of the ribs produces a complex interaction with the crossing jets. The jet–rib interaction produced two upward jet deflections for each inter-rib domain; these structures originate from the bottom wall and carry fluid to the upper wall where a periodic system of separation lines and impingements is observed. The turbulators tend to extend the positive effect of the impingement by breaking and deflecting the jets. The ribs insertion was responsible for an average heat-transfer enhancement of 25% on the bottom wall and of 16% on the upper wall.

4.5.4 Dimpled and Pin-Finned Channels

In the past decade, concavities or dimples on surfaces have been a focus for heat transfer enhancement especially due to the lower pressure drop for higher heat transfer enhancement obtained for these geometries. Bunker and Donnellan (2003) obtained heat transfer and friction coefficients measurements for fully developed, turbulent internal flows in circular tubes with six different concavity (dimple) surface array geometries. Two different concavity depths and three different concavity array densities were tested using tube bulk flow Reynolds numbers from 20,000 to 90,000. Their results showed that heat-transfer enhancements for dimpled internal surfaces of circular passages can reach factors of 2 or more when the relative dimple depth was greater than 0.3 and the dimple array density was about 0.5 or higher. Kovalenko and Khalatov (2003) studied dimpled rectangular channels with in-line and staggered dimple configurations in a Reynolds number range of 8,000–115,000. They saw the best heat-transfer enhancements in the range of 1.45–1.55 for the shallow dimples with a staggered configuration. Bunker (2007) demonstrated a new tip cap augmentation method to provide substantially increased convective heat flux on the internal cooled tip cap of a turbine blade. The design consists of several variations involving the fabrication or placement of arrays of discrete-shaped pins on the internal tip cap surface. They tested five tip cap surfaces including a smooth surface, two different heights of aluminum pin arrays, one more closely spaced pin array, and one pin array made of insulating material. Effective heat-transfer coefficients based on the original smooth surface area were increased by up to a factor of 2.5 primarily due to the added surface area of the pin array. However, factoring out this surface area effect showed that the local heat-transfer coefficient had also been increased by about 20%–30%, primarily over the base region of the tip cap itself with a negligible increase in tip turn pressure drop over that of a smooth surface. Chyu et al. (2007) experimentally examined the effects of imperfect manufacturing phenomena on the heat transfer and friction characteristics over pin-fin arrays with different pin inclinations. Their results showed a consistent trend that an increase in pin inclination reduces the levels of both heat transfer and friction loss. The magnitudes of heat-transfer coefficient and friction factor for the 40° inclined cases were about one-half the corresponding values of the 90° cases.

Taslim and Bethka (2007) investigated impingement on the leading-edge of an airfoil in the presence of cross-flows beyond the cross-flow created by the upstream jets (spent air). With measurements of heat-transfer coefficients on the airfoil nose area as well as the pressure and suction side areas. Based on their results, the presence of the external cross-flow reduced the impinging jet effectiveness both on the nose and side walls. Even for an axial to jet mass flow ratio as high as 5, the convective heat-transfer coefficient produced by the axial channel flow was less than that of the impinging jet without the presence of the external cross-flow. Zhou and Acharya (2009) presented mass/heat-transfer measurements in a square internal passage where one wall has a single dimple with four different types of dimple shapes: square, triangular, circular, and teardrop. They used the using the naphthalene sublimation method. The results both showed that the teardrop dimple provides the highest heat/mass transfer among the four dimples covered in this study. In the wake region, the triangular dimple shows the lowest Nu/Nu_0 values, while the other two dimple shapes (circular and square) show comparable Nu/Nu_0.

Siw et al. (2010) experimentally studied heat transfer and pressure characteristics in a rectangular channel with pin-fin arrays of partial detachment from one of the endwalls have been experimentally studied. The idea of implementing pin fins with detached spacing between the pin-tip and the adjacent endwall was to promote turbulent convection with separated shear layers induced near the pin-tips. Local heat-transfer data obtained over the entire wet surface suggested that the size of detached spacing, normalized by the pin diameter, i.e., C/D, had a profound influence on the overall heat-transfer characteristics in the channel. Compared to the baseline case of the fully attached pin-fin array, $S/D = X/D = 2.5$ and $H/D = 4$ (i.e., $C/D = 0$), the case for the staggered array with $C/D = 1$ resulted in the highest heat-transfer enhancement among all the cases studied, about 15% higher than the baseline case.

4.5.5 Combustor Liner Cooling and Effusion Cooling

Goldstein et al. (2004) investigated the local heat (mass) transfer characteristics of flow through two perforated plates placed, relative to each other, in either staggered, in-line or shifted in one direction. Hole length to diameter ratio of 1.5, hole pitch to diameter ratio of 3.0, and distance between the perforated plates of 1–3 hole diameters were used at hole Reynolds numbers of 3,000–14,000. Their results indicated that the overall averaged Sherwood number for the staggered arrangement and the shifted arrangement were approximately 70%–75% higher than that for the in-line arrangement and 3–3.5 times higher than that for the single layer flow (effusion only). For the small hole spacing, unlike the case with the large hole spacing, the staggered and the shifted hole arrangements had almost the same average values although the shifted case has slightly lower values due to the formation of low transfer regions. Rhee et al. (2004) and Cho et al. (2005) investigated the effect of rib arrangements on flow and heat/mass transfer characteristics for an impingement/effusion cooling system with initial crossflow using the same set-up as the one used by Goldstein et al. (2004).

Maurer et al. (2007) presented an experimental and numerical investigation to assess the thermal performance of V- and W-shaped ribs in a rectangular channel. The ribs were located on one channel sidewall in order to simulate a typical combustor liner cooling. Observing the overall thermal performance, W-shaped ribs with a rib pitch-to-height of $P/e = 10$ perform best. The difference to the next best configuration, which is a W-shaped ribbed channel with $P/e = 5$, is reduced for higher Reynolds numbers. All the W-rib configurations were better than the similar V-rib configurations. Esposito et al. (2007) experimentally tested two different styles of jet impingement geometries to be used in backside combustor cooling. In combustor liner cooling, the Reynolds numbers of the jets can be as high as 60,000. The higher jet Reynolds numbers lead to increased overall heat-transfer characteristics, but also an increase in cross-flow caused by spent air. Figure 4.194 shows the three different configurations tested. The first new geometry tested was a corrugated wall design. The corrugations in the wall allow spent air from upstream jets to exit the impingement array without interfering with the downstream jets. The second design with extended ports offers a higher cross-sectional area for cross-flow than the baseline and corrugated wall therefore reducing the overall cross-flow velocity. Also, the length of the impingement tube is an additional benefit that allows a more developed jet flow. This increased the peak jet velocity and further reduces cross-flow effects by increasing the core jet to cross-flow velocity ratio. A variation to the extended port design was also tested with variable extended port lengths, shown in Figure 4.194. This further increases the cross-flow area especially for the front nine rows. The length of the extended ports was linearly varied from the first to the last rows of jets. All ports were of uniform length in the spanwise direction.

Figure 4.195 then shows the heat-transfer distributions for all the jets comparing the different geometries. The extended port geometries significantly

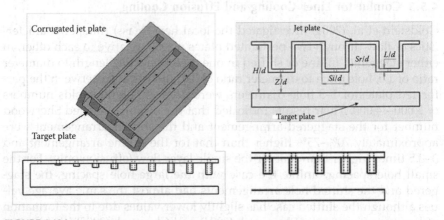

FIGURE 4.194
Different geometries tested by Esposito et al. Top left—corrugated; top right—side view of corrugated wall; bottom left—variable jet port lengths; bottom right—even length jet ports. (From Esposito, E. et al., Comparing extended port and corrugated wall jet impingement geometry for combustor liner backside cooling, ASME Paper GT2007-27390, 2007.)

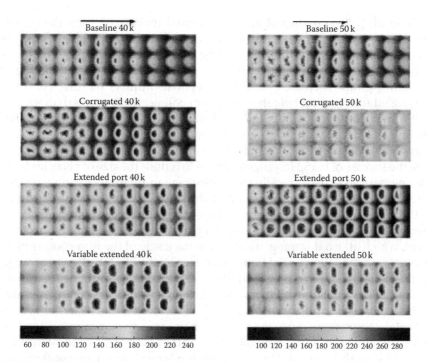

FIGURE 4.195
Detailed Nusselt numbers for different configurations at two different Reynolds numbers Esposito et al. (From Esposito, E. et al., Comparing extended port and corrugated wall jet impingement geometry for combustor liner backside cooling, ASME Paper GT2007-27390, 2007.)

improve heat transfer up to 40%–50% higher than baseline even at the 10th downstream row. Optimum extended port lengths are needed to allow the flow to develop prior to exiting the jet. Extended ports that are too short for a given Reynolds number can lead to a reduction in heat-transfer effectiveness, but if properly designed, enhancements of at least 20% in overall performance are possible for the downstream jets. It is clear that the use of corrugated walls or extended port geometry can produce significant benefits when the jets are packed in dense arrays.

Lauffer et al. (2007) used a model to represent a simplified combustor dome heat shield made out of perspex, and the distributions of the Nusselt number on the impingement target plate as well as on the side rims and along the central bolt recess of the heat shield were measured for different impingement Reynolds numbers. It was shown that the Nusselt number increases both locally and globally with increasing jet Reynolds number. However, for the impingement target wall, the overall averaged Nusselt number of the investigated realistic pattern was around 20% below the average Nusselt number that was expected by existing correlations for regular impingement fields.

Oh et al. (2008) conducted an experimental investigation on the cooling effectiveness of full-coverage film cooled wall with impingement jets. The main conclusions were that the cooling effectiveness increases due to the enhancement of both impingement array jet cooling and film cooling as the blowing ratio increases, i.e., the jet Reynolds number increases; and that the angled film cooling hole shows higher cooling effectiveness than the normal film cooling hole. Land and Thole (2008) investigated a double-walled cooling geometry with impingement and film-cooling for relieving sand ingested blockage effects. A number of parameters were simulated to investigate the success of using impingement jets to reduce the size of particles in the cooling passages that typically cause blockage in internal cooling passages. They concluded that the alignment of the impingement holes with respect to the film-cooling holes confirmed the beneficial effects of impingement in breaking up large particles into smaller particles. The staggered arrangement, which allowed impingement on the back side of the film-cooling plate, showed less reduction in the flow parameter relative to the aligned arrangement, where the impingement and film-cooling holes were aligned. Cardwell et al. (2008) evaluated sand blockages in double wall liners commonly used in the combustor and turbine for combined internal and external cooling of metal components. Specifically, sand blockages were evaluated through comparisons of measured flow rates for a particular pressure ratio across the liner. Results showed that the liner with the largest impingement flow area and least overlap between the impingement and film-cooling holes exhibited the lowest blocking overall at both ambient and heated conditions.

Behrendt et al. (2008) studied the effect of the cooling air properties on the cooling effectiveness of different advanced cooling concepts at realistic operating conditions. Quantitative data of the performance of advanced cooling concepts in a realistic environment in addition to existing data from scaled isothermal investigations is needed by combustor designers for further improvements. Their investigation revealed that the variation of the operating conditions led to a significant variation of the kinematic viscosity of the cooling air and thus contributing to the Reynolds number variation of the flow inside the cooling hole. Under high engine load, e.g., take-off and climb-out, the Reynolds effect on the discharge coefficient was negligible for most aero engine combustors.

Weaver et al. (2010) presented the development and validation of a testing method for microchannels as well as determine the effect of manufacturing roughness levels on these small channels. It was shown that at an average roughness height of 6.1 μm, which corresponded to 2.2% of the channel height, heat transfer was augmented by 1.1–1.2 while the friction factor was augmented significantly more by 2.1–2.6 over a smooth channel.

Hagari et al. (2010) investigated the heat-transfer performance of W-shaped ribs in a rectangular channel with typical geometries and flow conditions for a combustor liner cooling passage. Experiments were conducted with channel Reynolds number ranging from 40,000 to 550,000 with the ribs located on one side of the channel and the rib height-to-hydraulic diameter ratio (e/D_h) was 0.006–0.014, simulating the combustor liner cooling configurations. Heat-transfer enhancement factors ranged from 2.6 to 3.2 for the present rib configurations. It was found that sufficient level of heat transfer can be obtained even for considerably low blockage ratio such as $e/D_h < 0.01$ by applying the W-shaped ribs. Kunstmann et al. (2010) conducted an investigation to assess the thermal performance of 90° ribs, low and high W-shaped ribs, and combinations of low W-shaped ribs with high W-shaped ribs and with dimples in a rectangular channel with an aspect ratio (W/H) of 2:1. Their results showed that the highest heat-transfer enhancement rates were obtained by a combination of W-shaped ribs with $P/e = 10$ and $e/D_h = 0.06$ and W-shaped ribs with $P/e = 10$ and $e/D_h = 0.02$. The best thermal performance was achieved by regularly spaced lower W-shaped ribs and by a compound roughness of regularly spaced W-shaped ribs and dimples at Re below and above 300,000, respectively.

4.5.6 Innovative Cooling Approaches and Methods

Li et al. (2001) studied heat transfer enhancement due to mist added to steam in an impingement flow. A 150% enhancement with a mist concentration of 1.5% is typical in the stagnation region. The stagnation point enhancement was strongly influenced by heat flux, increasing from 40% at the highest flux to over 400% as the flux is reduced by a factor of 4. The heat transfer enhancement was modestly affected by steam velocity (Reynolds number). Taslim et al. (2001) and Talim and Setayeshgar (2001) reported on impingement cooling of a smooth as well as roughened airfoil leading edge. They examined sandpaper roughness and different conical bump and radial rib geometries in a test section with a circular nose, two tapered sidewalls, and a flat fourth wall on which the crossover jets were positioned. Circular and racetrack-shaped crossover jets, at 0° and 45° angles with the channel's radial axis, were compared. Results were also compared for leading-edge geometries with and without showerhead film holes. Taslim et al. (2003) studied three leading-edge surface geometries, consisting of a baseline smooth surface and two surfaces roughened with a combination of horseshoe and straight radial ribs, for impingement cooling. The smooth target surface produced the highest impingement heat transfer coefficients followed by the notched-horseshoe and horseshoe ribs. An overall increase of about 27% in heat removal was accomplished by roughening the leading-edge wall with these ribs. The increase was entirely attributed to the increase in the heat transfer area.

Kim et al. (2003) presented a comparison between three cooling schemes applicable to a serial cooled combustor liner configuration. They indicated that jet impingement provided the highest cooling, but also suffered from the largest pressure loss among the three studied cooling schemes. For given pumping power, the dimpled surface gave a superior cooling performance over jet array impingement and trip-strip surfaces. Nakamata et al. (2005) studied the effect of the spatial arrangement on the cooling performance of the integrated impingement and pin-fin cooling configuration using experiments and CFD analysis for some different configurations with various spatial arrangements of pins and holes. They indicated that a configuration with pins in the middle of an inline impingement array performed the best in terms of overall pressure drop and heat transfer enhancement. Maurer et al. (2008) conducted an experimental study to determine the heat transfer performance of advanced convective cooling techniques at the typical conditions found in a backside cooled combustion chamber. They indicate that for the design process of backside cooled combustion chamber walls, the range of the investigated test cases offered the possibility to find adequate convective cooling configurations for different design demands. In situations where the friction factor is the limiting parameter, walls with hemispheres proved to be an interesting alternative to ribbed channels whereas if heat transfer performance was the desired design parameter, W-shaped ribs represented the most efficient choice.

Jackson et al. (2009) studied a civil gas turbine HPT blade multi-pass system using a large-scale Perspex model and the heat transfer and pressure drop results compared to CFD predictions. The heat transfer test on the multipass model was replicated using CFD. The CFD accurately predicted the overall trends in the heat-transfer coefficient values. However, the absolute level of peak values looked slightly down, in the worst locations by up to 25%. Passage average heat-transfer coefficient values aligned well especially in the first and second pass, but in the third pass, the CFD-averaged heat-transfer coefficient values were under predicted by 20%. Poser et al. (2008) conducted transient heat transfer experiments in a model of a multipass gas turbine blade cooling circuit. The inner surface of the plexiglass model was coated with thermochromic liquid crystals in order to determine the internal heat transfer coefficients. The experiments were conducted with an engine-representative Reynolds number, Mach number, and heat flux direction as in a real blade. They evaluated their method and performed an uncertainty analysis that demonstrated robustness for the investigated geometries. Recently, LeBlanc et al. (2011) presented detailed heat-transfer distributions for a realistic turbine blade cooling circuit at engine representative Reynolds numbers. Different rib configurations were studied to determine the effect on heat transfer and pressure drop. The results showed a 50%–100% enhancement in heat transfer compared to smooth surface. The detailed distributions clearly showed the effect of coolant migration in these more complicated shaped channels with ribs and film cooling ejection holes. The results from the current study were compared with correlation-based predictions for the first pass rectangular channel for ribs and smooth surfaces. The correlation over-predicted the Nusselt numbers compared to the experimentally obtained Nusselt numbers. These results provided insight into using correlations developed for simplified geometries and idealistic flow conditions to predicting heat transfer for more realistic geometries.

References

Abuaf, N. and Kercher, D.M., 1992. Heat transfer and turbulence in a turbulated blade cooling circuit, *International Gas Turbine and Aeroengine Congress and Exposition*, June 1–4, Cologne, Germany, ASME Paper 92-GT-187.

Acharya, S., Dutta, S., Myrum, T.A., and Baker, R.S., 1993. Periodically developed flow and heat transfer in a ribbed duct. *International Journal of Heat and Mass Transfer*, 36(8), 2069–2082.

Acharya, S., Dutta, S., Myrum, T.A., and Baker, R.S., 1994. Turbulent flow past a surface mounted two-dimensional rib. *ASME Journal of Fluids Engineering*, 116(2), 238–246.

Acharya, S., Myrum, T., Qiu, X., and Sinha, S., 1997. Developing and periodically developed flow, temperature and heat transfer in a ribbed duct. *International Journal of Heat and Mass Transfer*, 40(2), 461–479.

Amaral, S., Verstraete, T., van den Braembussche, R., and Arts, T., 2008. Design and optimization of the internal cooling channels of HP turbine blade—Part I: Methodology. ASME Paper GT2008-51077.

Anderson, W.G., Hoff, S., Winstanley, D., Philips, J., and DelPorte, S., 1993. Heat pipe cooling of turboshaft engines, Presented at the *International Gas Turbine & Aeroengine Congress & Exposition*, May 24–27, Cincinnati, OH.

Armellini, A., Coletti, F., Arts, T., and Scholtes, C., 2008. Aero-thermal investigation of a rib-roughened trailing edge channel with crossing-jets. Part I: Flow field analysis. ASME Paper GT2008-50694.

Armstrong, J. and Winstanley, D., 1988. A review of staggered array pin fin heat transfer for turbine cooling applications. *ASME Journal of Turbomachinery*, 110, 94–103.

Arora, S.C. and Abdel-Messeh, W., 1989. Characteristics of partial length circular pin fins as heat transfer augmentors for airfoil internal cooling passages, *Gas Turbine and Aeroengine Congress and Exposition*, June 4–8, 1989, Toronto, Ontario, Canada, ASME Paper 89-GT-87.

Azad, G.M., Huang, Y., and Han, J.C., 2000a. Jet impingement heat transfer on pinned surfaces using a transient liquid crystal technique, *Proceedings of the 8th International Symposium on Transport Phenomena and Dynamics of Rotating Machinery*, March 2000, Honolulu, HI, Vol. II, pp. 731–738.

Azad, G.M., Huang, Y., and Han, J.C., 2000b. Jet impingement heat transfer on dimpled surfaces using a transient liquid crystal technique. *AIAA Journal of Thermophysics and Heat Transfer*, 14(2), 186–193.

Bailey, J.C. and Bunker, R.S., 2002. Local heat transfer and flow distributions for impinging jet arrays of dense and sparse extent. ASME Paper GT2002-30473.

Bailey, J.C. and Bunker, R., 2005. Turbine blade platform impingement cooling with effects of film extraction and roughness. ASME Paper GT2005-68415.

Behrendt, T., Gerendas, M., Lengyel, T., and Hassa, C., 2008. Characterization of advanced combustor cooling concepts under realistic operating conditions. ASME Paper GT2008-51191.

Bergholz, B., 2008. Cooling optimization of impingement heat transfer in turbine airfoils. ASME Paper GT2008-50704.

Bouchez, J.P. and Goldstein, R.J., 1975. Impingement cooling from a circular jet in a cross flow. *International Journal of Heat and Mass Transfer*, 18, 719–730.

Bradbury, L.J.S., 1965. The structure of a self-preserving turbulent plane jet. *Journal of Fluid Mechanics*, 23, 31–64.

Brigham, B.A. and VanFossen, G.J., 1984. Length to diameter ratio and row number effects in short pin fin heat transfer. *ASME Journal of Engineering for Gas Turbines and Power*, 106, 241–245.

Bruening, G.B. and Chang, W.C., 1999. Cooled cooling air systems for turbine thermal management, *Presented at the International Gas Turbine & Aeroengine Congress & Exhibition*, June 7–10, Indianapolis, IN, ASME Paper 99-GT-14.

Bunker, R., 2004. Latticework (vortex) cooling effectiveness. Part 1: Stationary channel experiments. ASME Paper GT2004-54157.

Bunker, R., 2007. The augmentation of internal blade tip-cap cooling by arrays of shaped pins. ASME Paper GT2007-27009.

Bunker, R.S. and Bailey, J.C., 2003. Heat transfer and friction in channels with very high blockage 45-deg staggered turbulators. ASME Paper GT2003-38611.

Bunker, R.S., Bailey, J.C., Lee, C.-P., Stevens, C.W., 2004. In-Wall Network (Mesh) cooling augmentation of gas turbine airfoils. ASME Paper GT2004-54260, 1007–1018.

Bunker, R.S. and Donnellan, K.F., 2003. Heat transfer and friction factors for flows inside circular tubes with concavity surfaces. ASME Paper GT2003-38053.

Bunker, R.S. and Metzger, D.E., 1990. Local heat transfer in internally cooled turbine airfoil leading edge regions. Part I: Impingement cooling without film coolant extraction. *ASME Journal of Turbomachinery*, 112, 451–458.

Byerley, A.R., Jones, T.V., and Ireland, P.T., 1992. Internal cooling passage heat transfer near the entrance to a film cooling hole: Experimental and computational results, *International Gas Turbine and Aeroengine Congress and Exposition*, June 1–4, Cologne, Germany.

Cardwell, N., Burd, S., and Thole, K., 2008. Investigation of sand blocking within impingement and film-cooling holes. ASME Paper GT2008-51351.

Casarsa, L. and Arts, T., 2002. Characterization of the velocity and heat transfer fields in an internal cooling channel with high blockage ratio. ASME Paper GT2002-30207.

Chakroun, W.M., Al-Fahed, S.F., and Abdel-Rehman, A.A., 1997. Heat transfer augmentation for air jet impinged on rough surface, *Presented at the International Gas Turbine & Aeroengine Congress & Exhibition*, June 2–5, Orlando, FL.

Chambers, A.C., Gillespie, D., Ireland, P.T., and Mitchell, M., 2006. Enhancement of impingement cooling in a high cross flow channel using shaped impingement cooling holes. ASME Paper GT2006-91229.

Chandra, P.R. and Cook, M.M., 1994. Effect of the number of channel ribbed walls on heat transfer and friction characteristics of turbulent flows. General papers in heat and mass transfer, ASME HTD-271, 201–209.

Chandra, P.R., Fontenot, M.L., and Han, J.C., 1997a. Effect of rib profiles on turbulent channel flow heat transfer. *Journal of Thermophysics and Heat Transfer*, 12(1), 116–118.

Chandra, P.R., Niland, M.E., and Han, J.C., 1997b. Turbulent flow heat transfer and friction in a rectangular channel with varying number of ribbed walls. *ASME Journal of Turbomachinery*, 119, 374–380.

Chanteloup, D. and Boelcs, A., 2003. Flow effects on the bend region heat transfer distribution of 2-pass internal coolant passages of gas turbine airfoils: Influence of film cooling extraction. ASME Paper GT2003-38702.

Chen, S., Li, P., Chyu, M., Cunha, F., and Abdel-Messeh, W., 2006. Heat transfer in an airfoil trailing edge configuration with shaped pedestals mounted internal cooling channel and pressure side cutback. ASME Paper GT2006-91019.

Cho, H.H., Hong, S.K., and Rhee, D.H., 2005. Effects of fin shapes and arrangements on heat transfer for impingement/effusion cooling with crossflow. ASME Paper GT2005-68684.

Cho, H.H., Moon, H.K., Rhee, D.H., and Lee, D.H., 2003. Effects of duct aspect ratios on heat/mass transfer with discrete V-shaped ribs. ASME Paper GT2003-38622.

Cho, H.H., Nam, Y.W., and Rhee, D.H., 2004. Local heat/mass transfer with various rib arrangements in impingement/effusion cooling system with crossflow. ASME Paper GT2004-53686.

Chupp, R.E., Helms, H.E., McFadden, P.W., and Brown, T.R., 1969. Evaluation of internal heat transfer coefficients for impingement cooled turbine airfoils. *AIAA Journal of Aircraft*, 6, 203–208.

Chyu, M.K., 1990. Heat transfer and pressure drop for short pin-fin arrays with pin-endwall fillet. *ASME Journal of Heat Transfer*, 112, 926–932.

Chyu, M.K. and Goldstein, 1991. Influence of an array of wall-mounted cylinders on the mass transfer from a flat surface. *International Journal of Heat and Mass Transfer*, 34(9), 2175–2186.

Chyu, M.K., Hsing, Y.C., and Natarajan, V., 1998a. Convective heat transfer of cubic fin arrays in a narrow channel. *ASME Journal of Turbomachinery*, 120, 362–367.

Chyu, M.K., Hsing, Y.C., Shih, T.I.P., and Natarajan, V., 1998b. Heat transfer contributions of pins and endwall in pin-fin arrays: Effects of thermal boundary condition modeling, *International Gas Turbine & Aeroengine Congress & Exhibition*, June 2–5, Stockholm, Sweden, ASME Paper 98-GT-175.

Chyu, M.K. and Natarajan, V., 1996. Heat transfer on the base surface of three-dimensional protruding elements. *International Journal of Heat and Mass Transfer*, 39(14), 2925–2935.

Chyu, M.K., Natarajan, V., and Metzger, D.E., 1992. Heat/mass transfer from pin-fin arrays with perpendicular flow entry. In *Fundamentals and Applied Heat Transfer Research for Gas Turbine Engines*, ASME HTD, Vol. 226. ASME, New York, pp. 31–39.

Chyu, M.K., Oluyede, E.O., and Moon, H.K., 2007. Heat transfer on convective surfaces with pin-fins mounted in inclined angles. ASME Paper GT2007-28138.

Chyu, M.K., Yu, Y., Ding, H., Downs, J.P., and Soechting, F.O., 1997. Concavity enhanced heat transfer in an internal cooling passage, Presented at the *International Gas Turbine & Aeroengine Congress & Exhibition*, June 2–5, Orlando, FL, ASME Paper 97-GT-437.

Coletti, F., Armellini, A., Arts, T., and Scholtes, C., 2008. Aero-thermal investigation of a rib-roughened trailing edge channel with crossing-jets. Part II: Heat transfer analysis. ASME Paper GT2008-50695.

Corman, J.C. and Paul, T.C., 1995. Power systems for the 21st century "H" gas turbine combined cycles. GE Power Systems, Schenectady, New York, GER-3935, pp. 1–12.

Dipprey, D.F. and Sabersky, R.H., 1963. Heat and momentum transfer in smooth and rough tubes in various Prandtl number. *International Journal of Heat and Mass Transfer*, 6, 329–353.

Downs, S.J. and James, E.H., 1987. Jet impingement heat transfer: A literature survey, *National Heat Transfer Conference*, August 9–12, Pittsburgh, PA, ASME Paper 87-HT-35.

Dutta, S., Dutta, P., Jones, R.E., and Khan, J.A., 1998a. Heat transfer coefficient enhancement with perforated baffles. *ASME Journal of Heat Transfer*, 120, 795–797.

Dutta, P., Dutta, S., and Khan, J.A., 1998b. Internal heat transfer enhancement by two perforated baffles in a rectangular channel, Presented at the *International Gas Turbine & Aeroengine Congress & Exhibition*, June 2–5, Stockholm, Sweden, ASME Paper 98-GT-55.

Ekkad, S.V. and Han, J.C., 1995. Local heat transfer measurements near a sharp 180° turn of a two-pass smooth square channel with a transient liquid crystal image technique. *Journal of Flow Visualization and Image Processing*, 2, 285–297.

Ekkad, S.V. and Han, J.C., 1997. Detailed heat transfer distributions in two-pass square channels with rib turbulators. *International Journal of Heat and Mass Transfer*, 40(11), 2525–2537.

Ekkad, S.V., Huang, Y., and Han, J.C., 1998. Detailed heat transfer distributions in two-pass smooth and turbulated square channels with bleed holes. *International Journal of Heat and Mass Transfer*, 41(13), 3781–3791.

Ekkad, S.V., Huang, Y., and Han, J.C., 1999. Impingement heat transfer on a target plate with film holes. *AIAA Journal of Thermophysics and Heat Transfer*, 13, 522–528.

Esposito, E., Ekkad, S., Kim, Y., and Dutta, P., 2007. Comparing extended port and corrugated wall jet impingement geometry for combustor liner backside cooling. ASME Paper GT2007-27390.

Fan, C.S. and Metzger, D.E., 1987. Effects of channel aspect ratio on heat transfer in rectangular passage sharp 180-deg turns, *Gas Turbine Conference and Exhibition*, May 31–June 4, Anaheim, CA, ASME Paper 87-GT-113.

Florschuetz, L.W., Berry, R.A., and Metzger, D.E., 1980. Periodic streamwise variations of heat transfer coefficients for inline and staggered arrays of circular jets with crossflow of spent air. *ASME Journal of Heat Transfer*, 102, 132–137.

Florschuetz, L.W., Metzger, D.E., and Su, C.C., 1984. Heat transfer characteristics for jet array impingement with initial crossflow. *ASME Journal of Heat Transfer*, 106, 34–41.

Florschuetz, L.W. and Su, C.C., 1987. Effects of crossflow temperature on heat transfer within an array of impinging jets. *ASME Journal of Heat Transfer*, 109, 74–82.

Florschuetz, L.W., Truman, C.R., and Metzger, D.E., 1981. Streamwise flow and heat transfer distributions for jet array impingement with crossflow. *ASME Journal of Heat Transfer*, 103, 337–342.

Gaddis, L., Wang, T., and Li, X., 2004. Mist/steam heat transfer of multiple rows of impinging jets. ASME Paper GT2004-54206.

Gau, C. and Lee, C.C., 1992. Impingement cooling flow structure and heat transfer along rib-roughened walls. *International Journal of Heat and Mass Transfer*, 35(11), 3009–3020.

Gee, D.L. and Webb, R.L., 1980. Forced convection heat transfer in helically rib-roughened tubes. *International Journal of Heat and Mass Transfer*, 23, 1127–1135.

Gillespie, D.R.H., Wang, Z., and Ireland, P.T., 1998. Full surface local heat transfer coefficient measurements in a model of an integrally cast impingement cooling geometry. *ASME Journal of Turbomachinery*, 120, 92–99.

Goldstein, R.J. and Behbahani, A.I., 1982. Impingement of a circular jet with and without cross flow. *International Journal of Heat and Mass Transfer*, 25, 1377–1382.

Goldstein, R.J., Behbahani, A.I., and Heppelmann, K.K., 1986. Streamwise distribution of the recovery factor and the local heat transfer coefficient to an impinging circular air jet. *International Journal of Heat and Mass Transfer*, 29, 1227–1235.

Goldstein, R., Cho, H.H., and Rhee, D.H., 2004. Effect of hole arrangements on local heat/mass transfer for impingement/effusion cooling with small hole spacing. ASME Paper GT2004-53685.

Goldstein, R.J., Chyu, M.K., and Hain, R.C., 1985. Measurement of local mass transfer on a surface in the region of the base of a protruding cylinder with a computer-controlled data acquisition system. *International Journal of Heat and Mass Transfer*, 28(5), 977–985.

Goldstein, R.J., Jabbari, M.Y., and Chen, S.B., 1994. Convective mass transfer and pressure loss characteristics of staggered short pin-fin arrays. *International Journal of Heat and Mass Transfer*, 37(Suppl. 1), 149–160.

Goldstein, R.J. and Seol, W.S., 1991. Heat transfer to a row of impinging circular air jets including the effect of entrainment. *International Journal of Heat and Mass Transfer*, 34, 2133–2147.

Goldstein, R.J., Sobolik, K.A., and Seol, W.S., 1990. Effect of entrainment on the heat transfer to a heated circular air jet impinging on a flat surface. *ASME Journal of Heat Transfer*, 112, 608–611.

Goldstein, R.J. and Timmers, J.F., 1982. Visualization of heat transfer from arrays of impinging jets. *International Journal of Heat and Mass Transfer*, 25, 1857–1868.

Glezer, B., Moon, H.K., and O'Connell, T., 1996. A novel technique for the internal blade cooling. ASME Paper 96-GT-l 81.

Guo, T., Wang, T., and Gaddis, J.L., 1999. Mist/steam cooling in a heated horizontal tube. Part 2: Results and modeling, Presented at the *International Gas Turbine & Aeroengine Congress & Exhibition*, June 7–10, Indianapolis, IN, ASME Paper 99-GT-145.

Hagari, T., Ishida, K., Oda, T., Douura, Y., and Kinoshita, Y., 2010. Heat transfer and pressure losses of W-shaped small ribs at high Reynolds numbers for combustor liner. ASME Paper GT2010-23197.

Haiping, C., Dalin, Z., and Taiping, H., 1997. Impingement heat transfer from rib roughened surface within arrays of circular jet: The effect of the relative position of the jet hole to the ribs, Presented at the *International Gas Turbine & Aeroengine Congress & Exhibition*, June 2–5, Orlando, FL, ASME Paper 97-GT-331.

Han, J.C., 1984. Heat transfer and friction in channels with two opposite rib-roughened walls. *ASME Journal of Heat Transfer*, 106, 774–781.

Han, J.C., 1988. Heat transfer and friction characteristics in rectangular channels with rib turbulators. *ASME Journal of Heat Transfer*, 110, 321–328.

Han, J.C., Chandra, P.R., and Lau, S.C., 1988. Local heat/mass transfer distributions around sharp 180 deg turns in two-pass smooth and rib-roughened channels. *ASME Journal of Heat Transfer*, 110, 91–98.

Han, J.C. and Dutta, S., 1995. Internal convection heat transfer and cooling: An experimental approach, *Lecture Series 1995-05 on Heat Transfer and Cooling in Gas Turbines*, May 8–12, von Karman Institute for Fluid Dynamics, Belgium, Europe.

Han, J.C., Glicksman, L.R., and Rohsenow, W.M., 1978. An investigation of heat transfer and friction for rib-roughened surfaces. *International Journal of Heat and Mass Transfer*, 21, 1143–1156.

Han, J.C., Huang, J.J., and Lee, C.P, 1993. Augmented heat transfer in square chan- nels with wedge-shaped and delta-shaped turbulence promoters. *Enhanced Heat Transfer*, 1(1), 37–52.

Han, J.C. and Park, J.S., 1988. Developing heat transfer in rectangular channels with rib turbulators. *International Journal of Heat and Mass Transfer*, 31(1), 183–195.

Han, J.C., Park, J.S., and Lei, C.K., 1989. Augmented heat transfer in rectangular chan- nels of narrow aspect ratios with rib turbulators. *International Journal of Heat and Mass Transfer*, 32(9), 1619–1630.

Han, J.C. and Zhang, P., 1989. Pressure loss distribution in three-pass rectangular channels with rib turbulators. *ASME Journal of Turbomachinery*, 111, 515–521.

Han, J.C. and Zhang, P., 1991. Effect of rib-angle orientation on local mass transfer dis- tribution in a three-pass rib-roughened channel. *ASME Journal of Turbomachinery*, 113, 123–130.

Han, J.C. and Zhang, Y.M., 1992. High performance heat transfer ducts with parallel broken and V-shaped broken ribs. *International Journal of Heat and Mass Transfer*, 35(2), 513–523.

Han, J.C., Zhang, Y.M., and Lee, C.P., 1991. Augmented heat transfer in square chan- nels with parallel, crossed, and V-shaped angled ribs. *ASME Journal of Heat Transfer*, 113, 590–596.

Han, J.C., Zhang, Y.M., and Lee, C.P, 1992. Influence of surface heat flux ratio on heat transfer augmentation in square channels with parallel, crossed, and V-shaped angled ribs. *ASME Journal of Turbomachinery*, 114, 872–880.

Hollworth, B.R. and Berry, R.D., 1978. Heat transfer from arrays of impinging jets with large jet-to-jet spacing. *ASME Journal of Heat Transfer*, 100, 352–357.

Hollworth, B.R. and Dagan, L., 1980. Arrays of impingement jets with spent fluid removal through vent holes on the target surface. Part I: Average heat transfer. *Journal of Engineering for Power*, 102, 994–999.

Hrycak, P., 1981. Heat transfer from a row of impinging jets to concave cylindrical surfaces. *International Journal of Heat and Mass Transfer*, 24, 407–419.

Huang, Y., Ekkad, S.V., and Han, J.C., 1996. Detailed heat transfer coefficient distribu- tions under an array of inclined impinging jets using a transient liquid crystal technique, *9th International Symposium on Transport Phenomena in Thermal Fluids Engineering (ISTP-9)*, June 25–28, Singapore.

Huang, Y., Ekkad, S.V., and Han, J.C., 1998. Detailed heat transfer distributions under an array of orthogonal impinging jets. *AIAA Journal of Thermophysics and Heat Transfer*, 12(1), 73–79.

Incropera, F.P. and DeWitt, D.P., 1996. *Fundamentals of Heat and Mass Transfer*, 4th edn. Wiley, New York.

Jackson, D., Ireland, P., and Cheong, B., 2009. Combined experimental and CFD study of a HP blade multi-pass cooling system. ASME Paper GT2009-60070.

Jubran, B.A., Hamdan, M.A., and Abdualh, R.M., 1993. Enhanced heat transfer, miss- ing pin, and optimization for cylindrical pin fin arrays. *ASME Journal of Heat Transfer*, 115, 576–583.

Kanokjaruvijit, K. and Martinez-Botas, R., 2004. Parametric effects on heat transfer of impingement on dimpled surface. ASME Paper GT2004-53142.

Kanokjaruvijit, K. and Martinez-Botas, R.F., 2005. Heat transfer and pressure loss investigation of dimple impingement. ASME Paper GT2005-68823.

Kercher, D.M. and Tabakoff, W., 1970. Heat transfer by a square array of round air jets impinging perpendicular to a flat surface including the effect of spent air. *ASME Journal of Engineering for Power*, 92, 73–82.

Kerrebrock, J.L. and Stickler, D.B., 1998. Vaporization cooling for gas turbines, the return-flow cascade, Presented at the *Gas Turbine & Aeroengine Congress & Exhibition*, June 2–5, Stockholm, Sweden, ASME Paper 98-GT-177.

Kieda, S., Torii, T., and Fujie, K., 1984. Heat transfer enhancement in a twisted tube having a rectangular cross-section with or without internal ribs. ASME Paper 84-HT-75.

Kim, Y.W., Leond, A., Mank, V., Hee-Kee, M., Kenneth, O.S., 2003. Comparison of trip-strips/impingement/dimple cooling concepts at high Reynolds numbers. ASME Paper GT2003-38935.

Koopman, R.N. and Sparrow, E.M., 1976. Local and average transfer coefficients due to an impinging row of jets. *International Journal of Heat and Mass Transfer*, 19, 673–683.

Korotky, G.J. and Taslim, M.E., 1998. Rib heat transfer coefficient measurements in a rib-roughened square passage. *ASME Journal of Turbomachinery*, 120, 376–385.

Kovalenko, G.V. and Khalatov, A.A., 2003. Fluid flow and heat transfer features at a cross-flow of dimpled tubes in a confined space. ASME Paper GT2003-38155.

Kukreja, R.T., Lau, S.C., and McMillin, R.D., 1993. Local heat/mass transfer distribution in a square channel with full and V-shaped ribs. *International Journal of Heat and Mass Transfer*, 36(8), 2013–2020.

Kumaran, T.K., Han, J.C., and Lau, S.C., 1991. Augmented heat transfer in a pin fin channel with short or long ejection holes. *International Journal of Heat and Mass Transfer*, 34(10), 2617–2628.

Kunstmann, S., von Wolfersdorf, J., and Ruedel, U., 2010. Heat transfer and pressure drop in combustor cooling channels with combinations of geometrical elements. ASME Paper GT2010-23234.

Land, C. and Thole, K., 2008. Considerations of a double-wall cooling design to reduce sand blockage. ASME Paper GT2008-50160.

Lau, S., Cervantes, J., Han, J.C., and Rudolph, R.J., 2006. Internal cooling near trailing edge of a gas turbine airfoil with cooling airflow through blockages with holes. ASME Paper GT2006-91230.

Lau, S.C., Han, J.C., and Batten, T., 1989a. Heat transfer, pressure drop, and mass flow rate in pin fin channels with long and short trailing edge ejection holes. *ASME Journal of Turbomachinery*, 111, 116–122.

Lau, S.C., Han, J.C., and Kim, Y.S., 1989b. Turbulent heat transfer and friction in pin fin channels with lateral flow ejection. *ASME Journal of Heat Transfer*, 111, 51–58.

Lau, S.C., Kim, Y.S., and Han, J.C., 1985. Effects of fin configuration and entrance length on local endwall heat/mass transfer in a pin fin channel. ASME Paper 85-WA/HT-62.

Lau, S.C., McMillin, R.D., and Kukreja, R.T., 1992. Segmental heat transfer in a pin fin channel with ejection holes. *International Journal of Heat and Mass Transfer*, 35(6), 1407–1417.

Lauffer, D., Weigand, B., von Wolfersdorf, J., Dahlke, S., and Liebe, R., 2007. Heat transfer enhancement by impingement cooling in a combustor liner heat shield. ASME Paper GT2007-27908.

Leblanc, C., Ekkad, S.V., Rajendran, V., and Lambert, T., 2011. Detailed heat transfer distributions in engine similar cooling channels for a turbine rotor blade with different rib orientations, *ASME Turbo Expo 2011*, Vancouver, British Columbia, Canada, June 2011, GT2011-45254.

Li, X., Gaddis, J.L., and Wang, T., 2001. Mist/steam cooling by a row of impinging jets. ASME Paper 2001-GT-0151.

Liou, T.M. and Hwang, J.J., 1993. Effect of ridge shapes on turbulent heat transfer and friction in a rectangular channel. *International Journal of Heat and Mass Transfer*, 36(4), 931–940.

Maurer, M., Ruedel, U., Gritsch, M., and von Wolfersdorf, J., 2008. Experimental study of advanced convective cooling techniques for combustor liners. ASME Paper GT2008-51026.

Maurer, M., von Wolfersdorf, J., and Gritsch, M., 2007. An experimental and numerical study of heat transfer and pressure losses of V- and W-shaped ribs at high Reynolds numbers. ASME Paper GT2007-27167.

McMillin, R.D. and Lau, S.C., 1994. Effect of trailing-edge ejection on local heat (mass) transfer in pin fin cooling channels in turbine blades. *ASME Journal of Turbomachinery*, 116, 159–164.

Metzger, D.E., Berry, R.A., and Bronson, J.P., 1982. Developing heat transfer in rectangular ducts with staggered arrays of short pin fins. *ASME Journal of Heat Transfer*, 104, 700–706.

Metzger, D.E. and Bunker, R.S., 1990. Local heat transfer in internally cooled turbine airfoil leading edge regions. Part II: Impingement cooling with film coolant extraction. *ASME Journal of Turbomachinery*, 112, 459–466.

Metzger, D.E., Chyu, M.K., and Bunker, R.S., 1988. The contribution of on-rib heat transfer coefficients to total heat transfer from rib-roughened surfaces. Personal communication.

Metzger, D.E. and Fan, C.S., 1992. Heat transfer in pin fin arrays with jet supply and large alternating wall roughness ribs. In *Fundamental and Applied Heat Transfer Research for Gas Turbine Engines*, HTD-Vol. 226, ASME, New York, pp. 23–30.

Metzger, D.E., Fan, S.C., and Haley, S.W., 1984. Effects of pin shape and array orientation on heat transfer and pressure loss in pin fin arrays. *ASME Journal of Engineering for Gas Turbines and Power*, 106, 252–257.

Metzger, D.E., Florschuetz, L.W., Takeuchi, D.I., Behee, R.D., and Berry, R.A., 1979. Heat transfer characteristics for inline and staggered arrays of circular jets with crossflow of spent air. *ASME Journal of Heat Transfer*, 101, 526–531.

Metzger, D.E. and Haley, S.W., 1982. Heat transfer experiments and flow visualization for arrays of short pin fins. ASME Paper 82-GT-138.

Metzger, D.E., Plevich, C.W., and Fan, C.S., 1984. Pressure loss through sharp 180 deg. turns in smooth rectangular channels. ASME Paper 84-GT-154.

Metzger, D.E., Shephard, W.B., and Haley, S.W., 1986. Row resolved heat transfer variations in pin-fin arrays including effects of non-uniform arrays and flow convergence, *International Gas Turbine Conference and Exhibit*, June 8–12, Dusseldorf, West Germany, ASME Paper 86-GT-132.

Metzger, D.E. and Vedula, R.P., 1987. Heat transfer in triangular channels with angled roughness ribs on two walls. *Experimental Heat Transfer*, 1, 31–44.

Moon, H.K., O'Connell, T., and Glezer, B., 1999. Channel height effect on heat transfer and friction in a dimpled passage, Presented at the *International Gas Turbine & Aeroengine Congress & Exhibition*, June 7–10, Indianapolis, IN, ASME Paper 99-GT-163.

Nakamata, C., Okita, Y., Mimura, F., Matsushita, M., Yamane, T., Fukuyama, Y., Matsuno, S., and Yoshida, T., 2005. Spatial arrangement dependence of cooling performance of an integrated impingement and pin fin cooling configuration. ASME Paper GT2005-68348.

Nikuradse, J., 1950. Laws for flow in rough pipes. NACA TM 1292.

Oh, S., Lee, D.H., Kim, K.M., Cho, H.H., and Kim, M.Y., 2008. Enhanced cooling effectiveness in full coverage film cooling system with impingement jets. ASME Paper GT2008-50784.

Olson, D.A., 1992. Heat transfer in thin, compact heat exchangers with circular, rectangular, or pin-fin flow passages. *ASME Journal of Heat Transfer*, 114, 373–382.

Park, J., Goodro, M., Ligrani, P., Fox, M., and Moon, H.K., 2006. Separate effects of Mach number and Reynolds number on jet array impingement heat transfer. ASME Paper GT2006-90628.

Park, J.S., Han, J.C., Huang, Y, Ou, S., and Boyle, R.J., 1992. Heat transfer performance comparisons of five different rectangular channels with parallel angled ribs. *International Journal of Heat and Mass Transfer*, 35(11), 2891–2903.

Peng, Y., 1983. Heat transfer and friction loss characteristics of pin fin cooling configuration. ASME Paper 83-GT-123.

Poser, R., Von Wolfersdorf, J., Lutum, E., and Semmler, K., 2008. Performing heat transfer experiments in blade cooling circuits using a transient technique with thermochromic liquid crystals, *ASME Turbo Expo 2008*, Berlin, Germany, June, GT2008-50364.

Rallabandi, A.P., Alkhamis, N., and Han, J.C., 2009a. Heat transfer and pressure drop measurements for a square channel with 45deg round edged ribs at high Reynolds numbers, *Proceedings of ASME Turbo-Expo 2009*, June 8–12, Orlando, FL, ASME Paper GT2009-59546.

Rallabandi, A.P., Yang, H., and Han, J.C., 2009b. Heat transfer and pressure drop correlations for 45° parallel rib roughened square channels at high Reynolds numbers. *ASME Journal of Heat Transfer*, 131(7), 071703.

Rau, G., Cakan, M., Moeller, D., and Arts, T., 1998. The effect of periodic ribs on the local aerodynamic and heat transfer performance of a straight cooling channel. *ASME Journal of Turbomachinery*, 120, 368–375.

Rhee, D.H., Nam, Y.W., and Cho, H.H., 2004. Local heat/mass transfer with various rib arrangements in impingement/effusion cooling system with crossflow, *ASME Turbo Expo 2004: Power for Land, Sea, and Air (GT2004)*, June 14–17, 2004 , Vienna, Austria, ASME Paper GT2004-53686, pp. 653–664 http://dx.doi.org/10.1115/GT2004-53686

Saha, K., Guo, S., Acharya, S., and Nakamata, C., 2008. Heat transfer and pressure measurements in a lattice-cooled trailing edge of a turbine airfoil. ASME Paper GT2008-51324.

Schabacker, J., Boles, A., and Johnson, B.V., 1998. PIV investigation of the flow characteristics in an internal coolant passage with two ducts connected by a sharp 180° bend. *International Gas Turbine and Aeroengine Congress and Exposition*, June 2–5, Stockholm, Sweden, ASME Paper 98-GT-544.

Segura, D. and Acharya, S., 2010. Internal cooling using novel swirl enhancement strategies in a slot shaped single pass channel. ASME Paper GT2010-23679.

Shen, J.R., Wang, Z., Ireland, P.T., Jones, T.V., and Byerley, A.R., 1996. Heat transfer enhancement within a turbine blade cooling passage using ribs and combinations of ribs with film cooling holes. *ASME Journal of Turbomachinery*, 118, 428–433.

Siw, S.C., Chyu, M.K., Shih, T., and Alvin, M.A., 2010. Effects of pin detached space on heat transfer and from pin fin arrays. ASME Paper GT2010-23227.

Son, C., Dailey, G., Gillespie, D.R.H., and Ireland, P.T., 2005. An investigation of the application of roughness elements to enhance heat transfer in an impingement cooling system. ASME Paper GT2005-68504.

Sparrow, E.M., Goldstein, R.J., and Rouf, M.A., 1975. Effect of nozzle-surface separation distance on impingement heat transfer for a jet in a crossflow. *ASME Journal of Heat Transfer*, 97, 528–533.

Sparrow, E.M. and Molki, M., 1982. Effect of a missing cylinder on heat transfer and fluid flow in an array of cylinders in cross-flow. *International Journal of Heat and Mass Transfer*, 25(4), 449–456.

Sparrow, E.M., Suopys, A.P., and Ansari, M.A., 1984. Effect of inlet, exit, and fin geometry on pin fins situated in a turning flow. *International Journal of Heat and Mass Transfer*, 27(7), 1039–1054.

Sparrow, E.M. and Wong, T.C., 1975. Impingement transfer coefficients due to initially laminar slot jets. *International Journal of Heat and Mass Transfer*, 18, 597–605.

Spence, R.B. and Lau, S.C., 1997. Heat transfer and friction in segmental turbine blade cooling channels. *AIAA Journal of Thermophysics and Heat Transfer*, 11(3), 486–488.

Steuber, G.D. and Metzger, D.E., 1986. Heat transfer and pressure loss performance for families of partial length pin fin arrays in high aspect ratio rectangular ducts, *8th International Heat Transfer Conference*, San Francisco, CA, Vol. 6, pp. 2915–2920.

Taslim, M.E. and Bethka, D., 2007. Experimental and numerical impingement heat transfer in an airfoil leading-edge cooling channel with crossflow. ASME Paper GT2007-28212.

Taslim, M.E. and Lengkong, A., 1998a. 45° round-corner rib heat transfer coefficient measurements in a square channel, *International Gas Turbine and Aeroengine Congress and Exhibition*, June 2–5, Stockholm, Sweden, ASME Paper 98-GT-176.

Taslim, M.E. and Lengkong, A., 1998b. 45 deg. staggered rib heat transfer coefficient measurements in a square channel. *ASME Journal of Turbomachinery*, 120, 571–580.

Taslim, M.E., Li, T., and Spring, S.D., 1997. Measurements of heat transfer coefficients and friction factors in rib-roughened channels simulating leading-edge cavities of a modern turbine blade. *ASME Journal of Turbomachinery*, 119, 601–609.

Taslim, M.E., Li, T., and Spring, S.D., 1998a. Measurements of heat transfer coefficients in rib-roughened trailing-edge cavities with crossover jets, Presented at the *Gas Turbine & Aeroengine Congress & Exhibition*, June 2–5, Stockholm, Sweden, ASME Paper 98-GT-435.

Taslim, M.E., Li, T., and Spring, S.D., 1998b. Measurements of heat transfer coefficients and friction factors in passages rib-roughened on all walls. *ASME Journal of Turbomachinery*, 120, 564–570.

Taslim, M.E., Liu, H., and Bakhtari, K., 2003. Experimental and numerical investigation of impingement on a rib-roughened leading-edge wall. ASME Paper GT2003-38118.

Taslim, M.E., Pan, Y., and Bakhtari, K., 2002. Experimental racetrack shaped jet impingement on a roughened leading-edge wall with film holes. ASME Paper GT2002-30477.

Taslim, M.E., Pan, Y., and Spring, S.D., 2001. An experimental study of impingement on roughened airfoil leading-edge walls with film holes. ASME Paper 2001-GT-0152.

Taslim, M.E. and Setayeshgar, L., 2001. Experimental leading-edge impingement cooling through racetrack crossover holes. ASME Paper 2001-GT-0153.

Taslim, M.E. and Spring, S.D., 1994. Effects of turbulator profile and spacing on heat transfer and friction in a channel. *AIAA Journal of Thermophysics and Heat Transfer*, 8(3), 555–562.

Taslim, M.E. and Wadsworth, C.M., 1997. An experimental investigation of the rib surface-averaged heat transfer coefficient in a rib-roughened square passage. *ASME Journal of Turbomachinery*, 119, 381–389.

Taylor, J.R., 1980. Heat transfer phenomena in gas turbines. ASME Paper 80-GT-172.

Trabolt, T.A. and Obot, N.T., 1987. Impingement heat transfer within arrays of circular jets. Part II: Effects of crossflow in the presence of roughness elements, Presented at the *Gas Turbine Conference & Exhibition*, May 31–June 4, Anaheim, CA.

Van Fossen, G.J., 1982. Heat transfer coefficients for staggered arrays of short pin fins. *ASME Journal of Engineering for Power*, 104(2), 268–274, ASME paper No. 81-GT-75.

Van Treuren, K.W., Wang, Z., Ireland, P.T., and Jones, T.V., 1994. Detailed measurements of local heat transfer coefficient and adiabatic wall temperature beneath an array of impingement jets. *ASME Journal of Turbomachinery*, 116, 369–374.

Verstraete, T., Amaral, S., van den Braembussche, R., and Arts, T., 2008. Design and optimization of the internal cooling channels of HP turbine blade—Part II: Optimization. ASME Paper GT2008-51080.

Viskanta, R., 1993. Heat transfer to impinging isothermal gas and flame jets. *Experimental Thermal and Fluid Science*, 6, 111–134.

Wang, Z., Ireland, P.T., and Jones, T.V., 1995. Detailed heat transfer coefficient measurements and thermal analysis at engine conditions of a pedestal with fillet radii. *ASME Journal of Turbomachinery*, 117, 290–295.

Wang, Z., Ireland, P.T., Kohler, S.T., and Chew, J.W., 1998. Heat transfer measurements to a gas turbine cooling passage with inclined ribs. *ASME Journal of Turbomachinery*, 120, 63–69.

Weaver, S., Barringer, M., and Thole, K.A., 2010. Micro channels with manufacturing roughness levels. ASME Paper GT2010-22976.

Webb, R.L., Eckert, E.R.G., and Goldstein, R.J., 1971. Heat transfer and friction in tubes with repeated-rib roughness. *International Journal of Heat and Mass Transfer*, 14(4), 601–617.

Womac, D.J., Incropera, F.P., and Ramadhyani, S., 1994. Correlating equations for impingement cooling of small heat sources with multiple circular liquid jets. *ASME Journal of Heat Transfer*, 116, 482–486.

Xing, Y. and Weigand, B., 2010. Experimental investigation on staggered impingement heat transfer on a rib roughened plate with different crossflow schemes. ASME Paper GT2010-22043.

Yamawaki, S., Yoshida, T., Taki, M., and Mimura, F., 1997. Fundamental heat transfer experiments of heat pipes for turbine cooling, Presented at the *International Gas Turbine & Aeroengine Congress & Exhibition*, June 2–5, Orlando, FL, ASME Paper 97-GT-438.

Zehnder, F., Schuler, M., Weigand, B., von Wolferdof, J., and Neumann, S.O., 2009. The effect of turning vanes on pressure loss and heat transfer of a ribbed rectangular two-pass internal cooling channel. ASME Paper GT2009-59482.

Zhang, Y.M., Azad, G.M., Han, J.C., and Lee, C.P., 2000. Turbulent heat transfer enhancement and surface heating effect in square channels with wavy, and twisted tape inserts with interrupted ribs. *Enhanced Heat Transfer*, 7, 35–49.

Zhang, Y.M., Gu, W.Z., and Han, J.C., 1994a. Augmented heat transfer in triangular ducts with full and partial ribbed walls. *AIAA Journal of Thermophysics and Heat Transfer*, 8(3), 574–579.

Zhang, Y.M., Gu, W.Z., and Han, J.C., 1994b. Heat transfer and friction in rectangular channels with ribbed or ribbed-grooved walls. *ASME Journal of Heat Transfer*, 116, 58–65.

Zhang, Y.M., Han, J.C., and Lee, C.P., 1997. Heat transfer and friction characteristics of turbulent flow in circular tubes with twisted-tape inserts and axial interrupted ribs. *Enhanced Heat Transfer*, 4, 297–308.

Zhou, F. and Acharya, S., 2009. Experimental and computational study of heat/mass transfer and flow structure for four dimple shapes in a square internal passage. ASME Paper GT2009-60240.

Zukauskas, A.A., 1972. Heat-transfer from tubes in cross flow. *Advances in Heat Transfer*, 8, 116–133.

Zuo, Z.J., Faghri, A., and Langston, L., 1997. A parametric study of heat pipe turbine vane cooling, Presented at the *International Gas Turbine & Aeroengine Congress & Exhibition*, June 2–5, Orlando, FL, ASME Paper 97-GT-443.

Zuo, Z.J., Faghri, A., and Langston, L., 1998. Numerical analysis of heat pipe turbine vane cooling. *Journal of Engineering for Gas Turbine and Power*, 120(4), 735–740.

5

Turbine Internal Cooling with Rotation

5.1 Rotational Effects on Cooling

In modern gas turbine blades, cooling air is circulated through internal cooling passages to remove heat from the blade. This cooling is necessary to maintain the blades' operating strength. The combined effects from flow turbulence, property variations, surface roughness, and geometry of channels affect coolant flow and its performance in a stator airfoil. Coolant flow analysis in rotor airfoils has another dimension added by rotational forces, and these rotational effects in a coolant passage of a rotor airfoil significantly affect the heat-transfer distribution. Inherent forces (Coriolis and rotational buoyancy forces) of rotation alter the flow field and thus affect heat transfer. Several studies have shown that coolant passing through a rotor blade cooling passage experiences strong Coriolis and rotational buoyancy forces. These forces generate secondary flow and distribute the core flow asymmetrically in the channel. In a serpentine multipass channel, flow separates and recirculates in the turn regions, and the flow development downstream of the sharp turn also has significant effects on the distribution of the local heat-transfer coefficient and on the overall channel heat transfer. This section focuses on the effects of rotation on heat transfer. In this section, rotating smooth, rotating ribbed, and rotating channels with impingement cooling are considered. Prior publications by different experimentalists did not anticipate the compressibility effects by rotation. However, recent numerical predictions indicate that at high rotation speeds, i.e., more than 10,000 rpm of aircraft engines, compressibility effects may become important. Aeroderivative land-based power generation units run at a slower speed; experimental results presented here can be directly applied.

5.2 Smooth-Wall Coolant Passage

5.2.1 Effect of Rotation on Flow Field

Heat transfer is a side effect of the flow field, and therefore a brief discussion of the effects of rotation on flow is presented first, followed by a detailed discussion on heat-transfer distributions. Flow in a rotating channel is significantly different from flow in a nonrotating stationary channel. The secondary flow in rotation redistributes velocity and also alters the random velocity fluctuation patterns in turbulent flows. Lezius and Johnston (1976) examined flow instability caused by rotation. A rotating high-aspect-ratio rectangular channel was used, and the working fluid was water. Figure 5.1 illustrates their observations. It can be seen that one side of the channel showed flow unstabilization (turbulence enhancement), and the other side showed flow stabilization (turbulence decay). Figure 5.2 shows the analyzed stability boundary for different rotation numbers. Ro is the rotation number, defined as $Ro = \omega D/V$, where ω is the rotation speed, D is the channel depth, and V is the average axial velocity. This figure shows that the lowest critical Reynolds number of 88.5 occurs at $Ro = 0.5$. This critical Reynolds number means that rotation favors the onset of turbulence, and at $Ro = 0.5$ turbulent flow can be initiated at a low Reynolds number of 88.5, compared to a critical Reynolds number of 400 at $Ro = 0.01$. Note that this experimental work of Lezius and Johnston (1976) was in an unheated channel, and therefore their results include only the Coriolis effect and not the rotational buoyancy effect. Figure 5.3 shows the measured velocity and eddy viscosity of Johnston et al. (1972). Results indicate that rotation increases flow

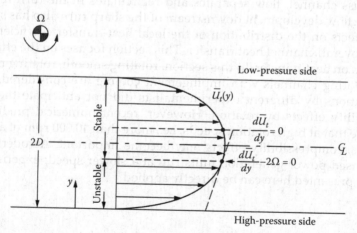

FIGURE 5.1
Stable and unstable regions in a rotating flow. The unstable side shows turbulence enhancement; the stable side shows a suppression of turbulence. (From Lezius, D.K. and Johnston, J.P., *J. Fluid Mech.*, 77(Part 1), 153, 1976. With permission.)

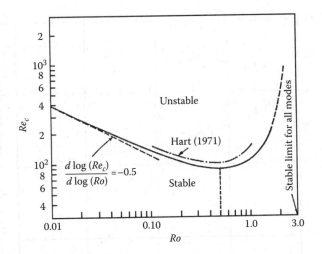

FIGURE 5.2
Critical Reynolds numbers for transition to turbulence in the presence of rotation. (From Lezius, D.K. and Johnston, J.P., *J. Fluid Mech.*, 77(Part 1), 153, 1976. With permission.)

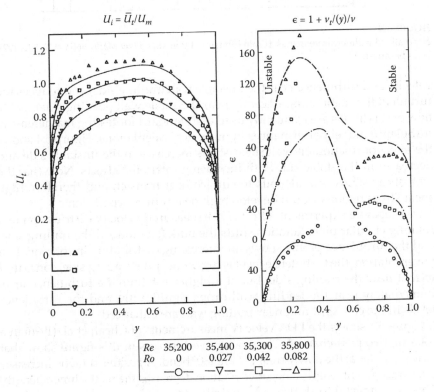

FIGURE 5.3
Velocity and eddy viscosity distribution in rotating channel flow. (From Lezius, D.K. and Johnston, J.P., *J. Fluid Mech.*, 77(Part 1), 153, 1976. With permission.)

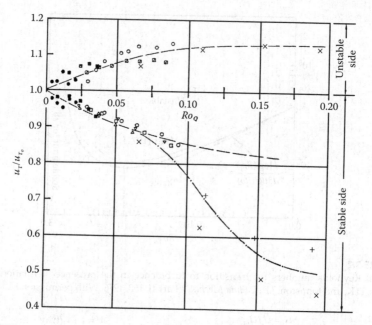

FIGURE 5.4
Normalized wall-shear velocities. (From Johnston, J.P. et al., *J. Fluid Mech.*, 56(Part 3), 533, 1972. With permission.)

velocity and turbulence near the unstable wall (trailing side) and reduces the turbulent fluctuations significantly near the stable wall (leading side). These flow observations are frequently used as a standard for testing mathematical turbulence model performances in numerical calculations. Figure 5.4 shows the effect of rotation on wall shear stress increases in the unstable side and decreases in the stable side with increasing rotational effects. Note that the wall shear stress is analogous to surface heat transfer, and therefore high shear stress regions would develop high heat-transfer coefficients.

In a separate experiment, Elfert (1993) measured velocity distribution in a rotating circular pipe. Rotation shifts the bulk flow toward the trailing side, and the turbulence profile shows an interesting distribution in rotation. In a lower rotation, the turbulence level is higher near the leading side, compared to that near the trailing side. But at a higher rotation, the turbulence level immediately near the leading wall drops significantly, and the turbulence levels in the core flow and near trailing wall are enhanced.

Figure 5.5 shows the LDA velocity measurements of Cheah et al. (1996) in a rotating two-pass channel. The velocity vector plots in this figure show that flow separates at the downstream of the 180° bend. A positive rotation increases the reattachment length; a negative rotation decreases the reattachment length. This experimental work also indicated that streamwise turbulence intensity at upstream of the bend ($z/D = -3.0$) was not much affected by rotation, whereas the core flow turbulence intensity at $z/D = 5.9$, which is located downstream of

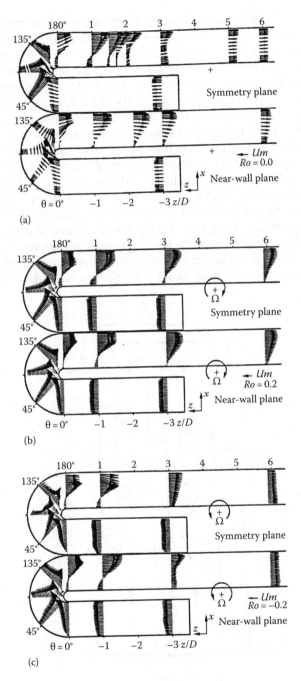

FIGURE 5.5
Velocity distribution in a rotating U-duct. (a) Stationary U-bend, (b) U-bend in positive rotation, and (c) U-bend in negative rotation. (From Cheah, S.C. et al., *ASME J. Turbomach.*, 118(3), 590, 1996. With permission.)

bend, increased with rotation. Measurements showed that cross stream fluctuations near the bend were comparable to those in a streamwise direction.

Tse and McGrath (1995) used laser-Doppler velocimetry to measure rotating flow at a Reynolds number of 25,000 and rotation number of 0.24. They observed that the flow was not fully developed, and higher streamwise flow is noted near the lower and upper sidewalls. The tangential component of the secondary flow indicated that the Coriolis-driven secondary flow was from the low-pressure to the high-pressure side in the middle part of the channel. Results also showed that flow was from high pressure to low pressure near the upper and lower sidewalls; an asymmetric distribution of the secondary flow in the turn region was noted. In the second pass with radial inward flow, the leading surface showed higher streamwise velocity than the trailing side, and the secondary flow in the core of the channel indicates flow from the trailing to the leading side. A stronger secondary flow exists near the upper and lower sidewalls that carry fluid from the leading to the trailing side.

Secondary flows in a two-pass channel are different for radial outflow and radial inflow passes. Figure 5.6 shows the schematic secondary flow and axial

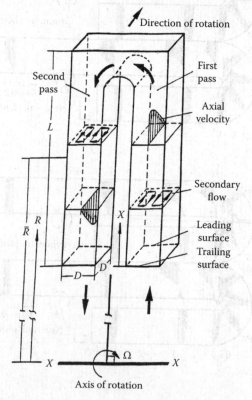

FIGURE 5.6

Conceptual view of a two-pass rotating coolant flow distribution. (From Han, J.C. et al., *ASME J. Heat Transfer*, 114(4), 850, 1993.)

flow distribution in a rotating two-pass channel. Flow is radial outward in the first pass and radial inward in the second pass. Since the direction of the Coriolis force is dependent on the direction of rotation and flow, the Coriolis force has a different direction in the two passes. The rotation direction remains the same for the two channels, but the direction of flow gets reversed from the first channel to the second in the 180° bend. Therefore, the direction of the Coriolis force is opposite in these two channels, and the resultant secondary flow is different. Figure 5.7 shows the combined effects of Coriolis and rotational buoyancy on flow distribution. For radial outward flow in the first channel, the Coriolis force shifts the core flow toward the trailing wall. If both trailing and leading walls are symmetrically heated, then faster-moving coolant near the trailing wall would be cooler than the slow-moving coolant near the leading wall. Rotational buoyancy is caused by a strong centrifugal force that pushes cooler heavier fluid away from the center of rotation. In the first channel, rotational buoyancy affects the flow in a fashion similar to the Coriolis force and causes a further increase in flow near the trailing wall of the first channel, whereas the Coriolis force favors the leading side of the second channel. The rotational buoyancy

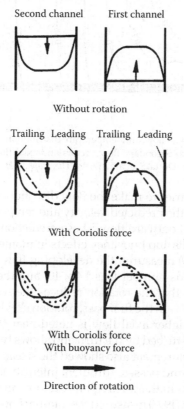

FIGURE 5.7
Conceptual view of effects of inertia, Coriolis, and rotational buoyancy on radially outward and inward flows. (From Han, J.C. et al., *ASME J. Heat Transf.*, 114(4), 850, 1993.)

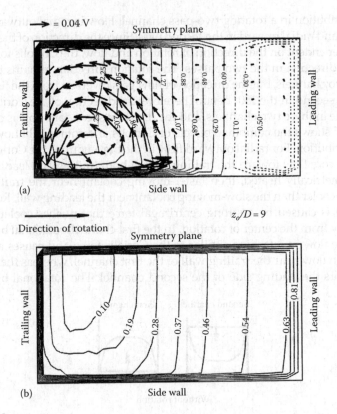

FIGURE 5.8
Predicted secondary flow, axial flow, and temperature distribution in a rotating square channel
with radial outward flow. (a) Secondary flow vectors and axial flow contours and (b) coolant
temperature contours. (From Dutta, S. et al., *AIAA J. Thermophys. Heat Transfer*, 9(2), 381, 1995a.)

in the second channel tries to make the flow distribution more uniform in the
duct. Figure 5.8 shows the predicted velocity and temperature distribution in a
square duct with radial outward flow. These predictions of Dutta et al. (1995b)
include both the Coriolis and buoyancy effects in momentum and turbulence.

Liou and Chen (1999) measured the developing flow in a two-pass channel
with LDV. The Reynolds number was 1.4×10^4 and the rotation number was
0.082. It was observed that the rotation reduces the separation bubble in the
after-turn region by 25%. In the first pass, rotation skews the streamwise mean
velocity profile, and higher axial flow is noted near the trailing side. In the
second pass, flow is disturbed by the turn and shows two peaks in the velocity
distribution. This measurement also showed the streamwise turbulence intensity in the first and second passes. Turbulent intensity is increased by rotation,
and this increase is less in the core region and more in the near walls.

Bons and Kerrebrock (1998) measured the internal flow in a simulated smooth-
wall turbine blade cooling passage using particle image velocimetry (PIV) for
both heated and nonheated cases. The centerline velocity vectors showed that

the streamwise velocity vectors were strong and not significantly affected by the secondary flow as such. However, the average flow velocity in this location is affected by rotation that is not apparent in the plot but discussed by them in a separate figure. The near-wall flow is strongly affected by the secondary flow and, as expected, flow vectors are tilted in the direction of the leading side.

The streamwise velocity distribution is strongly affected by the buoyancy effect, and the velocity profiles with heating and without heating show remarkable differences. The heated flow shows higher velocity near the trailing side. This trend in velocity shift is in agreement with the numerical predictions. There is an indication that flow separates near the leading side with the addition of heat. Unfortunately, details of the flow near wall could not be captured, but isolated vectors from a few selected locations indicate the presence of the reverse flow. Secondary flow magnitudes are higher at two locations in the channel, and core secondary flow velocity is less than these two peaks. Streamwise velocity vectors indicated a stagnation region near the wall at the leading side of the channel.

5.2.2 Effect of Rotation on Heat Transfer

Figure 5.9 shows a four-pass square cross-sectioned channel used by Wagner et al. (1991). This rotating test facility simulates the conditions in a serpentine

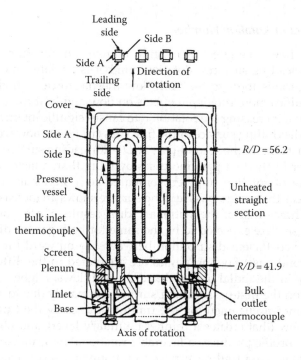

FIGURE 5.9
Serpentine coolant passage model used by Wagner et al. (From Wagner, J.H. et al., *ASME J. Turbomach.*, 113, 321, 1991. With permission.)

rotor blade coolant passage. The first, second, and third passes were instrumented to study heat transfer. However, the results presented were mostly for the first two passes. Results were presented in the nondimensional form of Nusselt number ratio (Nu/Nu_∞), where Nu_∞ is the Nusselt number for fully developed pipe flow. Wagner et al. (1991) rotated the test assembly at 1100 rpm or less, and they operated at approximately 10 atmospheric pressures (147 psi). The inlet temperature was typically 27°C, and the copper elements, which constitute heaters, were held at different constant temperatures: 49°C, 71°C, 93°C, and 116°C. Different heater temperatures showed different buoyancy effects in the heat transfer. Temperatures of the copper elements were measured with two thermocouples inserted in each element. This pioneering work of Wagner et al. (1991) quantified the effects of rotation on heat transfer. They showed that heat transfer is primarily dependent on rotation number ($Ro = \omega D/V$) and density ratio [$DR = (\rho_{in} - \rho_{wall})/\rho_{in}$]. The rotation number is the relative strength of the Coriolis force to the bulk inertia, and the density ratio is the relative strength of the rotational buoyancy force to the bulk inertia. Besides the rotational effects, the effects from the turn can significantly influence the heat-transfer results (Han et al., 1988). Therefore, interpretation of serpentine rotating channels should include the effects of turns in the near-turn regions.

5.2.2.1 Effect of Rotation Number

Figure 5.10 shows the effect of rotation number on heat-transfer ratio for trailing and leading surfaces. Rotation number is a relative measure of the rotational Coriolis force to the bulk flow inertial force. A higher rotation number signifies more rotational effects on flow and heat transfer. This figure shows that increasing the rotation rate causes significant increase in heat transfer on the trailing surface of the first pass ($x/D < 12$); however, comparatively less increase on the second-pass ($x/D > 20$) leading surface is observed. Heat transfer in the first pass increases by more than a factor of 3.5 at the largest value of rotation number, compared to a stationary heat-transfer value, whereas the second-pass leading surface shows an increase of approximately 1.5 times that of the stationary duct results. Since Coriolis effects on heat transfer are expected to be the same for both passes, differences in heat-transfer coefficient distribution between the outward (first pass) and inward (second pass) flowing passes are attributed to the different effects of buoyancy in the radial outward and inward flows. Higher heat-transfer coefficients on the first-pass trailing surface are mainly due to two reasons: (1) the coolant impingement on the trailing surface by the Coriolis-driven secondary flow (that creates a thinner boundary layer); and (2) the coolant temperature profile produces a favorable rotational buoyancy-induced flow near the destabilized trailing surface thin boundary layer. The decrease in heat-transfer coefficients near the inlet of the channel on the first-pass leading surfaces is due to a thicker stabilized boundary layer. The subsequent

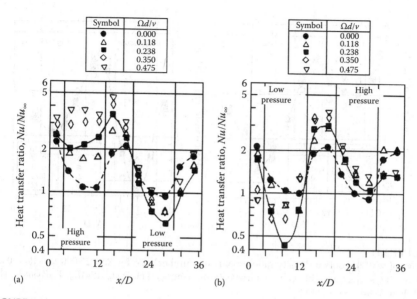

FIGURE 5.10
Effect of rotation number on Nusselt number ratio for $Re = 25,000$ and density ratio $= 0.13$. (a) Trailing surface and (b) Leading surface. (From Wagner, J.H. et al., *ASME J. Turbomach.*, 113, 321, 1991. With permission.)

increase in heat-transfer coefficients near the end of the channel is attributed to the stronger centrifugal buoyancy-induced, destabilized wall turbulence boundary layer. Effects of turn can also increase the heat transfer in the later part of the first pass. Since the coolant flows radial inward in the second pass (opposite to first pass), the Coriolis force acts in the opposite direction, which shifts the bulk flow toward the leading surface. This results in the cooler fluid moving faster near the leading surface and the warmer fluid moving slower near the trailing surface. Therefore, the heat-transfer coefficient on the leading surface is higher than on the trailing surface.

Figure 5.10 also shows the effect of rotation on the heat transfer from the first-pass leading side and the second-pass trailing side. The heat-transfer coefficients from these surfaces decrease with an increase in the rotation number. This decrease in the heat transfer is due to the decrease in axial flow and the stabilization of the near-wall flow on the leading side (Johnston et al., 1972). Moreover, the secondary flow in rotation carries the fluid along the other hot walls before reaching the leading wall. This also results in a reduced heat-transfer coefficient from the leading wall.

5.2.2.2 Effect of Density Ratio

Figure 5.11 shows the effects of density ratio, i.e., rotational buoyancy, on heat transfer. The inlet density ratio, $(\rho_{in} - \rho_{wall})/\rho_{in}$, was varied from 0.07 to 0.22. The Reynolds number, rotation number, and mean rotating radius were

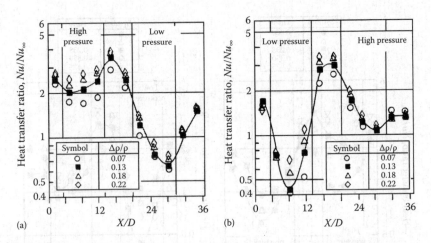

FIGURE 5.11
Effect of wall-to-coolant density ratio on Nusselt number ratio for *Re* = 25,000 and *Ro* = 0.24.
(a) Trailing surface and (b) Leading surface. (From Wagner, J.H. et al., *ASME J. Turbomach.*, 113,
321, 1991. With permission.)

held constant at 25,000, 0.24, and 49 hydraulic diameters, respectively. This
figure shows that increasing the inlet density ratio from 0.07 to 0.22 increases
the heat-transfer ratios in the first-pass trailing surface by as much as 50%,
and the heat-transfer coefficient in the first-pass leading surface increases by
as much as 100%. The first-pass leading surface shows an increase in heat
transfer with an increase in the density ratio. This increase in heat-transfer
coefficient of the first-pass leading wall, with an increase in the density ratio,
is attributed to a flow separation mechanism, as predicted by Prakash and
Zerkle (1992) and Dutta et al. (1995a). Their predictions showed that flow
separates at the leading surface of the first pass due to rotational buoyancy.
This flow separation increases turbulence and heat transfer at downstream
locations. Heat-transfer coefficient of the second-pass radial inward flow
increases with an increase in the density ratio.

5.2.2.3 Combined Effects of Rotation Number and Density Ratio

Figure 5.12 shows the heat-transfer results of the leading and trailing sides of
the first pass for different rotation numbers and density ratios. There are no
effects of density ratio on the heat-transfer ratio for a rotation number of 0.
This is because the gravitational buoyancy is negligible in this case, and rotational buoyancy is absent at 0 rotation number. Increasing the rotation number causes local increases in the heat transfer in the first-pass trailing surface.
Heat transfer from the first-pass leading surface decreases with increasing
rotation number and then increases again, especially with higher density
ratios. These heat-transfer distributions on the leading side of the first pass
were attributed to two factors by Wagner et al. (1992): first is the combination

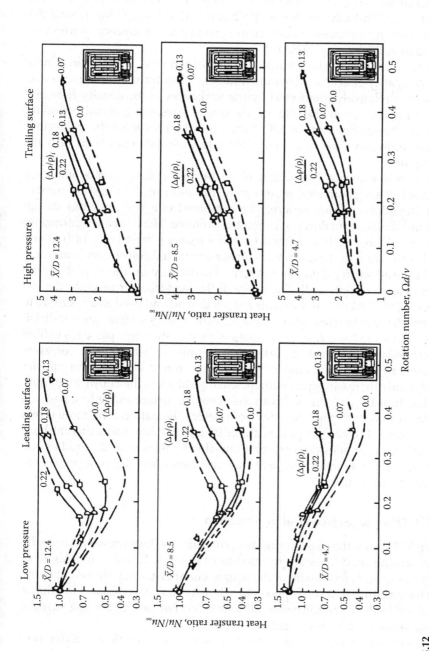

FIGURE 5.12

Effect of rotation number and wall-to-coolant density ratio on Nusselt number ratio in the first pass for $Re = 25,000$. (From Wagner, J.H. et al., *ASME J. Turbomach.*, 113, 321, 1991. With permission.)

of buoyancy forces and the stabilization of the near-wall flow; second is the Coriolis-driven secondary flow cells at larger rotation numbers. However, predictions of Prakash and Zerkle (1992) and Dutta et al. (1995a) showed that the increase in heat-transfer coefficient is related to a buoyancy-driven flow separation at the leading surface.

Figure 5.13 shows the heat transfer in the second pass, with different rotational buoyancy effects and rotation numbers. Heat-transfer ratio in the second pass is relatively unaffected by the variations in the density ratio and rotation number. Near the inlet of the second pass, the thermal boundary layers are thin because of the 180° turn effect. Moreover, the turn develops its own secondary flow that dominates over the rotational effects near the turn region.

Figure 5.14 shows a correlation of heat-transfer data with a buoyancy parameter. This parameter is similar to the ratio of the Grashof number to the Reynolds number square. The combined effect of the cross stream flows and the buoyant flows is difficult to evaluate. The buoyancy parameter (defined in Figure 5.14) is defined from the relative importance of free and forced convection in the analysis of stationary mixed convection heat transfer. The heat-transfer data show strong dependency on the flow direction (radial inward or outward). The range of the heat-transfer ratio for the last location in the first pass is shown as a shaded band and used as a reference to compare the results from the second and third passages. Heat-transfer distributions from the low-pressure surfaces of each of the three passages exhibit a similar relationship with the buoyancy parameter. Heat transfer decreases with increasing values of buoyancy parameter from 0.0 to 0.15. Heat transfer subsequently increases again with increasing values of the buoyancy parameter. The heat-transfer results from the pressure surface in the first pass are correlated well with the buoyancy parameter. The data can be plotted in a narrow band. The second pass with radial inward flow had different heat-transfer characteristics than the first and third passes that have radial outward flows. The second-pass heat transfer shows less dependency on the buoyancy parameter.

5.2.2.4 Effect of Surface-Heating Condition

Figure 5.15 shows the effects of heating condition on heat transfer. Three different heating conditions are referred to as cases A, B, and C by Han et al. (1993). For case A, four walls of the square channel are maintained at a uniform temperature that corresponds to a density ratio of 0.1. Case B maintains a uniform heat flux condition on all four sides. For case C, trailing surfaces are maintained hotter than leading surfaces.

Uniform wall temperature (case A) heat-transfer coefficient differences between the rotating and nonrotating conditions on the first-pass leading and trailing surfaces are due to both Coriolis-generated secondary cross stream flow vortices (as observed by Johnston et al., 1972) and the centrifugal

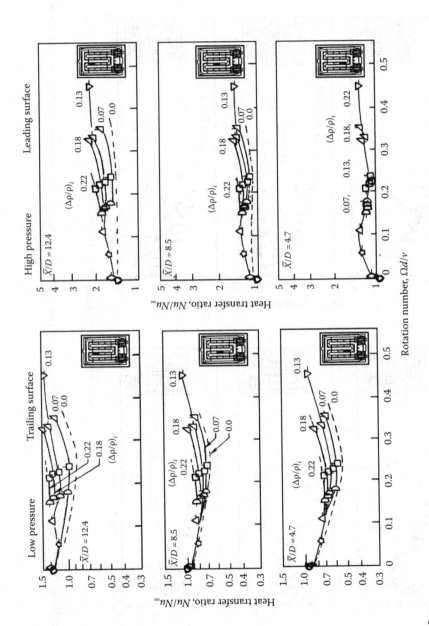

FIGURE 5.13
Effect of rotation number and wall-to-coolant density ratio on Nusselt number ratio in the second pass for *Re* = 25,000. (From Wagner, J.H. et al., *ASME J. Turbomach.*, 113, 321, 1991. With permission.)

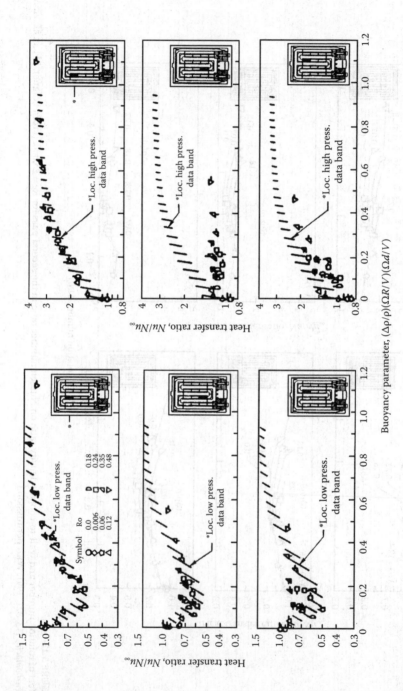

FIGURE 5.14

Comparison of Nusselt number ratios of the first, second, and third passes. (From Wagner, J.H. et al., *ASME J. Turbomach.*, 113, 321, 1991. With permission.)

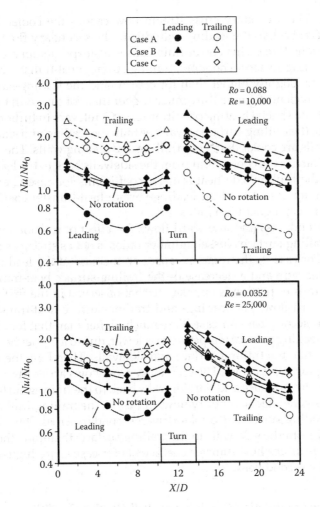

FIGURE 5.15
Effect of surface heating condition on Nusselt number ratio for heating cases A, B, and C of Han et al. (From Han, J.C. et al., *ASME J. Heat Transfer*, 114(4), 850, 1993.)

buoyancy (as observed by Wagner et al., 1991, 1992; Han and Zhang, 1992; Han et al., 1993). The observed heat-transfer coefficient patterns are the typical combined effects of Coriolis and buoyancy forces due to rotation on the leading and trailing surfaces of a square channel with radial outward flow. The first-pass trailing surface temperature is lower than the leading and side surfaces. Han et al. (1993) conjectured that the cooler fluid is accelerated due to centrifugal buoyancy, and a thinner boundary layer is created near the trailing surface.

The uniform heat flux condition (case B) shows that the first-pass trailing surface is cooler than the other sides because of higher heat-transfer coefficient. The secondary cross stream flow carries the cooler fluid from trailing walls back to the leading wall due to the secondary flow generated by the Coriolis force. This cooler fluid causes a steeper temperature profile near the leading surface. Therefore, trailing-surface heat-transfer coefficients for case B are slightly higher than for case A, and the leading-surface heat-transfer coefficients for case B are much higher than for case A in the rotating first pass. For a special heating condition (case C) relevant to turbomachinery application, the trailing surface is warmer than the leading and side surfaces. The sidewalls are cooler than both trailing and leading walls. The secondary cross flow carries the cooler fluid from the sidewalls back to the leading wall; therefore, the trailing-side heat-transfer coefficients for case C are slightly higher than for case A, and the leading-surface heat-transfer coefficients for case C are much higher than for case A.

In the second pass (radial inward flow), the effect of rotation on the leading- and trailing-surface Nusselt number ratios is not as strong as that in the first pass. For case A, the rotation causes an increase of the leading-surface heat-transfer ratio and a decrease of the trailing-surface heat-transfer ratio. This heat-transfer pattern is opposite to that observed in the first pass with radial outward flow. The leading- and trailing-surface heat-transfer coefficients in the second pass for case B are much higher than that for case A. For case B, the leading-surface temperature is lower, due to a higher heat-transfer coefficient, than the trailing surface and sidewalls. Therefore, the fluid near the leading surface is cooler. The secondary cross flow carries the cooler fluid from the leading surface and the sidewalls back to the trailing surface due to the Coriolis-generated secondary flow. In case C, the trailing wall is warmer than the leading surfaces. Coriolis-driven secondary cross flow carries the cooler fluid from the sidewalls to the trailing surface; therefore, the trailing- and leading-surface heat-transfer coefficients for case C are higher than for case A in the second pass.

5.2.2.5 *Effect of Rotation Number and Wall-Heating Condition*

Figure 5.16 shows the effects of rotation number and uneven wall heating on the heat-transfer coefficient. This figure shows the variations of Nusselt number ratio with rotation number at selected axial locations for three surface-heating conditions (cases A, B, and C). The experimental results from Wagner et al. (1991) for the uniform wall temperature, i.e., similar to case A, are also included for comparison. Results show that the first-pass trailing-surface Nusselt number ratio increases with an increase in the rotation number, whereas the first-pass leading-surface Nusselt number ratio decreases and then increases with an increasing rotation number for the case of uniform wall temperature. This figure also shows that the difference between the leading- and trailing-surface heat-transfer coefficients in the second pass is not as significant as that in the first pass. Moreover, the leading- and trailing-side heat-transfer coefficients in the second pass are relatively

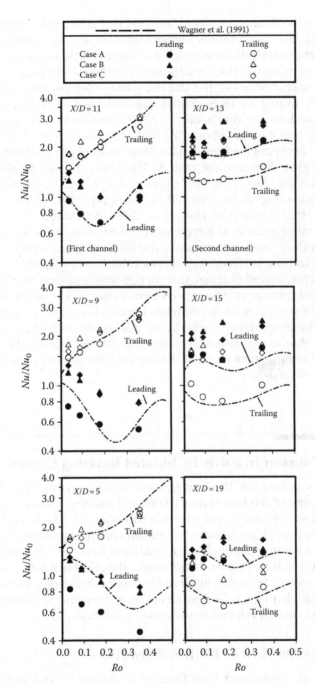

FIGURE 5.16
Effect of rotation number on Nusselt number ratio at selected axial locations. (From Han, J.C. et al., *ASME J. Heat Transfer*, 114(4), 850, 1993.)

insensitive to the rotation number compared to that in the first pass. It is possible that the 180° turn induces higher turbulent mixing, which dominates over the rotational effects at near vicinity. Moreover, the rotational buoyancy effects tend to minimize the Coriolis effects in the second pass.

As previously discussed, the uneven surface temperature on both the leading and trailing surfaces creates unequal local buoyancy forces that alter the heat-transfer coefficients. The trailing-surface Nusselt number ratios for the first pass are about 20% higher in case B than that for case A, whereas the first-pass leading-surface Nusselt number ratios for case B are 40%–80% higher than that for case A. The first-pass trailing-side Nusselt number ratios for case C are also slightly higher than that for case A, but the first-pass leading-surface Nusselt number ratios for case C are 40%–80% higher than that for case A. Han et al. (1993) concluded that in the first pass, the local uneven wall temperature interacts with the Coriolis-force-driven secondary flow and enhances the heat-transfer coefficients in both leading and trailing surfaces, with a noticeable increase in the leading side. However, in the second channel, the uneven wall temperature significantly enhances heat-transfer coefficients on both surfaces. The asymmetric heating condition generates an asymmetric temperature profile in the fluid. It seems that the temperature profile at the inlet to the second pass is already asymmetric due to the heating condition of the first pass. Therefore, the uneven wall temperature effect in the second pass is more significant than that in the first pass.

5.3 Heat Transfer in a Rib-Turbulated Rotating Coolant Passage

Periodic ribs placed on the heat-transfer surface disturb the boundary layer and augment the heat transfer from the surface of interest. However, ribs cause flow separation and increase turbulence in the boundary layer. Therefore, the flow characteristics in a ribbed channel are different from that observed in a smooth channel. Since rotational forces are coupled with the local velocity and temperature distributions, the effect of rotation in a ribbed channel is different from that observed in a smooth channel. Moreover, ribs placed at an angle to the bulk flow create secondary flow that interacts with the secondary flow developed in turn and by rotation.

5.3.1 Effect of Rotation on Rib-Turbulated Flow

Iacovides et al. (1998) used laser-Doppler anemometry and wall pressure measurements in a square section, rotating two-pass ribbed channel. The two passes are connected with a smooth U-bend. The ribbed-channel results are compared with the smooth-channel results. Ribbed-channel velocity

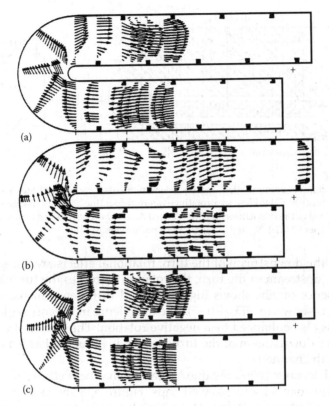

FIGURE 5.17
Comparison of velocity fields at positive and negative rotations with the stationary case. (a) Positive rotation, $Ro = 0.2$, (b) Stationary case, and (c) Negative rotation, $Ro = -0.2$. (From Iacovides, H. et al., *ASME J. Turbomach.*, 120(2), 386, 1998. With permission.)

profiles are significantly different from the smooth-channel results, especially because the separation bubble after the turn is smaller in a ribbed channel than in a smooth channel. Figure 5.17 shows the effect of positive and negative rotation on the ribbed channel, and rotating channel measurements are compared with the stationary channel. Results for the first pass, turn region, and second pass are presented. The velocity profiles are affected by the rotation direction in the first pass and early stages of the turn, but downstream of turn, the rotation direction does not have a significant effect. The turn effect is stronger than the rotational effects, and therefore immediate downstream of the turn, rotation effects (for the Ro number considered) do not influence the flow significantly. A detailed measurement indicates that rotational forces distort the velocity profile, but, surprisingly, the direction of rotation does not significantly alter the rotation flow profile.

This study also presented the turbulent intensities in the streamwise and cross stream directions in the first pass, second pass, and the turn region. As observed in the mean velocity distribution, the turn effects are

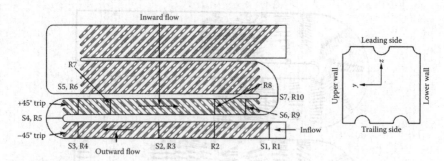

FIGURE 5.18

Serpentine channel configuration and rib orientations used by Tse and Steuber. (From Tse, D.G.N. and Steuber, G.D., Flow in a rotating square serpentine coolant passage with skewed trips, presented at the *International Gas Turbine and Aeroengine Congress and Exhibition*, Orlando, FL, ASME Paper 97-GT-529, June 2–5, 1997. With permission.)

strong in the downstream of the turn; rotational effects are not significant. However, upstream of the turn, the rotation increases the turbulent intensity. Presence of ribs shows high near-wall turbulence in the stationary channel (as shown in $z/D = -0.45$). The turbulence in the core region for this location is also enhanced by a negative rotation. The effects of rotation on turbulence downstream of the turn are not as influential as was observed in a smooth channel.

Tse and Steuber (1997) obtained rotating channel velocity data by LDV in a square channel with skewed trips. Figure 5.18 shows their multipass channel and rib orientations. Measuring locations are marked in this figure. They used a Reynolds number of 25,000 and a rotation number of 0.24 based on the channel hydraulic diameter. The ribs are at 45° angle of attack, and trailing and leading sides have a staggered but parallel rib arrangement. In a staggered rib arrangement, ribs are laid in an offset manner in two opposite walls; in a parallel rib arrangement, the ribs are inclined in the same direction to the flow. Secondary flow vectors in the first pass of the test section shows typical double vortex structure of a parallel skewed-ribbed channel.

Figure 5.19 shows the velocity distribution in a rotating channel's first and second turns. Unlike a smooth channel or a transverse rib channel, the secondary flow shows a large single vortex structure. A smaller vortex is noted at the corner of the channel. The strength of the primary vortex is much stronger than that in a smooth rotating channel. The secondary flow structures in the first turn and second turn are significantly different. These differences can be attributed to the relative position of the ribs, turn direction, and rotational effects. Bulk of the flow turns clockwise in the first turn, whereas the bulk flow turns counterclockwise in the second turn.

Hsieh et al. (1997) used LDV to measure the velocity in a two-pass rotating channel with transverse ribs. Their flow measurements in the first and the second pass show a decrease in the core flow velocity.

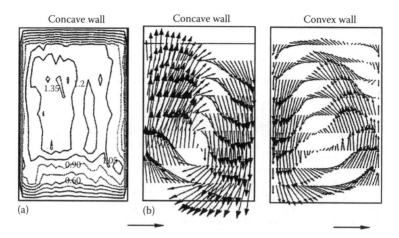

FIGURE 5.19

Secondary flows at the first and second bends of the serpentine channel of Tse and Steuber. (a) Convex wall rotation and (b) Concave wall rotation. (From Tse, D.G.N. and Steuber, G.D., Flow in a rotating square serpentine coolant passage with skewed trips, presented at the *International Gas Turbine and Aeroengine Congress and Exhibition*, Orlando, FL, ASME Paper 97-GT-529, June 2–5, 1997. With permission.)

The rotation-affected skew in the velocity profile is not as prominent as in a smooth channel. It can be argued that the presence of ribs increases turbulence mixing in the flow, and therefore it requires a greater force to distort the flow.

5.3.2 Effect of Rotation on Heat Transfer in Channels with 90° Ribs

Several different rib orientations are used in internal cooling. Figure 5.20 shows some typical arrangements used in internal ribbed passages. Unlike inclined ribs, the 90° orthogonal rib arrangement does not induce secondary flow of its own. The broken ribs create maximum flow disturbance and therefore show a higher heat-transfer coefficient than other rib arrangements. In this section, the heat-transfer patterns observed in orthogonal ribs are discussed. Wagner et al. (1992) experimented to determine the effects of buoyancy and Coriolis forces on heat transfer in a ribbed duct. Their experiments were conducted with a large-scale, multipass, heat-transfer model with both radial outward and inward flows. Ribs placed normal to the flow were used to produce rough walls. Figures 5.21 and 5.22 show the Nusselt number ratio (Nu/NU_∞) distribution for different rotation numbers and density ratios. As mentioned earlier, NU_∞ is the fully developed Nusselt number for fully developed turbulent smooth-pipe flow, rotation number ($Ro = \omega D/V$) is the relative strength of the Coriolis force to that of the bulk inertial force, and the density ratio [$DR = (\rho_{jn} - \rho_{Wall})/\rho_{in}$] is the measure of the buoyancy effects.

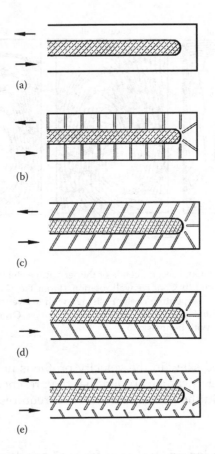

FIGURE 5.20
Different popular rib orientations to enhance heat-transfer coefficient in a two-pass chan-nel. (a) Smooth—no rib (From Han, J.C. et al., *ASME J. Heat Transfer*, 114(4), 850, 1993.), (b) 90° Transverse rib (From Parsons, J.A. et al., *Int. J. Heat Mass Transfer*, 37(9), 1411, 1994.), (c) 60° Parallel rib (Zhang et al., *ASME 93-GT-336. ASME J. Turbomach.*, 117(2), 272, 1995.), (d) 60° Parallel rib, and (e) 60° Staggered rib. (From Dutta, S. et al., *Int. J. Rotating Mach.*, 1(2), 129, 1995b.)

5.3.2.1 Effect of Rotation Number

Heat-transfer results from the high-pressure sides of two passes are shown in Figures 5.21 and 5.22. Results include a variation in rotation number and density ratio. Since the rotational forces are absent at no rotation ($Ro = 0$), no effect of density ratio on heat transfer is observed for a zero rotation number. Increasing the rotation number causes local increases in the heat transfer in the first-pass trailing surface by as much as 75% compared to the heat trans-fer in a stationary duct. The heat-transfer ratios for the high-pressure sur-faces in the first-pass trailing increase sharply with an increase in either the density ratio or the rotation number. The heat-transfer ratios in the second-pass leading are less affected by variations of either parameter. The increase

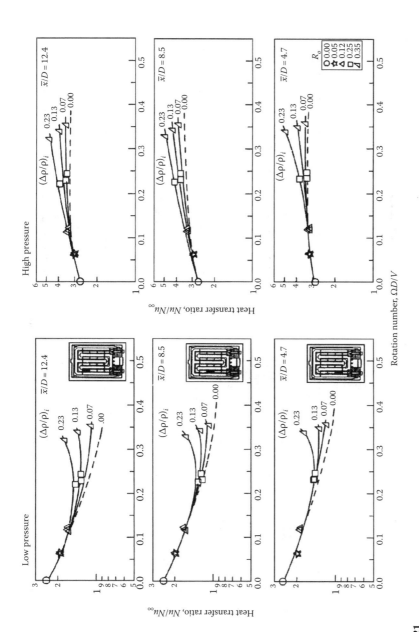

FIGURE 5.21

Effect of rotation number and coolant-to-wall density ratio on heat-transfer coefficient ratios in the 90° rib-roughened first pass with *Re* = 25,000. (From Wagner, J.H. et al., *ASME J. Turbomach.*, 114, 847, 1992. With permission.)

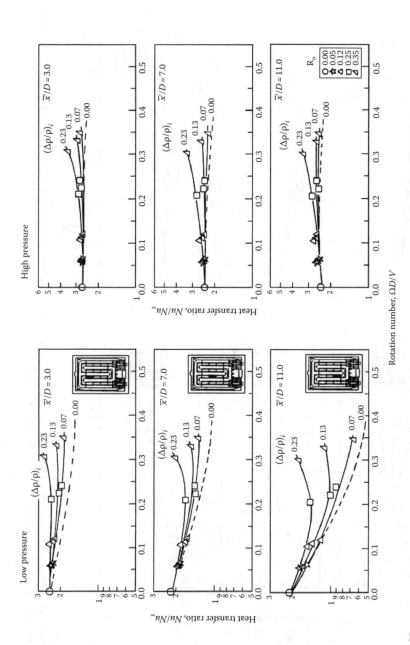

FIGURE 5.22

Effect of rotation number and coolant-to-wall density ratio on heat-transfer coefficient ratios in the 90° rib-roughened second pass with $Re = 25,000$. (From Wagner, J.H. et al., *ASME J. Turbomach.*, 114, 847, 1992. With permission.)

in heat-transfer ratio for the second pass is of the order of 30%–35% compared to the heat-transfer ratio in a stationary duct.

Figures 5.21 and 5.22 also show the heat transfer from low-pressure sides in the two passes. The results are more complex than that in the high-pressure surfaces. The heat-transfer ratio in the first-pass leading decreases with increasing rotation number for low values of rotation number ($Ro < 0.25$) and then, depending on density ratio, increases with increasing rotation for larger values of density ratio. The heat-transfer ratio increases with increases in the density ratio, and that is similar to the results obtained for the first-pass trailing surface. The effects of density ratio on the Nusselt number ratio are larger in the second-pass trailing with radial inward flow than in the first-pass leading with radial outward flow. Note that the estimated local density ratios in the second pass are less than half the inlet density ratios, because the air entering the second pass gets heated in the first pass.

Wagner et al. (1992) explained these uneven heat-transfer coefficient distributions on the low-pressure surfaces as a combined effect of buoyancy forces and the stabilization of the near-wall flow for low rotation numbers, the developing Coriolis-driven secondary flows, and the increases in flow reattachment lengths downstream of ribs for the larger values of the rotation number. It was postulated that the relatively small effects from variations in density ratio near the inlet of the second pass and the larger effects near the end of the second pass were due to the development of the near-wall thermal boundary layer. Near the inlet of the second pass, the thermal boundary layer was thin because of the strong turbulent mixing and secondary flow effects of the turn region. With an increase of distance from the turn, the turn-dominated secondary flows diminished, and the effects of buoyancy and Coriolis-driven secondary cross flow increased.

5.3.2.2 Effect of Wall-Heating Condition

Parsons et al. (1994) studied the influence of wall-heating condition on the local heat-transfer coefficient in a rotating, two-pass, square channel with 90° ribs on the leading and trailing walls. Like a smooth rotating channel, three thermal boundary conditions were studied: case A, all four walls at the same temperature; case B, all four walls at the same heat flux; and case C, trailing wall hotter than leading wall, and insulated sidewalls. Figure 5.23 shows the effect of rotation number on the Nusselt number ratio at six selected axial locations. For comparison, case A results for a smooth wall (Han et al., 1993) with uniform wall temperature are included. These results show that the first-pass trailing-wall Nusselt number ratio increases with an increasing rotation number for the smooth-wall channel, whereas the first-pass leading-wall Nusselt number ratio decreases and then increases with an increasing rotation number. The reverse is true for the second pass.

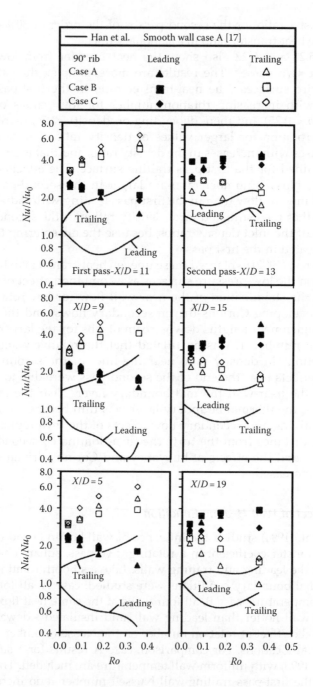

FIGURE 5.23
Effect of rotation number on Nusselt number ratio at selected axial locations for different heating cases. (From Parsons, J.A. et al., *Int. J. Heat Mass Transfer*, 37(9), 1411, 1994.)

Results also show that differences between the leading and trailing Nusselt number ratios decrease in the second pass, as compared to the first pass. A higher Nusselt number ratio difference in the first pass results because the inertia force is aided by the centrifugal buoyancy force in the first pass. The inertia force is partially counteracted by the buoyancy force in the second pass; this reduces differences between leading- and trailing-side Nusselt number ratios. The first-pass leading and second-pass trailing ribbed-wall Nusselt number ratios for cases B and C are up to 50% higher than those for case A. Results indicate that heating conditions significantly affect heat-transfer coefficient distribution. Compared to the smooth-wall results, ribs greatly enhance the wall Nusselt number ratios for the entire two-pass channel in both rotating and nonrotating conditions.

5.3.3 Effect of Rotation on Heat Transfer for Channels with Angled (Skewed) Ribs

Johnson et al. (1994a) used 45° angled ribs instead of orthogonal 90° ribs because the nonorthogonal ribs perform better in stationary channel heat-transfer enhancement. These nonorthogonal ribs are more frequently used in the turbine blade coolant channels. Their test section is shown in Figure 5.24 and its effect of rotation number on the local heat-transfer ratio is shown in Figure 5.25. Results show that the effect of rotation is more visible in the first

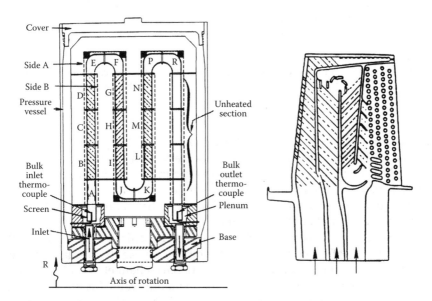

FIGURE 5.24
Multipass rotating channel with 45° ribs used by Johnson et al. (From Johnson, B.V. et al., *ASME J. Turbomach.*, 116(1), 113, ASME Paper 92-GT-191, 1994a. With permission.) to simulate a turbine rotor blade coolant passage.

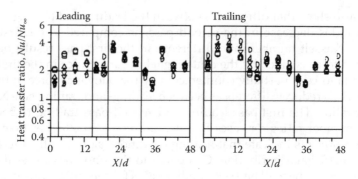

FIGURE 5.25
Effect of rotation number on ribbed channel heat-transfer ratio at $Re = 25,000$. (From Johnson, B.V. et al., *ASME J. Turbomach.*, 116(1), 113, ASME Paper 92-GT-191, 1994a.)

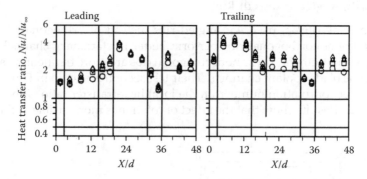

FIGURE 5.26
Effect of coolant-to-wall density ratio on ribbed channel heat-transfer coefficient ratio at $Re = 25,000$. (From Johnson, B.V. et al., *ASME J. Turbomach.*, 116(1), 113, ASME Paper 92-GT-191, 1994a. With permission.)

pass, and the following passes do not show much change in the presence of rotation. Figure 5.26 shows the buoyancy effect on the heat-transfer coefficient. The rotation and buoyancy effects are in general less for the ribbed channel compared to that in a smooth channel. Comparison of Figures 5.25 and 5.26 indicates that the rotational buoyancy has more effects at downstream passes than that of Coriolis forces.

Figures 5.27 and 5.28 compare the heat-transfer coefficients of smooth walls (open symbols), 90° ribbed walls (solid symbols), and 45° (half solid symbols) ribbed walls. Results show that, like a stationary channel, 90° ribbed walls have higher heat-transfer coefficients than smooth walls. Results also indicate that the 45° ribs perform better than the 90° ribs; the highest heat-transfer coefficients on low-pressure surfaces in all three passes are obtained with 45° angled ribs. However, the heat-transfer coefficients on high-pressure surfaces by 45° angled ribs are not so significantly better than 90° orthogonal ribs.

Symbol	ΔT_{in}(°C)	ΔT_{in}(°E)	$(\Delta \rho / \rho)_{in}$
	22.4	40	0.07
	44.7	80	0.13
	67.1	120	0.18
	89.1	160	0.22

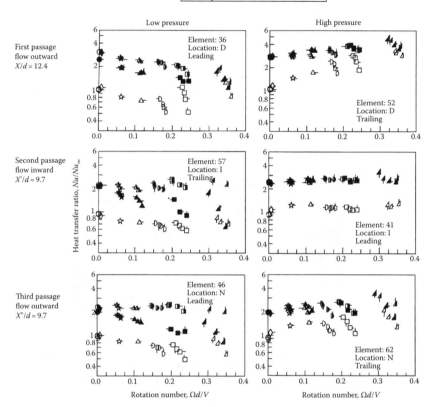

FIGURE 5.27

Comparison of heat-transfer coefficients based on rotation number from smooth channel (open symbols), 90° orthogonal ribs (solid symbols), and 45° inclined ribs (half solid symbols). (From Johnson, B.V. et al., *ASME J. Turbomach.*, 116(1), 113, ASME Paper 92-GT-191, 1994a. With permission.)

5.3.3.1 Effect of Angled Ribs and Heating Condition

Zhang et al. (1995) studied the influence of uneven wall temperature on the local heat-transfer coefficient in a rotating two-pass square channel with 60° rib turbulators. Their rib arrangement is shown in Figure 5.29. Figure 5.30 shows the effect of rotation number on the Nusselt number ratio at six selected axial locations of the two-pass square channel. The experimental results of Johnson et al. (1994a) for the case of uniform wall temperature and with 45° ribbed walls, and from Han et al. (1993) for the case of four walls at the same temperature and smooth walls, are included for comparison. The effect of rotation number on the 60° ribbed-wall Nusselt number ratio shows trends similar to those of the smooth wall, except that the 60° ribs significantly

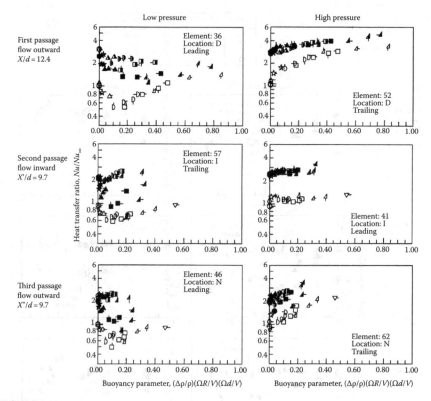

FIGURE 5.28

Comparison of heat-transfer coefficients based on buoyancy parameter from smooth channel (open symbols), 90° orthogonal ribs (solid symbols), and 45° inclined ribs (half solid symbols). (From Johnson, B.V. et al., *ASME J. Turbomach.*, 116(1), 113, ASME Paper 92-GT-191, 1994a. With permission.)

enhance the surface Nusselt number ratios in the entire two-pass channel. Ribbed-surface results show that the first-pass Nusselt number ratios on the leading side for cases B and C are higher than that for case A. The second-pass Nusselt number ratios on the trailing ribbed surface for cases B and C are also higher than for case A. However, the second-pass Nusselt number ratios on the leading ribbed surface for case C are lower than that for case A.

Figure 5.30 shows the comparison between Zhang et al.'s (1995) 60° rib and Johnson et al.'s (1994a) 45° rib results in conditions of uniform wall temperature. In the first pass, the 60° rib Nusselt number ratios on the leading surface agree with those of the 45° rib, while the 60° rib Nusselt number ratios on the trailing surface are significantly higher than those of the 45° rib. In the second pass, the 60° rib data on the leading surface are higher. However, the 60° rib data on the trailing surface are lower than those shown by Johnson et al. (1994a). The difference between these two studies is explained by Zhang et al. (1995) as follows: The 45° ribs of Johnson et al. (1994a) are semicircular

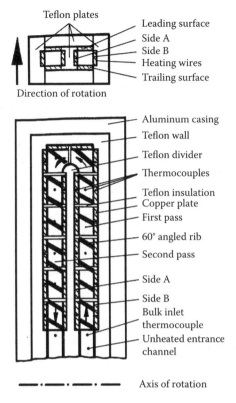

FIGURE 5.29
60° angled ribs used by Zhang et al. (From Zhang, Y.M. et al., *ASME J. Turbomach.*, 117(2), 272, 1995.)

in cross section and have a rib height ratio of $e/D = 0.10$. The 60° ribs of Zhang et al. (1995) are sharp-edged square in cross section, different from those of Johnson et al. (1994a), and have a rib height ratio of $e/D = 0.125$, i.e., larger than those of Johnson et al. (1994a). Rotation creates a thinner boundary layer near the first-pass trailing wall, and geometrical features like rib height, rib shape, and rib orientation have a more significant effect on the thinner trailing boundary layer than on the thicker leading-side boundary layer. Therefore, the 60° rib Nusselt number ratios on the ribbed trailing wall are higher than those of Johnson et al. (1994a), while the leading-wall Nusselt number ratios are about the same for the two studies. Similarly, in the second pass, the rotation induces a thinner boundary layer on the leading wall and a thicker boundary layer on the trailing wall. Therefore, due to the sharper and taller ribs of this study, the 60° rib data on the second-pass leading are higher than those of Johnson et al. (1994a). However, the 60° rib data on the trailing surface are lower than the 45° rib, particularly at the higher rotation numbers. This is due to the different combined effects of rotation, rib orientation, and second-pass entrance geometry.

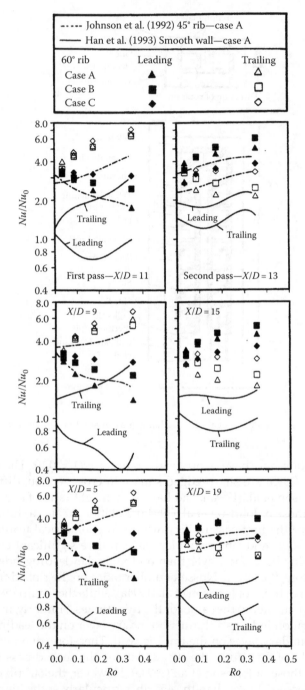

FIGURE 5.30
Effect of rotation number on Nusselt number ratio at selected axial locations for different heating conditions. (From Zhang, Y.M. et al., *ASME J. Turbomach.*, 117(2), 272, 1995.)

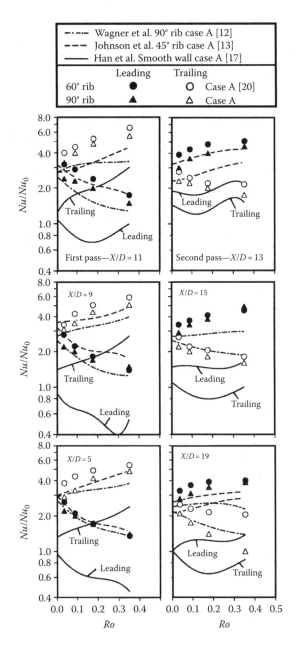

FIGURE 5.31

Effect of rotation number on Nusselt number ratio at selected axial locations for heating condition A of Parsons et al. (From Parsons, J.A. et al., *Int. J. Heat Mass Transfer*, 37(9), 1411, 1994.)

5.3.3.2 Comparison of Orthogonal and Angled Ribs

Parsons et al. (1994) studied the wall-heating condition on local heat transfer in a two-pass square channel with 90° ribs. Figure 5.31 shows the effect of rotation number on the Nusselt number ratio at six selected channel locations for case A, the uniform wall temperature heating condition. Experimental results from four previous investigations (Wagner et al., 1992; Han et al., 1993; Johnson et al., 1994a; Zhang et al., 1995) are included for comparison. Results show that the effect of rib angle is significant on the heat-transfer distribution. The 60° rib data are higher than the 90° rib data; similarly, the 45° rib data are higher than the 90° rib data. The combination of the inertial (streamwise) and the Coriolis (cross stream) forces produces a spiral fluid flow path. Parsons et al. (1994) postulate that the angled ribs are oriented in a more effective direction to the flow to trip the boundary layers than are the 90° transverse ribs. Therefore, angled ribs produce higher Nusselt number ratios than 90° ribs. This heat-transfer enhancement difference is more pronounced in the first-pass and second-pass trailing walls of the channel.

5.4 Effect of Channel Orientation with Respect to the Rotation Direction on Both Smooth and Ribbed Channels

Besides the effects of rotation on ribbed surfaces, effects of model orientation on heat-transfer distribution in rotating ducts are also important in turbomachinery applications. Figures 5.32 and 5.33 show the cooling-channel orientation with respect to the rotation direction. Since the turbine blade is curved, the rotor blade cooling passage can have a different model orientation with respect to the rotating plane. The schematic of the secondary vortices in Figure 5.33 shows that the distribution of flow is different for different model orientations.

5.4.1 Effect of Rotation Number

Johnson et al. (1994b) experimented to determine the effects of model orientation as well as buoyancy and Coriolis forces on heat transfer. The results from model orientations of 0° and 45° to the axis of rotation were compared. The experiments were conducted for passages with smooth surfaces and with 45° ribs. The results of Johnson et al. (1993) are shown in Figures 5.34 through 5.36. Figure 5.34 shows the effects of orientation and rotation direction for a smooth channel, and Figure 5.35 shows the effects on a ribbed channel. The first-pass leading wall is most significantly affected by rotation. Local Nusselt number distributions are shown in Figures 5.34 and 5.35. Figure 5.36 summarizes the rotational effects for different operating conditions.

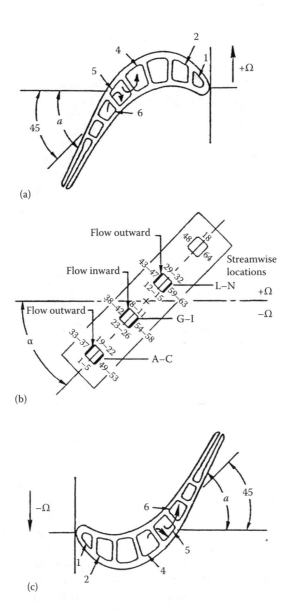

FIGURE 5.32
Model orientation and rotation directions used by Johnson et al. (a) Blade with forward-flowing coolant, (b) heat transfer model, and (c) blade with rearward-flowing coolant. (From Johnson, B.V. et al., *ASME J. Turbomach.*, 116, 738, ASME Paper 93-GT-305, 1994b. With permission.)

Figure 5.36 shows the effects of rotation number on the heat-transfer ratio for smooth and ribbed channels. At a typical flow condition, the heat transfer on the leading surfaces for outward flow in the first pass with smooth walls was twice as much for the model at 45° compared to the model at 0°. However, the differences for the other passages and with ribs

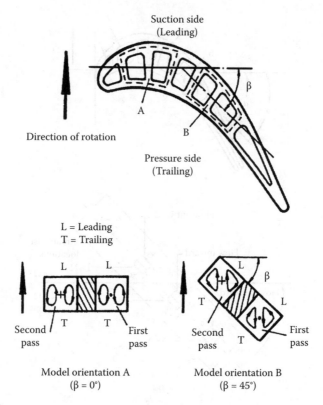

FIGURE 5.33
Schematic of cooling-channel orientations in an airfoil. (From Parsons, J.A. et al., *Int. J. Heat Mass Transfer*, 38(7), 1151, 1995.)

were less. In addition, the effects of buoyancy and Coriolis forces on heat transfer in the rotating passage were decreased with the model at 45° compared to the results at 0°. The heat transfer in the turn regions and immediately downstream of the turns in the second pass with flow inward and in the third pass with flow outward also depended on model orientation, with differences as large as 40%–50%. This figure also shows the heat-transfer ratio from the leading and trailing surfaces as a function of rotation number and the inlet density ratio. The smooth-wall results show less effect of inlet density ratio for the 45° model orientation than those for the 0° orientation.

5.4.2 Effect of Model Orientation and Wall-Heating Condition

Parsons et al. (1995) studied the influences of channel orientation and wall-heating condition on local heat transfer in a rotating, two-pass square channel with 60° and 90° ribs on the leading and trailing walls. Two thermal

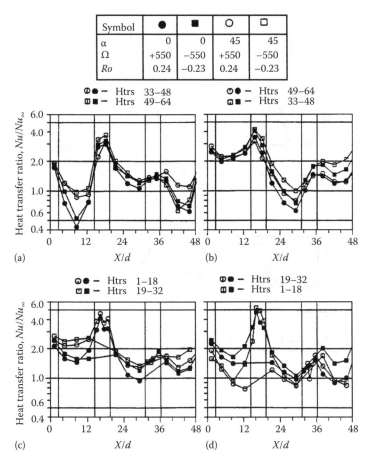

FIGURE 5.34
Effects of channel orientation and rotation direction on heat-transfer coefficient ratio for smooth-wall rotating channels at Re = 25,000, density ratio = 0.13. (a) Leading, (b) trailing, (c) side A, and (d) side B. (From Johnson, B.V. et al., *ASME J. Turbomach.*, 116, 738, ASME Paper 93-GT-305, 1994b. With permission.)

boundary conditions were studied. The first condition, case A, had all four walls at the same temperature; the second condition, case B, had all four walls at the same heat flux. Figures 5.37 and 5.38 show the results of cases A and B for different rib orientations and model orientations.

Figure 5.37 shows the heat-transfer results for case A with 60° angled ribs. The Nusselt number ratios for the first-pass trailing walls and second-pass leading walls for the twisted (model orientation of 45°) orientation are nearly equal to those Nusselt number ratios corresponding to the untwisted (model orientation of 0°) model orientation. The Nusselt number ratios for the twisted orientation of the first-pass leading wall are

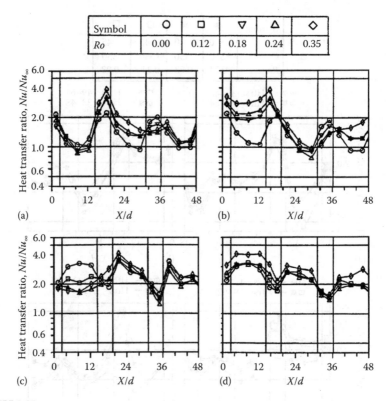

Symbol	O	□	▽	△	◇
Ro	0.00	0.12	0.18	0.24	0.35

FIGURE 5.35
Effects of rotation number on heat-transfer coefficient for smooth and ribbed rotating channels with 45° inclined channel orientation with *Re* = 25,000 and density ratio = 0.13. Smooth—(a) leading, (b) trailing. Skewed—(c) leading, and (d) trailing. (From Johnson, B.V. et al., *ASME J. Turbomach.*, 116, 738, ASME Paper 93-GT-305, 1994b. With permission.)

up to 120% higher and about 40% lower for the second-pass trailing wall, respectively, than those Nusselt number ratios for the untwisted orientation. The effect of the Coriolis force on the first-pass trailing walls and second-pass leading walls for the untwisted orientation is more, since these forces and cross stream flows are normally incident on these unstabilized surfaces. This produces more differences between leading and trailing Nusselt number ratios in each pass. However, the Coriolis effect is still present for the twisted orientation but reduced by being oblique at 45° to the first-pass trailing walls and second-pass leading walls. Therefore, the differences in Nusselt number ratios between leading and trailing walls are not as significant for the twisted orientation as for the untwisted orientation. In fact, the benefits of flow unstabilization and drawbacks of flow stabilization are shared by more than one surface in the 45° orientation. This sharing reduces the intensity of the corresponding effects in specified surfaces with twisted 45° model orientation.

Symbol flag	ȯ	O⊢	⊖	–O
$(\Delta\rho/\rho)_i$	0.07	0.13	0.18	0.22

Symbol	O	□	◇	▲	▼	★
α	0	0	0	45	45	45
Ω	Positive	Positive	Negative	Positive	Positive	Negative
Trips	Smooth	Skewed	Smooth	Smooth	Skewed	Smooth

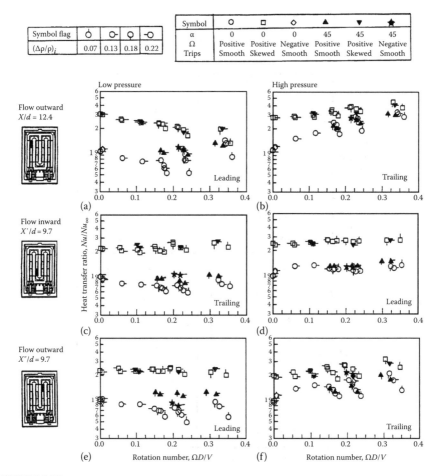

FIGURE 5.36

Effects of rotation number and inlet coolant-to-wall density ratio on Nusselt number ratio for different channel orientations at $Re = 25{,}000$. (a) First pass leading surface at $X/d = 12.4$, (b) first pass trailing surface at $X/d = 12.4$, (c) second pass trailing surface at $X/d = 9.7$, (d) second pass leading surface at $X/d = 9.7$, (e) third pass leading surface at $X/d = 9.7$, and (f) third pass trailing surface at $X/d = 9.7$. (From Johnson, B.V. et al., *ASME J. Turbomach.*, 116, 738, ASME Paper 93-GT-305, 1994b. With permission.)

Figure 5.38 shows the Nusselt number ratios at different rotation numbers for 90° ribs with heating condition of case B (uniform wall heat flux). Several Nusselt number ratios on both leading and trailing walls for the twisted model orientation for this heating condition are up to 100% higher than those Nusselt number ratios with untwisted model orientation. The most significant increase occurs on the first-pass leading wall. The effect of model orientation on the Coriolis force clearly increases heat transfer for the first-pass leading walls and the second-pass trailing walls. The first-pass trailing surface and the second-pass leading surface show

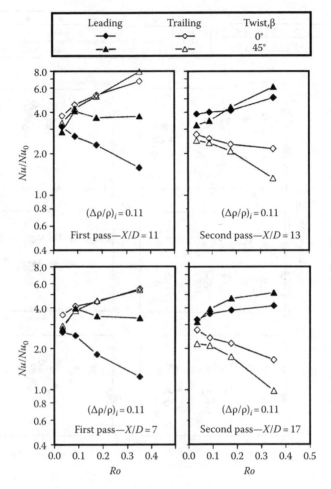

FIGURE 5.37
Effect of rotation number on Nusselt number ratio at selected axial locations for heating condition case A, two channel orientations, and 60° angled ribs. (From Parsons, J.A. et al., *Int. J. Heat Mass Transfer*, 38(7), 1151, 1995.)

increased Nusselt number ratios when the model twists (45° model orientation) under uneven wall temperatures. For the untwisted orientation, the flow in each of the two secondary cross stream flow vortices is heated in the boundary layers of three walls: the leading wall, trailing wall, and one of the two sidewalls. The temperature of the first-pass leading wall is higher than the other walls in the first pass with uniform wall heat flux heating condition. However, in the twisted-model orientation, each secondary cross stream flow vortex is heated either by the leading wall or by the trailing wall, and also by one of the two sidewalls. Thus, the twisted orientation secondary flow vortex next to the trailing wall and

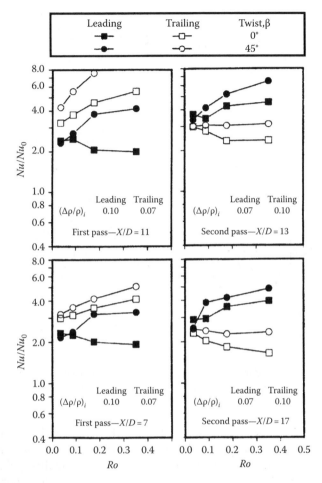

FIGURE 5.38
Effect of rotation number on Nusselt number ratio at selected axial locations for heating condition case B, two channel orientations, and 90° orthogonal ribs. (From Parsons, J.A. et al., *Int. J. Heat Mass Transfer*, 38(7), 1151, 1995.)

one sidewall is no longer as warm as the relatively warm leading wall, as the corresponding vortex did for the untwisted orientation. Therefore, the flow next to the trailing wall for the twisted orientation is relatively cooler than for the untwisted orientation. Thus heat transfer and Nusselt number ratios significantly increase for the first-pass trailing wall, and similarly for the second-pass leading wall in case B.

Figure 5.39 schematically shows the combined effects of rotation, rib orientation, and model orientation on flow distribution. Two different types of secondary flows in model orientation β = 0° (upper diagram in Figure 5.39)—one produced by the Coriolis force, the other by rib angle effects—combine to selectively enhance and reduce heat-transfer coefficients. They enhance

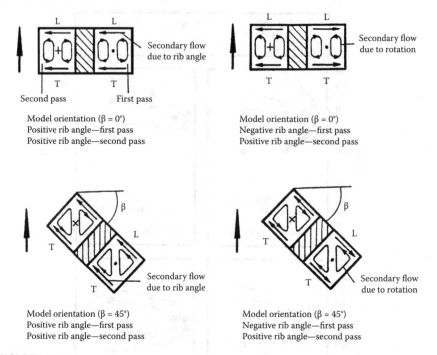

FIGURE 5.39
Conceptual view of secondary flow vortices induced by rotation, channel orientation, and rib angles. (From Parsons, J.A. et al., *Int. J. Heat Mass Transfer*, 38(7), 1151, 1995.)

heat transfer for one-half of each of the leading and trailing walls where they are constructively (same direction) oriented; and they reduce heat transfer for the other half of each of the leading and trailing walls where they are destructively (opposite direction) oriented. However, the two secondary flows for the rotating different model orientation (lower diagrams in Figure 5.39, β = 45°) combine to (1) constructively enhance heat transfer for the positive rib angle in the first pass, and (2) destructively reduce heat transfer for the negative rib angle in the first pass. Johnson et al. (1994a) used 45° negative ribs in the first pass, and Parsons et al. (1995) used 60° positive ribs in the first pass. The heat transfer on the first-pass leading wall increases for the 60° positive rib angle (used by Parsons et al., 1995) but does not change much for the 45° negative rib angle (used by Johnson et al., 1994a) by changing the model orientation from 0° to 45°. This is the effect of the combined Coriolis force and rib angle on the secondary cross flows.

Dutta and Han (1996) used high-performance ribs in a rotating two-pass square channel. Their rib configuration is shown in Figure 5.40. These broken V-ribs have been shown to be better performing in several rib configurations tested in a stationary channel. Three different channel orientations

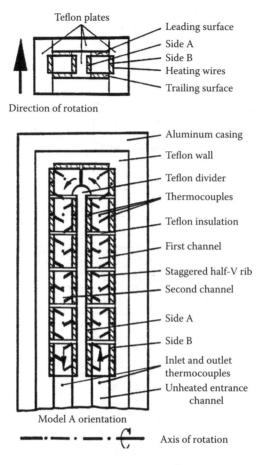

Direction of rotation

Model A orientation

FIGURE 5.40
Schematic of the broken rib arrangement used in a two-pass rotating channel by Dutta and Han for different channel orientation studies. (From Dutta, S. and Han, J.C., *ASME J. Heat Transfer*, 118, 578, 1996.)

were used, and these orientations are shown in Figure 5.41. These high-performance ribs have their own secondary flows. Figure 5.41 shows the schematics of the secondary flows developed by rib orientation and rotation. The channel orientation with respect to the rotation axis influences the secondary vortices. The secondary flow developed by ribs can interact with the secondary flow of rotation, and a new flow condition may be established. Figure 5.42 shows the heat-transfer comparison for two model orientations: 60° parallel ribs of Parsons et al. (1995) are compared with the high-performance ribs. This figure shows that the second-pass leading surface has higher Nusselt number ratios with model C than with model A. The broken V-shaped ribs perform better than the 90° and 60° ribs.

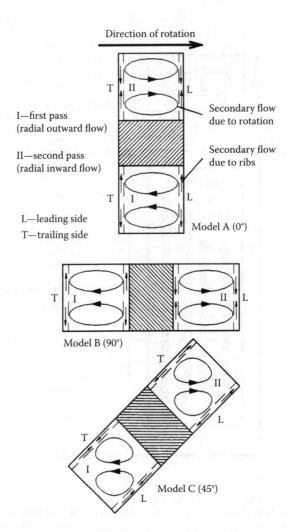

FIGURE 5.41

Three-channel orientations used by Dutta and Han. (From Dutta, S. and Han, J.C., *ASME J. Heat Transfer*, 118, 578, 1996.)

5.5 Effect of Channel Cross Section on Rotating Heat Transfer

5.5.1 Triangular Cross Section

So far only square-duct results are discussed. The following discussions are based on another cross-sectional geometry: triangular cross section. Clifford et al. (1984) studied the mean heat transfer in a triangular-sectioned rotating duct. Figure 5.43 shows their rotational triangular channel.

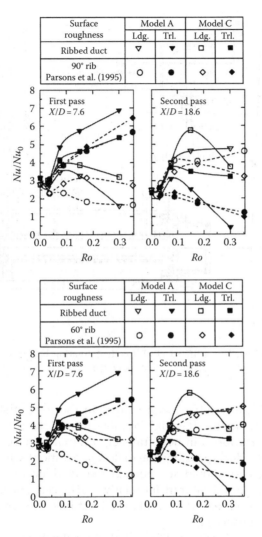

FIGURE 5.42
Variation in Nusselt number ratio with rotation number for different rib arrangements at different channel orientations. (From Dutta, S. and Han, J.C., *ASME J. Heat Transfer*, 118, 578, 1996.)

Figure 5.44 shows the typical distributions of local Nusselt number. The mean heat-transfer data presented by Clifford et al. (1984) showed little effect from rotational buoyancy. The zero rotational data are shown as the hatched band.

The data band for the stationary duct is narrower than for the corresponding distributions for the local Nusselt numbers at different rotational Rayleigh numbers. This dispersion in data is due to rotational buoyancy effects. The individual lines show changes in the heating rate and the

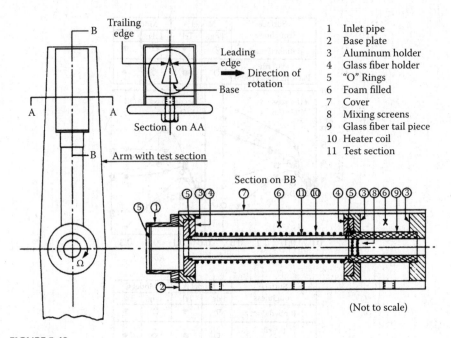

FIGURE 5.43
Triangular channel used by Clifford et al. for their rotating heat-transfer analysis. (From Clifford, R.J. et al., *ASME J. Eng. Gas Turbines Power*, 106, 661, 1984. With permission.)

rotational Rayleigh number mean values. In the immediate entry region, increasing the rotational Rayleigh number causes the expected reduction (adverse buoyancy force) in heat transfer, but that trend is reversed in the exit section of the duct (possibly due to flow separation). Clifford et al. (1984) exemplified the case shown at a normalized axial location of 0.2. A maximum increase in local heat transfer relative to the zero speed mean value of about 59% was observed, whereas at axial location of 0.8, a maximum reduction of 44% was observed.

Harasgama and Morris (1988) studied the influence of rotation on heat-transfer characteristics with different cross-sectioned ducts. They developed a correlation for duct average Nusselt number for radial outward flow as

$$Nu = 0.022\left(\frac{Ra}{Re^2}\right)^{-0.186} Re^{0.8}Ro^{0.33} \qquad (5.1)$$

where
    Ra is the rotational Rayleigh number
    Re is the through flow Reynolds number
    Ro is the inverse Rossby number ($\omega D/V$)

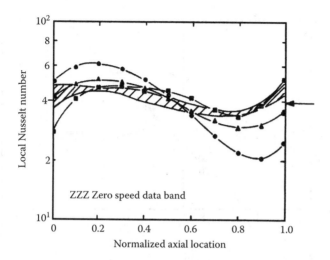

FIGURE 5.44

Circumferentially averaged Nusselt number distribution in a rotating triangular channel at $Re = 14{,}500$. (From Clifford, R.J. et al., *ASME J. Eng. Gas Turbines Power*, 106, 661, 1984. With permission.)

Figure 5.45 shows that the correlation is valid for mean duct heat-transfer coefficients in both triangular and circular ducts.

5.5.2 Rectangular Channel

Guidez (1989) did an experimental and theoretical study of a rotating rectangular channel. Figure 5.46 shows the channel shape and rotation directions. Rotation-induced secondary flow vortices are dependent on the channel cross section. Secondary flows developed by the turn distort the bulk flow distribution and affect the heat-transfer pattern. Figure 5.47 shows Nusselt number ratios in the pressure and suction sides for different rotation numbers. The leading-surface (suction side) results are nearly independent of the flow Reynolds number. The trailing side shows a considerable effect from the flow Reynolds number. The suction-side heat-transfer coefficient decreases with rotation, but reaches an equilibrium value at $Ro = 0.1$. The trailing side shows that the Nusselt number ratio increases with an increase

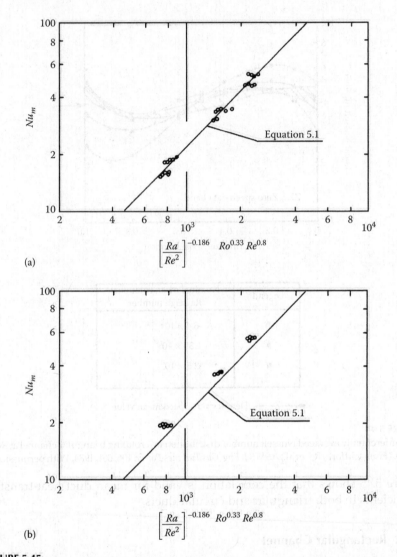

(a)

(b)

FIGURE 5.45
Comparison of circular and triangular rotating channel Nusselt numbers with correlation developed by Harasgama and Morris. (a) Correlation of circular duct, outward flow, leading side data and (b) Correlation of circular duct, outward flow, trailing side data. (From Harasgama, S.P. and Morris, W.D., *ASME J. Turbomach.*, 110, 44, 1988. With permission.)

in the rotation number. Figure 5.48 shows the effect of rotational buoyancy on heat transfer. Note that the shape of the curves changes curvature from the previous plot, i.e., Figure 5.47. Results of a circular channel are also plotted, and the trailing side shows mostly higher Nusselt number ratios with the rectangular channel used.

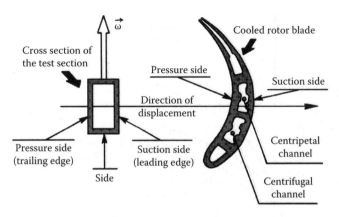

FIGURE 5.46
Schematics of rotor blade internal coolant passages and justification for a rotating rectangular channel analysis to simulate midchord region. (From Guidez, J., *ASME J. Turbomach.*, 111, 43, 1989. With permission.)

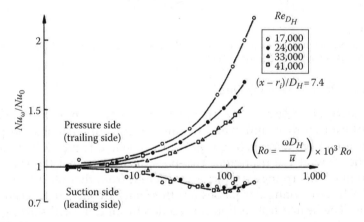

FIGURE 5.47
Variation of Nusselt number ratio with rotation number. (From Guidez, J., *ASME J. Turbomach.*, 111, 43, 1989. With permission.)

5.5.3 Circular Cross Section

Morris and Salemi (1992) and Morris and Chang (1997) show Nusselt number distribution in a rotating circular cross-sectional channel. Overall effects of Coriolis and buoyancy on the leading and trailing sides are shown in Figure 5.49. As observed with other cross sections, the trailing-edge heat transfer increases with increasing rotational effects, and the leading side shows lower heat-transfer coefficient than the stationary condition.

Figure 5.50 shows the effect of buoyancy on heat transfer. The rotation speed and bulk flow velocities are kept constant. The axial heat-transfer coefficient shows significant dependency on the wall-heating condition. A higher heat flux shows higher normalized Nusselt number.

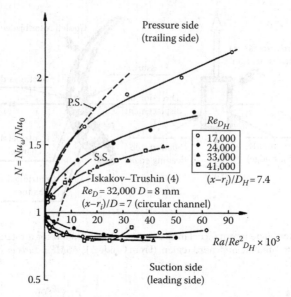

FIGURE 5.48
Variation of Nusselt number ratio with rotational Rayleigh number. (From Guidez, J., *ASME J. Turbomach.*, 111, 43, 1989. With permission.)

5.5.4 Two-Pass Triangular Duct

Dutta et al. (1994, 1996b,c) used a triangular duct to study smooth- and ribbed-wall heat-transfer distribution at different model angles. Air at atmospheric pressure and room temperature was used as coolant. Figure 5.51 shows the schematic view of the triangular two-pass duct. Note that the shape and orientation of this triangular duct are different from that used by Clifford et al. (1984). Two triangular passes are thermally isolated from each other by Teflon. The heated test section is preceded by an unheated starting duct, which has the same triangular shape of the test section and is made of Teflon.

Figure 5.52 shows the effect of rotation number at selected locations of the triangular duct. The wall-heating conditions are case A, all walls at the same temperature (the heated walls are at 40°C above ambient, giving a density ratio of 0.12); case B, all walls at uniform heat flux; and case C, enhanced heat-transfer surfaces warmer than other surfaces. The square-duct results for case A from Han et al. (1993) at available nearby locations are included for comparison. It is observed that wall-heating conditions (or centrifugal buoyancy effect) can significantly alter the Nusselt number ratios. For the locations in the first pass, the triangular-duct Nusselt number ratios for case A are mostly contained within the upper and lower limits imposed by the square duct. This is because in a triangular duct, there is less space for the fluid to form secondary flow by rotation. In the second pass, the leading side of the triangular duct shows a very high Nusselt number ratio compared to the square duct. An increase of up to 120% over the square-duct results is noted at the

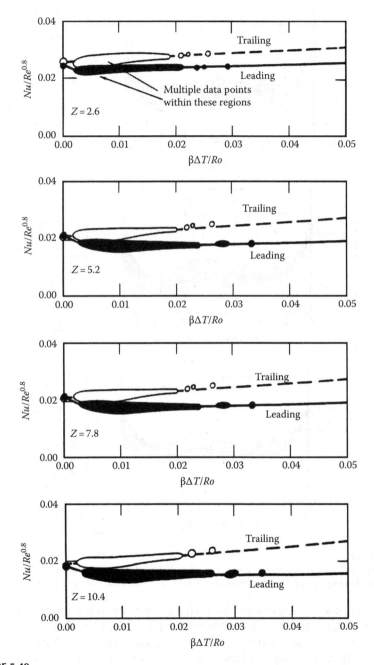

FIGURE 5.49
Overall rotational effects on the heat-transfer coefficient. (From Morris, W.D. and Salemi, R., *ASME J. Turbomach.*, 114(4), 858, 1992. With permission.)

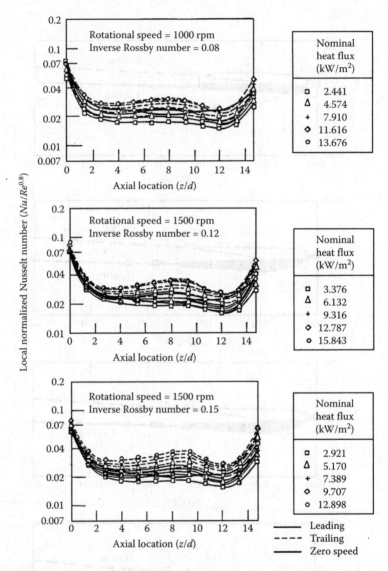

FIGURE 5.50

Normalized Nusselt number distribution with different heating conditions. (From Morris, W.D. and Chang, S.W., *Int. J. Heat Mass Transfer*, 40(15), 3703, 1997. With permission.)

second-pass leading surface. This is due to more intensive mixing and favorable secondary flow in the 180° bend for the triangular-duct configuration. Note that the average of trailing-surface and leading-surface Nusselt number ratios of the square duct and the triangular duct is about the same in the first pass. Harasgama and Morris (1988) had similarly observed almost identical mean duct heat transfer in rotating triangular and circular ducts. The other two heating conditions shown in Figure 5.52 show that the Nusselt number

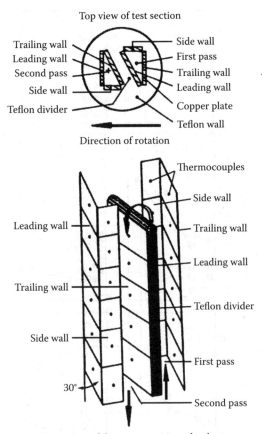

Top view of test section

Direction of rotation

Open view of the two-pass triangular duct

FIGURE 5.51
Schematic of the two-pass triangular channel used by Dutta et al. (From Dutta, S. et al., Effect of model orientation on local heat transfer in a rotating two-pass smooth triangular duct, *1994 Winter Annual Meeting*, Chicago, IL, ASME HTD Vol. 300, pp. 147–153, November 6–11, 1994.)

ratio difference between the leading and trailing sides for case B is slightly greater than for case A for the triangular duct. The Nusselt number ratio for case C is least among the group in leading surfaces of both passes.

The model orientation of the triangular duct is varied by turning the two-pass duct about its own axis. Figure 5.53 shows two model orientations with respect to the rotation plane. The two model orientations are referred to as model A and model B. Direction of rotation is also shown in this figure. The arrow represents the direction of the tangential velocity not the rotation vector.

Figure 5.54 shows the effect of rotation number Ro on the Nusselt number ratio (Nu/Nu_0) for model orientation A. The ribbed-duct results are compared with smooth-wall results. The first-pass trailing wall and second-pass

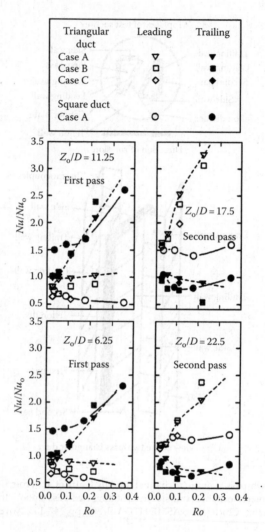

FIGURE 5.52
Effect of rotation number on Nusselt number ratio in a rotating two-pass triangular channel. (From Dutta, S. et al., Effect of model orientation on local heat transfer in a rotating two-pass smooth triangular duct, *1994 Winter Annual Meeting*, Chicago, IL, ASME HTD Vol. 300, pp. 147–153, November 6–11, 1994.)

leading wall show an increase in the Nusselt number ratio with an increase in the rotation number. Influence of rotation number on heat transfer from the sidewalls is mixed (increase and decrease). There is an increase in the heat transfer from the sidewalls at low rotation numbers from the corresponding no-rotation case. At higher rotation number ($Ro = 0.22$), the sidewall Nusselt number ratio decreases from the previous rotation number value. Inclusion of ribs enhances turbulence in the duct, and heat transfer from the smooth

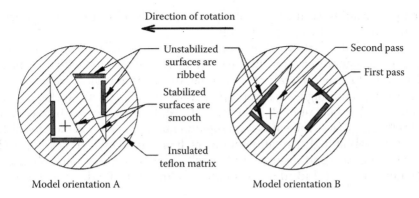

FIGURE 5.53
Two model orientations used by Dutta et al. in their two-pass ribbed triangular channel. (From Dutta, S. et al., *Int. J. Heat Mass Transfer*, 39(4), 707, 1996b.)

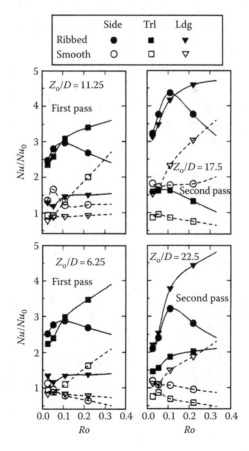

FIGURE 5.54
Nusselt number ratio variations with rotation number with model orientation A of Dutta et al. (From Dutta, S. et al., *Int. J. Heat Mass Transfer*, 39(4), 707, 1996b.)

first-pass leading wall and second-pass trailing wall of the ribbed duct is more than that with all smooth walls.

Figure 5.55 shows the rotation number effects on heat transfer for model orientation B. Smooth-wall results are included for comparison with the ribbed-channel data. Smooth-wall results show an increase in heat transfer from the first-pass trailing wall, second-pass leading wall, and sidewalls, with an increase in rotation number. Ribbed-duct results show that the heat transfer from those surfaces increases with rotation number at low rotation numbers but decreases at higher rotation numbers. It is argued by Dutta et al. (1995b) that the increase and decrease in Nusselt number ratio with an increase in rotation number is linked to the spiral path line of the fluid particles caused

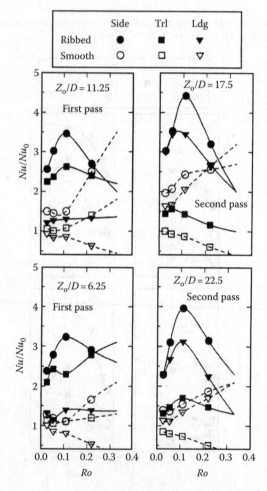

FIGURE 5.55

Nusselt number ratio variations with rotation number with model orientation B of Dutta et al. (From Dutta, S. et al., *Int. J. Heat Mass Transfer*, 39(4), 707, 1996b.)

by the rotation-induced secondary flow. The angle of approach of the flow with respect to the rib orientation is dependent on the rotation number, and perhaps the angle of approach is not favorable for heat-transfer enhancement beyond a certain number for partially unstabilized surfaces.

Figure 5.56 shows the effects of rotation number on the Nusselt number ratio for two model orientations. It is seen that the model orientation can significantly influence the heat-transfer pattern. In model A, the first-pass trailing wall and the second-pass leading wall are considered as fully unstabilized by rotation. Since the sidewalls are neither trailing wall nor leading wall in model A, sidewalls of both the passes in model A are partially unstabilized. With a similar analogy, all four ribbed walls are partially unstabilized

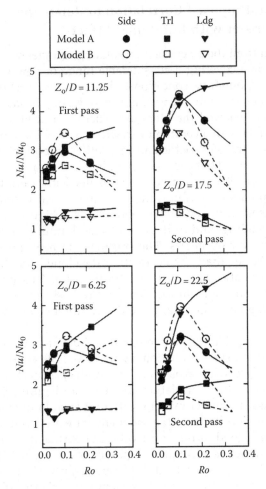

FIGURE 5.56
Effect of model orientation on rotation number dependency of rib-enhanced Nusselt number ratio in a two-pass triangular-sectioned rotating channel. (From Dutta, S. et al., *Int. J. Heat Mass Transfer,* 39(4), 707, 1996b.)

by rotation in model B. Results show that the effect of rotation number on the Nusselt number ratio for the partially unstabilized surfaces has a similarity. Nusselt number ratio from the partially unstabilized surfaces decreases at the highest rotation number experimented ($Ro = 0.22$), compared to the Nusselt number ratio of the previous rotation number ($Ro = 0.11$), whereas the Nusselt number ratio for the fully unstabilized walls increases with rotation number.

5.6 Different Proposed Correlation to Relate the Heat Transfer with Rotational Effects

As can be seen from the discussion on rotating channel heat transfer, much experimental data covering a considerable range of flow velocity, rotation speed, thermal boundary conditions, channel cross section, rib orientation, and channel orientation were published during the last 10 years. Dutta and Han (1997) reviewed the literature. There is a need to correlate these experimental data with a common variable or a combination of variables to estimate the heat-transfer coefficients in rotation. Important attempts to correlate rotating channel data are presented here.

Wagner et al. (1991) correlated their results with a buoyancy parameter. Their formulation of a buoyancy parameter was essentially a Gr/Re^2 formulation used in mixed convection analysis. Figure 5.57 shows their correlation of leading- and trailing-side Nusselt number ratios plotted for various buoyancy parameters. The trailing-side heat-transfer coefficient correlation is better than the correlation developed for the leading-side Nusselt number ratio. However, the trailing-side Nusselt number ratio profiles depend on the axial location. Figure 5.58 shows another compilation of rotating channel data, plotted with the buoyancy parameter by Mochizuki et al. (1994).

Since the leading-side correlation based on buoyancy parameter is not as good as the trailing-side correlation, Dutta et al. (1996a) developed a Gr/Re (instead of Gr/Re^2) correlation based on the Pohlhausen parameter for adverse pressure gradient. Their correlation, as shown in Figure 5.59, on the leading-surface Nusselt number ratio for different axial locations and different rotation conditions collected from independent experimental observations appears to fall in a good correlation. However, note that the experimentalists' selections of parameters make them almost nonoverlapping. This correlation of Nusselt number ratios is extended for ribbed channels. Figure 5.60 shows the correlation on the leading-side Nusselt number ratio for ribbed channels. Here also the correlation is good for a combination of two independent data sets, and the scatter is significantly less than that observed with a buoyancy parameter.

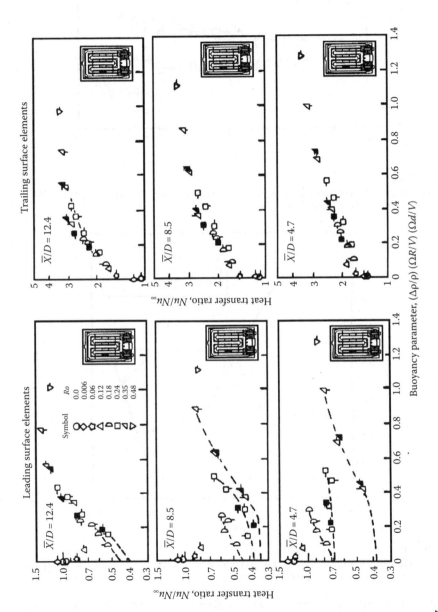

FIGURE 5.57
Heat-transfer coefficient ratio dependency on buoyancy parameter at $Re = 25,000$. (From Wagner, J.H. et al., *ASME J. Turbomach.*, 113, 321, 1991. With permission.)

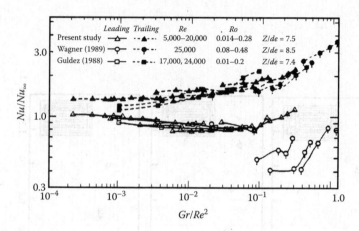

FIGURE 5.58
Nusselt number ratio variations with buoyancy parameter. (From Mochizuki, S. et al., *ASME J. Turbomach.*, 116, 133, 1994. With permission.)

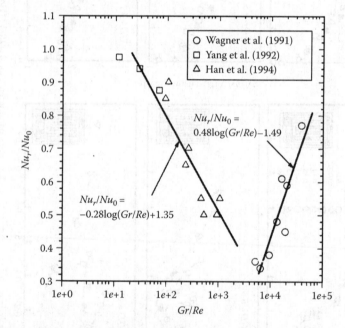

FIGURE 5.59
Leading-side Nusselt number ratio correlation with different adverse buoyancy levels. (From Dutta, S. et al., *ASME J. Heat Transfer*, 118, 977, 1996a.)

Another recent correlation attempt is shown in Figure 5.61. Morris and Rahmat-Abadi (1996) used the difference between local wall temperature and local bulk temperature as the local buoyancy effect rather than the overall buoyancy effect used in previous publications. They used their data from rotating ribbed circular pipe as the basis of the correlation. Their test sections A, B, and C

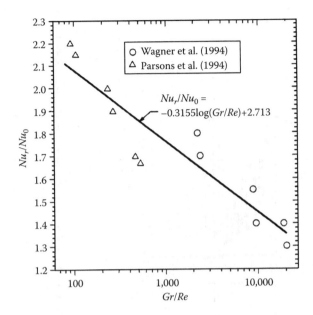

FIGURE 5.60
Ribbed leading-side Nusselt number correlation with different adverse buoyancy levels. (From Dutta, S. et al., *ASME J. Heat Transfer*, 118, 977, 1996a.)

had different rib pitch and height. The correlation in Figure 5.61 shows large scatters in data except for the test section B. Data from other publications are not included; therefore, validity of the correlation parameter is not fully tested. In another recent attempt, Kuo and Hwang (1996) correlated the Nusselt number ratio with a polynomial curve fitting of rotation and Rayleigh numbers, and non-dimensional axial locations. They suggested a correlation equation of the form

$$\frac{Nu}{Nu_0} = 1 + c_1 Ro \pm c_2 Ro \pm c_3 \frac{Ra}{Re^2} \qquad (5.2)$$

where coefficients c_1, c_2, and c_3 are positive functions of axial distance. The coefficient c_1 signifies the general enhancement of heat-transfer coefficient due to more mixing by the secondary flow. The coefficient c_2 reflects the relative position of a surface with respect to the Coriolis direction. Coefficient c_3 is attached to the buoyancy effect, and it changes sign depending on the inward or outward flow situation. Note that their correlation is linear in Ro and Ra/Re^2. Figure 5.62 shows their correlation after neglecting the buoyancy effect. The data of Kuo and Hwang (1996) reasonably follow the correlation, but experimental data of other researchers are not of much success.

Later, Hwang and Kuo (1997) obtained heat-transfer results from a multi-pass rotating channel. A similar correlation method based on a polynomial equation is developed for three passes. The first-pass correlation is discussed in the previous figure. The equations and correlation are plotted in

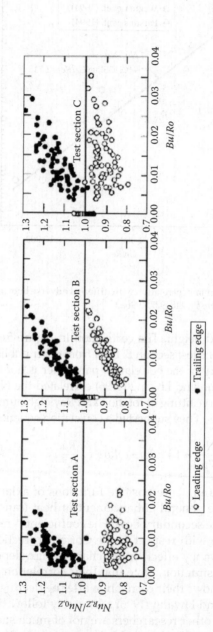

FIGURE 5.61

Nusselt number ratio dependency on local buoyancy effects. (From Morris, W.D. and Rahmat-Abadi, K.F., *Int. J. Heat Mass Transfer*, 39(11), 2253, 1996. With permission.)

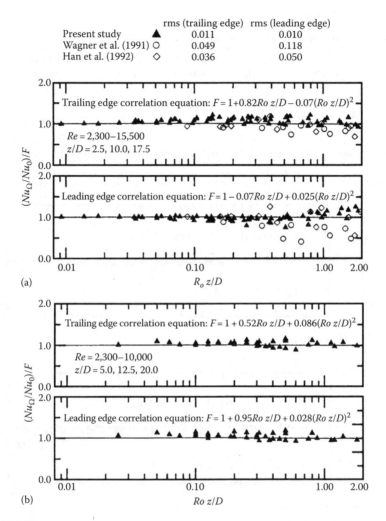

FIGURE 5.62
Leading- and trailing-side heat-transfer correlations developed by Kuo and Hwang. (a) Outward flow and (b) inward flow. (From Kuo, C.R. and Hwang, G.J., *ASME J. Heat Transfer*, 118, 23, 1996. With permission.)

Figure 5.63. Results indicate scatter of data at higher rotation speeds. The correlation for the second and third passes are given in the form of

$$\frac{Nu}{Nu_0} = 1 + c_1 \left(Ro\frac{z}{D} \right) \pm c_2 \left(Ro\frac{z}{d} \right)^2 \qquad (5.3)$$

where c_1 and c_2 are constants of correlation. Their values are given in Table 5.1.
    The second pass trailing and third pass leading show that the constants are smaller than the other two sides.

TABLE 5.1

Correlation Constants for Second and Third Passes

Constants	Second Pass Leading	Second Pass Trailing	Third Pass Leading	Third Pass Trailing
c_t	0.52	−0.05	−0.07	0.82
c_2	−0.07	0.03	0.025	−0.07

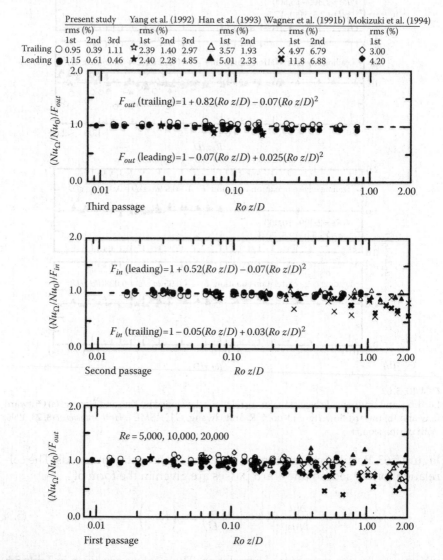

FIGURE 5.63
Heat-transfer coefficient correlation for multipass rotating channel. (From Hwang, G.J. and Kuo, C.R., *ASME J. Heat Transfer*, 119, 460, 1997. With permission.)

5.7 Heat-Mass-Transfer Analogy and Detail Measurements

Prior to the recent detailed heat-transfer coefficient measurement techniques, like liquid crystal and infrared thermography, detailed heat transfer is estimated from the mass-transfer results. Park and Lau (1998) and Park et al. (1998, 2000) used naphthalene sublimation to study the effects of channel orientation, rotational Coriolis force, and a sharp turn on the local mass-transfer distribution. Their experiment did not include the effect of rotational buoyancy. Figure 5.64 shows the schematic of the two-pass channel. The channel orientation with respect to the rotation direction is shown in the figure. Local measurement grid shows that the turn region has a higher resolution than the straight channels. Figure 5.65 shows the average mass-transfer ratios in the two-pass channel. The sharp turn enhances mass transfer. Rotation enhances mass transfer from the first-pass trailing wall. The enhancement is more for a normally oriented channel than for a diagonally oriented channel. Figure 5.66 shows the detailed mass-transfer distribution in a ribbed channel. Clearly the addition of ribs shows a

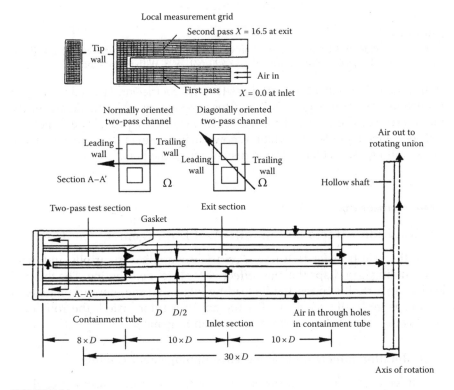

FIGURE 5.64
Schematic of the test channel used by Park and Lau for detailed mass-transfer analysis in a rotating environment. (From Park, C.W. and Lau, S.C., *ASME J. Heat Transfer*, 120, 624, 1998. With permission.)

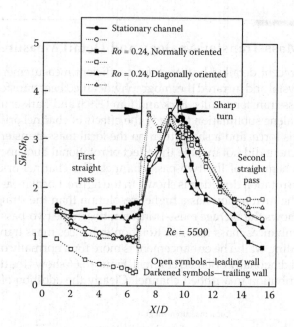

FIGURE 5.65
Effects of rotation and channel orientation on streamwise mass-transfer coefficient. (From Park, C.W. and Lau, S.C., *ASME J. Heat Transfer*, 120, 624, 1998. With permission.)

significant enhancement of local mass-transfer coefficient. The span-averaged results are shown in Figure 5.67, and these results provide the details of the interrib region. The channel length is not long enough to provide periodically a fully developed profile. The rotation shows an increasing mass-transfer pattern in the trailing wall with increasing distance from the inlet.

5.8 Rotation Effects on Smooth-Wall Impingement Cooling

Of all the heat-transfer techniques commonly applied in airfoils, impingement cooling is perhaps the most effective for localized cooling. The thermal load in an airfoil is highest at its leading edge, and therefore impingement cooling is generally used for this region. In a rotor airfoil, the rotational effects affect the cooling performance by impingement.

5.8.1 Rotation Effects on Leading-Edge Impingement Cooling

Epstein et al. (1985) studied the effects of rotation on impingement cooling. Figure 5.68 shows schematics of their test facility and impingement arrangement. Note that the impingement direction is not aligned with the rotation direction. Figures 5.69 and 5.70 show the local Nusselt numbers along the

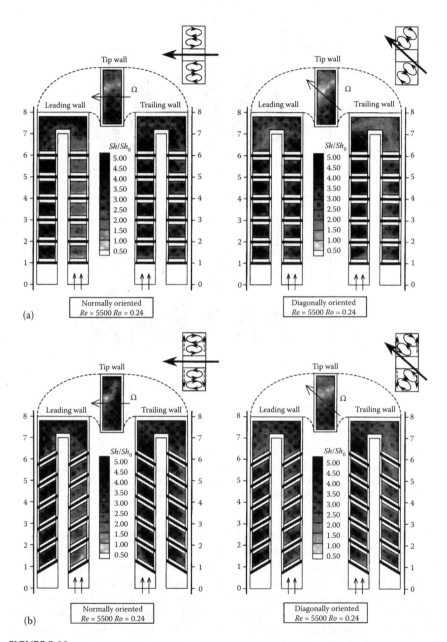

FIGURE 5.66
Local mass-transfer coefficient ratios in a ribbed two-pass square cross-sectioned channel for (a) $Re = 1 \times 10^4$, $Ro = 0.09$, and (b) $Re = 5.5 \times 10^3$, $Ro = 0.24$. (From Park, C.W. et al., *ASME J. Heat Transfer*, 122, 208, 2000. With permission.)

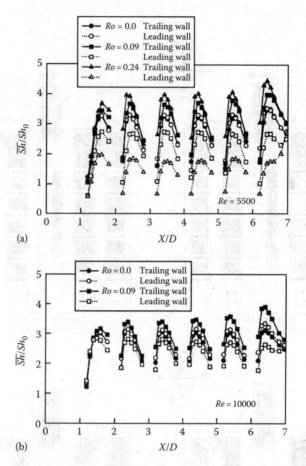

FIGURE 5.67
Effect of rotation on streamwise variation of mass-transfer coefficient ratio in a rib-roughened two-pass channel. (a) $Re = 5500$ and (b) $Re = 10000$. (From Park, C.W. et al., *AIAA J. Thermophys. Heat Transfer*, 12(1), 80, 1998. With permission.)

impingement channel. Reference 1 in the figures represents the rotating channel heat transfer as observed by Morris and Ayhan (1979), and data of Ref. 4 show the stationary channel impingement heat transfer as observed by Chupp et al. (1969). Results indicate that the rotation decreases the impingement heat transfer, but the effective heat transfer is better than a smooth rotating channel. The zero stagger of cooling jets in Figure 5.69 also shows lower Nusselt numbers compared to that with a stagger angle as shown in Figure 5.70. The high Nusselt number at the root of the channel with negative stagger was attributed by Epstein et al. (1985) to the design of the flow channel rather than to the flow physics of the staggered jets.

Mattern and Hennecke (1996) studied the impingement effects in the leading edge with a mass-transfer analogy. They used naphthalene sublimation

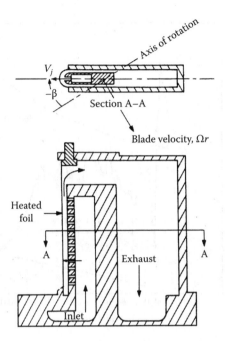

FIGURE 5.68
Schematic of a leading-edge impingement test facility used by Epstein et al. (From Epstein, A.H. et al., Rotational effects on impingement cooling, GTL Report No. 184, 1985. With permission.)

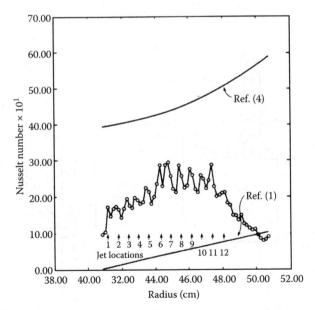

FIGURE 5.69
Radial variation of Nusselt number for zero stagger and $Ro_j = 0.53$. (From Epstein, A.H. et al., Rotational effects on impingement cooling, GTL Report No. 184, 1985. With permission.)

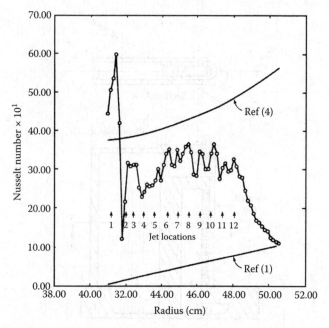

FIGURE 5.70
Radial variation of Nusselt number for −30° stagger and Ro_j = 0.53. (From Epstein, A.H. et al., Rotational effects on impingement cooling, GTL Report No. 184, 1985. With permission.)

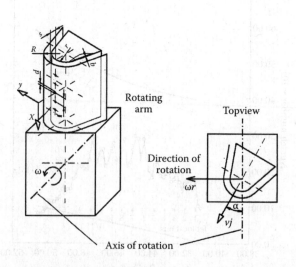

FIGURE 5.71
Schematic of the leading-edge impingement configuration used by Mattern and Hennecke. (From Mattern, C. and Hennecke, D.K., The influence of rotation on impingement cooling, presented at the *International Gas Turbine and Aeroengine Congress and Exhibition*, Birmingham, U.K., ASME Paper 96-GT-161, June 10–13, 1996. With permission.)

technique to get details of the surface mass transfer. Their experiments did not include the rotational buoyancy effect. In general, the rotation decreased the mass transfer from the impingement surface. Figure 5.71 shows the test model. Like the previous study, the jet direction has an offset angle with respect to the rotation direction. A maximum of 40% reduction in the mass transfer was noted by rotation.

Figure 5.72 shows the local mass transfer by impingement. The peaks correspond to the stagnation locations. Figure 5.73 shows the span-averaged Sherwood numbers in two stagger angles. It can be seen that the mass transfer decreases with an increase in the rotation effects for both stagger angles. The decrease is more for 90° stagger angle. The effect of stagger angle is shown in Figure 5.74. Results show that the effect of rotation is least on

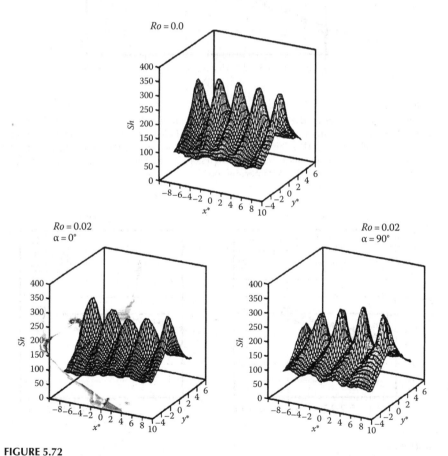

FIGURE 5.72
Effects of rotation on local mass-transfer coefficient at Re_j = 25,000, jet-to-impingement surface distance = 6.1 jet diameters, and jet hole spacing = 4 jet diameters. (From Mattern, C. and Hennecke, D.K., The influence of rotation on impingement cooling, presented at the *International Gas Turbine and Aeroengine Congress and Exhibition*, Birmingham, U.K., ASME Paper 96-GT-161, June 10–13, 1996. With permission.)

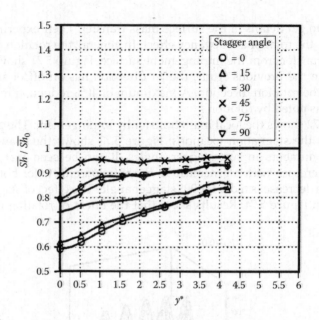

FIGURE 5.73
Effect of rotation number on the span-averaged Sherwood numbers with $Re_j = 25{,}000$, jet-to-impingement surface distance = 6.1 jet diameters, and jet hole spacing = 4 jet diameters. (From Mattern, C. and Hennecke, D.K., The influence of rotation on impingement cooling, presented at the *International Gas Turbine and Aeroengine Congress and Exhibition*, Birmingham, U.K., ASME Paper 96-GT-161, June 10–13, 1996. With permission.)

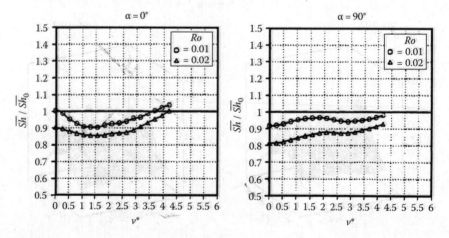

FIGURE 5.74
Relative changes in span-averaged mass-transfer coefficients as a function of stagger angle for $Re_j = 25{,}000$, $Ro = 0.02$, jet-to-impingement surface distance = 6.1 jet diameters, and jet hole spacing = 2 jet diameters. (From Mattern, C. and Hennecke, D.K., The influence of rotation on impingement cooling, presented at the *International Gas Turbine and Aeroengine Congress and Exhibition*, Birmingham, U.K., ASME Paper 96-GT-161, June 10–13, 1996. With permission.)

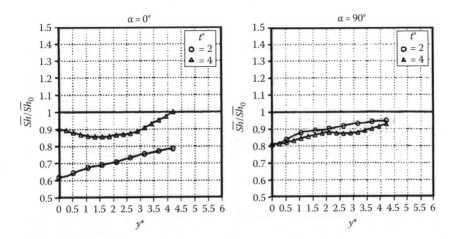

FIGURE 5.75
Relative change in span-averaged mass-transfer coefficients as a function of jet-to-jet spacing at Re_j = 25,000, Ro = 0.02, and jet-to-impingement surface distance = 6.1 jet diameters. (From Mattern, C. and Hennecke, D.K., The influence of rotation on impingement cooling, presented at the *International Gas Turbine and Aeroengine Congress and Exhibition*, Birmingham, U.K., ASME Paper 96-GT-161, June 10–13, 1996. With permission.)

α = 45°. However, all stagger angles show that rotation reduces the mass transfer compared to that with stationary channel mass transfer.

Figure 5.75 shows the effect of jet spacing on the mass transfer. The effect of jet spacing is less prominent in α = 90°. The closer spacing performs worse in α = 0°. Figure 5.76 shows the effect of jet-to-target-plate spacing. Results show that a smaller spacing increases mass transfer in the presence of rotation. The increase in mass transfer for a short distance is insignificant in α = 90°. In this configuration, the Coriolis effect deflects the jet. Since the target plate is close to the jet exit, the deflection is not significant. Coriolis effect in α = 0° is in the secondary flow, i.e., spent jets. For this small gap, the cross flow is expected to be high, and increased swirl in the cross flow increases the mass transfer.

A new jet impingement and swirl is investigated by Glezer et al. (1998). Figure 5.77 shows the schematic of the swirl tube test configuration. A preliminary test showed significant improvement in the heat-transfer performance; based on that study, a new airfoil is designed. Figure 5.78 shows the modified airfoil with swirling impingement in the leading edge. This new airfoil is tested in the hot-cascade test section. Results indicate that screw-shaped swirl cooling can significantly improve the heat-transfer coefficient over a smooth channel; this improvement is not significantly dependent on the temperature ratio and rotational forces. Moreover, it was concluded that optimization of the internal passage geometry in relation to location and size of the tangential slots is very important in achieving the best performance of the screw-shaped swirl in the leading-edge cooling.

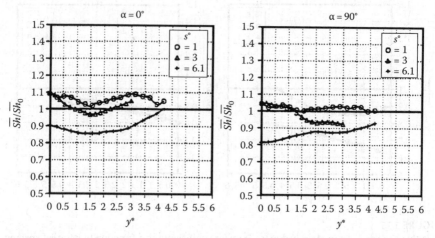

FIGURE 5.76
Relative changes in span-averaged mass-transfer coefficients as a function of jet-to-target plate distance for $Re_j = 25,000$, $Ro = 0.02$, and jet-to-jet spacing = 4 jet diameters. (From Mattern, C. and Hennecke, D.K., The influence of rotation on impingement cooling, presented at the *International Gas Turbine and Aeroengine Congress and Exhibition*, Birmingham, U.K., ASME Paper 96-GT-161, June 10–13, 1996. With permission.)

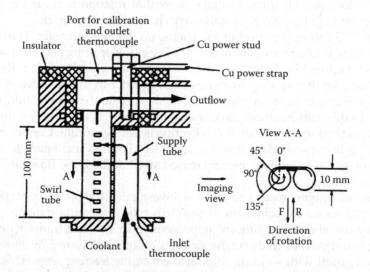

FIGURE 5.77
Schematic of swirl tube test facility used by Glezer et al. (From Glezer, B. et al., Heat transfer in a rotating radial channel with swirling internal flow, presented at the *International Gas Turbine and Aeroengine Congress and Exhibition*, Stockholm, Sweden, ASME Paper 98-GT-214, June 2–5, 1998. With permission.)

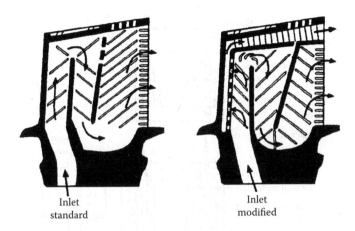

Inlet
standard

Inlet
modified

FIGURE 5.78
Standard and modified cooling arrangements by Glezer et al. (From Glezer, B. et al., Heat transfer in a rotating radial channel with swirling internal flow, presented at the *International Gas Turbine and Aeroengine Congress and Exhibition*, Stockholm, Sweden, ASME Paper 98-GT-214, June 2–5, 1998. With permission.)

5.8.2 Rotation Effect on Midchord Impingement Cooling

Parsons and Han (1998) used a rotating channel to study the effect of rotation on impingement cooling. However, their impingement configuration simulated the cooling requirements for the midchord region of an airfoil.

This test facility is similar to the rotating ribbed-channel configurations used by the same research group. The assembled impingement channel rotates in a horizontal plane. A schematic of the impingement airflow is shown in Figure 5.79. A central chamber serves as the pressure chamber, and impingement jets are released in either direction to impinge on two heated surfaces. Note that the impingement flow directions have different orientations with respect to the direction of rotation. Both pressure and heat-transfer data are obtained from this rotating channel facility.

Figure 5.80 shows the channel pressure distribution. From the differential pressures between the center channel and leading or trailing channels, the discharge mass flow rates in each jet location are estimated. In general, mass flow rate increases with an increase in the radius, i.e., a higher jet discharge is observed near the outlet. This type of mass distribution has been observed in other stationary impingement configurations. The amount of mass discharge is mostly unaffected by the rotation effects. Figure 5.81 shows the schematics of the Coriolis forces on the jet and cross flow. In a smooth rotating channel, the leading side of a radial outward flow has lower Nusselt numbers compared to that for the trailing edge. However, in the jet impingement configuration of Parsons and Han (1998), the leading-side heat-transfer coefficient is higher than the trailing side in the rotating condition. Figure 5.82 shows their Nusselt number distributions. Overall, the impingement effectiveness decreases in the presence of rotation, as indicated earlier by the direction

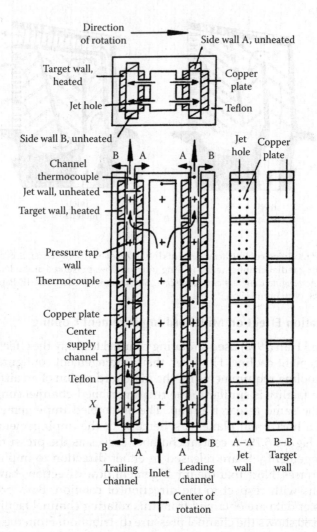

FIGURE 5.79
Schematic of the impingement test facility used by Parsons and Han (From Parsons, J.A. and Han, J.C., *Int. J. Heat Mass Transfer*, 41(13), 2059, 1998.)

of Coriolis, and centrifugal forces in both impingement chambers deflect impingement jets away from the target plates.

Figure 5.83 shows the decrease in the Nusselt number ratio in both leading and trailing sides with an increase in the rotation number. In a smooth channel without impingement cooling, the trailing-side heat transfer generally increases with rotation. But the configuration tested shows that the trailing side shows more decrease in the heat-transfer coefficient in the presence of rotation. It can be argued that the bulk cross flow and associated turbulence increase near the trailing wall, and the Coriolis and centrifugal forces deflect

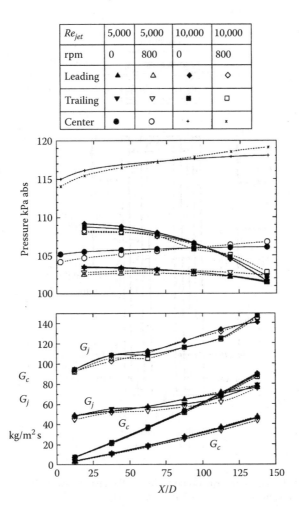

Re_{jet}	5,000	5,000	10,000	10,000
rpm	0	800	0	800
Leading	▲	△	◆	◇
Trailing	▼	▽	■	□
Center	●	○	+	×

FIGURE 5.80
Pressure and mass flux distributions for different test conditions of Parsons and Han. (From Parsons, J.A. and Han, J.C., *Int. J. Heat Mass Transfer*, 41(13), 2059, 1998.)

the impingement jets. These two effects, i.e., flow alterations and jet deflection, decrease the effectiveness of the impingement cooling in the trailing side.

In a separate study, Parsons et al. (1998) studied the rotating effects on impingement with all sides heated. In the all-side heated condition, the effects of buoyancy are stronger and heat-transfer coefficients from all four sides are obtained. They showed the stationary impingement heat-transfer results. Results indicate that the heat-transfer coefficient of the jet-orifice surface is highest followed by the target surface. The sidewalls show the lowest heat-transfer coefficient. It is interesting to note that only the one-side heated condition of Parsons and Han (1998) had a lower heat-transfer coefficient for the target surface compared to the present study.

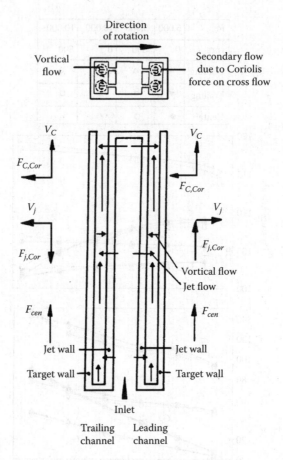

FIGURE 5.81
Schematics of the vortices and rotational forces for the test facility of Parsons and Han. (From Parsons, J.A. and Han, J.C., *Int. J. Heat Mass Transfer*, 41(13), 2059, 1998.)

Figure 5.84 shows the effect of rotation on the Nusselt number ratio for different walls. In general, the trailing side shows a decrease in the heat transfer with rotation. Both sidewalls of the trailing side show a significant reduction in the heat-transfer coefficient.

Akella and Han (1998) studied the effects of rotation on impingement for another variation of impingement configuration, as shown in Figure 5.85.

The difference from the earlier experiment by this group is that spent jets from the trailing channel are used as cooling jets for the leading channel. Therefore, the cross flow in the trailing side is radial outward; for the leading side, it is radial inward. In smooth-channel experiments, it has been shown that the trailing-side heat-transfer coefficient is higher for radial outward flow, and the leading side has a higher heat-transfer

Re_{jet}		10,000	10,000	5,000	5,000
rpm	0	400	800	400	800
Ro	0	0.0008	0.0015	0.0015	0.0028
Leading	—	◆	▲	●	■
Trailing	—	◇	△	○	□

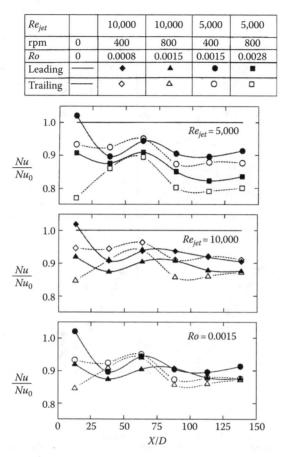

FIGURE 5.82
Effect of rotation on target wall Nusselt number ratio. (From Parsons, J.A. and Han, J.C., *Int. J. Heat Mass Transfer*, 41(13), 2059, 1998.)

coefficient for radial inward flow. Figure 5.86 shows the schematic of the Coriolis force in the forward and reverse rotation directions. The direction of the Coriolis force reverses direction with a reverse in direction of rotation.

Figure 5.87 shows the effect of rotation on Nusselt number distribution. Results indicate that irrespective of the direction of rotation, the Nusselt number decreases in the presence of rotation. The forward rotation has a slightly better performance over the reverse rotation. Figure 5.88 shows the effect of rotation number on the average Nusselt number ratio. The forward and reverse directions are not much different from each other, but the heat-transfer coefficient decreases linearly with increasing log of *Ro*.

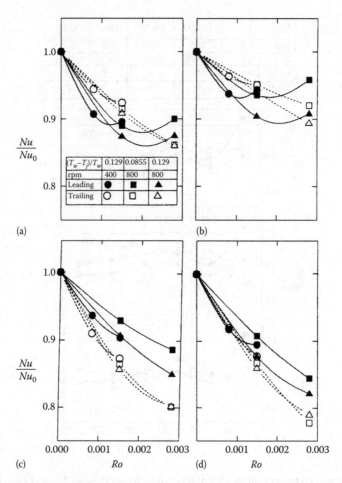

FIGURE 5.83
Effect of rotation number and coolant-to-wall temperature difference ratio on target wall Nusselt number ratio. (a) $X/D = 38$, (b) $X/D = 63$, (c) $X/D = 88$, and (d) $X/D = 113$. (From Parsons, J.A. and Han, J.C., *Int. J. Heat Mass Transfer*, 41(13), 2059, 1998.)

5.8.3 Effect of Film-Cooling Hole

Parsons and Han (1996) studied the effect of film coolant extraction from their rotating impingement channel. Figure 5.89 shows the schematic of the impingement with coolant extraction. The stationary channel results indicated that half extraction had a Nusselt number comparable to that with all extraction. The average Nusselt number variations with jet Reynolds number showed that coolant extraction increased the heat-transfer coefficient. Therefore, film coolant extraction is beneficial for the impingement cooling. It can be argued that the coolant extraction decreases the cross flow; therefore, the jet deflection by the cross flow is less in the presence of coolant extraction.

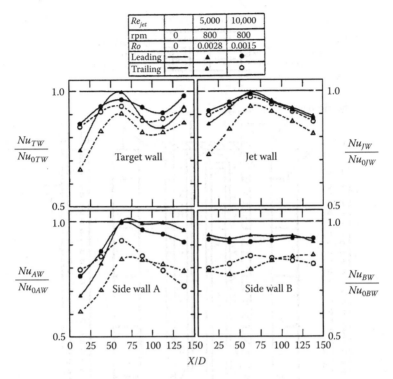

FIGURE 5.84
Effect of rotation on Nusselt number ratio for target, jet, and sidewalls. (From Parsons, J.A. and Han, J.C., *Int. J. Heat Mass Transfer*, 41(13), 2059, 1998.)

The effect of rotation on the impingement heat transfer with coolant extraction is plotted in Figure 5.90. It can be observed that rotation decreases the impingement heat-transfer coefficient. This figure summarizes the effect of rotation on heat transfer. The effect of the rotation number on Nusselt number ratio is seen at channel locations $x/D = 38$ and 113. At $x/D = 38$, the heat-transfer coefficient decreases by up to 20%; for $x/D = 113$, the heat-transfer coefficient decreases by 25%.

5.9 Rotational Effects on Rib-Turbulated Wall Impingement Cooling

Akella and Han (1999) included skew ribs in the target surfaces of their rotating impingement test facility. Rotation effects on impingement for this test channel have been discussed earlier. The target plates are rib turbulated, and two streamwise rows of jets are used for impingement. Conceptual flow

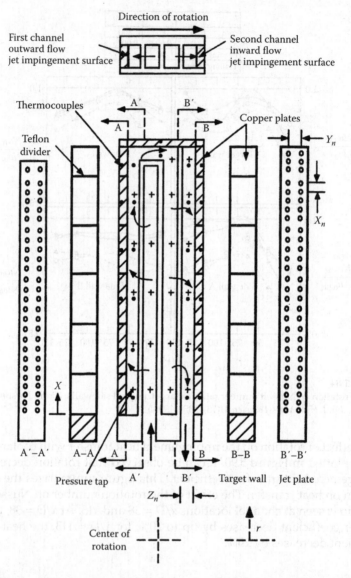

FIGURE 5.85
Schematic of the rotating channel with impingement heat transfer used by Akella and Han. (From Akella, K.V. and Han, J.C., *AIAA J. Thermophys. Heat Transfer*, 12(4), 582, 1998.)

fields of the ribbed channel and impingement are shown in Figures 5.91 and 5.92. The 45° angled ribs produce a secondary flow, and the secondary flow orientations are different for the first and second pass. The rotational forces influence the flow distribution. Figure 5.92 schematically shows the Coriolis and centrifugal forces on the flow. The Coriolis effects on the core flow, represented by a subscript c, and jet impinging flow, represented by a

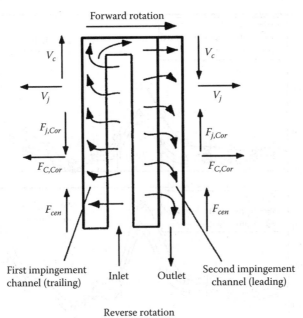

Forward rotation

First impingement channel (trailing)   Inlet   Outlet   Second impingement channel (leading)

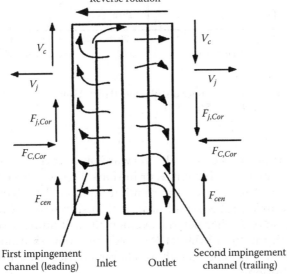

Reverse rotation

First impingement channel (leading)   Inlet   Outlet   Second impingement channel (trailing)

FIGURE 5.86
Conceptual view of flow and rotational forces on the test facility of Akella and Han. (From Akella, K.V. and Han, J.C., *AIAA J. Thermophys. Heat Transfer*, 12(4), 582, 1998.)

subscript j, are shown in this figure. Since flow directions change in the two passes, but rotation direction remains the same, the Coriolis forces are in different directions in the two passes.

Heat transfers in stationary, forward, and reverse rotation are shown in Figure 5.93. As in earlier impingement configurations, rotation reduces

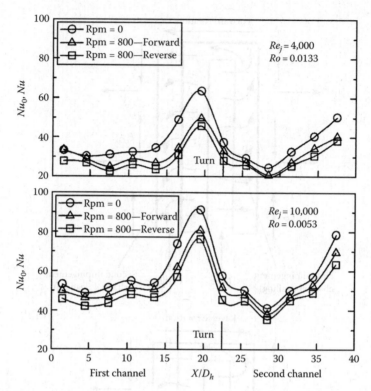

FIGURE 5.87
Effect of rotation on Nusselt number distribution. (From Akella, K.V. and Han, J.C., *AIAA J. Thermophys. Heat Transfer*, 12(4), 582, 1998.)

the heat-transfer coefficient. The forward and reverse rotations show similar heat-transfer coefficient distributions. Figure 5.94 summarizes the heat-transfer dependency on the rotation number. The decrease in average Nusselt number ratio has a linear relationship with the log of rotation number. In a smooth channel, the average Nusselt number shows a linear variation with the log of the jet Reynolds number. Addition of inclined ribs in the flow path changes those characteristics of heat-transfer coefficient distribution.

Figure 5.95 shows that the increase in Nusselt number by the addition of ribs is nonlinear with respect to the log of the jet Reynolds number. The enhancement is more at a higher Reynolds number. It can be argued that the cross flow developed by spent jets is stronger for a higher jet Reynolds number, and ribs are more effective in heat-transfer coefficient enhancement with a stronger cross flow. It also shows that rotation reduces heat transfer.

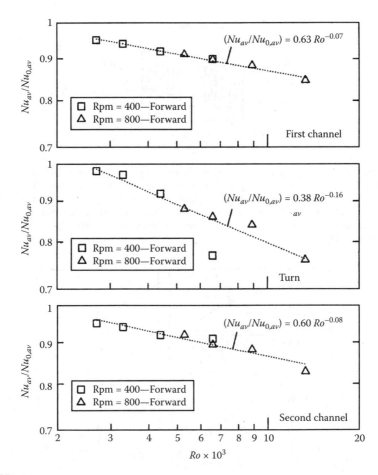

FIGURE 5.88
Effect of rotation number on channel-averaged Nusselt number ratio. (From Akella, K.V. and Han, J.C., *AIAA J. Thermophys. Heat Transfer*, 12(4), 582, 1998.)

5.10 New Information from 2000 to 2010

This section presents the recent (2000–2010) additions to heat-transfer analysis in triangular- or wedge-shaped rotating channels. The two narrow ends of an airfoil prevent the use of conventional rectangular or square cross-sectioned channels. The natural shape that fits this region is a triangular- or a wedge-shaped cooling channel. There have been a few publications on the triangular- or wedge-shaped cooling channel with and without ribs to simulate the leading edge and the trailing edge of a rotating airfoil. These studies with rotational effects show that heat-transfer patterns with this geometry

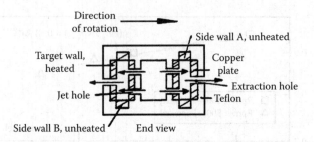

FIGURE 5.89
Schematic of the impingement arrangement used by Parsons and Han for studying impingement heat transfer in the presence of rotation and film extraction. (From Parsons, J.A. and Han, J.C., Rotation effect on jet impingement heat transfer in smooth rectangular channels with heated target walls and film coolant extraction, presented at the *International Mechanical Engineering Congress and Exhibition*, Atlanta, GA, ASME Paper 96-WA/HT-9, November 17–22, 1996.)

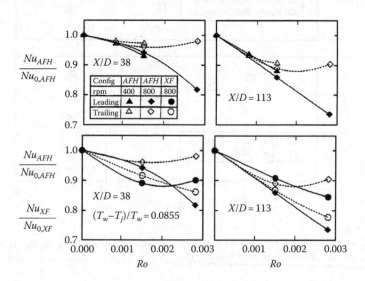

FIGURE 5.90
Effect of rotation number on target wall Nusselt number ratio. (From Parsons, J.A. and Han, J.C., Rotation effect on jet impingement heat transfer in smooth rectangular channels with heated target walls and film coolant extraction, presented at the *International Mechanical Engineering Congress and Exhibition*, Atlanta, GA, ASME Paper 96-WA/HT-9, November 17–22, 1996.)

are different from regular square or rectangular rotating channels; and heat-transfer pattern in these channels also get significantly affected by the rib orientations. In this section, recent publications for the effects of channel aspect ratio on the heat transfer from rotating channels are also presented. Several rib configurations and channel entrance conditions are integral parts of these studies. Recent studies focus on the coolant passages heat transfer under high rotation numbers to simulate actual turbine rotation conditions.

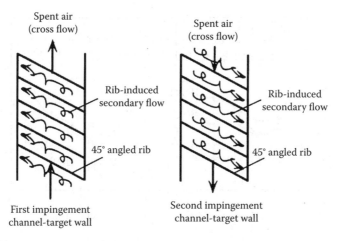

FIGURE 5.91
Schematic of the rib-induced secondary flow on stationary target walls. (From Akella, K.V. and Han, J.C., *AIAA J. Thermophys. Heat Transfer*, 13(3), 364, 1999.)

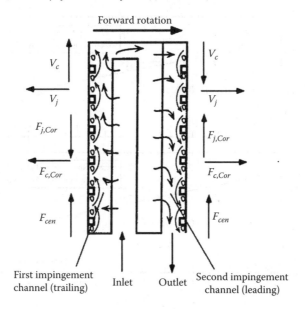

FIGURE 5.92
Schematic of the flow and rotational forces in rotating impingement channel of Akella and Han. (From Akella, K.V. and Han, J.C., *AIAA J. Thermophys. Heat Transfer*, 13(3), 364, 1999.)

5.10.1 Heat Transfer in Rotating Triangular Cooling Channels

The effect of ribs may be more pronounced in triangular channels and possibly suppressed the rotational effects. Lee et al. (2006) studied the heat transfer in a rotating equilateral triangular channel with three different rib arrangements, 45°, 90°, and 135° angles of attacks. The maximum rotation

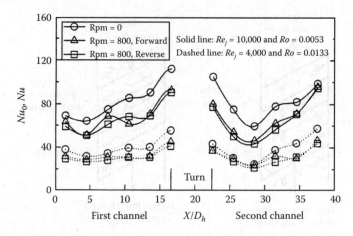

FIGURE 5.93
Effect of rotation on Nusselt number distribution on ribbed rotating channels with impingement. (From Akella, K.V. and Han, J.C., *AIAA J. Thermophys. Heat Transfer*, 13(3), 364, 1999.)

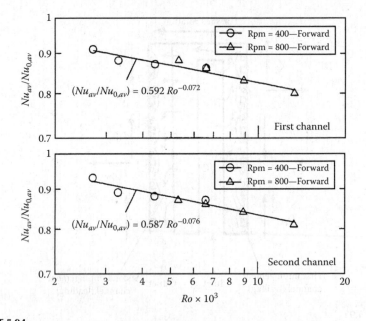

FIGURE 5.94
Effect of rotation on ribbed channel-averaged Nusselt number ratio. (From Akella, K.V. and Han, J.C., *AIAA J. Thermophys. Heat Transfer*, 13(3), 364, 1999.)

number tested was 0.1 and that did not change much of the heat-transfer distribution from the stationary conditions. The heat-transfer distribution along the channel flow direction differed among the surfaces due to the rib arrangements. The secondary flows induced by ribs were strong enough to dominate over the rotational effects. Later Haiyong et al. (2010) performed an

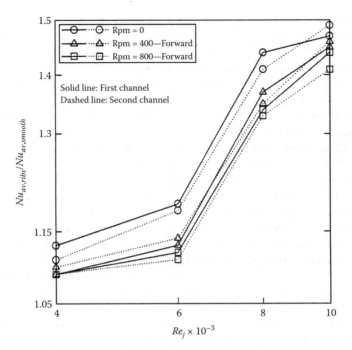

FIGURE 5.95
Effect of jet Reynolds number on ribbed and smooth surface heat-transfer enhancement ratios at different rotation speeds. (From Akella, K.V. and Han, J.C., *AIAA J. Thermophys. Heat Transfer*, 13(3), 364, 1999.)

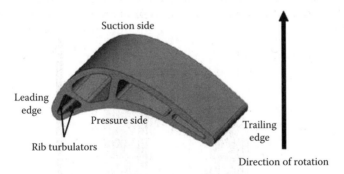

FIGURE 5.96
Triangular channel simulates the leading edge cooling channel of an airfoil. (From Liu, Y.H. et al., Heat transfer in leading edge triangular shaped cooling channels with angled ribs under high rotation numbers, ASME Paper, GT2008-50344, 2008a.)

experimental study on the trapezoidal duct with impingement jets to simulate the leading edge but no rotational effects were included.

It requires a higher rotation number than 0.1 to observe rotational effects on ribbed rotating triangular channels. Liu et al. (2008a) presented an experimental study of heat-transfer analysis in an equilateral triangular channel with rotation. The shape was selected to simulate the leading edge of an airfoil as shown in Figure 5.96. 45° angled ribs were used to enhance heat-transfer coefficient. The rib pitch-to-height (*P/e*) ratio was maintained at 8 and the rib height-to-channel hydraulic diameter was kept at 0.087.

Liu et al. (2008a) divided each of the leading and trailing walls in two heating regions and Figure 5.97 shows the heating regions and the angle to the rotation direction as used by them. The flow was radially outward. Ribs of square cross section of 1.59 mm × 1.59 mm were glued to the leading (marked L1 and L2) and on the trailing (T1 and T2) surfaces. The inner wall was left smooth. The ribs were aligned as 45° staggered V to the flow direction. The entrance to the test section was an unheated triangular channel of *L/D* = 2.09. So, the flow was not fully developed as it entered the heated test section. They show the Nusselt number ratios in the stationary channel for both ribbed and smooth channels. Results from Metzger and Vedula (1987) are included to validate the stationary channel results. The smooth channel shows that the Nusselt number ratio starts at 2.5 and decays to about 2. The higher values at the entrance of the smooth channel in this study signify the unheated entry to the test section. But the ribbed channel shows a continued increase in the Nusselt Number ratio along the flow direction. It is consistent with the observations made by Metzger and Vedula (60° angled ribs with *P/e* = 7.5) in their fully developed flow channel. Note that Metzger and Vedula (1987) used a heated entrance that was more than 10 hydraulic diameters long. Therefore, their smooth Nusselt number ratios matched the fully developed values.

Figure 5.98 shows the average Nusselt number ratio (*Nu* in rotation/*Nu* in stationary channel) variations with rotation number and buoyancy parameter. Both ribbed- and smooth-channel results are included. The suggested

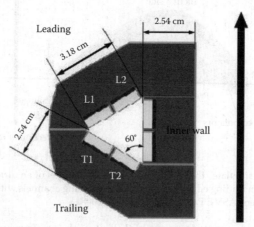

FIGURE 5.97
Cross-sectional view of the test section used by Liu et al. (From Liu, Y.H. et al., Heat transfer in leading edge triangular shaped cooling channels with angled ribs under high rotation numbers, ASME Paper, GT2008-50344, 2008a.)

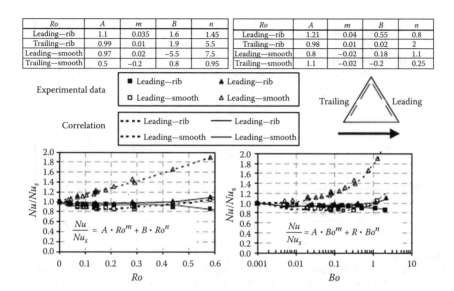

Ro	A	m	B	n	Ro	A	m	B	n
Leading—rib	1.1	0.035	1.6	1.45	Leading—rib	1.21	0.04	0.55	0.8
Trailing—rib	0.99	0.01	1.9	5.5	Trailing—rib	0.98	0.01	0.02	2
Leading—smooth	0.97	0.02	−5.5	7.5	Leading—smooth	0.8	−0.02	0.18	1.1
Trailing—smooth	0.5	−0.2	0.8	0.95	Trailing—smooth	1.1	−0.02	−0.2	0.25

FIGURE 5.98
Variation of average Nusselt number ratio with rotation number and buoyancy parameter. (From Liu, Y.H. et al., Heat transfer in leading edge triangular shaped cooling channels with angled ribs under high rotation numbers, ASME Paper, GT2008-50344, 2008a.)

correlations are within 6.8% of the experimental data. Four Reynolds numbers (10,000, 20,000, 30,000, and 40,000) and five rotation speeds with a maximum of 400 rpm were used to generate this set of data. Results show that rotation has significantly less effect in the presence of the ribs as was observed also by Lee et al. (2006). This is unlike what has been observed in rectangular channels where ribbed results were affected by rotation. Earlier, Sewall and Tafti (2005) performed LES simulations and have shown the changes in the separation region by the effects of rotation and buoyancy; but their channel was not triangular. Clearly the channel geometry with respect to the rotation direction as considered here has a different effect by rotation than that observed in square or rectangular channels. The smooth leading side does not show much degradation with rotation, but the smooth trailing side shows marked increase in the heat-transfer coefficient with rotation. The fitted curves show a nearly linear increase in Nusselt number ratio of the smooth trailing side with rotation number, but the dependency on the buoyancy parameter is significantly nonlinear.

Liu et al. (2009) expanded their heat-transfer analysis on equilateral triangular rotating channel with more rib configurations. Their objective was to simulate the leading-edge cooling chamber of a rotating airfoil. Three different rib orientations were used—orthogonal and two different staggered V-ribs. The schematic of the secondary flows induced by the two staggered rib orientation and rotation is shown in Figure 5.99.

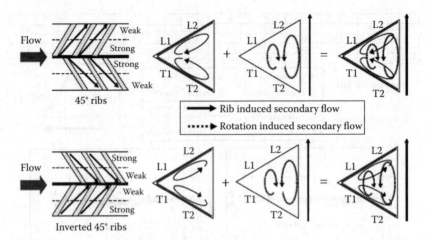

FIGURE 5.99

Conceptual view of the secondary flow in triangular channel as caused by inclined ribs and rotation. (From Liu, Y.H. et al., High rotation number effect on heat transfer in a triangular channel with 45-deg, inverted 45-deg and 90-deg ribs, ASME Paper, GT2009-59216, 2009.)

The stationary channel results with these three rib orientations are presented in Figure 5.100. The Nusselt number ratios indicate that inclined ribs enhance heat transfer but the enhancement location is strongly dependent on the rib orientation. As expected the perpendicular 90° rib does not show any preference on location and the Nusselt number ratios are improved in all sides with about the same enhancement factors. However, the inverted 45° rib improves heat transfer in the wider end of the triangle and the 45° rib enhances heat-transfer ratios in the narrow triangular tip region. Since there is no rotation, this difference was caused primarily by the secondary flow induced by the rib orientations. Results are consistent with observations made in rectangular channels with V-ribs. 45° ribs show enhancement in the base of the V-corner and inverted 45° ribs show enhancement at the other corners.

Figure 5.101 shows the effect of rotation on heat transfer in the triangular channel of Liu et al. (2009) to simulate the leading edge of an airfoil. Reynolds number was varied from 10,000 to 40,000 and rotation speed was varied from 0 to 400 rpm to generate these plots. Three axial locations were presented to show the effect of flow development along the cooling channel. The stationary channel results were presented earlier and here the Nusselt number ratios were calculated based on the stationary Nusselt numbers. The leading surface and trailing surface with respect to rotation were divided in two parts to provide more details and Nusselt number ratios were plotted with varying rotation numbers. The rib alignments and the conceptual secondary flow vortices have been shown earlier and are used to justify some of the observed patterns. The rotation-induced secondary flow interacts with the secondary flows from rib orientations and produces interesting Nusselt number ratio distributions.

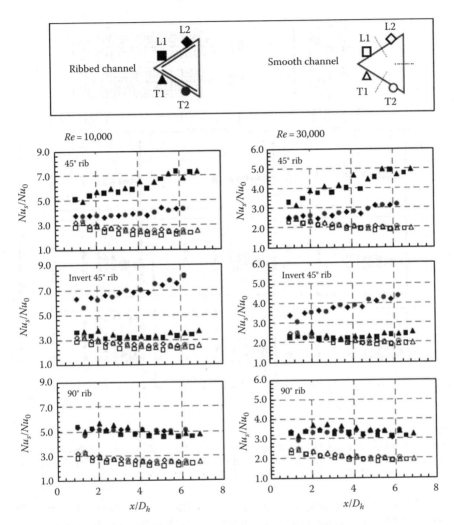

FIGURE 5.100

Nusselt number ratio distribution in the stationary triangular channel. (From Liu, Y.H. et al., High rotation number effect on heat transfer in a triangular channel with 45-deg, inverted 45-deg and 90-deg ribs, ASME Paper, GT2009-59216, 2009.)

The 90° rib is most influenced by the rotation numbers and the trend is similar to other rotating 90° ribbed-channel studies (Wagner et al., 1992). The trailing side shows Nusselt number ratio enhancement and the leading side shows a decrease with increasing rotation numbers. The effect of rotation becomes more prominent as the flow develops through the channel. The enhancement at the trailing side is more than the observed decrease in the leading side. The leading side at the wide end of the triangle remains about the same value as that of a stationary channel. The narrow end of the triangle, where leading side meets trailing side, shows an increasing trend of

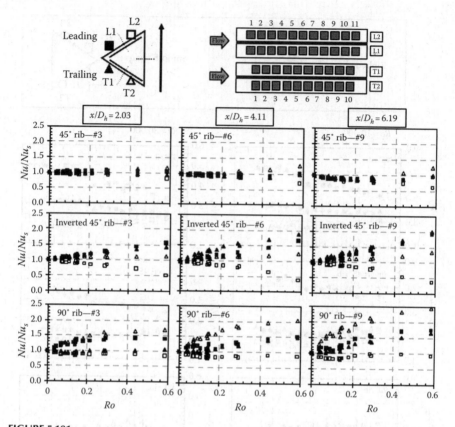

FIGURE 5.101
Nusselt number ratio variations with rotation number at three different streamwise locations. (From Liu, Y.H. et al., High rotation number effect on heat transfer in a triangular channel with 45-deg, inverted 45-deg and 90-deg ribs, ASME Paper, GT2009-59216, 2009.)

the Nusselt number ratio with an increase in the rotation number, especially at downstream locations of the channel. But at the entrance, the narrow end shows a reverse trend, the leading-side Nusselt number ratio is higher than the trailing side. It can be argued that the flow is separated and thus is in reverse direction past the ribs. This reverse flow is perhaps dominant in this region to reverse the rotation number effects from the effects observed in other locations.

The 45° orientation of the rib is not so affected by rotation. It seems that the secondary flow effects of rib orientation are preserved and not significantly altered by rotation because the Nusselt number ratios with the stationary channel results are nearly one at most rotation numbers. Only at the highest rotation number, the classic rotation-induced pattern is observed at the wider side of the triangle; that is, the trailing side shows enhancement and the leading side shows a decrease in the Nusselt number ratios. The narrow end of the triangle does not show much effect by rotation.

The inverted 45° orientation shows the enhancement and decrease in Nusselt number ratios more prominently than other rib orientations. The increase in the trailing side of the broad end of triangle is less prominent than the decrease in the leading side. The corner region, where leading side meets trailing side, shows an increase in Nusselt number ratio, especially at downstream locations. Note that this corner had low Nusselt numbers in stationary channel. So, it can be argued that rotation increases flow or turbulence in this region and thus heat-transfer coefficient is enhanced.

They also report that the 90° rib shows the highest friction factor because the ribs were orthogonal to the flow and it is consistent with rectangular channel (Taslim and Lengkong, 1998). The two types of 45° orientations do not show much difference in the friction factor ratios even if the secondary flows generated by them are in the opposite directions. This plot indicates that even though the ribs at an angle mix the core flow, the net pressure drop is less as compared to orthogonal 90° ribs, which does not promote core flow mixing.

Figure 5.102 shows the channel-averaged Nusselt number ratio variations with rotation number and buoyancy parameter. The paper shows the correlation coefficients that were developed based on curve fitting. The 45° rib shows the least effect of rotation on heat-transfer ratio. The most effect is seen on the 90° rib where both leading and trailing surfaces show an increase in the Nusselt number ratios with an increase in either rotation number or the buoyancy parameter. Note that 90° rib orientation used only one inlet density ratio. It is interesting to note that none of the configurations including the smooth channel showed much decrease in the Nusselt number ratio on the leading side. The heat transfer was enhanced in the trailing side in most cases. So, it can be said that in general rotation improved the overall heat transfer from this shaped cooling channel.

5.10.2 Heat Transfer in Rotating Wedge-Shaped Cooling Channels

Figure 5.103 shows the wedge-shaped cooling channel with turn exit at top reported by Wright et al. (2007). This wedge-shaped channel has a sharp entrance due to the change in geometry from rectangular to the trapezoidal as shown. The sidewall results compare favorably with previous sharp entrance results in a rectangular channel (Wright et al., 2005). The other two surfaces show significantly higher Nusselt number ratios and the effect of the turn enhances heat-transfer coefficient further in the narrow end of the channel. The bend at the end helps to increase the heat-transfer coefficient, and numerical work of Sleiti and Kapat (2004) provided some insight to the heat-transfer enhancement in the turn region in rotation.

Figure 5.104 shows an increase in heat-transfer coefficient with an increase in the rotation number. The sidewall is least affected, but unlike a square or rectangular channel, both leading and trailing surfaces show an enhancement in the heat-transfer coefficient with rotation. A correlation based on

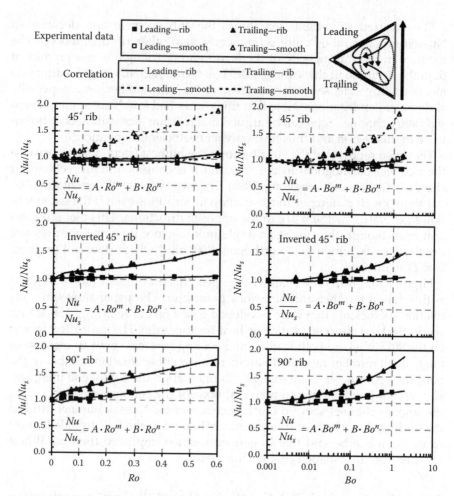

FIGURE 5.102

Streamwise-averaged Nusselt number variations with rotation number and buoyancy parameter. (From Liu, Y.H. et al., High rotation number effect on heat transfer in a triangular channel with 45-deg, inverted 45-deg and 90-deg ribs, ASME Paper, GT2009-59216, 2009.)

curve fitting shows that buoyancy parameter can capture the trend quite well in the given range.

The heat-transfer analysis in trailing edge remains incomplete without a discussion on coolant extraction or bleed because that is a common practice in most designs of high-temperature turbine. Among notable publications, Jeon et al. (2006) and Kim et al. (2007) studied the mass transfer in rotating channels with bleed in the presence of orthogonal and angled ribs. They investigated the effects of secondary flow due to rotation and angled rib turbulators on the heat/mass transfer in the square channels with channel rotation and bleed flow. The angle of attack of the angled ribs was 45° and the bleed holes were located between the rib turbulators on either the leading or

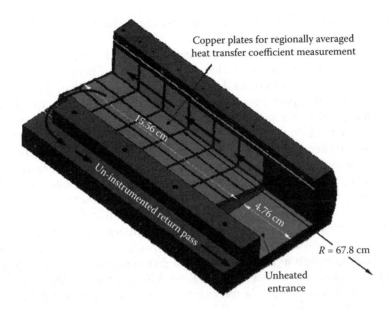

FIGURE 5.103
Wedge-shaped channel with turn exit at top as used by Wright et al. (From Wright, L. et al., Heat transfer in trailing edge wedge-shaped cooling channels under high rotation numbers, GT2007-27093, 2007.)

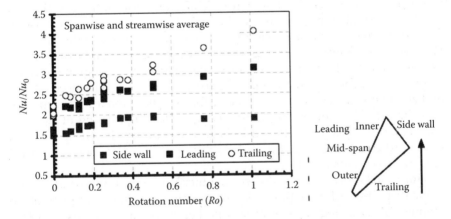

FIGURE 5.104
Effect of rotation number on the streamwise and spanwise averaged Nusselt number ratios. (From Wright, L. et al., Heat transfer in trailing edge wedge-shaped cooling channels under high rotation numbers, GT2007-27093, 2007.)

trailing surface. As the bleed ratio increased, the Sherwood number ratios decreased on both the bleeding and nonbleeding surfaces for the 45° angled ribs but increased on the bleeding surface for the 90° angled ribs. The drop in Sherwood number ratio with rotation in the leading surface with 45° ribs was less prominent when the bleed was on the trailing surface. But the opposite pattern was observed with 90° ribs. The leading surface benefitted from the leading-side bleed with orthogonal ribs.

Chang et al. (2007) presented another heat-transfer study done on trapezoidal duct with 45° staggered ribs with bleed. Local heat-transfer results showed the influences of sidewall bleeds and heat-transfer correlations with *Re*, *Ro*, and *buoyancy parameter* were presented. The heat load on the trailing edge is high and this region can have complex cooling arrangements and some stationary channel results can be found in Bunker et al. (2002) and Rigby and Bunker (2002). Their study included tapered channel, inclined ribs, and internal bleed arrangements but no rotational effects were included.

Liu et al. (2008b) studied the effect of trailing-edge bleed in rotating wedge-shaped channels with sharp entrance. Figure 5.105 shows the schematic of their channel configurations, and Figure 5.106 shows that the coolant extraction effectively changes the regional Reynolds numbers. The drop in Reynolds number was linear except for the end region. It is possible that the exit channel was not large enough to maintain uniform dump pressure for the extracted coolant. It is interesting to note that the Reynolds number profiles maintained the inlet ratios downstream of the entrance. For example, the end Reynolds number for Re_{inlet} of 10,000 was 1,635 as compared to 6,104 for Re_{inlet} of 40,000. The ratio is quite close to 4 similar to the corresponding ratio at the entrance.

Since coolant extraction adds another effect to the heat-transfer distribution, to understand the rotation effects, it is perhaps better to look at the stationary channel results first. Figure 5.107 shows the stationary channel results. The heat transfer in most locations decreases with an increase in *x*, axial distance, because coolant is extracted out of the cooling channel. However, the pattern of heat-transfer coefficient near the slot exit, marked as outer, is interesting. The heat-transfer coefficient in this region increases before decreasing downstream. The effect is more pronounced at the higher Reynolds numbers. It is argued that the flow makes a sharp turn to make an exit through the trailing-edge slots. This turning of the flow increases mixing and thus increases heat-transfer coefficient near the slots. It is interesting to note that the regions away from the slot show the lowest heat-transfer coefficients.

Figure 5.108 shows the effect of rotation number on the trapezoidal cooling channel's heat-transfer coefficient. Results are presented in three regions of the channel. The outer region is the narrow part of the channel, where the coolant exiting through the trailing-edge slots plays a major role. The mid-span is more like a standard rotating coolant channel and the inner part is a combination of side, and trailing and leading walls. The rotation number is varied by the combination of four Reynolds numbers and five rotation speeds. The Nusselt number ratios are local and calculated based on the local Reynolds numbers.

The outer section results show that the flow is strong in the region and perhaps parallel to direction to the rotation vector because of the coolant exit. As a result, the rotational effects are not significant between the two sides. However, an increase in the Nusselt number ratio is observed with

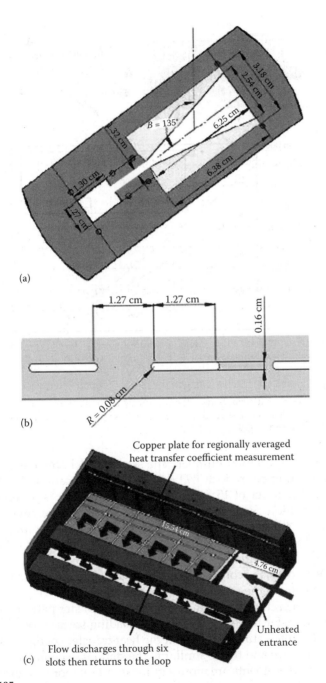

FIGURE 5.105
Trapezoidal channel is used here with trailing-edge extraction. Six slots were used as coolant extraction to simulate the trailing edge cooling. (a) Channel geometries, (b) slot geometries, and (c) schematic of the flow path. (From Liu, Y.H. et al., Heat transfer in trailing edge wedge shaped cooling channels with slot ejection under high rotation numbers, ASME Paper, GT2008-50343, 2008b.)

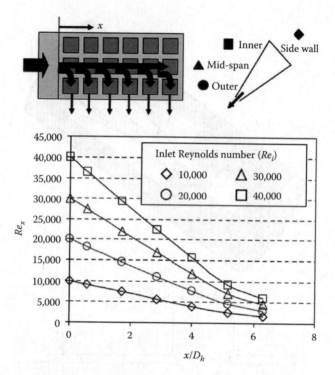

FIGURE 5.106

Local Reynolds number distribution in the stationary channel. (From Liu, Y.H. et al., Heat transfer in trailing edge wedge shaped cooling channels with slot ejection under high rotation numbers, ASME Paper, GT2008-50343, 2008b.)

an increase in the rotation number. The stationary channel also showed a higher Nusselt number ratio at $x/D = 6.3$ than other locations; but the maximum was of the order of 15, which is lower than what is observed in the presence of rotation. Since there are no local flow measurement data available in rotation, it can be argued that this is an effect of the pumping action by rotation. The coolant is centrifuged as well as pushed by Coriolis forces to the higher radius region and more coolant exits at the $x/D_h = 6.3$ than at $x/D_h = 1.7$. Moreover, the cooler coolant tends to be pushed to the further end (higher X/D_h) due to the rotational buoyancy effect.

The mid-span section shows the classical heat-transfer pattern of trailing-surface heat transfer to be higher than the leading surface for radially outward flow. It is interesting to note that the heat-transfer coefficients increase with radial distance and this is similar to the stationary channel characteristics, but the enhancements are more significant in rotation. For example, the lowest Reynolds number showed a Nusselt number ratio of 8 in a stationary channel and it is about 13 in the presence of rotation. The coolant is most likely being pushed to the further end by the centrifugal forces and cooler coolant is available at a higher x/D. The coolant changes direction from the

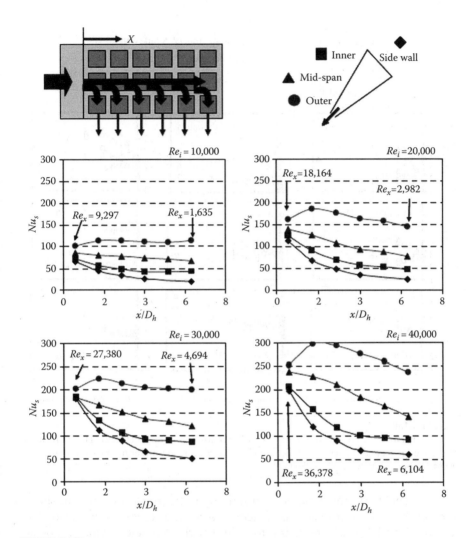

FIGURE 5.107
Nusselt number distributions in the stationary channel. (From Liu, Y.H. et al., Heat transfer in trailing edge wedge shaped cooling channels with slot ejection under high rotation numbers, ASME Paper, GT2008-50343, 2008b.)

radially outward to the direction of slot exit, thus reducing the effect of Coriolis forces on the flow distribution at higher radius locations.

The inner sidewall location has more space for the coolant to develop rotation-induced secondary flow and the separation between the trailing-surface and leading-surface heat-transfer coefficients are more significant in this region. Like a stationary channel, the sidewall shows the least magnitude of the heat-transfer coefficient among the three surfaces except at the $x/D = 1.7$, where the rotational effects reduced the heat-transfer coefficient of the leading surface for a rotation number up to 1. The trailing surface,

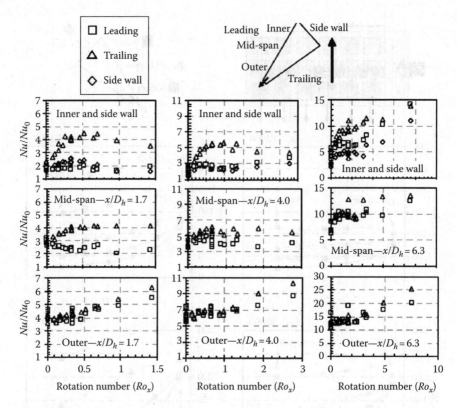

FIGURE 5.108
Effect of rotation number on the Nusselt number ratios. (From Liu, Y.H. et al., Heat transfer in trailing edge wedge shaped cooling channels with slot ejection under high rotation numbers, ASME Paper, GT2008-50343, 2008b.)

which benefits by the rotation in a radially outward flow situation, shows a drop in the heat-transfer coefficient beyond the rotation number of 1 in $x/D = 1.7$ and 4. It is interesting to note that the Nusselt number ratios for the leading and trailing surfaces at axial location $x/D = 4$ are similar to the stationary channel ratio of 3 at the highest rotation number corresponding to the lowest Reynolds number. The sidewall shows heat-transfer enhancement with rotation at $x/D = 4$ as compared to the stationary channel. The outer most location of $x/D = 6.3$ shows that all sides including the leading surface show heat-transfer enhancement with an increase in the rotation number. The Nusselt number ratio is significantly higher than the stationary channel ratio of 6.7. Clearly more coolant is being pushed to this region but the coolant is flowing through the core of the channel without benefiting the $x/D = 4$ location, which shows a drop in the heat-transfer coefficient of the trailing edge at the higher rotation numbers.

Rallabandi et al. (2010) studied the effect of full-length conducting, partial-length conducting, and nonconducting pins in a wedge-shaped channel with trailing-edge bleeding. The rotational effects were altered in different regions by the presence of the pins. Previously the flow pattern and heat transfer in wedge-shaped duct with pin fin are discussed in Andreini et al. (2004) in a stationary frame and rotation was not included in that study. Detailed analysis of the flow field pointed out the key flow features as horseshoe vortex, a stagnant wake behind the pin and a bulk flow acceleration due to convergent shape of the channel. Results revealed the presence of a weak jetlike flow field toward endwall surfaces caused by the strong recirculation behind each pin. Carcasci et al. (2003) studied the heat transfer and pressure drop in stationary wedge-shaped trailing edge with pins and longitudinal pedestal ribs. A combination of ribs and pins was recommended due to heat-transfer enhancement with less pressure drop penalty. Kulasekharan and Prasad (2008) studied the effects of coolant entry configurations in the trailing edge by stationary numerical models and found that straight flow to the trailing edge from inner chambers have higher pressure drop than the turned flow orientation.

Figure 5.109 shows the schematics of the test section and pin arrangements used by Rallabandi et al. (2010). Figure 5.110 shows the effect of local rotation

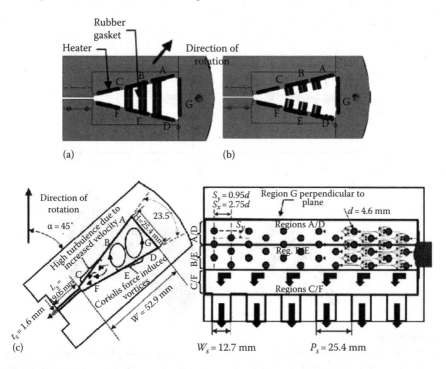

FIGURE 5.109
Schematics of the flow and pin arrangements used by Rallabandi et al. In (a) and (b) only one staggered row is shown for clarity. (a) Full length pins, (b) half-length pins, and (c) schematic of the coolant flow. (From Rallabandi, A.P. et al., Heat transfer in trailing edge wedge-shaped pin-fin channels with slot ejection under high rotation numbers, ASME GT2010-22832, 2010.)

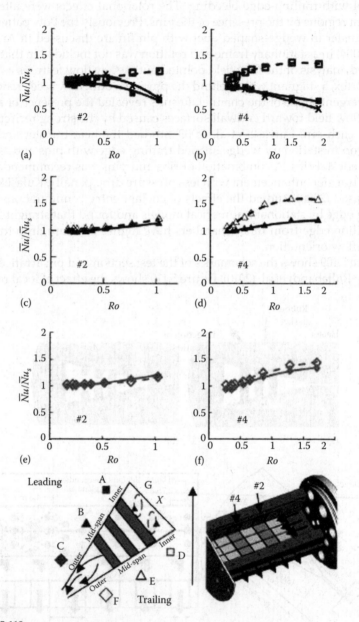

FIGURE 5.110
Effect of local rotation number on local Nusselt number distributions for full conductive pin-fin. (a) A, D, and G, (b) B and E, (c) C and F, (d) A, D, and G, (e) B and E, and (f) C and F. (From Rallabandi, A.P. et al., Heat transfer in trailing edge wedge-shaped pin-fin channels with slot ejection under high rotation numbers, ASME GT2010-22832, 2010.)

number on regional Nusselt number distribution. The narrow exit region shows that the walls are closer to each other and therefore the rotational effects are negligible as shown for regions C and F. Regions A, D, and G were strongly influenced by the Coriolis effects and show the classical characteristics of rotational effects—the trailing side shows heat-transfer coefficient enhancement over the leading surface in rotating radial outward flow. Regions B and E are connected by conductive copper pins but still the rotational effects are prominent at station #4, further from the inlet.

The effects of partial conductive pins are shown in Figure 5.111. Regions C and F are near the trailing-edge extraction and show similar heat-transfer coefficients at all rotation numbers. The rotational effects are stronger for the station #4 that is further from the inlet than station #2. The sidewall behaves more like a leading wall and shows a decrease in the Nusselt number with an increase in the rotation number. It is interesting to note that region G shows a decrease in Nusselt number with an increase in rotation number. This observation is opposite to what has been seen in smooth channel.

5.10.3 Effect of Aspect Ratio and Rib Configurations on Rotating Channel Heat Transfer

In this section, the effects of channel aspect ratio on the heat transfer from rotating channels are presented. Several rib configurations and channel entrance conditions were also integral parts of these studies.

Several experimental works were presented during 2000–2010 including the effect of aspect ratio on smooth and ribbed rotating channels. Azad et al. (2002) used a 2:1 aspect ratio channel with inclined ribs. Both parallel and cross ribbed configurations were used as shown in Figure 5.112. Three different Reynolds numbers, 5,000, 10,000, and 25,000, were used with corresponding rotation numbers of 0.21, 0.11, and 0.04. The rib height-to-hydraulic diameter ratio, e/D, was 0.094, and the pitch-to-height ratio, P/e, was 10. The inlet to coolant density ratio was 0.115. Two channel orientations, 90° and 135° with respect to the rotation direction, were used. The Nusselt number ratio is normalized based on Dittus-Boelter/McAdams correlation.

Figure 5.113 shows the schematic secondary flow generated by angled ribs. A double cell was noted for parallel rib and a single secondary flow vortex cell was noted in cross ribbed configuration in stationary channel. These secondary flows interacted with the secondary flows developed by rotation as discussed later.

Figure 5.114 shows the heat-transfer results from leading and trailing surfaces of the rotating channel of Azad et al. (2002). Both 90° and 135° results are presented. As expected the stationary channel results ($Ro = 0$) show significant improvement over the fully developed smooth results; Nusselt number ratio is more than 2 in most locations at lower Reynolds numbers. The first passage did not show the characteristic decrease in heat-transfer coefficient past the entrance. This behavior is attributed to the development

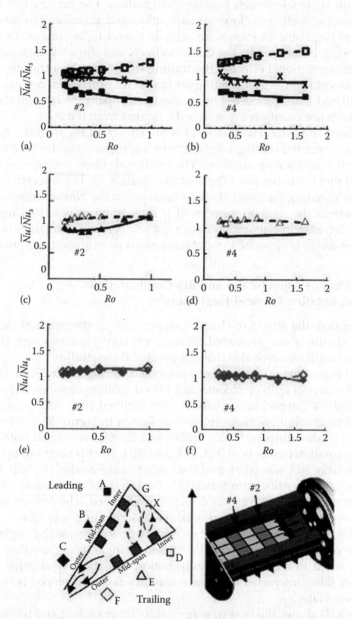

FIGURE 5.111

Effect of local rotation number on local Nusselt number distribution for partial-conductive pin fin. (a) A, D, and G, (b) B and E, (c) C and F, (d) A, D, and G, (e) B and E, and (f) C and F. (From Rallabandi, A.P. et al., Heat transfer in trailing edge wedge-shaped pin-fin channels with slot ejection under high rotation numbers, ASME GT2010-22832, 2010.)

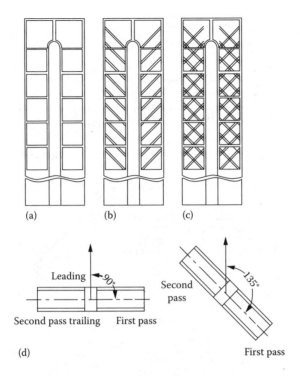

FIGURE 5.112
Rib configurations in a rotating 2:1 aspect ratio channel by Azad et al. Two-pass test sections with (a) smooth walls, (b) 45° parallel ribs on leading and trailing surfaces, (c) 45° cross ribs on leading and trailing surfaces, and (d) two different channel orientations. (From Azad, G.S. et al., *ASME J. Turbomach.*, 124, 251, 2002.)

of the secondary flow by the rib orientation. Note that there is no rib at the turn region and the Nusselt number ratio dropped because of that. The 90° orientation showed the most increase in the first-pass trailing wall, but this orientation also showed a reduced Nusselt number ratio in the second-pass trailing side. Since 135° orientation puts the leading and trailing surfaces at a nonorthogonal direction to the rotation, it was expected that the rotational effects would have been lower. But results indicate that the decrease in first-pass leading and second-pass trailing surfaces was comparable to the 90° orientation, but the increase in first-pass trailing and second-pass leading surfaces was not as good as the 90° results.

Figure 5.115 shows the effect of ribs is also present in the nonturbulated inner and outer walls. Nusselt number ratio is significantly higher than 1 in most locations. In general, rotation decreased the Nusselt number ratios in the first pass and increased in the second pass. The increase in Nusselt number ratio near the exit is perhaps caused by a sharp 90 turn at the exit of the heated test section. Overall the inner wall showed a higher Nusselt number ratio in the first pass than the outer wall and a lower Nusselt number ratio in the second pass. This pattern must have been caused by the rib generated

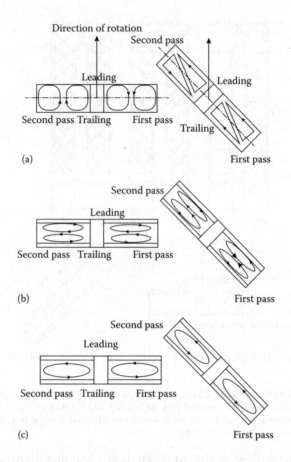

FIGURE 5.113
Conceptual secondary flow patterns are shown for parallel and cross ribbed configurations. (a) Rotation-induced double cell secondary flow, (b) 45° parallel rib-induced double cell secondary flow, (c) 45° cross rib-induced single cell secondary flow. (From Azad, G.S. et al., *ASME J. Turbomach.*, 124, 251, 2002.)

secondary flow because the pattern is consistent in $Ro = 0$. The rotation did not change the pattern, but the magnitude of increase and decrease changed because rotation had its own secondary flow effects.

Figure 5.116 shows the Nusselt number ratio distribution along the two-pass channel for three different Reynolds numbers in stationary and rotating configurations. Results show that unlike the parallel rib case, the Nusselt number ratio decreases with increasing x/D at the entrance of the first channel. The schematic vortex structure indicated that the cross ribbed configuration produced a single secondary vortex as compared to two counterrotating vortices in parallel rib case. This weaker secondary mixing is responsible for the lower Nusselt number ratio in stationary channel. Similar observations were made by Metzger and Vedula and Han et al. (1991). The rib orientation

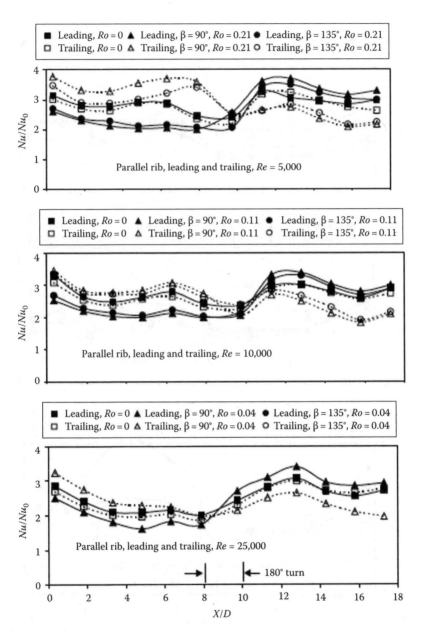

FIGURE 5.114
Leading- and trailing-wall heat-transfer results presented as Nusselt number ratios in parallel rib configuration. (From Azad, G.S. et al., *ASME J. Turbomach.*, 124, 251, 2002.)

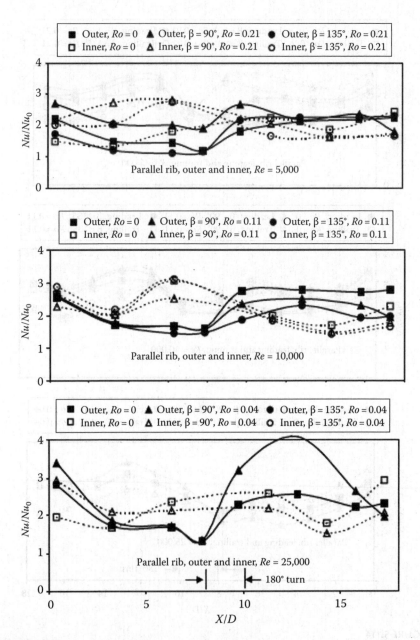

FIGURE 5.115
Nusselt number ratios in outer and inner nonribbed surfaces with 45° parallel rib. (From Azad, G.S. et al., *ASME J. Turbomach.*, 124, 251, 2002.)

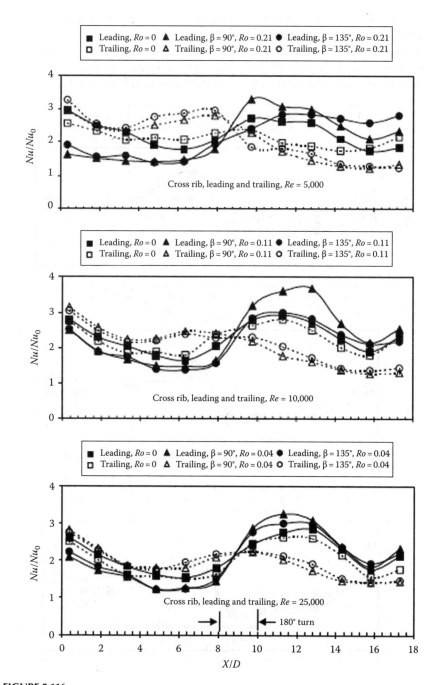

FIGURE 5.116
Nusselt number ratio along the rotating two-pass channel with cross inclined ribs. (From Azad, G.S. et al., *ASME J. Turbomach.*, 124, 251, 2002.)

gives a big enhancement to the Nusselt number ratio past the turn in the second-pass leading surface.

Figure 5.117 shows the Nusselt number ratio for 3 Reynolds numbers with cross rib configuration. In stationary channel, the Nusselt number ratio increases near the turn in the first pass due to the development of the

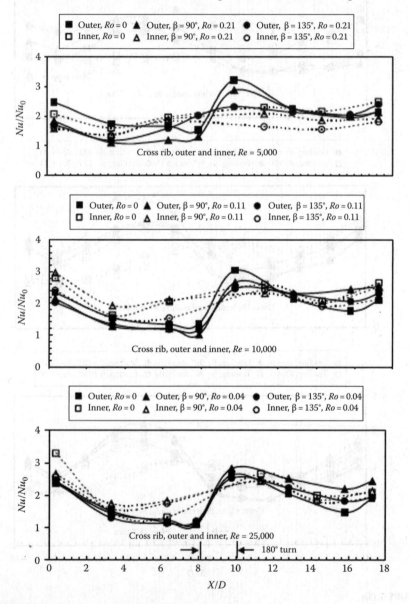

FIGURE 5.117
Nusselt number ratios for 45° cross ribs at outer and inner surfaces. (From Azad, G.S. et al., *ASME J. Turbomach.*, 124, 251, 2002.)

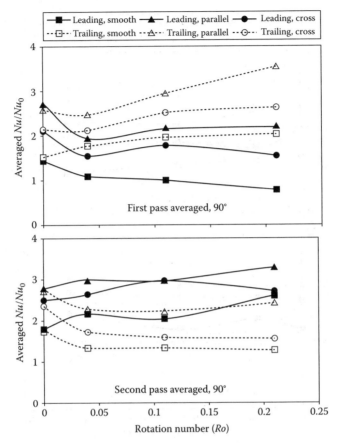

FIGURE 5.118
Surface-averaged Nusselt number ratios in leading and trailing walls at 90° channel orientation. (From Azad, G.S. et al., *ASME J. Turbomach.*, 124, 251, 2002.)

secondary flow by the rib orientation. In rotation, the turn region shows significant enhancement. At lower *Re*, the effect of rotation is more prominent in the first-pass outer wall, but the effect of rotation is not visible in this wall of second pass. The inner wall shows that Nusselt number decreases in 135° orientation. Interestingly, the high Re situation indicates more rotational effects in the second-pass outer wall than the lower *Re*, where rotation number was higher.

Figure 5.118 shows the surface-averaged Nusselt number ratios with 90° channel orientation. The parallel rib orientation performs better because it creates more mixing by the double cell vortex structure. Figure 5.119 shows the surface-averaged Nusselt number ratios with channel orientation of 135°. The parallel rib configuration clearly shows the better heat-transfer coefficients than the cross rib.

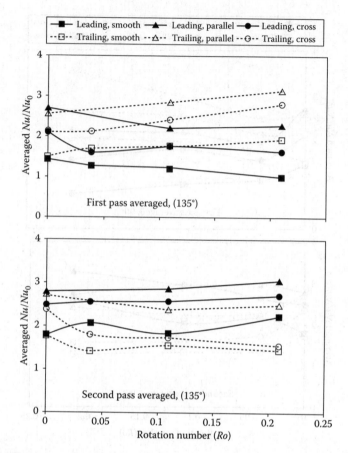

FIGURE 5.119
Surface-averaged Nusselt number ratios in leading and trailing walls at 135° channel orientation. (From Azad, G.S. et al., *ASME J. Turbomach.*, 124, 251, 2002.)

Figure 5.120 shows six different rib orientations used in a 4:1 rectangular channel with radial outward flow by Lee et al. (2003). V-shaped, and parallel and staggered angled ribs were used. Some rib configurations had a small gap at the corner and were referred to as ribs with gaps as shown in the figure. The Reynolds number was varied from 5,000 to 40,000 and rotation number was varied from 0 to 0.3. The inlet coolant-to-wall density ratio was maintained as 0.122. The channel was oriented 135° to the direction of rotation. The rib height-to-hydraulic diameter ratio was 0.078 and rib pitch-to-height ratio was 10.

Figure 5.121 shows the channel-averaged results for the stationary one-pass channel. The overall trend indicates that the ribbed results show decreasing Nusselt number ratio with increasing Reynolds numbers. Some previous studies are also included for comparison. The references are [4] Han et al. (1991), [23] Azad et al. (2002), [26] Al-Hadhrami et al. (2003), and [27] Griffith

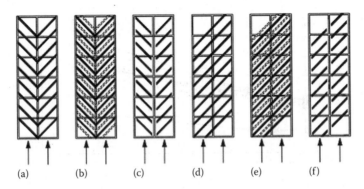

FIGURE 5.120
Schematic of the rib orientations used by Lee et al. in a 4:1 aspect ratio one-pass channel. (a) Parallel V-shaped ribs without gaps, (b) staggered V-shaped ribs without gaps, (c) parallel V-shaped ribs with gaps, (d) parallel angled ribs without gaps, (e) staggered angled ribs without gaps, and (f) parallel angled ribs with gaps. (From Lee, E. et al., Heat transfer in rotating rectangular channels (AR=4:1) with V-shaped and angled rib turbulators with and without gaps, GT2003-38900, 2003.)

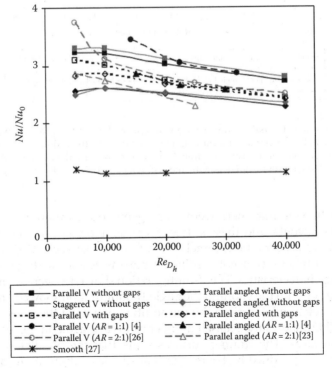

FIGURE 5.121
Channel-averaged Nusselt number ratios in stationary channel with and without different rib configurations. (From Lee, E. et al., Heat transfer in rotating rectangular channels (AR=4:1) with V-shaped and angled rib turbulators with and without gaps, GT2003-38900, 2003.)

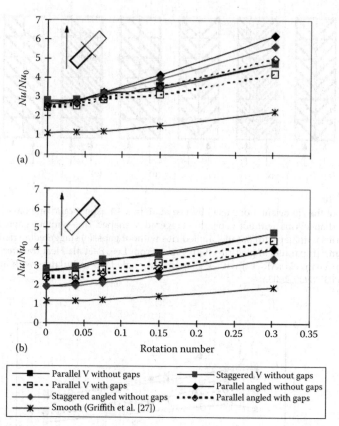

FIGURE 5.122
Streamwise-averaged Nusselt number ratio with and without ribs on the leading and trailing surfaces. (a) Leading-outer and trailing-outer surfaces and (b) leading-inner and trailing-inner surfaces. (From Lee, E. et al., Heat transfer in rotating rectangular channels (AR=4:1) with V-shaped and angled rib turbulators with and without gaps, GT2003-38900, 2003.)

et al. (2002). Note that there were a few geometrical differences in the rib shapes and layouts, and these studies were done on a square cross section channel or 2:1 aspect ratio channel. The difference between stagger and parallel configurations was not significant but the results with gaps showed interesting characteristics. The V-configuration showed a decrease in Nusselt number ratios with the gap, and the angled configuration showed an increase in the Nusselt number ratios with the addition of gaps at the sidewalls.

Figure 5.122 shows the Nusselt number ratios averaged on the leading and trailing surfaces. The trailing surfaces show that the parallel angled and staggered ribs produce better results than the V-configuration. But in the leading surfaces, the V-configuration without gaps performed better than the other rib orientations. There is no significant difference between parallel V and staggered V rib results.

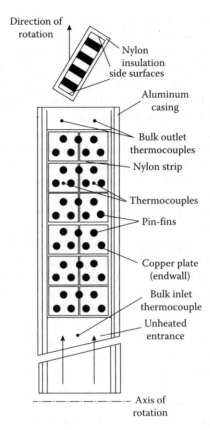

FIGURE 5.123
Schematic of the narrow test section used by Wright et al. (From Wright, L.M. et al., Effect of rotation on heat transfer in narrow rectangular cooling channels (AR = 8:1 and 4:1) with pin-fins, GT2003-38340, 2003.)

Wright et al. (2003) studied the rotational effects on narrow (aspect ratio of 8:1 and 4:1) channels with pin fins as heat-transfer augmentation. This narrow channel simulates the pin fins in narrow trailing edge of an airfoil. The channel is at 150° orientation with respect to the rotation direction as shown in Figure 5.123. Heat-transfer enhancement from both conducting (copper) and nonconducting (Plexiglas) pins were investigated. Due to changing aspect ratio of the channel, the height-to-diameter ratio of the pins varied from two for aspect ratio 4:1 to one for aspect ratio 8:1. A staggered array of pins with streamwise and spanwise spacing of two diameters is studied. The flow Reynolds number based on channel hydraulic diameter was varied from 5,000 to 20,000. The rotation number varied from 0 to 0.302, and inlet coolant-to-wall density ratio was 0.12.

The paper shows the Nusselt number ratio distribution in aspect ratio 4:1 with nonconducting pins. Pins extend completely across the channel from leading surface to the trailing surface. The results are compared with smooth-channel results with rotation. At lower rotation numbers, the effect of rotation is not noticeable in either smooth or pin-fin configurations. Note that the channel orientation is at an angle to the rotation direction as shown in the figure. At high rotation number, the leading surface shows the characteristic reduction in heat-transfer coefficient and the trailing surface shows an increase. The increase in trailing surface is more near the inlet, whereas the reduction in leading surface is more at the downstream locations.

The paper also shows the Nusselt number distribution with conducting pins in aspect ratio 4:1. Like nonconducting situation, rotation increases the overall Nusselt number ratio. The conducting pins show that the leading-surface Nusselt ratio did not decrease with an increase in rotation number. However, the trailing surface at farther location from the inlet shows a significant Nusselt number ratio enhancement with conductive pins.

Figure 5.124 shows the channel-averaged Nusselt number ratio as observed by Wright et al. (2003). The conducting pins show higher Nusselt number ratios. At high rotation number, the effect of pin conductivity becomes less significant and the enhancement by rotation is more prominent.

Fu et al. (2004, 2005) studied the heat-transfer coefficients in two-pass rotating rectangular channels with narrow aspect ratios of 1:4 and 1:2 as shown in Figures 5.125 and 5.126. The rib turbulators were placed at an angle of 45° to the direction of the flow. Four Reynolds numbers were considered varying from 5,000 to 40,000. The rotation number varied from 0.0 to 0.3. The ribs had a square 1.59 × 1.59 mm cross section. The rib height-to-hydraulic diameter ratios were 0.094 for $AR = 1:2$ and 0.078 for $AR = 1:4$ channel. The rib pitch-to-height ratio was 10 and the inlet coolant-to-wall density ratio was

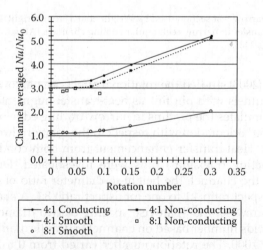

FIGURE 5.124
Channel-averaged Nusselt number ratio for conducting and nonconducting pins in rotating narrow channels. (From Wright, L.M. et al., Effect of rotation on heat transfer in narrow rectangular cooling channels (AR = 8:1 and 4:1) with pin-fins, GT2003-38340, 2003.)

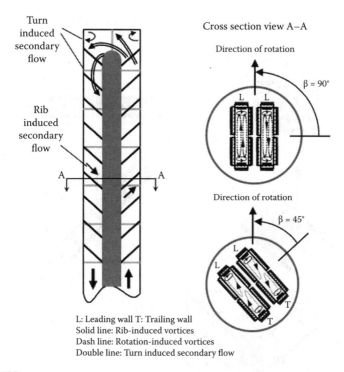

Turn induced secondary flow

Cross section view A–A

Direction of rotation

$\beta = 90°$

Rib induced secondary flow

Direction of rotation

$\beta = 45°$

L: Leading wall T: Trailing wall
Solid line: Rib-induced vortices
Dash line: Rotation-induced vortices
Double line: Turn induced secondary flow

FIGURE 5.125
Schematic of the two-pass narrow aspect ratio (AR = 1:4) channel with angled ribs as used by Fu et al. (From Fu, W.L. et al., Heat transfer in two-pass rotating rectangular channels (AR=1:2 and AR=1:4) with 45 deg angled rib turbulators, ASME Paper No. GT 2004-53261, 2004.)

maintained at 0.115. Two channel orientations, 90° and 45°, with respect to the rotation direction were considered.

Figure 5.127 shows the effect of buoyancy parameter on the Nusselt number ratio for different aspect ratio channels. The first-pass trailing surface shows the characteristic increase in the Nusselt number ratio with an increase in the buoyancy parameter, and the first-pass leading shows the decrease in the Nusselt number ratio. Figure 5.127b shows the Nusselt number ratios in the second pass. Interestingly, AR 1:4 and 1:2 channels indicate favorable Nusselt number ratios in the second pass. The rotational buoyancy enhances heat-transfer coefficient from both leading and trailing surfaces in these aspect ratios.

Figure 5.128 compares the Nusselt number ratios with different channel orientations and aspect ratios. The AR 1:4 and 1:2 aspect ratios are mostly suitable for the thicker part of the airfoil and they are oriented 45° to the rotation direction. The square cross section of AR 1:1 ids most likely be used in the middle part of the airfoil and the rotation orientation given is 90°. The other cross sections AR = 2:1 and 4:1 are more suitable for the trailing edge and the channel orientations considered for them is 135° to the rotation direction.

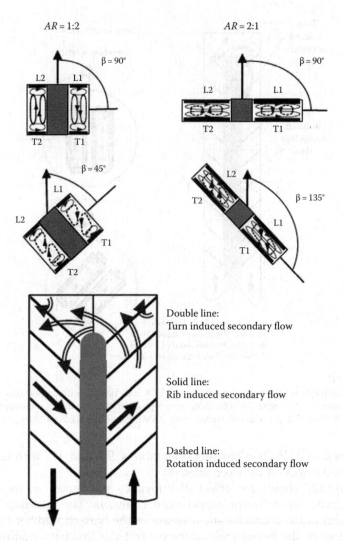

FIGURE 5.126
Schematic flow patterns from rib orientation and rotation. (From Fu, W.L. et al., Buoyancy effects on heat transfer in five different aspect-ratio rectangular channel with smooth walls and 45-degree ribbed walls, GT 2005-68493, 2005.)

Wright et al. (2004) used a 4:1 test channel to study some innovative rib configurations. The schematics of the test section and the rib arrangements are shown in Figures 5.129 and 5.130. In general, the square-sectioned ribs had $e/D = 0.078$ and $P/e = 10$. The channel is rotated at 135° to the direction of flow to simulate trailing portion of an airfoil. The Reynolds number is varied from 10,000 to 40,000 and the rotation number varied from 0 to 0.15. The inlet coolant-to-wall density ratio was maintained at 0.12. Results indicated that W-shaped configurations were better performers in heat-transfer enhancement.

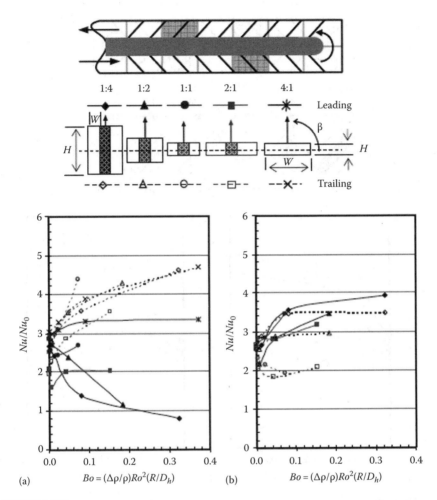

FIGURE 5.127
Effect of aspect ratio on the Nusselt number ratio. (From Fu, W.L. et al., Buoyancy effects on heat transfer in five different aspect-ratio rectangular channel with smooth walls and 45-degree ribbed walls, GT 2005-68493, 2005.) Comparisons are made with Fu et al. (From Fu, W.L. et al., Heat transfer in two-pass rotating rectangular channels (AR=1:2 and AR=1:4) with 45 deg angled rib turbulators, ASME Paper No. GT 2004-53261, 2004.), AR = 1:4 and 1:2; (From Griffith, T.S. et al., *ASME J. Heat Transfer*, 124, 617, 2002.), AR = 4:1, (From Azad, G.S. et al., *ASME J. Turbomach.*, 124, 251, 2002.), AR = 2:1, (From Al-Hadhrami, L. et al., *ASME J. Heat Transfer*, 125, 232, 2003.), AR = 1:1. (a) First pass and (b) second pass.

The Nusselt number distribution along the stationary channel with different rib configurations is plotted in the paper. The inclined ribs indicate the coolant to be forced from the outer surface to the inner surface. The Nusselt number ratio is significantly higher for the outer region. The other rib configurations do not prefer outer or inner regions. The W-shaped ribs show the most significant heat-transfer enhancement.

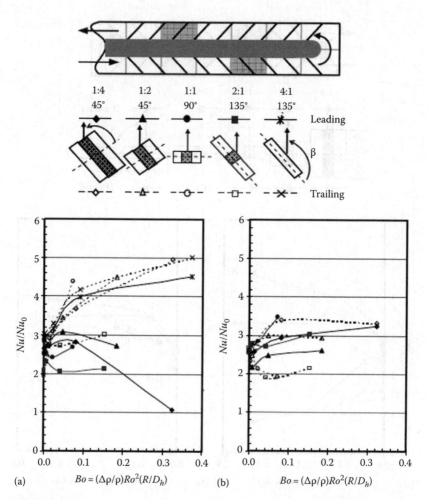

FIGURE 5.128
Channel orientation effect on the Nusselt number ratio with varying buoyancy parameter in different aspect ratio rotating channels. (a) First pass and (b) second pass. (From Fu, W.L. et al., Buoyancy effects on heat transfer in five different aspect-ratio rectangular channel with smooth walls and 45-degree ribbed walls, GT 2005-68493, 2005.)

The paper shows the rotating channel Nusselt number ratios with six different rib configurations at $Re = 20,000$ and $Ro = 0.075$. The angled ribs show the classic heat-transfer increase and decrease on trailing and leading surfaces at the inner region. The effect of rotation was not significant at the higher heat-transfer outer region. The V-shaped ribs show the minimum effect by rotation. Unlike V-shaped ribs, the W-shaped ribs show the leading- and trailing-side differences at downstream locations from the inlet.

Figure 5.131 shows the channel-averaged Nusselt number ratios and the friction factor ratios with different rib configurations. The stationary channel shows that heat transfer and friction factors are high for parallel W-rib.

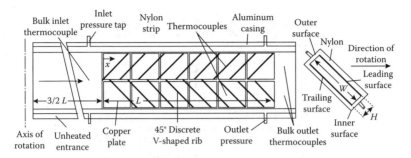

FIGURE 5.129
Schematic of the 4:1 test section used by Wright et al. (From Wright, L.M. et al., Thermal performance of angled, V-shaped, and W-shaped rib turbulators in rotating rectangular cooling channels (AR=4:1), GT2004-54073, 2004.)

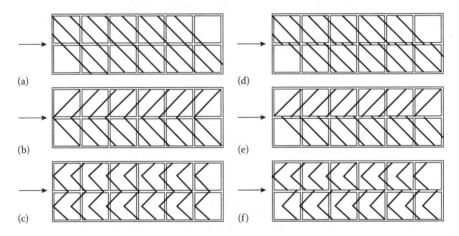

FIGURE 5.130
Different rib configurations used by Wright et al. (a) Angled ribs, (b) V-shaped ribs, (c) W-shaped ribs, (d) discrete angled ribs, (e) discrete V-shaped ribs, and (f) discrete W-shaped ribs. (From Wright, L.M. et al., Thermal performance of angled, V-shaped, and W-shaped rib turbulators in rotating rectangular cooling channels (AR=4:1), GT2004-54073, 2004.)

The discrete W showed comparable heat-transfer coefficients but the friction factor was lower than parallel W-ribs at higher *Re*. In rotating channel, the W-rib configurations maintained the higher heat-transfer patterns. The friction factor of discrete W-ribs was significantly lower that the parallel W-ribs, making this discrete W configuration more desirable for the rotating channels.

Figures 5.132 and 5.133 show the thermal performance (estimated from heat-transfer enhancement and pressure drop penalty) of different rib configurations used by Wright et al. (2004). Both stationary and rotating characteristics are presented. Discrete V- and discrete W-ribs provide best performance among the configurations used.

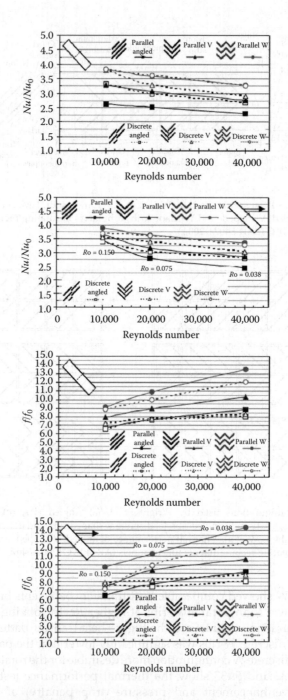

FIGURE 5.131
Channel-averaged Nusselt number ratios in rotating channels and corresponding friction factor ratios. (From Wright, L.M. et al., Thermal performance of angled, V-shaped, and W-shaped rib turbulators in rotating rectangular cooling channels (AR=4:1), GT2004-54073, 2004.)

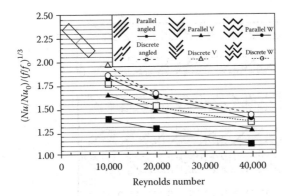

FIGURE 5.132
Thermal performance in nonrotating channels. (From Wright, L.M. et al., Thermal performance of angled, V-shaped, and W-shaped rib turbulators in rotating rectangular cooling channels (AR=4:1), GT2004-54073, 2004.)

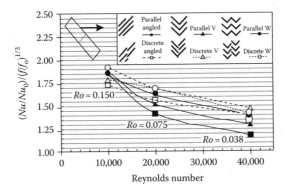

FIGURE 5.133
Thermal performance in rotating channels with different rib configurations. (From Wright, L.M. et al., Thermal performance of angled, V-shaped, and W-shaped rib turbulators in rotating rectangular cooling channels (AR=4:1), GT2004-54073, 2004.)

Liu et al. (2006) used a 1:2 rectangular channel and placed 45° ribs at different spacing. Figure 5.134 shows the rib configurations used. It also shows the effective increase in the surface area by the addition of the ribs. The rib pitch-to-height ratios were 10, 7.5, 5, and 3. Square ribs with a 1.59 mm × 1.59 mm cross section are used. The e/D ratio was 0.094. The channel was rotated at 550 rpm and Reynolds number was varied from 5,000 to 40,000; that generated rotation numbers from 0 to 0.2.

Figure 5.135 shows the effect of angled rib spacing on the first-pass Nusselt number ratio. The skewed rib results are compared with orthogonal ribs. The skewed angle ribs show that the heat-transfer coefficients continue to increase with closer spacing for $AR = 1:2$ channel.

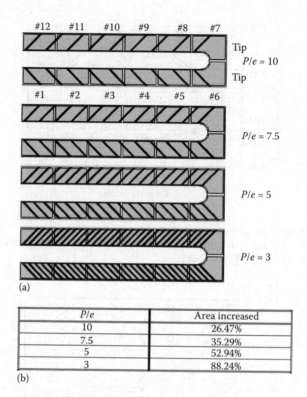

(a)

P/e	Area increased
10	26.47%
7.5	35.29%
5	52.94%
3	88.24%

(b)

FIGURE 5.134

(a) Different rib spacing used by Liu et al. with $AR = 1{:}2$. Area increased due to dense packing of ribs. (b) area increase by the rib spacing in the channel. (From Liu, Y.H. et al., Rib spacing effect on heat transfer and pressure loss in a rotating two-pass rectangular channel (AR=1:2) with 45-degree angled ribs, GT2006-90368, 2006.)

Figure 5.136 shows the overall pressure drop characteristics for different rib spacing. The friction factor ratio increased with a decrease in the rib spacing till $P/e = 5$. $P/e = 3$ showed a decrease in the friction factor ratio. This low spacing also showed a decrease in the friction factor ratio at higher rotation number (low Re). Other rib spacing showed an increase in the friction factor ratio with rotation.

Figures 5.137 and 5.138 show the Nusselt number ratios at selected locations in the first pass and second pass of the channel. The smooth-channel results indicate the classic pattern of higher Nusselt number ratio in the trailing side with respect to the leading side in the first pass. The ribbed channel also shows similar patterns, but the difference in the trailing and leading surfaces at higher Bo is more prominent than a smooth channel. The second-pass results with ribs show the trends of leading and trailing surfaces to be crossing over each other with increasing Bo. The ribbed leading surface shows heat-transfer enhancement with an increase in Bo for the second pass.

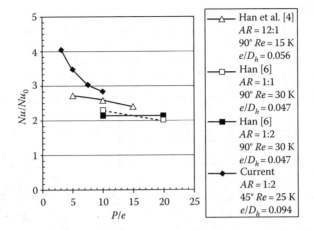

FIGURE 5.135
First-pass averaged Nusselt number ratios with different rib spacing. (From Liu, Y.H. et al., Rib spacing effect on heat transfer and pressure loss in a rotating two-pass rectangular channel (AR=1:2) with 45-degree angled ribs, GT2006-90368, 2006; From Han, J.C., *ASME J. Heat Transfer*, 110, 321, 1988; From Han, J.C. et al., *Int. J. Heat Mass Transfer*, 21, 1143, 1978.)

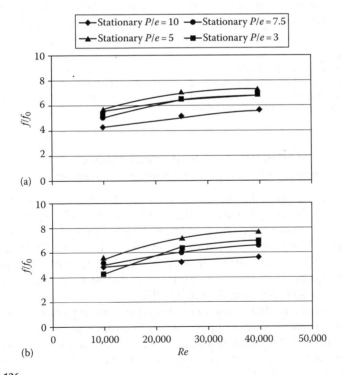

FIGURE 5.136
Overall friction factor ratios for (a) stationary and (b) rotating channels, Liu et al. (From Liu, Y.H. et al., Rib spacing effect on heat transfer and pressure loss in a rotating two-pass rectangular channel (AR=1:2) with 45-degree angled ribs, GT2006-90368, 2006.)

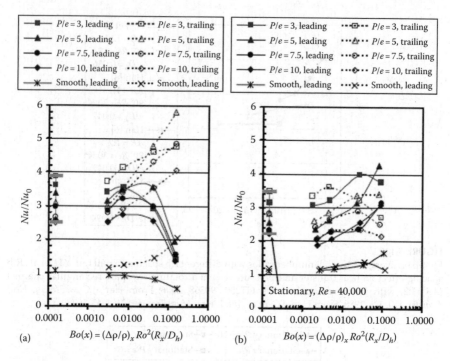

FIGURE 5.137
Nusselt number ratios at different buoyancy parameters with different rib spacings. (a) Leading and trailing, #4 and (b) Leading and trailing, #11. (From Liu, Y.H. et al., Rib spacing effect on heat transfer and pressure loss in a rotating two-pass rectangular channel (AR=1:2) with 45-degree angled ribs, GT2006-90368, 2006.)

5.10.4 Effect of High Rotation Number and Entrance Geometry on Rectangular Channel Heat Transfer

Liu et al. (2007) studied the effect of redirected sharp bend on the Nusselt number ratio. Figure 5.139 shows the schematic of their entrance and comparison with prior entrance shapes. The redirected sharp bends perform best in the group. The channel has $AR = 1:4$ applicable to the cooling channels in the leading edge of the airfoil. It is a two-pass rotating configuration. The Reynolds number is varied from 10,000 to 40,000, and the rotation numbers achieved with 400 rpm varied from 0 to 0.67.

Figure 5.140 shows the Nusselt number ratio with respect to the stationary channel in different buoyancy parameters. Buoyancy parameter is varied by rotation and flow Reynolds numbers. The region #1 is primarily dominated by the entrance effect, and the rotational buoyancy does not show much effect except for the higher Bo numbers. The buoyancy effect is more prominent downstream from the inlet.

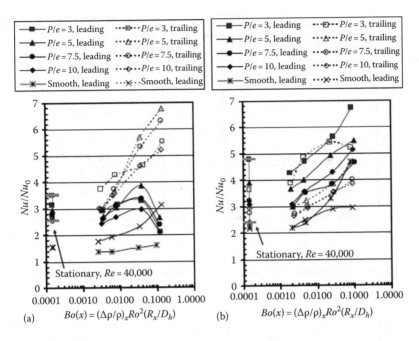

FIGURE 5.138
Nusselt number ratios at different buoyancy parameters with different rib spacings. (a) Leading and trailing, #6 and (b) leading and trailing, #7. (From Liu, Y.H. et al., Rib spacing effect on heat transfer and pressure loss in a rotating two-pass rectangular channel (AR=1:2) with 45-degree angled ribs, GT2006-90368, 2006.)

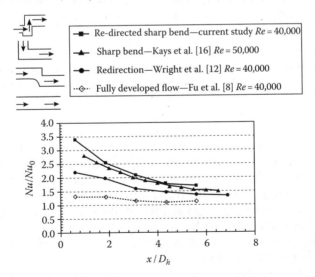

FIGURE 5.139
Schematic of the entrance configurations and stationary channel results by Liu et al. (From Liu, Y.H. et al., Heat transfer in a two-pass rectangular channel (AR = 1:4) under high rotation numbers, GT2007-27067, 2007.)

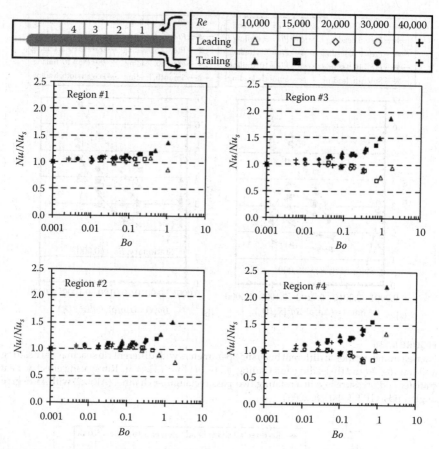

FIGURE 5.140
Nusselt number ratio (with stationary results) variation with buoyancy parameter in the radially outward flow of first pass, AR = 1:4. (From Liu, Y.H. et al., Heat transfer in a two-pass rectangular channel (AR=1:4) under high rotation numbers, GT2007-27067, 2007.)

The turn region Nusselt number ratios with the stationary channel Nu are presented in Figure 5.141. Interestingly both leading and trailing surfaces behave similarly in the presence of rotational buoyancy. The Nusselt number ratio increases irrespective of the turn location at higher Bo.

Like the turn region, the radial inward flow indicates an increase in the Nusselt number ratio with increasing Bo as shown in Figure 5.142. However, the increase in Nusselt number ratio is higher in the leading surface than in the trailing surface.

Figure 5.143 shows the effect of rotation number and buoyancy parameter on the Nusselt number ratio with respect to the stationary channel. Smooth unheated entrance results are plotted from Fu et al. (2005) for comparison.

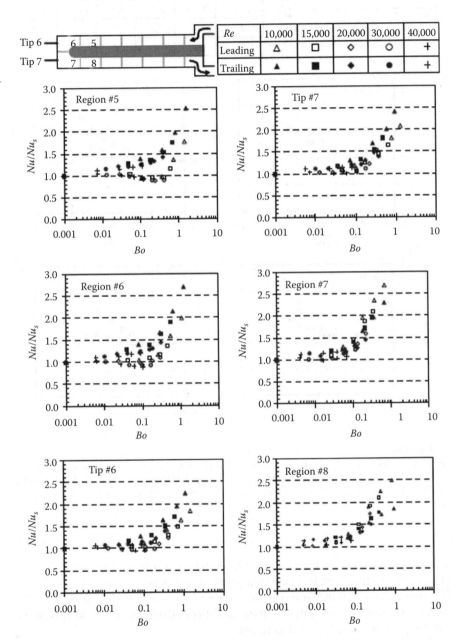

FIGURE 5.141
Nusselt number ratio (with stationary results) variation with buoyancy parameter in the turn region, $AR = 1{:}4$. (From Liu, Y.H. et al., Heat transfer in a two-pass rectangular channel (AR = 1:4) under high rotation numbers, GT2007-27067, 2007.)

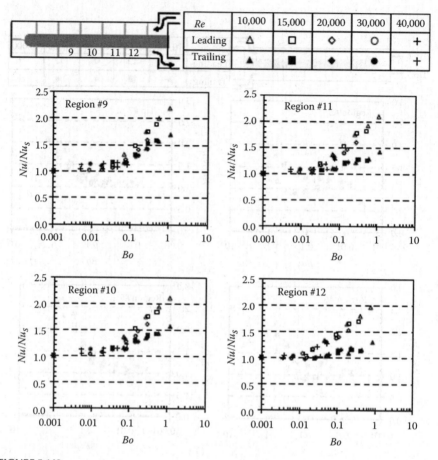

FIGURE 5.142

Nusselt number ratio (with stationary results) variation with buoyancy parameter in the radially inward flow of the second pass region, $AR = 1{:}4$. (From Liu, Y.H. et al., Heat transfer in a two-pass rectangular channel (AR=1:4) under high rotation numbers, GT2007-27067, 2007.)

The first-pass results clearly show the effect of entrance. The results of Liu et al. (2007) with a turn indicate that rotational effects are less than a smooth-channel entrance in the first pass. The second-pass results do not show much difference from the smooth entrance conditions. Correlations based on curve fitting are also included in this figure.

Huh et al. (2008) studied the effect of rib spacing on heat transfer in a rotating rectangular channel with $AR = 1{:}4$. The entrance had a sharp turn. Figure 5.144 shows the schematic of different rib spacing used by them. It was a two-pass channel with sharp turn at the entrance of the

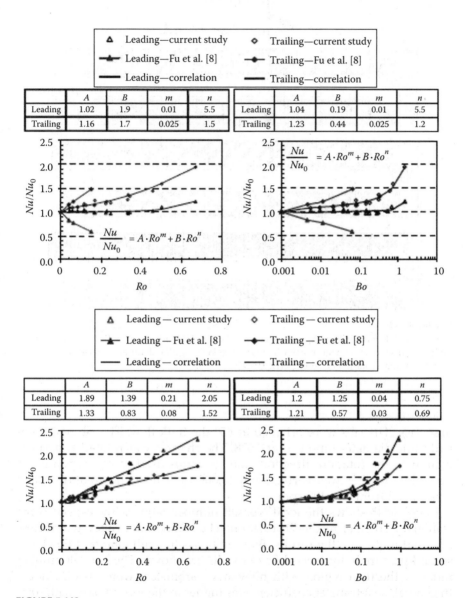

FIGURE 5.143
Average Nusselt number ratio with respect to the stationary channel in first and second pass, *AR* = 1:4. (From Liu, Y.H. et al., Heat transfer in a two-pass rectangular channel (AR = 1:4) under high rotation numbers, GT2007-27067, 2007; Fu, W.L. et al., Buoyancy effects on heat transfer in five different aspect-ratio rectangular channel with smooth walls and 45-degree ribbed walls, GT 2005-68493, 2005.)

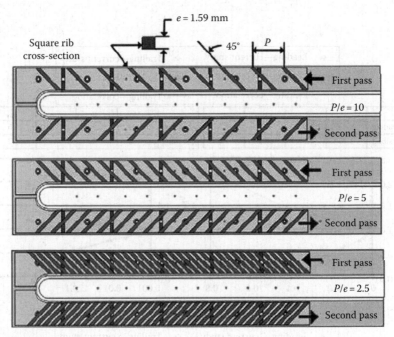

FIGURE 5.144

Schematic of different rib configurations used by Huh et al. with $AR = 1{:}4$. (From Huh, M. et al., Effect of rib spacing on heat transfer in a two-pass rectangular channel (AR=1:4) with a sharp entrance at high rotation numbers, GT2008-50311, 2008.)

first pass. The ribs were oriented at 45° the bulk flow. The rib height-to-hydraulic diameter ratio was at 0.078. The channel was rotated at perpendicular to the rotation direction, and P/e ratios used were 2.5, 5, and 10. Reynolds number was varied from 10,000 to 40,000, and rotation speed was varied from 0 to 400 rpm.

Figure 5.145 shows the local Nusselt number ratios with respect to the stationary channel results in rotation. The rotation buoyancy parameters are used to present the results. Both first-pass (radially outward flow) and second-pass (radially inward flow) results are shown. The Nusselt number ratios at the turn region with buoyancy parameters were also included. In general, heat-transfer coefficient was higher in the second pass with rotation. The tip turn region also benefitted in rotation for smooth as well as all rib spacing.

Figure 5.146 compares the Nusselt number ratios with respect to the stationary channel with Fu et al. (2004). Unlike the experimental setup of Huh et al. (2008), Fu et al. (2004) had smooth entrance to the heated section. Smooth-channel results clearly show that the turn at the entrance reduces

(continued)

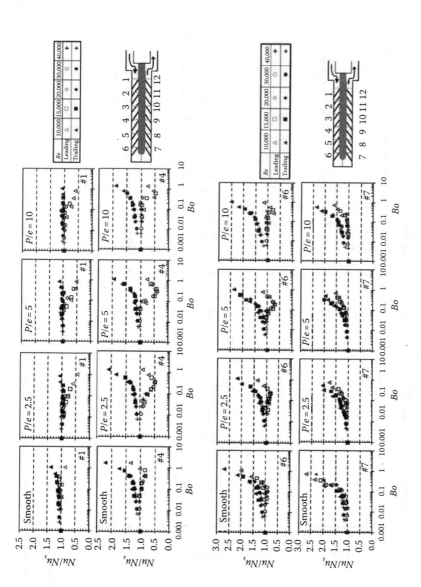

FIGURE 5.145
Nusselt number ratio with respect to the stationary channel at different buoyancy parameters with different rib spacing, *AR* = 1:4.

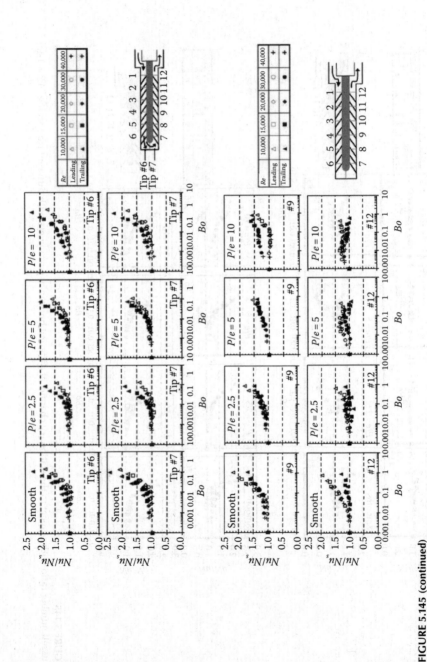

FIGURE 5.145 (continued)

Nusselt number ratio with respect to the stationary channel at different buoyancy parameters with different rib spacing, AR = 1:4. (From Huh, M. et al., Effect of rib spacing on heat transfer in a two-pass rectangular channel (AR=1:4) with a sharp entrance at high rotation numbers, GT2008-50311, 2008.)

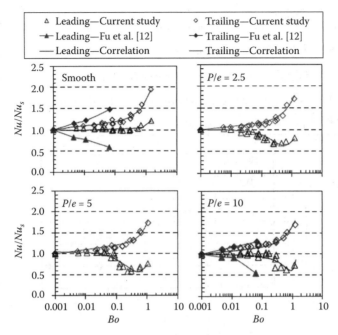

FIGURE 5.146
Comparison of Nusselt number ratios in leading and trailing sides of the first pass, $AR = 1{:}4$. (From Huh, M. et al., Effect of rib spacing on heat transfer in a two-pass rectangular channel (AR = 1:4) with a sharp entrance at high rotation numbers, GT2008-50311, 2008.)

the rotational effects as compared with the smooth-channel entrance. The decrease in Nusselt number ratio for the leading side with rib spacing of $P/e = 10$ is lower with turn at the entrance.

Huh et al. (2009) studied the entrance effects in a rotating smooth two-pass channel with a rectangular cross section of aspect ratio 2:1. The rotation numbers achieved in this study were higher than the rotation number presented earlier by this group. The rotation number and buoyancy parameter ranges were 0–0.45 and 0–0.8, respectively. Figure 5.147 shows the streamwise-averaged Nusselt number ratios for different entrance regions in a stationary channel. The presence of disturbances in the entrance increased the heat-transfer coefficient and that increase was sustained for the length of the channel studied.

Figures 5.148 and 5.149 show the heat-transfer enhancement at the tip region with rotation. Region 6 was noted to have more enhancement than region 7. Both surfaces showed increased Nusselt number ratios at rotation numbers higher than 0.2. Figure 5.150 shows the heat-transfer patterns

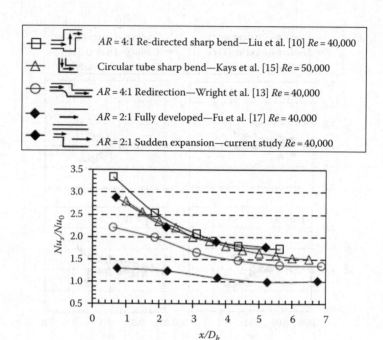

FIGURE 5.147
Comparison of stationary channel Nusselt number ratios for different entrance regions. (From Huh, M. et al., High rotation number effects on heat transfer in a rectangular (AR = 2:1) two pass channel, GT2009-59421, 2009.)

with varying buoyancy parameter. The curves are correlated with an empirical equation:

$$\frac{Nu}{Nu_s} = A \cdot Bo^a + B \cdot Bo^b.$$

The coefficients and exponents for different surfaces to be used in the above equation are available in Huh et al. (2009).

Huh et al. (2010) conducted heat-transfer experiments on a rotating rectangular channel with aspect ratio 2:1. The channel hydraulic diameter was 16.9 mm. Both smooth and ribbed configurations were studied. Ribs were applied on leading and trailing surfaces with $P/e = 10$ and $e/D = 0.094$. Ribs were at 45° to bulk flow. Channel was oriented at 90° and 135° to the direction of rotation. The Reynolds number was varied from 10,000 to 40,000. The rotation number range tested was 0–0.45, and the buoyancy parameter range was 0–0.85.

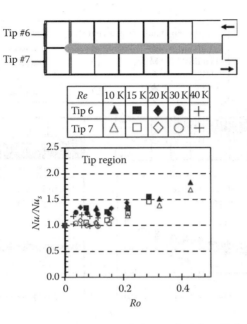

FIGURE 5.148
Nusselt number ratio with respect to the stationary channel at the tip region of the sharp turn, $AR = 2:1$. (From Huh, M. et al., High rotation number effects on heat transfer in a rectangular ($AR = 2:1$) two pass channel, GT2009-59421, 2009.)

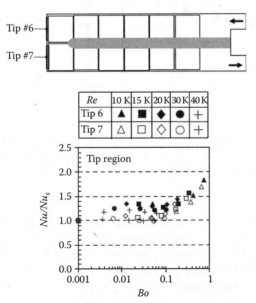

FIGURE 5.149
Effect of buoyancy parameter on heat-transfer enhancement at the tip region, $AR = 2:1$. (From Huh, M. et al., High rotation number effects on heat transfer in a rectangular ($AR = 2:1$) two pass channel, GT2009-59421, 2009.)

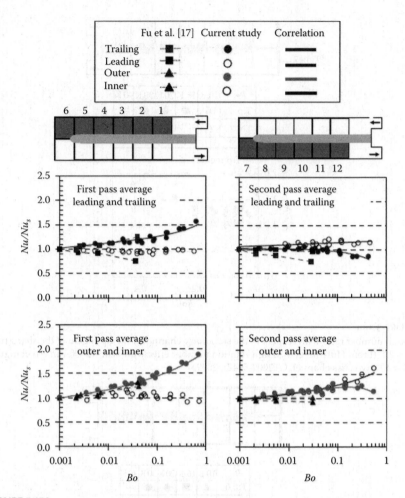

FIGURE 5.150

Effect of buoyancy parameter on the heat-transfer enhancement in the first and second pass of the rotating channel, $AR = 2:1$. (From Huh, M. et al., High rotation number effects on heat transfer in a rectangular (AR=2:1) two pass channel, GT2009-59421, 2009.)

Figure 5.151 shows the Nusselt number ratios for regions 4 and 10 in the two-pass channel. These two locations are selected to minimize the effect of entrance and the turn region on the heat-transfer pattern. Interestingly the leading smooth surface indicated an increase in the heat-transfer coefficient at 135° orientation in rotation. The outer smooth wall gets some of the trailing-side benefits in this tilted position and shows heat-transfer enhancement in rotation. The inner walls on both passes at 135° tilted orientation show significant reduction in heat-transfer coefficient. But the ribbed

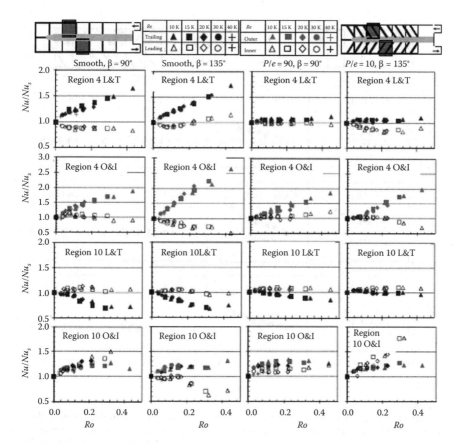

FIGURE 5.151
Effect of rotation number on heat transfer in leading, trailing, outer, and inner walls, $AR = 2:1$. (From Huh, M. et al., Influence of channel orientation on heat transfer in a two-pass smooth and ribbed rectangular channel (AR=2:1) under large rotation numbers, GT2010-22190, 2010.)

channels show that the same inner wall shows heat-transfer enhancement with rotation in this tilted position.

The effect of local buoyancy parameter on heat transfer in both smooth and ribbed channels is presented in Figure 5.152. The ribbed channel shows less effect by the buoyancy in the 90° channel orientation. The smooth channel in 135° orientation indicates all Nusselt number ratios to be similar to each other in region 6. Channel orientation 135° in ribbed rotating channel shows more difference in the leading and trailing sides than with channel orientation 90°.

Figure 5.153 shows the Nusselt number ratio with respect to the stationary channel results. This tip cap is smooth but gets affected by neighboring

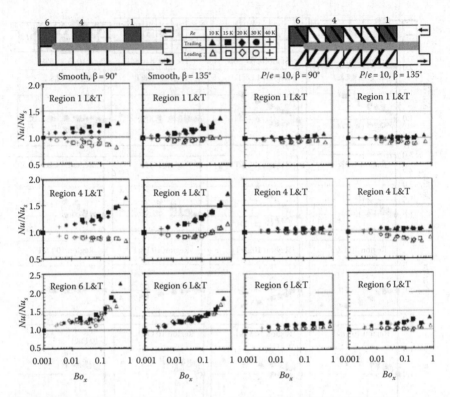

FIGURE 5.152
The effect of local buoyancy parameter on leading- and trailing-wall heat transfer in the first pass, $AR = 2{:}1$. (From Huh, M. et al., Influence of channel orientation on heat transfer in a two-pass smooth and ribbed rectangular channel (AR=2:1) under large rotation numbers, GT2010-22190, 2010.)

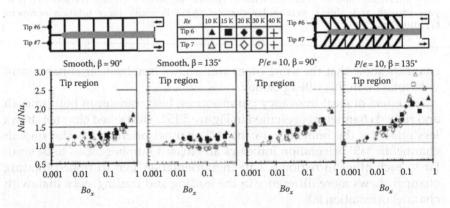

FIGURE 5.153
Effect of local buoyancy parameter on the heat transfer in the tip-cap region, $AR = 2{:}1$. (From Huh, M. et al., Influence of channel orientation on heat transfer in a two-pass smooth and ribbed rectangular channel (AR=2:1) under large rotation numbers, GT2010-22190, 2010.)

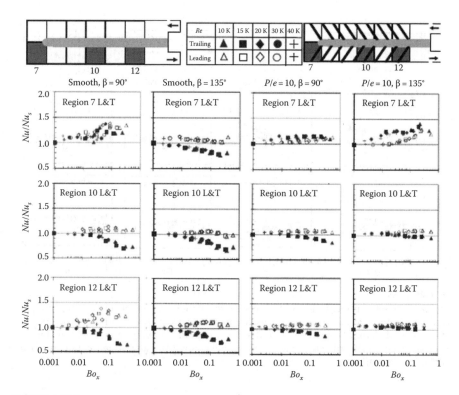

FIGURE 5.154
Effect of local buoyancy parameter on the leading- and trailing-wall heat transfer in the second pass, $AR = 2{:}1$. (From Huh, M. et al., Influence of channel orientation on heat transfer in a two-pass smooth and ribbed rectangular channel (AR=2:1) under large rotation numbers, GT2010-22190, 2010.)

trailing and leading surfaces when they are ribbed. The ribbed channel with 135° orientation shows the highest heat-transfer enhancement in rotation for the smooth tip-cap region.

Figure 5.154 shows the effect of local buoyancy parameter on the Nusselt number ratio with respect to the stationary channel. This plot shows the leading and trailing surface Nusselt number ratios for both smooth and ribbed channels and for two rotation channel orientations. The location 7, which is in the turn region, shows heat-transfer coefficient to increase with an increase in the local buoyancy parameter. The ribbed channel results downstream from the turn indicate reduced effects from rotational buoyancy.

Figures 5.155 and 5.156 show the Nusselt number ratio with respect to the stationary channel Nusselt number with varying buoyancy parameters. Both smooth and ribbed channel results are presented. Two channel orientations with respect to the rotating direction were considered. As expected, the effect of rotational buoyancy is less in the ribbed channel than that

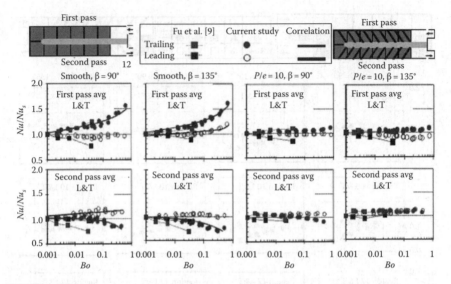

FIGURE 5.155
Average Nusselt number ratios with respect to the stationary channel on leading and trailing walls in the two-pass channel, Nusselt number ratio variations with buoyancy parameter, $AR = 2{:}1$. (From Huh, M. et al., Influence of channel orientation on heat transfer in a two-pass smooth and ribbed rectangular channel (AR = 2:1) under large rotation numbers, GT2010-22190, 2010.)

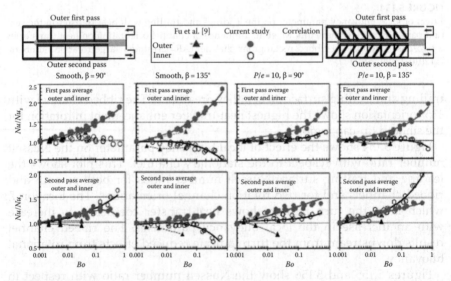

FIGURE 5.156
Average Nusselt number ratios with respect to the stationary channel on outer and inner walls in the two-pass channel. Nusselt number ratio variations with buoyancy parameter, $AR = 2{:}1$. (From Huh, M. et al., Influence of channel orientation on heat transfer in a two-pass smooth and ribbed rectangular channel (AR=2:1) under large rotation numbers, GT2010-22190, 2010.)

observed in the smooth channel. Curve-fitted correlations were formed based on curve fitted correlations that are available in Huh et al. (2010).

Huh et al. (2010) summarized the coefficients and exponents of these curve-fitted correlations in the paper.

References

Akella, K.V. and Han, J.C., 1998. Impingement cooling in rotating two-pass rectangular channels. *AIAA Journal of Thermophysics and Heat Transfer*, 12(4), 582–588.

Akella, K.V. and Han, J.C., 1999. Impingement cooling in rotating two-pass rectangular channels with ribbed walls. *AIAA Journal of Thermophysics and Heat Transfer*, 13(3), 364–371.

Al-Hadhrami, L., Griffith, T.S., and Han, J.C., 2003. Heat transfer in two-pass rotating rectangular channels (AR = 2) with five different orientations of 45° V-shaped rib turbulators. *ASME Journal of Heat Transfer*, 125, 232–242.

Al-Hadhrami, L. and Han, J.C., 2003. Effect of rotation on heat transfer in two-pass square channels with five different orientations of 45° angled rib turbulators. *International Journal of Heat and Mass Transfer*, 46, 653–669.

Andreini, A., Carcasci, C., and Magi, A., 2004. Heat transfer analysis of a wedge shaped duct with pin fin and pedestal arrays: A comparison between numerical and experimental results. GT2004-53319.

Azad, G.S., Uddin, M.J., Han, J.C., Moon, H.K., and Glezer, B., 2002. Heat transfer in a two-pass rectangular rotating channel with 45-deg angled rib turbulators. *ASME Journal of Turbomachinery*, 124, 251–259.

Bons, J.P. and Kerrebrock, J.L., 1998. Complementary velocity and heat transfer measurements in a rotating cooling passage with smooth walls, presented at the *International Gas Turbine and Aeroengine Congress and Exhibition*, June 2–5, 1998, Stockholm, Sweden, ASME Paper 98-GT-464.

Bunker, R.S., Wetzel, T.G., and Rigby, D.L., 2002. Heat transfer in a complex trailing edge passage for a high pressure turbine blade—Part1: Experimental measurements. GT2002-30212.

Carcasci, C., Facchini, B., and Innocenti, L., 2003. Heat transfer and pressure drop evaluation in the wedge-shaped trailing edge. ASME GT2003-38197.

Chang, S.W., Liou, T.M., Chiou, S.F., and Chang, S.F., 2007. High rotation number heat transfer of rotating trapezoidal duct with 45-deg staggered ribs and bleeds from apical side wall. ASME Paper GT2007-28174.

Cheah, S.C., Iacovides, H., Jackson, D.C., Ji, H., and Launder, B.E., 1996. LDA investigation of the flow development through rotating U-ducts. *ASME Journal of Turbomachinery*, 118(3), 590–595.

Chupp, R.E., Helms, H.E., McFadden, P.W., and Brown, T.R., 1969. Evaluation of internal heat transfer coefficients for impingement cooled turbine airfoils. *Journal of Aircraft*, 6, 203–208.

Clifford, R.J., Morris, W.D., and Harasgama, S.P., 1984. An experimental study of local and mean heat transfer in a triangular-sectioned duct rotating in the orthogonal mode. *ASME Journal of Engineering for Gas Turbines and Power*, 106, 661–667.

Dutta, S., Andrews, M.J., and Han, J.C., 1995a. On the simulation of turbulent heat transfer in a rotating duct. *AIAA Journal of Thermophysics and Heat Transfer*, 9(2), 381–382.

Dutta, S., Andrews, M.J., and Han, J.C., 1996a. On flow separation with adverse rotational buoyancy. *ASME Journal of Heat Transfer*, 118, 977–979.

Dutta, S. and Han, J.C., 1996. Local heat transfer in rotating smooth and ribbed two-pass square channels with three channel orientations. *ASME Journal of Heat Transfer*, 118, 578–584.

Dutta, S. and Han, J.C., 1997. Rotational effects on the turbine blade coolant passage heat transfer. *Annual Review of Heat Transfer*, 9, 269–314.

Dutta, S., Han, J.C., and Lee, C.P., 1994. Effect of model orientation on local heat transfer in a rotating two-pass smooth triangular duct, *1994 Winter Annual Meeting*, November 6–11, 1994, Chicago, IL, ASME HTD, Vol. 300, pp. 147–153.

Dutta, S., Han, J.C., and Lee, C.P., 1996b. Local heat transfer in a rotating two-pass ribbed triangular duct with two model orientations. *International Journal of Heat and Mass Transfer*, 39(4), 707–715.

Dutta, S., Han, J.C., and Zhang, Y.M., 1995b. Influence of rotation on heat transfer from a two-pass channel with periodically placed turbulence and secondary flow promoters. *International Journal of Rotating Machinery*, 1(2), 129–144.

Dutta, S., Han, J.C., Zhang, Y.M., and Lee, C.P., 1996c. Local heat transfer in a rotating two-pass triangular duct with smooth walls. *ASME Journal of Turbomachinery*, 118(3), 435–443, ASME Paper 94-GT-337.

Elfert, M., 1993. The effect of rotation and buoyancy on flow development in a rotating circular coolant channel, *2nd International Symposium on Engineering Turbulence Modeling and Measurements*, May 31–June 2, Florence, Italy.

Epstein, A.H., Kerrebrock, J.L., Koo, J.J., and Preiser, U.Z., 1985. Rotational effects on impingement cooling. GTL Report No. 184.

Fu, W.L., Wright, L.M., and Han, J.C., 2004. Heat transfer in two-pass rotating rectangular channels (AR = 1:2 and AR = 1:4) with 45 deg angled rib turbulators. ASME Paper No. GT 2004-53261.

Fu, W.L., Wright, L.M., and Han, J.C., 2005. Buoyancy effects on heat transfer in five different aspect-ratio rectangular channel with smooth walls and 45-degree ribbed walls. GT 2005-68493.

Glezer, B., Moon, H.K., Kerrebrock, J., Bons, J., and Guenette, G., 1998. Heat transfer in a rotating radial channel with swirling internal flow, presented at the *International Gas Turbine and Aeroengine Congress and Exhibition*, June 2–5, 1998, Stockholm, Sweden, ASME Paper 98-GT-214.

Griffith, T.S., Al-Hadhrami, L., and Han, J.C., 2002. Heat transfer in rotating rectangular channels (AR = 4) with angled ribs. *ASME Journal of Heat Transfer*, 124, 617–625.

Guidez, J., 1989. Study of the convective heat transfer in a rotating coolant channel. *ASME Journal of Turbomachinery*, 111, 43–50.

Haiyong, L., Songling, L., Hongfu, Q., and Cunliang, L., 2010. Experimental flow visualization study in a trapezoidal duct with impingement jets and cross flow near the leading edge. GT2010-23076.

Han, J.C., 1988. Heat transfer and friction characteristics in rectangular channels with rib turbulators. *ASME Journal of Heat Transfer*, 110, 321–328.

Han, J.C., Chandra, P.R., and Lau, S.C., 1988. Local heat/mass transfer distributions around sharp 180 deg turns in two-pass smooth and rib-roughened channels. *ASME Journal of Heat Transfer*, 110, 91–98.

Han, J.C., Glicksman L.R., and Rohsenow, W.M., 1978. An investigation of heat transfer and friction for rib-roughened surfaces. *International Journal of Heat and Mass Transfer*, 21, 1143–1156.

Han, J.C. and Zhang, Y.M., 1992. Effect of uneven wall temperature on local heat transfer in a rotating square channel with smooth walls and radial outward flow. *ASME Journal of Heat Transfer*, 114(4), 850–858.

Han, J.C., Zhang, Y.M., and Kalkuehler, K., 1993. Uneven wall temperature effect on local heat transfer in a rotating two-pass square channel with smooth walls. *ASME Journal of Heat Transfer*, 114(4), 850–858.

Han, J.C., Zhang, Y.M., and Lee, C.P., 1991. Augmented heat transfer in square channels with parallel, crossed, and V shaped angled ribs. *ASME Journal of Heat Transfer*, 113, 590–596.

Harasgama, S.P. and Morris, W.D., 1988. The influence of rotation on the heat transfer characteristics of circular, triangular, and square-sectioned coolant passages of gas turbine rotor blades. *ASME Journal of Turbomachinery*, 110, 44–50.

Hsieh, S.S., Chiang, M.H., and Chen, P.J., 1997. Velocity measurements and local heat transfer in a rotating ribbed two-pass square channel with uneven wall heat flux, presented at the *International Gas Turbine and Aeroengine Congress and Exhibition*, June 2–5, 1997, Orlando, FL, ASME Paper 97-GT-160.

Huh, M., Lei, J., and Han, J.C., 2010. Influence of channel orientation on heat transfer in a two-pass smooth and ribbed rectangular channel (AR = 2:1) under large rotation numbers. GT2010-22190.

Huh, M., Lei J., Liu, Y.H., and Han, J.C., 2009. High rotation number effects on heat transfer in a rectangular (AR = 2:1) two pass channel. GT2009-59421.

Huh, M., Liu, Y.H., Han, J.C., and Chopra, S., 2008. Effect of rib spacing on heat transfer in a two-pass rectangular channel (AR = 1:4) with a sharp entrance at high rotation numbers. GT2008-50311.

Hwang, G.J. and Kuo, C.R., 1997. Experimental studies and correlations of convective heat transfer in a radially rotating serpentine passage. *ASME Journal of Heat Transfer*, 119, 460–466.

Iacovides, H., Jackson, D.C., Ji, H., Kelemenis, G., Launder, B.E., and Nikas, K., 1998. LDA study of the flow development through an orthogonally rotating U-bend of strong curvature and rib roughened walls. *ASME Journal of Turbomachinery*, 120(2), 386–391.

Jeon, Y.H., Park, S.H., Kim, K.M., Lee, D.H., and Cho, H.H., 2006. Effects of bleed flow on heat/mass transfer in a rotating rib-roughened channel. ASME Paper, GT2006-91122.

Johnson, B.V., Wagner, J.H., Steuber, G.D., and Yeh, F.C., 1994a. Heat transfer in rotating serpentine passages with trips skewed to the flow. *ASME Journal of Turbomachinery*, 116(1), 113–123, ASME Paper 92-GT-191.

Johnson, B.V., Wagner, J.H., Steuber, G.D., and Yeh, F.C., 1994b. Heat transfer in rotating serpentine passages with selected model orientations for smooth or skewed trip walls. *ASME Journal of Turbomachinery*, 116, 738–744, ASME Paper 93-GT-305.

Johnston, J.P., Halleen, R.M., and Lezius, D.K., 1972. Effect of spanwise rotation on the structure of two-dimensional fully developed turbulent channel flow. *Journal of Fluid Mechanics*, 56(Part 3), 533–557.

Kim, K.M., Park, S.H., Jeon, Y.H., Lee, D.H., and Cho, H.H., 2007. Heat/mass transfer characteristics in angled ribbed channels with various bleed ratios and rotation numbers. GT2007-27166.

Kulasekharan, N. and Prasad, B.V.S.S.S., 2008. Effect of coolant entry orientation on flow and heat transfer in the trailing region channels of a gas turbine vane. GT2008-50951.

Kuo, C.R. and Hwang, G.J., 1996. Experimental studies and correlations of radially outward and inward air-flow heat transfer in a rotating square duct. *ASME Journal of Heat Transfer*, 118, 23–30.

Lee, D.H., Rhee, D.H., and Cho, H.H., 2006. Heat transfer measurements in a rotating equilateral triangular channel with various rib arrangements. ASME GT2006-90973.

Lee, E., Wright, L.M., and Han, J.C., 2003. Heat transfer in rotating rectangular channels (AR = 4:1) with V-shaped and angled rib turbulators with and without gaps. GT2003-38900.

Lezius, D.K. and Johnston, J.P., 1976. Roll-cell instabilities in rotating laminar and turbulent channel flows. *Journal of Fluid Mechanics*, 77(Part 1), 153–175.

Liou, T.M. and Chen, C.C., 1999. LDV study of developing flows through a smooth duct with a 180 deg straight-corner turn. *ASME Journal of Turbomachinery*, 121(1), 167–174.

Liu, Y.H., Huh, M., Han, J.C., and Chopra, S., 2007. Heat transfer in a two-pass rectangular channel (AR = 1:4) under high rotation numbers. GT2007-27067.

Liu, Y.H., Huh, M., Han, J.C., and Moon, H.K., 2009. High rotation number effect on heat transfer in a triangular channel with 45-deg, inverted 45-deg and 90-deg ribs. ASME Paper GT2009-59216.

Liu, Y.H., Huh, M., Rhee, D.H., Han, J.C., and Moon, H.K., 2008a. Heat transfer in leading edge triangular shaped cooling channels with angled ribs under high rotation numbers. ASME Paper GT2008-50344.

Liu, Y.H., Huh, M., Wright, L.M., and Han, J.C., 2008b. Heat transfer in trailing edge wedge shaped cooling channels with slot ejection under high rotation numbers. ASME Paper GT2008-50343.

Liu, Y.H., Wright, L.M., Fu, W.L., and Han, J.C., 2006. Rib spacing effect on heat transfer and pressure loss in a rotating two-pass rectangular channel (AR = 1:2) with 45-degree angled ribs. GT2006-90368.

Mattern, C. and Hennecke, D.K., 1996. The influence of rotation on impingement cooling, presented at the *International Gas Turbine and Aeroengine Congress and Exhibition*, June 10–13, 1996, Birmingham, U.K., ASME Paper 96-GT-161.

Metzger, D.E. and Vedula, R.P., 1987. Heat transfer in triangular channels with angled roughness ribs on two walls. *Experimental Heat Transfer*, 1, 31–44.

Mochizuki, S., Takamura, J., Yamawaki, S., and Yang, W.J., 1994. Heat transfer in serpentine flow passages with rotation. *ASME Journal of Turbomachinery*, 116, 133–140.

Morris, W.D. and Ayhan, T., 1979. Observations on the influence of rotation on heat transfer in the coolant channels of gas turbine rotor blades. *Proceedings of the Institute of Mechanical Engineers*, 193, 303–311.

Morris, W.D. and Chang, S.W., 1997. An experimental study of heat transfer in a simulated turbine blade cooling passage. *International Journal of Heat and Mass Transfer*, 40(15), 3703–3716.

Morris, W.D. and Rahmat-Abadi, K.F., 1996. Convective heat transfer in rotating ribbed tubes. *International Journal of Heat and Mass Transfer*, 39(11), 2253–2266.

Morris, W.D. and Salemi, R., 1992. An attempt to experimentally uncouple the effect of coriolis and buoyancy forces on heat transfer in smooth circular tubes which rotate in the orthogonal mode. *ASME Journal of Turbomachinery*, 114(4), 858–863.

Park, C.W. and Lau, S.C., 1998. Effect of channel orientation of local heat (mass) transfer distributions in a rotating two-pass square channel with smooth walls. *ASME Journal of Heat Transfer*, 120, 624–632.

Park, C.W., Lau, S.C., and Kukreja, R.T., 1998. Heat/mass transfer in a rotating two-pass channel with transverse ribs. *AIAA Journal of Thermophysics and Heat Transfer*, 12(1), 80–86.

Park, C.W. Yoon, C., and Lau, S.C., 2000. Heat (mass) transfer in a diagonally oriented rotating two-pass channel with rib-roughened walls. *ASME Journal of Heat Transfer*, 122, 208–211.

Parsons, J.A. and Han, J.C., 1996. Rotation effect on jet impingement heat transfer in smooth rectangular channels with heated target walls and film coolant extraction, presented at the *International Mechanical Engineering Congress and Exhibition*, November 17–22, 1996, Atlanta, GA, ASME Paper 96-WA/HT-9.

Parsons, J.A. and Han, J.C., 1998. Rotation effect on jet impingement heat transfer in smooth rectangular channels with heated target walls and radially outward cross flow. *International Journal of Heat and Mass Transfer*, 41(13), 2059–2071.

Parsons, J.A., Han, J.C., and Lee, C.P., 1998. Rotation effect on jet impingement heat transfer in smooth rectangular channels with four heated walls and radially outward crossflow. *ASME Journal of Turbomachinery*, 120, 79–85.

Parsons, J.A., Han, J.C., and Zhang, Y.M., 1994. Wall heating effect on local heat transfer in a rotating two-pass square channel with 90° rib turbulators. *International Journal of Heat and Mass Transfer*, 37(9), 1411–1420.

Parsons, J.A., Han, J.C., and Zhang, Y.M., 1995. Effects of model orientation and wall heating condition on local heat transfer in a rotating two-pass square channel with rib turbulators. *International Journal of Heat and Mass Transfer*, 38(7), 1151–1159.

Prakash, C. and Zerkle, R., 1992. Prediction of turbulent flow and heat transfer in a radially rotating square duct. *Journal of Turbomachinery*, 114, 835–846.

Rallabandi, A.P., Liu, Y.H., and Han, J.C., 2010. Heat transfer in trailing edge wedge-shaped pin-fin channels with slot ejection under high rotation numbers. ASME GT2010-22832.

Rigby, D.L. and Bunker, R.S., 2002. Heat transfer in a complex trailing edge passage for a high pressure turbine blade—Part2: Simulation results. GT2002-30213.

Sewall, E.A. and Tafti, D.K., 2005. Large eddy simulation of flow and heat transfer in the developing flow region of a rotating gas turbine blade internal cooling duct with Coriolis and buoyancy forces. GT2005-68519.

Sleiti, A.K. and Kapat, J.S., 2004. Fluid flow and heat transfer in rotating curved duct at high rotation and density ratios. GT2004-53028.

Taslim, M.E. and Lengkong, A., 1998. 45 deg. staggered rib heat transfer coefficient measurements in a square channel. *ASME Journal of Turbomachinery*, 120, 571–580.

Tse, D.G.N. and McGrath, D.B., 1995. A combined experimental/computational study of flow in turbine blade cooling passage. Part I: Experimental study, presented at the *International Gas Turbine and Aeroengine Congress and Exposition*, June 5–8, 1995, Houston, TX, ASME Paper 95-GT-355.

Tse, D.G.N. and Steuber, G.D., 1997. Flow in a rotating square serpentine cool-
    ant passage with skewed trips, presented at the *International Gas Turbine and
    Aeroengine Congress and Exhibition*, June 2–5, 1997, Orlando, FL, ASME Paper
    97-GT-529.
Wagner, J.H., Johnson, B.V., Graziani, R.A., and Yeh, F.C., 1992. Heat transfer in
    rotating serpentine passages with trips normal to the flow. *ASME Journal of
    Turbomachinery*, 114, 847–857.
Wagner, J.H., Johnson, B.V., and Kopper, F.C., 1991. Heat transfer in rotating serpen-
    . tine passages with smooth walls. *ASME Journal of Turbomachinery*, 113, 321–330.
Wright, L.M., Fu, W.L., and Han, J.C., 2004. Thermal performance of angled, V-shaped,
    and W-shaped rib turbulators in rotating rectangular cooling channels (AR =
    4:1). GT2004-54073.
Wright, L.M., Lee, E., and Han, J.C., 2003. Effect of rotation on heat transfer in narrow
    rectangular cooling channels (AR = 8:1 and 4:1) with pin-fins. GT2003-38340.
Wright, L.M., Lee, E., and Han, J.C., 2005. Influence of entrance geometry on heat
    transfer in rotating rectangular cooling channels (AR = 4:1) with angled ribs.
    ASME Journal of Heat Transfer, 127(4), 378–387.
Wright, L., Liu, Y.H., Han, J.C., and Chopra, S., 2007. Heat transfer in trailing edge
    wedge-shaped cooling channels under high rotation numbers. GT2007-27093.
Zhang, Y.M., Han, J.C., Parsons, J.A., and Lee, C.P., 1995. Surface heating effect on
    local heat transfer in a rotating two-pass square channel with 60° angled rib
    turbulators. *ASME Journal of Turbomachinery*, 117(2), 272–280, April 1995, ASME
    93-GT-336.

6

Experimental Methods

6.1 Introduction

Experimental techniques are important for obtaining measurements on heat-transfer surfaces. Cooling designs depend on experimental correlations that cover the entire range of parameters that affect that particular design. Design correlations have been developed by researchers who conducted experiments for actual components or for simplistic models. Researchers opt to use experimental methods that are most applicable for the test model and the kind of results they expect to obtain. Depending on the experimental measurement tool and the available capability, experiments can be run at low temperature or high temperature. Also, some experiments can only provide either heat-transfer coefficient measurements or film effectiveness measurements, whereas some techniques can provide both for film-cooling situations. Also, researchers have used mass-transfer measurements to obtain heat-transfer results with heat-/mass-transfer analogy. Optical, surface visualization techniques are also popular. In addition to surface heat-transfer measurements, it is also important to characterize the flow behavior. There are several flow field and thermal field measurement techniques. We have attempted to discuss each experimental methodology in this chapter so that the reader can understand how the data presented in the earlier chapters were acquired.

6.2 Heat-Transfer Measurement Techniques

6.2.1 Introduction

Actual turbine tests incur instrumentation and operational costs that are tremendous. Experimenters have simulated flow conditions similar to that in actual turbines and measured heat-transfer coefficients or wall temperatures for low-temperature experiments. The data thus obtained under low-temperature conditions are typically scaled to engine flow conditions by designers in their design codes. Typically, results are expressed in non-dimensional parameters such as Nusselt number, heat-transfer coefficient ratio, or temperature ratios.

6.2.2 Heat Flux Gages

Thin-film heat flux gages have been used by researchers for measuring heat-transfer rates on actual gas turbine blades under engine simulated conditions. Thin-film heat gages provide relatively accurate information for actual turbine rigs, which are difficult to instrument with standard thermocouples. The thin-film heat flux gages can easily be attached to a heat-transfer surface flush on the wall and connected to a data acquisition unit. The responses of these elements are very fast, of the order of 10^{-8} s. These thin-film transducers measure both dc and ac components of the heat flux. Figure 6.1 shows the thin-film gage circuitry and typical surface temperature and heat flux outputs out of a heat flux gage. The gages were used by Schultz and Jones (1973), Oldfield et al. (1978), and Doorly and Oldfield (1986) for transient turbine cascade heat-transfer measurements. Figure 6.2 shows

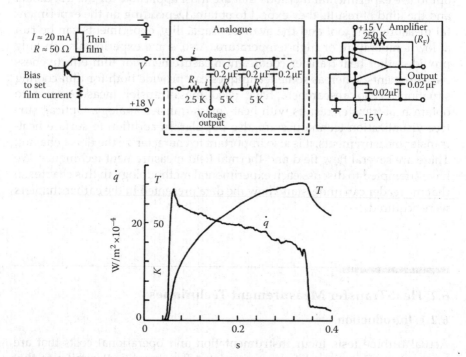

FIGURE 6.1
Early thin-film gage circuitry and typical surface temperature and heat-transfer outputs. (From Schulz, D.L. and Jones, T.V., Heat-transfer measurements in short-duration hypersonic facilities, AGARD Report AD0758590, 1973.)

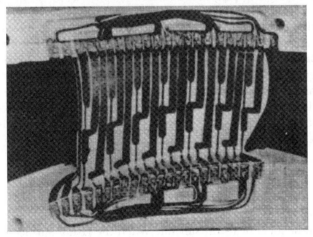

FIGURE 6.2

Photograph of a glass ceramic blade instrumented with thin-film gages. (From Oldfield, M.L.G. et al., A study of passage flow through a cascade of turbine blades using image plane holographic interferometry, AGARD CP 399, Paper 33, 1986.)

a complex three-dimensional (3-D) glass ceramic blade instrumented with thin-film gages.

The gages used by Guenette et al. (1989) consist of two thin-film nickel temperature transducers mounted on either side of a 25 μm thick polyimide insulator. The heat flux gage signals are transmitted, amplified, filtered, and

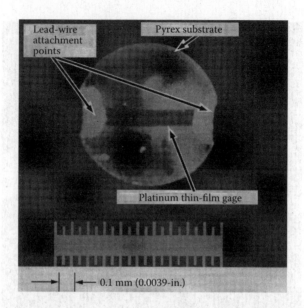

FIGURE 6.3
Photograph of a thin-film heat flux gage used by Dunn. (Dunn, M.G., Turbine heat flux measurements: Influence of slot injection on vane trailing edge heat transfer and influence of rotor on vane heat transfer, ASME Paper 84-GT-175, 1984.)

then recorded on a 12 bit A/D system at 200,000 Hz sampling rate per channel. Figure 6.3 shows a photograph of the thin-film platinum heat flux gage used by Dunn et al. (1986). Both Dunn et al. (Dunn, 1984; Dunn et al., 1986) and Guenette et al. (1989) have extensively used thin-film heat flux gages for heat-transfer measurements in turbine rigs. Ashworth et al. (1985) and Doorly and Oldfield (1985) also used thin-film heat flux gages for unsteady heat-transfer measurements. The steady-state heat-transfer results are presented in terms of the dimensionless Nusselt number as

$$Nu = \frac{\dot{q}m^c}{(T_0 - T_m)k} \qquad (6.1)$$

where $\dot{q}_m$ is the local heat-transfer rate when the rising surface temperature is T_m. Figure 6.4 shows the transient heat flux responses for several gages on a turbine blade in a cascade. The movement of the wake passage across the blade passage is captured by the heat flux responses at different axial locations on the blade surface. Abhari and Epstein (1994) measure the unsteady heat-transfer responses of heat flux gages placed on a film-cooled rotor blade. Yang and Diller (1995) also used thin-film heat flux gages to measure heat transfer inside a turbine blade tip clearance gap.

Heat flux gages can operate under realistic conditions and provide information for real engine airfoils. However, the instrumentation is extensive to

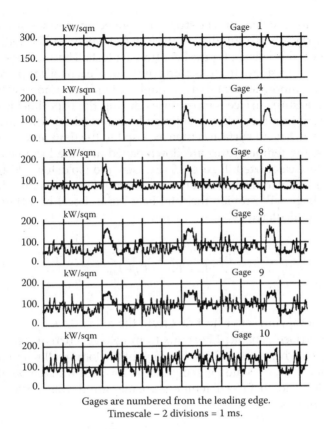

Gages are numbered from the leading edge.
Timescale – 2 divisions = 1 ms.

FIGURE 6.4
Heat transfer measured by several heat flux gages at different locations on a blade. (From Ashworth, D.A. et al., *ASME J. Eng. Gas Turbines Power*, 107, 1022, 1985.)

generate the information from the heat flux gages. Also, the information from this technique is of low resolution and can only be used as a test of the design.

6.2.3 Thin-Foil Heaters with Thermocouples

Thermocouple techniques provide direct temperature measurements on heat-transfer surfaces. Under real engine conditions, it is possible to measure actual wall temperatures and study the various effects on the metal component. However, instrumentation for actual turbine rigs can be difficult and expensive. Several cheaper, alternative tests provide reliable data for designers. Results at lower temperatures are scaled to actual conditions. Typically, experiments provide results in the form of heat-transfer coefficients and, additionally, film effectiveness values for film-cooling situations.

The test plate is typically made of low-conducting material to avoid heat losses from the noncontact side of the plate. The test surface is instrumented with thin stainless-steel foil strips. Each foil strip is relatively small in width compared to its span length. The strips are connected in series by copper bus bars. This surface, when supplied with electrical power and heated, serves as a constant heat flux surface for the heat-transfer tests. Thermocouples are soldered on the underside of the foils along the span of each foil strip. Depending on spatial resolution, several thermocouples are placed underneath the foil surface to measure surface temperatures during the experiment. Electrical power is input into the foils, providing a high-resistance, low-current heating device. The heated surface is then exposed to the flow, and on achieving steady state, the temperatures are downloaded and the local heat-transfer coefficients are calculated. Figure 6.5 (Eriksen and Goldstein, 1974) shows a typical instrumented thin-foil heater–thermocouple test section. Circuit current and voltage will provide the total heat input. Local heat-transfer coefficient can be calculated as

$$h = \frac{q''}{T_w - T_\infty} = \frac{q''_{gen} - q''_{loss}}{T_w - T_{aw}} \tag{6.2}$$

where

q'' is a net local convective heat flux from the foil surface
q''_{gen} is the surface-generated heat flux from voltage–current measurement
q''_{loss} is the local heat loss and is a function of the local wall temperature
T_w is the local steady-state foil temperature (or local wall temperature)
T_{aw} is the local adiabatic wall temperature measured when airflow is on but the foil heat is off

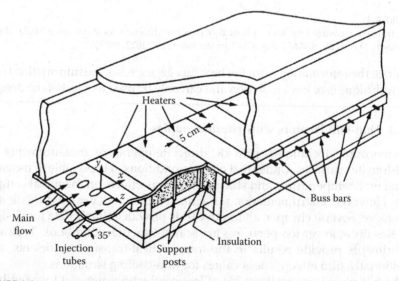

FIGURE 6.5
Heater-foil-thermocouple-instrumented flat-plate test section. (From Eriksen, V.L. and Goldstein, R.J., *ASME J. Heat Transfer*, 96, 239, 1974.)

Loss tests found total heat loss from the test blade for a no-flow condition. This method provides low-resolution, sufficiently accurate heat-transfer coefficient measurements on surfaces that can be suitably instrumented with heater foils. Leiss (1975), Han et al. (1993), and others have used this technique for surface heat-transfer coefficient measurement.

The same surface can be used for film effectiveness measurements in film-cooling situations. In film-cooling situations, the test section is instrumented downstream of the injection holes with thin-foil strips and thermocouples. The heat-transfer coefficients are measured with mainstream and coolant air at the same temperature, which is ambient temperature, and the test surface is heated. The T_{aw} term is equal to the wall temperature when both the mainstream and coolant flows are on and the surface is unheated. The film effectiveness measurements are made with the similar test section, with the mainstream at ambient temperature, coolant heated, and the surface unheated. The local wall temperature measured is now the adiabatic wall temperature, as the surface is unheated and well insulated. The coolant temperature is measured before the air is injected. The local wall temperature is a mixture temperature of the coolant and mainstream, and thus the film effectiveness is defined as

$$\eta = \frac{T_{aw} - T_\infty}{T_c - T_\infty} \tag{6.3}$$

where
T_∞ is the ambient air temperature
T_c is the coolant temperature before injection
T_{aw} is the local measured steady-state adiabatic wall temperature

Several researchers (Mick and Mayle, 1988; Mehendale and Han, 1992; Ou et al., 1994) have used this methodology. Some studies have just concentrated on measuring either heat-transfer coefficients or film effectiveness. In the case of heat-transfer coefficients, the test setup is similar to that described earlier. However, for film-cooling effectiveness measurement only, the surface need not be instrumented with thin foils. Researchers have used well-insulated test plates (made of styrofoam or urethane) with embedded thermocouples for film effectiveness measurements as shown in Figure 6.6.

Han (1988) used thin electric foil heaters and imbedded thermocouples for heat-transfer measurements in internal ribbed channels. The sketch of the test channel indicating the distributions of electric heaters is shown in Figure 6.7. Each wall of the test channel is made of nonconductive material. The nonconductive wall was a combination of a thick wood board and a thick plexiglass plate. Then, 0.0025 cm thick stainless-steel foils were cemented to the inner face of the wooden board of each side. Each wall heater was individually controlled using a variac transformer. Thermocouples are

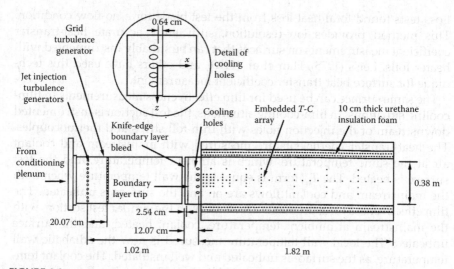

FIGURE 6.6
Turbine blade instrumented with heater foils and thermocouples. (From Ou, S. et al., *ASME J. Turbomach.*, 116, 721, 1994.)

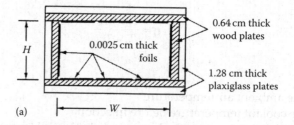

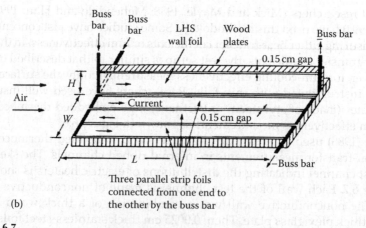

FIGURE 6.7
Test surface with heater foils and thermocouples. (a) Cross-section view and (b) heater strips shown on test plate. (From Han, J.C., *ASME J. Heat Transfer*, 110, 321, 1988.)

soldered to the underside of the stainless-steel foils at fixed locations, and the wire is pulled out through the wooden and plexiglass plates and connected to data loggers. The thermocouples will measure the wall temperature (assuming that the foil will indicate the same temperature on either side) of the channel. The local heat-transfer coefficient is then calculated using the equation

$$h = \frac{q''}{T_w - T_b} = \frac{q''_{gen} - q''_{loss}}{T_w - T_b} \qquad (6.4)$$

This technique provides low resolution and robust measurements, and is the cheapest method for performing convective heat-transfer measurements. However, test surface instrumentation is very critical. Heat losses and thermocouple accuracy are important in estimating uncertainty in measurements. Depending on test section instrumentation, uncertainty levels can vary from ±5% to ±10%.

6.2.4 Copper Plate Heaters with Thermocouples

For regionally averaged heat-transfer measurements, some researchers have used copper plates with embedded thermocouples along the test plate. The copper strips were heated with thin heaters underneath to provide a locally constant heat flux. The measured wall temperatures were considered as regionally averaged, as copper is an excellent conductor of heat. The heat-transfer coefficient was then calculated, as in the case of the thin-foil strips from the input heat flux, measured wall temperatures, and the local air temperatures. Kercher and Tabakoff (1970), Fan and Metzger (1987), Wagner et al. (1991), Han et al. (1992), and Han et al. (1992) have all used copper plates and presented regional-averaged measurements inside stationary and rotating internal cooling channels (Chapters 4 and 5). Kercher and Tabakoff (1970) used this technique to measure heat-transfer coefficients underneath an array of impinging jets. Figure 6.8 shows their impingement test facility. The test surface shows the various copper plates and the electrical power input. Each heater plate had an independently controlled power source, with heater wire supplying the current to the copper plate. There were thermocouples embedded inside these copper plates to measure temperatures of the copper plates during the experiment. The local heat flux is known from the heat supplied to the copper plates, and the wall temperature is measured with the embedded thermocouples. Since copper plate has high thermal conductivity, the heater plate temperature is already a regionally averaged value of the temperature of the entire copper plate. The Nusselt number obtained from Equation 6.4 is thus a regionally averaged value.

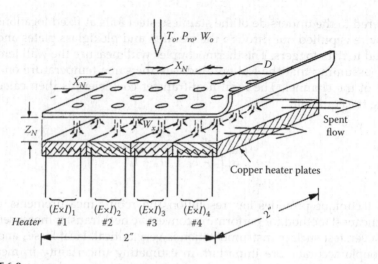

FIGURE 6.8
Copper plate heaters and thermocouple-instrumented test surface. (From Kercher, D.M. and Tabakoff, W., *ASME J. Eng. Power*, 92, 73, 1970.)

6.2.5 Transient Technique

The test surface is imbedded with thermocouples and routed into an automatic data acquisition system. The surface is then exposed to hot mainstream, and the transient responses of the individual thermocouples are stored. The transient temperature data are then converted into the local heat-transfer coefficient using the lumped-parameter equation (Lander et al., 1972). The test surface has to be a thin-walled metal to satisfy the lumped-parameter assumption. The lumped-parameter approximation is usually valid for a small value of the Biot number ($Bi = h_f l/k$). The wall temperatures are measured at several locations, and the heat-transfer coefficient is calculated from the equation

$$\frac{T_w(\text{steady state}) - T_w(t)}{T_w(\text{steady state}) - T_w(\text{initial})} = e^{-at} \quad \text{where } a = \frac{h_f}{\rho l c_p} \tag{6.5}$$

The wall temperatures are represented at different times: The initial wall temperature is before the transient, the subscript t indicates the wall temperature at any time t, and, finally, the steady state indicates the wall temperature at the end of the transient. The terms ρ and l indicate the density and thickness of the test wall. The heat-transfer coefficient h_f is determined by a curve-fitting program and is based on the temperature difference between the final and initial values of the wall (Abuaf et al., 1997).

In film-cooling situations, coolant is at a lower temperature and is injected into the boundary layer on the surface. The film effectiveness is calculated from

$$\eta = \frac{T_{rec} - T_{aw}}{T_{rec} - T_{co}} \qquad (6.6)$$

where

T_{aw} is the wall temperature measured at the end of the transient test when steady conditions are reached

T_{co} is the coolant exit temperature

The local recovery temperature T_{rec} is calculated using a typical Mach number distribution inside the test channel. If the Mach number is low, then the recovery temperature is just the measured air temperature. Leiss (1975) used this technique on a flat surface with film cooling, using single copper strips and heating the surface with ambient mainstream air. Abuaf et al. (1997) recently used this technique in a warm wind tunnel facility and made heat-transfer measurements on cylindrical test models and thin-walled airfoils with film cooling.

This technique also provides low-resolution information on the test surface. However, instrumentation and time of operation are less. Uncertainty levels can vary from ±2% to ±17% for film effectiveness measurements and averaged around ±8% for heat-transfer coefficient measurements.

6.3 Mass-Transfer Analogy Techniques

6.3.1 Introduction

Several researchers have used mass-transfer analogy experiments to measure heat-transfer coefficients. Mass-transfer analogy experiments provide relatively high-resolution results with no heat loss, axial conduction effects, as in heat-transfer experiments. Several mass-transfer methodologies have been used by researchers in experiments. We provide a brief discussion of all the techniques in this section.

6.3.2 Naphthalene Sublimation Technique

Naphthalene sublimation has been used to determine local/average mass-transfer coefficients by several researchers. The mass-transfer coefficients are determined by measuring the mass of naphthalene sublimed by either forced or natural convection. A naphthalene surface is cast in the particular test geometry, and the mass is lost by sublimation due to diffusion or convection, as in the case of heat lost in heat-transfer experiments. The analogy between mass transfer and heat transfer is directly applied by accounting for differences in properties and assuming the turbulent transport and boundary conditions to be similar in both cases. The local mass-transfer coefficient at any measurement point was determined from the rate of mass transfer per unit surface area, the local naphthalene vapor density at the measurement point, and the local bulk naphthalene vapor density:

$$h_m = \frac{\dot{m}''}{\rho_w - \rho_\infty} \tag{6.7}$$

where
   h_m is the mass transfer coefficient
   ρ_w is the local density of naphthalene at the surface
   ρ_∞ is the density of naphthalene in the mainstream

For experiments in a large wind tunnel, $\rho_\infty \sim 0$. However, for internal-cooling experiments, $\rho_\infty \sim \rho b$, which is the local bulk naphthalene vapor. The rate of mass transfer per unit surface area at the measurement point was evaluated from the density of solid naphthalene, the measured change of elevation at the measurement point ΔZ, and the duration of the test run Δt:

$$\dot{m}'' = \rho_s \frac{\Delta Z}{\Delta t} \tag{6.8}$$

Vapor pressure at the wall was calculated from the equation $\log_{10}(\rho_w) = A - B/T_w$, where A and B are constant, and T_w is the absolute wall temperature measured using thermocouples. Wall vapor density ρ_w is calculated using the perfect gas law. The local mass transfer dimensionless Sherwood number is calculated using the equation

$$Sh = h_m \frac{D}{\bar{D}} \tag{6.9}$$

where
   D is the hydraulic diameter of the channel or can be replaced with a suitable length scale for external flow conditions
   $\bar{D}$ is the binary diffusion coefficient given as $\bar{D} = \upsilon/Sc$, where υ is the kinematic viscosity of air and Sc is the Schmidt number ($Sc = 2.5$ for naphthalene)

The heat-mass transfer analogy states that the Sherwood number is analogous to the Nusselt number, $Sh \sim Nu$. Ratios of Sherwood number for similar conditions can be directly estimated as a ratio of Nusselt numbers for similar flow conditions.

Goldstein and Spores (1988) and Goldstein et al. (1995) used this method for studying turbine blade endwall heat/mass transfer. Goldstein and Chen (1991) measured turbine blade heat/mass transfer. Figure 6.9 shows the test blade indicating the naphthalene-coated surface, and also the probe traversing the airfoil surface measuring the mass transfer. Recently, Haring et al. (1995) also presented heat-/mass-transfer measurements on turbine airfoils using this technique.

Goldstein and Taylor (1982) used this technique for measuring heat-/ mass-transfer coefficients dowstream of film-cooling holes. They presented detailed measurements for a flat surface with film cooling, which was not possible with other techniques at that time. Figure 6.10 shows the flat plate with film-cooling holes and the indicated surface measurement region coated with naphthalene.

For film-cooling tests, to determine the impermeable wall effectiveness, two tests are run. In the first test, the mainstream and coolant are pure air, and the surface is coated with naphthalene. The measured wall thicknesses can be directly used to calculate the mass-transfer coefficients as described earlier. In the second test, the injectant is a naphthalene-saturated air such

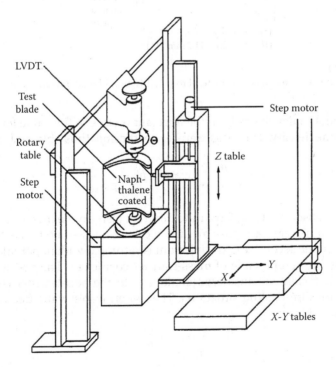

FIGURE 6.9
Naphthalene-cast airfoil and probe traversing system. (From Chen, P.H. and Goldstein, R.J., Convective transport phenomena on a turbine blade, *Proceedings of the Third International Symposium on Transport Phenomena in Thermal Control*, Taipei, Taiwan, 1988.)

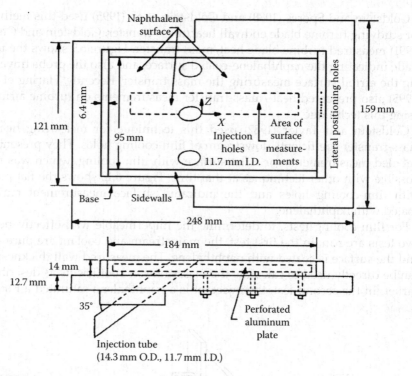

FIGURE 6.10
Naphthalene-cast film-cooled flat plate. (From Goldstein, R.J. and Taylor, J.R., *ASME J. Heat Transfer*, 104, 715, 1982.)

that the naphthalene density in the injectant is close to the value for the wall naphthalene density. The mass-transfer coefficients thus obtained are

$$h_m = \frac{\dot{m}_1''}{\rho_{v,w1} - \rho_\infty} \quad \text{and} \quad h_m' = \frac{\dot{m}_2''}{\rho_{v,w2} - \rho_\infty} \tag{6.10}$$

where the subscripts 1 and 2 indicate different tests. In the first test, the injectant and mainstream are pure air, and a certain mass-transfer coefficient is obtained. In the second test, the injectant is saturated with naphthalene, thus providing a different value of mass-transfer coefficient. The first test mass-transfer coefficient is considered analogous to the heat-transfer coefficient obtained for film-cooling situations. The impermeable wall effectiveness is then obtained from

$$\eta_{iw} = 1 - \frac{h_m'}{h_m} \tag{6.11}$$

The impermeable wall effectiveness is then converted to the isothermal wall film-cooling effectiveness as described in the next section. This technique

has been used by Cho and Goldstein (1995) and Goldstein et al. (1999) for
film-cooling measurements.

The naphthalene sublimation technique has also been used extensively
for internal cooling experiments as described earlier. Figure 6.11 shows the
sketch of the test section used by Han et al. (1988). The locations for measuring naphthalene depths are also shown in the figure. Mattern and Hennecke
(1996), Park and Lau (1998), and Acharya et al. (1998) have also used this
technique for measuring internal heat-/mass-transfer distributions for rotating channels.

Han et al. (1988) indicated that the local naphthalene vapor density at a
measurement point can be found to change by 6% for a 0.56°C variation in
the naphthalene surface temperature. It is necessary that the surface temperature at any measurement point be measured accurately. The overall uncertainty in the local Sherwood number was found to be around ±8%.

6.3.3 Foreign-Gas Concentration Sampling Technique

A mass transfer analogy is applied for measuring only film-cooling effectiveness. Instead of heating or cooling the injectant and measuring surface
temperatures, a fluid at mainstream temperature containing a foreign gas is
injected. The local impermeable wall effectiveness, η, is defined as

$$\eta = \frac{C_{iw} - C_\infty}{C_2 - C_\infty} \tag{6.12}$$

If there is no foreign gas in the mainstream, then $\eta = C_{iw}/C_2$, where C_{iw} is
the mass-fraction of foreign gas at an impermeable wall, and C_2, C_∞ are the
mass fraction of foreign gas present in the coolant fluid and the mainstream,
respectively. The analogy of mass transfer to heat transfer holds if the turbulent Lewis number and molecular Lewis number are unity. When the analogy holds, the relation between the adiabatic film-cooling effectiveness and
the impermeable wall effectiveness is given by

$$\eta_T = \frac{\eta}{(c_{p\infty}/c_{p2}) + (1 - (c_{p\infty}/c_{p2}))\eta} \tag{6.13}$$

where $c_{p\infty}$ and c_{p2} are the specific heats of the mainstream and the coolant
fluids in the heat-transfer situation. This relation is verified by stating that if
$c_{p2} = c_{p\infty}$ then $\eta_T = \eta$. Pedersen et al. (1977) and Ito et al. (1978) have described
this technique in detail. They measured the gas sampling for foreign-gas
content using gas chromatography. The air was sampled along the test wall
at discrete locations, and the film effectiveness values were estimated. This
technique provides film effectiveness only. This is a very useful technique as it

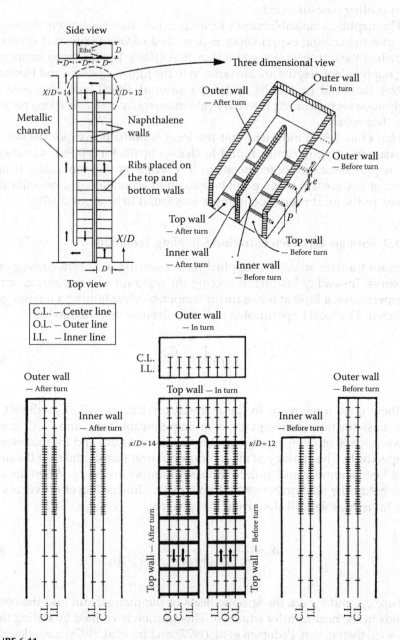

FIGURE 6.11

Sketch of two-pass test section used by Han et al. (1988) for naphthalene sublimation measurements. (From Han, J.C., *ASME J. Heat Transfer*, 110, 321, 1988.)

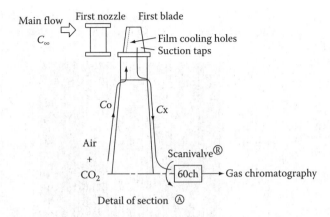

FIGURE 6.12
Schematic of test setup used by Takeishi et al. (1990) for gas chromatography. (From Takeishi, K. et al., *ASME J. Turbomach.*, 112, 488, 1990.)

can be used to simulate actual density ratios that occur inside real gas turbine engines without having to operate under high-temperature environments.

Other studies that have used this technique for film effectiveness measurements are by Goldstein and Chen (1985) and Takeishi et al. (1990). Figure 6.12 shows the schematic layout of the turbine rig used by Takeishi et al. (1990). The sampled gas is routed into the gas chromatograph for determining the cooling effectiveness. The uncertainty of the measured impermeable wall film effectiveness is about 6% for effectiveness values greater than 0.1. The uncertainty levels could be higher for lower effectiveness values. The main disadvantage of this technique is that the measurements provide only film effectiveness values.

6.3.4 Swollen-Polymer Technique

A swollen-polymer technique was presented for mass-transfer coefficient measurements by McLeod and Todd (1973). The mass-transfer test surface consists of a coating of a polymer that is swollen in a reversible manner in an organic fluid of suitable volatility. Exposure of the swollen polymer to airstreams causes the swelling agent to evaporate, resulting in a change in the polymer coating thickness. The rate of change of the polymer thickness at any location is proportional to the local mass-transfer coefficient as described earlier for the naphthalene sublimation technique. The time of exposure is controlled so that the concentration of the swelling agent in the polymer and its vapor pressure over the polymer surface are maintained constant, and lateral diffusion of the swelling agent within the polymer is insignificant. Hay et al. (1985) used laser holographic interferometry to measure the thickness changes of the polymer at various locations. This method is described in detail by Kapur and McLeod (1976). The swollen-polymer

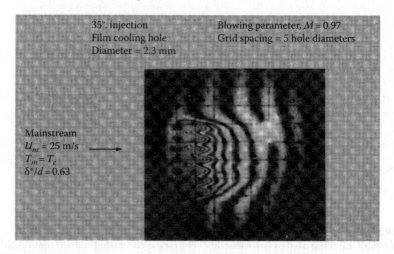

FIGURE 6.13
Typical interference patterns depicting contours of iso-mass transfer coefficients. Interference fringes depicting contours of constant mass transfer coefficient (or of analogous heat transfer coefficient) for injection through a row of holes, pitch/diameter = 3. (From Hay, N. et al., *ASME J. Eng. Gas Turbines Power*, 107, 105, 1985.)

technique analysis for determining the mass-transfer coefficients is similar to that of the naphthalene sublimation technique. The heat-mass-transfer analogy is employed for estimating the heat-transfer coefficients. The knowledge of the physical properties of the polymer and of the swelling agent, the geometry of the optical setup, and the order of the fringes at the locations are required. Figure 6.13 shows typical interference fringes depicting contours of constant mass-transfer coefficient downstream of a row of film-cooling holes. For identification of the order of all the fringes on the interferogram, it is also necessary to determine the way the order of the fringes changes in any direction. The "live" fringes are constructed, and their movements are observed in a prescribed manner described by Hay et al. (1985).

This methodology provides fairly detailed heat-transfer measurements. The uncertainty in heat-transfer coefficient measurement is estimated to be around ±3%. The number of contours in the holograms increases or decreases depending on the operating temperature and test duration. The test surface preparation is very critical in this technique in the choice of the right polymer and swelling agent and also the duration of the test.

6.3.5 Ammonia–Diazo Technique

Ammonia and diazo technique is a surface-flow visualization technique. The surface of the test piece is covered with a diazo-coated paper. When pure ammonia gas is passed over the surface, it reacts with the diazo coating, leaving a trace on the paper as it is transported over the surface. Soechting et al.

(1987) were the first to make quantitative effectiveness measurements using the ammonia and diazo technique. Friedrichs et al. (1996) were the first to calibrate the traces of the ammonia gas on diazo paper to quantitative measurements.

The measurement technique depends on the mass-transfer principle, which states that the traces on the diazo-coated surface are dependent on the surface concentration of ammonia and water vapor in the coolant gas. Prior to the experiment, the test surface is coated with the diazo film. The cooling air is then seeded with ammonia gas and water vapor. The test surface was then exposed to the coolant for 1 and 2 min. The amount of ammonia concentration and relative humidity can be varied to achieve the desired darkness traces. The image is then fixed by exposure to light to prevent further reaction. To determine quantitative data, the relationship between the darkness of the trace and surface concentration of coolant has to be determined. Calibration of the trace depends on the ammonia concentration, humidity, and exposure time. Also, an increase in temperature results in a lighter image.

A reference experiment is performed parallel to the actual experiment to avoid calibration errors. A calibration strip is produced by mixing the coolant gas mixture with free-stream air from the wind tunnel in known ratios. To quantify the darkness distribution, both the calibration strip and the exposed test surface are digitized simultaneously using an optical scanner. The analysis of the calibration strip provides a relationship between the darkness of the trace and the relative concentration of the coolant. Since only the coolant is seeded with ammonia and water vapor, the adiabatic film-cooling effectiveness can be determined from the equation

$$\eta = \frac{C_{iw}}{C_{jet}} = C_{rel} \tag{6.14}$$

where
   C_{iw} is the impermeable wall concentration
   C_{jet} is the coolant jet condition
   C_{rel} is the concentration relative to the coolant in the plenum

This technique is an excellent surface flow visualization tool and also a quantitative method for film effectiveness. It provides only film effectiveness measurements; another test with a different method needs to be run to obtain the heat-transfer coefficient measurements.

6.3.6 Pressure-Sensitive Paint Techniques

Pressure-sensitive paint (PSP) techniques are based on oxygen-quenched photoluminescence. It is a property of some compounds to emit light after being illuminated by a suitable light source. The intensity of light depends on the partial pressure of oxygen and directly relates to the pressure of the surrounding gas containing oxygen. The PSP technique uses a mass-concentration principle for measuring film effectiveness. Zhang et al. (1999) presented film effectiveness results on a turbine vane using the PSP technique. They used compressed air for the mainstream and heated nitrogen gas for the coolant. The nitrogen gas is heated to about the same temperature as the mainstream to eliminate any temperature contamination effects. The PSP is sensitive to temperature variations also. If the mainstream is air (79% N_2) and the coolant is 100% N_2, then the film effectiveness is expressed in terms of oxygen concentrations, the accuracy of the IR camera is reduced, thus making the application more difficult. It is a nonintrusive measurement technique.

Infrared thermography (IRT) can be used in several ways to obtain heat-transfer information:

1. Direct temperature measurements can be obtained by taking IR images of the hot surface. Typically, turbine vanes in high-temperature environments have been tested by several experimenters. The information thus obtained has little value in a global sense and is limited to that particular condition or component. Results in the form of heat-transfer coefficient or adiabatic wall temperature are more valuable for cooling designers.

2. IRT has been used by several researchers for adiabatic wall temperature maps. The test surface, made of low thermal conductivity insulation to reduce conduction losses, is instrumented with stainless-steel heater foils with embedded thermocouples measuring discrete location temperatures. The test surface is then painted black, with known emissivity of around 0.95. The IR camera system provides a two-dimensional (2-D) distribution of the surface temperature. The image is digitized into an array of pixel values. The thermocouples placed on the foil surface provide an in situ method of calibration for the IR signal. This helps in improving the accuracy of the system. Since conduction losses are assumed negligible, the measured surface temperatures are considered adiabatic wall temperatures. The film effectiveness is then defined as

$$\eta = \frac{T_{aw} - T_\infty}{T_c - T_\infty} \qquad (6.15)$$

Figure 6.14 shows the typical test setup for IRT (Scherer et al., 1991). For improved accuracy, IR imaging is used with a cold window and a heated/unheated surface with emissivity close to 1.0. Blair and Lander (1975) introduced this technique for film effectiveness measurements in gas turbine applications. Figure 6.15 shows the sketch of the leading-edge model installation used by them and a typical iso-effectiveness for the airfoil leading-edge configuration. Other researchers such as Gritsch et al. (1998)

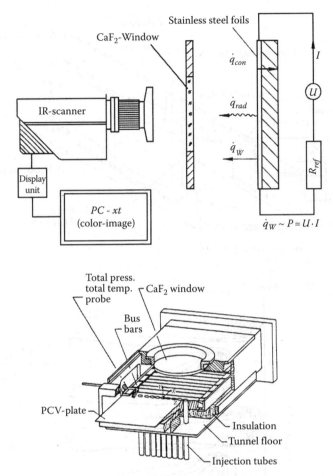

FIGURE 6.14
IRT setup used by Scherer et al. (Scherer, V. et al., Jets in a crossflow: Effects of hole spacing to diameter ratio on the spatial distribution of heat transfer, ASME Paper 91-GT-356, 1991.)

and Kohli and Bogard (1998) have since used IRT for adiabatic wall temperature measurements.

Also, Epstein et al. (1985) used an IR measurement technique for impingement-cooled channels under rotation. Bons and Kerrebrock (1998) and Glezer et al. (1998) used a similar technique for measuring heat transfer in rotating channels. Uncertainty levels can be as small as ±1% for accurate IR cameras. The high-resolution wall temperature measurements are directly obtained from the IR-sensitive camera. The technique is very useful for application in high-temperature environments. However, any optical technique requires that the test surface be visually tracked during the experiment. At high temperatures, the windows for viewing the test surface have been made of quartz or zinc-salinide, which are expensive.

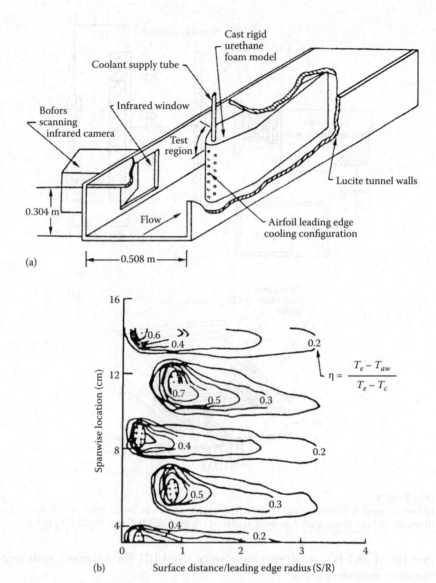

FIGURE 6.15
Leading-edge model setup used by Blair and Lander (1975) and iso-effectiveness distributions.
(a) Test section view of leading edge and (b) iso effectiveness contours on model. (From Blair,
M.F. and Lander, R.D., *ASME J. Heat Transfer*, 97, 539, 1975.)

6.3.7 Thermographic Phosphors

Noel et al. (1990) used thermographic phosphors (TP) to measure temperatures of engine components. The remote-temperature measurement
technique exploits the temperature dependence of the characteristic
decay time of the laser-induced fluorescence of the TP. Figure 6.16 shows

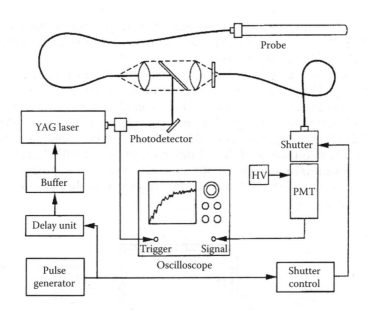

FIGURE 6.16
Experimental setup for thermographic phosphorescence method used by Noel et al. (From Noel, B.W. et al., Evaluating thermographic phosphors in an operating turbine engine, ASME Paper 90-GT-266, 1990.)

the experiment setup used by Noel et al. The YAG laser excites the TP on the turbine vanes by way of the fiber optics to the probe. The return signal is detected by the photomultiplier tube (PMT) and analyzed in the waveform-processing oscilloscope. Figure 6.17 shows the comparison between thermocouple-measured data and TP-measured data. The data

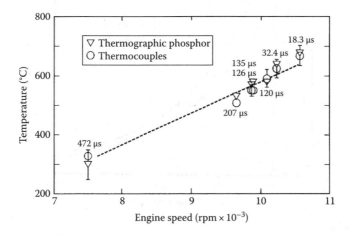

FIGURE 6.17
Comparison of TP measurement and thermocouple data. (From Noel, B.W. et al., Evaluating thermographic phosphors in an operating turbine engine, ASME Paper 90-GT-266, 1990.)

are plotted for increasing engine rpm. Their work showed the feasibility for using TP as a nonintrusive method to measure turbine component temperatures during operation.

Chyu and Bizzak (1993) developed a laser-induced fluorescence (LIF) method for 2-D surface temperature measurement. The system exploits the temperature sensitivity of both the fluorescence intensity and the lifetime of the 512 nm emission triplet of europium-doped lanthanum oxysulfide ($La_2O_2S:Eu^{3+}$). A pulsed laser is used to excite a phosphor-coated surface, and the resulting fluorescence of the 512 nm emission, along with that of the relatively temperature-independent 620 nm emission line, is acquired. The ratio of the intensities of the 512 and 620 nm emissions is then correlated with temperature.

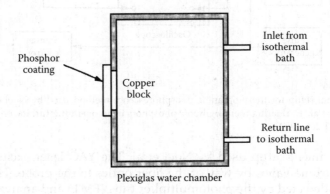

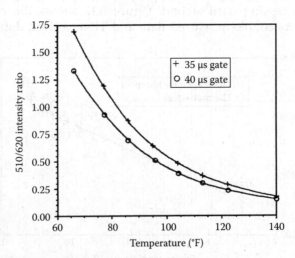

FIGURE 6.18
Phosphor calibration setup and calibration curve. (From Chyu, M.K. and Bizzak, D.J., Measurement of surface temperature using a laser-induced fluorescence thermal imaging system, ASME Paper 93-GT-214, 1993.)

The primary components of the LIF thermal imaging system are the neodymium-doped yttrium-aluminum-garnet (Nd:YAG) pulsed laser and the image-intensified CCD camera. First, a phosphor coating is applied to the surface. Care should to be taken to ensure that the coating is not thick enough to affect surface-heat transfer. During temperature measurement, the phosphor coating on the test surface is excited by the 355 nm tripled output of the laser. The image acquisition and the laser pulse are coordinated via an advance synchronization signal. More details are provided by Chyu and Bizzak (1993). A calibration curve is developed for the phosphorescence intensity versus the surface temperature. Figure 6.18 shows the calibration setup, which has an isothermal bath operated with water that keeps the phosphor coating at a uniform temperature. The intensity versus the surface temperature is drawn as a calibration curve, as shown in Figure 6.18. The LIF technique is a relatively new technique compared to the other techniques presented in this chapter. It has been used sparingly in combustion-related temperature measurements. The uncertainty in surface temperature measurement can be approximately 0.7°C. The surface needs to be coated with the phosphor, and viewability is also an issue. This technique has not yet been used significantly in heat-transfer measurements.

6.4 Liquid Crystal Thermography

Liquid crystals are referred to as thermochromic since they reflect different colors selectively when subjected to temperature changes. At any particular temperature, liquid crystals reflect a single wavelength of light. The colors can be calibrated to particular temperatures since the transition of colors is sharp and precise. Liquid crystal color range can be factory set, wherein the appearance of red, green, and blue colors is calibrated to fixed temperatures. The bandwidth from the appearance of red to the appearance of blue can also be fixed. If the bandwidth is greater than 5°C, then the liquid crystals are called wide band. However, if the bandwidth is less than 2°C, then the liquid crystals are called narrow-band liquid crystals.

6.4.1 Steady-State Yellow-Band Tracking Technique

Simonich and Moffat (1984) and Hippensteele et al. (1983) used 35 mm cameras or eyeballing to track the color bands. They instituted experiments with heater foil surfaces and liquid crystal sheets. Figure 6.19 shows the test surface instrumented by Hippensteele et al. (1983) for liquid crystal color measurement. The early technique involved setting a uniform heat flux surface and tracking the yellow band of the liquid crystal. A fixed grid was mapped on the test surface, and the local heat flux for

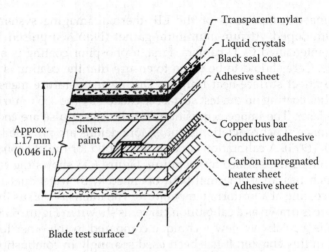

FIGURE 6.19
Cross section of the liquid crystal/heater sheet composite used by Hippensteele et al. (From Hippensteele, S.A. et al., *ASME J. Heat Transfer*, 105, 184, 1983.)

appearance of yellow band at every location was estimated. The local heat flux was estimated by subtracting conduction heat loss from the input power, and the heat-transfer coefficient at every location was determined using the equation

$$h = \frac{q''_{conv}}{T_w - T_\infty} \tag{6.16}$$

where
T_w is the yellow-band temperature
T_∞ is the mainstream temperature

This technique was popular for qualitative and quantitative surface-heat-transfer measurement before improved computer technology provided accurate image-processing systems. There are several studies that presented heat-transfer measurements inside internal channels also using this technique.

Hippensteele et al. (1983) indicate that the uncertainty in the measurement of heat-transfer coefficient using this technique can be about ±8%, or even less if the heater sheet is carefully selected such that uniform heat flux is provided and minimal heat losses due to conduction and radiation are estimated.

6.4.2 Steady-State HSI Technique

Color can be represented in terms of intensities of the three primary colors: red, green, and blue (RGB) or as hue, saturation, and intensity (HSI). A typical

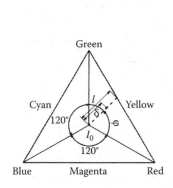

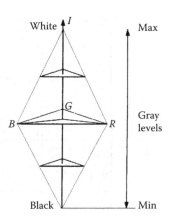

Cross-sectional view of HSI triangle model
(I axis runs perpendicular to the page through I_0)

$$S = (\delta / l) \times 255$$
$$H = (\varphi / 360) \times 255$$

3-Dimensional view shows intensity axis
$$I = (R + G + B)/3$$

FIGURE 6.20
Triangular model representation of HSI implementation system. (From Camci, C. et al., *ASME J. Turbomach.*, 114, 765, 1992.)

triangular model of the HSI representation system is shown in Figure 6.20. The color can be converted from RGB to HSI values. The hue signal of the color is typically monotonic with temperatures over the major range of liquid crystal color display. Camci et al. (1992) discuss the conversion of RGB color display to HSI display in detail. Figure 6.23 also shows the conversion equations from the triangular model. The expressions are simplified and given in simple trigonometric functions by Pritchard (1977). The intensity is the average of the three luminance values of the R, G, and B components in the color signal. The expressions are

$$I = \frac{R+G+B}{3}$$

$$S = 1 - \frac{Min(R,G,B)}{I}$$

$$H = \frac{\pi}{2} - \arcsin \left[\frac{\left(0.5 \cdot \left((R-G)+(R-B)\right)\right)}{\sqrt{\left((R-G)^2 + (R-G) \cdot (G-B)\right)}} \right] \qquad (6.17)$$

$$H = \frac{H \cdot 180}{\pi}$$

(If $H > 180$ then $H = 360 - H$)

$$H = \frac{H}{360} \cdot 255$$

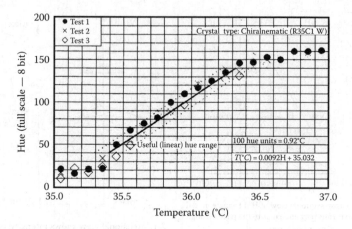

FIGURE 6.21

Hue versus temperature distributions for a typical pixel. (From Camci, C. et al., *ASME J. Turbomach.*, 114, 765, 1992.)

The range of HSI values will be from 0 to 255 (8 bit scale). The local hue value can be calibrated against the local temperature and shows a monotonic distribution. Figure 6.21 shows a typical hue versus temperature distribution for a liquid crystal band for a typical pixel (Camci et al., 1992).

The HSI steady-state technique is just an advanced method of the aforementioned technique. The color analysis for the HSI technique is performed using an image-processing system. The color range parameter hue is calibrated with temperature during a calibration experiment. The hue calibration provides a direct measurement of color versus local wall temperature. A wide-band liquid crystal sheet is used over the constant heat flux surface. Once the surface heat flux is set, a set color pattern is obtained on the surface. The image is stored and processed for local hue values at every pixel location. The local hue values are then converted to local wall temperatures from the calibration curve. This calibration curve is valid for the test surface under calibration light conditions and camera angles only. If there is any change in lighting or camera angle, the hue versus temperature curve needs to be recalibrated. Once the surface temperatures are calculated from the hue values, heat-transfer coefficients are obtained from Equation 6.16 using the known heat flux and mainstream air temperature information. This technique was used extensively by Camci et al. (1992). Figure 6.22 shows the experimental setup used by Camci et al. (1992) for heat transfer from an impinging jet. The hue versus temperature calibration curve, as shown in Figure 6.21, represents the relationship between color and temperature for all the pixel points on the test surface. Based on the hue distribution, the surface temperature distribution can be determined from the calibration curve. This technique has a strong advantage over the yellow-band tracking method. However, the technique still requires accurate estimation of heat losses and also a method to estimate the liquid crystal bandwidth before actually running

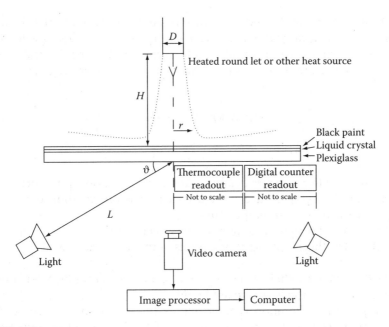

FIGURE 6.22
Experimental setup of a heated round jet used by Camci et al. (From Camci, C. et al., *ASME J. Turbomach.*, 114, 765, 1992.)

the experiment. Because of these reasons, transient liquid crystal techniques are favored and considered more accurate.

Camci et al. (1992) indicated that the overall uncertainty of convective heat-transfer coefficient is estimated to be around ±5%. However, the calibration of hue versus temperature is very sensitive to light conditions (illumination intensity and angle), camera angle, and settings. This causes very significant complications, in that an effort should be made to ensure that the camera and light conditions are similar for all tests so that the same calibration curve can be used. In the event something changes, an entire new calibration curve has to be generated. Since this is still a steady-state technique, the time of duration of each test is around 2–3 h. The results are high-resolution measurements, and an effort should be made to calculate the steady-state heat losses correctly.

6.4.3 Transient HSI Technique

The color recognition scheme for the transient HSI technique is similar to that described in the earlier section. In this case, the RGB attributes for the color are determined using three 8 bit video A/D convertors. The real-time conversions from RGB to HSI attributes are determined as described in the equations given earlier.

The test surface wall is typically made of plexiglass so that the wall can be assumed to be a semi-infinite solid. This allows for the assumption that the penetration depth of the thermal pulse is small compared to the thickness of the wall. The local wall temperature rise for an impulsively started heat-transfer experiment can be related to the time and thermophysical properties of the body, and the convective heat-transfer coefficient is given as

$$\frac{T_w - T_i}{T_\infty - T_i} = 1 - \exp(\beta^2) erfc(\beta) \quad \text{where } \beta = h\left(\frac{t}{\rho C k}\right)^{1/2} \tag{6.18}$$

where
   T_w is the wall temperature at any instant in time at every pixel location
   T_i is the initial temperature of the test surface before the transient
   T_∞ is the oncoming air temperature
   t is the time at which the temperature is a certain T_w
   ($\rho C k$) are the thermophysical properties of the test surface wall

The heat-transfer coefficient at every pixel location can thus be determined if all other values in the equation are known.

However, the calibration is intensive as the entire surface temperature distribution needs to be captured at any particular instant of time. Camci et al. (1993) came upon a solution wherein they mixed three narrow-band liquid crystals and sprayed them on the surface. Each liquid crystal has a bandwidth of 1°C. Figure 6.23 shows the variation of hue of color with respect to temperature for all three liquid crystals. Figure 6.23 shows the variation of local intensity with respect to temperature for all three liquid crystals. During a transient experiment, the wall temperature is determined by matching both the local intensity and hue to the correct scale and then determining the actual liquid crystal range that will eventually determine the actual wall temperature (T_w). Then based on the transient time and other known parameters, the local heat-transfer coefficient h can thus be determined from Equation 6.18.

Several advances have been made in this technique by Wang et al. (1993, 1994). The methodology of obtaining data using image-processing systems is well explained in both the aforementioned papers. Image-processing tools are widely available for obtaining color information. These techniques thus help determine more accurate surface heat-transfer information for complicated geometries. Wang et al. (1990) have also attempted to use this technique for measuring convective heat transfer over rough surfaces. Van Treuren et al. (1994) used the same methodology described by Wang et al. (1994) to determine both the convective heat-transfer coefficient and the adiabatic wall temperatures at all locations for impinging jet heat transfer (Chapter 4, on Impingement Heat Transfer).

The overall uncertainty in heat-transfer coefficient measurements with this technique is estimated in the range between ±5% and 6%. Calibration curve problems still exist with light and camera conditions. However, test times are

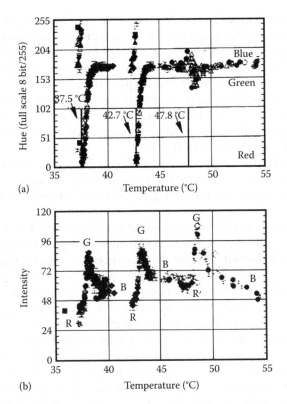

(a)

(b)

FIGURE 6.23
Hue and intensity calibration with respect to temperature for a mixture of three liquid crystals. (From Camci, C. et al., *ASME J. Heat Transfer*, 115, 311, 1993.)

less than a minute, and heat losses are not considered an issue. Wang et al. (1993, 1994) indicated that they found no dependency of the color versus temperature curve with lighting and camera angles. They showed repeatability of the curves for different light and camera conditions. However, they used a full-intensity history and a scaling analysis to convert color signals to temperature measurements.

6.4.4 Transient Single-Color Capturing Technique

This technique measures the time of appearance of a particular color band at every pixel location on the test surface during the transient experiment, compared to the HSI technique wherein the time is kept constant and the wall temperature variations are measured from the calibration curve. This technique measures the transient time for any pixel location to change to a known prescribed temperature. The liquid crystal band typically changes from colorless, to red, to green, and to blue. It is easy to use an image-processing system to determine the color change temperature for any transition.

The appearance of green, when the color changes from red to green, is also the disappearance of the red color. The image-processing system can be programmed to look for the appearance of green (or disappearance of red) during the transient test and measure the time since the start of the transient test. In this test, the locations that change color at about the same instant of time, in general, have the same heat-transfer coefficients. The calibration is not as rigorous as the HSI technique. The calibration helps determine the chosen transition color band region and helps obtain the temperature under the laboratory experiment conditions.

A test is suddenly initiated by diverting a heated flow onto the liquid-crystal-coated test surface. Typically, the test surface is made of low conductivity, thermal diffusivity material (e.g., plexiglass). At any surface point, the wall temperature is represented by the solution for the semi-infinite solid, 1 D transient conduction problem as in Equation 6.18. The wall temperature (T_w) here is the known prescribed color change temperature. Bunker and Metzger (1990) used the appearance of green as the prescribed temperature in their impingement-cooling experiments. According to the authors, the red-to-green transition of the liquid crystal represents a strong intensity that makes it more accurate to track. The time of color change (t) varies for each location depending on the local heat-transfer coefficient. The measurement of required times to reach the prescribed liquid crystal color (or temperature) is the essence of the experiment. The equation assumes that the oncoming mainstream imposes a step change in air temperature on the surface. However, in actual practice, the wall surface does not experience a pure step change in air temperature because of the transient heating of upstream regions. Therefore, Equation 6.18 is changed to represent the response of this changing mainstream temperature. The measured temperature response is incorporated as a superposed set of elemental steps in T_∞, arranged to represent the actual rise in temperature. Metzger and Larson (1986) show that the wall temperature response can be represented as

$$T_w - T_i = \sum_{j=1}^{N} U(t - \tau_j)\Delta T_\infty \qquad (6.19)$$

where

$$U(t - \tau_j) = 1 - \exp\left[\frac{h^2}{k^2}\alpha(t - \tau_j)\right] erfc\left[\frac{h}{k}\sqrt{\alpha(t - \tau_j)}\right]$$

The superposition can be imposed by measuring the T_∞ variations with time (τ). The step changes in τ and ΔT_∞ are then input into the wall temperature solution. The heat-transfer coefficient (h) can then be determined from the aforementioned equations. This technique has been significantly used

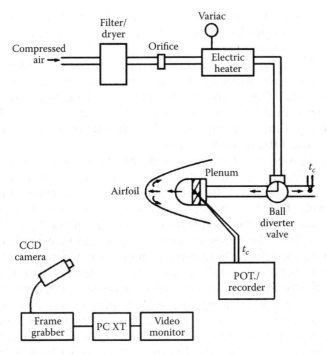

FIGURE 6.24
Test setup used by Bunker and Metzger. (From Bunker, R.S. and Metzger, D.E., *ASME J. Turbomach.*, 112, 451, 1990.)

by several researchers for measuring heat-transfer coefficients over external flow surfaces like flat plates, internal turbulated channels (Ekkad and Han, 1997; Ekkad et al., 1998b), and turbine blade tip heat-transfer studies (Kim et al., 1995). Recently, some researchers have started using mesh heaters that provide a true step change in mainstream temperature. That will eliminate the need for using superposition as described earlier (Chen et al., 1998); Figure 6.24 shows the test setup used by Bunker and Metzger (1990) for a leading-edge impingement heat-transfer study.

Metzger and Larson (1986) used phase change paints instead of liquid crystals. Phase change paints sublimate from solid coating to no coating during a transient test. Metzger and Larson (1986) painted a surface with these melting-point surface phase change paints and determined when each point on the surface changes phase during the transient test. The temperature of the phase change is factory set. The same theory of semi-infinite solid model was assumed, and the mainstream temperature was represented as a series of time steps to determine the local heat-transfer coefficients.

The liquid crystal technique can also be used in reverse mode wherein the surface is preheated and then suddenly cooled transiently. In that case, the surface is heated to a temperature above the liquid crystal color range and suddenly exposed to a cooler mainstream at constant temperature. The color change times at every pixel are measured, and the local heat-transfer coefficient is calculated from Equation 6.18. The need for mainstream temperature superposition is not required as in the case of the hot-mainstream problem. The only difficulty in this technique is to preheat the test surface uniformly so that the entire test surface is at a uniform initial temperature before the transient. Du et al. (1997) used this technique by preheating a turbine blade and measuring heat-transfer coefficients on the blade surface. The results from their experiments compared with earlier thermocouple data for the same blade under the same conditions.

Vedula and Metzger (1991) enhanced the above single color capturing technique to determine both heat-transfer coefficient and film effectiveness for film-cooling situations. In three temperature convection situations, as in film cooling, both heat transfer and adiabatic aspects of the problem can be established by any two points on the line determined by measurement of local quantities. The aforementioned method can be extended to three-temperature situations by determining the local film temperature and the heat-transfer coefficient simultaneously. In three-temperature film-cooling situations, the heat-transfer surface comes in contact with a mixture of the mainstream and coolant temperature. This mixture quality varies from immediately downstream of coolant injection all the way to a location far downstream of injection. Figure 6.25 shows an illustration of the film-cooling problem. To determine the actual heat-transfer coefficient, it is imperative

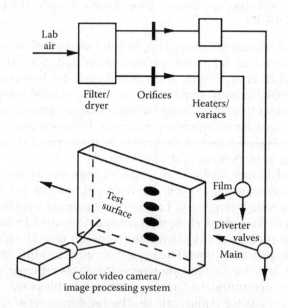

FIGURE 6.25
Illustration of a typical film-cooling problem. (From Vedula, R.J. and Metzger, D.E., A method for the simultaneous determination of local effectiveness and heat transfer distributions in three-temperature convection situations, ASME Paper 91-GT-345, 1991.)

to know the actual film temperature. For this reason, Vedula and Metzger (1991) determined two sets of equations that completely define the problem. They suggested that the heat-transfer coefficient and T_f can be determined by solving two equations of the form of Equation 6.18. These two equations can be obtained from a single transient test with two surface temperature indications at different times during the transient test or from two separate related transient tests.

Vedula and Metzger (1991) provide an example wherein two color change times are obtained from a single transient test at every location. The transient test has two unknowns (T_f and h). If during the transient the liquid crystal coating indicates one surface temperature (T_{w1}) at time t_1 corresponding to some color change and displays another surface temperature (T_{w2}) at time t_2, then h and T_f can be determined from the simultaneous solution of the two equations:

$$\frac{T_{w1} - T_i}{T_f - T_i} = 1 - \exp\left(\frac{h^2 \alpha t_1}{k^2}\right) erfc\left(\frac{h\sqrt{\alpha t_1}}{k}\right)$$

$$\frac{T_{w2} - T_i}{T_f - T_i} = 1 - \exp\left(\frac{h^2 \alpha t_2}{k^2}\right) erfc\left(\frac{h\sqrt{\alpha t_2}}{k}\right)$$

$$(6.20)$$

Alternatively, two separate tests, 1 and 2 with the same flows but different temperatures, can be conducted with both the tests measuring the indication of the same color change temperature (T_w). In that case, the film temperature will be different for both tests. Vedula and Metzger (1991) ran two tests, with one that had the mainstream and coolant running at about the same temperatures and another test that had the mainstream hot and the coolant at room temperature. Since they did not have a true step change in air temperatures for both mainstream and coolant, they used the superposition principle and introduced time-step integration as described earlier. For this they defined film effectiveness (η) as

$$\eta = \frac{T_f - T_c}{T_\infty - T_c} \qquad (6.21)$$

where T_∞ and T_c are the mainstream and coolant temperatures upstream of injection. This is based on the assumption that the T_f is the same as the adiabatic wall temperature as there is no heat transfer into the wall during the transient test. Using the definition of film effectiveness, Equation 6.19 is written as

$$T_w - T_i = \sum_{j=1}^{N} U(t - \tau_j) \Delta T_f \qquad (6.22)$$

where

$$U(t - \tau_j) = 1 - \exp\left[\frac{h^2}{k^2}\alpha(t - \tau_j)\right]erfc\left[\frac{h}{k}\sqrt{\alpha(t - \tau_j)}\right]$$

Since T_f can be written in the form of film effectiveness, mainstream temperature, and coolant temperature, the aforementioned equation becomes

$$T_w - T_i = \sum_{j=1}^{N} U(t - \tau_j)\left[(1 - \eta)\Delta T_\infty + \eta\Delta T_c\right] \tag{6.23}$$

To obtain both unknowns, h and η, two similar related transient tests are run. Ekkad et al. (1997) used the same technique for measuring both heat-transfer coefficients and film effectiveness over a flat surface with compound-angle film injection.

To avoid using superposition of the varying mainstream, several researchers have chosen to use a preheated test surface and suddenly impose a cool mainstream for the transient. This technique was used by Ekkad et al. (1998a) and Du et al. (1998) to measure both the heat-transfer coefficient and the film effectiveness simultaneously. Figure 6.26 shows the experimental setup and liquid-crystal-coated blade used by Du et al. (1998). They preheated a turbine blade model using an enclosure heater. They used four camera locations to map the entire blade surface. They controlled heater inputs at various locations to obtain a fairly uniform initial surface temperature. They ran two separate tests to obtain both heat-transfer coefficient and film effectiveness. They used the semi-infinite solid assumption solution as shown by Equation 6.18. They replaced the local air temperature with the local film temperature. In the first test, they ran both the mainstream and the coolant at room temperature and suddenly exposed the heated surface to the mainstream and coolant. Since the coolant and mainstream are at the same temperature, the film temperature is assumed to be the same as the coolant or mainstream temperature, in which case the heat-transfer coefficient can be determined from the first test. In the second test, the mainstream is kept cool, and the coolant is heated to about the initial temperature of the test surface. The times of color change are measured, and the known heat-transfer coefficient from the first test is used to determine the film effectiveness from the equation

$$\eta = \frac{T_w - T_{i2}}{T_{c2} - T_{\infty 2}}\frac{1}{F(h)} + \frac{T_{i2} - T_{\infty 2}}{T_{c2} - T_{\infty 2}} \tag{6.24}$$

where

$$F(h) = 1 - \exp\left(\frac{h^2\alpha t_2}{k^2}\right)erfc\left(\frac{h\sqrt{\alpha t_2}}{k}\right)$$

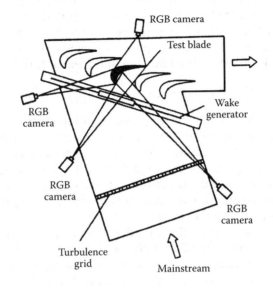

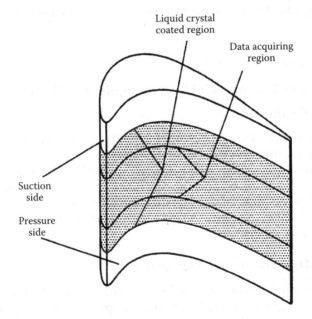

FIGURE 6.26
Test setup and liquid-crystal-coated blade used by Du et al. (From Du, H. et al., *ASME J. Turbomach.*, 120, 808, 1998.)

The subscript 2 indicates the second test. The mainstream and coolant flow conditions are identical for the two tests to obtain consistent results.

Yu and Chyu (1998) provided yet another complication to the aforementioned technique wherein there are two rows of injection at different temperature. In that case, the film effectiveness contribution of each hole injection is required. They determined that the film temperature T_f can be written as

$$T_f = T_{f1} + T_{f2} - T_\infty = (1 - \eta_2 - \eta_2)T_\infty + \eta_1 T_{c1} + \eta_2 T_{c2} \tag{6.25}$$

Liquid crystal techniques have gained popularity with researchers due to the capability of application to complex geometries. Another advantage of the liquid crystal technique has been the tremendous improvements in image-processing technology. Liquid crystal color change measurements can be easily obtained using commercial software and hardware. However, the major disadvantage has been the limitation of liquid crystals only in experiments at low temperatures.

This technique avoids calibrating the entire liquid crystal band and tracks only a single event in the liquid crystal display band. This helps avoid the problems in the earlier techniques of camera and light angles. The uncertainty in heat-transfer coefficient measurements is indicated as ±6%–8% and effectiveness measurements as ±9%–12%. The accuracy in this technique strongly depends on the accuracy of the time of color change at each pixel.

6.5 Flow and Thermal Field Measurement Techniques

6.5.1 Introduction

In addition to heat-transfer measurement, flow field velocity and thermal measurements enhance the understanding of heat-transfer phenomena. Since all the earlier sections in this chapter typically dealt with surface heat-transfer characteristics, it is also important to understand the flow field and thermal field away from the surface that cause the surface heat-transfer characteristics. Flow field and thermal field measurements are typically obtained using several techniques.

6.5.2 Five-Hole Probe/Thermocouples

Miniature five-hole probes are used in measuring three mean velocity components at a single location. A schematic of the miniature five-hole probe is shown in Figure 6.27. The diameter of the probe tip is 1.22 mm, where the central tube is surrounded by four other tubes. The end of each tube is tapered at a 45° angle with respect to the mouth of the central tube. The probe is calibrated in a wind tunnel before actual application. A traversing mechanism is used to place the probe at the required location for measurements.

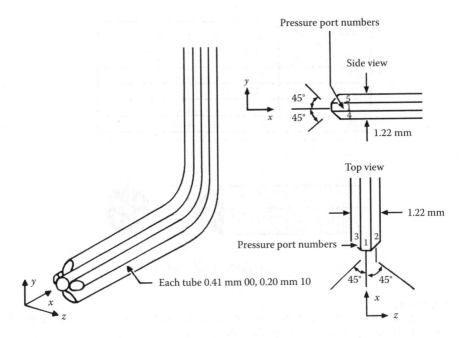

FIGURE 6.27
Schematic diagram of the miniature five-hole probe. (From Ligrani, P.M. et al., *J. Phys. Eng.: Sci. Instrum.*, 22, 868, 1989.)

Measured pressures from each probe port are stored through pressure transducers into the computer at each traverse location. Ligrani et al. (1989) provided a detailed description of the usage of five-hole probes. From calibration, the yaw, pitch, total, and total minus static pressure coefficients are calculated. The total velocity magnitude V is determined using

$$V = \left[\frac{2C_{pts}(P_1 - \bar{P})}{\rho}\right]^{1/2} \tag{6.26}$$

where C_{pts} is the total minus static pressure coefficient. The total pressure relative to atmospheric pressure P_0 at each location is determined using $P_0 - P_t = (P_0 - \bar{P}) - C_{pts}(P_1 - \bar{P})$. The three velocity components are subsequently given by

$$V_\theta = V\cos y \cos p$$

$$V_r = V \sin y \tag{6.27}$$

$$V_z = V \cos y \sin p$$

where y and p are the yaw and pitch angles of the probe.

Probe Pressure Port Perturbation	$\alpha = 0°/\beta = 0°$ RMS Error	$\alpha = 20°/\beta = -20°$ RMS Error
±0%	±0°	±0°
±1%/4%	±0.5°/± 0.5°	±1.0°/± 0.7°
±1%/2%	±1°/± 1°	±2°/± 1.5°
±1%	±2°/± 2°	±7.5°/± 16°

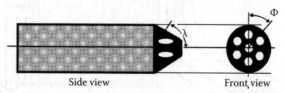

Side view   Front view

Port numbering convention :
$$\begin{matrix} 6 & 1 & 2 \\ & 7 & \\ 5 & 4 & 3 \end{matrix}$$

FIGURE 6.28
General schematic of the seven-hole probe. (From Takahashi, T.T., Measurement of air flow characteristics using seven-hole cone probes, AIAA Paper 97-0600, 1997.)

Takahashi (1997) used a seven-hole cone probe for 3-D mean velocity distributions including high flow angularity measurements. Figure 6.28 shows the general schematic of the seven-hole probe. Takahashi developed a simple, very efficient algorithm to compute flow properties from the measured probe tip pressures. Multihole probes have been used by fluid dynamicists for a long time. Prandtl and Tietjens (1934) were the earliest to use a multihole probe.

6.5.3 Hot-Wire/Cold-Wire Anemometry

Hot-wire anemometry systems are an effective tool for measuring fluctuating velocity and temperature components in nonisothermal boundary layer flows. The technique for using multiple wire overheats for separating concomitant velocity and temperature fluctuations was developed by Corrsin (1947). Blair and Bennett (1984) describe the development of an instrumentation system designed for the simultaneous measurement of two components of instantaneous velocity (u and v) and temperature (T) in low-speed boundary layers. They provide a detailed description on the design of hot-wire system instrumentation and selection. They designed a three-sensor probe for simultaneous measurement of u, v, and T. The three-sensor probe consisted of a vertical X wire, with a third wire mounted equidistant between the wires. The third wire was placed parallel to one of the wires in the X wire. Figure 6.29 shows the triple-wire

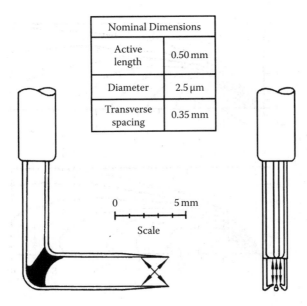

Nominal Dimensions	
Active length	0.50 mm
Diameter	2.5 µm
Transverse spacing	0.35 mm

FIGURE 6.29
Diagram of the triple-wire probe and sensor configuration. (From Blair, M.F. and Bennett, J.C., Hot wire measurements of velocity and temperature fluctuations in a heated turbulent boundary layer, ASME Paper 84-GT-M 234, 1984.)

probe with sensor configuration used by Blair and Bennett (1984). Hot wires (single, X, and triple) have been used to measure fluctuating velocity and temperature components by several investigators. It is common to use hot wires for measuring turbulence intensity inside wind tunnels. Kim and Simon (1987) and Zhou and Wang (1995) have also used hot-wire probes similar to the one shown earlier for measuring free-stream turbulence and thermal structures in boundary layers. The temperature probe is also referred to as a cold wire, as it measures the temperature component.

Thole and Bogard (1994) and Kohli and Bogard (1998) provided temperature measurements only using the cold-wire probe inside flows that are highly turbulent, such as film cooling. It provides detailed temperature measurements in mixing flows and can resolve mean and fluctuating components of the temperature field. The near-wall temperature distributions can be used to estimate film effectiveness for an adiabatic test surface. Figure 6.30 shows the mean temperature contours along a film hole center-line as presented by Kohli and Bogard (1998).

6.5.4 Laser Doppler Velocimetry

The laser doppler velocimetry (LDV) method is a high-resolution nonintrusive technique. Full-field flow measurements can be obtained at any plane

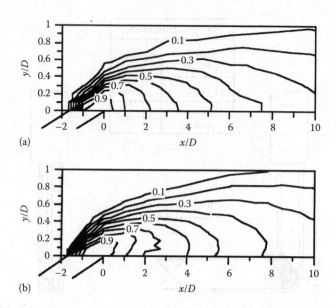

FIGURE 6.30

Mean temperature contours measured by Kohli and Bogard (1998) downstream of film holes. (a) Mean temperature contours (Θ) on jet centerline; $M = 0.4$, $DR = 1.05$, $I = 0.156$. (b) Mean temperature contours (Θ) on jet centerline from Schmidt et al.; $M = 0.5$, $DR = 1.6$, $I = 0.16$. (From Kohli, A. and Bogard, D.G., *ASME J. Turbomach.*, 120, 86, 1998.)

of the flow. It can provide all turbulence quantities and Reynolds stresses. The technique is applicable for film-cooling flows and can provide detailed flow structures of the jet coming out of the hole. Pietrzyk et al. (1990) provided hydrodynamic measurements for film cooling on a flat plate using a two-component LDV system. They provided near-field mean velocity vectors, turbulent shear stress contours, and turbulent quantities from their measurements. The LDV system is described in detail by Pietrzyk et al. (1990). Thole et al. (1998) recently provided detailed flow measurements for film injection through several hole shapes. They showed the reduced turbulence generated by shaping film holes, which eventually leads to lower heat-transfer coefficients and higher film effectiveness. Figure 6.31 shows the mean velocity measurements for three different film hole geometries made by Thole et al. (1998).

Rivir et al. (1987) used laser sheet light for visualization of film-cooling flows. Illumination times of 10 ns were used for the still photographs. LDA measurement systems have also been used for flow field measurements inside rotating channels. Cheah et al. (1994) provided velocity, turbulence intensity, and Reynolds shear stress measurements inside stationary and rotating smooth channels with a U bend. Iocovides et al. (1996) provided similar results inside rotating channels with a U bend and rib-roughened walls. Figure 6.32 shows the comparison of rotational effect on mean velocity

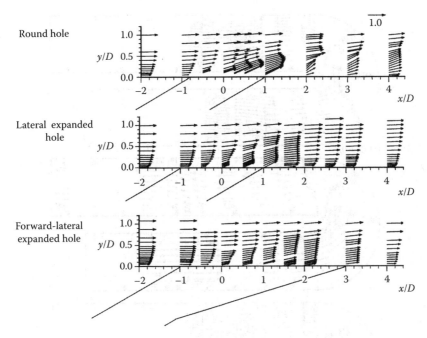

Round hole

Lateral expanded hole

Forward-lateral expanded hole

FIGURE 6.31
Mean velocity vectors for round, laterally expanded, and forward-laterally expanded holes. (From Thole, K. et al., *ASME J. Turbomach.*, 120, 327, 1998.)

fields for the ribbed U-bend channel. Results clearly show the effect of the Coriolis component on the velocity profiles along the length of the channel. Tse and Steuber (1997) provided velocity data in rotating coolant passages with skewed ribs using an LDA system.

6.5.5 Particle Image Velocimetry

In this technique, a laser sheet of the seeded flow is captured in an image. The pulsed laser of short duration is optically spread into a sheet. Two laser pulses separated by a certain time difference depending on the flow speed are generated. The CCD camera takes a double-exposure image. The particle image displacement is determined from the autocorrelation of the image intensity distribution. The technique has been used for flow measurements in internal rib-turbulated channels by Schabaker et al. (1998). This technique has difficulty in resolving highly 3-D flows. Figure 6.33 shows the PIV setup used by Schabaker et al. (1998). They measured the flow field in the turn region of a two-pass channel with 180° turn. Recently, Son et al. (1999) presented PIV measurements in the turn region of a two-pass channel. Figure 6.34 shows the mean flow field around the turn region and the streamwise turbulence intensity distributions.

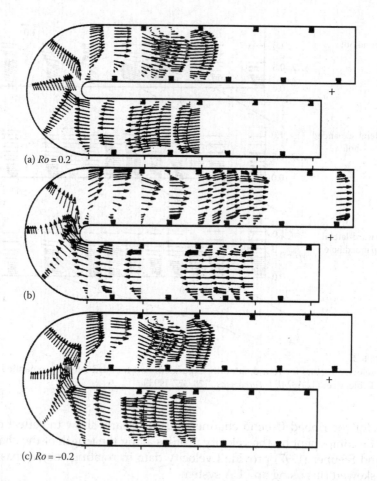

FIGURE 6.32
Comparison of rotational effect on mean velocity fields for ribbed U bend. (a) Positive rotation ($Ro = 0.2$), (b) stationary case, and (c) negative rotation ($Ro = -0.2$). (From Iacovides, H. et al., *ASME J. Turbomach.*, 120, 386, 1998.)

Treml and Lawless (1998) used a PIV system to measure the 2-D velocity field in a turbine passage with vane–rotor interactions. Also, Bons and Kerrebrock (1998) provided velocity measurements inside a rotating cooling passage with smooth walls using the PIV system. They used a stationary camera and captured images from a rotating frame. The image-acquisition system is synchronized so that the laser fires and the camera forms an image only when the rotating test section is located in the camera's field of view. Gogineni et al. (1996) made PIV measurements on periodically forced flat-plate film cooling. However, highly 3-D flow measurements could have significant errors because the out-of-plane motion cannot be corrected completely for PIV systems.

The PIV system

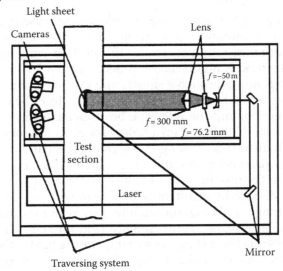

FIGURE 6.33
PIV setup used by Schabaker et al. (From Schabaker, J. et al., PIV investigation of the flow characteristics in an internal coolant passage with two ducts connected by a sharp 180° bend, ASME Paper 98-GT-544, 1998.)

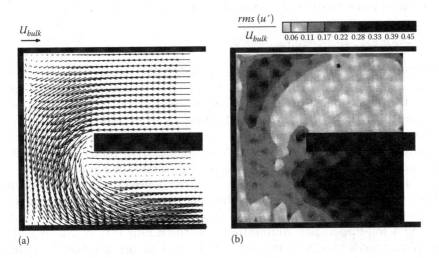

(a)   (b)

FIGURE 6.34
Detailed mean flow field and streamwise turbulent intensity by Son et al. (1999) using PIV. (a) PIV result of mean flow field ($Re = 30,000$). (b) PIV result of streamwise turbulence intensity. (From Son, S. et al., *ASME J. Heat Transfer*, 121 509, 1999.)

6.5.6 Laser Holographic Interferometry

A laser holographic interferometry technique was used by Liou and Hwang (1992) for temperature measurements inside rib-roughened channels. The temperature measurements are for the local air temperature in the channel. The method is nonintrusive. A beam from a 3 W argon-ion laser was split after passing through a shutter and expanded into two beams of 150 mm diameter with plane wave fronts. The object beam passes through the test section, and the reference beam bypasses it. The two expanded beams are then interfered on a hologram plate. The phase hologram represented a diffraction grating in which the comparison beam was stored. The comparison beam described the state in which the ribbed walls were not heated. Because the ribs were heated, the expanded object beam gets distorted due to the resultant refractive index field. The distorted beam passes through the hologram and interferes with the comparison reference beam. The instantaneous interference field formed behind the hologram plate represents the instantaneous temperature field of the channel. Figure 6.35 shows some examples of the holographic interferograms in periodically fully developed flows in ribbed channels. Each fringe shift represents a typical temperature difference. Liou and Hwang (1992) provide more details in their paper. This technique is not very popular. However, Hay et al. (1985) have used a similar technique to measure surface mass transfer as described earlier in the swollen-polymer technique. Oldfield et al. (1986) presented some holographic interferometry results inside a passage flow through a cascade of turbine blades. Figure 6.36 shows the interference patterns on the endwalls of a turbine blade cascade.

6.5.7 Surface Visualization

Surface visualization tools are valuable in helping visualize the fluid flow near a solid surface. Flow visualization can be a very effective and simple tool for providing a great deal of information. It can also be used concurrently with other quantitative techniques to provide a better understanding of the problem. Some of the techniques presented earlier can easily double as both qualitative and quantitative tools. For example, liquid crystal techniques can provide both qualitative and quantitative measurements. Langston and Boyle (1982) presented a surface-streamline flow visualization technique. In this technique, a matrix of ink dots is marked around the test surface to act as flow tracers. The dots are made using a felt-tipped pen containing permanent blue ink. A drafting film is placed underneath an ink dot matrix. Just before the flow is turned on, the dotted area is sprayed with oil of wintergreen so that the entire dotted area is covered with a thin but continuous liquid film. The matrix of dots becomes blurred as the ink dots dissolve and diffuse into the film of oil. Once the air is turned on, the oil of wintergreen film is then

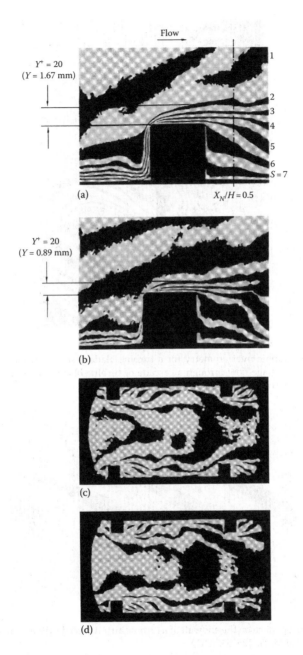

(a)

(b)

(c)

(d)

FIGURE 6.35
Examples of holographic interferograms of periodically fully developed flows. (a) $Re = 6,400$, $Pi/H = 10$, $H/De = 0.081$ (×6 magnification of near rib region). (b) $Re = 10,200$, $Pi/H = 10$, $H/De = 0.081$ (×6 magnification of near rib region). (c) $Re = 8,300$, $Pi/H = 10$, $H/De = 0.081$, 0 s (×1 full field). (d) $Re = 8,300$, $Pi/H = 10$, $H/De = 0.081$, 0.033 s (×1 full field). (From Liou, T.M. and Hwang, J.J., *ASME J. Heat Transfer*, 114, 56, 1992.)

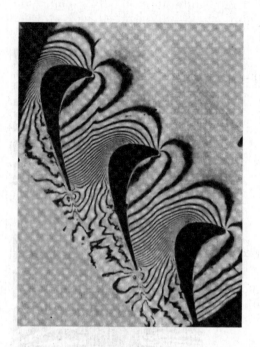

FIGURE 6.36
Example of holographic interferometry for a turbine blade cascade. (From Oldfield, M.L.G. et al., A study of passage flow through a cascade of turbine blades using image plane holographic interferometry, AGARD CP 399, Paper 33, 1986.)

FIGURE 6.37
Ink-dot streakline pattern on the endwall of a turbine airfoil cascade. (From Langston, L.S. and Boyle, M.T., *J. Fluid Mech.*, 125, 53, 1982.)

acted upon by wall shear forces and flows in response to them. The ink then acts as a tracer, with each dot producing a streakline. The oil then evaporates leaving permanent ink traces of the surface streaklines. This technique is a very popular method for surface flow visualization and has been extensively used by researchers. Figure 6.37 shows the ink-dot

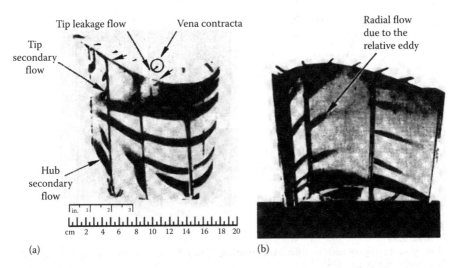

FIGURE 6.38
Flow visualization on a turbine rotor blade using the ammonia/ozalid paper. (a) Suction surface and tip and (b) pressure surface and tip. (From Dring, R.P. and Joslyn, H.D., *ASME J. Eng. Power*, 103(2), 400, 1981.)

streakline pattern on the endwall of a turbine airfoil cascade as obtained by Langston and Boyle (1981).

Dring and Joslyn (1981) used an ammonia/ozalid paper surface indicator technique for surface flow visualization to obtain airfoil and endwall flow visualization. A trace amount of anhydrous ammonia is seeped onto the test surface through a static pressure tap location. Ozalid paper is attached smoothly to the surface immediately downstream of the pressure taps. The ammonia is swept across and leaves a dark blue trace in the flow direction on the ozalid paper. A high-intensity photograph of the trace while the flow is active will avoid the possibility of overexposing the ozalid paper to ammonia. Figure 6.38 shows the flow visualization on a turbine rotor blade. The surface streaks clearly indicate the three dimensionality of the flows along airfoils.

Dring and Joslyn (1981) also presented a smoke flow visualization technique for turbine and compressor passages. A smoke generator was used to inject smoke into the passages at various locations. A mineral oil is pumped through a probe tip where an electrical overheat produces a dense white vapor at the injection site. A high-speed camera can be employed to capture the smoke traces. Figure 6.39 presents the flow visualization in turbine blade passages. The wake passage through the blade passage is well captured by the flow visualization technique.

Goldstein and Chen (1984) used a mixture of fine carbon powder and oil spread on contact paper for surface flow visualization. Figure 6.40 shows the streamline tracks on the endwall of the gas turbine blade.

Theory   Experiment

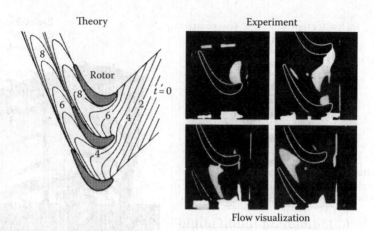

Flow visualization

FIGURE 6.39
Flow visualization of turbine rotor blade passage. (From Dring, R.P. and Joslyn, H.D., *ASME J. Eng. Power*, 103(2), 400, 1981.)

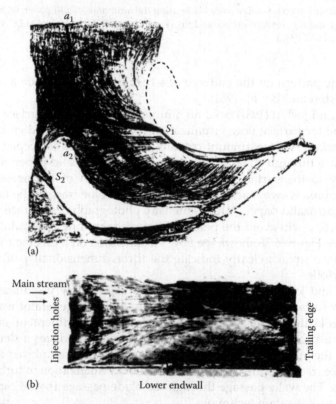

FIGURE 6.40
(a) Streamline streaks on endwall of a gas turbine blade. (b) Streamline tracks on convex surface of gas turbine blade. (From Goldstein, R.J. and Chen, H.P., *ASME J. Eng. Gas Turbines Power*, 107, 117, 1985.)

6.6 New Information from 2000 to 2010

Since 2000, similar experimental methods have been used to determine turbine blade surface heat transfer coefficients and film cooling effectiveness, as well as turbine blade internal coolant passage heat transfer. There are so many good papers using the aforementioned experimental methods available in the open literature. Only a few papers are selected to be included in this section. However, the detailed descriptions on the corresponding experimental method are referred to the previous sections in this chapter.

6.6.1 Transient Thin-Film Heat Flux Gages

Transient response of heat flux to the order of 10^{-8} s can be measured using thin-film heat flux gages. Due to its fast response, this technique is frequently used only in short duration experiments. This technique gives reliable data for heat transfer coefficients with low uncertainty. Transient response due to unsteady flow can be investigated using this technique due to its fast response. Another major advantage is that if properly mounted, the thin-film heat flux gages are robust and can be used at high operating temperatures up to 800°C. The primary disadvantage of this technique is that it gives only point measurements. Thin-film heat flux gages have been used extensively to measure the fast time response of the fluctuating heat flux on the test surface under harsh operating conditions. The thin-film sensor is manufactured from a highly conductive material such as platinum or copper alloys. The gages are available as button type and continuous strip-type gages (Figure 6.41). The thickness of the thin-film sensor is typically expressed in fractions of a micrometer that gives it a frequency response of as high as 50 kHz. The thin-film sensors are manufactured by depositing or painting the material on a substrate in very fine layers and curing them in an autoclave at high temperatures. Refer to Section 6.2.2 for the details.

Since 2000, the following selected research groups continuously using the transient thin-film heat flux gages for turbine blade/vane surface heat-transfer measurements are listed here for reference: For example, Ohio State University (Bergholz et al., 2000; Dunn and Haldeman, 2003; Haldeman et al., 2006; Mathison et al., 2010); VKI (Didier et al., 2002; Iliopoulou et al., 2004); Oxford (Ainsworth et al., 2004; Jones et al., 2004; O'Dowd et al., 2009). Interested readers can see these papers for detailed information. The following provides fundamental background on how to determine time-dependent heat-transfer coefficient in unsteady flow using thin-film heat flux gages.

The determination of the heat-transfer coefficient requires the prior knowledge of three components as shown in Equation 6.28: the fluid temperature,

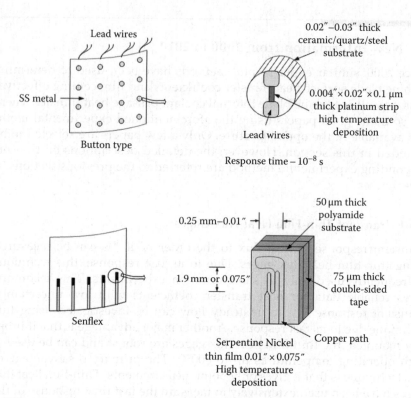

FIGURE 6.41
Sketches of typical button type and continuous strip-type gages.

the wall temperature, and the heat flux at the wall. The fluid temperature can be measured using a thermocouple inserted into the mainstream:

$$h = \frac{q''(t)}{T_w(t) - T_\infty} \tag{6.28}$$

The wall temperature is measured using the principle of a variable resistance thermometer. Temperature changes in the material of the thin-film sensor will cause its resistance to change as a linear approximation given as $R(t) = aT(t) + b$, where a and b are constants. The film sensor is connected as part of a Wheatstone bridge circuit operating under a constant current mode (Figure 6.42). A Wheatstone bridge is a divided electrical circuit used to measure the dynamic change in resistance of the thin-film sensor. Using V (voltage) = I (current) × R (resistance), the fluctuating voltage output $V(t)$ from this circuit is proportional to the fluctuations in wall temperature with time $T(t)$ due to the linear approximation between film resistance $R(t)$ and its temperature $T(t)$. A calibration curve between the resistance and the temperature

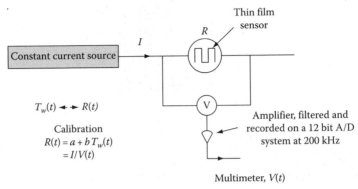

Wheatstone bridge circuit

FIGURE 6.42
Sketch of typical thin-film sensor operation principle.

will give the wall temperature. An amplifier may be connected to boost the output voltage from the bridge circuit. A suitable gain may also be applied. Several thin-film heat flux gages are nowadays also combined with a separate resistance-temperature sensing element to measure surface temperature mounted on the same substrate.

In order to convert the measured fluctuating voltage signals to heat flux, appropriate analytical models are used to determine the wall heat flux history. As the wall heat flux changes, the wall temperature will also change rapidly. The substrate for the thin film is designed such that the temperature on the back side of the substrate does not change even if the wall temperature changes, for the duration of the experiment. If the substrate thickness is large and thermal conductivity is low, the wall heat flux can be prevented from conducting to the back side and the semi-infinite model can be assumed for the substrate. The one-dimensional (1-D) unsteady conduction equation given in Equation 6.29 can be solved with the wall temperature history and the constant initial temperature on the substrate back side (Figure 6.43). The semi-infinite model assumption is critical for the accurate determination of the heat flux:

$$\frac{\partial^2 T(x,t)}{\partial x^2} = \frac{1}{\alpha(x)} \frac{\partial T(x,t)}{\partial t} \tag{6.29}$$

where α is the thermal diffusivity of the substrate. The initial and boundary conditions necessary to solve this equation are as follows:

Initial condition:

$$T(x,0) = T_i \quad (x \geq 0)$$

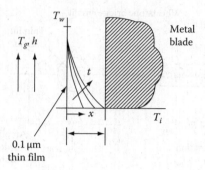

FIGURE 6.43
Sketch of typical thin-film sensor 1-D unsteady conduction model.

Boundary condition:

$$T(\infty,t) = T_i \quad (t \geq 0) \quad \text{and} \quad T(0,t) = T_w \quad (t \geq 0) \tag{6.30}$$

The analytical solution to the aforementioned equation can be given as

$$\frac{T - T_i}{T_\infty - T_i} \, erf\left(\frac{x}{2\sqrt{\alpha t}}\right) \tag{6.31}$$

The resulting surface heat flux obtained from Fourier's law of conduction can be expressed as

$$q_w''(t) = \frac{k_{substrate}\left(T_w(t) - T_\infty\right)}{\sqrt{\pi \alpha t}} \tag{6.32}$$

Equation 6.32 gives the wall heat flux on the test surface from the thin-film heat flux gage. Schultz and Jones (1973) proposed a criterion to determine the minimum substrate thickness needed to satisfy the semi-infinite model assumption. The criterion states that at a thickness of $x = 3.648\sqrt{\alpha t}$ the wall heat flux ratio $q(x)/q_w(x = 0)$ is less than 1%. In some cases, multiple substrates with different properties may be used for mounting a heat flux gage (Iliopoulou et al., 2004). In such cases, constant heat flux and constant temperature boundary conditions may be assumed at their interfaces and Equation 6.29 can be solved numerically.

The gages are embedded into the test surface into grooves such that the film is flush with the test surface. Gages not inserted flush with the wall will detect unwanted flow disturbances caused due to the projecting surfaces and hence will mask the true data. Before inserting into the test wall though, the gages have to be mounted on an insulating substrate such as quartz or ceramic. Thus, the semi-infinite model can be applied to calculate the heat flux from the thin-film sensor. Use of insulating materials such as plastics is avoided as they do not hold their properties at high working temperatures. Quartz and ceramic are selected as their thermal property variation at high temperatures is small and is well documented. They can be also machined and cast in complicated shapes. The low thermal diffusivity of the substrate will ensure that the change in its temperature is small for the duration of the experiment. This is particularly important when the gage is mounted on a metal wall. If the test surface is itself made from insulating material, the substrate will then become redundant.

In some experiments such as heat flux investigation on a rotating turbine blade, the application may require that the thin-film sensor and substrate be exposed to large stresses. In these situations, the thin-film sensor may be directly mounted onto a metal surface for increased strength or the substrate size may be reduced to minimize stress concentration on the test surface by groove cutting for heat flux gage insertion (Bergholz et al., 2000; Didier et al., 2002). However, the semi-infinite assumption may not be true in these cases. Appropriate corrections for heat conduction into the metal surface may be used to offset this setback. Several researchers have introduced numerical models to correct the data from the sensor for heat conduction. Duration of the experiment can also be decreased to minimize heat conduction errors.

6.6.2 Advanced Liquid Crystal Thermography

Liquid crystals are thermochromic, i.e., they can change color with temperature. Thus, a surface coated with thermochromic liquid crystals (TLC) will show a color change when exposed to certain temperatures. Due to this advantage of a tangible color response, TLCs have found applications in a broad range of areas. Gas turbine blades, electronic component cooling, boiling heat transfer, etc., are some areas of research performed using TLCs. Even some medical tools have utilized their benefits in measuring body temperature. The liquid crystal technique (LCT) requires the use of three major components: the liquid-crystal-coated test surface, a white light source/s to illuminate the surface, and a camera to capture the local distribution of color on the test surface. There are several techniques using liquid crystals for measuring surface temperature distributions: steady-state yellow-band tracking technique, steady-state HIS technique, transient HIS technique, and transient single-color capturing technique. Refer to Section 6.4 for the details.

Since 2000, the following selected research groups continuously using the liquid crystal technique for turbine airfoil surface heat transfer and blade tip film cooling measurements are listed here for reference: For example,

Oxford (Guo et al., 2000; Chambers et al., 2002; Ling et al., 2003; Talib et al., 2003); NASA (Thurman and Poinsatte, 2000); Texas A&M (Ekkad and Han, 2000; Kwak and Han, 2003a,b; Kwak et al., 2003). Interested readers can see these papers for detailed information. The following provides fundamental background on how to obtain accurate data using the liquid crystal techniques.

Liquid crystals are composed of chiral or twisted molecular structures that become optically active when they change phase. Below the event temperature (temperature at which the first color change starts), the liquid crystals are in a solid phase and appear transparent. The liquid crystals change phase from solid to liquid state over a range of temperature. Based on the relative solid–liquid content on the phase change curve, the two-phase solution refracts light and appears to change color. When the temperature of the liquid-crystal-coated surface reaches its event temperature, melting of the liquid crystals on the surface is initiated and they will start reflecting visible light at a certain wavelength when illuminated with white light. At this temperature, the liquid crystal will change color from clear to red. As the temperature increases further, more quantities of the unmelted liquid crystals enter the liquid phase and start reflecting color wavelengths proportional to the mass of the melted liquid crystals. Thus, the reflected wavelength or color changes with increasing temperature till the clearing point temperature when the liquid crystal goes into the liquid phase. At this stage, the TLC becomes transparent again with no change in color with further increases in temperature.

The color change is reversible and on cooling the color sequence is reversed. Upon re-heating, color changes occur at the same temperature and hence are repeatable. This fact is important as the experiments performed using this technique give reproducible results. Generally, liquid crystals are viewed against a nonreflective black background mainly to avoid the transmitted light from adversely affecting the interpretation of the reflected light. There are two categories in which TLCs can be classified: cholesteric and chiral-nematic. Due to their inherently oily form, the true liquid crystals are micro-encapsulated in tiny protective capsules and are suspended in an aqueous binder material. This makes the liquid crystals easier to work with and also chemical contamination is avoided.

The range of temperature when the TLC changes color to red till the beginning of blue is called the bandwidth. TLCs are grouped by means of a two color/temperature designator to describe its relationship between color and temperature. A typical example would be "R35C5W," which is a commonly used TLC formulation. In this case, "R35C" means that the start of the red color or the event temperature is at 35°C while "5W" indicates that the start of the blue color is 5°C above the event temperature. Thus, the bandwidth for this type of liquid crystal is 5°C. TLCs are available either as narrow band with a bandwidth between 0.5°C and 2°C or as wide band with bandwidth ranging from 5°C to 30°C. Depending on the application, a TLC with an appropriate event temperature and bandwidth can be selected.

Selection of a narrow band or a wide band TLC depends on the kind of application to be studied. Very accurate temperature measurements can be obtained using a narrow band TLC as the color change is very sharp at a certain temperature. Also, image processing is easier as a single color can be tracked for different input heat fluxes. The chief advantage of using a wide band liquid crystal is that the entire isotherm pattern of the surface can be mapped in a single image. A single image can also capture large variations in surface temperature at high spatial resolution. Careful calibration of the color change to temperature response is needed to obtain high accuracy.

The color reflected from the TLC is dependent on the light source used to illuminate the test surface. A sufficiently bright and stable white light source is needed. When the test surface is curved, more than one light source might be needed to ensure uniform light intensity over the test surface. The light sources selected should not have infrared or ultraviolet light in its output spectrum. Presence of infrared light in the light source output can cause unwanted heating of the TLC-coated test surface through radiation. This will give inaccurate measurements of the surface temperature. Ultraviolet radiation, on the other hand, can cause rapid deterioration of the liquid crystal response, resulting in an inconsistent response of color change to temperature. Typically, the use of fluorescent lights should also be avoided as they emit light periodically but at high frequencies. This periodicity is not observable by the human eye but it gets magnified when viewed using a camera with a relatively small exposure time. This periodic nature results in inconsistent intensity distribution with time on the test surface.

Calibration of the TLC can be either performed in situ or in a separate calibration facility. In situ calibration is recommended as the color perceived from a true TLC is dependent on the lighting and viewing arrangement at a given temperature. Thus, due to the difficulty in having two exactly same setups for measurement and calibration, in situ calibration is preferred. It should be also noted that the color play of the TLC is dependent on the viewing angle. If the viewing angle is very large, the observed color may be different as to the color seen from a normal viewing angle even if the surface is at the same temperature. However, the error induced when the viewing angle is less than ±30° from the normal angle is small and hence can be neglected. Hence, additional calibration should be performed for large viewing angles to account for any discrepancy when the surface to be viewed is not normal to the camera. Calibration is performed by applying the TLC to an isothermal surface whose temperature can be controlled. An example of such a surface would be a copper plate with an embedded thermocouple with a heater attached on the other side. The high conductivity of copper ensures that the surface temperature on the TLC-coated surface is uniform. By controlling the heat input to the heater, successive increments in temperature can be obtained and the color response can be recorded. This process is repeated from the event temperature till the clearing point temperature. The hue of the color when plotted against temperature is typically linear from the start of red till the beginning of blue with the slope changing after that.

The color image of the surface is typically taken with an RGB CCD camera. If detailed measurements of the entire surface are to be performed, the camera should give a sufficiently good image resolution. The camera resolution necessary depends on the size of the test surface and the smallest surface feature that has to be captured in detail. The pixel size on the test surface should be at least less than half the size of the smallest feature on the test surface. More pixels will give an image with better resolution. However, high-resolution cameras are expensive and frequently a compromise has to be reached between cost and the camera pixel resolution. For example, an image resolution of 0.5 mm/pixel will be able to resolve a hole of diameter 2 mm on the test surface in sufficient detail in most cases. However, if the pixel resolution is 2 mm/pixel, only one pixel will be available to cover the entire hole, which as a result will appear blurred in the image captured from the camera. Also, the camera chosen should be able to take several images of the test surface in a relatively short time, so that the color or hue magnitudes at each pixel can be averaged to get the true surface color distribution. Several hundred images may have to be averaged to reduce noise from the camera CCD.

TLCs are available as a sprayable liquid or as a coated (printed) sheet. Sprayable liquid crystal is typically used when the surface is not flat. A coating of a nonreflective black paint on the test surface is necessary before spraying the TLC for good visualization and to accurately interpret the color response. Sufficient care should be taken while spraying in order to obtain a uniform layer thickness of the black paint and liquid crystal. A coated sheet consists of a thin film of liquid crystal sandwiched between a transparent substrate (sheet) and a black background. They are usually made by printing an ink containing microencapsulated TLC onto the reverse side of the substrate. A black ink is then applied on top of the dry TLC coating and color change effects are viewed from the uncoated side. A coated sheet has a much longer life as compared to a surface that is sprayed with TLC as the transparent coating prevents the direct contact of the TLC and air and thus protects it from dust.

Due to their property of changing color with temperature, TLCs have been successfully used to measure heat-transfer coefficients on a surface. Steady-state as well as transient techniques have been utilized. The following section discusses the steady-state methods available for measuring heat transfer along with some applications in research.

6.6.3 Infrared Thermography

All objects absorb and emit thermal radiation and the amount an object will absorb or emit depends on the temperature of that object. From spectral blackbody emissive power versus wavelength over a wide range of temperatures, infrared radiation stretches from a wavelength of about 1 μm (near-infrared) to 200 μm (far-infrared). The radiated infrared energy will be zero when the object is at absolute zero temperature and increases with increased temperature. The heat radiated by a black body per unit time is given by the Stefan–Boltzmann law of thermal radiation as $\dot{Q} = \sigma A T^4$. The net radiation heat flow between two black bodies at different temperatures of T_1 and T_2 is given by $\dot{Q} = \sigma A \left(T_1^4 - T_2^4 \right)$. The net radiation heat flow between two real bodies that emit less radiation than a black body can be expressed by introducing a parameter known as emissivity. The emissivity of an object is a measure of the objects efficiency in emitting radiation as compared to a black body:

$$\dot{Q} = \varepsilon \sigma A \left(T_1^4 - T_2^4 \right) \tag{6.33}$$

Cameras that measure the infrared radiation utilize the aforementioned equation to measure the surface temperature of an object. If the object surface emissivity is known, the infrared sensor in the camera composed of quantum detectors measures the net radiation heat flux from the object surface, thus calculating its surface temperature. The quantum detectors convert the photon energy of the particular range of wavelength into an electrical signal by releasing electrons. The electrical output generated by these detectors is very small and can be overshadowed by the background noise generated by the device. Hence, to improve the signal-to-noise ratio from these detectors, they have to be cooled to cryogenic temperatures. Using an array of sensors, the IR camera can resolve the surface into several pixels and produce a monochromatic image of the temperature distribution on the surface. To get a better visual perception of the temperature distribution on the surface, false color is often assigned to different intensities of the monochrome image to convert it to color.

The IRT technique is a powerful method that can give detailed distributions of the local heat transfer coefficients on the surface. It has several inherent advantages over other optical techniques such as liquid crystal thermography:

1. The IR method is not bounded by an operating temperature range as in the case of liquid crystals.

2. Typically, only black paint needs to be sprayed on the test surface for IRT. In case of liquid crystal thermography, the liquid crystal also needs to be painted on the test surface. Thus, IR technique is free from errors due to nonuniform paint distribution, thickness of coating, etc.

3. For transient tests, the IR method can give the initial temperature of the test surface which is difficult to obtain using liquid crystals.

4. IR cameras do not require a sophisticated illumination system as in the case of LCT.

5. Most IR cameras are precalibrated for temperature, which reduces the time and cost for detailed calibration for hue against temperature required for wide band liquid crystals.

A major disadvantage of the IRT technique is the high cost and sophistication of the IR camera itself. In addition, windows made from special materials such as zinc selenide, transparent for infrared radiation, have to be fabricated for image capture.

For steady-state IR camera technique, the surface film cooling effectiveness can be determined by using Equation 6.15 as explained in previous section. For steady-state IR camera technique, the heat-transfer coefficients are obtained on the surface by heating it with a heater such as a thin-foil heater. By applying a suitable voltage across the heater, the desired temperature distribution can be obtained. A nonreflective black paint with known emissivity is spray painted on the test surface on top of the heater. The black paint will result in a uniform high emissivity on the test surface. The test surface is placed in the test section or wind tunnel and the mainstream flow is turned on. Heat is applied to the heater such that a temperature difference of about 30°C can be obtained. Once temperature equilibrium conditions are obtained on the test surface, an image of the surface is captured by the IR camera through a zinc selenide window. The power input to the heater and the mainstream flow rate and temperature are measured. With the knowledge of heat loss from the heater to ambient surroundings, the net heat input can be calculated. The heat-transfer coefficients for each pixel can then be calculated using

$$h = \frac{q''}{T_{w,IR} - T_\infty} = \frac{q''_{gen} - q''_{loss}}{T_{w,IR} - T_\infty} \tag{6.34}$$

For transient IR camera technique, similar to transient liquid crystal technique in principle, refer to Section 6.4 for the details. Using IR camera for local surface temperature distribution instead of using CCD camera for liquid crystals, all equations from (6.19) through (6.26) can be applied to determine local heat-transfer coefficient and film-cooling effectiveness distributions. Since 2000, the following selected research groups continuously using the IR camera technique for turbine airfoil surface heat transfer coefficient, film cooling effectiveness, and internal cooling channel heat transfer measurements are listed here for reference: For example, Moon and Jaiswal (2000); Boyle et al. (2000); Bogard and Cutbrith (2003); Ekkad et al. (2004); Brauckmann and von Wolfersdorf (2005); Lynch and Thole (2007); Somawardhana and Borgard (2007); Chang et al. (2010); Thrift et al. (2010). Interested readers can see these papers for detailed information.

6.6.4 Pressure-Sensitive Paint

PSP is a photo-luminescent material that emits light with intensity proportional to the surrounding partial pressure of oxygen. Any pressure variation on the PSP-coated surface causes emitting light intensity to change due to an oxygen-quenching process. A higher pressure will result in low intensity and vice versa. The intensity emitted by PSP is also dependent on the temperature to a smaller extent. Hence, accurate measurement of pressure requires knowledge of the surface temperature distribution. Performing all experiments under the same thermal conditions also helps in reducing uncertainty due to temperature variation. A calibration performed (at same temperature level as in real experiments) for intensity ratio to give pressure ratio gives pressure information. A major advantage of this technique is that detailed contour plots of local static pressure and film-cooling effectiveness can be obtained. Due to the absence of a temperature difference between the test surface and mainstream and coolant, this measurement technique provides for robust measurement near sharp edges such as film-cooling-hole break outs due to avoidance of heat conduction errors. Figure 6.44 shows the film-cooling effectiveness measurement using heat (thermal) and mass (PSP) transfer analogy by Han and Rallabandi (2010). Figure 6.45 shows flat plate film cooling using PSP technique and comparison with other measurement methods (Han and Rallabandi, 2008).

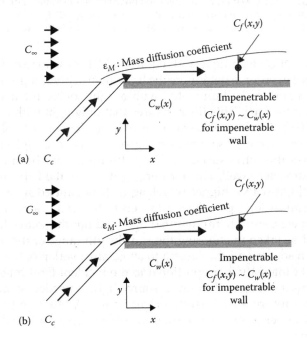

FIGURE 6.44

Measurement of film-cooling effectiveness using the heat (a) and mass (b) transfer analogy. (From Han, J.C. and Rallabandi, A.P., *Front Heat Mass Transfer*, 1, 1, 2010.)

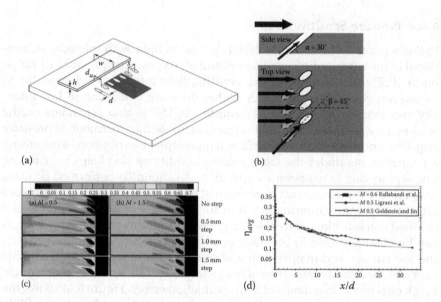

FIGURE 6.45

Flat-plate film cooling using PSP measurement method from Rallabandi et al. (a) Various upstream step parameters tested. (b) Definition of compound angle. (c) Parametric effect of upstream step on cylindrical compound angled holes. (d) Benchmarking of data with open literature: Ligrani et al. (1994b) and Goldstein and Jin (2001). (From Rallabandi, A.P. et al., Effect of upstream step on flat plate film cooling effectiveness using PSP, *Proceedings of ASME 2008 Summer Heat Transfer Conference*, Jacksonville, FL, Paper No. 2008-56194, August 10–14, 2008.)

This is an optical technique and requires a CCD camera to capture images of the surface in order to determine the light intensity emitted from the paint. The painted surface is illuminated with a light source fitted with a blue-green band pass filter. This filter ensures that only light with a wavelength between 450 nm (blue) and 520 nm (green) is incident on the PSP-coated test surface. The chosen light source should give uniform and high-intensity distribution over the entire surface. LED light sources have been found to suit this application very well. This incident light excites the PSP surface, which then emits light with a higher wavelength. This emitted light is generally red (>590 nm). A long pass filter attached to the CCD camera will ensure that the camera sees only this emitted light and not the source light. Special care should be taken in choosing the wavelength range of the filters for the illumination source and the camera to avoid any overlap of the ranges. The emitted light intensity is proportional to the incident light intensity on the surface. Thus a strong and uniform source light is needed so that the CCD camera can capture a sufficiently bright image. A black and white image is recorded under gray scale settings which can encapsulate this intensity information.

The test surface is first cleaned and painted with PSP. Several coats of this paint may have to be applied in order to get a sufficient response in intensity for small pressure changes. About seven to eight coats may be needed depending on the type of paint used. Spraying the surface with more than eight coats can cause flaking of the paint and hence should be avoided. An air brush should be used for painting as it ensures a uniform distribution of the paint over the surface.

To obtain the intensity ratio from PSP, three kinds of images are required: a reference image (with illumination, no mainstream flow, surrounding pressure uniform at 1 atm), an air image (with illumination and mainstream flow), and a black image (no illumination and no mainstream flow) to remove noise effects due to the camera.

Oxygen partial pressure information is obtained from the intensity ratio and calibration curve. This oxygen partial pressure information can be directly converted into static pressure distribution for the case with air coolant injection. Intensity ratio for air and air/nitrogen mixture is calculated using (6.36):

$$\frac{I_{ref} - I_{blk}}{I_{air} - I_{blk}} = func\left((P_{O_2})_{air}\right) \quad \text{or} \quad func(P) \tag{6.35}$$

where
  I denotes the intensity obtained for each pixel for reference (*ref*), black (*blk*), and air (*air*) images
  $func(P)$ is the relation between intensity ratio and pressure ratio obtained after calibrating the PSP
  $(P_{O_2})_{air}$ are the partial pressures of oxygen on the test surface for air images

The PSP should be calibrated for the range of pressure measurement in the test section. Calibration of the PSP can be either performed in a pressure chamber or in situ. When performed in a pressure chamber such as a vacuum chamber, a test piece painted with PSP is constructed and placed in the chamber. Figure 6.46b depicts a schematic of such a calibration setup. One wall of this chamber should be made from a transparent material such as plexiglass in order to provide optical access to the PSP-coated test piece. This chamber should be connected to a pump and a pressure gage for adjusting and measuring the pressure inside the chamber, respectively. While performing the calibration, it should be ensured that the pressure inside the chamber is uniform, i.e., no flow should be occurring internally. The effect of temperature on PSP can also be measured by attaching a heater underneath the test piece. A thermocouple placed on the test piece surface can be used to indicate the surface temperature.

In situ calibration of PSP can be performed by subjecting the test surface to different pressures and recording the intensities at several pressures using a camera and also measuring the static pressure using pressure taps distributed at several locations. Thus a relationship between the static pressure

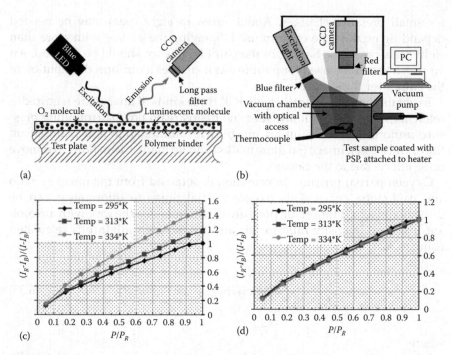

FIGURE 6.46

PSP working principle and typical calibration curve. (a) PSP: principle of operation, (b) PSP: calibration setup, (c) PSP: calibration results, T_R = 295 K, and (d) PSP: calibration results, T_R maintained at test temperature. (From Mhetras, S.P. et al., Effect of flow parameter variations on full coverage film-cooling effectiveness for a gas turbine blade, *Proceedings of ASME Turbo-Expo 2007*, Montreal, Quebec, Canada, ASME GT2007-27071, May 14–17, 2007.)

and the emitted intensity can be directly established. An advantage of using in situ calibration is that both the calibration and the tests can be performed under the same thermal conditions, thus minimizing the errors due to temperature variation.

There is a temperature dependency of PSP. However, if the intensity is normalized by that of the reference image (at 1 atm), the calibration curves at different temperatures fall into one curve. Figure 6.46 shows a typical calibration curve of intensity ratio versus pressure ratio (Mhetras et al., 2007). During testing, it must be ascertained that temperatures of mainstream air, coolant, and test section are same while taking reference, air, and nitrogen images to minimize uncertainty.

An image-processing tool should be used for analyzing the images acquired by the camera. Images obtained from the camera can be saved as TIF images. Several images should be captured for each test case, and the average pixel intensity should be calculated from these images. These intensity magnitudes can be then converted to partial pressures of oxygen using an appropriate calibration curve and then to surface pressure. The results obtained for each pixel can be then plotted as contour plots.

For film-cooling effectiveness measurements, four kinds of images are required: a reference image (with illumination, no mainstream flow, surrounding pressure uniform at 1 atm), an air image (with illumination and mainstream flow, air used as coolant), an air/nitrogen image (with illumination and mainstream flow, nitrogen gas used as coolant), and a black image (no illumination and no mainstream and coolant flow) to remove noise effects due to the camera. The intensity ratio for air and air/nitrogen mixture is calculated using Equations 6.35 and 6.36, respectively,

$$\frac{I_{ref} - I_{blk}}{I_{mix} - I_{blk}} = func\left((P_{O_2})_{mix}\right) \tag{6.36}$$

where
  I denotes the intensity obtained for each pixel for reference (*ref*), black (*blk*), and air/nitrogen (*mix*) images
  $(P_{O_2})_{mix}$ are the partial pressures of oxygen on the test surface for air/nitrogen mixture images

The film-cooling effectiveness can be expressed as a ratio of oxygen concentrations, or partial pressures measured by PSP and can be calculated using the following equation:

$$\eta = \frac{Co_{air} - Co_{mix}}{Co_{air}} = \frac{(P_{O_2})_{air} - (P_{O_2})_{mix}}{(P_{O_2})_{air}} \tag{6.37}$$

where Co_{air} and Co_{mix} are the oxygen concentrations, respectively, for mainstream air and air/nitrogen mixture on the test surface. By assuming the molecular weights of air and nitrogen as same, effectiveness can be expressed as a ratio of partial pressures of oxygen due to proportionality between concentration and partial pressure. Figure 6.47 shows light and camera setup to capture full blade coverage film effectiveness using PSP technique by Mhetras et al. (2007). Figure 6.48 shows film-cooled blade, and Figure 6.49 shows effect of blowing ratio on full-blade film-cooling effectiveness distributions on both pressure and suction surfaces using PSP measurement technique by Mhetras et al. (2007).

Since 2000, the aforementioned equations have been used to determine the local film-cooling effectiveness when nitrogen is injected through the film-cooling holes, by converting the intensities recorded to partial pressures of oxygen over the tested surface. Several selected papers have been published using this methodology for airfoil film cooling (Mhetras et al., 2007), blade tip film cooling (Ahn et al., 2004), endwall film cooling (Gao et al., 2008), as well as rotating blade leading-edge film cooling as shown in Figure 6.50 (Ahn et al., 2005) and rotating blade platform film cooling as shown in Figure 6.51 (Suryanyanan et al., 2006). In order to simulate realistic

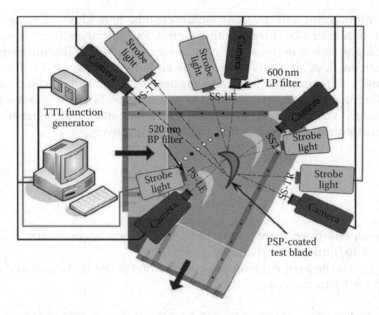

FIGURE 6.47
Light and camera setup to capture full-blade coverage film cooling using PSP technique by Mhetras et al. (From Mhetras, S.P. et al., Effect of flow parameter variations on full coverage film-cooling effectiveness for a gas turbine blade, *Proceedings of ASME Turbo-Expo 2007*, Montreal, Quebec, Canada, ASME GT2007-27071, May 14–17, 2007.)

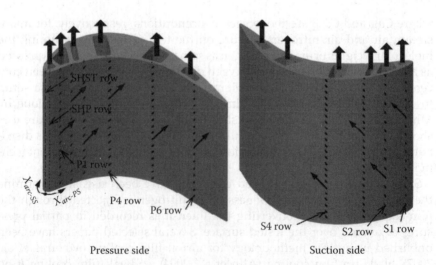

FIGURE 6.48
Bottom half of 3-D blade cut at mid-span. (From Mhetras, S.P. et al., Effect of flow parameter variations on full coverage film-cooling effectiveness for a gas turbine blade, *Proceedings of ASME Turbo-Expo 2007*, Montreal, Quebec, Canada, ASME GT2007-27071, May 14–17, 2007.)

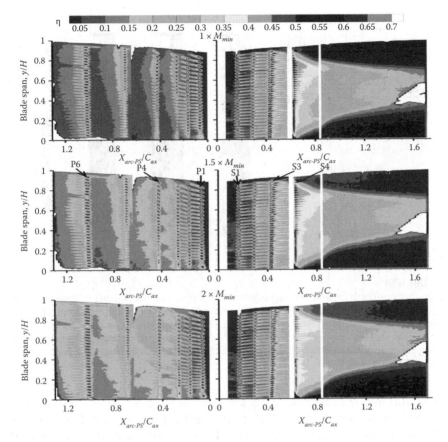

FIGURE 6.49

Film-cooling effectiveness distributions for three blowing ratios at $M_{min} = 0.36$ without showerhead ejection using PSP technique. (From Mhetras, S.P. et al., Effect of flow parameter variations on full coverage film-cooling effectiveness for a gas turbine blade, *Proceedings of ASME Turbo-Expo 2007*, Montreal, Quebec, Canada, ASME GT2007-27071, May 14–17, 2007.)

coolant-to-mainstream density ratios for turbine film cooling, a coolant with a heavier density such as carbon dioxide, a mixture of SF6 and argon, is injected through the film-cooling holes. In this case, Equation 6.37 has been modified to include the coolant-to-mainstream molecular weight ratio for local film-cooling calculations (Charbonnier et al., 2009; Narzary et al., 2009, 2010; Han and Rallabandi, 2010). Interested readers can see these papers for detailed information.

6.6.5 Temperature-Sensitive Paint

Temperature-sensitive paint (TSP) is similar in principle and operation to the PSP technique. When the paint is excited by incident light of a particular wavelength, it emits light at a different wavelength which can then

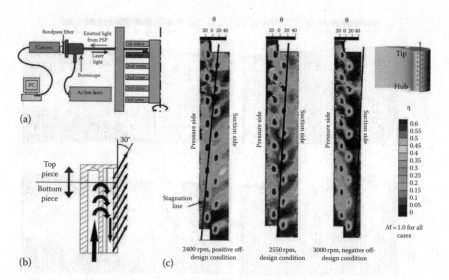

FIGURE 6.50
Film-cooling effectiveness on leading edge under rotating conditions using PSP measurement technique. (a) Experimental setup: PSP camera and light access through borescope. (b) Blade internal passage details. (c) Film cooling effectiveness measurements on blade surface. (From Ahn, J.Y. et al., Film cooling effectiveness on the leading edge of a rotating film-cooled blade using pressure sensitive paint, *Proceedings of ASME Turbo-Expo 2005*, Reno, NV, Paper No. GT 2005-68344, June 6–9, 2005.)

be quantified by capturing images of the test surfaces. Thus, TSP is also a photo-luminescent material that emits light proportional to the temperature of the coated surface. A typical TSP-coated surface contains an oxygen impermeable binder and luminescent temperature-sensitive molecules. When light is incident on this test surface, these molecules, which are initially in a ground state, absorb the incident photons, resulting in a boost in their energy level. This energy can either be dissipated through the emission of radiation or through radiationless deactivation. Fluorescence is produced when the molecules lose energy by emitting photons by radiationless deactivation also known as thermal quenching. Increase in temperature of the luminescent molecules increases the likelihood of thermal quenching. Photon emission occurs at a higher wavelength and, upon emission, the molecules return to their original ground state. Figure 6.52 shows the TSP working principle and basic setup for measurement by Narzary et al. (2009).

As in PSP technique, the test surface is coated with paint, and the light intensity from the captured images is calibrated against temperature as opposed to pressure. The illumination and image-capture systems for TSP and PSP techniques are essentially the same. Different optical filters on the illumination source and camera might be needed depending on the TSP

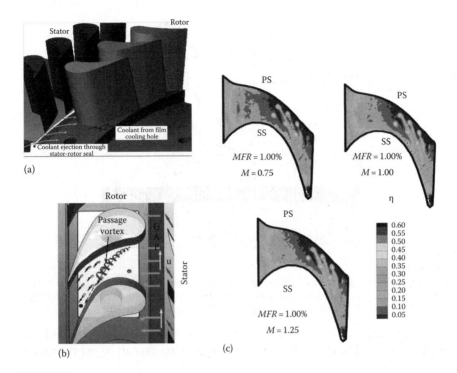

FIGURE 6.51
Film-cooling effectiveness on rotor endwall under rotating conditions using PSP measurement technique. (a) Rotor-stator configuration with inclined slot. (b) Schematic of expected fluid mechanics in endwall film cooling. (c) Measured film cooling effectiveness values on endwall under rotation. (From Suryanarayanan, A. et al., Film cooling effectiveness on a rotating blade platform, *Proceedings of ASME Turbo-Expo 2006*, Barcelona, Spain, Paper No. GT2006-90034, May 8–11, 2006.)

formulation used and their absorption and emission spectra. If a colored illumination source such as LED light is used, then a filter is not necessary for the light. TSP is generally insensitive to changes in surface pressure as the polymer binder used is not permeable to oxygen. For PSP though, the polymer binder is oxygen permeable to allow for oxygen quenching. Hence, PSP needs to be calibrated for pressure as well as temperature.

For TSP measurements, the measured intensity data are normalized using a reference condition. This reference condition will depend on the environment and conditions in which the experiment is performed. For most experiments, the reference condition is usually set as the room temperature before the experiment is started. The primary reason for normalizing the intensity is to eliminate errors due to inconsistencies in spatial illumination, paint thickness and distribution, camera sensitivity, etc. By taking two images for the test surface under the same lighting and image acquisition conditions but one at a known temperature and the other at an unknown temperature,

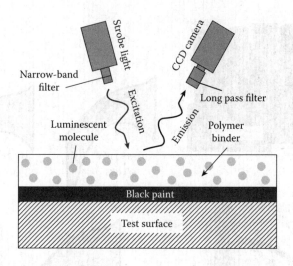

FIGURE 6.52
TSP working principle and basic TSP setup for measurement by Narzary et al. (From Narzary, D.P. et al., Influence of coolant density on turbine blade platform film-cooling, *Proceedings of ASME Turbo-Expo 2009*, Orlando, FL, ASME GT-2009-59342, June 8–12, 2009.)

the errors due to the aforementioned inconsistencies can be avoided after normalization. The intensity ratio after normalization can be expressed as

$$\frac{I(T) - I_{blk}}{I(T)_{ref} - I_{blk}} = f(T) \tag{6.38}$$

The intensity of the test surface under no lighting is subtracted from the image intensities measured during testing to remove unwanted noise in the data. To determine the functional relationship between intensity and temperature, a calibration must be performed. A small copper piece fitted with a thermo-couple to measure its temperature is coated with TSP. A heater attached on the other side of the TSP-coated surface helps in regulating the temperature. This copper block is then positioned underneath the camera and the illu-mination source. The temperature is controlled using the heater through the range in which temperature data will be recorded and the intensity is recorded through images taken with the camera. A reference condition is set and using the aforementioned equation, the relationship between tempera-ture and intensity is determined. It should be noted that a pressurized cham-ber used when calibrating PSP is not necessary for calibration of TSP as TSP is virtually insensitive to pressure. A typical calibration curve for TSP is shown in Figure 6.53 by Narzary et al. (2009). After the wall temperature measured by TSP, Equation 6.15 can be used to calculate the film-cooling effectiveness.

For film-cooling study, Equation 6.15 can be used to determine the local film-cooling effectiveness when coolant is injected through the film-cooling holes, by converting the intensities recorded to temperature distribution

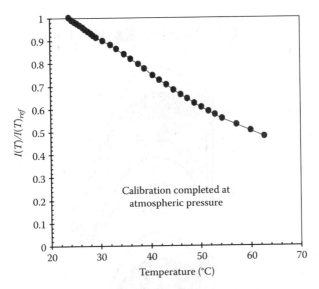

FIGURE 6.53
TSP calibration by Narzary et al. (From Narzary, D.P. et al., Influence of coolant density on turbine blade platform film-cooling, *Proceedings of ASME Turbo-Expo 2009*, Orlando, FL, ASME GT-2009-59342, June 8–12, 2009.)

over the tested surface. Figure 6.54 shows the effect of free stream turbulence intensity on endwall film-cooling effectiveness with purge flow using the TSP measurement technique by Narzary et al. (2009). Several selected papers have been published using this methodology for film-cooling effectiveness calculations (Liu et al., 2003; Kunze et al., 2009; Narzary et al., 2009; Russin et al., 2009). Interested readers can see these papers for detailed information.

6.6.6 Flow and Thermal Field Measurements

Flow and Thermal Field Measurements with Film Cooling: Since 2000, the following selected research groups continuously using the miniature five-hole probes/thermocouples, hot/cold-wires, LDV, and PIV measurement techniques for flow and thermal field measurements with film cooling are listed here for reference: For example, flow and thermal field measurements within a contoured endwall passage with and without slot bleed injection or film cooling (Bud and Simon, 2000; Bud et al., 2000; Oke and Simon, 2000; Oke et al., 2000; Piggush and Simon, 2005); flow and thermal field measurements within a stagnation region of a film-cooled turbine vane (Cutbirth and Bogard, 2001; Polanka et al., 2001); film cooling flow field measurements with jet pulsing and periodic wakes (Couthard et al., 2006; Womack et al., 2007); PIV measurements of film cooling in an adverse pressure gradient flow (Jessen et al., 2010); PIV measurements on flat-plate film cooling with free-stream turbulence (Wright et al., 2010); velocity measurements around

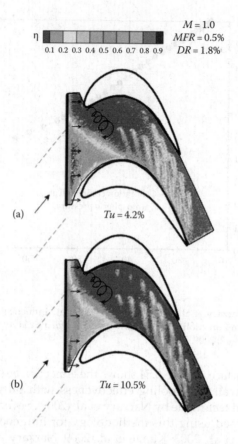

FIGURE 6.54
Adiabatic effectiveness distribution at two different free stream turbulence intensities using TSP measurement technique. (a) Tu = 4.2% and (b) Tu = 10.5%. (From Narzary, D.P. et al., Influence of coolant density on turbine blade platform film-cooling, *Proceedings of ASME Turbo-Expo 2009*, Orlando, FL, ASME GT-2009-59342, June 8–12, 2009.)

film-cooling holes with deposition (Haase and Bons, 2010). Interested readers can see these papers for detailed information.

Flow Field Measurements in a Rotating Coolant Passage: Since 2000, the following selected research groups continuously using the LDV and PIV measurement techniques for flow field measurements in a rotating coolant passage are listed here for reference: For example, flow characteristics in a rotating two-pass square duct with angled ribs (Liou and Dai, 2003; Liou et al., 2004); flow and thermal filed measurements of a row of cooling jets impinging on a rotating concave surface (Iacovides et al., 2004); PIV measurements in a rotating two-pass square cooling system with ribbed walls (Elfert et al., 2008, 2010). Interested readers can see these papers for detailed information.

6.7 Closure

A variety of experimental techniques have been presented in this chapter as an overview to the results presented in the earlier chapters. Experimenters have used different measurement techniques for obtaining results based on the requirements, test apparatus, parameter selection, cost, and time. Each experimental technique has its own advantages and disadvantages versus other techniques. This chapter provides a brief overview of all the heat-transfer and flow measurement techniques.

References

Abhari, R.S. and Epstein, A.H., 1994. An experimental study of film cooling in a rotating transonic turbine. *ASME Journal of Turbomachinery*, 116, 63–70.

Abuaf, N., Bunker, R., and Lee, C.P., 1997. Heat transfer and film cooling effectiveness in a linear airfoil cascade. *ASME Journal of Turbomachinery*, 119, 302–309.

Ahn, J.Y., Mhetras, S., and Han, J.C., 2004. Film cooling effectiveness on a gas turbine blade tip and shroud using pressure sensitive paint, *Proceedings of ASME Turbo-Expo 2004*, Vienna, Austria, June 14–17, 2004, GT 2004-53249.

Ahn, J.Y., Schobeiri, M.T., Han, J.C., and Moon, H.K., 2005. Film cooling effectiveness on the leading edge of a rotating film-cooled blade using pressure sensitive paint, *Proceedings of ASME Turbo-Expo 2005*, Reno, NV, June 6–9, 2005, Paper No. GT 2005-68344.

Ainsworth, R., Thorpe, S., and Allan, W., 2004. Unsteady heat transfer measurements from transonic turbine blades at engine representative conditions in a transient facility. ASME Paper GT2004-53835.

Ashworth, D.A., LaGraff, J.E., Schultz, D.L., and Grindrod, K.J., 1985. Unsteady aerodynamic and heat transfer processes in a transonic turbine stage. *ASME Journal of Engineering for Gas Turbines and Power*, 107, 1022–1030.

Bergholz, R.F., Steuber, G.D., and Dunn, M.G., 2000. Rotor/stator heat transfer measurements and CFD predictions for short-duration turbine rig tests. ASME Paper 2000-GT-0208.

Blair, M.F. and Bennett, J.C., 1984. Hot wire measurements of velocity and temperature fluctuations in a heated turbulent boundary layer. ASME Paper 84-GT-M 234.

Blair, M.F. and Lander, R.D., 1975. New techniques for measuring film cooling effectiveness. *ASME Journal of Heat Transfer*, 97, 539–543.

Bogard, D.G. and Cutbrith, J.M., 2003. Effects of coolant density ratio on film cooling performance on a vane. ASME Paper GT2003-38582.

Bons, J.P. and Kerrebrock, J.L., 1998. Complementary velocity and heat transfer measurements in a rotating cooling passage with smooth walls. ASME Paper 98-GT-464.

Boyle, R.J., Spuckler, C.M., Lucci, B.L., and Camperchioli, W.P., 2001. Infrared low-temperature turbine vane rough surface heat transfer measurements. *ASME Journal of Turbomachinery*, 123, 168–177.

Brauckmann, D. and von Wolfersdorf, J., 2005. Application of steady state and transient IR-thermography measurements to film cooling experiments for a row of shaped holes. ASME Paper GT-2005-68035.

Bunker, R.S. and Metzger, D.E., 1990. Local heat transfer in internally cooled turbine airfoil leading edge regions. Part I: Impingement cooling without film extraction. *ASME Journal of Turbomachinery*, 112, 451–458.

Camci, C., Kim, K., and Hippensteele, S.A., 1992. A new hue capturing technique for the quantitative interpretation of liquid crystal images used in convective heat transfer studies. *ASME Journal of Turbomachinery*, 114, 765–775.

Camci, C., Kim, K., Hippensteele, S.A., and Poinsatte, P.E., 1993. Evaluation of a hue capturing based transient liquid crystal method for high-resolution mapping of convective heat transfer on curved surfaces. *ASME Journal of Heat Transfer*, 115, 311–318.

Chambers, A.C., Gillespie, D., Ireland, P.T., and Daily, G.M., 2002. A novel transient liquid crystal technique to determine heat transfer coefficient distributions and adiabatic wall temperature in a three temperature problem. ASME Paper GT2002-30532.

Chang, S.W., Liou, T.M., and Chen, W.G., 2010. Heat transfer in a rotating rectangular channel with two opposite walls roughened with spherical protrusions at high rotation numbers. ASME Paper GT2010-22609.

Cheah, S.C., Iacovides, H., Jackson, D.C., Ji, H., and Launder, B.E., 1994. LDA investigation of the flow development through rotating U-ducts. *ASME Journal of Turbomachinery*, 118, 590–596.

Chen, P.H. and Goldstein, R.J., 1992. Convective transport phenomena on the suction surface of a turbine blade including the influence of secondary flows near the endwall. *Journal of Turbomachinery*, 114, 776–787.

Chen, P.H., Ai, D., and Lee, S.H., 1998. Effect of compound angle injection on flat-plate film cooling through a row of conical holes. ASME Paper 98-GT-459.

Cho, H.H. and Goldstein, R.J., 1995. Heat (mass) transfer and film cooling effectiveness with injection through discrete holes. Part I: Within holes and on the back surface. *ASME Journal of Turbomachinery*, 117, 451–460.

Chyu, M.K. and Bizzak, D.J., 1993. Measurement of surface temperature using a laser-induced fluorescence thermal imaging system. ASME Paper 93-GT-214.

Corrsin, S., 1947. Extended applications of the hot-wire anemometer. *Review of Scientific Instruments*, 18, 469–471.

Cutbirth, J.M. and Bogard, D.G., 2001. Thermal field and flow visualization within the stagnation region of a film cooled turbine vane. ASME Paper 2001-GT-0401.

Didier, F., Denos, R., and Arts, T., 2002. Unsteady rotor heat transfer in a transonic turbine stage. ASME Paper GT2002-30195.

Doorly, D.J. and Oldfield, M.L.G., 1985. Simulation of the effects of shock-waves passing on a turbine rotor blade. *ASME Journal of Engineering for Gas Turbines and Power*, 107, 998–1006.

Doorly, J.E. and Oldfield, M.L.G., 1986. New heat transfer gauges for use on multilayered substrates, Presented at *ASME. Gas Turbine Conference*, Dusseldorf, Germany, June 1986.

Dring, R.P. and Joslyn, H.D., 1981. Measurement of turbine rotor blade flows. *ASME Journal of Engineering for Power*, 103(2), 400–405.

Du, H., Ekkad, S.V., and Han, J.C., 1997. Effect of unsteady wake with trailing edge coolant ejection on detailed heat transfer coefficient distributions for a gas turbine blade. *ASME Journal of Heat Transfer*, 119, 242–248.

Du, H., Han, J.C., and Ekkad, S.V., 1998. Effect of unsteady wake on detailed heat transfer coefficient and film effectiveness distributions for a gas turbine blade. *ASME Journal of Turbomachinery*, 120, 808–817.

Dunn, M.G., 1984. Turbine heat flux measurements: Influence of slot injection on vane trailing edge heat transfer and influence of rotor on vane heat transfer. ASME Paper 84-GT-175.

Dunn, M.G. and Haldeman, C.W., 2003. Heat transfer measurements and predictions for the vane and blade of a rotating high-pressure turbine stage. ASME Paper GT2003-38726.

Dunn, M.G., Martin, H.L., and Stanek, M.J., 1986. Heat flux and pressure measurements and comparison with prediction for a low aspect ratio turbine stage. *ASME Journal of Turbomachinery*, 108, 108–115.

Ekkad, S.V., Du, H., and Han, J.C., 1998a. Detailed film cooling measurements on a cylindrical leading edge model: Effect of free-stream turbulence and density ratio. *ASME Journal of Turbomachinery*, 120, 799–807.

Ekkad, S.V. and Han, J.C., 1997. Detailed heat transfer distributions in two-pass square channels with rib turbulators. *International Journal of Heat and Mass Transfer*, 40(11), 2525–2537.

Ekkad, S.V. and Han, J.C., 2000. Liquid crystal thermography for turbine heat transfer and cooling measurement. *Measurement Science and Technology*, 11(7), 957–968.

Ekkad, S.V., Huang, Y., and Han, J.C., 1998b. Detailed heat transfer distributions in two-pass smooth and turbulated square channels with bleed holes. *International Journal of Heat and Mass Transfer*, 41, 3781–3791.

Ekkad, S.V., Ou, S., and Rivir, R.B., 2004. A transient infrared thermography method for simultaneous film effectiveness and heat transfer coefficient measurements from a single test. *ASME Journal of Turbomachinery*, 126, 597–603, October 2004.

Ekkad, S.V., Zapata, D., and Han, J.C., 1997. Film effectiveness over a flat surface with air and CO_2 through compound angle holes using a transient liquid crystal image method. *ASME Journal of Turbomachinery*, 119, 587–593.

Elfert, M., Schroll, M., and Foerster, W., 2010. PIV-measurement of secondary flow in a rotating two-pass cooling system with an improved sequencer technique. ASME Paper GT2010-23510.

Elfert, M., Voges, M., and Klinner, J., 2008. Detailed flow investigation using PIV in a rotating square-sectioned two-pass cooling system with ribbed walls. ASME Paper GT2008-51183.

Epstein, A.H., Kerrebrock, J.L., Koo, J.J., and Preiser, U.Z., 1985. Rotating effects on impingement cooling. GTL Report No. 184, MIT.

Eriksen, V.L. and Goldstein, R.J., 1974. Heat transfer and film cooling following injection through inclined circular holes. *ASME Journal of Heat Transfer*, 96, 239–245.

Fan, C.S. and Metzger, D.E., 1987. Effect of channel aspect ratio on heat transfer in rectangular passage sharp 180-deg turns. ASME Paper 87-GT-113.

Friedrichs, S., Hodson, H.P., and Dawes, W.N., 1996. Distribution of film-cooling effectiveness on a turbine endwall measured using the ammonia and diazo technique. *ASME Journal of Turbomachinery*, 118, 613–621.

Gao, Z., Narzary, D.P., and Han, J.C., 2008. Turbine blade platform film cooling with typical stator-rotor purge flow and discrete-hole film cooling, *Proceedings of ASME Turbo-Expo 2008*, Berlin, Germany, June 9–13, 2008, ASME GT2008-50286.

Glezer, B., Moon, H.K., Kerrebrock, J., Bons, J., and Guenette, G., 1998. Heat transfer in a rotating radial channel with swirling internal flow. ASME Paper 98-GT-214.

Gogineni, S.P., Trump, D.D., Rivir, R.B., and Pestian, P.J., 1996. PIV measurements of periodically forced flat plate film cooling flows with high free-stream turbulence. ASME Paper 96-GT-236.

Goldstein, R.J. and Chen, H.P., 1984. Film cooling on a gas turbine blade near the end wall, *American Society of Mechanical Engineers, 29th International Gas Turbine Conference and Exhibit*, Amsterdam, the Netherlands.

Goldstein, R.J. and Chen, H.P., 1985. Film cooling on a gas turbine near the endwall. *ASME Journal of Engineering for Gas Turbines and Power*, 107, 117–122.

Goldstein, R.J., Jin, P., and Olson, R.L., 1999. Film cooling effectiveness and mass/heat transfer coefficient downstream of one row of discrete holes. *ASME Journal of Turbomachinery*, 121, 225–232.

Goldstein, R.J. and Spores, R.A., 1988. Turbulent transport on the endwall in the region between adjacent turbine blades. *ASME Journal of Heat Transfer*, 110, 862–869.

Goldstein, R.J. and Taylor, J.R., 1982. Mass transfer in the neighborhood of jets entering a crossflow. *ASME Journal of Heat Transfer*, 104, 715–721.

Goldstein, R.J., Wang, H.P., and Jabbari, M.Y., 1995. The influence of secondary flows near the endwall and boundary layer disturbance on convective transport from a turbine blade. *ASME Journal of Turbomachinery*, 117, 657–663.

Gritsch, M., Schulz, A., and Wittig, S., 1998. Adiabatic wall effectiveness measurements of film cooling holes with expanded exits. *ASME Journal of Turbomachinery*, 120, 549–556.

Guenette, G.R., Epstein, A.H., Giles, M.B., Hanes, R., and Norton, R.J.G., 1989. Fully scaled transonic turbine rotor heat transfer measurements. *ASME Journal of Turbomachinery*, 111, 1–7.

Guo, S.M., Lai, C.C., Jones, T.V., Oldfield, M.L.G., Lock, G.D., and Rawlinson, A.J., 2000. Influence of surface roughness on heat transfer and effectiveness for a fully film cooled nozzle guide vane measured by wide band liquid crystals and direct heat flux gauges. ASME Paper 2000-GT-0204.

Haase, K. and Bons, J., 2010. Velocity measurements around film cooling holes with deposition. ASME Paper GT2010-22358.

Haldeman, C., Mathison, R., Dunn, M., Southworth, S., Harral, J., and Heitland, G., 2006. Aerodynamic and heat flux measurements in a single stage fully cooled turbine—Part I: Experimental approach. ASME Paper GT2006-90966.

Han, J.C., 1988. Heat transfer and friction characteristics in rectangular channels with rib turbulators. *ASME Journal of Heat Transfer*, 110, 321–328.

Han, J.C., Chandra, P.R., and Lau, S.C., 1988. Local heat/mass transfer distributions around sharp 180° turns in two-pass smooth and rib-roughened channels. *ASME Journal of Heat Transfer*, 110, 91–98.

Han, J.C. and Rallabandi, A.P., 2010. Turbine blade film cooling using PSP technique. *Frontiers in Heat and Mass Transfer*, 1(1), 1–21, ISSN 2151-8629.

Han, J.C. and Zhang, P., 1991. Effect of rib-angle orientation on local mass transfer distribution in a three-pass rib-roughened channel. *ASME Journal of Turbomachinery*, 113, 123–130.

Han, J.C., Zhang, P., and Lee, C.P., 1992. Influence of surface heat flux ratio on heat transfer augmentation in square channels with parallel, crossed, and V-shaped angled ribs. *ASME Journal of Turbomachinery*, 114, 872–880.

Han, J.C., Zhang, L., and Ou, S., 1993. Influence of unsteady wake on heat transfer coefficients from a gas turbine blade. *ASME Journal of Heat Transfer*, 115, 904–911.

Haring, M., Bolcs, A., Harasgama, S.P., and Richter, J., 1995. Heat transfer measurements on turbine airfoils using the naphthalene sublimation technique. *ASME Journal of Turbomachinery*, 117, 432–437.

Hay, N., Lampard, D., and Saluja, C.L., 1985. Effect of cooling films on the heat transfer coefficient on a flat plate with zero mainstream pressure gradient. *ASME Journal of Engineering for Gas Turbines and Power*, 107, 105–110.

Hibbs, R., Acharya, S., Chen, Y., Nikitopoulos, D., Myrum, T., 1998. Heat transfer in a two-pass internally ribbed turbine blade coolant channel with cylindrical vortex generators. *ASME Journal of Turbomachinery*, 120, 724–734.

Hippensteele, S.A., Russell, L.M., and Stepka, F.S., 1983. Evaluation of a method for heat transfer measurements and thermal visualization using a composite of a heater element and liquid crystals. *ASME Journal of Heat Transfer*, 105, 184–189.

Iacovides, H., Jackson, D.C., Ji, H., Kelemenis, G., Launder, B.E., and Nikas, K., 1998. LDA study of the flow development through an orthogonally rotating u-bend of strong curvature and rib-roughened walls. *ASME Journal of Turbomachinery*, 120, 386–391.

Iacovides, H., Launder, B.E., Kounadis, D., Li, J., and Xu, Z., 2004. Experimental study of the flow and thermal development of a row of cooling jets impinging on a rotating concave surface. ASME Paper GT2004-53244.

Iliopoulou, V., Denos, R., Billiard, N., and Arts, T., 2004. Time-averaged and time-resolved heat flux measurements on a turbine stator blade using two-layered thin-film gauges. ASME Paper GT2004-53437.

Ito, S., Goldstein, R.J., and Eckert, E.R.G., 1978. Film cooling of a gas turbine blade. *ASME Journal of Engineering for Power*, 100, 476–481.

Jessen, W., Konopka, M., and Schroeder, W., 2010. Particle-image velocimetry measurements of film cooling in an adverse pressure gradient flow. ASME Paper GT2010-22411.

Jones, T., Anthony, R., and LaGraff, J., 2004. High frequency surface heat flux imaging of bypass transition. ASME Paper GT2004-54162.

Kapur, D.N. and McLeod, N., 1976. Vapor pressure determination for certain high-boiling liquids by holography. *Industrial Engineering in Chemistry*, 15, 50–54.

Kercher, D.M. and Tabakoff, W., 1970. Heat transfer by a square array of round air jets impinging perpendicular to a flat surface, including the effect of spent air. *ASME Journal of Engineering for Power*, 92, 73–82.

Kim, Y.W., Downs, J.P., Soechting, F.O., Abdel-Messeh, W., Steuber, G.D., and Tanrikut, S., 1995. A summary of the cooled turbine blade tip heat transfer and film effectiveness investigations performed by Dr. D.E. Metzger. *ASME Journal of Turbomachinery*, 117, 1–11.

Kim, J. and Simon, T.W., 1987. Measurements of the turbulent transport of heat and momentum in convexly curved boundary layers: Effects of curvature, recovery, and free-stream turbulence. ASME Paper 87-GT-199.

Kohli, A. and Bogard, D.G., 1998. Fluctuating thermal field in the near-hole region for film cooling flows. *ASME Journal of Turbomachinery*, 120, 86–91.

Kunze, M., Vogeler, K., Brown, G., Prakash, C., and Landis, K., 2009. Aerodynamic and endwall film cooling investigations of a gas turbine nozzle guide vane applying temperature-sensitive paint. ASME Paper GT-2009-59412.

Kwak, J.S., Ahn, J.Y., Han, J.C., Lee, C.P., Bunker, R.S., Boyle, R.J., and Gaugler, R.E., 2003. Heat transfer coefficients on the squealer-tip and near-tip regions of a gas turbine blade with single or double squealer. *ASME Journal of Turbomachinery*, 125, 778–787.

Kwak, J.S. and Han, J.C., 2003a. Heat transfer coefficient and film cooling effectiveness on a gas turbine blade tip. *ASME Journal of Heat Transfer*, 125, 494–502.

Kwak, J.S. and Han, J.C., 2003b. Heat transfer coefficient and film cooling effectiveness on the squealer tip of a gas turbine blade. *ASME Journal of Turbomachinery*, 125, 648–657.

Lander, R.D., Fish, R.W., and Suo, M., 1972. External heat-transfer distribution on a film cooled turbine vane. *AIAA Journal of Aircraft*, 9, 707–714.

Langston, L.S. and Boyle, M.T., 1982. A new surface-streamline flow-visualization technique. *Journal of Fluid Mechanics*, 125, 53–57.

Leiss, C., 1975. Experimental investigation of film cooling with injection from a row of holes for the application to gas turbine blades. *ASME Journal of Engineering for Power*, 97, 21–27.

Ligrani, P.M., Singer, B.A., and Baun, L.R., 1989. Miniature five-hole pressure probe for measurement of three-mean velocity components in low-speed flows. *Journal of Physics Engineering: Scientific Instrumentation*, 22, 868–876.

Ling, J.P., Turner, L., and Ireland, P.T., 2003. A technique for processing transient heat transfer, liquid crystal experiments in the presence of lateral conduction. ASME Paper GT2003-38446.

Liou, T.M. and Dai, G.Y., 2003. Pressure and flow characteristics in a rotating two-pass square duct with 45-deg angled ribs. ASME Paper GT2003-38346.

Liou, T.M. and Hwang, J.J., 1992. Turbulent heat transfer augmentation and friction in periodic fully developed channel flows. *ASME Journal of Heat Transfer*, 114, 56–64.

Liou, T.M., Li, Y.C., and Hwang, Y.S., 2004. Flowfield and pressure measurements in a rotating two-pass duct with staggered rounded ribs skewed 45-deg to the flow. ASME Paper GT2004-53173.

Liu, Q., Douglass, C.J., Kapat, J.S., and Qiu, J., 2003. Applicability of temperature sensitive paints for measurement of surface temperature distribution. ASME Paper GT2003-38591.

Lynch, S. and Thole, K., 2007. The effect of combustor-turbine interface gap leakage on the endwall heat transfer for a nozzle guide vane. ASME Paper GT2007-27867.

Mathison, R., Haldeman, C., and Dunn, M., 2010. Aerodynamics and heat transfer for a cooled one and one-half stage high-pressure turbine: Part I: Vane inlet temperature profile generation and migration. ASME Paper GT2010-22716.

Mattern, Ch. and Hennecke, D.K., 1996. The influence of rotation on impingement cooling. ASME Paper 96-GT-161.

McLeod, N. and Todd, R.B., 1973. The experimental determination of wall fluid mass-transfer coefficients using plasticised polymer surface coatings. *International Journal of Heat and Mass Transfer*, 16, 485–503.

Mehendale, A.B. and Han, J.C., 1992. Influence of high mainstream turbulence on leading edge film cooling heat transfer. *ASME Journal of Turbomachinery*, 114, 707–715.

Metzger, D.E. and Larson, D.E., 1986. Use of melting point surface coatings for local convection heat transfer measurements in rectangular channel flows with 90-deg turns. *ASME Journal of Heat Transfer*, 108, 48–54.

Mhetras, S.P., Han, J.C., and Rudolph, R., 2007. Effect of flow parameter variations on full coverage film-cooling effectiveness for a gas turbine blade, *Proceedings of ASME Turbo-Expo 2007*, Montreal, Quebec, Canada, May 14–17, 2007, ASME GT2007-27071.

Mick, W.J. and Mayle, R.E., 1988. Stagnation film cooling and heat transfer, including its effect within the hole pattern. *ASME Journal of Turbomachinery*, 110, 66–72.

Moon, H.K., Jaiswal, R., 2000. Cooling effectiveness measurements with thermal radiometry in a turbine cascade, *ASME Turbo Expo 2000*, Munich, Germany.

Narzary, D.P., Liu, K.C., and Han, J.C., 2009. Influence of coolant density on turbine blade platform film-cooling, *Proceedings of ASME Turbo-Expo 2009*, Orlando, FL, June 8–12, 2009, ASME GT-2009-59342.

Narzary, D.P., Liu, K.C., Rallabandi, A.P., and Han, J.C., 2010. Influence of coolant density on turbine blade film-cooling using pressure sensitive paint technique, *Proceedings of ASME Turbo-Expo 2010*, Glasgow, U.K., June 14–18, 2010, ASME GT-2010-22781.

Noel, B.W., Borella, H.M., Lewis, W., Turley, W.D., Beshears, D.L., Capps, G.J., Cates, M.R., Muhs, J.D., and Tobin, K.W., 1990. Evaluating thermographic phosphors in an operating turbine engine. ASME Paper 90-GT-266.

O'Dowd, D., Zhang, Q., Ligrani, P., He, L., and Friedrichs, S., 2009. Comparison of heat transfer measurement techniques on a transonic turbine blade tip. ASME Paper GT2009-59376.

Oke, R.A. and Simon, T.W., 2000. Measurements in film cooling with lateral injection: Adiabatic effectiveness values and temperature fields. ASME Paper 2000-GT-0597.

Oke, R.A., Simon, T.W., Burd, S.W., and Vahlberg, R., 2000. Measurements in a turbine cascade over a contoured endwall: Discrete hole injection of bleed flow. ASME Paper 2000-GT-0214.

Oldfield, M.L.G., Bryanston-Cross, P.J., Nicholson, J.H., and Scrivener, C.T.J., 1986. A study of passage flow through a cascade of turbine blades using image plane holographic interferometry. AGARD CP 399, Paper 33.

Oldfield, M.L.G., Jones, T.V., and Schultz, D.L., 1978. On-line computer for transient turbine cascade instrumentation, *IEEE Transactions on Aerospace Electronic Systems* AES-14(5).

Ou, S., Han, J.C., Mehendale, A.B., and Lee, C.P., 1994. Unsteady wake over a linear turbine blade cascade with air and CO_2 film injection. Part I: Effect on heat transfer coefficients. *ASME Journal of Turbomachinery*, 116, 721–729.

Park, C.W. and Lau, S.C., 1998. Effect of channel orientation of local heat (mass) transfer distributions in a rotating two-pass square channel with smooth walls. *ASME Journal of Heat Transfer*, 120, 624–632.

Pedersen, D.R., Eckert, E.R.G., and Goldstein, R.J., 1977. Film-cooling with large density differences between the mainstream and secondary fluid measured by the heat-mass transfer analogy. *ASME Journal of Heat Transfer*, 99, 620–627.

Pietrzyk, J.R., Bogard, D.G., and Crawford, M.E., 1990. Effects of density ratio on the hydrodynamics of film cooling. *ASME Journal of Turbomachinery*, 112, 437–443.

Piggush, J.D. and Simon, T., 2005. Flow measurements in a first stage nozzle cascade having endwall contouring, leakage and assembly features. ASME Paper GT2005-68340.

Polanka, M.D., Cutbirth, J.M., and Bogard, D.G., 2001. Three component velocity field measurements in the stagnation region of a film cooled turbine vane. ASME Paper 2001-GT-0402.

Prandtl, L. and Tietjens, O.G., 1934. *Applied Hydro- and Aeromechanics*. McGraw-Hill, New York.

Pritchard, D.H., 1977. US color television fundamentals: A review. *IEEE Transactions, Consumer Electronics*, CE-23, 467–478.

Rivir, R.B., Roquemore, W.M., and McCarthy, J.W., 1987. Visualization of film cooling flows using laser sheet light. AIAA Paper 87-1914.

Russin, R.A., Alfred, D., and Wright, L.M., 2009. Measurement of detailed heat transfer coefficient and film cooling effectiveness distributions using PSP and TSP. ASME Paper GT2009-59975.

Schabaker, J., Bolcs, A., and Johnson, B.V., 1998. PIV investigation of the flow characteristics in an internal coolant passage with two ducts connected by a sharp 180° bend. ASME Paper 98-GT-544.

Scherer, V., Wittig, S., Morad, K., and Mikhael, N., 1991. Jets in a crossflow: Effects of hole spacing to diameter ratio on the spatial distribution of heat transfer. ASME Paper 91-GT-356.

Schultz, D.L. and Jones, T.V., 1973. Heat transfer measurements in short duration hypersonic facilities, *NATO Advanced Group for Aerospace Research and Development AG-165*, AGARD Report AD0758590.

Simonich, J.C. and Moffat, R.J., 1984. Liquid crystal visualization of surface heat transfer on a concavely curved turbulent boundary layer. *ASME Journal of Engineering for Gas Turbines and Power*, 106, 619–627.

Soechting, F.O., Landis, K.K., and Dobrowolski, R., 1987. Development of low-cost test techniques for advanced film cooling technology. AIAA Paper 87–1913.

Somawardhana, R. and Borgard, D.G., 2007. Effects of surface roughness and near hole obstructions on film cooling effectiveness. ASME Paper GT2007-28004.

Son, S., Kihm, K.D., Han, J.C., and Shon, D.K., 1999. Flow field measurements in a two-pass channel using particle image velocimetry. *ASME Journal of Heat Transfer*, 121(3), 509 (picture gallery).

Suryanarayanan, A., Mhetras, S.P., Schobeiri, M.T., and Han, J.C., 2006. Film cooling effectiveness on a rotating blade platform, *Proceedings of ASME Turbo-Expo 2006*, Barcelona, Spain, May 8–11, 2006, Paper No. GT2006-90034.

Takahashi, T.T., 1997. Measurement of air flow characteristics using seven-hole cone probes. AIAA Paper 97-0600.

Takeishi, K., Matsuura, M., Aoki, S., and Sato, T., 1990. An experimental study of heat transfer and film cooling on low aspect ratio turbine nozzles. *ASME Journal of Turbomachinery*, 112, 488–496.

Talib, A.A., Ireland, P.T., Mullender, A.J., and Neely, A.J., 2003. A novel liquid crystal image processing technique using multiple gas temperature steps to determine heat transfer coefficient distribution and adiabatic wall temperature. ASME Paper GT2003-38198.

Thole, K. and Bogard, D.G., 1994. Simultaneous temperature and velocity measurements. *Measurement Science and Technology*, 5, 435–439.

Thole, K., Gritsch, M., Schulz, A., and Wittig, S., 1998. Flowfield measurements for film cooling holes with expanded exits. *ASME Journal of Turbomachinery*, 120, 327–336.

Thrift, A., Thole, K.A., and Hada, S., 2010. Effects of an axisymmetric contoured endwall on a nozzle guide vane: Adiabatic heat transfer measurements. ASME Paper GT-2010-22970.

Thurman, D. and Poinsatte, P., 2000. Experimental heat transfer and bulk air temperature measurements for multipass internal cooling model with ribs and bleed. ASME Paper 2000-GT-0233.

Treml, K. and Lawless, P.B., 1998. Particle image velocimetry of vane-rotor interaction in a turbine stage. AIAA Paper 98-3599.

Tse, D.G.N. and Steuber, G.D., 1997. Flow in a rotating square serpentine coolant passage with skewed trips. ASME Paper 97-GT-529.

Van Treuren, K.W., Wang, Z., Ireland, P.T., and Jones, T.V., 1994. Detailed measurements of local heat transfer coefficient and adiabatic wall temperature beneath an array of impingement jets. *ASME Journal of Turbomachinery*, 116, 369–374.

Vedula, R.J. and Metzger, D.E., 1991. A method for the simultaneous determination of local effectiveness and heat transfer distributions in three-temperature convection situations. ASME Paper 91-GT-345.

Wagner, J.H., Johnson, B.V., and Hajek, T.J., 1991. Heat transfer in rotating passages with smooth walls and radial outward Flow. *ASME Journal of Turbomachinery*, 113, 42–51.

Wang, Z., Ireland, P.T., and Jones, T.V., 1990. A technique for measuring convective heat transfer at rough surfaces. ASME Paper 90-GT-300.

Wang, Z., Ireland, P.T., and Jones, T.V., 1993. An advanced method of processing liquid crystal video signals from transient heat transfer experiments. ASME Paper 93-GT-282.

Wang, Z., Ireland, P.T., Jones, T.V., and Davenport, R., 1994. A color image processing system for transient liquid crystal heat transfer experiments. ASME Paper 94-GT-290.

Womack, K.M., Volino, R., and Schultz, M.P., 2007. Measurements in film cooling flows with periodic wakes. ASME Paper GT2007-27917.

Wright, L., McClain, S.T., and Clemenson, M., 2010. Effect of freestream turbulence intensity on film cooling jet structure and surface effectiveness using PIV and PSP. ASME Paper GT2010-23054.

Yang, T.T. and Diller, T.E., 1995. Heat transfer and flow for a grooved turbine blade tip in a transonic cascade. ASME Paper 95-WA/HT-29.

Yu, Y. and Chyu, M.K., 1998. Influence of gap leakage downstream of the injection holes on film cooling performance. *ASME Journal of Turbomachinery*, 120, 541–548.

Zhang, L., Baltz, M., Padupatty, R., and Fox, M., 1999. Turbine nozzle film cooling study using the pressure sensitive paint (PSP) technique. ASME Paper 99-GT-196.

Zhou, D. and Wang, T., 1995. Effects of elevated free-stream turbulence on flow and thermal structures in transitional boundary layers. *ASME Journal of Turbomachinery*, 117, 407–412.

7

Numerical Modeling

7.1 Governing Equations and Turbulence Models

7.1.1 Introduction

Experimental techniques and results of gas turbine–related heat transfer were discussed in previous chapters. It can be seen that experimental heat transfer covers more than half a century of research, and it developed significant new design improvements to increase efficiency and durability of turbine components that are exposed to high operating temperatures. Numerical model development by Crawford (STAN series) and Patankar (SIMPLE velocity and pressure interlinkage method) opened new avenues for numerical flow and heat-transfer research. Numerical predictions provide the details that are difficult to obtain by experimental means. Moreover, the increase in computation power in desktop computers has made it economical to optimize the design parameters based on numerical analyses. Turbulence models developed by Launder, Spalding, Satyanarayana, Rodi, Speziale, and others improved the quality of the numerical prediction significantly. In this chapter, different aspects of turbine heat-transfer prediction—namely, heat transfer in airfoil, endwall, and tip gap—are discussed. Like experimental heat-transfer analysis, predictions have two broad classifications: the external and internal cooling arrangements. Most external heat-transfer predictions are done with the time-averaged turbulence model of Baldwin–Lomax and two equation models, e.g., k-ε and k-ω turbulence models, whereas internal cooling arrangements are simulated with more computation-intensive turbulence models. Details of these models are discussed in turbulence- and flow-modeling texts (Wilcox, 1993; Tannehill et al., 1997). A brief description of these models is given here. Several numerical models are used to capture the turbulence effects in a gas turbine. Based on the number of equations to be solved, turbulence modeling approaches can be broadly classified as one-equation, two-equation, and second-moment closure models. Most common models are based on a two-equation turbulence model, namely, k-ε model, low Reynolds number k-ε model, two-layer k-ε model, and k-ε model. Other popular models are

Baldwin–Lomax model, algebraic closure model, and second-moment closure model. The basic equations solved in these models are given here.

Detailed derivations of these models require a discussion of turbulence; derivations are available in books dedicated for turbulence modeling. Note that several constants in these models depend on the model developer and the problem being solved. Since most turbulence models are engineering problem solvers, i.e., constants are set to predict the experimental data without a scientific derivation, model users select model constants that perform well for their flow conditions.

7.1.2 Governing Equations

The governing equations solved for all turbulence models are the continuity equation

$$\nabla \cdot (\rho \overline{V}) = 0 \tag{7.1}$$

and the momentum equation

$$\nabla \cdot (\rho v_i \overline{V} - \mu_{eff} \nabla v_i) = -\frac{\partial p}{\partial x_i} + S_{vi} \tag{7.2}$$

where

$\overline{V}$ is the time-averaged velocity vector (u, v, w)

v_i represents the velocity component in the i direction

S_{vi} represents the source term for the momentum transport equation in the i direction

The effective viscosity μ_{eff} includes laminar viscosity μ and turbulent eddy viscosity μ_t, as

$$\mu_{eff} = \mu + \mu_t \tag{7.3}$$

The temperature transport equation can be written as

$$\nabla \cdot (\rho T \overline{V} - \alpha_{eff} \nabla T) = S_T \tag{7.4}$$

where S_T is the source term for energy transport, and the effective diffusion coefficient is given as

$$\alpha_{eff} = \frac{\mu}{Pr} + \frac{\mu_t}{Pr_t} \tag{7.5}$$

where

Pr is the molecular

Pr_t are the turbulent Prandtl numbers

7.1.3 Turbulence Models

Turbulence models are developed based on time-averaged statistical models. The local velocities are defined as a time-averaged component and a time-dependent fluctuating component:

$$U(\text{local velocity}) = u(\text{time averaged}) + u'(\text{fluctuating component}) \qquad (7.6)$$

Most turbulence models solve steady-state turbulence by time averaging the second moments of the fluctuating velocity component. In general, the turbulent kinetic energy k is the sum of turbulent energies and is defined as

$$k = \frac{1}{2}[u_i' u_i'] \qquad (7.7)$$

7.1.3.1 Standard k-ε Model

This two-equation model is the most widely used turbulence model for engineering simulations. The high Reynolds number k-ε model is based on the Boussinesque approximation of the Reynolds turbulent stresses. The turbulent eddy diffusivity is expressed in terms of turbulence parameters k and ε. Two additional scalar transport equations, one for the turbulent kinetic energy k and the other for the turbulence dissipation ε, are solved to model the turbulence effects. The eddy viscosity of this model is obtained as

$$\mu_t = \rho C_\mu \frac{k^2}{\varepsilon} \qquad (7.8)$$

The k and ε transport equations are

$$\nabla \cdot \left(\rho k \bar{V} - \frac{\mu_t}{\sigma_k} \nabla k \right) = P - \rho\varepsilon \qquad (7.9)$$

$$\nabla \cdot \left(\rho \varepsilon \bar{V} - \frac{\mu_t}{\sigma_\varepsilon} \nabla \varepsilon \right) = C_1 P \frac{\varepsilon}{k} - C_2 \rho \frac{\varepsilon^2}{k} \qquad (7.10)$$

where P is the usual Reynolds stress turbulence production term given as

$$P = \mu_t \left[2 \left[\left(\frac{\partial u}{\partial x} \right)^2 + \left(\frac{\partial v}{\partial y} \right)^2 + \left(\frac{\partial w}{\partial z} \right)^2 \right] \right.$$

$$\left. + \left(\frac{\partial u}{\partial y} + \frac{\partial v}{\partial x} \right)^2 + \left(\frac{\partial u}{\partial z} + \frac{\partial w}{\partial x} \right)^2 + \left(\frac{\partial w}{\partial y} + \frac{\partial v}{\partial z} \right)^2 \right] \qquad (7.11)$$

The near-wall region is solved by a wall function. The near-wall k and ε values are algebraically assigned in this model. The near-wall turbulent eddy viscosity is defined as

$$\mu_{eff\ near\ wall} = \frac{\mu y^+}{2.5\ln(9y^+)} \quad \text{where } y^+ = \frac{\rho y}{\mu}\left(\sqrt{C_\mu k}\right)^{0.5} \tag{7.12}$$

The near-wall diffusion coefficient for energy transport equation is given as

$$\frac{\mu_{eff\ near\ wall}}{Pr_t} = \frac{\mu}{Pr_t}\frac{y^+}{2.5\ \ln(9y^+) + P_{fn}}$$

$$\text{where } P_{fn} = \frac{9(Pr/Pr_t - 1)}{(Pr/Pr_t)^{1/4}} \tag{7.13}$$

The wall function of the standard k-ε model limits the resolution of the near-wall details. The nearest grid point for a standard model is placed outside the laminar sublayer of the turbulent boundary layer. Therefore, this model cannot capture the transition effects and also the details of a thin boundary layer that prevails on the leading edge of an airfoil.

7.1.3.2 Low-Re k-ε Model

This model does not restrict the near-wall grid distribution. Additional functions, as given in Tannehill et al. (1997), are included in the turbulence model. The model equations for the low-*Re* k-ε model follow.

The eddy viscosity of this model is obtained as

$$\mu_t = \rho C_\mu f_\mu \frac{k^2}{\varepsilon} \tag{7.14}$$

The k and ε transport equations are

$$\nabla\cdot\left(\rho k \bar{V} - \frac{\mu_t}{\sigma_k}\nabla k\right) = P - \rho\varepsilon - 2\mu\frac{k}{y^2} \tag{7.15}$$

$$\nabla\cdot\left(\rho\varepsilon\bar{V} - \frac{\mu_t}{\sigma_\varepsilon}\nabla\varepsilon\right) = C_1 P\frac{\varepsilon}{k} - C_2 f_1\rho\frac{\varepsilon^2}{k} - \left(-2\mu\frac{\varepsilon}{y^2}e^{-y^+/2}\right) \tag{7.16}$$

where the additional functions are evaluated as

$$f_\mu = 1 - e^{-0.0155y^+} \quad f_1 = 1 - 0.22e^{-(Re_T/6)^2} \tag{7.17}$$

This model allows a higher resolution at near-wall locations. However, computation time is increased due to the greater number of grids and additional functions in the model.

7.1.3.3 Two-Layer k-ε Model

This is another approach to avoid the wall function at near-wall locations. The core flow region is solved with standard k-ε model, and in the near-wall region the turbulent dissipation is specified in terms of the turbulent kinetic energy. Therefore, in essence, the model combines a one-equation model for the near-wall region and a two-equation model for the away-from-the-wall region. The dissipation and eddy viscosity in the near-wall locations are calculated as

$$\varepsilon = \frac{k^{3/2}}{l_\varepsilon} \quad \text{and} \quad \mu_t = \rho C_\mu \sqrt{k} l_\mu \tag{7.18}$$

where the dissipation length scale l_ε and viscosity length scale l_μ are defined from the local turbulent and distance-from-wall conditions. The constants and model equations for these length scales vary widely depending on the application and model developer. For example, Iacovides (1998) uses the following equations for the near-wall turbulence model:

$$l_\varepsilon = 2.55y\left[1 - \exp(-0.236y^*)\right]$$

$$l_\mu = 2.55y\left[1 - \exp(-0.016y^*)\right] \quad \text{where } y^* = \frac{y^{k1/2}}{\upsilon} \tag{7.19}$$

7.1.3.4 k-ω Model

This is a two-equation turbulence model proposed by Wilcox (1993). The turbulence transport equations involved are the turbulent kinetic energy k and a specific dissipation rate ω. This model does not need a separate near-wall treatment for the transport equations. The dissipation and specific dissipation of k are related as $\varepsilon = \beta^* k \omega$. The eddy viscosity of this model is formulated as

$$\mu_t = \rho \alpha^* \frac{k}{\omega} \quad \text{or} \quad \mu_t = \rho \frac{C_\mu}{\beta^*} \frac{k}{\omega} \tag{7.20}$$

The k and ω transport equations are

$$\nabla \cdot \left(\rho k \bar{V} - \frac{\mu_t}{\sigma_k^*} \nabla k \right) = P - \rho \beta^* \omega k \tag{7.21}$$

$$\nabla \cdot \left(\rho \omega \bar{V} - \frac{\mu_t}{\sigma_\omega} \nabla \omega \right) = \alpha P \frac{\omega}{k} - \beta \rho \omega^2 \tag{7.22}$$

where P is the usual Reynolds stress turbulence production term. The other related terms are

$$\alpha^* = \frac{1/40 + Re_T/6}{1 + Re_T/6} \qquad \alpha = \frac{5}{9} \frac{0.1 + Re_T/2.7}{1 + Re_T/2.7} \frac{1}{\alpha^*}$$

$$\beta^* = 0.09 \frac{5/18 + \left(Re_T/8 \right)^4}{\left(1 + Re_T/8 \right)^4} \qquad \beta = 3/40 \tag{7.23}$$

and the turbulent Reynolds number is defined as

$$Re_T = \frac{\rho k}{\omega \mu} \tag{7.24}$$

The near-wall boundary implementation technique is described in Wilcox (1993).

7.1.3.5 Baldwin–Lomax Model

In this model, the turbulence characteristics are solved with Reynolds averaged Navier–Stokes (RANS) equations, and the inner region is resolved with the use of vorticity. Unlike the k-ε model, this is a one-equation turbulence model. The computation time required for this model is less compared to a two-equation model, and recent publications indicate satisfactory results for external flows. The eddy viscosity in the inner layer is defined as

$$\mu_t = \rho l^2 \left| \omega \right| \tag{7.25}$$

where

$$\left| \omega \right| = \sqrt{\left(\frac{\partial v}{\partial x} - \frac{\partial u}{\partial y} \right)^2 + \left(\frac{\partial w}{\partial y} - \frac{\partial v}{\partial z} \right)^2 + \left(\frac{\partial u}{\partial z} - \frac{\partial w}{\partial x} \right)^2} \tag{7.26}$$

$$\text{and } l = ky(1 - e^{-(y^+/a^+)})$$

7.1.3.6 Second-Moment Closure Model

In this model, separate transport equations for the second moments—namely, auto- and cross-correlations of the time-dependent velocity fluctuations—are solved. Note that in the k-ε model, the turbulent kinetic energy is a combined effect of these turbulent correlations. Typically, with this model, six additional transport equations are required to be solved in a three-dimensional (3-D) turbulent flow. The eddy diffusivity in the momentum transport equation is replaced by the source terms developed from the turbulent Reynolds stress tensor $R^{ij} = u'^i u'^j$, the transport equation for the Reynolds stress tensor is given as (Chen et al., 1999)

$$\frac{\partial R^{ij}}{\partial t} + u_m \frac{\partial R^{ij}}{\partial x_m} = P^{ij} + D_u^{ij} + D_p^{ij} + D_v^{ij} + \phi^{ij} - \varepsilon^{ij} \tag{7.27}$$

where
  P^{ij} is the production term
  D_u^{ij} is diffusion by u
  D_p^{ij} is diffusion by p'
  D_v^{ij} is viscous diffusion
  ϕ^{ij} is the pressure strain
  ε^{ij} is viscous dissipation

Detailed expressions for these individual terms are given in Chen et al. (1999). These expressions are long and involve empirical functions.

7.1.3.7 Algebraic Closure Model

This model is a compromise between the second-moment closure model and low-resolution two-equation turbulence model. Instead of solving iterative transport equations for turbulent fluctuations, algebraic equations are used at each node. The algebraic closure for turbulent stresses (Launder, 1971) can be given as

$$\frac{\overline{u_i u_{j'}}}{k}(P - \varepsilon) = P_{ij} + \phi_{ij} - \varepsilon_{ij} \tag{7.28}$$

where
  P_{ij} is a production term
  ϕ_{ij} is a pressure strain term
  ε_{ij} is the dissipation term

The closure of the pressure strain term is given by Launder et al. (1975) as

$$\phi_{ij} = -C_{s1}\frac{\varepsilon}{k}\left(\overline{u_i u_j} - \frac{2}{3}\delta_{ij}k\right) - C_{s2}\left(P_{ij} - \frac{2}{3}P\delta_{ij}\right)$$

$$+ C_{slw}\frac{\varepsilon}{k}\left(\overline{u_i u_j} - \frac{2}{3}\delta_{ij}k\right)f\left(\frac{l}{x_w}\right) + C_{s2w}(P_{ij} - D_{ij})f\left(\frac{l}{x_w}\right) \tag{7.29}$$

The term $f(l/x_w)$ represents a near-wall function that accounts for near-wall viscous damping effects. This term is expressed as

$$f\left(\frac{l}{x_w}\right) = \frac{1}{C_w}\frac{k^{3/2}}{\varepsilon}\left(\frac{1}{x_1^m} + \frac{1}{x_2^m} + \frac{1}{y_1^m} + \frac{1}{y_2^m}\right)^{1/m} \tag{7.30}$$

where xs and ys are the distance measured from four surrounding walls.

The P_{ij} and D_{ij} terms in x–y coordinates can be represented as

$$P_{xy} = -\overline{v'^2}\frac{\partial u}{\partial y} - \overline{u'^2}\frac{\partial v}{\partial x} \qquad D_{xy} = -\overline{u'^2}\frac{\partial u}{\partial y} - \overline{v'^2}\frac{\partial v}{\partial x}$$

$$P_{xx} = -2\overline{u'^2}\frac{\partial u}{\partial x} - 2\overline{u'v'}\frac{\partial u}{\partial y} \qquad D_{xx} = -2\overline{u'^2}\frac{\partial u}{\partial x} - 2\overline{u'v'}\frac{\partial v}{\partial x} \tag{7.31}$$

$$P_{yy} = -2\overline{v'^2}\frac{\partial v}{\partial y} - 2\overline{u'v'}\frac{\partial v}{\partial x} \qquad D_{yy} = -2\overline{v'^2}\frac{\partial v}{\partial y} - 2\overline{u'v'}\frac{\partial u}{\partial y}$$

The Reynolds stresses can now be expanded as

$$\frac{\overline{u'^2}}{k} = \frac{\Lambda'}{\varepsilon}\left(-2\overline{u'^2}\frac{\partial u}{\partial x} - 2\overline{u'v'}\frac{\partial u}{\partial y} - \frac{2}{3}P\right)$$

$$+ \frac{\Lambda''}{\varepsilon}\left(-2\overline{u'^2}\frac{\partial u}{\partial x} - 2\overline{u'v'}\frac{\partial u}{\partial y} - 2\overline{u'^2}\frac{\partial u}{\partial x} - 2\overline{u'v'}\frac{\partial v}{\partial x}\right) + \frac{2}{3}$$

$$\frac{\overline{v'^2}}{k} = \frac{\Lambda'}{\varepsilon}\left(-2\overline{v'^2}\frac{\partial v}{\partial y} - 2\overline{u'v'}\frac{\partial v}{\partial x} - \frac{2}{3}P\right)$$

$$\tag{7.32}$$

$$+ \frac{\Lambda''}{\varepsilon}\left(-2\overline{v'^2}\frac{\partial v}{\partial y} - 2\overline{u'v'}\frac{\partial v}{\partial x} - 2\overline{v'^2}\frac{\partial v}{\partial y} - 2\overline{u'v'}\frac{\partial u}{\partial y}\right) + \frac{2}{3}$$

$$\frac{\overline{u'v'}}{k} = \frac{\Lambda'}{\varepsilon}\left(-\overline{v'^2}\frac{\partial u}{\partial y} - \overline{u'^2}\frac{\partial v}{\partial x}\right)$$

$$+ \frac{\Lambda''}{\varepsilon}\left(-\overline{v'^2}\frac{\partial u}{\partial y} - \overline{u'^2}\frac{\partial v}{\partial x} - \overline{u'^2}\frac{\partial u}{\partial y} - \overline{v'^2}\frac{\partial v}{\partial x}\right)$$

where

$$\Lambda' = \frac{1-C_{s2}}{(P/\varepsilon)-1+C_{sl}-C_{slw}f(l/x_w)} \quad \Lambda'' = \frac{C_{s2w}f(l/x_w)\Lambda'}{1-C_{s2}} \tag{7.33}$$

7.2 Numerical Prediction of Turbine Heat Transfer

7.2.1 Introduction

The accurate prediction of heat transfer in high-pressure turbine stages has long been recognized as a key to improve gas turbine performance and engine life. Development of faster computers with more memory makes it feasible to compute more details of heat-transfer analysis. The main difficulties in predictions are proper turbulence models and rotor–stator interlinkages. Simoneau and Simon (1993) summarized the convective heat-transfer prediction in a gas path.

7.2.2 Prediction of Turbine Blade/Vane Heat Transfer

There are many two-dimensional (2-D) viscid and inviscid predictions of airfoils. Among significant contributors, Abhari et al. (1991) examined the time-resolved aerodynamics and heat transfer in a transonic turbine rotor. Figure 7.1 shows the grid used by them. The calculation domain was time

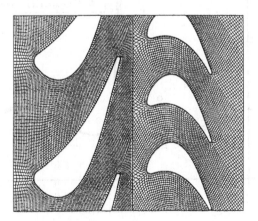

FIGURE 7.1
O-type grids used in unsteady-nozzle and blade calculation by Abhari et al. (From Abhari, R.S. et al., Comparison of time-resolved turbine rotor blade heat transfer measurements and numerical calculations, *ASME International Gas Turbine and Aeroengine Congress and Exposition*, Orlando, FL, ASME Paper 91-GT-268, June 3–6, 1991.)

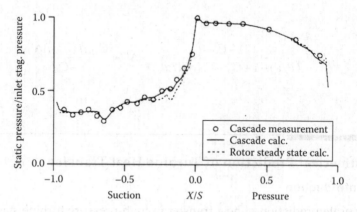

FIGURE 7.2
Static pressure predictions in a cascade. (From Abhari, R.S. et al., Comparison of time-resolved turbine rotor blade heat transfer measurements and numerical calculations, *ASME International Gas Turbine and Aeroengine Congress and Exposition*, Orlando, FL, ASME Paper 91-GT-268, June 3–6, 1991.)

dependent; and laminar, turbulent, steady, and unsteady heat-transfer predictions were compared with experimental data. Figure 7.2 shows the static pressure profile on a rotor cascade. Comparison of numerical and experimental work indicates that the quality of pressure prediction is acceptable. Two solution methods, cascade calculation and the steady-state rotor predictions, do not show any significant differences. Figure 7.3 shows their

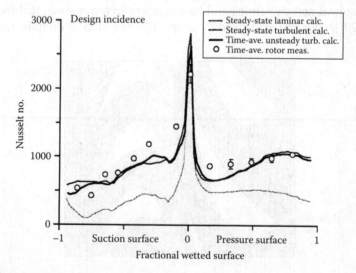

FIGURE 7.3
Comparison of measured heat-transfer coefficient with fully laminar and fully turbulent steady-state calculations. (From Abhari, R.S. et al., Comparison of time-resolved turbine rotor blade heat transfer measurements and numerical calculations, *ASME International Gas Turbine and Aeroengine Congress and Exposition*, Orlando, FL, ASME Paper 91-GT-268, June 3–6, 1991.)

heat-transfer prediction. Their predictions are compared with experimental data. Results indicate that the quality of heat-transfer prediction is inferior to the pressure profile that was shown in Figure 7.2. The near-surface effects dominate the surface heat transfer, and clearly numerical models could not capture the turbulence effects near the airfoil surface. Both laminar and turbulent predictions show a high Nusselt number at the stagnation point, and the laminar predictions are lower on the airfoil surface. There is relatively little difference between the steady-state turbulent and the time-averaged unsteady calculations. In general, the measured Nusselt number is greater than or equal to the predicted Nusselt numbers. Abhari et al. (1991) did not include the transition to turbulent boundary layer in their model. Predictions were done either with fully turbulent flow assumptions or with fully laminar flow. Biswas and Fukuyama (1993) calculated boundary layer transition with an improved low-Reynolds number turbulence model. In another study, Yang and Luo (1996) predicted flow transition with a k-ω model.

Figure 7.4 shows the measured and predicted unsteady Nusselt numbers. Along both the suction and pressure surfaces, the flow is unsteady in both measurement and prediction as they are in qualitative agreement. The difference between measurement and prediction is more in the leading edge, where predictions show a much larger degree of unsteadiness than in the experiment. The unsteadiness in the flow is primarily due to wake impingement and moving shock patterns. Among other contributors, Harasgama et al. (1993) used the 2-D boundary layer to predict heat transfer of turbine airfoil. Hurst et al. (1995) calculated heat transfer in transonic and supersonic flows. Yamamoto et al. (1995) used a quasi-unsteady study to predict wake interaction of rotor stator cascades.

Dunn et al. (1992) reported the time-averaged Stanton numbers and the surface-pressure distribution for the first-stage vane row and the first-stage blade row. The first-stage vane Stanton number distributions are compared with predictions obtained using a quasi-, 3-D Navier–Stokes solution and STAN5. Figure 7.5 shows the measured and predicted Stanton numbers for the vane at 50% span for a Reynolds number of 250,000. There is little difference among numerical predictions with different models at this Reynolds number. A lower Reynolds number showed more variations in prediction with different numerical models. However, the stagnation results are predicted reasonably well with all models. The figure shows that on the pressure surface, all of the predictions are in good agreement with each other but underpredict the Stanton number significantly. The trailing side of the suction surface is predicted reasonably well by all models.

Figure 7.6 shows the heat-transfer coefficient for a von Karman Institute blade. Unlike the prediction of Dunn et al. (1992), this figure shows that predictions by Harasgama et al. (1993) are higher on the pressure side, and predictions differ significantly from each other on the suction side. Flow conditions are closer to the engine conditions, with a free-stream turbulence

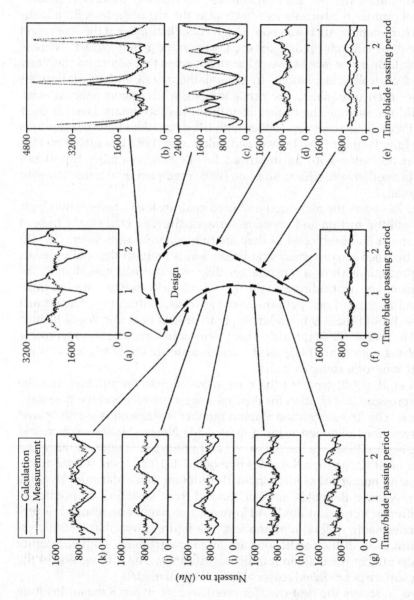

FIGURE 7.4

Time-resolved predictions of unsteady wake effects on heat-transfer coefficient. (From Abhari, R.S. et al., Comparison of time-resolved turbine rotor blade heat transfer measurements and numerical calculations, *ASME International Gas Turbine and Aeroengine Congress and Exposition*, Orlando, FL, ASME Paper 91-GT-268, June 3–6, 1991.)

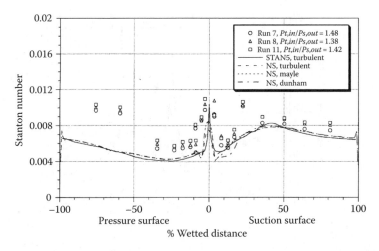

FIGURE 7.5
Stanton number distributions on the first vane with $Re = 250,000$. (From Dunn, M.G. et al., Time-averaged heat transfer and pressure measurements and comparison with prediction for a two-stage turbine, *ASME International Gas Turbine and Aeroengine Congress and Exposition*, Cologne, Germany, ASME Paper 92-GT-194, June 1–4, 1992.)

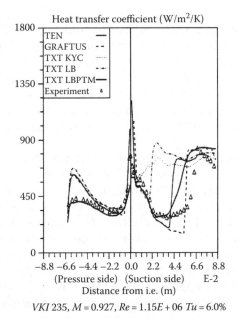

VKI 235, M = 0.927, Re = 1.15E + 06 Tu = 6.0%

FIGURE 7.6
Heat-transfer coefficient predictions by Harasgama et al. (From Harasgama, S.P. et al., Calculation of heat transfer to turbine blading using two-dimensional boundary layer methods, *ASME International Gas Turbine and Aeroengine Congress and Exposition*, Cincinnati, OH, ASME Paper 93-GT-79, May 24–27, 1993.)

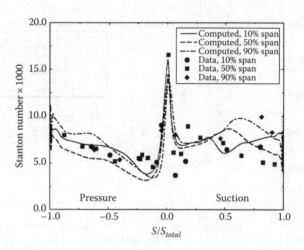

FIGURE 7.7
Comparison of predicted Stanton numbers with C grids. (From Ameri, A.A. and Steinthorsson, E., Prediction of unshrouded rotor blade tip heat transfer, *ASME International Gas Turbine and Aeroengine Congress and Exposition*, Houston, TX, ASME Paper 95-GT-142, June 5–8, 1995.)

level of 6%, and the flow Mach number is close to unity. Recent predictions by Ameri and Steinthorsson (1995, 1996) are shown in Figure 7.7. This plot also includes experimental Stanton number measurement by Dunn and Kim (1992). Note that predictions for the 90% span location show best agreement with experimental data. Predictions at other span locations did not show such good agreement between experiment and numerical prediction. Ameri and Steinthorsson (1995) used a multiblock grid with a TRAF3D computation model.

Boyle and Jackson (1997) predicted heat transfer in different nozzle geometries. Boyle (1994) used the Navier–Stokes analysis to predict the change in turbine efficiency resulting from the changes in blade surface roughness or incidence flow angles. The results of a midspan Navier–Stokes analysis are combined with those from a quasi-3-D flow analysis to determine turbine performance over a range of incidence flow angles. The validity of the approach is verified by comparisons with experimental data for a turbine with both smooth and rough blades tested over a wide range of blade incidence flow angles. Their analysis overpredicts the efficiency at positive incidences but is in reasonably good agreement with the experimental data for negative incidences. Among other contributors, Saxer and Felici (1996) predicted 3-D hot streak migration and shock interaction in turbine stages.

Luo and Lakshminarayana (1997) predicted the boundary layer development and heat transfer on transonic nozzle vanes. They used a low-Reynolds number k-ε model. Their predictions indicate that the predictions are

TABLE 7.1

Functions Used in Low-*Re* k-ε Model

Model	f_μ	f_1	f_2
CH	$1 - \exp(-0.0115y^+)$	1.0	$1 - 0.22\exp(-Re_t^2/36)$
LB	$[1 - \exp(-0.0165Re_y)]^2$ $(1 + 20.5/Re_t)$	$1 + (0.06/f_\mu)^3$	$1 - \exp(-Re_t^2)$
FLB	$0.4f_w/\sqrt{(Re_t)} +$ $\left[1 - 0.4f_w/\sqrt{(Re_t)}\right]$ $\left[1 - \exp\left(-Re_y/42.63\right)\right]^3$	1.0	$\left[1 - 0.22\exp\left(-Re_t^2/36\right)\right]t_w^2$

significantly dependent on the model selected. Prediction by the FLB model is the best among the models tested. These models differ in the formulation of f_μ, f_1, and f_2. These functions are given in Table 7.1, where $Re_y = \sqrt{ky/\upsilon}$; and $Re_t = k^2/\upsilon\varepsilon$ are the turbulent Reynolds numbers. Prediction and experimental data show qualitative agreement, but quantitative agreement is not satisfactory.

7.2.3 Prediction of the Endwall Heat Transfer

Ameri and Arnone (1994, 1996) have used the Baldwin–Lomax turbulence model to predict heat transfer of a rotor blade and endwall. Figure 7.8 shows

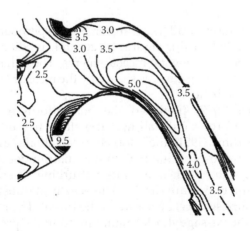

FIGURE 7.8
Endwall Stanton numbers (×1000) for constant surface temperature condition. (From Ameri, A.A. and Arnone, A., Prediction of turbine blade passage heat transfer using a zero and a two-equation turbulence model, *ASME International Gas Turbine and Aeroengine Congress and Exposition*, The Hague, the Netherlands, ASME Paper 94-GT-122, June 13–16, 1994.)

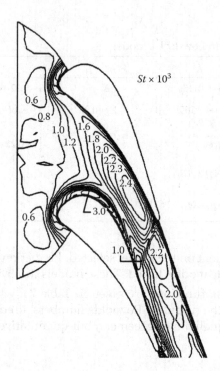

$St \times 10^3$

FIGURE 7.9

Hub endwall Stanton number distribution for stator of a transonic turbine. (From Boyle, R.J. and Giel, P.W., Prediction of nonuniform inlet temperature effects on vane and rotor heat transfer, *ASME International Gas Turbine and Aeroengine Congress and Exhibition*, Orlando, FL, ASME Paper 97-GT-133, June 2–5, 1997.)

their prediction of the endwall heat transfer. Both experimental and numerical results are plotted in this figure. In general, predictions show good agreement with the experimental measurements. The enhancement of heat transfer close to the leading edge, as shown by the measurement, is captured by the numerical prediction.

Boyle and Giel (1997) predicted the nonuniform inlet temperature effects on vane and rotor endwall heat transfer. They used RVC3D code as described in Chima (1991) and adopted an algebraic turbulence model. Figures 7.9 and 7.10 show the Stanton number distribution for hub and casing endwalls of a stator blade of a transonic turbine. Overall heat-transfer patterns in hub and casing are similar. The peak heat transfer is in a region close to the pressure surface, upstream of the throat. The hub heat transfer is influenced by the passage shock emanating from the pressure-side trailing edge. This shock effect is diffused before reaching the suction surface. The exit flow is subsonic. There is no evidence of a passage shock in the casing heat-transfer results.

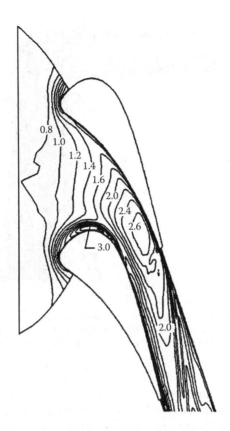

FIGURE 7.10
Casing endwall Stanton number distribution for stator of a transonic turbine. (From Boyle, R.J. and Giel, P.W., Prediction of nonuniform inlet temperature effects on vane and rotor heat transfer, *ASME International Gas Turbine and Aeroengine Congress and Exhibition*, Orlando, FL, ASME Paper 97-GT-133, June 2–5, 1997.)

The endwall flow is strongly affected by the compressibility effects. Koiro and Lakshminarayana (1998) used a low-Reynolds number k-ε model to predict this effect. They compared predicted secondary flow with the measured secondary flow in the endwall region. Their prediction evaluated a multigrid approach and showed that the convergence can be achieved faster with multigrid.

7.2.4 Prediction of Blade Tip Heat Transfer

Ameri and Steinthorsson (1995) predicted the unshrouded rotor blade tip flow and associated heat transfer. Figure 7.11 shows the flow predictions in the tip gap region for 3-D airfoil simulation. These flow predictions are relative to the blade tip. From this flow prediction, the suction-side tip leakage

FIGURE 7.11
Leakage flow predictions through the tip gap region. (From Ameri, A.A. and Steinthorsson, E., Prediction of unshrouded rotor blade tip heat transfer, *ASME International Gas Turbine and Aeroengine Congress and Exposition*, Houston, TX, ASME Paper 95-GT-142, June 5–8, 1995.)

vortex and tip separation vortex on the blade tip can be identified. Presence of these vortices has a large effect on the heat transfer at the blade tip region. Figure 7.12 shows the Stanton number distribution. A fair agreement between experimental and computation results is observed. This plot also includes heat-transfer calculations based on fully developed and developing flow conditions. It can be seen that fully developed and developing heat-transfer coefficient correlations are not capable of predicting the heat-transfer pattern. Therefore, it is necessary to calculate the actual geometry to estimate the heat-transfer coefficients.

Ameri et al. (1998) simulated the tip flow and heat transfer on the first stage E^3 turbine. They considered three different tip configurations: flat tip, 2% tip recess, and 3% tip recess. Figure 7.13 shows the Stanton number distribution. For the flat tip case, the heat transfer is higher near the pressure side and lower near the suction side. For the 2% and 3% recess cases, the heat transfer on the bottom of the cavity could be higher than the flat tip case, and the heat transfer on the pressure-side rim is comparable to the flat tip case but is higher on the suction-side rim. In addition, a 2-D cavity problem was also solved. Figure 7.14 shows a 2-D cavity. The Nusselt number distribution along the rim, sidewalls, and the bottom of the 2-D cavity is also shown in Figure 7.14. Experimental data from Metzger et al. (1989) are included for comparison. Generally a good agreement is observed between

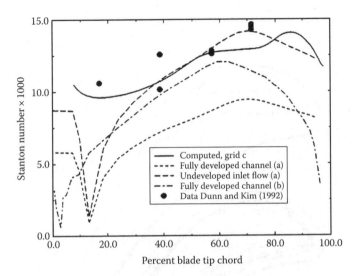

FIGURE 7.12
Comparison of the predicted mean camber line tip heat transfer with the fully developed channel flow correlation and assumed developing flow heat transfer. (From Ameri, A.A. and Steinthorsson, E., Prediction of unshrouded rotor blade tip heat transfer, *ASME International Gas Turbine and Aeroengine Congress and Exposition*, Houston, TX, ASME Paper 95-GT-142, June 5–8, 1995.)

the prediction and experimental measurement. A peak in the heat-transfer coefficient in the upper portion of the downstream sidewall is due to flow separation and reattachment in that region.

7.3 Numerical Prediction of Turbine Film Cooling

7.3.1 Introduction

The free surface boundary of a film coolant makes it difficult to model. Though the cooling effect is on the solid surface, the spread rate and the secondary flow in a coolant ejected from the film hole remain to be solved. To capture the flow details in a film, it is required to adopt advance turbulence models. On the other hand, an airfoil can have many holes at various locations that interact with high free-stream turbulence, unsteady flows, endwall secondary flows, accelerating or decelerating flows, and blade curvature. Therefore, an optimum balance is always sought to capture the effective heat-transfer coefficients without going through too many computation-intensive turbulence parameters.

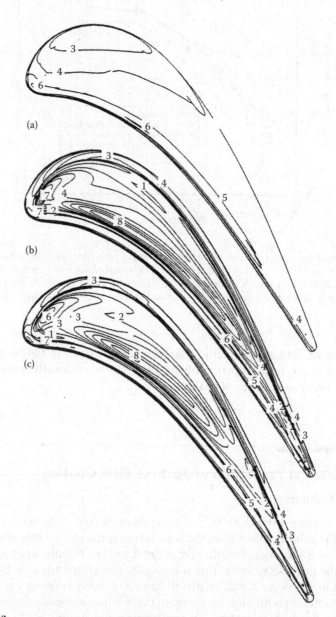

FIGURE 7.13
Heat-transfer distribution (1000 × Stanton number) on the cavity floor and rim for (a) no recess as well as (b) 2% and (c) 3% tip recess. (From Ameri, A.A. et al., *ASME J. Turbomach.*, 120, 753, 1998.)

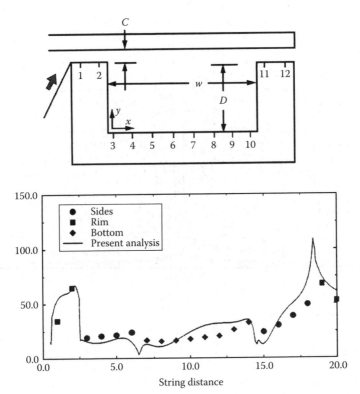

FIGURE 7.14
Comparison of Nusselt numbers along the rim, side walls, and the bottom of a 2-D blade tip cavity modeled by Ameri et al. (From Ameri, A.A. et al., *ASME J. Turbomach.*, 120, 753, 1998.)

7.3.2 Prediction of Flat-Surface Film Cooling

Leylek and Zerkle (1994) predicted the discrete-jet film cooling with the k-ε turbulence model. Large-scale computation analyses are performed and results were compared with experiments to understand coolant jet and cross-flow interaction in a discrete-jet film cooling. Predictions explain important aspects of film cooling, such as the development of complex flow within the film hole and the known counterrotating vortex structure in the cross-stream. Figure 7.15 shows a typical vortex structure in the ejected flow and cross-flow interaction. The strength of the secondary flow is directly related to the blowing ratio. The net aerodynamic effect of the interacting vortices is to bring the individual cores laterally close together and to lift the entire coolant film vertically away from the surface. This lift of coolant away from the heat-transfer surface deteriorates the film protection performance. Another undesirable action of this vortex pair is to bring the hot cross-flow down under the film layer. A kidney-shaped cross section of the coolant jet is the final consequence of all the coolant cross-flow interactions.

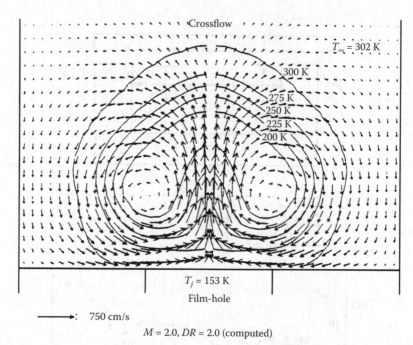

FIGURE 7.15
Velocity vectors and temperature contours at the cross section of a film coolant flow. (From Leylek, J.H. and Zerkle, R.D., *ASME J. Turbomach.*, 116, 358, 1994.)

Figure 7.16 shows the comparison of adiabatic effectiveness of discrete-hole film cooling. The comparison of prediction and experimental observation improved with increasing blowing ratios. The predictions show that the coolant jet stays attached to the surface and therefore cannot capture the discontinuity in the effectiveness as experimentally measured near the ejection hole. Figure 7.17 shows the details of the velocity vector in the film hole. Considerable flow separation and irregularities are noted in the film hole. This figure clearly shows that the flow is not fully developed in the film hole.

Walters and Leylek (1997) and Brittingham and Leylek (1997) did a detailed analysis of film-cooling flow physics. They used the RAMPANT software package from Fluent, Inc. A second-order discretization scheme was used with standard k-ε model with wall functions. Figure 7.18 shows their prediction of temperature contours. The diffusion of coolant as it travels downstream is shown in this plot. It is noted that the downstream behavior is highly sensitive to the near-field interaction that serves to locate the coolant at some given position away from the wall. Figure 7.19 shows the velocity magnitude contours in the jet exit plane. The distribution of flow variables at the jet exit is influenced by two primary mechanisms. Firstly, the counterrotating flow causes higher-momentum coolant fluid to exit from the upstream half of the film hole exit plane. Secondly, impingement of the oncoming

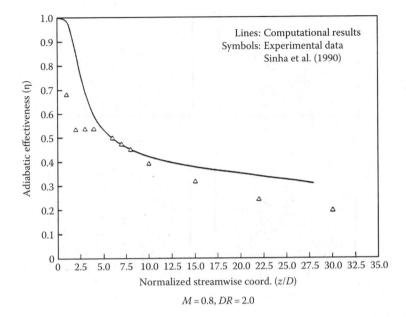

FIGURE 7.16
Prediction of adiabatic effectiveness of film cooling. (From Leylek, J.H. and Zerkle, R.D., *ASME J. Turbomach.*, 116, 358, 1994.)

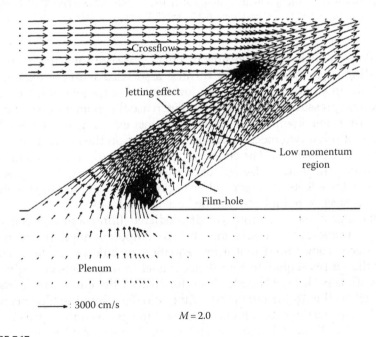

FIGURE 7.17
Computed velocity vectors inside a film hole. (From Leylek, J.H. and Zerkle, R.D., *ASME J. Turbomach.*, 116, 358, 1994.)

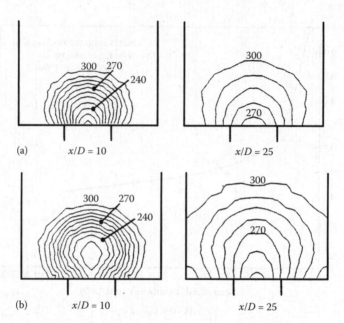

FIGURE 7.18

Temperature contours of film cooling as predicted by Walters and Leylek. (a) $M = 0.5$ and (b) $M = 1$. (From Walters, D.K. and Leylek, J.H., A detailed analysis of film-cooling physics. Part 1: Streamwise injection with cylindrical holes, *ASME International Gas Turbine and Aeroengine Congress and Exhibition*, Orlando, FL, ASME Paper 97-GT-269, June 2–5, 1997.)

cross-flow on the issuing jet results in a high-pressure zone upstream of the jet leading edge. While the flow turning associated with the exiting coolant results in a low-pressure region downstream of the jet trailing edge, the consequent pressure gradient serves to increase the momentum of the fluid exiting from the downstream portion of the jet exit. In general, the relative strength of these two mechanisms is dependent on the blowing ratio, density ratio, and geometry. The effect of the film hole L/D ratio is significant in determining the relative strength of the two mechanisms discussed. As L/D decreases, the effects of the separation region have less time to attenuate and therefore exert more influence on the jet exit conditions.

Neelakantan and Crawford (1995) predicted film-cooling effectiveness and heat transfer due to streamwise and compound-angle injection on a flat surface. They found that a one-equation turbulence model performs better than a two-equation turbulence model in their film-cooling predictions with the TEXSTAN code. They found that the entrainment fraction correlates well with the momentum flux ratio. For any given geometry, the entrainment fractions for a greater hole spacing are seen to be greater. This indicates that the model correctly captures the increasing dilution of the film coolant with increased hole spacing. They showed that drag coefficients are parallel to each other for different hole spacings for the inline round holes.

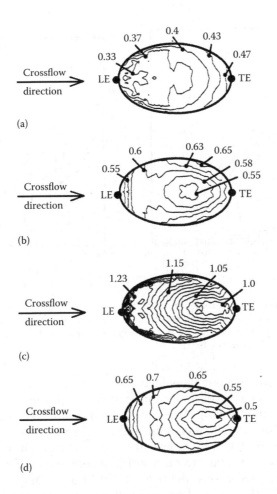

FIGURE 7.19
Velocity magnitude contours in the jet exit plane as predicted by Walters and Leylek. (a) $L/D = 3.5$, $M = 0.5$, (b) $L/D = 3.5$, $M = 1$, (c) $L/D = 3.5$, $M = 2$, (d) $L/D = 1.75$, $M = 1$. (From Walters, D.K. and Leylek, J.H., A detailed analysis of film-cooling physics. Part 1: Streamwise injection with cylindrical holes, *ASME International Gas Turbine and Aeroengine Congress and Exhibition,* Orlando, FL, ASME Paper 97-GT-269, June 2–5, 1997.)

But a similar pattern could not be seen in compound-angle holes. This can be explained by the fact that the one-equation turbulence model does not capture the additional mixing resulting from the compound angles. For a given geometry, the drag coefficients for the greater hole spacing are seen to plot below those for the lesser hole spacings. This is expected, since for the higher hole spacing, the same quantity of injected fluid is dispersed over a greater spanwise area.

Berhe and Patankar (1996, 1997) did a hydrodynamic study of discrete-hole film cooling using a standard k-ε model. Figure 7.20 compares their prediction of turbulent intensity with experimental observation. The shapes of

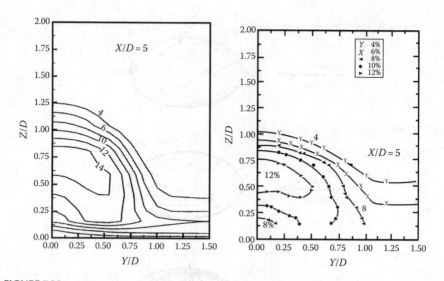

FIGURE 7.20
Comparison of predicted turbulence intensity with experimental observations. (From Berhe, M.K. and Patankar, S.V., Computation of discrete-hole film cooling: A hydrodynamic study, *ASME International Gas Turbine and Aeroengine Congress and Exhibition*, Orlando, FL, ASME Paper 97-GT-80, June 2–5, 1997.)

the corresponding contour lines in the two figures are comparable. There are differences in the vertical location of the contour lines. In the centerline plane, the computed contour lines are located 0.05D–0.2D higher than the corresponding experimental contour lines. In the midpitch plane, the computed contours are 0.1D–0.2D lower than the corresponding experimental lines. Their prediction shows a more uniform core in the ejection hole.

Among other predictions, Bohn et al. (1997) and Fougeres and Heider (1994) reported 3-D conjugate flow and heat-transfer analysis of a film-cooled guide vane. Fukuyama et al. (1995) predicted vane surface film-cooling effectiveness. Hyams and Leylek (1997) predicted details of film-cooling physics. McGovern and Leylek (1997) did a film-cooling analysis with compound-angle injection. Giebert et al. (1997) predicted film cooling from holes with expanded exits.

7.3.3 Prediction of Leading-Edge Film Cooling

Lin et al. (1997) computed the leading-edge film cooling with injection through rows of compound-angle holes. They used a low-Reynolds number k-ω turbulence model in a cell-centered finite volume code called CFL3D. Figure 7.21 shows their adiabatic effectiveness predictions. It can be observed that there is a region between the symmetry plane holes where adiabatic effectiveness reaches a very low value. This low adiabatic effectiveness is due to the horseshoe-like vortices entraining hot gases from the mainstream

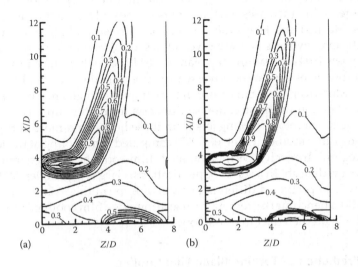

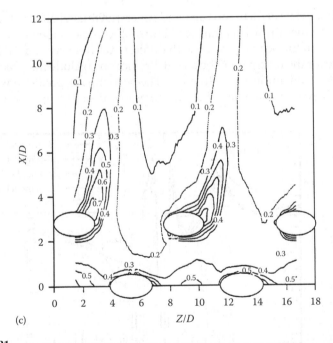

FIGURE 7.21
Adiabatic effectiveness near leading-edge film-cooling holes: (a) computed and averaged, (b) computed and unaveraged, and (c) measured. (From Lin, Y.L. et al., Computation of leading-edge film cooling with injection through rows of compound-angle holes, *ASME International Gas Turbine and Aeroengine Congress and Exhibition*, Orlando, FL, ASME Paper 97-GT-298, June 2–5, 1997.)

that got underneath the cooling jets. Note that flow separation due to jet-to-jet interaction increases adiabatic effectiveness. This interjet action helps the jets attached and keeps the surface cool. The second observation is that adiabatic effectiveness is low between the second row of film-cooling holes. Cooling jets from the symmetry plane holes lift off when they approach the second row holes due to the adverse pressure gradient induced by the cooling jets. Moreover, large curvature in the computation geometry causes high flow acceleration in the streamwise direction for both mainstream and the cooling jets. Therefore, cooling jets from these holes bend more abruptly.

Among other studies, He et al. (1995) computed film cooling at the leading-edge region. Irmisch (1994) simulated film-cooling aerodynamics with unstructured mesh. In a series of publications, Garg and Gaugler (1993, 1994, 1996, 1997a,b) predicted different aspects of film cooling. Martin and Thole (1997) did a benchmark prediction on leading-edge film cooling, and Walters et al. (1995) predicted jet in cross-flow.

7.3.4 Prediction of Turbine Blade Film Cooling

Garg and Gaugler (1997a,b) and Garg and Ameri (1998) used different turbulence models to predict airfoil heat transfer. Figure 7.22 compares their predictions for an airfoil with film-cooling holes with different turbulence models. Presented results are for the midspan of the airfoil. Predictions include the effects of film-cooling holes on surface heat transfer. The suction-side predictions compare better with experimental data than do the pressure-side comparisons. There is

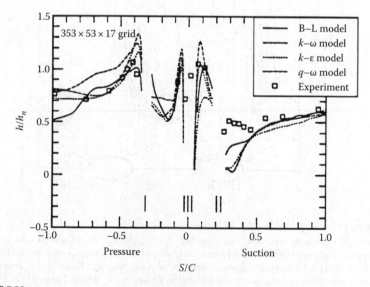

FIGURE 7.22
Effect of turbulence model on heat-transfer coefficient predictions. (From Garg, V.K. and Ameri, A.A., *Numer. Heat Transfer, Part A*, 32, 347, ASME Paper 97-GT-24, 1998.)

good agreement among different turbulence models as well as with the experimental observation in the downstream side of the suction surface. However, the pressure surface shows variations in the predictions with different numerical models. Overall, the q-w model and the k-ε model predict similar results, and the k-ω model predicts higher h values.

In a recent publication, Garg (1997) predicted film-cooled rotating blade heat transfer. Advantages of numerical prediction are clearly shown in Figure 7.23. Detailed Nusselt number contours are achievable by prediction, and flow features are relatively easy to change compared to an experimental setup. Once the numerical model is properly set up and basic geometry is properly mapped by grids, simple modification of the geometry and flow conditions is fairly easy in numerical modeling. However, reliability of predictions should always be verified with reliable experimental observations. Unavailability of detailed heat-transfer results by experimental methods prevents a comparison of prediction with experimental observation at this time. This figure compares two different turbulence models: q-w and Baldwin–Lomax models. The Nusselt number profiles on the pressure surface are similar, but those on the suction surface show large variations, and model-dependent predictions are observed. Most differences between model predictions are observed near the leading edge, hub and tip regions, and near the trailing edge. The q-w model predicts much higher heat loads than the B–L model. Due to the absence of film cooling over the leading edge, this part has a high heat load.

7.4 Numerical Prediction of Turbine Internal Cooling

7.4.1 Introduction

Several numerical studies are available on stationary-channel heat transfer. The flow and heat transfer related to internal cooling are affected by flow separation and rotation in the channel. The stationary-channel analyses were mostly focused on turbulence model development, whereas most studies useful for gas turbine heat transfer are based on rotating-channel studies.

7.4.2 Effect of Rotation

One of the earlier predictions on rotating channel flow was done by Majumdar et al. (1977). Figure 7.24 shows the comparison of their predictions with experimental measurements. Their standard k-ε model predictions compare favorably with the experimental flow measurement. However, the prediction of surface heat transfer is not so successful with the standard k-ε model. Prakash and Zerkle (1992) showed that the boundary layer separates

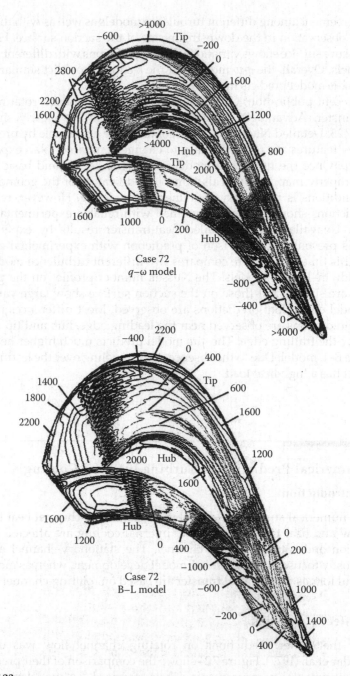

FIGURE 7.23
Nusselt number contours on a rotor blade predicted by two different turbulence models. (From Garg, V.K., Comparison of predicted and experimental heat transfer on a film cooled rotating blade using a two-equation turbulence model, *ASME International Gas Turbine and Aeroengine Congress and Exhibition*, Orlando, FL, ASME Paper 97-GT-220, June 2–5, 1997.)

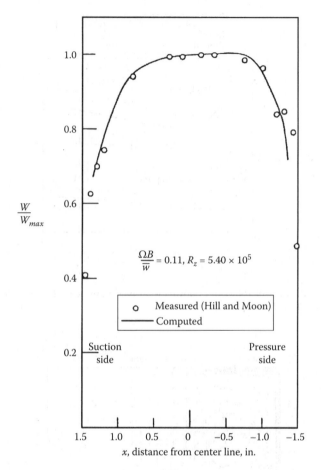

FIGURE 7.24
Prediction of axial flow velocity in an orthogonally rotating channel. (From Majumdar, A.K. et al., *ASME J. Fluids Eng.*, 99, 148, 1977.)

in adverse rotational buoyancy. Later, Tekriwal (1994) used an extended k-ε turbulence model, and results show that predictions are significantly dependent on the grid distribution and near-wall y^+.

After systematically investigating several prospective turbulence models, Dutta et al. (1996b) applied a modified k-ε model developed by Howard et al. (1980) and showed satisfactory predictions on the radial outward flow. Figure 7.25 shows the comparison of numerical predictions with experimental data for a given rotation number and with different buoyancy effects. Dutta et al. (1996b) included both Coriolis and rotational buoyancy effects in their prediction and also used the proper inlet condition of the experiment to get a satisfactory prediction of the heat-transfer coefficients. The secondary flows predicted by the numerical model are shown in Figure 7.26. This figure shows vector plots of secondary flow and contours of the axial flow

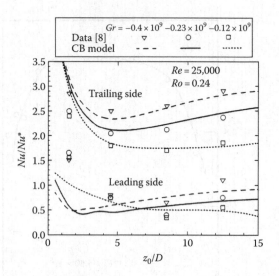

FIGURE 7.25
Prediction of the effect of rotational buoyancy on heat-transfer coefficient. (From Dutta, S. et al., *Int. J. Heat Mass Transfer*, 39(12), 2505, 1996b.)

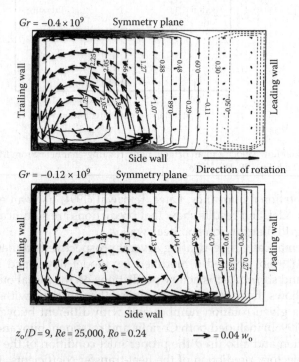

FIGURE 7.26
Effect of rotational buoyancy on the secondary flow. (From Dutta, S. et al., *Int. J. Heat Mass Transfer*, 39(12), 2505, 1996b.)

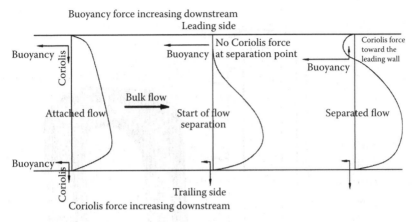

FIGURE 7.27
Schematic of the flow separation mechanism at the leading side of a rotating channel with radial outward flow. (From Dutta, S. et al., *Int. J. Heat Mass Transfer*, 39(12), 2505, 1996b.)

distribution for the highest and lowest Grashof numbers. In rotation the core flow shifts toward the trailing side, and the flow with stronger rotational buoyancy (more negative Grashof number) separates at the leading wall. This flow separation mechanism is shown in Figure 7.27.

Since Coriolis force is zero at the no-flow stagnation location, this force cannot create a separated boundary layer. However, once the separated bubble is formed, the Coriolis force acts toward the leading wall to collapse this bubble. Therefore, the balance between Coriolis and rotational forces determines the size of the separation bubble.

7.4.3 Effect of 180° Turn

The coolant channel used for internal cooling has a serpentine flowpath. Therefore, there are several turns present in the flowpath. Chen et al. (1999) have studied the effects of turn and rotation on the heat-transfer coefficient distribution; Figure 7.28 shows their grid. Note that their model used an arbitrary combination of embedded, overlapped, or matched grids using a chimera domain decomposition approach. In this approach, the solution domain is first decomposed into a number of smaller blocks that facilitate efficient adaptation of different block geometries, flow solvers, and boundary conditions for calculations involving complex configurations and flow conditions. Chen et al. (1999) used RANS with a finite analytic method, and their near-wall model used a second-moment closure.

Velocity vectors indicate that the axial velocity shifts toward the outer surface in the bend but returns quickly to a mostly flat profile in the second pass. There is no significant flow reversal in a stationary channel in the bend region for the two-pass configuration considered. However, the flow

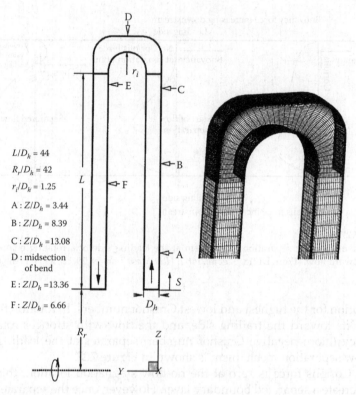

$L/D_h = 44$

$R_r/D_h = 42$

$r_i/D_h = 1.25$

A : $Z/D_h = 3.44$

B : $Z/D_h = 8.39$

C : $Z/D_h = 13.08$

D : midsection of bend

E : $Z/D_h = 13.36$

F : $Z/D_h = 6.66$

FIGURE 7.28

The computation geometry and details of the chimera grid used in the turn. (From Chen, H.C. et al., Computation of flow and heat transfer in rotating two-pass square channels by a Reynolds stress model, *ASME International Gas Turbine and Aeroengine Congress and Exhibition*, Indianapolis, IN, ASME Paper 99-GT-174, June 7–10, 1999.)

separates in the rotating channel, and the buoyancy effects affect the extent of this boundary layer separation. A higher density ratio indicates a longer flow separation bubble upstream of the turn.

The Nusselt number in the second-moment closure does not change significantly with grid refinement, but the two-layer model shows improvement in the turn region with grid refinement. Figure 7.29 shows the temperature distribution and the secondary flow in the nonrotating channel. Figure 7.29a indicates small secondary corner vortices in the first pass. Figure 7.29b shows the flow and temperature predictions in the bend. The centrifugal force and associated pressure gradients produce two counterrotating vortices that convect fluid from the core to the outer surface. This in turn generated secondary flow decays in the second pass, as shown in Figure 7.29c and d. The left column of the figure shows the isothermal contours in the two-pass channel. In the first pass, the cooler fluid is located in the core region, whereas after the bend, the cooler fluid is pushed toward the outer surface by the centrifugal force.

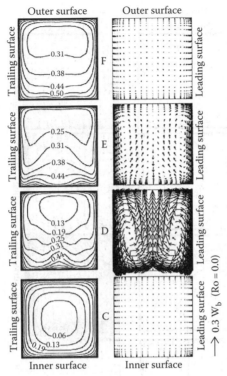

FIGURE 7.29
Dimensionless temperature $[\theta = (T - T_0)/(T_W - T_0)]$ contours and secondary flows for nonrotating channel. (From Chen, H.C. et al., Computation of flow and heat transfer in rotating two-pass square channels by a Reynolds stress model, *ASME International Gas Turbine and Aeroengine Congress and Exhibition*, Indianapolis, IN, ASME Paper 99-GT-174, June 7–10, 1999.)

Figure 7.30 shows the secondary flow velocity vectors and the isothermal contours for the rotating case at selected locations of the rotating channel. Locations a–f correspond to $Z/D = 3.44, 8.39, 13.08$ in the first pass, and 13.36 and 6.66 in the second pass, respectively. In the first pass, the Coriolis force pushes the cold fluid from the core toward the trailing surface and then returns along the side walls. This secondary flow also distorts the axial velocity profile as reported earlier in the straight-channel discussions. Figure 7.30d shows the secondary flow in the turn region. This asymmetric secondary flow indicates that the flow and heat-transfer simulations in this configuration have a strong 3-D effect.

Figure 7.31 shows the surface heat-transfer coefficient predictions for both stationary and rotating channels. Two surfaces, leading and trailing, are shown for the rotating channel. In the stationary channel, the heat transfer is high near the inlet due to the thinner thermal boundary layer.

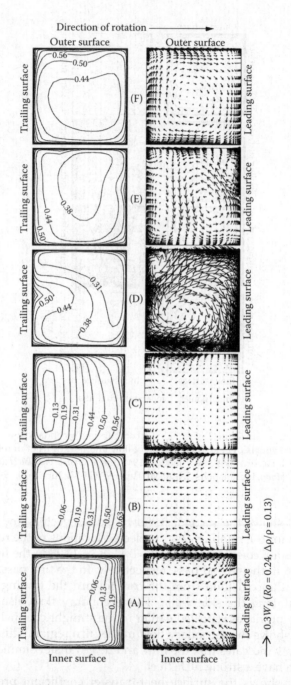

FIGURE 7.30
Dimensionless temperature [$\theta = (T - T_0)/(T_W - T_0)$] contours and secondary flows for nonrotating channel. (From Chen, H.C. et al., Computation of flow and heat transfer in rotating two-pass square channels by a Reynolds stress model, *ASME International Gas Turbine and Aeroengine Congress and Exhibition*, Indianapolis, IN, ASME Paper 99-GT-174, June 7–10, 1999.)

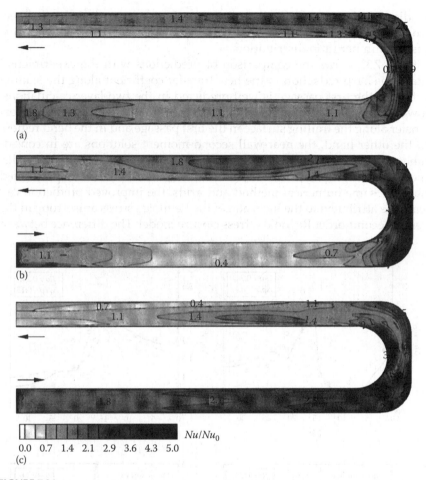

FIGURE 7.31
Detailed Nusselt number predictions by Chen et al. (a) Leading surface and trailing surface ($Ro = 0.0$), (b) Leading surface ($Ro = 0.24$, $\Delta\rho/\rho = 0.13$), (c) Trailing surface ($Ro = 0.24$, $\Delta\rho/\rho = 0.13$). (From Chen, H.C. et al., Computation of flow and heat transfer in rotating two-pass square channels by a Reynolds stress model, *ASME International Gas Turbine and Aeroengine Congress and Exhibition*, Indianapolis, IN, ASME Paper 99-GT-174, June 7–10, 1999.)

In the downstream, the heat-transfer coefficients decrease and asymptotically attain the fully developed value. The heat transfers in the bend and the outer surface of the second passage are high due to the secondary flows induced by the high-pressure gradient in the bend. For the rotating channel, the Nusselt number ratio reaches a minimum in the middle of the first pass of the leading surface and then increases along the outer surface in the bend. For the first-pass trailing surface, the Nusselt number increases in the streamwise direction and reaches a maximum in the bend region. This increase and decrease in the heat-transfer coefficient are discussed in more detail in the experimental rotating-channel section. However, the numerical prediction provides the details of the local Nusselt number ratio distribution.

Figure 7.32 shows the comparison of predictions with the experimental data. The sharp reduction in the heat-transfer coefficient along the leading surface in the first passage is well predicted in the two-layer calculations. However, the two-layer k-ε model failed to capture the steep increase in heat transfer along the trailing surface in the first passage and in the bend region. On the other hand, the near-wall second-moment solutions are in considerably better agreement with the experimental data on all four sidewalls. Since both the two-layer and second-moment calculations were performed using the same numerical method and grids, the improved prediction can clearly be attributed to the inclusion of the Reynolds stress anisotropy in the present second-order Reynolds stress closure model. The difference between

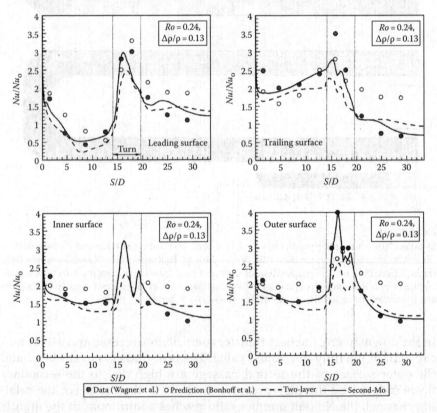

FIGURE 7.32
Comparison of predictions by standard k-ε model and second-moment turbulence model with experimental data. (From Chen, H.C. et al., Computation of flow and heat transfer in rotating two-pass square channels by a Reynolds stress model, *ASME International Gas Turbine and Aeroengine Congress and Exhibition*, Indianapolis, IN, ASME Paper 99-GT-174, June 7–10, 1999.)

prediction and experimental measurements could be due to the sudden contraction entrance condition in Wagner et al. (1991) that differs significantly from the fully developed flow condition used in the present calculations.

7.4.4 Effect of Transverse Ribs

Prakash and Zerkle (1995) used a ribbed channel to simulate the rotational effects on the separated flow. Predictions show that flow separates and reattaches in the interrib region. Predictions also show that both ribbed side and smooth side have higher heat-transfer coefficients than all the smooth-channel conditions. It is observed that the core flow is mostly undisturbed in the ribbed stationary channel, and the secondary flow by rotational forces is present in a ribbed rotating channel. They also predicted the heat-transfer coefficients in rotating channel conditions and have shown that the overall enhancement by rotation is less in a ribbed channel than in a smooth channel.

Prediction of convective heat-transfer coefficient is directly linked to the quality of flow predictions. Acharya et al. (1993) used the standard high-Reynolds number k-ε model and nonlinear k-ε model to predict the details of turbulent flow in a nonrotating, periodically rib-roughened channel. They showed their predictions of the turbulent shear stress at different locations in the interrib region. The two numerical models used in their study underpredicted turbulence in flow and surface heat-transfer coefficients. Among other contributors, Liou and Chen (1995) computed nonrotating, periodic flow and heat transfer with various rib shapes. Jang et al. (2000a) computed flow and heat transfer in a nonrotating two-pass square duct with 90° ribs by a near-wall second-moment turbulence closure model.

Iacovides (1997, 1998) studied velocity and heat transfer in rotating channels with orthogonal ribs. Figure 7.33 shows the velocity predictions in a staggered rib arrangement. The heat-transfer coefficients are also shown in Figure 7.33. Heat-transfer predictions by two turbulence models—low-Reynolds number differential stress model (DSM) and effective viscosity model (EVM)—are compared with experimental data. The DSM model better predicts the experimental data. The rotating ribbed-channel flow predictions are shown in Figure 7.34. Predictions capture the flow reasonably well for both positive and negative rotation directions.

7.4.5 Effect of Angled Ribs

Stephens and Shih (1997) computed the compressible flow and heat transfer in a rotating channel with inclined ribs and a 180° bend. They used a low-Reynolds number version of the k-ω turbulence model. Figure 7.35 shows their prediction of the secondary flow in a two-pass rotating smooth channel. The Coriolis force forms two symmetric counterrotating vortices at the inlet of the smooth duct. With radial outward flow, this secondary-flow vortex pair flows from the trailing side to the leading side along the two sidewalls, transporting

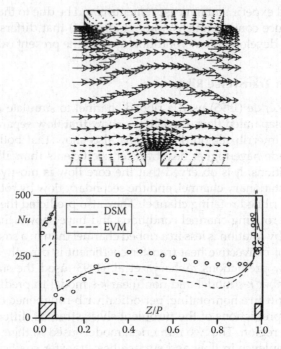

FIGURE 7.33
Mean flow prediction along the midplane of a stationary channel with staggered ribs and prediction of heat-transfer coefficient in the interrib region. (From Iacovides, H., *Int. J. Heat Fluid Flow*, 19, 393, 1998.)

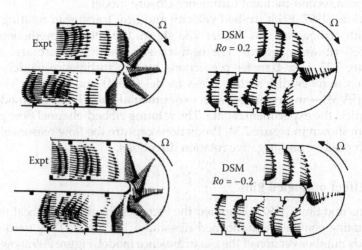

FIGURE 7.34
Prediction of mean flow in the turn region of a ribbed channel. (From Iacovides, H., The computation of turbulent flow through stationary and rotating U-bends with rib-roughened surfaces, *10th International Conference on Numerical Methods in Laminar and Turbulent Flows*, Swansea, U.K., 1997.)

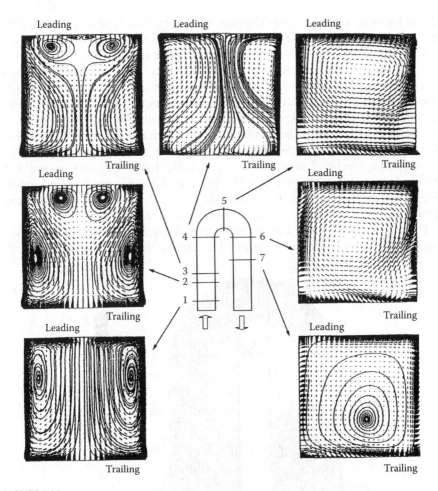

FIGURE 7.35
Velocity vectors at several locations of a rotating smooth channel as predicted by Stephens and Shih. (From Stephens, M.A. and Shih, T.I.-P., Computation of compressible flow and heat transfer in a rotating duct with inclined ribs and a 180-degree bend, *ASME International Gas Turbine and Aeroengine Congress and Exhibition*, Orlando, FL, ASME Paper 97-GT-192, June 2–5, 1997.)

cooler air from near the center of the channel to the walls. With higher temperature and hence lower density near the leading wall, centrifugal buoyancy tends to decelerate the flow, causing flow separation with reversed flow on the leading wall. Two events are observed with the flow separation. First, the radial outward flow next to the trailing wall increases in speed because of the reduced effective cross-sectional area. Second, additional secondary flow vortices form; the number of secondary flow vortices increases from two to four and then to six with the new counterrotating vortex pair forming next to the leading wall. The additional secondary flow in the separated region has a reverse direction because the bulk flow is reversed. Accordingly, in the down-leg part of the two-pass channel, where the flow is radial inward, centrifugal buoyancy accelerates instead of decelerating the lower density fluid.

This figure also shows that the rotational flow is significantly disturbed by the 180° turn. Secondary flow vortices in the bend are asymmetric instead of symmetric. The asymmetry is due to interactions between the secondary flows formed by Coriolis and centrifugal buoyancy in the upstream radial outward flow and the secondary flow formed in the bend. Downstream of the bend, flow is dominated by one large vortex that flows from the concave to the convex side along the leading face. This single vortex secondary flow developed in the bend persists at a considerable distance downstream of the bend. Among other contributors, Stephens et al. (1996) computed nonrotating heat transfer in an inclined rib condition. Bonhoff et al. (1997) predicted heat transfer in a rotating two-pass channel with skewed ribs.

Figure 7.36 shows the leading-surface and Figure 7.37 shows the trailing-surface Nusselt number ratios in a ribbed channel. The inclined rib

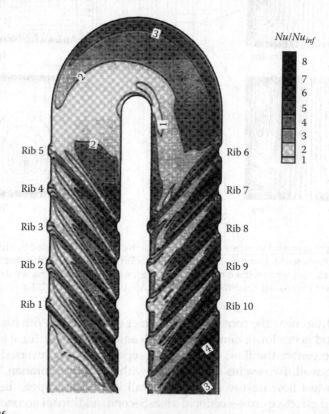

FIGURE 7.36
Nusselt number predictions in a rotating two-pass channel with inclined ribs (leading side). (From Stephens, M.A. and Shih, T.I.-P., Computation of compressible flow and heat transfer in a rotating duct with inclined ribs and a 180-degree bend, *ASME International Gas Turbine and Aeroengine Congress and Exhibition*, Orlando, FL, ASME Paper 97-GT-192, June 2–5, 1997.)

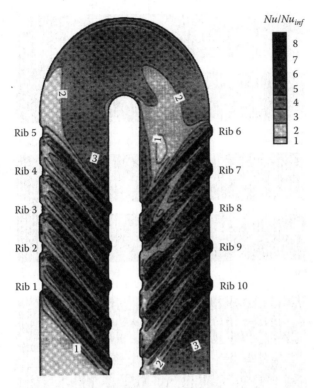

FIGURE 7.37
Nusselt number predictions in a rotating two-pass channel with inclined ribs (trailing side). (From Stephens, M.A. and Shih, T.I.-P., Computation of compressible flow and heat transfer in a rotating duct with inclined ribs and a 180-degree bend, *ASME International Gas Turbine and Aeroengine Congress and Exhibition*, Orlando, FL, ASME Paper 97-GT-192, June 2–5, 1997.)

arrangement shows an asymmetric heat-transfer coefficient distribution. The inclined ribs develop a secondary flow of their own, and that interacts with the secondary flow developed in rotation. The turn region does not have any ribs, and therefore the heat-transfer coefficients in the turn region are less compared to the interrib region. In the leading wall of the radial outward flow (Figure 7.36), heat transfer is highest at the upstream side of the rib, and in the interrib region next to the inner sidewall. For the bend part of the leading wall heat transfer is highest near the concave side. For the radial inward flow, heat transfer is highest on the upstream side of the rib and in the interrib region next to the outer sidewall. For the trailing wall (Figure 7.37), the overall heat transfer is higher than that on the leading wall, especially in the radial outward flow and the bend part.

Jang et al. (2000b) computed flow and heat transfer in two-pass, nonrotating square channels with 60° ribs by using a near-wall second-order Reynolds stress closure model. Figure 7.38 shows the predicted and measured Nusselt number ratio comparison. It shows that the second-moment Reynolds stress

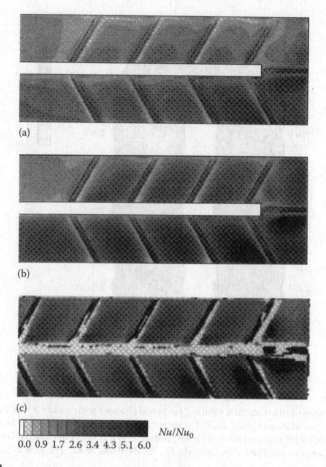

(a)

(b)

(c)

Nu/Nu_0

0.0 0.9 1.7 2.6 3.4 4.3 5.1 6.0

FIGURE 7.38

Comparison between predicted and measured Nusselt number ratio in a nonrotating two-pass square channel with 60° angled ribs: (a) two-layer model, (b) second-moment model, and (c) Ekkad and Han (1997). (From Jang, Y.J. and Chen, H.C., Computation of flow and heat transfer in two-pass channels with 60° ribs, *38th Aerospace Sciences Meeting and Exhibit*, Reno, NV, AIAA Paper 2000-1036, January 10–13, 2000b.)

model provides better heat-transfer prediction than the two-layer k-ε model. Jang et al. (2000c) also predicted flow and heat transfer in a rotating square channel with 45° angled ribs by Reynolds stress model. Figure 7.39 shows the predicted Nusselt number ratio distribution, and Figure 7.40 shows the predicted and measured Nusselt number ratio comparison. The heat-transfer enhancement due to the angled rib is clearly seen; the increased heat transfer on the trailing surface and the decreased heat transfer on the leading surface due to rotation are also clearly shown in Figures 7.39 and 7.40. Figure 7.40 shows the improved heat-transfer prediction compared with the experimental data by using a near-wall second-moment closure model.

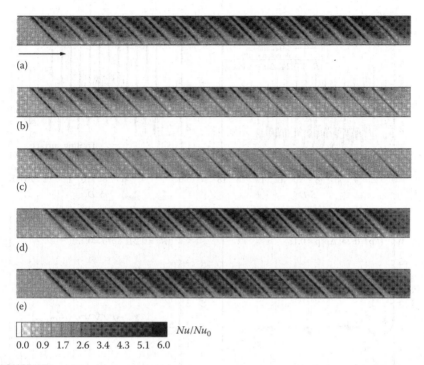

Nu/Nu_0

0.0 0.9 1.7 2.6 3.4 4.3 5.1 6.0

FIGURE 7.39
Nusselt number predictions in a rotating square channel with 45° angled ribs. (a) Leading and trailing surface ($Ro = 0.0$), (b) leading surface ($Ro = 0.12$, $\Delta\rho/\rho = 0.13$), (c) leading surface ($Ro = 0.24$, $\Delta\rho/\rho = 0.13$), (d) trailing surface ($Ro = 0.12$, $\Delta\rho/\rho = 0.13$), and (e) trailing surface ($Ro = 0.24$, $\Delta\rho/\rho = 0.13$). (From Jang, Y.J. et al., Flow and heat transfer in a rotating square channel with 45° angled ribs by Reynolds stress turbulence model, *ASME International Gas Turbine and Aeroengine Congress and Exhibition*, Munich, Germany, ASME Paper 2000-GT-0229, May 8–11, 2000c.)

7.4.6 Effect of Rotation on Channel Shapes

Dutta et al. (1996a) used a rotation-modified k-ε turbulence model to predict heat-transfer coefficients in different aspect ratio rotating channels. Figure 7.41 shows secondary flow distribution in rotating channels of different aspect ratios. This plot shows that the effect of buoyancy is less significant on the low aspect ratio channels ($AR = 1:2$ and $1:4$) compared to that on higher aspect ratio channels ($AR = 1:1$ and $2:1$). The aspect ratios 1:1 and 2:1 show that secondary flows near the middle of the channels are moving from trailing to the leading surfaces at $z/D = 10$. From previous predictions and reasoning based on the Coriolis force, the secondary flow in a rotating channel is expected to move from the leading to the trailing surface. It can be argued that the higher buoyancy forces active in the present rotating flow are responsible for this discrepancy between prior and present predictions.

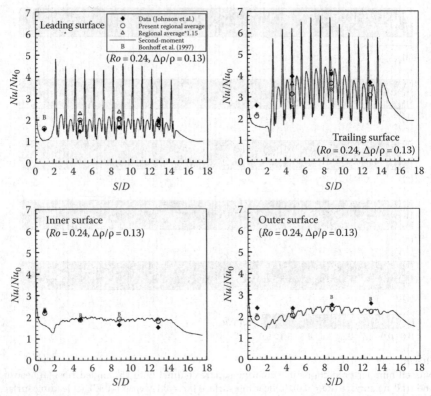

FIGURE 7.40

Comparison between predicted and measured Nusselt number ratio in a rotating square channel with 45° angled ribs. (From Jang, Y.J. et al., Flow and heat transfer in a rotating square channel with 45° angled ribs by Reynolds stress turbulence model, *ASME International Gas Turbine and Aeroengine Congress and Exhibition*, Munich, Germany, ASME Paper 2000-GT-0229, May 8–11, 2000c.)

Figure 7.42 shows the heat-transfer predictions for different aspect ratio channels. This figure shows the effect of rotation number on Nusselt number ratio at selected axial locations. Experimental data are plotted for several aspect ratios. Results indicate that the trend in the existing experimental data may be used to estimate the coolant channel performance in the real operating condition. The trailing wall of the 1:2 aspect ratio channel shows the highest Nusselt number ratios. However, the variations in this surface Nusselt number ratio are not smooth with rotation number. The trailing-surface heat-transfer coefficient is in general higher than the leading-surface Nusselt number ratio. This enhancement of trailing-side heat transfer is due to redistribution of flow and turbulence in the channel cross section. Coriolis force with rotation increases the axial flow velocity near the trailing wall compared to that near the leading wall. Moreover,

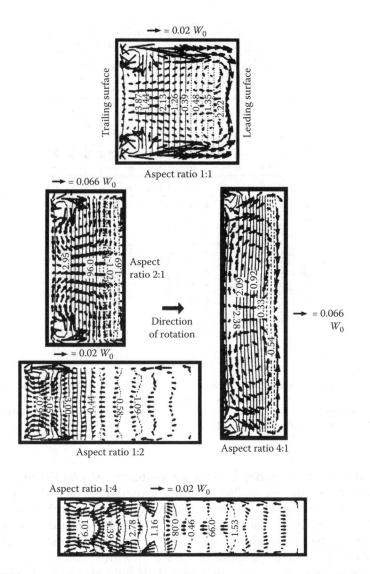

FIGURE 7.41
Secondary flow vectors and axial flow contours in different aspect ratio rotating channels. (From Dutta, S. et al., Prediction of turbulent flow and heat transfer in rotating square and rectangular smooth channels, *International Gas Turbine and Aeroengine Congress and Exhibition*, Birmingham, U.K., ASME Paper 96-GT-234, June 10–13, 1996a.)

the secondary flow redistributes the turbulence favoring heat-transfer enhancement from the trailing wall.

Dutta et al. (1996a,c) predicted heat transfer and secondary flows in triangular channels and rectangular channels in rotation. They used a standard k-ε model available in the finite element-based software package FIDAP.

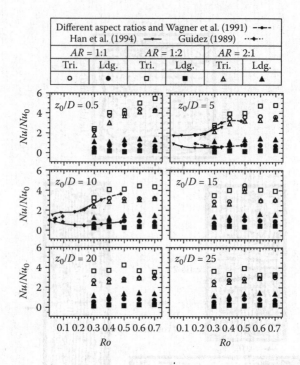

FIGURE 7.42
Effect of rotation number on Nusselt number ratio for different aspect ratio channels. (From Dutta, S. et al., Prediction of turbulent flow and heat transfer in rotating square and rectangular smooth channels, *International Gas Turbine and Aeroengine Congress and Exhibition*, Birmingham, U.K., ASME Paper 96-GT-234, June 10–13, 1996a.)

A sand roughness model is used for heat-transfer predictions in a rib-roughened channel. Figure 7.43 shows the secondary flow distribution in triangular channels for two different channel orientations with respect to the rotation direction. The secondary flows and radial outward flow distributions in the channel are different for the two model orientations. This figure also shows the tentative regions of stabilization and unstabilization of flow in rotation. The heat-transfer surfaces may be fully unstabilized or partially unstabilized by rotation.

7.4.7 Effect of Coolant Extraction

Rigby et al. (1997) used the k-ω model to predict the heat transfer in channels with film coolant extraction. Figure 7.44 shows their heat-transfer predictions along with selected experimental observations. The heat transfer is low just downstream of the rib, and then it rises sharply as the region of reattachment is approached, followed by a drop near the upstream side of the downstream rib. In addition to the disturbance caused by the ribs, a sharp increase in heat

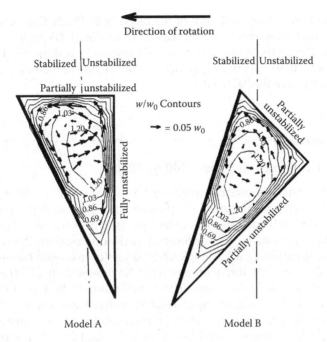

FIGURE 7.43
Prediction of secondary flow vectors and axial flow contours in rotating triangular channels
with different model orientations. (From Dutta, S. et al., *Int. J. Heat Mass Transfer*, 39(4), 707, 1996c.)

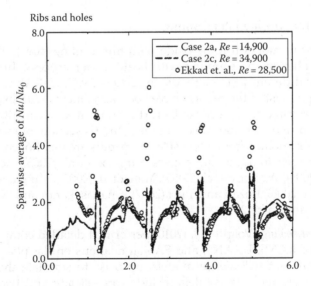

FIGURE 7.44
Prediction of heat-transfer coefficient in a ribbed channel with film extraction. (From Rigby, D.L.
et al., Numerical prediction of heat transfer in a channel with ribs and bleed, *ASME International Gas
Turbine and Aeroengine Congress and Exhibition*, Orlando, FL, ASME Paper 97-GT-431, June 2–5, 1997.)

transfer is observed downstream of the bleed hole. Predictions are in excellent agreement with experimental data of Ekkad et al. (1996) in the interrib region. On top of the ribs, predictions significantly underpredict the heat-transfer coefficient. This underprediction may be due to the lower resolution and the turbulence model used.

7.5 New Information from 2000 to 2010

Since 2000, similar numerical methods have been used to determine turbine blade surface heat-transfer coefficients and film-cooling effectiveness, as well as turbine blade internal coolant passage heat transfer. There are so many good papers using the aforementioned numerical methods available in the open literature. In addition, RANS (Reynolds averaged Navier–Stokes) and URANS (Unsteady Reynolds averaged Navier–Stokes), LES (Large Eddy Simulations) and DES (Detached Eddy Simulation), DNS (Direct Numerical Simulations), and CHT (Conjugate Heat Transfer) have recently been used for complex film-cooling flow and internal-cooling heat-transfer predictions. Only a few papers related to RANS, URANS, LES, DNS, and CHT are selected to be mentioned in this section. However, the detailed descriptions on the corresponding numerical modeling are referred to the original papers cited in this chapter.

7.5.1 CFD for Turbine Film Cooling

It is difficult to model turbine blade with film cooling due to the complicated flow phenomena coupling with heat-transfer process. Turbulence is resolved by different CFD methods, including RANS, URANS, LES, DNS, etc. Although most of the models have good agreements with experimental results, the accuracy still needs to be further improved. Generally, unsteady models perform better than steady models. In this section, numerical simulations of film cooling from a few selected papers are mentioned. Interested readers can refer to additional papers: Sargison et al. (2003), Acharya and Muldoon (2004), Whitney et al. (2005), Yang et al. (2005), Wang and Li (2006), Shih and Na (2006), Rozati and Tafti (2007), Dhanasekaran and Wang (2009), Sreedharan and Tafti (2010).

Flat plate film cooling: Voigt et al. (2010) performed detailed comparison and validation of RANS, URANS, and SAS simulations on flat plate film cooling. Five different turbulence models are used to simulate the flat plate film-cooling process. The models include three steady and two unsteady methods. The steady RANS methods are the shear stress transport (SST) model of Menter; the Reynolds stress model of Speziale, Sarkar, and Gatski; and a k-ε explicit algebraic Reynolds stress model. The unsteady models are a URANS formulation of the SST model and a scale-adaptive simulation (SAS). The solver used in this study is the commercial code ANSYS CFX 11.0. The results are compared to available experimental data. These data include velocity and turbulence intensity fields in several planes. Results show that the steady RANS approach has difficulties with predicting the flow field due to the highly 3-D unsteadiness. The URANS and SAS simulations, on the other hand, show good agreements with the experimental data. The deviation from the experimental data in velocity values in the steady cases is about 20%, whereas the error in the unsteady cases is below 10%.

Shape-hole film cooling: Leedom and Acharya (2008) presented LES of film-cooling flow fields from cylindrical, laterally diffused, and console shaped holes. The simulations include the coolant delivery tube and the feeding plenum chamber and are performed for a specific mass flow rate of coolant per unit width of blade. A cross-flow inlet is used on the plenum and the resulting asymmetric flow characteristics. The importance of the choice of boundary condition for the feeding plenum chamber (even with low plenum velocities) was documented. The shaping of the film-cooling holes is clearly an important consideration in the distribution and coverage of the coolant film. The console shape demonstrates reduced vertical penetration and improved coverage of the surface by the coolant film. The results show that the console's performance is superior to the laterally diffused and cylindrical holes in terms of jet penetration into the cross-flow. The cylindrical hole shows higher penetration and mixing in the jet and wake regions, while the console case shows that most of the coolant jet penetrates to less than $4d$ and covers the wall without mixing much with the cross-flow. For the laterally diffused case, the jetting effect is essentially eliminated, and the exit profile appears more uniform. However, higher jet-penetration and mixing is achieved in the vicinity of the console-hole mergers. The turbulence generated in the near field of film cooling jets was found to be highly anisotropic.

Leading edge region film cooling: Computational studies are carried out by Sreedharan and Tafti (2009), using LES to investigate the effect of coolant to mainstream blowing ratio in a leading edge region of a film cooled vane. The three row film-cooled leading edge vane geometry is modeled as a symmetric semi-cylinder with a flat after-body. One row of coolant holes is located along the stagnation line and the other two rows of coolant holes are located at each side of the stagnation line. The coolant is injected at 45° to the vane surface with 90° compound angle injection. The coolant to mainstream density ratio is set to unity and the free-stream Reynolds number based on leading edge diameter is 32,000. Blowing ratios ($B.R.$) of 0.5, 1.0, 1.5, and 2.0 are investigated. It is found that the stagnation cooling jets penetrate much further into the mainstream, both in the normal and lateral directions, than the off-stagnation jets for all blowing ratios. Jet dilution is characterized by turbulent diffusion and entrainment. The strength of both mechanisms increases with blowing ratio. The lateral and normal penetration of the stagnation coolant jet is much stronger than the off-stagnation jets. Increasing the blowing ratio from 0.5 to 2.0, the stagnation jet penetrates farther into the mainstream. At *B.R.* = 2.0, due to increased jet penetration into the mainstream and enhanced turbulent diffusion, there is little but uniform protection in the stagnation region. The adiabatic effectiveness in the stagnation region initially increases with blowing ratio but then generally decreases as the blowing ratio increases further. At *B.R.* = 0.5, the off-stagnation jet has a high effectiveness immediately downstream of injection but which quickly decays due to the small coolant mass and jet dilution. As the blowing ratio increases, an increase in jet penetration and subsequent dilution lowers the effectiveness immediately downstream of injection, but increases further downstream as the injected coolant mass increases with blowing ratio.

Deposition and erosion on leading edge region with film cooling: Sreedharan and Tafti (2009) performed a numerical study to investigate deposition and erosion of Syngas ash in the leading edge region of a turbine vane. LES is used to model the flow field of the coolant jet-mainstream interaction and syngas ash particles are modeled using a Lagrangian framework. Ash particle sizes of 5 and 7 µm are considered. Under the conditions of the current simulations, both ash particles have Stokes numbers less than unity of $O(1)$ and hence are strongly affected by the flow and thermal field generated by the coolant interaction with the mainstream. Because of this, the stagnation coolant jets are quite successful in pushing the particles away from the surface and minimizing deposition and erosion in the stagnation region. Overall, about 10% of the 5 µm particles versus 20% of the 7 µm particles are deposited on the surface at *B.R.* = 0.5. An increase to *B.R.* = 2 increases deposition of the 5 µm particles to 14% while decreasing deposition of the 7 µm particles to 15%. Erosive ash particles of 5 µm size increase from 5% of the total to 10% as the blowing ratio increases from 0.5 to 2.0, whereas 7 µm erosive particles remain nearly constant at 15%. Overall, for particles of size 5 µm, there is a combined increase in deposition and erosive particles from 16% to 24% as the blowing ratio increases from 0.5 to 2.0. The 7 µm particles, on the other hand, decrease from 35% to about 30% as the blowing ratio increases from 0.5 to 2.

Pulsed film cooling: Muldoon and Acharya (2007) presented DNS of a pulsed film-cooling jet to examine if pulsations of the coolant jet can enhance film-cooling effectiveness. Calculations are performed for a cylindrical jet inclined at 30°. The jet pulsation is defined by the duty cycle (*DC*) and the Strouhal number (*St*), both of which are varied in the present study. Calculations are done for a baseline steady blowing ratio of 1.5. Improvement in the overall film-cooling effectiveness was found with pulsing relative to the unpulsed flow at the same peak blowing ratio (*M* = 1.5). The range of nondimensional pulsing frequencies studied varied from 0.004 to 0.32, and best results were obtained at 0.32. When compared with the steady case on the same effective coolant flow rate (or mean blowing ratio) basis, pulsing at a *DC* of 0.5 showed comparable effectiveness and at a duty cycle of 0.25 showed reduced effectiveness. In many realistic configurations, it is not possible to lower the blowing ratio, and pulsing may be the only effective way to enhance film-cooling effectiveness. Further, the recirculation region behind the jet is greatly reduced with pulsing leading to improved film-cooling effectiveness. The present study only examined a small number of pulsed cases and there is a need to optimize the forcing parameters involved.

Endwall film cooling: Hada and Thole (2006) reported the computational study of the effect of upstream slot and mid-passage gap on turbine vane endwall film cooling. Yang et al. (2007) performed CFD prediction of film cooling and heat transfer from a rotating blade platform with stator-rotor purge and discrete film-hole flows in a 1½ turbine stage. Lynch et al. (2010) reported computational predictions of heat transfer and film cooling for a turbine blade with non-axisymmetric endwall contouring. Lawson and Thole (2010) performed simulations of mutiphase particle deposition on endwall film cooling. The aforementioned endwall film cooling studies used ANSYS FLUENT with RANS model to do the CFD analysis and Gambit to do meshing. They also present reasonable good comparisons with the existing experimental data.

7.5.2 CFD for Turbine Internal Cooling

RANS and LES are the most commonly used simulation methods for turbine blade internal flow and heat-transfer predictions. The RANS method becomes complex due to the requirement to model turbulent stresses and fluxes for including the effect of turbulent eddies of various sizes. The effective viscosity models employed tend to estimate lower the absence of isotropy in the Reynolds stress field. Hence, in flows such as that in serpentine ducts with considerable Coriolis and buoyancy forces, not accounting the non-isotropy of turbulence could lead to unexpected prediction errors. LES is a method of numerically solving partial differential equations that govern turbulent fluid flows. In this method, the larger eddies in the turbulent flow are explicitly resolved and the extra amount of detail can be studied. Note that RANS can provide only averaged solutions, whereas LES can give instantaneous solutions as well. In the LES approach, the effects of small-scale eddies need to be specifically modeled. The models used for sub-grid stresses and fluxes are in general simpler than that used in RANS. DNS is a deterministic approach, i.e., every flow is solved in a detailed fashion. Continuity, momentum, and energy equations are valid at every solution point unlike many other simplified and averaged solution methods that seek to find a suitable approximation to actual flow conditions. The extremely small grid spacing and time increments makes this type of simulations extremely expensive in terms of time and computational resources. However, some sort of averaging such as ensemble or time averaging is required to obtain useful information from this method.

Only a few papers related to RANS and LES are selected to be mentioned in this section. However, the detailed descriptions on the corresponding numerical modeling are referred to the original papers cited in this chapter. Interested readers can refer to additional papers: Al-Qahtani et al. (2001, 2002), Abdon and Sunden (2001), Su et al. (2003), Funazaki and Hachiya (2003), Acharya and Saha (2003), Jia and Sunden (2004), Tyagi and Acharya (2004), Tafti and Sewall (2004), Raisee and Iacovides (2004), Sleiti and Kapat (2004), Tafti et al. (2005), Leylek et al. (2005), Luo and Razinsky (2007), Schuler et al. (2010).

Rotating multi-pass rectangular channel with ribs: Su et al. (2004) employ Reynolds stress turbulence model to predict the flow and heat transfer in turbine blade cooling passages and study the effect of blade rotation, coolant-to-wall density ratio, rib shape, channel aspect ratio, and channel orientation on the flow turbulence and enhancement of surface heat transfer in the blade-cooling passages. The authors conclude the multi-block RANS method CFD simulation of the 3-D flow and heat transfer in the coolant passages, in stationary and rotating passages, by stating the relative impact of the aspect ratio, density ratio, and rib orientation on the flow and heat-transfer distributions. The second moment solutions display large anisotropy in turbulent stress and heat flux distributions. With rotation, the Coriolis and buoyancy forces result in strong non-isotropic turbulence flows. Without any rotation, the bend region displays two counter-rotating vortices. The $AR = 1:2$ nonrotating duct shows the largest heat-transfer rate. Here, the counter-rotating vortices nearly fill the entire duct area. However, note that for aspect ratio of $AR = 1:4$, the enhancement in the heat transfer and generated vortices are seen only along the leading and trailing surfaces. For 90° ribbed channels, soon after the bend, during rotation, vortices grow near the leading surface. The Nusselt number ratios decrease with increasing Reynolds numbers for all cases of aspect ratios. The friction factor is observed to decrease when Reynolds number is varied from 10,000 to 100,000. Rotation results in higher heat transfer on the first pass trailing surface, and reduced heat transfer on the leading side. After the bend, when the flow is directed inward, this trend is opposite. The effect of rotation on Nusselt numbers is pronounced only in low aspect ratio channels.

Squire duct with 90° ribs: Tyagi and Acharya (2005) present the LES of flow and heat transfer in rotating square duct with normal rib turbulators. The unsteady temperature field in periodic domain is computed directly. The authors observe that the large-scale vortices play a major role in the mixing of the core fluid and the near-wall heated fluid. The temperature field is mostly controlled by the large-scale mixing. The flow is largely turbulent and contains low-frequency flows. The authors present the 3-D energy spectrum of the instantaneous flow field. For the length scale $l/D = 0.1$, the corresponding wave number generates the peak energy. The authors explain that the vortex shedding behind the ribs are responsible for the large spike in the energy spectrum, and note that the time variation of the flow rate is attributed to the variation dominated by the vortex shedding frequency.

Squire duct with 45° ribs: Viswanathan and Tafti (2006) present LES of fully developed flow and heat transfer in a Rotating Duct with Ribs oriented at 45°. The authors observe in the simulation that the heat-transfer augmentation depends mostly on span-wise momentum in the case of 45° ribs, unlike the case in 90° ribs where stream-wise momentum is predominant. The authors conclude that the 90° ribbed ducts have heat-transfer augmentation strongly dependent on the stream-wise momentum, whereas in skewed ribs (45° or 60°), the span-wise momentum has an upper hand in determining the overall heat transfer. The Coriolis forces do not contribute much to the span-wise momentum. Hence, the rotation does not have major effect on the heat transfer at the ribbed walls at the trailing and leading sides, and they exhibit patterns similar to that seen in a stationary duct.

Channel with dimples and protrusions: Elyyan and Tafti (2007) simulate the flow and heat transfer in a channel with dimples and protrusions on opposite walls. LES is conducted for three Reynolds numbers, $Re = 220, 940$, and 9300, ranging from laminar, weakly turbulent, to fully turbulent, respectively. The authors aim to deliver a detailed analysis of the flow pattern in different types of flow, and obtain the effect of these features on the heat-transfer augmentation on either surface. Additionally, the estimation of heat transfer along with pressure loss aids in the heat-transfer enhancement assessment. The authors conclude that the heat-transfer augmentation on the dimpled surface can be attributed mainly to the turbulence generated in the separated shear layer at the upstream rim of the dimple. These turbulences impinge on the downstream sides of the dimples and result in higher heat-transfer rates. All the turbulences created, within and at the rim of the dimples, effectively enhance the heat-transfer rate of the flat portion downstream of the dimple where the flow reattaches. However, the heat-transfer augmentation on the protrusions are more due to impingement rather than turbulence. The flow acceleration over the protrusions creates higher heat-transfer rates. The form drag and friction drag contribute to the overall pressure loss. But, with dimples and protrusions, the overall pressure loss is very small. Hence, these types of surface modifications can be employed in situations where low pressure loss is desirable, even at the cost of low heat-transfer rates. Elyyan and Tafti (2010) continued a similar analysis, but on a rotating channel with dimples on two walls, and obtained a fairly similar result. They note that irrespective of the not-so-impressive cooling performance of dimpled channels, for applications that require very low pressure drop levels, these types of configurations can be used.

7.5.3 CFD for Conjugate Heat Transfer and Film Cooling

Turbine internal cooling and film cooling are the major cooling techniques that applied to turbine blades. The cooling gas is extracted from the compressor and pushed into the turbine cooling channel in order to cool the turbine from inside. Moreover, this coolant is also used for external cooling by extracting them from film-cooling holes. Film-cooling holes partially cover the surface area of the blade, thus protecting it from hot combustion gases. In both types of cooling, heat is removed by means of both convection and conduction. Moreover, convection and conduction are effected from each other. Conjugate heat transfer is basically termed as the interaction between the convection heat transfer from the surrounding fluid and the conduction heat transfer through the solid body. The conduction heat transfer is affected from the convection heat transfer of the surrounding fluid. Thus, they should be solved simultaneously. This coupling of solid to fluid is usually done by using the same wall temperature for the adjacent fluid block and solid block. The conjugate CFD methods provide good predictions for heat-transfer analysis in turbine blades. Eliminating the heat-transfer coefficient calculation by utilizing the relationship between solid and fluid interface, conjugate methods can provide faster solutions. There are many commercially available softwares (i.e., ANSYS Fluent, ANSYS CFX, Star CCM, etc.). However, the accuracy of conjugate methods must be compared with experimental data and still remains to be improved.

Only a few papers related to conjugate heat transfer and film cooling are selected to be mentioned in this section. However, the detailed descriptions on the corresponding numerical modeling are referred to the original papers cited in this chapter. Interested readers can refer to additional papers: Takahashi et al. (2000, 2001), Kusterer et al. (2003), Heidmann et al. (2003), Dulikravich et al. (2003), Bohn and Tummers (2003), Gatta et al. (2004), Sugimoto et al. (2004), Luo and Razinsky (2006), Na et al. (2007), Starke et al. (2008), Bohn and Krewinkel (2009), Sierra et al. (2009), Ceccherini et al. (2009), Dhiman and Yavuzkurt (2010), Mangani et al. (2010), Dyson et al. (2010), Jung et al. (2010).

Effect of Biot number on TBC-coated flat plate with internal and film cooling: Biot number, the ratio of conduction resistance to convection resistance, is one of the main parameters in conjugate heat transfer. Biot number has a significant effect on temperature and heat flux distributions over the solid body. Shih et al. (2009) performed an extensive study on the effects of Biot number on temperature and heat-flux distributions in a TBC-coated flat plate cooled by rib-enhanced internal cooling. Their flow domain contains two parallel plates with internal coolant flow in between plates and external hot gas flow from one side. The hot gas side of the plate is TBC-coated in order to reduce the heat transfer from surrounding fluid to the solid body. TBC has two layers where the outer one, which is directly in contact with hot gas flow, is a ceramic top coat and the other one is metallic bond coat. They are constructed on the super alloy plate. They used ANSYS Fluent-UNS version 6.3.26 with steady Reynolds averaged Navier–Stokes (SRANS) method to do the analysis. The turbulence effects are given by two-equation realizable k-ε model. Instead of wall functions, they used wall integration method to simulate the wall effects. Their final grid has more than three million cells and the cells near the walls are finer. This mesh and the mentioned CFD method give reasonable temperature, heat flux, and Biot number predictions.

Conjugate convective-conductive behavior of a rib-roughened channel: Conjugate heat-transfer behavior of a ribbed cooling channel when conduction heat transfer is not negligible is studied by Fedrizzi and Arts (2004). They investigated two different materials: copper and stainless steel. Typical 45° ribbed walls are constructed. Infrared thermography is used to evaluate temperature distribution on the channel inner surface. On the other hand, the heat flux on the surface is evaluated computationally. Using the measured temperature distribution, the numerical code ANSYS FLUENT is used to do the CFD analysis and Gambit to do meshing. The grid-independent study is also conducted. Computational procedure includes estimating the heat flux by using the surface temperature of the plate and the temperature of the bulk flow that obtained experimentally. In this study, it is assumed that heat flux supply to the slab is constant, the channel and endwalls are perfectly insulated, and the metal temperature distribution is periodic in the axial direction due to ribs. It should be noted that the mesh is finer on and around the rib where flow behavior gets complex due to separation and vortices. It is concluded that copper is not suitable for computation of heat flux distribution. However, the pattern of heat flux distribution of steel seems reasonable. The heat flux is highest at the immediate upstream and on the top of the rib. This is attributed to a well-known pattern of strong vortices on the immediate upstream of the rib due to flow separation on the top of the rib. A very similar pattern is observed for the Nusselt number distribution.

Effect of cooling-hole inner wall oxidation on TBC-coated flat plate film cooling: It is known that oxidation causes some restrictions on the flow domain in effusion cooling. Bohn and Krewinkel (2009) conducted a study in order to analyze these effects by means of computational methods in conjugate heat transfer. It consists of film-cooling holes in staggered array on a TBC-coated flat plate. The exit of the holes has an expansion angle in order to make it possible for coolant fluid to cover more surface area during film cooling. The oxidation layer causes the coolant jet to have a slightly higher temperature at the exit of the effusion hole. Thus, it is expected to have a slightly lower film-cooling effectiveness. The cooling-hole inner wall oxidation causes a slight decrease in film cooling effectiveness around upstream region of cooling hole. The effectiveness on downstream is quite similar. The computational is an implicit finite volume method combined with a multi-block technique. The governing equations in solid body are reduced to the Fourier equation which is directly coupled to the fluid flow region. The coupling is achieved by using the same temperature at the adjacent fluid and solid cells due to same amount of heat flux passing through those cells. Thus, there is no need to define a heat-transfer coefficient and heat flux, which is an important advantage of conjugate method. Algebraic eddy-viscosity turbulence model, which is known to give comparable results to the k-ε and k-ω methods for investigated effusion cooled geometries, is used. The authors also stated that the results of this turbulence model sufficiently agree with experimental data. In total, it has two million grid points. Moreover, conjugate heat-transfer technique requires fine mesh near the wall, where the coupling will be implemented, and accordingly the mesh structured finer near the walls. As a result, the walls and the oxidation layer have significantly finer mesh cells compared to that of main flow domain.

Turbine vane model with internal and film cooling: Ledezma et al. (2008) conducted a computational study on conjugate heat transfer behavior of internal and film cooling on a high-pressure turbine vane model. They investigated two different geometries: one without film cooling and the other with film cooling. They utilized k-ε and k-ω models as well as Omega Reynolds Stress and shear stress transport models. They also compared the accuracy of their results. They did an extensive study of grid-independent solution by considering wall function and wall integration methods. It is apparent that the wall integration method has finer mesh at near-wall region compared to that of wall function method. For near-wall regions, both methods have tetrahedral elements while for the main flow region they have prismatic elements. For solid fluid interfaces, nonconformal interface is used. They have used four different types of mesh with wall integration and wall function methods with different number of nodes on each. It is known that higher number of nodes will result in more accurate solution up to a level and then the solution will be grid independent. They concluded that wall integration method gives more accurate solution compared to wall function method because the mesh near the wall region are finer for that kind of mesh type.

Effect of hot streak on TBC-coated turbine blade heat transfer: He and Oldfield (2009) conducted a study on modeling of unsteady conjugate heat transfer. They wrote a code for unsteady conjugate analysis. Their code mainly includes three parts for convective heat transfer, conduction heat transfer, and solid–fluid interface. To determine the unsteady temperature distribution on the solid body, they have based their code on the following equation. $T = To + \sum_{n=1}^{N} An \cos nwt + Bn \sin nwt$, where w is frequency, t is the time step, and A and B are harmonic coefficients. Periodically unsteady flow disturbances are generated by the following harmonic unsteady total temperature term. $Tt = \overline{T}(1 + A_T \cos(wt + 2\Pi y/Y_{HS}))$. The authors conducted several different simulations in order to make sure that the code is giving accurate results compared to reference data. They performed the conjugate simulation of a turbine blade subject to hot streak with two internal cooling passages. The blades are covered with a TBC of 5 mm thickness. The inner cooling channel temperature of the blades is set to a constant value, and accordingly the temperature ratio of hot gas to the coolant in the cooling channel is generated for the flow domain. Hot streak has an angle of 60° between the axial directions. To understand the effect of unsteadiness on time-averaged values, the authors compared the results of this simulation with different hot streak amplitudes. It is shown that, as expected, as the amplitude increases, the difference between the steady-state solution and time-averaged solution increases. As the time changes, the temperature distribution on the leading edge significantly changes due to the unsteadiness of the hot streak. The authors studied the leading-edge temperature and stream trace distribution for two cases. In one case, the hot streak is impinging on the leading edge, whereas in the other case the cold fluid was impinging on the leading edge. The impingement capability of hot and cold portions was investigated. The stream traces showed a higher incidence when hot streak hit the leading edge compared to that of cold streak. Moreover, it seems that the cold portion of the fluid can get closer to the leading edge compared to hot streak. This is attributed to the different amount of momentums that hot and cold fluids carry.

7.5.4 CFD for Turbine Heat Transfer

Turbine blade surface heat transfer: There are continued efforts to predict turbine blade and vane surface heat-transfer coefficients without film cooling. For example, Boyle et al. (2000) predicted turbine vane rough surface heat transfer and compared with measured data; Garg and Ameri (2001) used two-equation turbulence models for heat-transfer prediction on a transonic turbine blade; Boyle and Geil (2001) predicted the effects of relaminization on turbine blade heat transfer; and Giel et al. (2004) predicted the effects of freestream turbulence on turbine blade heat transfer.

Effect of endwall contouring on heat transfer: Recently, endwall contouring has been proposed to reduce the secondary flow losses and decrease the associated heat-transfer coefficients. For example, Shih et al. (2000) presented control of secondary flows in a turbine nozzle guide vane by endwall contouring; Lin et al. (2000) predicted the effects of gap leakage on fluid flow in a contoured turbine nozzle guide vane; Zess and Thole (2001) reported computational design and experimental evaluation of using a leading edge fillet on a gas turbine vane; Shih and Lin (2002) presented controlling secondary-flow structure by leading-edge airfoil fillet and inlet swirl to reduce aerodynamic loss and surface heat transfer; Saha and Acharya (2006) predicted turbulent flow and heat transfer through a 3-D non-axisymmetric blade passage; and Shyam et al. (2010) predicted unsteady tip and endwall heat transfer in a highly loaded transonic turbine stage.

Turbine blade tip heat transfer and film cooling: Recently, there are many CFD predictions on turbine blade tip region heat transfer. For example, Yang et al. (2002) predicted flow and heat transfer for a turbine blade with a flat-tip and a squealer-tip. Acharya et al. (2002) reported numerical simulations of film cooling on the tip of a gas turbine blade; Acharya et al. (2003) conducted numerical study of flow and heat transfer on a blade tip with different leakage reduction strategies; Thole et al. (2003) reported predictions of cooling from dirt purge holes along the tip of a turbine blade; Polanka et al. (2003) presented turbine tip and shroud heat transfer and loading including unsteady effects; Yang et al. (2004) conducted CFD simulations of film cooling and heat transfer with different film-hole arrangement on the plane and squealer tip of a gas turbine blade; Janke et al. (2006) used CFD-based design for passive shroud cooling concepts for HP turbines; and Ameri et al. (2007) computed unsteady turbine blade and tip heat transfer due to wake passing.

References

Abdon, A. and Sunden, B., 2001. Numerical simulation of turbulent impingement cooling. ASME Paper 2001-GT-0150.

Abhari, R.S., Guenette, G.R., Epstein, A.H., and Giles, M.B., 1991. Comparison of time-resolved turbine rotor blade heat transfer measurements and numerical calculations, *ASME International Gas Turbine and Aeroengine Congress and Exposition*, June 3–6, 1991, Orlando, FL, ASME Paper 91-GT-268.

Acharya, S., Dutta, S., Myrum, T.A., and Baker, R.S., 1993. Periodically developed flow and heat transfer in a ribbed duct. *International Journal of Heat and Mass Transfer*, 36(8), 2069–2082.

Acharya, S. and Muldoon, F., 2004. Direct numerical simulation of a film cooling jet. ASME Paper GT2004-53502.

Acharya, S. and Saha, A., 2003. Flow and heat transfer in an internally ribbed ducts with rotation: An assessment of LES and URANS. ASME Paper GT2003-38619.

Acharya, S., Yang, H., Prakash, C., and Bunker, R.S., 2002. Numerical simulation of film cooling on the tip of a gas turbine blade. ASME Paper GT2002-30553.

Acharya, S., Yang, H., Prakash, C., and Bunker, R.S., 2003. Numerical study of flow and heat transfer on a blade tip with different leakage reduction strategies. ASME Paper GT2003-38617.

Al-Qahtani, M., Chen, H.C., and Han, J.C., 2002. A numerical study of flow and heat transfer in rotating rectangular channels (AR = 4) with 45-deg rib turbulators by Reynolds stress turbulence model. ASME Paper GT2002-30216.

Al-Qahtani, M., Jang, Y.J., Chen, H.C., and Han, J.C., 2001. Prediction of flow and heat transfer in rotating two-pass rectangular channels with 45-deg rib turbulators. ASME Paper 2001-GT-0187.

Ameri, A.A. and Arnone, A., 1994. Prediction of turbine blade passage heat transfer using a zero and a two-equation turbulence model, *ASME International Gas Turbine and Aeroengine Congress and Exposition*, June 13–16, 1994, The Hague, the Netherlands, ASME Paper 94-GT-122.

Ameri, A.A. and Arnone, A., 1996. Transition modeling effects on turbine rotor blade heat transfer predictions. *ASME Journal of Turbomachinery*, 118, 307–313.

Ameri, A., Rigby, D., Steinthorsson, E., Heidmann, J., and Fabian, J., 2007. Unsteady turbine blade and tip heat transfer due to wake passing. ASME Paper GT2007-27550.

Ameri, A.A. and Steinthorsson, E., 1995. Prediction of unshrouded rotor blade tip heat transfer, *ASME International Gas Turbine and Aeroengine Congress and Exposition*, June 5–8, 1995, Houston, TX, ASME Paper 95-GT-142.

Ameri, A.A. and Steinthorsson, E., 1996. Analysis of gas turbine rotor blade tip and shroud heat transfer, *ASME International Gas Turbine and Aeroengine Congress and Exhibition,* June 10–13, 1996, Birmingham, U.K., ASME Paper 96-GT-189.

Ameri, A.A., Steinthorsson, E., and Rigby, D.L., 1998. Effects of squealer tip on rotor heat transfer and efficiency. *ASME Journal of Turbomachinery,* 120, 753–758.

Berhe, M.K. and Patankar, S.V., 1996. A numerical study of discrete-hole film cooling, *ASME International Mechanical Engineering Congress and Exhibition,* November 17–22, 1996, Atlanta, GA, ASME Paper 96-WA/HT-8.

Berhe, M.K. and Patankar, S.V., 1997. Computation of discrete-hole film cooling: A hydrodynamic study, *ASME International Gas Turbine and Aeroengine Congress and Exhibition,* June 2–5, 1997, Orlando, FL, ASME Paper 97-GT-80.

Biswas, D. and Fukuyama, Y., 1993. Calculation of transitional boundary layers with an improved low-Reynolds number version of the K-6 turbulence model, *ASME International Gas Turbine and Aeroengine Congress and Exposition,* May 24–27, 1993, Cincinnati, OH, ASME Paper 93-GT-73.

Bohn, D.E., Becker, V.J., and Kusterer, K.A., 1997. 3-D conjugate flow and heat transfer calculations of a film-cooled turbine guide vane at different operation conditions, *ASME International Gas Turbine and Aeroengine Congress and Exhibition,* June 2–5, 1997, Orlando, FL, ASME Paper 97-GT-23.

Bohn, D. and Krewinkel, R., 2009a. Conjugate simulation of the effects of oxide formation in effusion cooling holes on cooling effectiveness. ASME Paper GT2009-59081.

Bohn, D. and Krewinkel, R., 2009b. Conjugate calculation of effusion cooling with realistic cooling hole geometries. ASME Paper GT2009-59082.

Bohn, D. and Tummers, C., 2003. Numerical 3-D conjugate flow and heat transfer investigation of a transonic convection-cooled thermal barrier coated turbine guide vane with reduced cooling fluid mass flow. ASME Paper GT2003-38431.

Bonhoff, B., Tomm, U., Johnson, B.V., and Jennisons, I., 1997. Heat transfer predictions for rotating U-shaped coolant channels with skewed ribs and with smooth walls, *ASME International Gas Turbine and Aeroengine Congress and Exhibition,* June 2–5, 1997, Orlando, FL, ASME Paper 97-GT-l 62.

Boyle, R.J., 1994. Prediction of surface roughness and incidence effects on turbine performance. *ASME Journal of Turbomachinery,* 116, 745–751.

Boyle, R.J. and Giel, P.W., 1997. Prediction of nonuniform inlet temperature effects on vane and rotor heat transfer, *ASME International Gas Turbine and Aeroengine Congress and Exhibition,* June 2–5, 1997, Orlando, FL, ASME Paper 97-GT-133.

Boyle, R.J. and Geil, P.W., 2001. Prediction of relaminization effects on turbine blade heat transfer. ASME Paper 2001-GT-0162.

Boyle, R.J. and Jackson, R., 1997. Heat transfer predictions for two turbine nozzle geometries at high Reynolds and Mach numbers. *ASME Journal of Turbomachinery,* 119, 270–283.

Boyle, R.J., Lucci, B.L., and Spuckler, C.M., 2000. Comparison of predicted and measured turbine vane rough surface heat transfer. ASME Paper 2000-GT-0217.

Brittingham, R.A. and Leylek, J.H., 1997. A detailed analysis of film cooling physics. Part IV: Compound-angle injection with shaped holes, *ASME International Gas Turbine and Aeroengine Congress and Exhibition,* June 2–5, 1997, Orlando, FL, ASME Paper 97-GT-272.

Ceccherini, A., Facchini, B., Tarchi, L., Toni, L., and Coutandin, D., 2009. Combined effect of slot injection, effusion array and dilution hole on the cooling performance of a real combustor liner. ASME Paper GT2009-60047.

Chen, H.C., Jang, Y.J., and Han, J.C., 1999. Computation of flow and heat transfer in rotating two-pass square channels by a Reynolds stress model, *ASME International Gas Turbine and Aeroengine Congress and Exhibition*, June 7–10, 1999, Indianapolis, IN, ASME Paper 99-GT-174.

Chima, R.V., 1991. Viscous three-dimensional calculations of transonic fan performance, *AGARD Propulsion and Energetics Symposium on Computational Fluid Mechanics for Propulsion*, NASA TM 103800, May 27–31, 1991, San Antonio, TX.

Dhanasekaran, T.S. and Wang, T., 2009. Simulation of mist film cooling on rotating gas turbine blades. ASME Paper GT2009-59424.

Dhiman, S. and Yavuzkurt, S., 2010. Film cooling calculations with an iterative conjugate heat transfer approach using empirical heat transfer coefficient corrections. ASME Paper GT2010-22958.

Dulikravich, G.S., Yoshimura, S., Egorov, I., and Dennis, B.H., 2003. Optimization of a large number of coolant passages located close to the surface of a turbine blade. ASME Paper GT2003-38051.

Dunn, M.G. and Kim, J., 1992. Turbine blade platform, blade tip and shroud heat transfer, *4th International Symposium on Transport Phenomena and Dynamics of Rotating Machinery*, Maui, HI.

Dunn, M.G., Kim, J., Civinskas, K.C., and Boyle, R.J., 1992. Time-averaged heat transfer and pressure measurements and comparison with prediction for a two-stage turbine, *ASME International Gas Turbine and Aeroengine Congress and Exposition*, June 1–4, 1992, Cologne, Germany, ASME Paper 92-GT-194.

Dutta, S., Andrews, M.J., and Han, J.C., 1996a. Prediction of turbulent flow and heat transfer in rotating square and rectangular smooth channels, *International Gas Turbine and Aeroengine Congress and Exhibition*, June 10–13, 1996a, Birmingham, U.K., ASME Paper 96-GT-234.

Dutta, S., Andrews, M.J., and Han, J.C., 1996b. Prediction of turbulent heat transfer in rotating smooth square ducts. *International Journal of Heat and Mass Transfer*, 39(12), 2505–2514.

Dutta, S., Han, J.C., and Lee, C.P., 1996c. Local heat transfer in a rotating two-pass ribbed triangular duct with two model orientations. *International Journal of Heat and Mass Transfer*, 39(4), 707–715.

Dyson, T., Bogard, D., Kohli, A., and Piggush, J., 2010. Overall effectiveness for a film cooled turbine blade leading edge with varying hole pitch. ASME Paper GT2010-23707.

Ekkad, S.V., Huang, Y., and Han, J.C., 1996. Detailed heat transfer distributions in two-pass smooth and turbulated square channels with bleed holes, *1996 National Heat Transfer Conference*, Houston, TX, Vol. 8, pp. 133–140.

Elyyan, M. and Tafti, D.K., 2007. LES investigation of flow and heat transfer in a channel with dimples and protrusions. ASME Paper GT2007-27811.

Elyyan, M. and Tafti, D.K., 2010. Investigation of Coriolis forces effect of flow structure and heat transfer distribution in a rotating dimpled channel. ASME Paper GT2010-22657.

Fougeres, J.M. and Heider, R., 1994. Three-dimensional Navier–Stokes prediction of heat transfer with film cooling, *International Gas Turbine and Aeroengine Congress and Exposition*, June 13–16, 1994, The Hague, the Netherlands, ASME Paper 94-GT-14.

Fedrizzi, R. and Arts, T., 2004. Investigation of the conjugate convective-conductive thermal behavior of a rib-roughened internal cooling channel. ASME Paper GT2004-53046.

Fukuyama, Y., Otomo, F., Sato, M., Kobayashi, Y., and Matsuzaki, H., 1995. Prediction of vane surface film cooling effectiveness using compressible Navier–Stokes procedure and k-F turbulence model with wall function, *International Gas Turbine and Aeroengine Congress and Exposition*, June 5–8, 1995, Houston, TX, ASME Paper 95-GT-25.

Funazaki, K.I. and Hachiya, H., 2003. Systematic numerical studies on heat transfer and aerodynamic characteristics of impingement cooling devices combined with pins. ASME Paper GT2003-38256.

Garg, V.K., 1997. Comparison of predicted and experimental heat transfer on a film cooled rotating blade using a two-equation turbulence model, *ASME International Gas Turbine and Aeroengine Congress and Exhibition*, June 2–5, 1997, Orlando, FL, ASME Paper 97-GT-220.

Garg, V.K. and Ameri, A.A., 1998. Comparison of two-equation turbulence models for prediction of heat transfer on film-cooled turbine blades. *Numerical Heat Transfer, Part A*, 32, 347–355. ASME Paper 97-GT-24.

Garg, V.K. and Ameri, A.A., 2001. Two-equation turbulence models for prediction of heat transfer on a transonic turbine blade. ASME Paper 2001-GT-0165.

Garg, V.K. and Gaugler, R.E., 1993. Heat transfer in film cooled turbine blades, *International Gas Turbine and Aeroengine Congress and Exposition*, May 24–27, 1993, Cincinnati, OH, ASME Paper 93-GT-81.

Garg, V.K. and Gaugler, R.E., 1994. Prediction of film cooling on gas turbine airfoils, *International Gas Turbine and Aeroengine Congress and Exposition*, June 13–16,1994, The Hague, the Netherlands, ASME Paper 94-GT-16.

Garg, V.K. and Gaugler, R.E., 1996. Leading edge film cooling effects on turbine blade heat transfer. *Numerical Heat Transfer, Part A*, 30, 165–187. ASME Paper 95-GT-275.

Garg, V.K. and Gaugler, R.E., 1997a. Effect of coolant temperature and mass flow on film cooling of turbine blades. *International Journal of Heat and Mass Transfer*, 40(2), 435–444. ASME Paper 95-WA/HT-l.

Garg, V.K. and Gaugler, R.E., 1997b. Effect of velocity and temperature distribution at the hole exit on film cooling of turbine blades. *ASME Journal of Turbomachinery*, 119, 343–349.

Gatta, S.D., Adami, P., Martelli, F., and Montomoli, F., 2004. Conjugate heat transfer modelling in film cooled blades. ASME Paper GT2004-53177.

Giebert, D., Gritsch, M., Schulz, A., and Wittig, S., 1997. Film-cooling from holes with expanded exits: A comparison of computational results with experiments, *ASME International Gas Turbine and Aeroengine Congress and Exhibition*, June 2–5, 1997, Orlando, FL, ASME Paper 97-GT-163.

Giel, P., Boyle, R.J., and Ames, F.E., 2004. Prediction for the effects of freestream turbulence on turbine blade heat transfer. ASME Paper GT2004-54332.

Hada, S. and Thole, K., 2006. Computational study of a mid-passage gap and upstream slot on vane endwall film-cooling. ASME Paper GT2006-91067.

Harasgama, S.P., Tarada, F.H., Baumann, R., Crawford, M.E., and Neelakantan, S., 1993. Calculation of heat transfer to turbine blading using two-dimensional boundary layer methods, *ASME International Gas Turbine and Aeroengine Congress and Exposition*, May 24–27, 1993, Cincinnati, OH, ASME Paper 93-GT-79.

He, L. and Oldfield, M.L.G., 2009. Unsteady conjugate heat transfer modeling. ASME Paper GT2009-59174.

He, P., Salcudean, M., and Gartshore, I.S., 1995. Computations of film cooling for the leading edge region of a turbine blade model, *International Gas Turbine and Aeroengine Congress and Exposition*, June 5–8, 1995, Houston, TX, ASME Paper 95-GT-20.

Heidmann, J.D., Steinthorsson, E., Divo, E.A., Kassab, A.J., and Rodriguez, F., 2003. Conjugate heat transfer effects on a realistic film-cooled turbine vane. ASME Paper GT2003-38553.

Howard, J.H.G., Patankar, S.V., and Bordynuik, R.M., 1980. Flow prediction in rotating ducts using Coriolis-modified turbulence models. *ASME Journal of Fluids Engineering*, 102, 456–461.

Hurst, C., Schulz, A., and Wittig, S., 1995. Comparison of calculated and measured heat transfer coefficients for transonic and supersonic boundary-layer flows. *ASME Journal of Turbomachinery*, 117, 248–254.

Hyams D.G. and Leylek, J.H., 1997. A detailed analysis of film cooling physics. Part III: Streamwise injection with shaped holes, *ASME International Gas Turbine and Aeroengine Congress and Exhibition*, June 2–5, 1997, Orlando, FL, ASME Paper 97-GT-271.

Iacovides, H., 1997. The computation of turbulent flow through stationary and rotating U-bends with rib-roughened surfaces, *10th International Conference on Numerical Methods in Laminar and Turbulent Flows*, Swansea, U.K.

Iacovides, H., 1998. Computation of flow and heat transfer through rotating ribbed passages. *International Journal of Heat and Fluid Flow*, 19, 393–400.

Irmisch, S., 1994. Simulation of film-cooling aerodynamics with a 2D Navier–Stokes solver using unstructured meshes, *International Gas Turbine and Aeroengine Congress and Exposition*, June 5–8, 1994, Houston, TX, ASME Paper 95-GT-24.

Jang, Y.J., Chen, H.C., and Han, J.C., 2000a. Numerical prediction of flow and heat transfer in a two-pass square channel with 90° ribs, *Proceedings of the 8th International Symposium on Transport Phenomena and Dynamics of Rotating Machinery*, March 27–30, Honolulu, HI, Vol. I, pp. 580–587.

Jang, Y.J., Chen, H.C., and Han, J.C., 2000b. Computation of flow and heat transfer in two-pass channels with 60° ribs, *38th Aerospace Sciences Meeting and Exhibit*, January 10–13, Reno, NV, AIAA Paper 2000–1036.

Jang, Y.J., Chen, H.C., and Han, J.C., 2000c. Flow and heat transfer in a rotating square channel with 45° angled ribs by Reynolds stress turbulence model, *ASME International Gas Turbine and Aeroengine Congress and Exhibition*, May 8–11, Munich, Germany, ASME Paper 2000-GT-0229.

Janke, E., Haselbach, F., Whitney, C., Hodson, H., Kanjirakkad, V., and Thomas, R.L., 2006. Passive shroud cooling concepts for HP turbines: CFD based design approach. ASME Paper GT2006-91194.

Jia, R. and Sunden, B., 2004. A low-re RSTM model for computations of heat transfer and fluid flow for impingement and convective cooling. ASME Paper GT2004-53459.

Jung, E.Y., Lee, D.H., Oh, S.H., Kim, K.M., and Cho, H.H., 2010. Total cooling effectiveness on a staggered full-coverage film cooling plate with impinging jet. ASME Paper GT2010-23725.

Koiro, M. and Lakshminarayana, B., 1998. Simulation and validation of Mach number effects on secondary flow in a transonic turbine cascade using a multigrid, k-ε solver. *ASME Journal of Turbomachinery*, 120, 285–297.

Kusterer, K., Bohn, D., and Ren, J., 2003. Conjugate heat transfer analysis for film cooling configurations with different hole geometries. ASME Paper GT2003-38369.

Launder, B.E., 1971. An improved algebraic stress model of turbulence. Imperial College Mechanical Engineering Report, TM TN/A19, London, U.K.

Launder, B.E., Reece, G.J., and Rodi, W., 1975. Progress in the development of a Reynolds-stress turbulence closure. *Journal of Fluid Mechanics*, 68, 537.

Lawson, S.A. and Thole, K.A., 2010. Simulations of multi-phase particle deposition on endwall film-cooling. ASME Paper GT2010-22376.

Ledezma, G., Laskowski, G.M., and Tolpadi, A.K., 2008. Turbulence model assessment for conjugate heat transfer predictions in a high-pressure turbine vane model. ASME Paper GT2008-50498.

Leedom, D. and Acharya, S., 2008. Large eddy simulations of film cooling flow fields from cylindrical and shaped holes. ASME Paper GT2008-51009.

Leylek, J.H., York, W., and Holloway, D.S., 2005. Prediction of heat transfer in a ribbed channel: Evaluation of unsteady RANS methodology. ASME Paper GT2005-68821.

Leylek, J.H. and Zerkle, R.D., 1994. Discrete-jet film cooling: A comparison of computational results with experiments. *ASME Journal of Turbomachinery*, 116, 358–363.

Lin, Y.L., Shih, T., Chyu, M.K., and Bunker, R.S., 2000. Effects of gap leakage on fluid flow in a contoured turbine nozzle guide vane. ASME Paper 2000-GT-0555.

Lin, Y.L., Stephens, M.A., and Shih, T.I.-P., 1997. Computation of leading-edge film cooling with injection through rows of compound-angle holes, *ASME International Gas Turbine and Aeroengine Congress and Exhibition*, June 2–5, 1997, Orlando, FL, ASME Paper 97-GT-298.

Liou, T.M. and Chen, S.H., 1995. Computation of spatially periodic turbulent fluid flow and heat transfer in a channel with various rib shapes, *ASME International Gas Turbine and Aeroengine Congress and Exposition*, June 5–8, 1995, Houston, TX, ASME Paper 95-GT-23.

Luo, J. and Lakshminarayana, B., 1997. Numerical simulation of turbine blade boundary layer and heat transfer and assessment of turbulence models. *ASME Journal of Turbomachinery*, 119, 794–801.

Luo, J. and Razinsky, E., 2006. Conjugate heat transfer analysis of a cooled turbine vane using the V2F turbulence model. ASME Paper GT2006-91109.

Luo, J. and Razinsky, E., 2007. Analysis of turbulent flow in 180-deg turning ducts with and without guide vanes. ASME Paper GT2007-28173.

Lynch, S., Thole, K.A., Kohli, A., and Lehane, C., 2010. Computational predictions of heat transfer and film cooling for a turbine blade with non-axisymmetric endwall contouring. ASME Paper GT2010-22984.

Majumdar, A.K., Pratap, V.S., and Spalding, D.B., 1977. Numerical computation of flow in rotating ducts. *ASME Journal of Fluids Engineering*, 99, 148–153.

Mangani, L., Cerutti, M., Maritano, M., and Spel, M., 2010. Conjugate heat transfer analysis of NASA C3X film cooled vane with an object-oriented CFD code. ASME Paper GT2010-23458.

Martin, C.A. and Thole, K.A., 1997. A CFD benchmark study: Leading edge film-cooling with compound angle injection, *ASME International Gas Turbine and Aeroengine Congress and Exhibition*, June 2–5, 1997, Orlando, FL, ASME Paper 97-GT-297.

McGovern, K.T. and Leylek, J.H., 1997. A detailed analysis of film cooling physics. Part II: Compound-angle injection with cylindrical holes, *ASME International Gas Turbine and Aeroengine Congress and Exhibition*, June 2–5, 1997, Orlando, FL, ASME Paper 97-GT-270.

Metzger, D.E., Bunker, R.S., and Chyu, M.K., 1989. Cavity heat transfer on a transverse grooved wall in a narrow channel. *Journal of Heat Transfer*, 111, 73–79.

Muldoon, F. and Acharya, S., 2007. Direct numerical simulations of pulsed film cooling. ASME Paper GT2007-28156.

Na, S., Williams, B., Dennis, R., Bryden, M., and Shih, T., 2007. Internal and film cooling of a flat plate with conjugate heat transfer. ASME Paper GT2007-27599.

Neelakantan, S. and Crawford, M.E., 1995. Prediction of film cooling effectiveness and heat transfer due to streamwise and compound angle injection on flat surfaces, *International Gas Turbine and Aeroengine Congress and Exposition*, June 5–8, 1995, Houston, TX, ASME Paper 95-GT-l 51.

Polanka, M.D., White, A.L., Praisner, T., and Clark, J.P., 2003. Turbine tip and shroud heat transfer and loading: Part B: Comparisons between prediction and experiment including unsteady effects. ASME Paper GT2003-38916.

Prakash, C. and Zerkle, R., 1992. Prediction of turbulent flow and heat transfer in a radially rotating square duct. *ASME Journal of Turbomachinery*, 114(4), 835–846.

Prakash, C. and Zerkle, R., 1995. Prediction of turbulent flow and heat transfer in a ribbed rectangular duct with and without rotation. *ASME Journal of Turbomachinery*, 117, 255–261.

Raisee, M. and Iacovides, H., 2004. Turbulent flow and heat transfer in stationary and rotating cooling passages with inclined ribs on opposite walls. ASME Paper GT2004-53245.

Rigby, D.L., Steinthorsson, E., and Ameri, A.A. 1997. Numerical prediction of heat transfer in a channel with ribs and bleed, *ASME International Gas Turbine and Aeroengine Congress and Exhibition*, June 2–5, 1997, Orlando, FL, ASME Paper 97-GT-431.

Rozati, A. and Tafti, D.K., 2007a. Large eddy simulation of leading edge film cooling. Part I: Computational domain and effect of coolant pipe inlet condition. ASME Paper GT2007-27689.

Rozati, A. and Tafti, D.K., 2007b. Large eddy simulation of leading edge film cooling. Part II: Heat transfer and effect of blowing ratio. ASME Paper GT2007-27690.

Saha, A. and Acharya, S., 2006. Computations of turbulent flow and heat transfer through a three-dimensional non-axisymmetric blade passage. ASME Paper GT2006-90390.

Sargison, J.E., Oldfield, M., Lock, G.D., Rawlinson, A., and Guo, S., 2003. Performance prediction of a converging slot-hole film cooling geometry. ASME Paper GT2003-38144.

Saxer, A.P. and Felici, H.M., 1996. Numerical analysis of 3-D unsteady hot streak migration and shock interaction in a turbine stage. *ASME Journal of Turbomachinery*, 118, 268–277.

Schuler, M., Dreher, H.-M., Neumann, S.O., Elfert, M., and Weigand, B., 2010. Numerical predictions of the effect of rotation on fluid flow and heat transfer in an engine-similar two-pass internal cooling channel with smooth and ribbed walls. ASME Paper GT2010-22870.

Shih, T., Chi, X., Bryden, K.M., Alsup, C., and Dennis, R.A., 2009. Effects of biot number on temperature and heat-flux distributions in a TBC-coated flat plate cooled by rib-enhanced internal cooling. ASME Paper GT2009-59726.

Shih, T. and Lin, Y.L., 2002. Controlling secondary-flow structure by leading-edge airfoil fillet and inlet swirl to reduce aerodynamic loss and surface heat transfer. ASME Paper GT2002-30529.

Shih, T., Lin, Y.L., and Simon, T.W., 2000. Control of secondary flows in a turbine nozzle guide vane by endwall contouring. ASME Paper 2000-GT-0556.

Shih, T. and Na, S., 2006. Increasing adiabatic film-cooling effectiveness by using upstream ramp. ASME Paper GT2006-91163.

Shyam, V., Ameri, A., and Chen, J.P., 2010. Analysis of unsteady tip and end-wall heat transfer in a highly loaded transonic turbine stage. ASME Paper GT2010-23694.

Sierra, F.Z., Bolaina, C., Kubiak, J., Narzary, D., Han, J.C., and Nebradt, J., 2009. Heat transfer and thermal mechanical stress distributions in gas turbine blades. ASME Paper GT2009-59194.

Simoneau, R.J. and Simon, F.F., 1993. Progress towards understanding and predicting convection heat transfer in the turbine gas path. *International Journal of Heat and Fluid Flow*, 14(2), 106–127.

Sleiti, A.K. and Kapat, J.S., 2004. Effect of Coriolis and centrifugal forces on turbulence and heat transfer at high rotation and buoyancy numbers in a rib-roughened internal cooling channel. ASME Paper GT2004-52018.

Sreedharan, S.H. and Tafti, D.K., 2009a. Effect of blowing ratio in the near stagnation region of a three-row leading edge film cooling geometry using large eddy simulations. ASME Paper GT2009-59325.

Sreedharan, S.H. and Tafti, D.K., 2009b. Effect of blowing ratio on syngas flyash particle deposition on a three-row leading edge film cooling geometry using large eddy simulations. ASME Paper GT2009-59326.

Sreedharan, S.S. and Tafti, D.K., 2010. Composition dependent model for the prediction of syngas ash deposition with application to a leading edge turbine vane. ASME Paper GT2010-23655.

Starke, C., Janke, E., Hofer, T., and Lengani, D., 2008. Comparison of a conventional thermal analysis of a turbine cascade to a full conjugate heat transfer computation. ASME Paper GT2008-51151.

Stephens, M.A., Chyu, M.K., Shin, T.I.-P., and Civinskas, K.C., 1996. Computation of convective heat transfer in a square duct with inclined ribs of rounded cross-section, *ASME International Mechanical Engineering Congress and Exhibition*, November 17–22, 1996, Atlanta, GA, ASME Paper 96-WA/HT-12.

Stephens, M.A. and Shih, T.I.-P., 1997. Computation of compressible flow and heat transfer in a rotating duct with inclined ribs and a 180-degree bend, *ASME International Gas Turbine and Aeroengine Congress and Exhibition*, June 2–5, 1997, Orlando, FL, ASME Paper 97-GT-192.

Su, G., Chen, H.C., Han, J.C., and Heidmann, J.D., 2004. Computation of flow and heat transfer in two-pass rotating rectangular channels (AR = 1:1, AR = 1:2, AR = 1:4) with 45-deg angled ribs by a Reynolds stress turbulence model. ASME Paper, GT2004-53662.

Su, G., Teng, S., Chen, H.C., and Han, J.C., 2003. Computation of flow and heat transfer in rotating rectangular channels (AR = 4) with V-shaped ribs by a Reynolds stress turbulence model. ASME Paper GT2003-38348.

Sugimoto, T., Bohn, D., Kusterer, K., and Tanaka, R., 2004. Conjugate calculations for a film-cooled blade under different operating conditions. ASME Paper GT2004-53719.

Tafti, D. and Sewall, E., 2004. Large eddy simulation of the developing region of a rotating ribbed internal turbine blade cooling channel. ASME Paper GT2004-53833.

Tafti, D., Viswanathan, A., and Abdel-Wahab, S., 2005. Large eddy simulation of flow and heat transfer in an internal cooling duct with high blockage ratio 45-deg staggered ribs. ASME Paper GT2005-68086.

Takahashi, T., Watanabe, K., and Takahashi, T., 2000. Thermal conjugate analysis of a first stage blade in a gas turbine. ASME Paper 2000-GT-0251.

Takahashi, T., Watanabe, K., and Takahashi, T., 2001. Transient analyses of conjugate heat transfer of a first stage rotor blade in start-up and shut-down. ASME Paper 2001-GT-0171.

Tannehill, J.C., Anderson, D.A., and Pletcher, R.H., 1997. *Computational Fluid Mechanics and Heat Transfer*, 2nd edn. Taylor & Francis Group, Washington, DC.

Tekriwal, P., 1994. Heat transfer predictions in rotating radial smooth channel: Comparative study of K-E models with wall function and low-RE model, *ASME International Gas Turbine and Aeroengine Congress and Exposition*, June 13–16, 1994, The Hague, the Netherlands, ASME Paper 94-GT-196.

Thole, K., Hohlfeld, E., Christophel, J., and Couch, E., 2003. Predictions of cooling from dirt purge holes along the tip of a turbine blade. ASME Paper GT2003-38251.

Tyagi, M. and Acharya, S., 2004. Large eddy simulations of flow and heat transfer in rotating ribbed duct flows. ASME Paper GT2004-53924.

Tyagi, M. and Acharya, S., 2005. Large eddy simulation of turbulent flows in complex and moving rigid geometries using the immersed boundary method. *International Journal for Numerical Methods in Fluids*, 48, 691–722, doi: 10.1002/fld.937.

Viswanathan, A. and Tafti, D., 2006. Large eddy simulation of fully developed flow and heat transfer in a rotating duct with 45-degree ribs. ASME Paper GT2006-90229.

Voigt, S., Noll, B., and Aigner, M., 2010. Comparison and validation of RANS, URANS and SAS simulations of flat plate film-cooling. ASME Paper GT2010-22475.

Wagner, J.H., Johnson, B.V., and Kopper, F.C., 1991. Heat transfer in rotating serpentine passages with smooth walls. *ASME Journal of Turbomachinery*, 113, 321–330.

Walters, D.K. and Leylek, J.H., 1997. A detailed analysis of film-cooling physics. Part 1: Streamwise injection with cylindrical holes, *ASME International Gas Turbine and Aeroengine Congress and Exhibition*, June 2–5, 1997, Orlando, FL, ASME Paper 97-GT-269.

Walters, D.B., McGovern, K.T., Butkiewicz, J.J., and Leylek, J.H., 1995. A systematic computational methodology applied to a jet-in-crossflow. Part 2: Unstructured/adaptive grid approach, *ASME International Mechanical Engineering Congress and Exposition*, November 12–17, 1995, San Francisco, CA, ASME Paper 95-WA/HT-2.

Wang, T. and Li, X., 2006. Simulation of mist film cooling at gas turbine operating conditions. ASME Paper GT2006-90742.

Whitney, C., Martini, P., Schulz, A., and Bauer, H.J., 2005. Detached eddy simulation of film cooling performance on the trailing edge cut-back of gas turbine airfoils. ASME Paper GT2005-68084.

Wilcox, D.C., 1993. *Turbulence Modeling for CFD*. DCW Industries, La Canada, CA.

Yang, H., Acharya, S., Ekkad, S.V., Prakash, C., and Bunker, R., 2002a. Flow and heat transfer predictions for a flat-tip turbine blade. ASME Paper GT2002-30190.

Yang, H., Acharya, S., Ekkad, S.V., Prakash, C., and Bunker, R., 2002b. Numerical simulation of flow and heat transfer past a turbine blade with a squealer-tip. ASME Paper GT2002-30193.

Yang, H., Chen, H.C., and Han, J.C., 2004. Numerical prediction of film cooling and heat transfer with different film-hole arrangements on the plane and squealer tip of a gas turbine blade. ASME Paper GT2004-53199.

Yang, H., Chen, H.C., Han, J.C., and Moon, H.K., 2005. Numerical prediction of film cooling and heat transfer on the leading edge of a rotating blade with two rows holes in a 1-1/2 turbine stage at design and off design conditions. ASME Paper GT2005-68335.

Yang, H., Gao, Z., Chen, H.C., Han, J.C., and Schobeiri, M.T., 2007. Prediction of film cooling and heat transfer from a rotating blade platform with stator–rotor purge and discrete film-hole flows in a 1-1/2 turbine stage. ASME Paper GT2007-27069.

Yamamoto, A., Murao, R., Suzuki, Y., and Aoi, Y., 1995. A quasi unsteady study on wake interaction of turbine stator and rotor cascades. *ASME Journal of Turbomachinery*, 117, 553–560.

Yang, R.J. and Luo, W.J., 1996. Turbine blade heat transfer prediction in flow transition using k-w two-equation model. *Journal of Thermophysics and Heat Transfer*, 10(4), 613–620.

Zess, G.A. and Thole, K., 2001. Computational design and experimental evaluation of using a leading edge fillet on a gas turbine vane. ASME Paper 2001-GT-0404.

8

Final Remarks

8.1 Turbine Heat Transfer and Film Cooling

Turbine blade external heat transfer and film cooling have been studied for many years. Recent research focuses on a high free-stream turbulence effect on turbine stator heat transfer with and without film cooling, upstream unsteady wake effect on downstream rotor blade heat transfer with and without film cooling, and surface roughness/TBC spallation effect on blade heat transfer with and without film cooling, under engine Mach and Reynolds number flow conditions. To optimize the film-cooling performance, effects of film hole size, length, spacing, shape, orientation on turbine blade, and surface heat-transfer distributions need to be considered. Recent research also focuses on providing highly detailed heat-transfer coefficient and film-cooling effectiveness distributions on turbine blades under unsteady high-turbulence engine flow conditions. It is well known that blade life can be reduced by half if the blade metal temperature prediction is off by 50°F. Therefore, highly accurate, high-resolution data are extremely important for engine designers. It is also important to note that turbine designers have used the existing/modified computational fluid dynamics (CFD) codes to predict film-cooling performance with complex film hole geometry under engine flow conditions, including the effect of rotor/stator interaction and blade rotation.

8.2 Turbine Internal Cooling with Rotation

Turbine vane internal heat transfer has been studied for many years, and many semi-empirical correlations are available for cooling designers. Recent research focuses on rotor blade coolant passage heat transfer, with and without bleed, through film holes under realistic engine flow Reynolds number and engine rotation number conditions. Rotation induces Coriolis and centrifugal forces, which produce cross-stream secondary flow in the rotating coolant passage. Therefore, heat-transfer coefficients in a rotor coolant passage are quite different from those in a nonrotating frame. As mentioned earlier, recent research also focuses on both highly accurate and highly detailed local heat-transfer coefficient distributions in rotor coolant passages under high coolant flow Reynolds numbers and conditions of high rotation numbers and high buoyancy parameters. To optimize the cooling effect, impingement cooling in the blade leading edge, rib-turbulated serpentine cooling in the blade midspan, and pin-fin cooling in the blade trailing edge have been used with and without incorporating film cooling under rotating conditions. Innovative concepts for blade internal cooling are of interest for turbine-cooling designers. As mentioned earlier, the existing or modified CFD codes become a useful tool for coolant passage heat-transfer predictions and coolant flow management/optimization.

8.3 Turbine Edge Heat Transfer and Cooling

To date, most available experimental data are for the main body of turbine airfoil external heat transfer, film cooling, and internal heat-transfer studies. Satisfying the even higher rotor inlet temperature requirement makes turbine airfoil edge cooling an urgent issue for this century's gas turbines. Edge cooling heat transfer includes turbine vane endwall heat transfer and film cooling under realistic engine Mach and Reynolds numbers, airfoil trailing-edge region heat transfer with internal coolant injection along the trailing-edge base, and turbine rotor blade tip region heat transfer with and without tip hole cooling under engine flow conditions. Highly accurate and highly detailed local heat-transfer data in these turbine edge regions would be useful in preventing blade failure due to local hot spots. Flow visualizations/measurements and CFD predictions would provide valuable information for designing effective cooled blades for advanced gas turbines.

8.4 New Information from 2000 to 2010

With the continued demand for power, there is a constant demand to increase the efficiency of power generation by natural-gas-fired gas turbines. Moreover, there is a need for the turbines to operate cleanly, lowering emissions and pollution. Therefore, the opportunity exists for new fuels that are potentially cleaner fuels and lead to increased engine efficiency. Syngas (produced from coal) is a viable alternative fuel as coal is America's most abundant fossil fuel, and it makes economic sense to utilize this resource. Furthermore, hydrogen (derived from coal gasification) is another possible fuel leading to more efficient and clean turbines. While it is important to take advantage of these coal-derived syngas and high hydrogen content (HHC) fuels, it is also vital to recognize, understand, and overcome the drawbacks of these fuels. These coal gasified fuels produce high percentage of water vapor (steam) and increase heavy heat load to the turbine components that require efficient cooling technology development. These coal-based fuels may introduce impurities into the mainstream gas, which can have a harmful impact on turbine components. The impurities can cause corrosion and erosion on the surface of the airfoils or the impurities may be deposited on the components. All of these scenarios increase the surface roughness of the components, increase heat transfer, increase aerodynamic losses, decrease film-cooling effectiveness, and decrease the performance of the turbine. Recent research focuses on how to address and access new turbine hot-gas-path working environments and how to model and solve new challenging aerodynamics, heat transfer, and cooling problems. It suggests that the advanced cooling technology and durable thermal barrier coatings still play most critical roles for developments of future coal-based gas turbines with near-zero emissions.

8.5 Closure

Gas turbine heat transfer and cooling is in the category of single-phase forced convection. Single-phase convection heat-transfer rates in gas turbines are very difficult to predict because of complex flow through complex blade geometry. Today, researchers are combining advanced measurement techniques and advanced CFD codes to understand complex flow physics and heat-transfer behaviors in turbines. This will allow them to develop highly sophisticated and highly efficient cooling technology for this century's gas turbine engines. Someday turbine-cooling designers should be able to easily predict the correct turbine blade surface temperature distributions for a given blade geometry and flow condition. How much research is needed to correctly predict flow and heat-transfer distributions from rotor internal coolant passages with impingement/ribs/pins, through film-cooling holes, to external hot-gas path with complex three-dimensional unsteady flow including endwall flows and tip leakage flows? How can researchers, particularly university researchers, contribute to reach that dream?

Index